Lecture Notes in Artificial Intelligence 6322

Edited by R. Goebel, J. Siekmann, and W. Wahlster

Subseries of Lecture Notes in Computer Science

José Luis Balcázar
Francesco Bonchi Aristides Gionis
Michèle Sebag (Eds.)

Machine Learning and Knowledge Discovery in Databases

European Conference, ECML PKDD 2010
Barcelona, Spain, September 20-24, 2010
Proceedings, Part II

Springer

Series Editors

Randy Goebel, University of Alberta, Edmonton, Canada
Jörg Siekmann, University of Saarland, Saarbrücken, Germany
Wolfgang Wahlster, DFKI and University of Saarland, Saarbrücken, Germany

Volume Editors

José Luis Balcázar
Universidad de Cantabria
Departamento de Matemáticas, Estadística y Computación
Avenida de los Castros, s/n, 39071 Santander, Spain
E-mail: joseluis.balcazar@unican.es

Francesco Bonchi
Aristides Gionis
Yahoo! Research Barcelona
Avinguda Diagonal 177, 08018 Barcelona, Spain
E-mail: {bonchi, gionis}@yahoo-inc.corp

Michèle Sebag
TAO, CNRS-INRIA-LRI, Université Paris-Sud
91405, Orsay, France
E-mail: sebag@lri.fr

Cover illustration: Decoration detail at the Park Güell, designed by Antoni Gaudí, and one of the landmarks of modernist art in Barcelona.

Library of Congress Control Number: 2010934301

CR Subject Classification (1998): I.2, H.3, H.4, H.2.8, J.1, H.5

LNCS Sublibrary: SL 7 – Artificial Intelligence

ISSN 0302-9743
ISBN-10 3-642-15882-X Springer Berlin Heidelberg New York
ISBN-13 978-3-642-15882-7 Springer Berlin Heidelberg New York

springer.com

Printed in Germany

Typesetting: Camera-ready by author, data conversion by Scientific Publishing Services, Chennai, India
Printed on acid-free paper 06/3180

Preface

The European Conference on Machine Learning and Principles and Practice of Knowledge Discovery in Databases, ECML PKDD 2010, was held in Barcelona, September 20–24, 2010, consolidating the long junction between the European Conference on Machine Learning (of which the first instance as European workshop dates back to 1986) and Principles and Practice of Knowledge Discovery in Data Bases (of which the first instance dates back to 1997). Since the two conferences were first collocated in 2001, both machine learning and data mining communities have realized how each discipline benefits from the advances, and participates to defining the challenges, of the sister discipline. Accordingly, a single ECML PKDD Steering Committee gathering senior members of both communities was appointed in 2008.

In 2010, as in previous years, ECML PKDD lasted from Monday to Friday. It involved six plenary invited talks, by Christos Faloutsos, Jiawei Han, Hod Lipson, Leslie Pack Kaelbling, Tomaso Poggio, and Jürgen Schmidhuber, respectively. Monday and Friday were devoted to workshops and tutorials, organized and selected by Colin de la Higuera and Gemma Garriga. Continuing from ECML PKDD 2009, an industrial session managed by Taneli Mielikainen and Hugo Zaragoza welcomed distinguished speakers from the ML and DM industry: Rakesh Agrawal, Mayank Bawa, Ignasi Belda, Michael Berthold, José Luis Flórez, Thore Graepel, and Alejandro Jaimes. The conference also featured a discovery challenge, organized by András Benczúr, Carlos Castillo, Zoltán Gyöngyi, and Julien Masanès.

From Tuesday to Thursday, 120 papers selected among 658 submitted full papers were presented in the technical parallel sessions. The selection process was handled by 28 area chairs and the 282 members of the Program Committee; additional 298 reviewers were recruited. While the selection process was made particularly intense due to the record number of submissions, we heartily thank all area chairs, members of the Program Committee, and additional reviewers for their commitment and hard work during the short reviewing period. The conference also featured a demo track, managed by Ulf Brefeld and Xavier Carreras; 12 demos out of 24 submitted ones were selected, attesting to the high impact technologies based on the ML and DM body of research.

Following an earlier tradition, seven ML and seven DM papers were distinguished by the program chairs on the basis of their exceptional scientific quality and high impact on the field, and they were directly published in the Machine Learning Journal and the Data Mining and Knowledge Discovery Journal, respectively. Among these papers, some were selected by the Best Paper Chair Hiroshi Motoda, and received the Best Paper Awards and Best Student Paper Awards in Machine Learning and in Data Mining, sponsored by Springer.

A topic widely explored from both ML and DM perspectives was graphs, with motivations ranging from molecular chemistry to social networks. The point of matching or clustering graphs was examined in connection with tractability and domain knowledge, where the latter could be acquired through common patterns, or formulated through spectral clustering. The study of social networks focused on how they develop, overlap, propagate information (and how information propagation can be hindered). Link prediction and exploitation in static or dynamic, possibly heterogeneous, graphs, was motivated by applications in information retrieval and collaborative filtering, and in connection with random walks.

Frequent itemset approaches were hybridized with constraint programming or statistical tools to efficiently explore the search space, deal with numerical attributes, or extract locally optimal patterns. Compressed representations and measures of robustness were proposed to optimize association rules. Formal concept analysis, with applications to pharmacovigilance or Web ontologies, was considered in connection with version spaces.

Bayesian learning features new geometric interpretations of prior knowledge and efficient approaches for independence testing. Generative approaches were motivated by applications in sequential, spatio-temporal or relational domains, or multi-variate signals with high dimensionality. Ensemble learning was used to support clustering and biclustering; the post-processing of random forests was also investigated.

In statistical relational learning and structure identification, with motivating applications in bio-informatics, neuro-imagery, spatio-temporal domains, and traffic forecasting, the stress was put on new learning criteria; gradient approaches, structural constraints, and/or feature selection were used to support computationally effective algorithms.

(Multiple) kernel learning and related approaches, challenged by applications in image retrieval, robotics, or bio-informatics, revisited the learning criteria and regularization terms, the processing of the kernel matrix, and the exploration of the kernel space. Dimensionality reduction, embeddings, and distance were investigated, notably in connection with image and document retrieval.

Reinforcement learning focussed on ever more scalable and tractable approaches through smart state or policy representations, a more efficient use of the available samples, and/or Bayesian approaches.

Specific settings such as ranking, multi-task learning, semi-supervised learning, and game-theoretic approaches were investigated, with some innovative applications to astrophysics, relation extraction, and multi-agent systems. New bounds were proved within the active, multi-label, and weighted ensemble learning frameworks.

A few papers aimed at efficient algorithms or computing environments, e.g., related to linear algebra, cutting plane algorithms, or graphical processing units, were proposed (with available source code in some cases). Numerical stability was also investigated in connection with sparse learning.

Among the applications presented were review mining, software debugging/ process modeling from traces, and audio mining.

To conclude this rapid tour of the scientific program, our special thanks go to the local chairs Ricard Gavaldà, Elena Torres, and Estefania Ricart, the Web and registration chair Albert Bifet, the sponsorship chair Debora Denato, and the many volunteers that eagerly contributed to make ECML PKDD 2010 a memorable event.

Our last and warmest thanks go to all invited speakers and other speakers, to all tutorial, workshop, demo, industrial, discovery, best paper, and local chairs, to the area chairs and all reviewers, to all attendees — and overall, to the authors who chose to submit their work to the ECML PKDD conference, and thus enabled us to build up this memorable scientific event.

July 2010

José L Balcázar
Francesco Bonchi
Aristides Gionis
Michèle Sebag

Organization

Program Chairs

José L Balcázar
Universidad de Cantabria and
Universitat Politècnica de Catalunya, Spain
http://personales.unican.es/balcazarjl/

Francesco Bonchi
Yahoo! Research
Barcelona, Spain
http://research.yahoo.com

Aristides Gionis
Yahoo! Research
Barcelona, Spain
http://research.yahoo.com

Michèle Sebag
CNRS
Université Paris Sud, Orsay Cedex, France
http://www.lri.fr/ sebag/

Local Organization Chairs

Ricard Gavaldà	Universitat Politècnica de Catalunya
Estefania Ricart	Barcelona Media
Elena Torres	Barcelona Media

Organization Team

Ulf Brefeld	Yahoo! Research
Eugenia Fuenmayor	Barcelona Media
Mia Padullés	Yahoo! Research
Natalia Pou	Barcelona Media

Workshop and Tutorial Chairs

Gemma C. Garriga	University of Paris 6
Colin de la Higuera	University of Nantes

Best Papers Chair

Hiroshi Motoda	AFOSR/AOARD and Osaka University

Industrial Track Chairs

Taneli Mielikainen	Nokia
Hugo Zaragoza	Yahoo! Research

Demo Chairs

Ulf Brefeld	Yahoo! Research
Xavier Carreras	Universitat Politècnica de Catalunya

Discovery Challenge Chairs

András Benczúr	Hungarian Academy of Sciences
Carlos Castillo	Yahoo! Research
Zoltán Gyöngyi	Google
Julien Masanès	European Internet Archive

Sponsorship Chair

Debora Donato	Yahoo! Labs

Web and Registration Chair

Albert Bifet	University of Waikato

Publicity Chair

Ricard Gavaldà	Universitat Politècnica de Catalunya

Steering Committee

Wray Buntine
Walter Daelemans
Bart Goethals
Marko Grobelnik
Katharina Morik
Joost N. Kok
Stan Matwin
Dunja Mladenic
John Shawe-Taylor
Andrzej Skowron

Area Chairs

Samy Bengio
Bettina Berendt
Paolo Boldi
Wray Buntine
Toon Calders
Luc de Raedt
Carlotta Domeniconi
Martin Ester
Paolo Frasconi
Joao Gama
Ricard Gavaldà
Joydeep Ghosh
Fosca Giannotti
Tu-Bao Ho
George Karypis
Laks V.S. Lakshmanan
Katharina Morik
Jan Peters
Kai Puolamäki
Yucel Saygin
Bruno Scherrer
Arno Siebes
Soeren Sonnenburg
Alexander Smola
Einoshin Suzuki
Evimaria Terzi
Michalis Vazirgiannis
Zhi-Hua Zhou

Program Committee

Osman Abul
Gagan Agrawal
Erick Alphonse
Carlos Alzate
Massih Amini
Aris Anagnostopoulos
Annalisa Appice
Thierry Artières
Sitaram Asur
Jean-Yves Audibert
Maria-Florina Balcan
Peter Bartlett
Younes Bennani
Paul Bennett
Michele Berlingerio
Michael Berthold
Albert Bifet
Hendrik Blockeel
Mario Boley
Antoine Bordes
Gloria Bordogna
Christian Borgelt
Karsten Borgwardt
Henrik Boström
Marco Botta
Guillaume Bouchard
Jean-Francois Boulicaut
Ulf Brefeld
Laurent Brehelin
Bjoern Bringmann
Carla Brodley
Rui Camacho
Stéphane Canu
Olivier Cappé
Carlos Castillo
Jorge Castro
Ciro Cattuto
Nicolò Cesa-Bianchi
Nitesh Chawla
Sanjay Chawla
David Cheung
Sylvia Chiappa
Boris Chidlovski
Flavio Chierichetti
Philipp Cimiano
Alexander Clark
Christopher Clifton
Antoine Cornuéjols
Fabrizio Costa
Bruno Crémilleux
James Cussens
Alfredo Cuzzocrea

Florence d'Alché-Buc
Claudia d'Amato
Gautam Das
Jeroen De Knijf
Colin de la Higuera
Krzysztof Dembczynski
Ayhan Demiriz
Francois Denis
Christos Dimitrakakis
Josep Domingo Ferrer
Debora Donato
Dejing Dou
Gérard Dreyfus
Kurt Driessens
John Duchi
Pierre Dupont
Saso Dzeroski
Charles Elkan
Damien Ernst
Floriana Esposito
Fazel Famili
Nicola Fanizzi
Ad Feelders
Alan Fern
Daan Fierens
Peter Flach
George Forman
Vojtech Franc
Eibe Frank
Dayne Freitag
Elisa Fromont
Patrick Gallinari
Auroop Ganguly
Fred Garcia
Gemma Garriga
Thomas Gärtner
Eric Gaussier
Floris Geerts
Matthieu Geist
Claudio Gentile
Mohammad Ghavamzadeh
Gourab Ghoshal
Chris Giannella
Attilio Giordana
Mark Girolami
Shantanu Godbole
Bart Goethals
Sally Goldman
Henrik Grosskreutz
Dimitrios Gunopulos
Amaury Habrard
Eyke Hüllermeier
Nikolaus Hansen
Iris Hendrickx
Melanie Hilario
Alexander Hinneburg
Kouichi Hirata
Frank Hoeppner
Jaakko Hollmen
Tamas Horvath
Andreas Hotho
Alex Jaimes
Szymon Jaroszewicz
Daxin Jiang
Felix Jungermann
Frederic Jurie
Alexandros Kalousis
Panagiotis Karras
Samuel Kaski
Dimitar Kazakov
Sathiya Keerthi
Jens Keilwagen
Roni Khardon
Angelika Kimmig
Ross King
Marius Kloft
Arno Knobbe
Levente Kocsis
Jukka Kohonen
Solmaz Kolahi
George Kollios
Igor Kononenko
Nick Koudas
Stefan Kramer
Andreas Krause
Vipin Kumar
Pedro Larrañaga
Mark Last
Longin Jan Latecki
Silvio Lattanzi

Anne Laurent
Nada Lavrac
Alessandro Lazaric
Philippe Leray
Jure Leskovec
Carson Leung
Chih-Jen Lin
Jessica Lin
Huan Liu
Kun Liu
Alneu Lopes
Ramón López de Màntaras
Eneldo Loza Mencía
Claudio Lucchese
Elliot Ludvig
Dario Malchiodi
Donato Malerba
Bradley Malin
Giuseppe Manco
Shie Mannor
Stan Matwin
Michael May
Thorsten Meinl
Prem Melville
Rosa Meo
Pauli Miettinen
Lily Mihalkova
Dunja Mladenic
Ali Mohammad-Djafari
Fabian Morchen
Alessandro Moschitti
Ion Muslea
Mirco Nanni
Amedeo Napoli
Claire Nedellec
Frank Nielsen
Siegfried Nijssen
Richard Nock
Sebastian Nowozin
Alexandros Ntoulas
Andreas Nuernberger
Arlindo Oliveira
Balaji Padmanabhan
George Paliouras
Themis Palpanas
Apostolos Papadopoulos
Andrea Passerini
Jason Pazis
Mykola Pechenizkiy
Dmitry Pechyony
Dino Pedreschi
Jian Pei
Jose Peña
Ruggero Pensa
Marc Plantevit
Enric Plaza
Doina Precup
Ariadna Quattoni
Predrag Radivojac
Davood Rafiei
Chedy Raissi
Alain Rakotomamonjy
Liva Ralaivola
Naren Ramakrishnan
Jan Ramon
Chotirat Ratanamahatana
Elisa Ricci
Bertrand Rivet
Philippe Rolet
Marco Rosa
Fabrice Rossi
Juho Rousu
Céline Rouveirol
Cynthia Rudin
Salvatore Ruggieri
Stefan Rüping
Massimo Santini
Lars Schmidt-Thieme
Marc Schoenauer
Marc Sebban
Nicu Sebe
Giovanni Semeraro
Benyah Shaparenko
Jude Shavlik
Fabrizio Silvestri
Dan Simovici
Carlos Soares
Diego Sona
Alessandro Sperduti
Myra Spiliopoulou

Gerd Stumme
Jiang Su
Masashi Sugiyama
Johan Suykens
Domenico Talia
Pang-Ning Tan
Tamir Tassa
Nikolaj Tatti
Yee Whye Teh
Maguelonne Teisseire
Olivier Teytaud
Jo-Anne Ting
Michalis Titsias
Hannu Toivonen
Ryota Tomioka
Marc Tommasi
Hanghang Tong
Luis Torgo
Fabien Torre
Marc Toussaint
Volker Tresp
Koji Tsuda
Alexey Tsymbal
Franco Turini
Antti Ukkonen
Matthijs van Leeuwen
Martijn van Otterlo
Maarten van Someren
Celine Vens
Jean-Philippe Vert
Ricardo Vilalta
Christel Vrain
Jilles Vreeken
Christian Walder
Louis Wehenkel
Markus Weimer
Dong Xin
Dit-Yan Yeung
Cong Yu
Philip Yu
Chun-Nam Yue
Francois Yvon
Bianca Zadrozny
Carlo Zaniolo
Gerson Zaverucha
Filip Zelezny
Albrecht Zimmermann

Additional Reviewers

Mohammad Ali Abbasi
Zubin Abraham
Yong-Yeol Ahn
Fabio Aiolli
Dima Alberg
Salem Alelyani
Aneeth Anand
Sunil Aryal
Arthur Asuncion
Gowtham Atluri
Martin Atzmueller
Paolo Avesani
Pranjal Awasthi
Hanane Azzag
Miriam Baglioni
Raphael Bailly
Jaume Baixeries
Jorn Bakker
Georgios Balkanas
Nicola Barbieri
Teresa M.A. Basile
Luca Bechetti
Dominik Benz
Maxime Berar
Juliana Bernardes
Aurélie Boisbunon
Shyam Boriah
Zoran Bosnic
Robert Bossy
Lydia Boudjeloud
Dominique Bouthinon
Janez Brank
Sandra Bringay
Fabian Buchwald
Krisztian Buza
Matthias Böck

José Caldas
Gabriele Capannini
Annalina Caputo
Franco Alberto Cardillo
Xavier Carreras
Giovanni Cavallanti
Michelangelo Ceci
Eugenio Cesario
Pirooz Chubak
Anna Ciampi
Ronan Collobert
Carmela Comito
Gianni Costa
Bertrand Cuissart
Boris Cule
Giovanni Da San Martino
Marco de Gemmis
Kurt De Grave
Gerben de Vries
Jean Decoster
Julien Delporte
Christian Desrosiers
Sanjoy Dey
Nicola Di Mauro
Joshua V. Dillon
Huyen Do
Stephan Doerfel
Brett Drury
Timo Duchrow
Wouter Duivesteijn
Alain Dutech
Ilenia Epifani
Ahmet Erhan Nergiz
Rémi Eyraud
Philippe Ezequel
Jean Baptiste Faddoul
Fabio Fassetti
Bruno Feres de Souza
Remi Flamary
Alex Freitas
Natalja Friesen
Gabriel P.C. Fung
Barbara Furletti
Zeno Gantner
Steven Ganzert
Huiji Gao
Ashish Garg
Aurelien Garivier
Gilles Gasso
Elisabeth Georgii
Edouard Gilbert
Tobias Girschick
Miha Grcar
Warren Greiff
Valerio Grossi
Nistor Grozavu
Massimo Guarascio
Tias Guns
Vibhor Gupta
Rohit Gupta
Tushar Gupta
Nico Görnitz
Hirotaka Hachiya
Steve Hanneke
Andreas Hapfelmeier
Daniel Hsu
Xian-Sheng Hua
Yi Huang
Romain Hérault
Leo Iaquinta
Dino Ienco
Elena Ikonomovska
Stéphanie Jacquemont
Jean-Christophe Janodet
Frederik Janssen
Baptiste Jeudy
Chao Ji
Goo Jun
U Kang
Anuj Karpatne
Jaya Kawale
Ashraf M. Kibriya
Kee-Eung Kim
Akisato Kimura
Arto Klami
Suzan Koknar-Tezel
Xiangnan Kong
Arne Koopman
Mikko Korpela
Wojciech Kotlowski

Alexis Kotsifakos
Petra Kralj Novak
Tetsuji Kuboyama
Matjaz Kukar
Sanjiv Kumar
Shashank Kumar
Pascale Kuntz
Ondrej Kuzelka
Benjamin Labbe
Mathieu Lajoie
Hugo Larochelle
Agnieszka Lawrynowicz
Gregor Leban
Mustapha Lebbah
John Lee
Sau Dan Lee
Gayle Leen
Florian Lemmerich
Biao Li
Ming Li
Rui Li
Tiancheng Li
Yong Li
Yuan Li
Wang Liang
Ryan Lichtenwalter
Haishan Liu
Jun Liu
Lei Liu
Xu-Ying Liu
Corrado Loglisci
Pasquale Lops
Chuan Lu
Ana Luisa Duboc
Panagis Magdalinos
Sebastien Mahler
Michael Mampaey
Prakash Mandayam
Alain-Pierre Manine
Patrick Marty
Jeremie Mary
André Mas
Elio Masciari
Emanuel Matos
Andreas Maunz
John McCrae
Marvin Meeng
Wannes Meert
Joao Mendes-Moreira
Aditya Menon
Peter Mika
Folke Mitzlaff
Anna Monreale
Tetsuro Morimura
Ryoko Morioka
Babak Mougouie
Barzan Mozafari
Igor Mozetic
Cataldo Musto
Alexandros Nanopoulos
Fedelucio Narducci
Maximilian Nickel
Inna Novalija
Benjamin Oatley
Marcia Oliveira
Emauele Olivetti
Santiago Ontañón
Francesco Orabona
Laurent Orseau
Riccardo Ortale
Aomar Osmani
Aline Paes
Sang-Hyeun Park
Juuso Parkkinen
Ioannis Partalas
Pekka Parviainen
Krishnan Pillaipakkamnatt
Fabio Pinelli
Cristiano Pitangui
Barbara Poblete
Vid Podpecan
Luigi Pontieri
Philippe Preux
Han Qin
Troy Raeder
Subramanian Ramanathan
Huzefa Rangwala
Guillaume Raschia
Konrad Rieck
François Rioult

Ettore Ritacco
Mathieu Roche
Christophe Rodrigues
Philippe Rolet
Andrea Romei
Jan Rupnik
Delia Rusu
Ulrich Rückert
Hiroshi Sakamoto
Vitor Santos Costa
Kengo Sato
Saket Saurabh
Francois Scharffe
Leander Schietgat
Jana Schmidt
Constanze Schmitt
Christoph Scholz
Dan Schrider
Madeleine Seeland
Or Sheffet
Noam Shental
Xiaoxiao Shi
Naoki Shibayama
Nobuyuki Shimizu
Kilho Shin
Kaushik Sinha
Arnaud Soulet
Michal Sramka
Florian Steinke
Guillaume Stempfel
Liwen Sun
Umar Syed
Gabor Szabo
Yasuo Tabei
Nima Taghipour
Hana Tai
Frédéric Tantini
Katerina Tashkova
Christine Task
Alexandre Termier
Lam Thoang Hoang
Xilan Tian
Xinmei Tian
Gabriele Tolomei
Aneta Trajanov
Roberto Trasarti
Abhishek Tripathi
Paolo Trunfio
Ivor Tsang
Theja Tulabandhula
Boudewijn van Dongen
Stijn Vanderlooy
Joaquin Vanschoren
Philippe Veber
Sriharsha Veeramachaneni
Sebastian Ventura
Alessia Visconti
Jun Wang
Xufei Wang
Osamu Watanabe
Lorenz Weizsäcker
Tomas Werner
Jörg Wicker
Derry Wijaya
Daya Wimalasuriya
Adam Woznica
Fuxiao Xin
Zenglin Xu
Makoto Yamada
Liu Yang
Xingwei Yang
Zhirong Yang
Florian Yger
Reza Bosagh Zadeh
Reza Zafarani
Amelia Zafra
Farida Zehraoui
Kai Zeng
Bernard Zenko
De-Chuan Zhan
Min-Ling Zhang
Indre Zliobaite

Sponsors

We wish to express our gratitude to the sponsors of ECML PKDD 2010 for their essential contribution to the conference: the French National Institute for Research in Computer Science and Control (INRIA), the Pascal2 European Network of Excellence, Nokia, Yahoo! Labs, Google, KNIME, Aster data, Microsoft Research, HP, MODAP (Mobility, Data Mining, and Privacy) a Coordination Action type project funded by EU, FET OPEN, the Data Mining and Knowledge Discovery Journal, the Machine Learning Journal, LRI (Laboratoire de Recherche en Informatique, Université Paris-Sud -CNRS), ARES (Advanced Research on Information Security and Privacy) a national Spanish project, the UNESCO Chair in Data Privacy, Xerox, Universitat Politecnica de Catalunya, IDESCAT (Institut d'Estadistica de Catalunya), and the Ministerio de Ciencia e Innovacion (Spanish government).

Table of Contents – Part II

Regular Papers

Bayesian Knowledge Corroboration with Logical Rules and User Feedback

Gjergji Kasneci, Jurgen Van Gael, Ralf Herbrich, and Thore Graepel

Microsoft Research Cambridge,
7 J J Thomson Avenue, Cambridge CB3 0FB, UK
{gjergjik,v-juvang,rherb,thoreg}@microsoft.com

Abstract. Current knowledge bases suffer from either low coverage or low accuracy. The underlying hypothesis of this work is that user feedback can greatly improve the quality of automatically extracted knowledge bases. The feedback could help quantify the uncertainty associated with the stored statements and would enable mechanisms for searching, ranking and reasoning at entity-relationship level. Most importantly, a principled model for exploiting user feedback to learn the truth values of statements in the knowledge base would be a major step forward in addressing the issue of knowledge base curation.

We present a family of probabilistic graphical models that builds on user feedback and logical inference rules derived from the popular Semantic-Web formalism of RDFS [1]. Through internal inference and belief propagation, these models can learn both, the truth values of the statements in the knowledge base and the reliabilities of the users who give feedback. We demonstrate the viability of our approach in extensive experiments on real-world datasets, with feedback collected from Amazon Mechanical Turk.

Keywords: Knowledge Base, RDFS, User Feedback, Reasoning, Probability, Graphical Model.

1 Introduction

1.1 Motivation

Recent efforts in the area of Semantic Web have given rise to rich triple stores [6,11,14], which are being exploited by the research community [12,13,15,16,17,18]. Appropriately combined with probabilistic reasoning capabilities, they could highly influence the next wave of Web technology. In fact, Semantic-Web-style knowledge bases (KBs) about entities and relationships are already being leveraged by prominent industrial projects [7,8,9].

A widely used Semantic-Web formalism for knowledge representation is the Resource Description Framework Schema (RDFS) [1]. The popularity of this formalism is based on the fact that it provides an extensible, common syntax for data transfer and allows the explicit and intuitive representation of knowledge in form of entity-relationship (ER) graphs. Each edge of an ER graph can be

J.L. Balcázar et al. (Eds.): ECML PKDD 2010, Part II, LNAI 6322, pp. 1–18, 2010.

thought of as an RDF *triple*, and each node as an RDFS *resource*. Furthermore, RDFS provides light-weight reasoning capabilities for inferring new knowledge from the one represented explicitly in the KB.

The triples contained in RDFS KBs are often subject to uncertainty, which may come from different sources:

Extraction & Integration Uncertainty: Usually, the triples are the result of information extraction processes applied to different Web sources. After the extraction, integration processes are responsible for organizing and storing the triples into the KB. The mentioned processes build on uncertain techniques such as natural language processing, pattern matching, statistical learning, etc.

Information Source Uncertainty: There is also uncertainty related to the Web pages from which the knowledge was extracted. Many Web pages may be unauthoritative on specific topics and contain unreliable information. For example, contrary to Michael Jackson's Wikipedia page, the Web site `michaeljacksonsightings.com` claims that Michael Jackson is still alive.

Inherent Knowledge Uncertainty: Another type of uncertainty is the one that is inherent to the knowledge itself. For example, it is difficult to say when the great philosophers Plato or Pythagoras were exactly born. For Plato, Wikipedia offers two possible birth dates 428 BC and 427 BC. These dates are usually estimated by investigating the historical context, which naturally leads to uncertain information.

Leveraging user feedback to deal with the uncertainty and curation of data in knowledge bases is acknowledged as one of the major challenges by the community of probabilistic databases [32]. A principled method for quantifying the uncertainty of knowledge triples would not only build the basis for knowledge curation but would also enable many inference, search and recommendation tasks. Such tasks could aim at retrieving relations between companies, people, prices, product types, etc. For example, the query that asks how Coca Cola, Pepsi and Christina Aguilera are related might yield the result that Christina Aguilera performed in Pepsi as well as in Coca Cola commercials. Since the triples composing the results might have been extracted from blog pages, one has to make sure that they convey reliable information. In full generality, there might be many important (indirect) relations between the query entities, which could be inferred from the underlying data. Quantifying the uncertainty of such associations would help ranking the results in a useful and principled way.

Unfortunately, Semantic-Web formalisms for knowledge representation do not consider uncertainty. As a matter of fact, knowledge representation formalisms and formalisms that can deal with uncertainty are evolving as separate fields of AI. While knowledge representation formalisms (e.g., Description Logics [5], frames [3], KL-ONE [4], RDFS, OWL [2], etc.) focus on expressiveness and borrow from subsets of first-order logics, techniques for representing uncertainty focus on modeling possible world states, and usually represent these by probability distributions. We believe that these two fields belong together and that a targeted effort has to be made to evoke the desired synergy.

1.2 Related Work

Most prior work that has dealt with user feedback, has done so from the viewpoint of user preferences, expertise, or authority (e.g., [34,35,36]). We are mainly interested in the truth values of the statements contained in a knowledge base and in the reliability of users who give feedback. Our goal is to learn these values jointly, that is, we aim to learn from the feedback of multiple users at once.

There are two research areas of AI which provide models for dealing with reasoning over KBs: (1) logical reasoning and (2) probabilistic reasoning. Logical reasoning builds mainly on first-order logic and is best at dealing with relational data. Probabilistic reasoning emphasizes the uncertainty inherent in data.

There have been several proposals for combining techniques from these two areas. In the following, we discuss the strengths and weaknesses of the main approaches.

***Probabilistic Database Model* (PDM).** The PDM [31,32,33] can be viewed as a generalization of the relational model which captures uncertainty with respect to the existence of database tuples (also known as tuple semantics) or to the values of database attributes (also known as attribute semantics).

In the tuple semantics, the main assumption is that the existence of a tuple is independent of the existence of other tuples. Given a database consisting of a single table, the number of possible worlds (i.e. possible databases) is 2^n, where n is the maximum number of the tuples in the table. Each possible world is associated with a probability which can be derived from the existence probabilities of the single tuples and from the independence assumption.

In the attribute semantics, the existence of tuples is certain, whereas the values of attributes are uncertain. Again, the main assumption in this semantics is that the values attributes take are independent of each other. Each attribute is associated with a discrete probability distribution over the possible values it can take. Consequently, the attribute semantics is more expressive than the tuple-level semantics, since in general tuple-level uncertainty can be converted into attribute-level uncertainty by adding one more (Boolean) attribute. Both semantics could also be used in combination, however, the number of possible worlds would be much larger, and deriving complete probabilistic representations would be very costly. So far, there exists no formal semantics for continuous attribute values [32]. Another major disadvantage of PDMs is that they build on rigid and restrictive independence assumptions which cannot easily model correlations among tuples or attributes [26].

***Statistical Relational Learning* (SRL).** SRL models [28] are concerned with domains that exhibit uncertainty and relational structure. They combine a subset of relational calculus (first-order logic) with probabilistic graphical models, such as Bayesian or Markov networks to model uncertainty. These models can capture both, the tuple and the attribute semantics from the PDM and can represent correlations between relational tuples or attributes in a natural way [26].

More ambitious models in this realm are *Markov Logic Networks* [23,24], *Multi-Entity Bayesian Networks* [29] and *Probabilistic Relational Models* [27]. Some of these models (e.g., [23,24,29]) aim at exploiting the whole expressive power of first-order logic. While [23,24] represent the formalism of first-order logic by factor graph models, [27] and [29] deal with Bayesian networks applied to first-order logic. Usually, inference in such models is performed using standard techniques such as belief propagation or Gibbs sampling. In order to avoid complex computations, [22,23,26] propose the technique of *lifted inference*, which avoids materializing all objects in the domain by creating all possible groundings of the logical clauses. Although lifted inference can be more efficient than standard inference on these kinds of models, it is not clear whether they can be trivially lifted (see [25]). Hence, very often these models fall prey to high complexity when applied to practical cases.

More related to our approach is the work by Galland et al. [38], which presents three probabilistic fix-point algorithms for aggregating disagreeing views about knowledge fragments and learning their truth values as well as the trust in the views. However, as admitted by the authors, their algorithms cannot be used in an online fashion, while our approach builds on a Bayesian framework and is inherently flexible to online updates. Furthermore, [38] does not deal with the problem of logical inference, which is a core ingredient of our approach. In our experiments, we show that our approach outperforms all algorithms from [38] on a real-world dataset (provided by the authors of [38]).

Finally, a very recent article [39] proposes a supervised learning approach to the mentioned problem. In contrast to our approach, the solution proposed in [39] is not fully Bayesian and does not deal with logical deduction rules.

1.3 Contributions and Outline

We argue that in many practical cases the full expressiveness of first-order logic is not required. Rather, reasoning models for knowledge bases need to make a trade-off between expressiveness and simplicity. Expressiveness is needed to reflect the domain complexity and allow inference; simplicity is crucial in anticipation of the future scale of Semantic-Web-style data sources [6].

In this paper, we present a Bayesian reasoning framework for inference in triple stores through logical rules and user feedback. The main contributions of this paper are:

- A family of probabilistic graphical models that exploits user feedback to learn the truth values of statements in a KB. As users may often be inconsistent or unreliable and give inaccurate feedback across knowledge domains, our probabilistic graphical models jointly estimate the truth values of statements and the reliabilities of users.
- The proposed model uses logical inference rules based on the proven RDFS formalism to propagate beliefs about truth values from and to derived statements. Consequently, the model can be applied to any RDF triple store.
- We present the superiority of our approach in comparison to prior work on real-world datasets with user feedback from Amazon Mechanical Turk.

In Section 2, we describe an extension of the RDFS formalism, which we refer to as RDFS#. In Section 3, we introduce the mentioned family of probabilistic graphical models on top of the RDFS# formalism. Section 4 is devoted to experimental evaluation and we conclude in Section 5.

2 Knowledge Representation with RDFS

Semantic-Web formalisms for knowledge representation build on the entity-relationship (ER) graph model. ER graphs can be used to describe the knowledge from a domain of discourse in a structured way. Once the elements of discourse (i.e., entities or so-called *resources* in RDFS) are determined, an ER graph can be built. In the following, we give a general definition of ER graphs.

Definition 1 (Entity-Relationship Graph). *Let* Ent *and* $Rel \subseteq Ent$ *be finite sets of entity and relationship labels respectively. An entity-relationship graph over* Ent *and* Rel *is a multigraph* $G = (V, l_{Ent}, E_{Rel})$ *where* V *is a finite set of nodes,* $l_{Ent} : V \rightarrow Ent$ *is an injective vertex labeling function, and* $E_{Rel} \subseteq l_{Ent}(V) \times Rel \times l_{Ent}(V)$ *is a set of labeled edges.*

The labeled nodes of an ER graph represent entities (e.g., people, locations, products, dates, etc.). The labeled edges represent relationship instances, which we refer to as *statements* about entities (e.g., <*AlbertEinstein*, *hasWon*, *NobelPrize*>). Figure 1 depicts a sample ER subgraph from the YAGO knowledge base.

Table 1. Correspondence of ER and RDFS terminology

Element of discourse	ER term	RDFS term
$c \in Ent$	entity	resource
$r \in Rel$	relationship (type)	property
$f \in E_{Rel}$	relationship instance / fact	statement / RDF triple / fact

One of the most prominent Semantic-Web languages for knowledge representation that builds on the concept of ER graphs is the Resource Description Framework Schema (RDFS) [1]. Table 1 shows the correspondence between ER and RDFS terminology.

RDFS is an extensible knowledge representation language recommended by the World Wide Web Consortium (W3C) for the description of a domain of discourse (such as the Web). It enables the definition of domain resources, such as individuals (e.g. *AlbertEinstein, NobelPrize, Germany,* etc.), classes (e.g. *Physicist, Prize, Location,* etc.) and relationships (or so-called properties, e.g. *type, hasWon, locatedIn,* etc.). The basis of RDFS is RDF which comes with three basic symbols: URIs (Uniform Resource Identifiers) for uniquely addressing resources, literals for representing values such as strings, numbers, dates, etc., and blank nodes for representing unknown or unimportant resources.

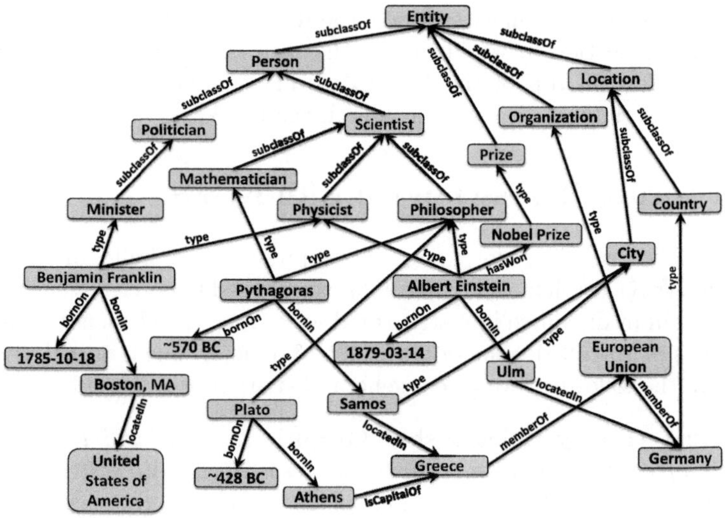

Fig. 1. Sample ER subgraph from the YAGO knowledge base

Another important RDF construct for expressing that two entities stand in a binary relationship is a statement. A statement is a triple of URIs and has the form <*Subject*, *Predicate*, *Object*>, for example <*AlbertEinstein*, *bornIn*, *Ulm*>. An RDF statement can be thought of as an edge from an ER graph, where the *Subject* and the *Object* represent entity nodes and the *Predicate* represents the relationship label of the corresponding edge. Consequently, a set of RDF statements can be viewed as an ER graph. RDFS extends the set of RDF symbols by new URIs for predefined class and relation types such as *rdfs:Resource* (the class of all resources), *rdfs:subClassOf* (for representing the subclass-class relationship), etc.

RDFS is popular because it is a light-weight modeling language with practical logical reasoning capabilities, including reasoning over properties of relationships (e.g., reflexivity, transitivity, domain, and range). However, in the current specification of RDFS, reflexivity and transitivity are defined only for *rdfs:subClassOf*, *rdfs:subPropertyOf*, and the combination of the relationships *rdf:type+rdfs:subClassOf*. The more expressive Web Ontology Language (OWL) [2], which builds on RDFS, allows the above properties to be defined for arbitrary relationships, but its expressive power makes consistency checking undecidable. The recently introduced YAGO model [14] permits the definition of arbitrary acyclic transitive relationships but has the advantage that it still remains decidable. Being able to define transitivity for arbitrary relationships can be a very useful feature for ontological models, since many practically relevant relationships, such as *isA, locatedIn, containedIn, partOf, ancestorOf, siblingOf*, etc., are transitive. Hence, in the following, we will consider a slightly different variant of RDFS.

Let RDFS#[1] denote the RDFS model, in which blank nodes are forbidden and the reasoning capabilities are derived from the following rules. For all $X, Y, Z \in Ent, R, R' \in Rel$ with $X \neq Y, Y \neq Z, X \neq Z, R \neq R'$:

1. $<X, \mathit{type}, Y> \wedge <Y, \mathit{subClassOf}, Z> \rightarrow <X, \mathit{type}, Z>$
2. $<X, R, Y> \wedge <Y, R, Z> \wedge <R, \mathit{type}, \mathit{TransitiveRelation}> \rightarrow <X, R, Z>$
3. $<R, \mathit{subPropertyOf}, R'> \wedge <X, R, Y> \rightarrow <X, R', Y>$
4. $<R, \mathit{hasDomain}, \mathit{Dom}> \wedge <X, R, Y> \rightarrow <X, \mathit{type}, \mathit{Dom}>$
5. $<R, \mathit{hasRange}, \mathit{Ran}> \wedge <X, R, Y> \rightarrow <Y, \mathit{type}, \mathit{Ran}>$

Theorem 1 (Tractability of Inference). *For any* RDFS# *knowledge base $\mathcal{K}$, the set of all statements that can be inferred by applying the inference rules can be computed in polynomial time in the size of $\mathcal{K}$ (i.e., number of statements in $\mathcal{K}$). Furthermore, consistency can be checked in polynomial time.*

The proof of the theorem is a straight-forward extension of the proof of tractability for RDFS entailment, when blank nodes are forbidden [37].

We conclude this section by sketching an algorithm to compute the deductive closure of an RDFS# knowledge base $\mathcal{K}$ with respect to the above rules. Let $\mathcal{F}_\mathcal{K}$ be the set of all statements in $\mathcal{K}$. We recursively identify and index all pairs of statements that can lead to a new statement (according to the above rules) as shown in Algorithm 1.

For each pair of statements (f, f') that imply another statement $\tilde{f}$ according to the RDFS# rules, Algorithm 1 indexes $(f, f', \tilde{f})$. In case $\tilde{f}$ is not present in $\mathcal{F}_\mathcal{K}$ it is added and the algorithm is ran recursively on the updated set $\mathcal{F}_\mathcal{K}$.

Algorithm 1. InferFacts($\mathcal{F}_\mathcal{K}$)

for all pairs $(f, f') \in \mathcal{F}_\mathcal{K} \times \mathcal{F}_\mathcal{K}$ **do**
  if $f \wedge f' \rightarrow \tilde{f}$ and $(f, f', \tilde{f})$ is not indexed **then**
    index $(f, f', \tilde{f})$
    $\mathcal{F}_\mathcal{K} = \mathcal{F}_\mathcal{K} \cup \{\tilde{f}\}$
    InferFacts($\mathcal{F}_\mathcal{K}$)
  end if
end for

3 A Family of Probabilistic Models

Using the language of graphical models, more specifically directed graphical models or Bayesian networks [40], we develop a family of Bayesian models each of which jointly models the truth value for each statement and the reliability for each user. The Bayesian graphical model formalism offers the following advantages:

[1] Read: RDFS sharp.

- Models can be built from existing and tested modules and can be extended in a flexible way.
- The conditional independence assumptions reflected in the model structure enable efficient inference through message passing.
- The hierarchical Bayesian approach integrates data sparsity and traces uncertainty through the model.

We explore four different probabilistic models each incorporating a different body of domain knowledge. Assume we are given an RDFS# KB $\mathcal{K}$. Let $\mathcal{F}_{\mathcal{K}} = \{f_1, ..., f_n\}$ be the set of all statements contained in and deducible from $\mathcal{K}$. For each statement $f_i \in \mathcal{F}_{\mathcal{K}}$ we introduce a random variable $t_i \in \{T, F\}$ to denote its (unknown) truth value. We denote by $y_{ik} \in \{T, F\}$ the random variable that captures the feedback from user k for statement f_i. Let us now explore two different priors on the truth values t_i and two user feedback models connecting for y_{ik}.

3.1 Fact Prior Distributions

Independent Statements Prior. A simple baseline prior assumes independence between the truth values of statements, $t_i \sim \text{Bernoulli}(\alpha_t)$. Thus, for $\mathbf{t} \in \{T, F\}^n$, the conditional probability distribution for the independent statements prior is

$$p(\mathbf{t}|\alpha_t) = \prod_{i=1}^{n} p(t_i|\alpha_t) = \prod_{i=1}^{n} \text{Bernoulli}(t_i; \alpha_t). \tag{1}$$

This strong independence assumption discards existing knowledge about the relationships between statements from RDFS#. This problem is addressed by the *Deduced Statements Prior.*

Deduced Statements Prior. A more complex prior will incorporate the deductions from RDFS# into a probabilistic graphical model. First, we describe a general mechanism to turn a logical deduction into a probabilistic graphical model. Then, we show how this can be used in the context of RDFS#.

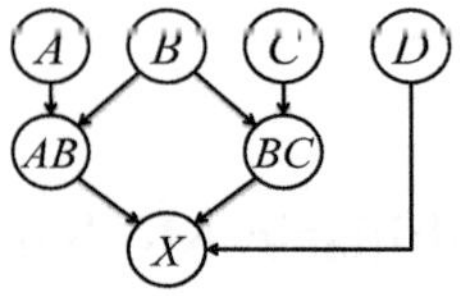

Fig. 2. A graphical model illustrating the logical derivation for the formula $X = (A \wedge B) \vee (B \wedge C) \vee D$

Let X denote a variable that can be derived from $A \wedge B$ or $B \wedge C$, where the premises A, B, and C are known. Let D denote all unknown derivations of X. The truth of X can be expressed in disjunctive normal form: $X = (A \wedge B) \vee (B \wedge C) \vee D$.

This can automatically be turned into the graphical model shown in Figure 2. For each conjuctive clause, a new variable with corresponding conditional probability distribution is introduced, e.g.,

$$p(AB|A,B) = \begin{cases} 1 & \text{if } A \wedge B \\ 0 & \text{otherwise} \end{cases} \tag{2}$$

This simplifies our disjunctive normal form to the expression $X = AB \vee BC \vee D$. Finally, we connect X with all the variables in the disjunctive normal form by a conditional probability:

$$p(X|AB,BC,D) = \begin{cases} 1 & \text{if } AB \vee BC \vee D \\ 0 & \text{otherwise} \end{cases} \tag{3}$$

This construction can be applied to all the deductions implied by RDFS#. After computing the deductive closure of the KB (see Algorithm 1), for each statement $f_i \in \mathcal{F}_\mathcal{K}$, all pairs of statements that imply f_i can be found; we denote this set by $\mathcal{D}_i$. An additional binary variable $\tilde{t}_i \sim \text{Bernoulli}(\alpha_t)$ is introduced to account for the possibility that our knowledge base does not contain *all* possible deductions of statement f_i. The variable $\tilde{t}_i$ is added to the probabilistic graphical model similar to the variable D in the example above. Hence, we derive the following conditional probability distribution for the prior on statements

$$p(\mathbf{t}|\alpha_t) = \prod_{i=1}^{n} \sum_{\tilde{t}_i \in \{T,F\}} p(t_i|\tilde{t}_i, \mathcal{D}_i, \alpha_t) p(\tilde{t}_i|\alpha_t), \tag{4}$$

where Equations (2) and (3) specify the conditional distribution $p(t_i|\tilde{t}_i, \mathcal{D}_i, \alpha_t)$.

3.2 User Feedback Models

The proposed user feedback model jointly models the truth values t_i, the feedback signals y_{ik} and the user reliabilities. In this section we discuss both a one-parameter and a two-parameter per user model for the user feedback component. Note that not all users rate all statements: this means that only a subset of the y_{ik} will be observed.

1-Parameter Model. This model represents the following user behavior. When user k evaluates a statement f_i, with probability u_k he will report the real truth value of f_i and with probability $1 - u_k$ he will report the opposite truth value. Figure 4 represents the conditional probability table for $p(y_{ik}|u_k, t_i)$. Consider the set $\{y_{ik}\}$ of observed true/false feedback labels for the statement-user pairs. The conditional probability distribution for $\mathbf{u} \in [0,1]^m$, $\mathbf{t} \in \{T,F\}^n$ and $\{y_{ik}\}$ in the 1-parameter model is

$$p(\{y_{ik}\}, \mathbf{u}|\mathbf{t}, \alpha_u, \beta_u) = \prod_{\text{users } k} p(u_k|\alpha_u, \beta_u) \prod_{\text{statements } i \text{ by } k} p(y_{ik}|t_i, u_k). \tag{5}$$

2-Parameter Model. This model represents a similar user behavior as above, but this time we model the reliability of each user k with two parameters $u_k \in [0, 1]$ and $\bar{u}_k \in [0, 1]$, one for true statements and one for false statements. Figure 5 represents the conditional probability table for $p(y_{ik}|u_k, \bar{u}_k, t_i)$. The conditional probability distribution for the 2-parameter model is

$$p(\{y_{ik}\}, \mathbf{u}, \bar{\mathbf{u}}|\mathbf{t}, \alpha_u, \beta_u, \alpha_{\bar{u}}, \beta_{\bar{u}}) = \prod_{\text{users } k} p(u_k|\alpha_u, \beta_u) p(\bar{u}_k|\alpha_{\bar{u}}, \beta_{\bar{u}}) \prod_{\text{statements } i \text{ by } k} p(y_{ik}|t_i, u_k, \bar{u}_k). \quad (6)$$

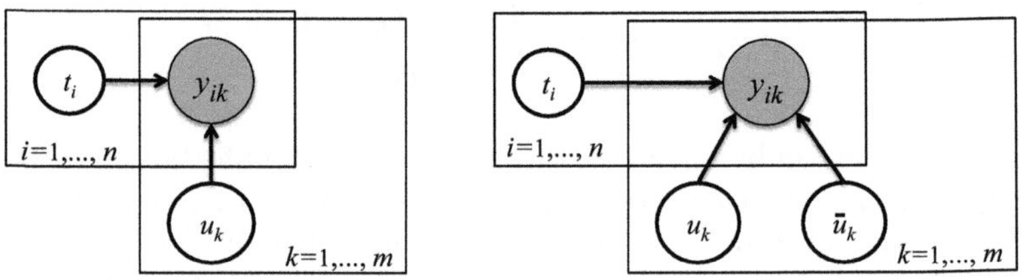

Fig. 3. The graphical models for the user feedback components. Left, the 1-parameter feedback model and right, the 2-parameter feedback model.

y_{ik} \ t_i	T	F
T	u_k	$1 - u_k$
F	$1 - u_k$	u_k

Fig. 4. The conditional probability distribution for feedback signal y_{ik} given reliability u_k and truth t_i

y_{ik} \ t_i	T	F
T	u_k	$1 - \bar{u}_k$
F	$1 - u_k$	$\bar{u}_k$

Fig. 5. The conditional probability distribution for feedback signal y_{ik} given reliabilities $u_k, \bar{u}_k$ and truth t_i

In both models, the prior belief about u_k (and $\bar{u}_k$ in the 2-parameter model) is modeled by a Beta(α_u, β_u) (and Beta$(\alpha_{\bar{u}}, \beta_{\bar{u}})$) distribution, which is a conjugate prior for the Bernoulli distribution.

Table 2 depicts four different models, composed using all four combinations of statement priors and user feedback models. We can write down the full joint probability distribution for the I1 model as

$$p(\mathbf{t}, \{y_{ik}\}, \mathbf{u}|\alpha_t, \alpha_u, \beta_u) = \left(\prod_i^n \text{Bernoulli}(t_i; \alpha_t)\right) \left(\prod_{\text{users } k} p(u_k|\alpha_u, \beta_u) \prod_{\text{statements } i \text{ by } k} p(y_{ik}|t_i, u_k)\right). \quad (7)$$

The joint distribution for I2, D1 and D2 can be written down similarly by combining the appropriate equations above.

Table 2. The four different models

Model Name	Composition
I1	independent priors & 1-parameter feedback model
I2	independent priors & 2-parameter feedback model
D1	deduced statements priors & 1-parameter feedback model
D2	deduced statements priors & 2-parameter feedback model

3.3 Discussion

Figure 6 illustrates how the 1-parameter feedback model, D1, can jointly learn the reliability of two users and the truth values of two statements, f_i and f_j, on which they provide feedback. Additionally, it can also learn the truth value of the statement f_l, which can be derived from $f_i \wedge f_j$. An additional variable $\tilde{t}_l$ is added to account for any deductions which might not be captured by the KB. Note that the model in Figure 6 is loopy but still satisfies the acyclicity required by a directed graphical model.

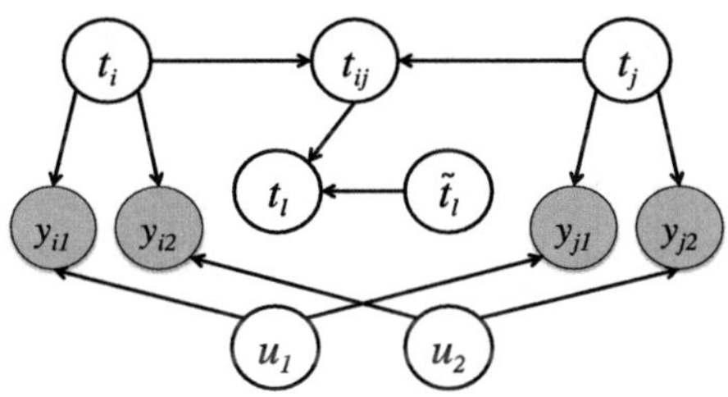

Fig. 6. Illustration of a small instance of the D1 model. Note how user feedback is propagated through the logical relations among the statements.

Given a probabilistic model we are interested in computing the posterior distribution for the statement truth variables and user reliabilities: $p(\mathbf{t}|\{y_{ik}\}, \alpha_t, \alpha_u, \beta_u)$ and $p(\mathbf{u}|\{y_{ik}\}, \alpha_t, \alpha_u, \beta_u)$. Both computations involve summing (or integrating) over all possible assignments for the unobserved variables

$$p(\mathbf{u}|\{y_{ik}\}, \alpha_t, \alpha_u, \beta_u) \propto \sum_{t_1 \in \{T,F\}} \cdots \sum_{t_n \in \{T,F\}} p(\mathbf{t}, \{y_{ik}\}, \mathbf{u}|\alpha_t, \alpha_u, \beta_u). \quad (8)$$

As illustrated in Figure 6, the resulting graphical models are loopy. Moreover deep deduction paths may lead to high treewidth graphical models making exact computation intractable. We chose to use an approximate inference scheme based on message passing known as expectation propagation [30,21].

From a computational perspective, it is easiest to translate the graphical models into factor graphs, and describe the message passing rules over them. Table 3 summarizes how to translate each component of the above graphical

Table 3. Detailed semantics for the graphical models. The first column depicts the Bayesian network dependencies for a component in the graphical model, the second column illustrates the corresponding factor graph, and the third column gives the exact semantics of the factor. The function $\llbracket t \rrbracket$ maps T and F to 1 and 0, respectively.

Bayes Net	Factor Graph	Factor Semantics
	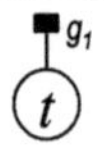	$g_1(t) = \pi^{\llbracket t \rrbracket}(1-\pi)^{(1-\llbracket t \rrbracket)}$
	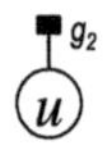	$g_2(u) = \frac{\Gamma(\alpha)\Gamma(\beta)}{\Gamma(\alpha+\beta)} u^{\alpha-1}(1-u)^{\beta-1}$
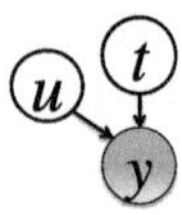	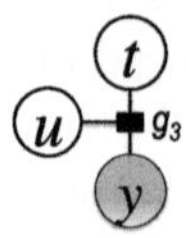	$g_3(y,t,u) = (u^{\llbracket y \rrbracket}(1-u)^{(1-\llbracket y \rrbracket)})^{\llbracket t \rrbracket} \cdot (u^{(1-\llbracket y \rrbracket)}(1-u)^{\llbracket y \rrbracket})^{(1-\llbracket t \rrbracket)}$
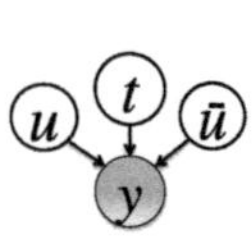	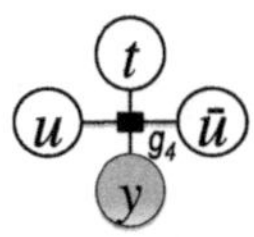	$g_4(y,t,u,\bar{u}) = (u^{\llbracket y \rrbracket}(1-u)^{(1-\llbracket y \rrbracket)})^{\llbracket t \rrbracket} \cdot (\bar{u}^{(1-\llbracket y \rrbracket)}(1-\bar{u})^{\llbracket y \rrbracket})^{(1-\llbracket t \rrbracket)}$
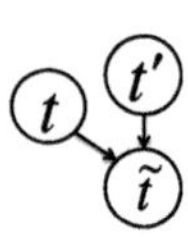	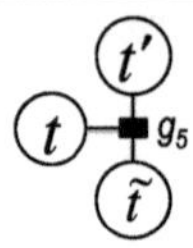	$g_5(t,t',\tilde{t}) = \llbracket t \rrbracket\llbracket t' \rrbracket\llbracket \tilde{t} \rrbracket + (1-\llbracket t \rrbracket)\llbracket t' \rrbracket(1-\llbracket \tilde{t} \rrbracket) + \llbracket t \rrbracket(1-\llbracket t' \rrbracket)(1-\llbracket \tilde{t} \rrbracket) + (1-\llbracket t \rrbracket)(1-\llbracket t' \rrbracket)(1-\llbracket \tilde{t} \rrbracket)$
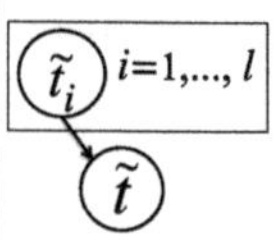	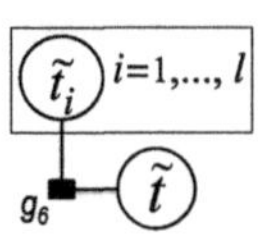	$l = 2 : g_6(\tilde{t}_1,\tilde{t}_2,\tilde{t}) = \llbracket \tilde{t}_1 \rrbracket\llbracket \tilde{t}_2 \rrbracket\llbracket \tilde{t} \rrbracket + (1-\llbracket \tilde{t}_1 \rrbracket)\llbracket \tilde{t}_2 \rrbracket\llbracket \tilde{t} \rrbracket + \llbracket \tilde{t}_1 \rrbracket(1-\llbracket \tilde{t}_2 \rrbracket)\llbracket \tilde{t} \rrbracket + (1-\llbracket \tilde{t}_1 \rrbracket)(1-\llbracket \tilde{t}_2 \rrbracket)(1-\llbracket \tilde{t} \rrbracket)$

models into a factor graph. We rely on Infer.NET [10] to compute a schedule for the message passing algorithms and to execute them. The message passing algorithms run until convergence. The complexity of every iteration is linear in the number of nodes in the underlying factor graph.

4 Experimental Evaluation

For the empirical evaluation we constructed a dataset by choosing a subset of 833 statements about prominent scientists from the YAGO knowledge base [14]. Since the majority of statements in YAGO are correct, we extended the extracted subset by 271 false but semantically meaningful statements[2] that were randomly

[2] E.g., the statement <*AlbertEinstein*, *bornIn*, *Berlin*> is meaningful although false, whereas <*Berlin*, *bornIn*, *AlbertEinstein*> is not semantically meaningful.

generated from YAGO entities and relationships, resulting in a final set of 1,104 statements. The statements from this dataset were manually labeled as true or false, resulting in a total of 803 true statements and 301 false statements.

YAGO provides transitive relationships, such as *locatedIn*, *isA*, *influences*, etc. Hence, we are in the RDFS# setting. We ran Algorithm 1 to compute the closure of our dataset with respect to the transitive relationships. This resulted in 329 pairs of statements from which another statement in the dataset could be derived.

For the above statements we collected feedback from Amazon Mechanical Turk (AMTurk). The users were presented with tasks of at most 5 statements each and asked to label each statement in a task with either true or false. This setup resulted in 221 AMTurk tasks to cover the 1,104 statements in our dataset. Additionally, the users were offered the option to use any external Web sources when assessing a statement. 111 AMTurk users completed between 1 and 186 tasks. For each task we payed 10 US cents. At the end we collected a total number of 11,031 feedback labels.

4.1 Quality Analysis

First we analyze the quality of the four models, I1, I2, D1, D2. As a baseline method we use a "voting" scheme, which computes the probability of a statement f being true as

$$p(f) = \frac{1 + \#\text{ of true votes for } f}{2 + \#\text{ of votes for } f}.$$

We choose the negative log score (in bits) as our accuracy measure. For a statement f_i with posterior p_i the negative log score is defined as

$$\text{nls}(p_i, t_i) := \begin{cases} -\log_2(p_i) & \text{if ground truth for } f_i \text{ is true} \\ -\log_2(1 - p_i) & \text{if ground truth for } f_i \text{ is false} \end{cases} \tag{9}$$

The negative log score represents how much information in the ground truth is captured by the posterior; when $p_i = [\![t_i]\!]$ the log score is zero. To illustrate the learning rate of each model, in Figure 7 we show aggregate negative log scores for nested subsets of the feedback labels. For each of the subsets, we use all 1,104 statements of the dataset.

Figure 7 shows that for smaller subsets of feedback labels the simpler models perform better and have lower negative log scores. However, as the number of labels increases, the two-parameter models become more accurate. This is in line with the intuition that simpler (i.e., one-parameter) models learn quicker (i.e., with fewer labels). Nonetheless, observe that with more labels, the more flexible (i.e., 2-parameter) models achieve lower negative log scores. Finally, the logical inference rules reduce the negative log scores by about 50 bits when there are no labels. Nevertheless, when the amount of labels grows, the logical inference rules hardly contribute to the decrease in negative log score. All models consistently outperform the voting approach.

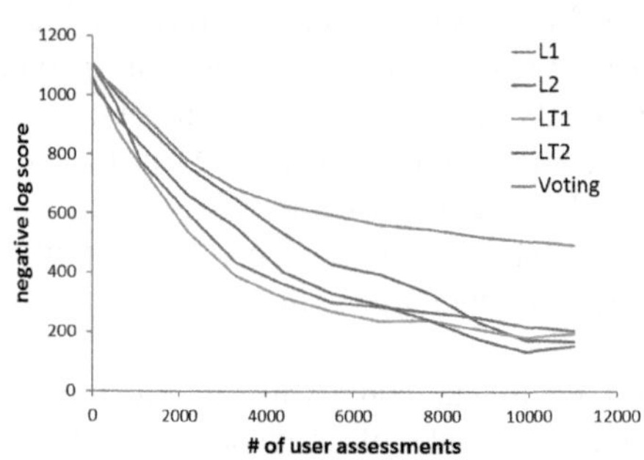

Fig. 7. The negative log score for the different models as a function of the number of user assessments

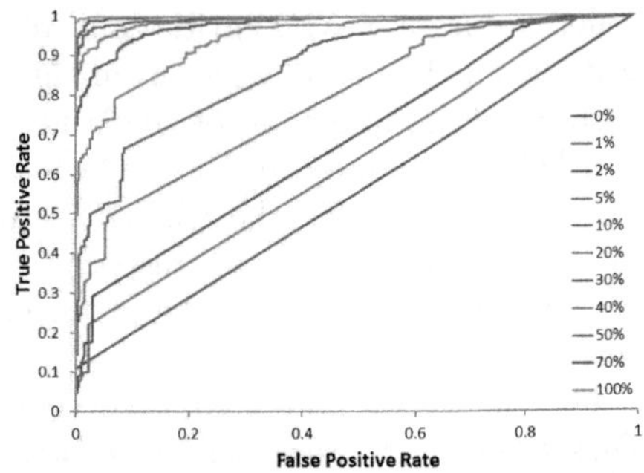

Fig. 8. The ROC curves for the D1 model, for varying numbers of user assessments

We computed ROC curves for model D1 for different nested subsets of the data. In Figure 8, when all labels are used, the ROC curve shows almost perfect true positive and false positive behavior. The model already performs with high accuracy for 30% of the feedback labels. Also, we get a consistent increase in AUC as we increase the number of feedback signals.

4.2 Case Studies

Our probabilistic models have another big advantage: the posterior probabilities for truths and reliabilities have clear semantics. By inspecting them we can discover different types of user behavior.

When analyzing the posterior probabilities for the D2 model, we found that the reliability of one of the users was 89% when statements were true, while it was only 8% when statements were false. When we inspected the labels that were generated by the user we found that he labelled 768 statements, out of which 693 statements were labelled as "true". This means that he labelled 90% of all statements that were presented to him as "true", whereas in our dataset only about 72% of all statements are true. Our model suggests that it is more likely that this user was consciously labelling almost all statements as true. Similarly we found users who almost always answered "false" to the statements that were presented. In Figure 9, the scatter plot for the mean values of u and $\bar{u}$ across all users gives evidence for the existence of such a biased behavior. The points in the lower-right and in the upper-left part of the plot represent users who report statements mainly as true and false, respectively.

Interestingly enough, we did not find any users who were consistently reporting the opposite truth values compared to their peers. We would have been able to discover this type of behavior by the D2 model. In such a case, a good indication would be reliabilities below 50%.

The previous analysis also hints at an important assumption of our model: only because most of our users are providing correct feedback, it is impossible for

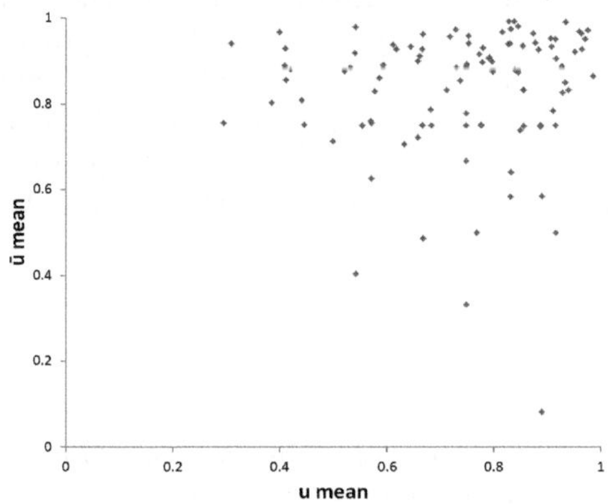

Fig. 9. Scatter plot of u versus $\bar{u}$ for the D2 model. Each dot represents a user.

malicious behavior to go undetected. If enough reliable users are wrong about a statement, our model can converge on the wrong belief.

4.3 Comparison

In addition, we evaluated the D1 model on another real-world dataset that was also used by the very recent approach presented in [38]. The authors of [38] present three fixed-point algorithms for learning truth values of statements by aggregating user feedback. They report results on various datasets one of which is a sixth-grade biology test dataset. This test consists of 15 yes-no questions which can be viewed as statements in our setting. The test was taken by 86 participants who gave a total of 1,290 answers, which we interpret as feedback labels. For all algorithms presented in [38], the authors state that they perform similarly to the voting baseline. The voting baseline yields a negative log score of 8.5, whereas the D1 model yields a much better negative log score of $3.04e-5$.

5 Conclusion

We presented a Bayesian approach to the problem of knowledge corroboration with user feedback and semantic rules. The strength of our solution lies in its capability to jointly learn the truth values of statements and the reliabilities of users, based on logical rules and internal belief propagation. We are currently investigating its application to large-scale knowledge bases with hundreds of millions of statements or more. Along this path, we are looking into more complex logical rules and more advanced user and statement features to learn about the background knowledge of users and the difficulty of statements. Finally, we are exploring active learning strategies to optimally leverage user feedback in an online fashion.

In recent years, we have witnessed an increasing involvement of users in annotation, labeling, and other knowledge creation tasks. At the same time,

Semantic Web technologies are giving rise to large knowledge bases that could facilitate automatic knowledge processing. The approach presented in this paper aims to transparently evoke the desired synergy from these two powerful trends, by laying the foundations for complex knowledge curation, search and recommendation tasks. We hope that this work will appeal to and further benefit from various research communities such as AI, Semantic Web, Social Web, and many more.

Acknowledgments

We thank the Infer.NET team, John Guiver, Tom Minka, and John Winn for their consistent support throughout this project.

References

1. W3C RDF: Vocabulary Description Language 1.0: RDF Schema, `http://www.w3.org/TR/rdf-schema/`
2. W3C: OWL Web Ontology Language, `http://www.w3.org/TR/owl-features/`
3. Minsky, M.: A Framework for Representing Knowledge. MIT-AI Laboratory Memo 306 (1974), `http://web.media.mit.edu/~minsky/papers/Frames/frames.html`
4. Brachman, R.J., Schmolze, J.: An Overview of the KL-ONE Knowledge Representation System. Cognitive Science 9(2) (1985)
5. Baader, F., Calvanese, D., McGuinness, D.L., Nardi, D., Patel-Schneider, P.F.: The Description Logic Handbook. Cambridge University Press, Cambridge (2003)
6. W3C SweoIG: The Linking Open Data Community Project, `http://esw.w3.org/topic/SweoIG/TaskForces/CommunityProjects/LinkingOpenData`
7. Wolfram Alpha: A Computational Knowledge Engine, `http://www.wolframalpha.com/`
8. EntityCube, `http://entitycube.research.microsoft.com/`
9. True Knowledge, `http://www.trueknowledge.com/`
10. Infer.NET, `http://research.microsoft.com/en-us/um/cambridge/projects/infernet/`
11. Bizer, C., Lehmann, J., Kobilarov, G., Auer, S., Becker, C., Cyganiak, R., Hellmann, S.: DBpedia: A Nucleus for a Web of Open Data. In: Aberer, K., Choi, K.-S., Noy, N., Allemang, D., Lee, K.-I., Nixon, L.J.B., Golbeck, J., Mika, P., Maynard, D., Mizoguchi, R., Schreiber, G., Cudré-Mauroux, P. (eds.) ASWC 2007 and ISWC 2007. LNCS, vol. 4825, pp. 722–735. Springer, Heidelberg (2007)
12. Lehmann, J., Schüppel, J., Auer, S.: Discovering Unknown Connections - The DBpedia Relationship Finder. In: 1st Conference on Social Semantic Web (CSSW 2007) pp. 99–110. GI (2007)
13. Suchanek, F.M., Sozio, M., Weikum, G.: SOFIE: Self-Organizing Flexible Information Extraction. In: 18th International World Wide Web conference (WWW 2009), pp. 631–640. ACM Press, New York (2009)
14. Suchanek, F.M., Kasneci, G., Weikum, G.: Yago: A Core of Semantic Knowledge. In: 16th International World Wide Web Conference (WWW 2007), pp. 697–706. ACM Press, New York (2007)

15. Kasneci, G., Suchanek, F.M., Ifrim, G., Ramanath, M., Weikum, G.: NAGA: Searching and Ranking Knowledge. In: 24th International Conference on Data Engineering (ICDE 2008), pp. 953–962. IEEE, Los Alamitos (2008)
16. Kasneci, G., Ramanath, M., Sozio, M., Suchanek, F.M., Weikum, G.: STAR: Steiner-Tree Approximation in Relationship Graphs. In: 25th International Conference on Data Engineering (ICDE 2009), pp. 868–879. IEEE, Los Alamitos (2009)
17. Kasneci, G., Shady, E., Weikum, G.: MING: Mining Informative Entity Relationship Subgraphs. In: 18th ACM Conference on Information and Knowledge Management (CIKM 2009), pp. 1653–1656. ACM Press, New York (2009)
18. Preda, N., Kasneci, G., Suchanek, F.M., Yuan, W., Neumann, T., Weikum, G.: Active Knowledge: Dynamically Enriching RDF Knowledge Bases by Web Services. In: 30th ACM International Conference on Management Of Data (SIGMOD 2010). ACM Press, New York (2010)
19. Wu, F., Weld, D.S.: Autonomously Semantifying Wikipedia. In: 16th ACM Conference on Information and Knowledge Management (CIKM 2007), pp. 41–50. ACM Press, New York (2007)
20. Weld, D.S., Wu, F., Adar, E., Amershi, S., Fogarty, J., Hoffmann, R., Patel, K., Skinner, M.: Intelligence in Wikipedia. In: 23rd AAAI Conference on Artificial Intelligence (AAAI 2008), pp. 1609–1614. AAAI Press, Menlo Park (2008)
21. Minka, T.P.: A Family of Algorithms for Approximate Bayesian Inference. Massachusetts Institute of Technology (2001)
22. Poole, D.: First-Order Probabilistic Inference. In: 8th International Joint Conference on Artificial Intelligence (IJCAI 2003), pp. 985–991. Morgan Kaufmann, San Francisco (2003)
23. Domingos, P., Singla, P.: Lifted First-Order Belief Propagation. In: 23rd AAAI Conference on Artificial Intelligence (AAAI 2008), pp. 1094–1099. AAAI Press, Menlo Park (2008)
24. Domingos, P., Richardson, M.: Markov Logic Networks. Machine Learning 62(1-2), 107–136 (2006)
25. Jaimovich, A., Meshi, O., Friedman, N.: Template Based Inference in Symmetric Relational Markov Random Fields. In: 23rd Conference on Uncertainty in Artificial Intelligence (UAI 2007), pp. 191–199. AUAI Press (2007)
26. Sen, P., Deshpande, A., Getoor, L.: PrDB: Managing and Exploiting Rich Correlations in Probabilistic Databases. Journal of Very Large Databases 18(5), 1065–1090 (2009)
27. Friedman, N., Getoor, L., Koller, D., Pfeffer, A.: Learning Probabilistic Relational Models. In: 16th International Joint Conference on Artificial Intelligence (IJCAI 1999), pp. 1300–1309. Morgan Kaufmann, San Francisco (1999)
28. Getoor, L.: Tutorial on Statistical Relational Learning. In: Kramer, S., Pfahringer, B. (eds.) ILP 2005. LNCS (LNAI), vol. 3625, pp. 415–415. Springer, Heidelberg (2005)
29. Da Costa, P.C.G., Ladeira, M., Carvalho, R.N., Laskey, K.B., Santos, L.L., Matsumoto, S.: A First-Order Bayesian Tool for Probabilistic Ontologies. In: 21st International Florida Artificial Intelligence Research Society Conference (FLAIRS 2008), pp. 631–636. AAAI Press, Menlo Park (2008)
30. Frey, B.J., Mackay, D.J.C.: A Revolution: Belief Propagation in Graphs with Cycles. In: Advances in Neural Information Processing Systems, vol. 10, pp. 479–485. MIT Press, Cambridge (1997)

31. Antova, L., Koch, C., Olteanu, D.: 10^{10^6} Worlds and Beyond: Efficient Representation and Processing of Incomplete Information. In: 23rd International Conference on Data Engineering (ICDE 2007), pp. 606–615. IEEE, Los Alamitos (2007)
32. Dalvi, N.N., Ré, C., Suciu, D.: Probabilistic Databases: Diamonds in the Dirt. Communications of ACM (CACM 2009) 52(7), 86–94 (2009)
33. Agrawal, P., Benjelloun, O., Sarma, A.D., Hayworth, C., Nabar, S.U., Sugihara, T., Widom, J.: Trio: A System for Data, Uncertainty, and Lineage. In: 32nd International Conference on Very Large Data Bases (VLDB 2006), pp. 1151–1154. ACM Press, New York (2006)
34. Osherson, D., Vardi, M.Y.: Aggregating Disparate Estimates of Chance. Games and Economic Behavior 56(1), 148–173 (2006)
35. Jøsang, A., Marsh, S., Pope, S.: Exploring Different Types of Trust Propagation. In: Stølen, K., Winsborough, W.H., Martinelli, F., Massacci, F. (eds.) iTrust 2006. LNCS, vol. 3986, pp. 179–192. Springer, Heidelberg (2006)
36. Kelly, D., Teevan, J.: Implicit Feedback for Inferring User Preference: A Bibliography. SIGIR Forum 37(2), 18–28 (2003)
37. Horst, H.J.T.: Completeness, Decidability and Complexity of Entailment for RDF Schema and a Semantic Extension Involving the OWL Vocabulary. Journal of Web Semantics: Science, Services and Agents on the World Wide Web 3(2-3), 79–115 (2005)
38. Galland, A., Abiteboul, S., Marian, A., Senellart, P.: Corroborating Information from Disagreeing Views. In: 3rd ACM International Conference on Web Search and Data Mining (WSDM 2010), pp. 1041–1064. ACM Press, New York (2010)
39. Raykar, V.C., Yu, S., Zhao, L.H., Valadez, G.H., Florin, C., Bogoni, L., Moy, L.: Learning From Crowds. Journal of Machine Learning Research 11, 1297–1322 (2010)
40. Pearl, J.: Probabilistic Reasoning in Intelligent Systems: Networks of Plausible Inference. Morgan Kaufmann, San Francisco (1997)

Learning an Affine Transformation for Non-linear Dimensionality Reduction

Pooyan Khajehpour Tadavani[1] and Ali Ghodsi[1,2]

[1] David R. Cheriton School of Computer Science
[2] Department of Statistics and Actuarial Science
University of Waterloo
Waterloo, Ontario, Canada

Abstract. The foremost nonlinear dimensionality reduction algorithms provide an embedding only for the given training data, with no straightforward extension for test points. This shortcoming makes them unsuitable for problems such as classification and regression. We propose a novel dimensionality reduction algorithm which learns a parametric mapping between the high-dimensional space and the embedded space. The key observation is that when the dimensionality of the data exceeds its quantity, it is always possible to find a linear transformation that preserves a given subset of distances, while changing the distances of another subset. Our method first maps the points into a high-dimensional feature space, and then explicitly searches for an affine transformation that preserves local distances while pulling non-neighbor points as far apart as possible. This search is formulated as an instance of semi-definite programming, and the resulting transformation can be used to map out-of-sample points into the embedded space.

Keywords: Machine Learning, Dimensionality Reduction, Data Mining.

1 Introduction

Manifold discovery is an important form of data analysis in a wide variety of fields, including pattern recognition, data compression, machine learning, and database navigation. In many problems, input data consists of high-dimensional observations, where there is reason to believe that the data lies on or near a low-dimensional manifold. In other words, multiple measurements forming a high-dimensional data vector are typically indirect measurements of a single underlying source. Learning a suitable low-dimensional manifold from high-dimensional data is essentially the task of learning a model to represent the underlying source. This type of dimensionality reduction[1] can also be seen as the process of deriving a set of degrees of freedom which can be used to reproduce most of the variability of the data set.

[1] In this paper the terms 'manifold learning' and 'dimensionality reduction' are used interchangeably.

J.L. Balcázar et al. (Eds.): ECML PKDD 2010, Part II, LNAI 6322, pp. 19–34, 2010.

Several algorithms for dimensionality reduction have been developed based on eigen-decomposition. Principal components analysis (PCA) [4] is a classical method that provides a sequence of the best linear approximations to the given high-dimensional observations. Another classical method is multidimensional scaling (MDS) [2], which is closely related to PCA. Both of these methods estimate a linear transformation from the training data that projects the high-dimensional data points to a low-dimensional subspace. This transformation can then be used to embed a new test point into the same subspace, and consequently, PCA and MDS can easily handle out-of-sample examples.

The effectiveness of PCA and MDS is limited by the linearity of the subspace they reveal. In order to resolve the problem of dimensionality reduction in nonlinear cases, many nonlinear techniques have been proposed, including kernel PCA (KPCA) [5], locally linear embedding (LLE) [6], Laplacian Eigenmaps [1], and Isomap [12]. It has been shown that all of these algorithms can be formulated as KPCA [3]. The difference lies mainly in the choice of kernel. Common kernels such as RBF and polynomial kernels generally perform poorly in manifold learning, which is perhaps what motivated the development of algorithms such as LLE and Isomap.

The problem of choosing an appropriate kernel remained crucial until more recently, when a number of authors [9,10,11,13,14] have cast the manifold learning problem as an instance of semi-definite programming (SDP). These algorithms usually provide a faithful embedding for a given training data; however, they have no straightforward extension for test points[2]. This shortcoming makes them unsuitable for supervised problems such as classification and regression.

In this paper we propose a novel nonlinear dimensionality reduction algorithm: *Embedding by Affine Transformation (EAT).* The proposed method learns a parametric mapping between the high-dimensional space and the embedding space, which unfolds the manifold of data while preserving its local structure. An intuitive explanation of the method is outlined in Section 2. Section 3 presents the details of the algorithm followed by experimental results in Section 4.

2 The Key Intuition

Kernel PCA first implicitly projects its input data into a high-dimensional feature space, and then performs PCA in that feature space. PCA provides a linear and distance preserving transformation - i.e., all of the pairwise distances between the points in the feature space will be preserved in the embedded space. In this way, KPCA relies on the strength of kernel to unfold a given manifold and reveal its underlying structure in the feature space.

We present a method that, similar to KPCA, maps the input points into a high-dimensional feature space. The similarity ends here, however, as we explicitly search for an affine transformation in the feature space that preserves only the local distances while pulling the non-neighbor points as far apart as possible.

[2] An exception is kernel PCA with closed-form kernels; however, closed-form kernels generally have poor performance even on the training data.

In KPCA, the choice of kernel is crucial, as it is assumed that mapping the data into a high-dimensional feature space can flatten the manifold; if this assumption is not true, its low-dimensional mapping will not be a faithful representation of the manifold. On the contrary, our proposed method does not expect the kernel to reveal the underlying structure of the data. The kernel simply helps us make use of "the blessing of dimensionality". That is, when the dimensionality of the data exceeds its quantity, a linear transformation can span the whole space. This means we can find a linear transformation that preserves the distances between the neighbors and also pulls the non-neighbors apart; thus flattening the manifold. This intuition is depicted in Fig. 1.

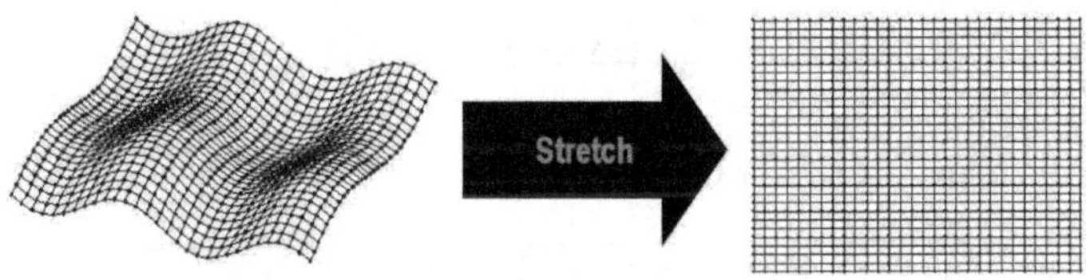

Fig. 1. A simple 2D manifold is represented in a higher-dimensional space. Stretching out the manifold allows it to be correctly embedded in a 2D space.

3 Embedding by Affine Transformation (EAT)

We would like to learn an Affine transformation that preserves the distances between neighboring points, while pulling the non-neighbor points as far apart as possible. In a high dimensional space, this transformation looks locally like a rotation plus a translation which leads to a local isometry; however, for non-neighbor points, it acts as a scaling.

Consider a training data set of n d-dimensional points $\{\mathbf{x}_i\}_{i=1}^n \subset \mathbb{R}^d$. We wish to learn a $d \times d$ transformation matrix $\mathbf{W}$ which will unfold the underlying manifold of the original data points, and embed them into $\{\mathbf{y}_i\}_{i=1}^n$ by:

$$\mathbf{y} = \mathbf{W}^\mathsf{T}\mathbf{x} \tag{1}$$

This mapping will not change the dimensionality of the data; $\mathbf{y}$ has the same dimensionality as $\mathbf{x}$. Rather, the goal is to learn $\mathbf{W}$ such that it preserves the local structure but stretches the distances between the non-local pairs. After this projection, the projected data points $\{\mathbf{y}_i\}_{i=1}^n$ will hopefully lie on or close to a linear subspace. Therefore, in order to reduce the dimensionality of $\{\mathbf{y}_i\}_{i=1}^n$, one may simply apply PCA to obtain the ortho-normal axes of the linear subspace. In order to learn $\mathbf{W}$, we must first define two disjoint sets of pairs:

$$\mathcal{S} = \{(i,j)|\ \mathbf{x}_i \text{ and } \mathbf{x}_j \text{ are neighbors}\}$$
$$\mathcal{O} = \{(i,j)|\ \mathbf{x}_i \text{ and } \mathbf{x}_j \text{ are non-neighbors}\}$$

The first set consists of pairs of neighbor points in the original space, for which their pairwise distances should be preserved. These pairs can be identified, for example, by computing a neighborhood graph using the K-Nearest Neighbor (KNN) algorithm. The second set is the set of pairs of non-neighbor points, which we would like to pull as far apart as possible. This set can simply include all of the pairs that are not in the first set.

3.1 Preserving Local Distances

Assume for all (i,j) in a given set $\mathcal{S}$, the target distances are known as τ_{ij}. we specify the following cost function, which attempts to preserve the known squared distances:

$$\sum_{(i,j)\in\mathcal{S}} \left(\|\mathbf{y}_i - \mathbf{y}_j\|^2 - \tau_{ij}^2\right)^2 \tag{2}$$

Then we normalize it to obtain[3]:

$$Err = \sum_{(i,j)\in\mathcal{S}} \left(\|\frac{\mathbf{y}_i - \mathbf{y}_j}{\tau_{ij}}\|^2 - 1\right)^2 \tag{3}$$

By substituting (1) into (3), we have:

$$\sum_{(i,j)\in\mathcal{S}} \left(\left(\frac{\mathbf{x}_i - \mathbf{x}_j}{\tau_{ij}}\right)^\top \mathbf{W}\mathbf{W}^\top \left(\frac{\mathbf{x}_i - \mathbf{x}_j}{\tau_{ij}}\right) - 1\right)^2 = \sum_{(i,j)\in\mathcal{S}} \left(\boldsymbol{\delta}_{ij}^\top \mathbf{A}\boldsymbol{\delta}_{ij} - 1\right)^2 \tag{4}$$

where $\boldsymbol{\delta}_{ij} = \frac{\mathbf{x}_i - \mathbf{x}_j}{\tau_{ij}}$ and $\mathbf{A} = \mathbf{W}\mathbf{W}^\top$ is a positive semidefinite (PSD) matrix. It can be verified that:

$$\boldsymbol{\delta}_{ij}^\top \mathbf{A}\boldsymbol{\delta}_{ij} = vec(\mathbf{A})^\top vec(\boldsymbol{\delta}_{ij}\boldsymbol{\delta}_{ij}^\top) = vec(\mathbf{A})^\top vec(\boldsymbol{\Delta}_{ij}) \tag{5}$$

where $vec()$ simply rearranges a matrix into a vector by concatenating its columns, and $\boldsymbol{\Delta}_{ij} = \boldsymbol{\delta}_{ij}\boldsymbol{\delta}_{ij}^\top$. For a symmetric matrix $\mathbf{A}$ we know:

$$vec(\mathbf{A}) = \mathbf{D}_d vech(\mathbf{A}) \tag{6}$$

where $vech(\mathbf{A})$ is the half-vectorization operator, and $\mathbf{D}_d$ is the unique $d^2 \times \frac{d(d+1)}{2}$ duplication matrix. Similar to the $vec()$ operator, the half-vectorization operator rearranges a matrix into a vector by concatenating its columns; however, it stacks the columns from the principal diagonal downwards in a column vector. In other words, a symmetric matrix of size d will be rearranged to a column vector of size d^2 by the $vec()$ operator, whereas $vech()$ will stack it into a column vector

[3] (2) corresponds to the assumption that noise is additive while (3) captures a multiplicative error, i.e. if $\|\mathbf{y}_i - \mathbf{y}_j\|^2 = \tau_{ij}^2 + \varepsilon_{ij}$, where ε_{ij} is additive noise, then clearly $\varepsilon_{ij} = \left(\|\mathbf{y}_i - \mathbf{y}_j\|^2 - \tau_{ij}^2\right)^2$; however, if $\|\mathbf{y}_i - \mathbf{y}_j\|^2 = \tau_{ij}^2 + \varepsilon_{ij} \times \tau_{ij}^2$, then $\varepsilon_{ij} = \left(\|\frac{\mathbf{y}_i - \mathbf{y}_j}{\tau_{ij}}\|^2 - 1\right)^2$. The latter one makes the summation terms comparable.

of size $\frac{d(d+1)}{2}$. This can significantly reduce the number of unknown variables, especially when d is large. D_d is a unique constant matrix. For example, for a 2×2 symmetric matrix $\mathbf{A}$ we have:

$$vec(\mathbf{A}) = \begin{bmatrix} 1\,0\,0 \\ 0\,1\,0 \\ 0\,1\,0 \\ 0\,0\,1 \end{bmatrix} vech(\mathbf{A})$$

Since both $\mathbf{A}$ and $\boldsymbol{\Delta}_{ij}$ are symmetric matrices, we can rewrite (5) using $vech()$ and reduce the size of the problem:

$$\boldsymbol{\delta}_{ij}^{\mathsf{T}} \mathbf{A} \boldsymbol{\delta}_{ij} = vech(\mathbf{A})^{\mathsf{T}} \mathbf{D}_d^{\mathsf{T}} \mathbf{D}_d vech(\boldsymbol{\Delta}_{ij}) = vech(\mathbf{A})^{\mathsf{T}} \boldsymbol{\xi}_{ij} \quad (7)$$

where $\boldsymbol{\xi}_{ij} = \mathbf{D}_d^{\mathsf{T}} \mathbf{D}_d vech(\boldsymbol{\Delta}_{ij})$. Using (7), we can reformulate (4) as:

$$Err = \sum_{(i,j)\in\mathcal{S}} (vech(\mathbf{A})^{\mathsf{T}} \boldsymbol{\xi}_{ij} - 1)^2 = vech(\mathbf{A})^{\mathsf{T}} \mathbf{Q} vech(\mathbf{A}) - 2vech(\mathbf{A})^{\mathsf{T}} \mathbf{p} + |\mathcal{S}| \quad (8)$$

where $\mathbf{Q} = \sum_{(i,j)\in\mathcal{S}} \boldsymbol{\xi}_{ij} \boldsymbol{\xi}_{ij}^{\mathsf{T}}$ and $\mathbf{p} = \sum_{(i,j)\in\mathcal{S}} \boldsymbol{\xi}_{ij}$, and $|\mathcal{S}|$ in (8) denotes the number of elements in $\mathcal{S}$, which is constant and can be dropped from the optimization. Now, we can decompose the matrix $\mathbf{Q}$ using the singular value decomposition technique to obtain:

$$\mathbf{Q} = \mathbf{U} \boldsymbol{\Lambda} \mathbf{U}^{\mathsf{T}} \quad (9)$$

If $rank(\mathbf{Q}) = r$, then $\mathbf{U}$ is a $\frac{d(d+1)}{2} \times r$ matrix with r orthonormal basis vectors. We denote the null space of $\mathbf{Q}$ by $\overline{\mathbf{U}}$. Any vector of size $\frac{d(d+1)}{2}$, including vector $vech(\mathbf{A})$, can be represented using the space and the null space of $\mathbf{Q}$:

$$vech(\mathbf{A}) = \mathbf{U}\boldsymbol{\alpha} + \overline{\mathbf{U}}\boldsymbol{\beta} \quad (10)$$

$\boldsymbol{\alpha}$ and $\boldsymbol{\beta}$ are vectors of size r and $\left(\frac{d(d+1)}{2} - r\right)$ respectively. Since $\mathbf{Q}$ is the summation of $\boldsymbol{\xi}_{ij}\boldsymbol{\xi}_{ij}^{\mathsf{T}}$, and $\mathbf{p}$ is the summation of $\boldsymbol{\xi}_{ij}$, it is easy to verify that $\mathbf{p}$ is in the space of $\mathbf{Q}$ and therefore $\overline{\mathbf{U}}^{\mathsf{T}}\mathbf{p} = 0$. Substituting (9) and (10) in (8), the objective function can be expressed as:

$$vech(\mathbf{A})^{\mathsf{T}} \left(\mathbf{Q} vech(\mathbf{A}) - 2\mathbf{p}\right) = \boldsymbol{\alpha}^{\mathsf{T}} \left(\boldsymbol{\Lambda}\boldsymbol{\alpha} - 2\mathbf{U}^{\mathsf{T}}\mathbf{p}\right) \quad (11)$$

The only unknown variable in this equation is $\boldsymbol{\alpha}$. Hence, (11) can be solved in closed form to obtain:

$$\boldsymbol{\alpha} = \boldsymbol{\Lambda}^{-1}\mathbf{U}^{\mathsf{T}}\mathbf{p} \quad (12)$$

Interestingly, (11) does not depend on $\boldsymbol{\beta}$. This means that the transformation $\mathbf{A}$ which preserves the distances in $\mathcal{S}$ (local distances) is not unique. In fact, there is a family of transformations in the form of (10) that preserve local distances for any value of $\boldsymbol{\beta}$. In this family, we can search for the one that is both positive semi-definite, and increases the distances of the pairs in the set $\mathcal{O}$ as much as possible. The next section shows how the freedom of vector $\boldsymbol{\beta}$ can be exploited to search for a transformation that satisfies these conditions.

3.2 Stretching the Non-local Distances

We define the following objective function which, when optimized, attempts to maximize the squared distances between the non-neighbor points. That is, it attempts to maximize the squared distances between $\mathbf{x}_i$ and $\mathbf{x}_j$ if $(i,j) \in \mathcal{O}$.

$$Str = \sum_{(i,j)\in\mathcal{O}} \frac{||\mathbf{y}_i - \mathbf{y}_j||^2}{\tau_{ij}^2} \tag{13}$$

Similar to the cost function *Err* in the previous section, we have:

$$\begin{aligned} Str &= \sum_{(i,j)\in\mathcal{O}} \left(\left(\frac{\mathbf{x}_i - \mathbf{x}_j}{\tau_{ij}} \right)^{\mathsf{T}} \mathbf{W}\mathbf{W}^{\mathsf{T}} \left(\frac{\mathbf{x}_i - \mathbf{x}_j}{\tau_{ij}} \right) \right) \\ &= \sum_{(i,j)\in\mathcal{O}} \boldsymbol{\delta}_{ij}^{\mathsf{T}} \mathbf{A} \boldsymbol{\delta}_{ij} = \sum_{(i,j)\in\mathcal{O}} vech(\mathbf{A})^{\mathsf{T}} \boldsymbol{\xi}_{ij} = vech(\mathbf{A})^{\mathsf{T}} \mathbf{s} \end{aligned} \tag{14}$$

where $\mathbf{s} = \sum_{(i,j)\in\mathcal{O}} \boldsymbol{\xi}_{ij}$. Then, the optimization problem is:

$$\max_{\mathbf{A}\succeq 0} \quad vech(\mathbf{A})^{\mathsf{T}} \mathbf{s} \tag{15}$$

Recall that $vech(\mathbf{A}) = \mathbf{U}\boldsymbol{\alpha} + \overline{\mathbf{U}}\boldsymbol{\beta}$, and $\boldsymbol{\alpha}$ is already determined from (12). So the problem can be simplified as:

$$\max_{\mathbf{A}\succeq 0} \quad \boldsymbol{\beta}^{\mathsf{T}} \overline{\mathbf{U}}^{\mathsf{T}} \mathbf{s} \tag{16}$$

Clearly if $\mathbf{Q}$ is full rank, then the matrix $\overline{\mathbf{U}}$ (i.e. the null space of $\mathbf{Q}$) does not exist and therefore, it is not possible to stretch the non-local distances. However, it can be shown that if the dimensionality of the data is more than its quantity, $\mathbf{Q}$ is always rank deficient, and $\overline{\mathbf{U}}$ exists. The rank of $\mathbf{Q}$ is at most $|\mathcal{S}|$, which is due to the fact that $\mathbf{Q}$ is defined in (8) as a summation of $|\mathcal{S}|$ rank-one matrices. Clearly, the maximum of $|\mathcal{S}|$ is the maximum possible number of pairs i.e. $\frac{n\times(n-1)}{2}$; however the size of $\mathbf{Q}$ is $\frac{d\times(d+1)}{2}$.

$\mathbf{Q}$ is rank deficient when $d \geq n$. To make sure that $\mathbf{Q}$ is rank deficient, one can project the points into a high-dimensional space, by some mapping $\phi()$; however, performing the mapping is typically undesirable (e.g. the features may have infinite dimension), so we employ the well-known *kernel trick* [8], using some kernel $\mathcal{K}(\mathbf{x}_i, \mathbf{x}_j)$ function that computes the inner products between the feature vectors without explicitly constructing them.

3.3 Kernelizing the Method

In this section, we show how to extend our method to non-linear mappings of data. Conceptually, the points are mapped into a feature space by some non-linear mapping $\phi()$, and then the desired transformation is learned in that space. This can be done implicitly through the use of kernels.

The columns of the linear transformation $\mathbf{W}$ can always be re-expressed as linear combinations of the data points in the feature space, $\mathbf{W} = \mathbf{X}\boldsymbol{\Omega}$. Therefore, we can rewrite the squared distance as:

$$\begin{aligned} |\mathbf{y}_i - \mathbf{y}_j|^2 &= (\mathbf{x}_i - \mathbf{x}_j)^\mathsf{T}\mathbf{W}\mathbf{W}^\mathsf{T}(\mathbf{x}_i - \mathbf{x}_j) \\ &= (\mathbf{x}_i - \mathbf{x}_j)^\mathsf{T}\mathbf{X}\boldsymbol{\Omega}\boldsymbol{\Omega}^\mathsf{T}\mathbf{X}^\mathsf{T}(\mathbf{x}_i - \mathbf{x}_j) \\ &= (\mathbf{x}_i^\mathsf{T}\mathbf{X} - \mathbf{x}_j^\mathsf{T}\mathbf{X})\boldsymbol{\Omega}\boldsymbol{\Omega}^\mathsf{T}(\mathbf{X}^\mathsf{T}\mathbf{x}_i - \mathbf{X}^T\mathbf{x}_j) \\ &= (\mathbf{X}^\mathsf{T}\,\mathbf{x}_i - \mathbf{X}^\mathsf{T}\mathbf{x}_j)^\mathsf{T}\boldsymbol{\mathcal{A}}(\mathbf{X}^\mathsf{T}\mathbf{x}_i - \mathbf{X}^\mathsf{T}\mathbf{x}_j) \end{aligned} \tag{17}$$

where $\boldsymbol{\mathcal{A}} = \boldsymbol{\Omega}\boldsymbol{\Omega}^\mathsf{T}$. We have now expressed the distance in terms of a matrix to be learned, $\boldsymbol{\mathcal{A}}$, and the inner products between the data points which can be computed via the kernel, $\mathbf{K}$.

$$\begin{aligned} |\mathbf{y}_i - \mathbf{y}_j|^2 &= (\mathcal{K}(\mathbf{X}, \mathbf{x}_i) - \mathcal{K}(\mathbf{X}, \mathbf{x}_j))^\mathsf{T}\boldsymbol{\mathcal{A}}(\mathcal{K}(\mathbf{X}, \mathbf{x}_i) - \mathcal{K}(\mathbf{X}, \mathbf{x}_j)) \\ &= (\mathbf{K}_i - \mathbf{K}_j)^\mathsf{T}\boldsymbol{\mathcal{A}}(\mathbf{K}_i - \mathbf{K}_j) \end{aligned} \tag{18}$$

where $\mathbf{K}_i = \mathcal{K}(\mathbf{X}, \mathbf{x}_i)$ is the i_{th} column of the kernel matrix $\mathbf{K}$. The optimization of $\boldsymbol{\mathcal{A}}$ then proceeds just as in the non-kernelized version presented earlier, by substituting $\mathbf{X}$ and $\mathbf{W}$ by $\mathbf{K}$ and $\boldsymbol{\Omega}$ respectively.

3.4 The Algorithm

The training procedure of Embedding by Affine Transformation (EAT) is summarized in Alg.1. Following it, Alg.2 explains how out-of-sample points can be mapped into the embedded space. In these algorithms, we suppose that all training data points are stacked into the columns of a $d \times n$ matrix $\mathbf{X}$. Likewise, all projected data points $\{\mathbf{y}_i\}_{i=1}^n$ are stacked into the columns of a matrix $\mathbf{Y}$ and the $d' \times n$ matrix $\mathbf{Z}$ denotes the low-dimensional representation of the data. In the last line of Alg.1, the columns of $\mathbf{C}$ are the eigenvectors of $\mathbf{Y}\mathbf{Y}^\mathsf{T}$ corresponding to the top d' eigenvalues which are calculated by PCA.

Alg. 1. EAT - Training

Input: $\mathbf{X}$, and d'
Output: $\mathbf{Z}$, and linear transformations $\mathbf{W}$ (or $\boldsymbol{\Omega}$) and $\mathbf{C}$

1: Compute a neighborhood graph and form the sets $\mathcal{S}$ and $\mathcal{O}$
2: Choose a kernel function and compute the kernel matrix $\mathbf{K}$
3: Calculate the matrix $\mathbf{Q}$, and the vectors $\mathbf{p}$ and $\mathbf{s}$, based on $\mathbf{K}$, $\mathcal{S}$ and $\mathcal{O}$
4: Compute $\mathbf{U}$ and $\boldsymbol{\Lambda}$ by performing SVD on $\mathbf{Q}$ such that $\mathbf{Q} = \mathbf{U}\boldsymbol{\Lambda}\mathbf{U}^\mathsf{T}$
5: Let $\boldsymbol{\alpha} = \boldsymbol{\Lambda}^{-1}\mathbf{U}^\mathsf{T}\mathbf{p}$
6: Solve the SPD problem $\max_{\mathbf{A}\succeq 0}(\boldsymbol{\beta}^\mathsf{T}\overline{\mathbf{U}}^\mathsf{T}\mathbf{s})$, where $vech(\mathbf{A}) = \mathbf{U}\boldsymbol{\alpha} + \overline{\mathbf{U}}\boldsymbol{\beta}$
8: Decompose $\mathbf{A} = \mathbf{W}\mathbf{W}^\mathsf{T}$ (or in the kernelized version $\mathcal{A} = \boldsymbol{\Omega}\boldsymbol{\Omega}^\mathsf{T}$)
6: Compute $\mathbf{Y} = \mathbf{W}^\mathsf{T}\mathbf{X} = \boldsymbol{\Omega}^\mathsf{T}\mathbf{K}$
7: Apply PCA to $\mathbf{Y}$ and obtain the final embedding $\mathbf{Z} = \mathbf{C}^\mathsf{T}\mathbf{Y}$

After the training phase of EAT, we have the desired transformation $\mathbf{W}$ for unfolding the latent structure of the data. We also have $\mathbf{C}$ from PCA, which is used to reduce the dimensionality of the unfolded data. As a result, we can embed any new point $\mathbf{x}$ by using the algorithm shown in Alg.2.

Alg. 2. EAT - Embedding

Input: out-of-sample example $\mathbf{x}_{d\times 1}$, and the transformations $\mathbf{W}$ (or $\boldsymbol{\Omega}$) and $\mathbf{C}$
Output: vector $\mathbf{z}_{d'\times 1}$ which is a low-dimensional representation of $\mathbf{x}$

1: Compute $\mathbf{K_x} = \mathcal{K}(., \mathbf{x})$
2: Let $\mathbf{y} = \mathbf{W}^\mathsf{T}\mathbf{x} = \boldsymbol{\Omega}^\mathsf{T}\mathbf{K_x}$
3: Compute $z = \mathbf{C}^\mathsf{T}\mathbf{y}$

---

4 Experimental Results

In order to evaluate the performance of the proposed method, we have conducted several experiments on synthetic and real data sets.

To emphasize the difference between the transformation computed by EAT and the one that PCA provides, we designed a simple experiment on a synthetic data set. In this experiment we consider a three-dimensional V-shape manifold illustrated in the top-left panel of Fig. 2. We represent this manifold by 1000 uniformly distributed sample points, and divide it into two subsets: a training set of 28 well-sampled points, and a test set of 972 points. EAT is applied to the training set, and then the learned transformation is used to project the test set. The result is depicted in the top-right panel of Fig. 2. This image illustrates $\mathbf{Y} = \mathbf{W}^\mathsf{T}\mathbf{X}$, which is the result of EAT in 3D before applying PCA. It shows that the third dimension carries no information, and the unfolding happens before PCA is applied to reduce the dimensionality to 2D.

The bottom-left and bottom-right panels of Fig. 2 show the results of PCA and KPCA, when applied to the whole data set. PCA computes a global distance preserving transformation, and captures the directions of maximum variation in the data. Clearly, in this example, the direction with the maximum variation is not the one that unfolds the V-shape. This is the key difference between the functionality of PCA and EAT. Kernel PCA does not provide a satisfactory embedding either. Fig. 2 shows the result that is generated by an RBF kernel; we experimented KPCA with a variety of popular kernels, but none were able to reveal a faithful embedding of the V-shape.

Unlike kernel PCA, EAT does not expect the kernel to reveal the underlying structure of data. When the dimensionality of data is higher than its quantity, a linear transformation can span the whole space. This means we can always find $\mathbf{W}$ to flatten the manifold.

When the original dimensionality of data is high ($d > n$, e.g. for images), EAT does not need a kernel in principal; however, using a linear kernel reduces the

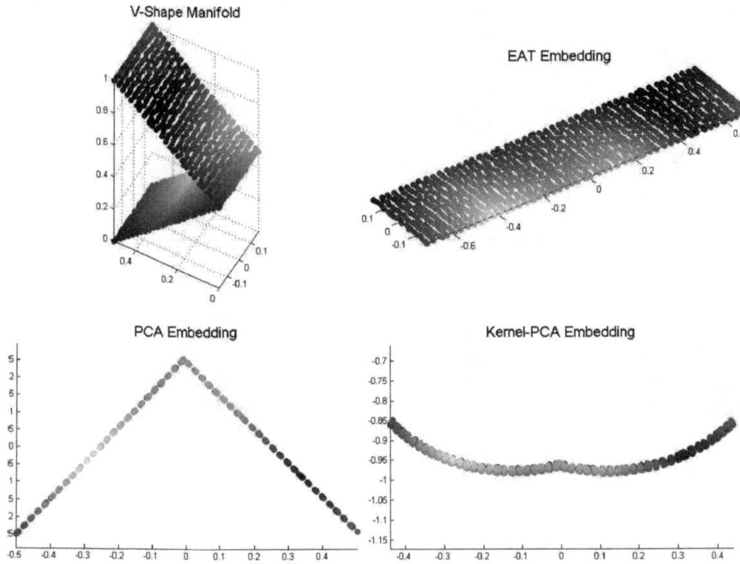

Fig. 2. A V-shape manifold, and the results of EAT, PCA and kernel PCA

computational complexity of the method[4]. In all of the following experiments, we use a linear kernel when the original dimensionality of data is high (e.g. for images), and RBF in all other cases. In general EAT is not that sensitive to the type of kernel. We will discuss the effect of kernel type and its parameter(s) later in this section.

The next experiment is on a Swiss roll manifold, depicted in the bottom-left panel of Fig. 3. Although Swiss roll is a three-dimensional data set, it tends to be one of the most challenging data sets due to its complex global structure. We sample 50 points for our training set, and 950 points as an out-of-sample test set. The results of Maximum Variance Unfolding (MVU), Isomap, and EAT [5] are presented in the first row of Fig. 3. The second row shows the projection of the out-of-sample points into a two-dimensional embedded space. EAT computes a transformation that maps the new data points into the low-dimensional space. MVU and Isomap, however, do not provide any direct way to handle out-of-sample examples. A common approach to resolve this problem is to learn a non-parametric model between the low and high dimensional spaces.

In this approach, a high-dimensional test data point $\mathbf{x}$ is mapped to the low dimensional space in three steps: (i) the k nearest neighbors of $\mathbf{x}$ among the train-

[4] In the kernelized version, $\mathbf{W}$ is $n \times n$ but in the original version it is $d \times d$. Thus, Computing $\mathbf{W}$ in the kernelized form is less complex when $d > n$.

[5] In general, Kernel PCA fails to unfold the Swiss roll data set. LLE generally produces a good embedding, but not on small data sets (e.g. the training set in this experiment). For this reason we do not demonstrate their results.

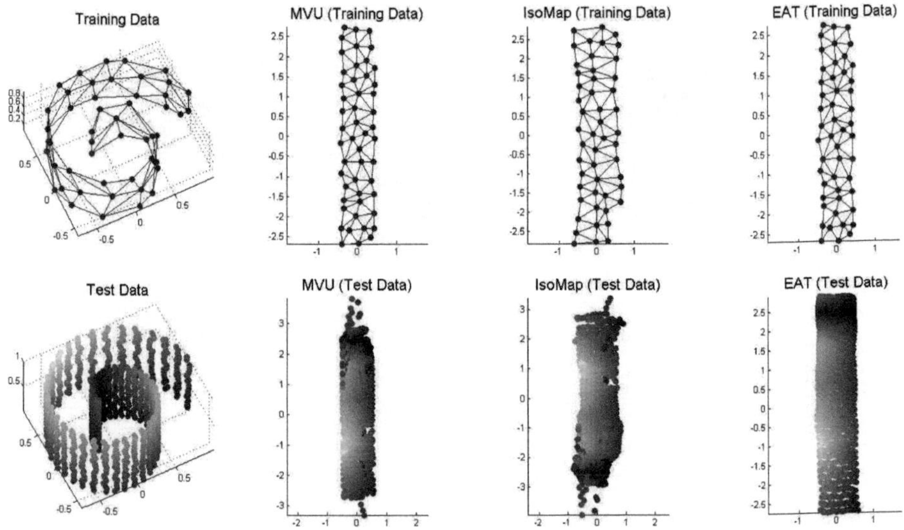

Fig. 3. A Swiss roll manifold, and the results of different dimensionality reduction methods: MVU, Isomap, and EAT. The top row demonstrates the results on the training set, and the bottom row shows the results of the out-of-sample test set.

ing inputs (in the original space) are identified; (ii) the linear weights that best reconstruct $\mathbf{x}$ from its neighbors, subject to a sum-to-one constraint, are computed; (iii) the low-dimensional representation of $\mathbf{x}$ is computed as the weighted sum (with weights computed in the previous step) of the embedded points corresponding to those k neighbors of $\mathbf{x}$ in the original space. In all of the examples in this paper, the out-of-sample embedding is conducted using this non-parametric model except for EAT, PCA, and Kernel PCA which provide parametric models. It is clear that the out-of-sample estimates of MVU and Isomap are not faithful to the Swiss roll shape, especially along its border.

Now we illustrate the performance of the proposed method on some real data sets. Fig. 4 shows the result of EAT when applied to a data set of face images. This data set consists of 698 images, from which we randomly selected 35 as the training set and the rest are used as the test data. Training points are indicated with a solid blue border.

The images in this experiment have three degrees of freedom: pan, tilt, and brightness. In Fig. 4, the horizontal and vertical axes appear to represent the pan and tilt, respectively. Interestingly, while there are no low-intensity images among the training samples, darker out-of-sample points appear to have been organized together in the embedding. These darker images still maintain the correct trends in the variation of pan and tilt across the embedding. In this example, EAT was used with a linear kernel.

In another experiment , we conducted an experiment on a subset of the Olivetti image data set [7]. Face images of three different persons are used as

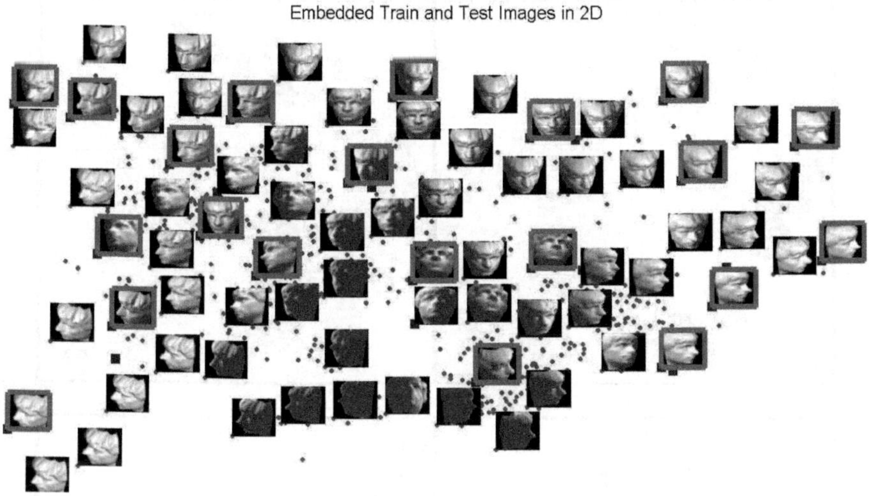

Fig. 4. The result of manifold learning with EAT (using a linear kernel) on a data set of face images

the training set, and images of a fourth person are used as the out-of-sample test examples. The results of MVU, LLE, Isomap, PCA, KPCA, and EAT are illustrated in Fig. 5. Different persons in the training data are indicated by red squares, green triangles and purple diamonds. PCA and Kernel PCA do not provide interpretable results even for the training set. The other methods, however, separate the different people along different chains. Each chain shows a smooth change between the side view and the frontal view of an individual.

The key difference between the algorithms is the way they embed the images of the new person (represented by blue circles). MVU, LLE, Isomap, PCA, and Kernel PCA all superimpose these images onto the images of the most similar individual in the training set, and by this, they clearly lose a part of information. This is due to the fact, that they learn a non-parametric model for embedding the out-of-samples. EAT, however, embeds the images of the new person as a separate cluster (chain), and maintains a smooth gradient between the frontal and side views.

Finally, we attempt to unfold a globe map (top-left of Fig. 6) into a faithful 2D representation. Since a complete globe is a closed surface and thus cannot be unfolded, our experiment is on a half-globe. A regular mesh is drawn over the half-globe (top-right of Fig. 6), and 181 samples are taken for the training set. EAT is used to unfold the sampled mesh and find its transformation (bottom-right of Fig. 6).

Note that it is not possible to unfold a globe into a 2D space while preserving the original local distances; in fact, the transformation with the minimum preservation error is the identity function. So rather than preserving the local distances, we define Euclidean distances based on the latitude and longitude of

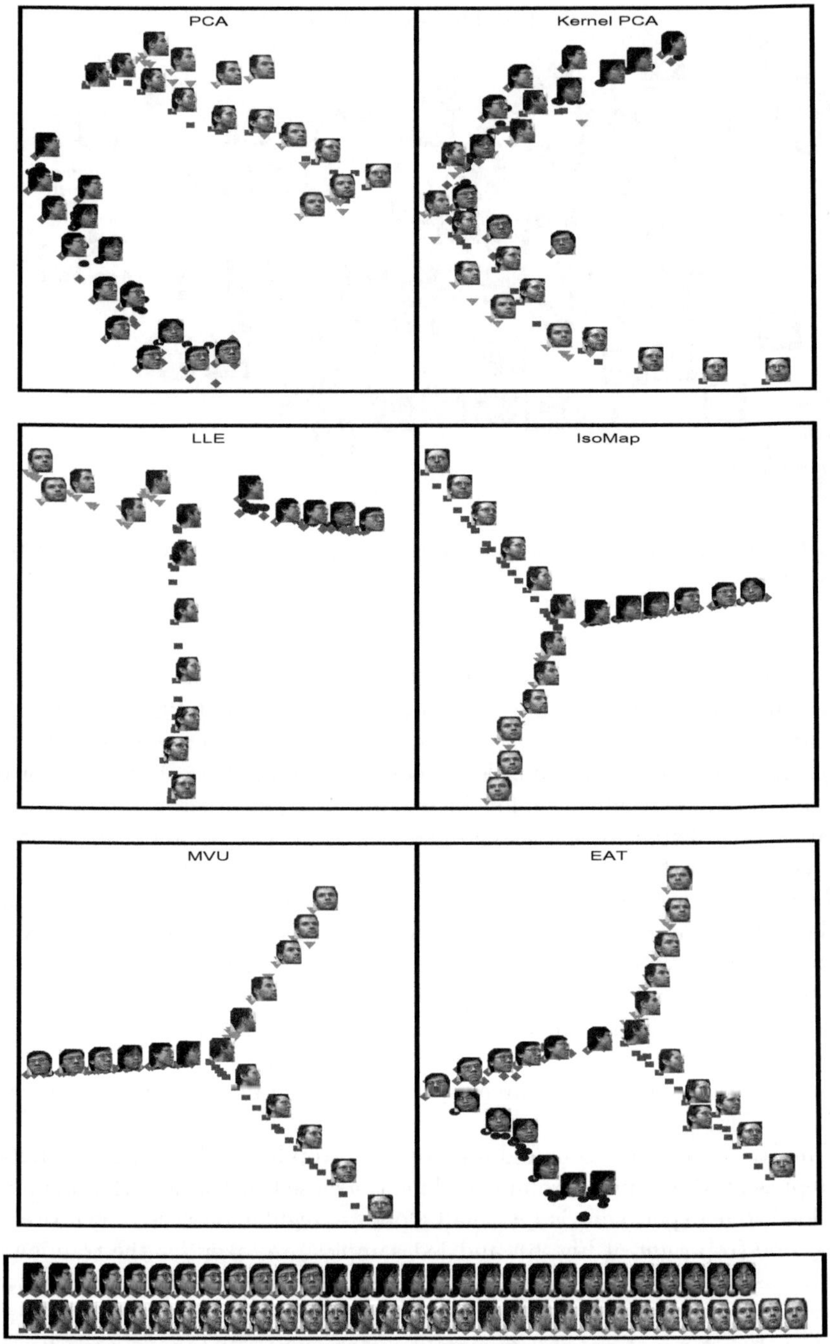

Fig. 5. The results of different dimensionality reduction techniques on a data set of face photos. Each color represents the pictures of one of four individuals. The blue circles show the test data (pictures of the fourth individual).

the training points along the surface of the globe; then the 2D embedding becomes feasible. This is an interesting aspect of EAT: it does not need to operate on the original distances of the data, but can instead be supplied with arbitrary distance values (as long as they are compliant with the desired dimensionality reduction of the data).

Our out-of-sample test set consists of 30,000 points as specified by their 3D position with respect to the center of the globe. For this experiment we used an RBF kernel with $\sigma = 0.3$. Applying the output transformation of EAT results in the 2D embedding shown in the bottom-left of Fig. 6; color is used to denote elevation in these images. Note that the pattern of the globe does not change during the embedding process, which demonstrates that the representation of EAT is faithful. However, the 2D embedding of the test points is distorted at the sides, which is due to the lack of information from the training samples in these areas.

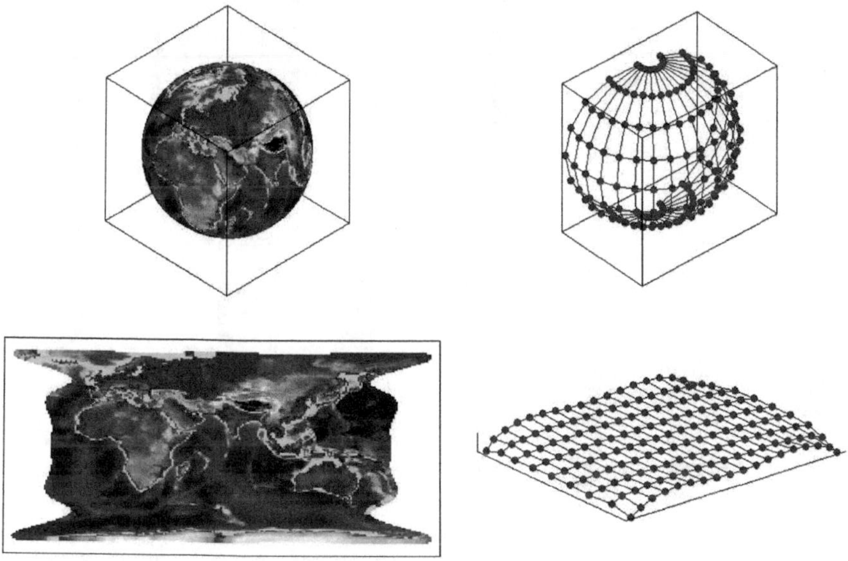

Fig. 6. Unfolding a half-globe into a 2D map by EAT. A half-sphere mesh is used for training. Color is used to denote elevation. The out-of-sample test set comprises 30,000 points from the surface of Earth.

4.1 The Effect of Type and Parameters of the Kernels

The number of bases corresponding to a particular kernel matrix is equal to the rank of that matrix. If we use a full rank kernel matrix (i.e. $rank(\mathbf{K}) = n$), then the number of bases is equal to the number of data points and a linear transformation can span the whole space. That is, it is always possible to find a transformation $\mathbf{W}$ that perfectly unfolds the data as far as the training data points are concerned. For example, an identity kernel matrix can perfectly unfold any training data set; but it will fail to map out-of-sample points correctly, because it cannot measure the similarity between the out-of-sample points and the training examples. In other words, using a full rank kernel is a sufficient condition in order to faithfully embed the training points. But if the correlation between the kernel and the data is weak (an extreme case is using the identity matrix as a kernel), EAT will not perform well for the out-of-sample points.

We define $r = \frac{rank(\mathbf{K})}{n}$. Clearly $r = 1$ indicates a full rank matrix and $r < 1$ shows a rank deficient kernel matrix $\mathbf{K}$. The effect of using different kernels on the Swiss roll manifold (bottom-left of Fig. 3) are illustrated in Fig. 7.

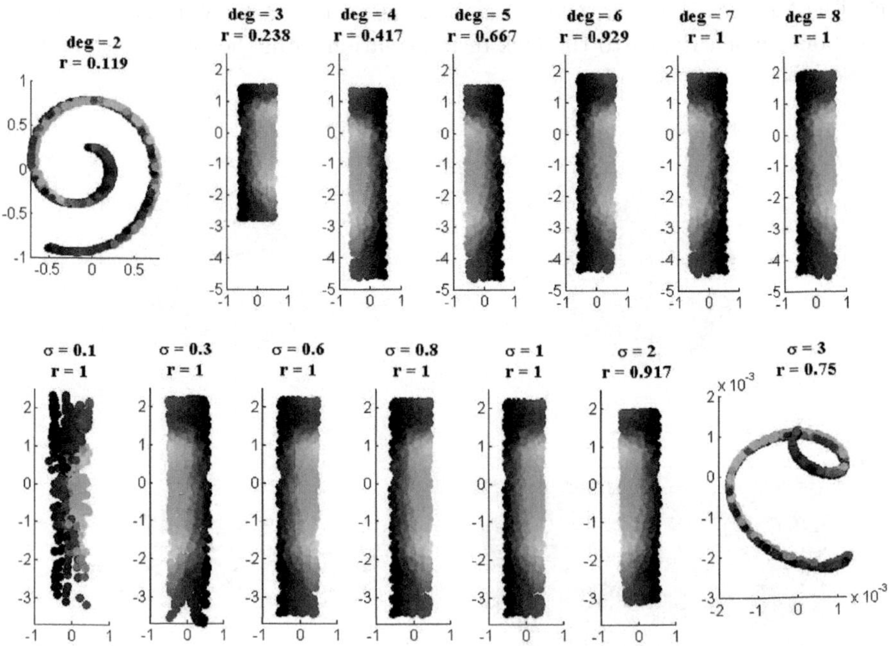

Fig. 7. The effect of using different kernels for embedding a Swiss-roll manifold. Polynomials of different degrees are used in the first row, and in the second row RBF kernels with different σ values map the original data to the feature space.

Two different kernels are demonstrated. In the first row polynomial kernels of different degrees are used, and the second row shows the result of RBF kernels which have different values for their variance parameter σ. The dimensionality of the feature spaces of the low degree polynomial kernels ($deg = 2, 3$) is not high enough; thus they do not produce satisfactory results. Similarly, in the experiment with RBF kernels, when σ is high, EAT is not able to find the desired affine transformation in the feature space to unfold the data (e.g. the rightmost-bottom result).

The leftmost-bottom result is generated by an RBF kernel with a very small value assigned to σ. In this case, the kernel is full rank and consequently $r = 1$. The training data points are mapped perfectly as expected but EAT fails to embed the out-of-sample points correctly. Note that with such a small σ the resulted RBF kernel matrix is very close to the identity matrix, so over-fitting will happen in this case. Experiments with a wide variety of other kernels on different data sets show similar results. Based on these experiments, we suggest that an RBF kernel can be used for any data set. The parameter σ should be selected such that (i) the kernel matrix is full rank or close to full rank ($r \approx 1$), and (ii) the resulting kernel is able to measure the similarity between non-identical data points (σ is not too small). This method is not sensitive to type of kernel. For an RBF kernel a wild range of values for σ can be safely used, as long as the conditions (i) and (ii) are satisfied. When the dimensionality of the original data is more than or equal to the number of the data points, there is no need for a kernel, but one may use a simple linear kernel to reduce the computational complexity[6].

5 Conclusion

We presented a novel dimensionality reduction method which, unlike other prominent methods, can easily embed out-of-sample examples. Our method learns a parametric mapping between the high and low dimensional spaces, and is performed in two steps. First, the input data is projected into a high-dimensional feature space, and then an affine transformation is learned that maps the data points from the feature space into the low dimensional embedding space. The search for this transformation is cast as an instance of semi-definite programming (SDP), which is convex and always converges to a global optimum. However, SDP is computationally intensive, which can make it inefficient to train EAT on large data sets.

Our experimental results on real and synthetic data sets demonstrate that EAT produces a robust and faithful embedding even for very small data sets. It also shows that it is successful at projecting out-of-sample examples. Thus, one approach for handling large data sets with EAT would be to downsample the data by selecting a small subset as the training input and embedding the rest of the data as test examples.

Another feature of EAT is that it treats the distances between the data points in three different ways. One can preserve a subset of the distances (set $\mathcal{S}$), stretch another subset (set $\mathcal{O}$) and leave the third set (pairs that are not in $\mathcal{S}$ and $\mathcal{O}$) unspecified. This is in contrast with methods like MVU that preserve local distances but stretch any non-local pairs. This property means that EAT could be useful for semi-supervised tasks where only partial information about similarity and dissimilarity of points is known.

[6] An RBF kernel can be used for this case as well.

References

1. Belkin, M., Niyogi, P.: Laplacian eigenmaps and spectral techniques for embedding and clustering. In: Proceedings NIPS (2001)
2. Cox, T., Cox, M.: Multidimensional Scaling, 2nd edn. Chapman Hall, Boca Raton (2001)
3. Ham, J., Lee, D., Mika, S., Schölkopf, B.: A kernel view of the dimensionality reduction of manifolds. In: International Conference on Machine Learning (2004)
4. Jolliffe, I.: Principal Component Analysis. Springer, New York (1986)
5. Mika, S., Schölkopf, B., Smola, A., Müller, K.R., Scholz, M., Rätsch, G.: Kernel PCA and de-noising in feature spaces. In: Kearns, M.S., Solla, S.A., Cohn, D.A. (eds.) Proceedings NIPS 11. MIT Press, Cambridge (1999)
6. Roweis, S., Saul, L.: Nonlinear dimensionality reduction by locally linear embedding. Science 290, 2323–2326 (2000)
7. Samaria, F., Harter, A.: Parameterisation of a Stochastic Model for Human Face Identification. In: Proceedings of 2nd IEEE Workshop on Applications of Computer Vision (1994)
8. Schölkopf, B., Smola, A.: Learning with Kernels. MIT Press, Cambridge (2002)
9. Shaw, B., Jebara, T.: Minimum volume embedding. In: Meila, M., Shen, X. (eds.) Proceedings of the Eleventh International Conference on Artificial Intelligence and Statistics, San Juan, Puerto Rico, March 21-24. JMLR: W&CP, vol. 2, pp. 460–467 (2007)
10. Shaw, B., Jebara, T.: Structure preserving embedding. In: Bottou, L., Littman, M. (eds.) Proceedings of the 26th International Conference on Machine Learning, pp. 937–944. Omnipress, Montreal (June 2009)
11. Song, L., Smola, A.J., Borgwardt, K.M., Gretton, A.: Colored maximum variance unfolding. In: NIPS (2007)
12. Tenenbaum, J.: Mapping a manifold of perceptual observations. Advances in Neural Information Processing Systems 10, 682–687 (1998)
13. Weinberger, K.Q., Saul, L.K.: Unsupervised learning of image manifolds by semidefinite programming. In: Proc. of the IEEE Conf. on Computer Vision and Pattern Recognition (CVPR 2004), vol. II, pp. 988–995 (2004)
14. Weinberger, K., Sha, F., Zhu, Q., Saul, L.: Graph Laplacian regularization for large-scale semidefinite programming. In: Advances in Neural Information Processing Systems, vol. 19, p. 1489 (2007)

NDPMine: Efficiently Mining Discriminative Numerical Features for Pattern-Based Classification*

Hyungsul Kim, Sangkyum Kim, Tim Weninger, Jiawei Han, and Tarek Abdelzaher

Department of Computer Science,
University of Illinois at Urbana-Champaign,
Urbana IL 61801, USA
{hkim21,kim71,weninge1,hanj,zaher}@illinois.edu

Abstract. Pattern-based classification has demonstrated its power in recent studies, but because the cost of mining discriminative patterns as features in classification is very expensive, several efficient algorithms have been proposed to rectify this problem. These algorithms assume that feature values of the mined patterns are binary, i.e., a pattern either exists or not. In some problems, however, the number of times a pattern appears is more informative than whether a pattern appears or not. To resolve these deficiencies, we propose a mathematical programming method that directly mines discriminative patterns as numerical features for classification. We also propose a novel search space shrinking technique which addresses the inefficiencies in iterative pattern mining algorithms. Finally, we show that our method is an order of magnitude faster, significantly more memory efficient and more accurate than current approaches.

Keywords: Pattern-Based Classification, Discriminative Pattern Mining, SVM.

1 Introduction

Pattern-based classification is a process of learning a classification model where patterns are used as features. Recent studies show that classification models which make use of pattern-features can be more accurate and more understandable than the original feature set [2,3]. Pattern-based classification has been adapted to work on data with complex structures such as sequences [12,9,14,6,19], and graphs [16,17,15], where discriminative frequent patterns are taken as features to build high quality classifiers.

These approaches can be grouped into two settings: binary or numerical. Binary pattern-based classification is the well-known problem setting in which the feature

* Research was sponsored in part by the U.S. National Science Foundation under grants CCF-0905014, and CNS-0931975, Air Force Office of Scientific Research MURI award FA9550-08-1-0265, and by the Army Research Laboratory under Cooperative Agreement Number W911NF-09-2-0053 (NS-CTA). The views and conclusions contained in this document are those of the authors and should not be interpreted as representing the official policies, either expressed or implied, of the Army Research Laboratory or the U.S. Government. The U.S. Government is authorized to reproduce and distribute reprints for Government purposes notwithstanding any copyright notation here on. The second author was supported by the National Science Foundation OCI-07-25070 and the state of Illinois. The third author was supported by a NDSEG PhD Fellowship.

J.L. Balcázar et al. (Eds.): ECML PKDD 2010, Part II, LNAI 6322, pp. 35–50, 2010.

space is $\{0, 1\}^d$, where d is the number of features. This means that a classification model only uses information about whether an interesting pattern exists or not. On the other hand, a numerical pattern-based classification model's feature space is $\mathbb{N}^d$, which means that the classification model uses information about how many times an interesting pattern appears. For instance, in the analysis of software traces loops and other repetitive behaviors may be responsible for failures. Therefore, it is necessary to determine the number of times a pattern occurs in traces.

Pattern-based classification techniques are prone to a major efficiency problem due to the exponential number of possible patterns. Several studies have identified this issue and offered solutions [3,6]. However, to our knowledge there has not been any work addressing this issue in the case of numerical features.

Recently a boosting approach was proposed by Saigo *et al.* called *gBoost* [17]. Their algorithm employs a linear programming approach to boosting as a base algorithm combined with a pattern mining algorithm. The linear programming approach to boosting algorithm (LPBoost) [5] is shown to converge faster than ADABoost [7] and is proven to converge to a global solution. *gBoost* works by iteratively growing and pruning a search space of patterns via branch and bound search. In work prior to *gBoost* [15] by the same authors, the search space is erased and rebuilt during each iteration. However, in their most recent work, the constructed search space is reused in each iteration to minimize computation time; the authors admit that this approach would not scale but were able to complete their case study with 8GB main memory.

The high cost of finding numerical features along with the accuracy issues of binary-only features motivates us to investigate an alternative approach. What we wish to develop is a method which is both efficient and able to mine numerical features for classification. This leads to our proposal of a *numerical direct pattern mining* approach, **NDPMine**. Our approach employs a mathematical programming method that directly mines discriminative patterns as numerical features. We also address the fundamental problem of iterative pattern mining algorithms, and propose a novel search space shrinking technique to prune memory space without removing potential features. We show that our method is an order of magnitude faster, significantly more memory efficient and more accurate than current approaches.

The structure of this paper is as follows. In Section 2 we provide a brief background survey and discuss in further detail the problems that *NDPMine* claims to remedy. In Section 3 we introduce the problem setting. Section 4 describes our discriminative pattern mining approach, pattern search strategy and search space shrinking technique. The experiments in Section 5 compare our algorithm with current methods in terms of efficiency and accuracy. Finally, Section 6 contains our conclusions.

2 Background and Related Work

The first pattern-based classification algorithms originated from the domain of association rule mining in which CBA [11] and CMAR [10] used the **two-step** pattern mining process to generate a feature set for classification. Cheng *et al.* [2] showed that, within a large set of frequent patterns, those patterns which have higher *discriminative* power, *i.e.* higher information gain and/or Fisher score, are useful in classification. With this intuition, their algorithm (*MMRFS*) selects patterns for inclusion in the feature set based on the information gain or Fisher score of each pattern. The following year, Cheng *et al.* [3] showed that they could be more efficient if they performed pattern-based classification by a **direct process** which directly mines discriminative patterns (*DDPMine*). A separate algorithm by Fan *et al.* (called M^bT) [6], developed at the same time as *DDPMine*, uses a decision tree-like approach, which recursively splits the training instances by picking the most discriminative patterns.

As alluded to earlier, an important problem with the many approaches is that the feature set used to build the classification model is entirely binary. This is a significant drawback because many datasets rely on the number of occurrences of a pattern in order to train an effective classifier. One such dataset comes from the realm of software behavior analysis in which patterns of events in software traces are available for analysis. Loops and other repetitive behaviors observed in program traces may be responsible for failures. Therefore, it is necessary to mine not only the execution patterns, but also the number of occurrences of the patterns. Lo *et al.* [12] proposed a solution to this problem (hereafter called SoftMine) which mines *closed unique iterative patterns* from normal and failing program traces in order to identify software anomalies. Unfortunately, this approach employs the less efficient two-step process which exhaustively enumerates a huge number of frequent patterns before finding the most discriminative patterns.

Other approaches have been developed to address specific datasets. For time series classification, Ye and Keogh [19] used patterns called shapelets to classify time-series data. Other algorithms include DPrefixSpan [14] which classifies action sequences, XRules [20] which classifies trees, and gPLS [16] which classifies graph structures.

Table 1. Comparison of related work

	Binary	**Numerical**
Two-step	MMRFS	SoftMine, Shapelet
Direct	DDPMine, M^bT, gPLS DPrefixSpan, gBoost	**NDPMine**

Table 1 compares the aforementioned algorithms in terms of the pattern's feature value (binary or numerical) and feature selection process (two-step or direct). To the best of our knowledge there do not exist any algorithms which mine patterns as numerical features in a direct manner.

3 Problem Formulation

Our framework is a general framework for numerical pattern-based classification. We, however, confine our algorithm for structural data classification such as sequences, trees, and/or graphs in order to present our framework clearly. There are several pattern definitions for each structural data. For example, for sequence datasets, there are sequential patterns, episode patterns, iterative patterns, and unique iterative patterns [12]. The pattern definition which is better for classification depends on each dataset, and thus, we assume that the definition of a pattern is given as an input. Let $D = \{x_i, y_i\}_{i=1}^n$ be a dataset, containing structural data, where x_i is an object and y_i is its label. Let P be the set of all possible patterns in the dataset.

We will introduce several definitions, many of which are frequently used in pattern mining papers.

A pattern p in P is a sub-pattern of q if q contains p. If p is a sub-pattern of q, we say q is a super-pattern of p. For example, in a sequence, a sequential pattern $\langle A, B\rangle$ is a *sub-pattern* of a sequential pattern $\langle A, B, C\rangle$ because we can find $\langle A, B\rangle$ within $\langle A, B, C\rangle$.

The number of occurrences of a given pattern p in a data instance x is denoted by $occ(p, x)$. For example, if we count the number of non-overlapped occurrences of a pattern, the number of occurrences of a pattern $p = \langle A, B\rangle$ in a data instance $x = \langle A, B, C, D, A, B\rangle$ is 2, and $occ(p, x) = 2$. Since the number of occurrences of a pattern in a data depends on a user's definition, we assume that the function *occ* is given as an input.

The *support* of a pattern p in D is denoted by $sup(p, D)$, where $sup(p, D) = \sum_{x_i \in D} occ(p, x_i)$. A pattern p is *frequent* if $sup(p, D) \geq \theta$, where θ is a minimum support threshold.

A function f on P is said to posses the *apriori property* if $f(p) \leq f(q)$ for any pattern p and all its sub-patterns q.

With these definitions, the problem we present in this paper is as follows: Given a dataset $D = \{x_i, y_i\}_{i=1}^n$, and an occurrence function *occ* with the *apriori property*, we want to find a good feature set of a small number of discriminative patterns $F = \{p_1, p_2, \ldots, p_m\} \subseteq P$ so that we map D into $\mathbb{N}^m$ space to build a classification model. The training dataset in $\mathbb{N}^m$ space for building a classification model is denoted by $D' = \{x'_i, y_i\}_{i=1}^n$, where $x'_{ij} = occ(p_j, x_i)$.

4 NDPMine

From the discussion in Section 1, we see the need for a method which efficiently mines discriminative numerical features for pattern-based classification. This section describes such a method called NDPMine (**N**umerical **D**iscriminative **P**attern Mining).

4.1 Discriminative Pattern Mining with LP

For direct mining of discriminative patterns two properties are required: (1) a measure for discriminative power of patterns, (2) a theoretical bound of the measure for pruning search space. Using information gain and Fisher score, *DDPMine* successfully showed the theoretical bound when feature values of patterns are binary. However, there are no theoretical bounds for information gain and Fisher score when feature values of patterns are numerical. Since standard statistical measures for discriminative power are not suitable in our problem, we take a different approach: model-based feature set mining. Model-based feature set mining find a set of patterns as a feature set while building a classifier. In this section, we will show that *NDPMine* has the two properties required for direct mining of discriminative patterns by formulating and solving an optimization problem of building a classifier.

To do that, we first convert a given dataset into a high-dimensional dataset, and learn a hyperplane as a classification boundary.

Definition 1. *A pattern and class label pair* (p, c) *is called class-dependent pattern, where* $p \in P$ *and* $c \in C = \{-1, 1\}$. *Then, the value of a class-dependent pattern* (p, c) *for data instance* x *is denoted by* $s_c(p, x)$, *where* $s_c(p, x) = c \cdot occ(p, x)$.

Since there are $2|P|$ class-dependent patterns, we have $2|P|$ values for an object x in D. Therefore, by using all class-dependent patterns, we can map x_i in D into $\boldsymbol{x}'_i$ in $\mathbb{N}^{2|P|}$ space, where $x'_{ij} = s_{c_j}(p_j, x'_i)$. One way to train a classifier in high dimensional space is to learn a classification hyperplane (*i.e.*, a bound with maximum margin) by formulating and solving an optimization problem. Given the training data $D' = \{x'_i, y_i\}_{i=1}^n$, the optimization problem is formulated as follows:

$$\begin{aligned} &\max_{\boldsymbol{\alpha},\rho} \rho \\ &\text{s.t.} \quad \sum_{(p,c)\in P\times C} y_i \alpha_{p,c} s_c(p, x'_i) \geq \rho, \quad \forall i \\ &\qquad \sum_{(p,c)\in P\times C} \alpha_{p,c} = 1, \quad \alpha_{p,c} \geq 0, \end{aligned} \tag{1}$$

where $\boldsymbol{\alpha}$ represents the classification boundary, and ρ is the margin between two classes and the boundary.

Let $\tilde{\boldsymbol{\alpha}}$ and $\tilde{\rho}$ be the optimal solution for (1). Then, the prediction rule learned from (1) is $f(\boldsymbol{x}') = sign(\boldsymbol{x}' \cdot \tilde{\boldsymbol{\alpha}})$, where $sign(v) = 1$ if $v \geq 0$ and -1, otherwise. If $\exists (p, c) \in P \times C : \alpha_{p,c} = 0$, $f(\boldsymbol{x}')$ is not affected by the dimension of the class-dependent pattern (p, c). Let $F = \{p | \exists c \in C, \tilde{\alpha}_{p,c} > 0\}$. If using F instead of P in (1), we will have the same prediction rule. In other words, only the small number of patterns in F, we can learn the same classification model as the one learned by P. With this observation, we want to mine such a pattern set (equivalently: a feature set) F to build a classification model.

In addition, we want F to be as small as possible. In order to obtain a relatively small feature set, we need to obtain a very sparse vector $\boldsymbol{\alpha}$, where only few dimensions are non-zero values. To obtain a sparse weight vector $\boldsymbol{\alpha}$, we adopt the formulation from LPBoost [5].

$$\begin{aligned} &\max_{\boldsymbol{\alpha},\boldsymbol{\xi},\rho} \rho - \omega \textstyle\sum_{i=1}^n \xi_i \\ &\text{s.t.} \quad \sum_{(p,c)\in P\times C} y_i \alpha_{p,c} s(\boldsymbol{x}'_i; p, c) + \xi_i \geq \rho, \quad \forall i \\ &\qquad \sum_{(p,c)\in P\times C} \alpha_{p,c} = 1, \quad \alpha_{p,c} \geq 0 \\ &\qquad \xi_i \geq 0, \quad i = 1, \ldots, n, \end{aligned} \tag{2}$$

where ρ is a soft-margin, $\omega = \frac{1}{\nu \cdot n}$, and ν is a parameter for misclassification cost. The difference between the two formulas is that (2) allows mis-classifications of the training instances to cost ω, where (1) does not. To allow mis-classifications, (2) introduces slack variables $\boldsymbol{\xi}$, and makes $\boldsymbol{\alpha}$ sparse in its optimal solution [5]. Next, we do not know all patterns in P unless we mine all of them, and mining all patterns in P is intractable.

Therefore, we cannot solve (2) directly. Fortunately, such a linear optimization problem can be solved by *column generation*, a classic optimization technique [13]. The column generation technique, also called the cutting-plane algorithm, starts with an empty set of constraints in the dual problem and iteratively adds the most violated constraints. When there are no more violated constraints, the optimal solution under the set of selected constraints is equal to the optimal solution under all constraints. To use the column generation technique in our problem, we give the dual problem of (2) as shown in [5].

$$\begin{aligned} \min_{\boldsymbol{\mu},\gamma} \ & \gamma \\ \text{s.t.} \ & \sum_{i=1}^{n} \mu_i y_i s_c(p, \boldsymbol{x}'_i) \leq \gamma, \quad \forall (p,c) \in P \times C \\ & \sum_{i=1}^{n} \mu_i = 1, \quad 0 \leq \mu_i \leq \omega, \quad i = 1, \ldots, n, \end{aligned} \tag{3}$$

where $\boldsymbol{\mu}$ can be interpreted as a weight vector for the training instances.

Each constraint $\sum_{i=1}^{n} \mu_i y_i s_c(p, \boldsymbol{x}'_i) \leq \gamma$ in the dual (3) corresponds to a class-dependent pattern (p, c). Thus, the column generation finds a class-dependent pattern at each iteration whose corresponding constraint is violated the most. Let $H^{(k)}$ be the set of class-dependent patterns found so far at the kth iteration. Let $\boldsymbol{\mu}^{(k)}$ and $\gamma^{(k)}$ be the optimal solution for k^{th} restricted problem:

$$\begin{aligned} \min_{\boldsymbol{\mu}^{(k)},\gamma^{(k)}} \ & \gamma^{(k)} \\ \text{s.t.} \ & \sum_{i=1}^{n} \mu_i^{(k)} y_i s_c(p, x'_i) \leq \gamma^{(k)}, \quad \forall (p,c) \in H^{(k)} \\ & \sum_{i=1}^{n} \mu_i^{(k)} = 1, \quad 0 \leq \mu_i^{(k)} \leq \omega, \quad i = 1, \ldots, n \end{aligned} \tag{4}$$

After solving the kth restricted problem, we search a class-dependent pattern (p^*, c^*) whose corresponding constraint is violated the most by the optimal solution $\gamma^{(k)}$ and $\boldsymbol{\mu}^{(k)}$, and add (p^*, c^*) to $H^{(k)}$.

Definition 2. *For a given* (p, c)*, let* $v = \sum_{i=1}^{n} \mu_i^{(k)} y_i s_c(p, x'_i)$*. If* $v \leq \gamma^{(k)}$*, the corresponding constraint of* (p, c) *is not violated by* $\gamma^{(k)}$ *and* μ^k *because* $\sum_{i=1}^{n} \mu_i^{(k)} y_i s_c(p, x'_i) = v \leq \gamma^{(k)}$*. If* $v > \gamma^{(k)}$*, then we say the corresponding constraint of* (p, c) *is violated by* $\gamma^{(k)}$ *and* $\boldsymbol{\mu}^{(k)}$*, and the margin of the constraint is defined as* $v - \gamma^{(k)}$*.*

In this view, (p^*, c^*) is the class-dependent pattern with the maximum margin. Now, we define our measure for discriminative power of class-dependent patterns.

Definition 3. *We define a* gain function *for a given weight* $\boldsymbol{\mu}$ *as follows:*

$$gain(p, c; \boldsymbol{\mu}) = \sum_{i=1}^{n} \mu_i y_i s_c(p, x'_i).$$

Algorithm 1. Discriminative Pattern Mining

$H^{(0)} \leftarrow \emptyset$
$\gamma^{(0)} \leftarrow 0$
$\mu_i^{(0)} = 1/n \quad \forall i = 1, \ldots, n$
for $k = 1, \ldots$ **do**
  $(p^*, c^*) = \operatorname{argmax}_{(p,c)\in P\times C}\, gain(p, c; \boldsymbol{\mu}^{(k-1)})$
  if $gain(p^*, c^*; \boldsymbol{\mu}^{(k-1)}) - \gamma^{(k-1)} < \epsilon$ **then**
    break
  end if
  $H^{(k)} \leftarrow H^{(k-1)} \cup \{(p^*, c^*)\}$
  Solve the kth restricted problem (4) to get $\gamma^{(k)}$ and $\boldsymbol{\mu}^{(k)}$
end for
Solve (5) to get $\tilde{\boldsymbol{\alpha}}$
$F \leftarrow \{p | \exists c \in C, \tilde{\alpha}_{p,c} > 0\}$

For given $\gamma^{(k)}$ and $\boldsymbol{\mu}^{(k)}$, choosing the constraint with maximum margin is the same as choosing the constraint with maximum gain. Thus, we search for a class-dependent pattern with maximum gain in each iteration until there are no more violated constraints. Let k^* be the last iteration. Then, we can get the optimal solution $\tilde{\rho}$ and $\tilde{\boldsymbol{\alpha}}$ for (2) by solving the following optimization problem and setting $\tilde{\alpha}_{(p,c)} = 0$ for all $(p,c) \notin H^{(k^*)}$.

$$\begin{aligned} \min_{\boldsymbol{\alpha},\boldsymbol{\xi},\rho} \; & -\rho + \omega \textstyle\sum_{i=1}^{n} \xi_i \\ \text{s.t.} \; & \sum_{(p,c)\in H^{(k^*)}} y_i \alpha_{p,c} s(\boldsymbol{x}'_i; p, c) + \xi_i \geq \rho, \quad \forall i \\ & \sum_{(p,c)\in H^{(k^*)}} \alpha_{p,c} = 1, \quad \alpha_{p,c} \geq 0 \\ & \xi_i \geq 0, \quad i = 1, \ldots, n \end{aligned} \tag{5}$$

The difference is that now we have the training instances in $|H^{(k^*)}|$ dimensions, not in $2|P|$ dimensions. Once we have $\tilde{\boldsymbol{\alpha}}$, as explained before, we can make a feature set F such that $F = \{p | \exists c \in C, \tilde{\alpha}_{p,c} > 0\}$. As a summary, the main algorithm of *NDPMine* is presented in Algorithm 1.

4.2 Optimal Pattern Search

As in *DDPMine* and other direct mining algorithms, our search strategy is a branch-and-bound approach. We assume that there is a canonical search order for P such that all patterns in P are enumerated without duplication. Many studies have been done for canonical search orders for most of structural data such as sequence, tree, and graph. Most of the pattern enumeration methods in these canonical search orders create the next pattern by extending the current pattern. Our aim is to find a pattern with maximum gain. Thus, for efficient search, it is important to prune the unnecessary or unpromising search space. Let p be the current pattern. Then, we compute the maximum gain bound for all super-patterns of p and decide whether we can prune the branch or not based on the following theorem.

Algorithm 2. Branch-and-bound Pattern Search

```
Global variables: maxGain, maxPat
procedure search_optimal_pattern(μ, θ, D)
  maxGain ← 0
  maxPat ← ∅
  branch_and_bound(∅, μ, θ, D)

function branch_and_bound(p, μ, θ, D)
  for q ∈ {extended patterns of p in the canonical order} do
    if sup(q, D) ≥ θ then
      for c ∈ {−1, +1} do
        if gain(q, c; μ) > maxGain then
          maxGain ← gain(q, c; μ)
          maxPat ← (q, c)
        end if
      end for
      if gainBound(p; μ) > maxGain then
        branch_and_bound(q, μ, θ, D)
      end if
    end if
  end for
```

Theorem 1. *If $gainBound(p;\boldsymbol{\mu}) \le g^*$ for some g^*, then $gain(q,c;\boldsymbol{\mu}) \le g^*$ for all super-patterns q of p and all $c \in C$, where*

$$gainBound(p;\boldsymbol{\mu}) = \max\left(\sum_{\{i|y_i=+1\}} \mu_i \cdot occ(p,x_i'),\ \sum_{\{i|y_i=-1\}} \mu_i \cdot occ(p,x_i') \right)$$

Proof. We will prove it by contradiction. Suppose that there is a super-pattern q of p such that $gain(q,c;\boldsymbol{\mu}) > gainBound(p;\boldsymbol{\mu})$. If $c = 1$,

$$\begin{aligned} gain(q,c;\boldsymbol{\mu}) &= \sum_{i=1}^{n} \mu_i y_i s_c(q,x_i') = \sum_{i=1}^{n} \mu_i y_i occ(q,x_i') \\ &= \sum_{\{i|y_i=1\}} \mu_i occ(q,x_i') - \sum_{\{i|y_=-1\}} \mu_i occ(q,x_i') \\ &\le \sum_{\{i|y_i=1\}} \mu_i occ(q,x_i') \le \sum_{\{i|y_i=1\}} \mu_i occ(p,x_i') \\ &\le gainBound(p;\boldsymbol{\mu}) \end{aligned}$$

Therefore, it is a contradiction. Likewise, if $c = -1$, we can derive a similar contradiction. Note that $occ(q,x_i') \le occ(p,x_i')$ because occ has *apriori property*.

If the maximum gain among the ones observed so far is greater than $gainBound(p;\boldsymbol{\mu})$, we can prune the branch of a pattern p. The optimal pattern search algorithm is presented in Algorithm 2.

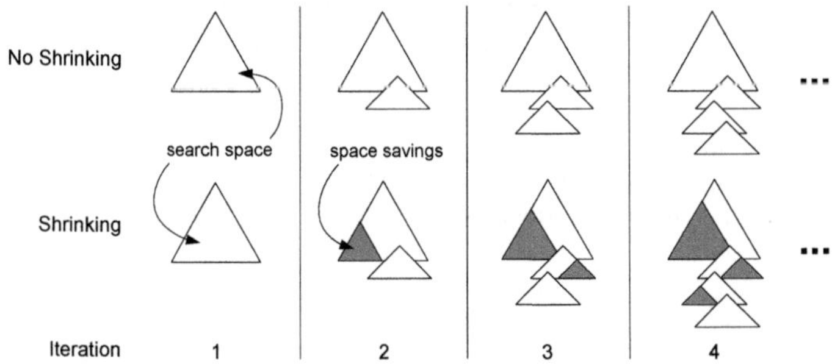

Fig. 1. Search Space Growth with and without Shrinking Technique. Dark regions represent shrinked search space (memory savings).

4.3 Search Space Shrinking Technique

In this section, we explain our novel search space shrinking technique. Mining discriminative patterns instead of frequent patterns can prune more search space by using a bound function $gainBound$. However, this requires an iterative procedure like in *DDPMine*, which builds a search space tree again and again. To avoid the repetitive searching, *gBoost* [17] stores the search space tree of previous iterations in main memory. The search space tree keeps expanding as iteration goes because it needs to mine different discriminative patterns. This may work for small datasets on a machine with enough main memory, but is not scalable.

In this paper, we also store the search space of previous iterations, but introduce search space shrinking technique to resolve the scalability issue.

In each iteration k of the column generation, we look for a pattern whose gain is greater than $\gamma^{(k-1)}$, otherwise the termination condition will hold. Thus, if a pattern p cannot have greater gain than $\gamma^{(k-1)}$, we do not need to consider p in the kth iteration and afterwards because $\gamma^{(k)}$ is non-decreasing by the following theorem.

Theorem 2. *$\gamma^{(k)}$ is non-decreasing as k increases.*

Proof. In each iteration, we add a constraint that is violated by the previous optimal solution. Adding more constraints does not decrease the value of objective function in a minimization problem. Thus, $\gamma^{(k)}$ is not decreasing.

Definition 4. $maxGain(p) = \max_{\boldsymbol{\mu},c} gain(p, c; \boldsymbol{\mu})$, *where* $c \in C$, *and* $\forall i\ 0 \leq \mu_i \leq \omega$.

If there is a pattern p such that $maxGain(p) \leq \gamma^{(k)}$, we can safely remove the pattern from main memory after the kth iteration without affecting the final result of *NDPMine*. By removing those patterns, we shrink the search space in main memory after each iteration. Also, since $\gamma^{(k)}$ increases during each iteration, we remove more patterns as k increases. This memory shrinking technique is illustrated in Figure 2.

In order to compute $maxGain(p)$, we could consider all the possible values of $\boldsymbol{\mu}$ by using linear programming. However, we can compute $maxGain(p)$ efficiently by using the greedy algorithm *greedy_maxGain* presented in Algorithm 3.

Algorithm 3. Greedy Algorithm for $maxGain$

Global Parameter: ω

function $greedy\_maxGain(p)$

$maxGain^+ \leftarrow greedy\_maxGainSub(p, +1)$
$maxGain^- \leftarrow greedy\_maxGainSub(p, -1)$
if $maxGain^+ > maxGain^-$ **then**
    return $maxGain^+$
else
    return $maxGain^-$
end if

function $greedy\_maxGainSub(p, c)$

$maxGain \leftarrow 0$
$weight \leftarrow 1$
$X \leftarrow \{x'_1, x'_2, \ldots, x'_n\}$
while $weight > 0$ **do**
    $x'_{best} = \text{argmax}_{x'_i \in X}\, y_i \cdot s_c(p, x'_i)$
    if $weight \geq \omega$ **then**
        $maxGain \leftarrow maxGain + \omega \cdot y_{best} \cdot s_c(p, x'_{best})$
        $weight \leftarrow weight - \omega$
    else
        $maxGain \leftarrow maxGain + weight \cdot y_{best} \cdot s_c(p, x'_{best})$
        $weight \leftarrow 0$
    end if
    $X \leftarrow X - \{x'_{best}\}$
end while
return $maxGain$

Theorem 3. *The greedy algorithm $greedy\_maxGain(p)$ gives the optimal solution, which is equal to $maxGain(p)$*

Proof. Computing $maxGain(p)$ is very similar to continuous knapsack problem (or fractional knapsack problem) – one of the classic greedy problems. We can think our problem as follows: Suppose that we have n items, each with weight of 1 pound and a value. Also, we have a knapsack with capacity of 1 pound. We can have fractions of items as we want, but not more than ω. The only difference from continuous knapsack problem is that we need to have the knapsack full, and the values of items can be negative. Therefore, the optimality of the greedy algorithm for continuous knapsack problem shows the optimality of $greedy\_maxGain$.

5 Experiments

The major advantages of our method is that it is accurate, efficient in both time and space, produces a small number of expressive features, and operates on different data types. In this section, we evaluate these claims by testing the accuracy, efficiency and expressiveness on two different data types: sequences and trees. For comparison-sake we re-implemented the two baseline approaches described in Section 5.1. All experiments are done on a 3.0GHz Pentium Core 2 Duo computer with 8GB main memory.

5.1 Comparison Baselines

As described in previous sections, *NDPMine* is the only algorithm that uses the direct approach to mine numerical features, therefore we compare *NDPMine* to the two-step process of mining numerical features in computation time and memory usage. Since we have two different types of datasets, sequences and trees, we re-implemented the two-step *SoftMine* algorithm by Lo *et al.* [12] which is only available for sequences. By showing the running time of *NDPMine* and *SoftMine*, we can appropriately compare the computational efficiency of direct and two-step approaches.

In order to show the effectiveness of the numerical feature values used by *NDPMine* over the effectiveness of binary feature values, we re-implemented the binary *DDPMine* algorithm by Cheng *et al.* [3] for sequences and trees. *DDPMine* uses the sequential covering method to avoid forming redundant patterns in a feature set. In the original *DDPMine* algorithm [3], both the Fisher score and information gain were introduced as the measure for discriminative power of patterns; however, for fair comparison of the effectiveness with *SoftMine*, we only use the Fisher score in *DDPMine*.

By comparing the accuracy of both methods, we can appropriately compare the numerical features mined by *NDPMine* with the binary features mined by *DDPMine*.

In order to show the effectiveness of the memory shrinking technique, we implemented our framework in two different versions, one with memory shrinking technique and the other without it.

5.2 Experiments on Sequence Datasets

Sequence data is a ubiquitous data structure. Examples of sequence data include text, DNA sequences, protein sequences, web usage data, and software execution traces. Among several publicly available sequence classification datasets, we chose to use software execution traces from [12]. These software trace datasets contained sequences of nine different software traces. More detail description of the software execution trace datasets is available in [12]. The goal of this classification task was to determine whether a program's execution trace (represented as an instance in the dataset) contains a failure or not. For this task, we needed to define what constitutes a pattern in a sequence and how to count the number of occurrences of a pattern in a sequence. We defined a pattern and the occurrences of a pattern the same as in [12].

5.3 Experiments on Tree Datasets

Datasets in tree structure are also widely available. Web documents in XML are good examples of tree datasets. XML datasets from [20] are one of the commonly used datasets in tree classification studies. However, we collected a very interesting tree dataset for authorship classification. In information retrieval and computational linguistics, authorship classification is one of the classic problems. Authorship classification aims to classify the author of a document. In order to attempt this difficult problem with our *NDPMine* algorithm, we randomly chose 4 authors – Jack Healy, Eric Dash, Denise Grady, and Gina Kolata – and collected 100 documents for each author from *NYTimes.com*. Then, using the Stanford parser [18], we parsed each sentence into a tree of *POS*(Part of Speech) tags. We assumed that these trees reflected the author's writing style and thus could be used in authorship classification. Since a document consisted of multiple sentences, each document was parsed into a set of labeled trees where its author's name was used as its class label for classification. We used induced subtree patterns as features in classification. The formal definition of induced subtree patterns can be found in [4]. We defined the number of occurrences of a pattern in a document is the number of sentences in the document that contained the pattern. We mined frequent induced subtree patterns with several pruning techniques similar to *CMTreeMiner* [4], the-state-of-art tree mining algorithm. Since the goal of this classification task was to determine the author of each document, all pairs of authors and their documents were combined to make two-class classification dataset.

5.4 Parameter Selection

Besides the definition of a pattern and the occurrence counting function for a given dataset, *NDPMine* algorithm needs two parameters as input: (1) the minimum support threshold θ, and (2) the misclassification cost parameter ν. The θ parameter was given as input. The ν parameter was tuned in the same way as SVM tunes its parameters: using cross-validation on the training dataset.

DDPMine and *SoftMine* are dependent on two parameters: (1) the minimum support threshold θ, and (2) the sequential coverage threshold δ. Because we were comparing these algorithms to *NDPMine* in accuracy and efficiency, for sequence and tree datasets, we selected parameters which were best suited to each task.

First, we fixed $\delta = 10$ for the sequence datasets as suggested in [12], and $\delta = 20$ for the tree datasets. Then, we found the appropriate minimum support θ in which *DDPMine* and *SoftMine* performed their best. Thus, we set $\theta = 0.05$ for the sequence datasets and $\theta = 0.01$ for the tree datasets.

5.5 Computation Efficiency Evaluation

We discussed in Section 1 that some pattern-based classification models can be inefficient because they use the two-step mining process. We compared the computation efficiency of the two-step mining algorithm *SoftMine* with *NDPMine* as θ varies. The sequential coverage threshold is fixed to the value from Section 5.4. Due to the limited space, we only show the running time for each algorithm on the *schedule* dataset and the $\langle$D. Grady, G. Kolata$\rangle$ dataset in Figure 2. Other datasets showed similar results.

We see from the graphs in Figure 2 that *NDPMine* outperforms *SoftMine* by an order of magnitude. Although the running times are similar for larger values of θ, the results show that the direct mining approach used in *NDPMine* is computationally more efficient than the two-step mining approach used in *SoftMine*.

5.6 Memory Usage Evaluation

As discussed in Section 3, *NDPMine* uses memory shrinking technique which prunes the search space in main memory during each iteration. We evaluated the effectiveness of this technique by comparing the memory usage of *NDPMine* with the memory shrinking technique to *NDPMine* without the memory shrinking technique. Memory usage is evaluated in terms of the number of the size (in megabytes) of the memory heap.

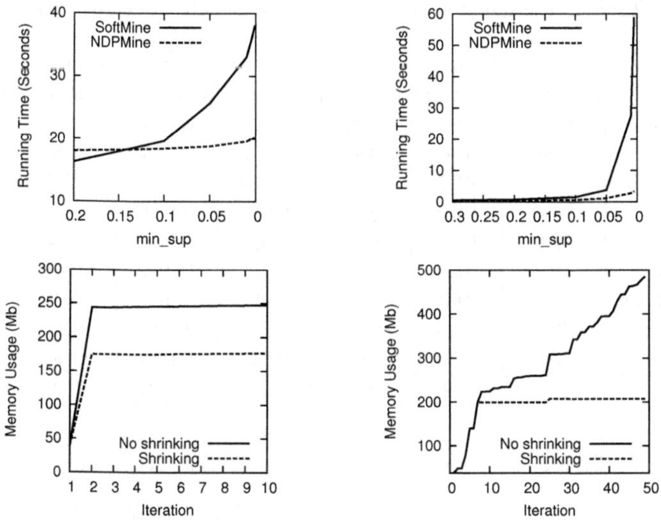

(a) Sequence (b) Tree

Fig. 2. Running Time and Memory Usage

Figure 2 shows the memory usage time for each algorithm on the *schedule* dataset and the ⟨D. Grady, G. Kolata⟩ dataset. We set $\theta = 0$ in order to use as much memory as possible. We see from the graphs in Figure 2 that *NDPMine* with memory shrinking technique is more memory efficient than *NDPMine* without memory shrinking. Although the memory space expands roughly at the same rate initially, the search space shrinking begins to save space as soon as $\gamma^{(k)}$ increases. The difference between the sequence dataset and the tree dataset in Figure 2 is because the search spaces of the tree datasets are much larger than the search spaces of the sequence datasets.

5.7 Accuracy Evaluation

We discussed in Section 1 that some pattern-based classification algorithms can only mine binary feature values, and therefore may not be able to learn an accurate classification model. For evaluation purposes, we compared the accuracy of the classification model learned with features from *NDPMine* to the classification model learned with features from *DDPMine* and *SoftMine* for the sequence and tree datasets. After the feature set was formed, an SVM (from the LIBSVM [1] package) with linear kernel was used to learn a classification model. The accuracy of each model was also measured by 5-fold cross validation. Table 2 shows the results for each algorithm in the sequence datasets. Similarly, Table 3 shows the results in the tree datasets. The accuracy is defined as the number of true positives and true negatives over the total number of examples, and determined by 5-fold cross validation.

In the sequence dataset, the pattern search space is relatively small and the classification tasks are easy. Thus, Table 2 shows marginal improvements. However, for the tree dataset, which has larger pattern search space, and enough difficulty for classification, our method shows the improvements clearly.

Table 2. The summary of results on software behavior classification

	Accuracy			Running Time		Number of Patterns	
Software	*DDPMine*	*SoftMine*	*NDPMine*	*SoftMine*	*NDPMine*	*SoftMine*	*NDPMine*
x11	93.2	**100**	**100**	0.002	0.008	17.0	6.6
cvs_omission	**100**	**100**	**100**	0.008	0.014	88.8	3.0
cvs_ordering	96.4	**96.7**	96.1	0.025	0.090	103.2	24.2
cvs_mix	96.4	94.2	**97.5**	0.020	0.061	34.6	10.6
tot_info	**92.8**	91.2	92.7	0.631	0.780	136.4	25.6
schedule	92.2	**92.5**	90.4	25.010	24.950	113.8	16.2
print_tokens	96.6	**100**	99.6	11.480	24.623	76.4	27.4
replace	85.3	**90.8**	90.0	0.325	1.829	51.6	15.4
mysql	100	95.0	**100**	0.024	0.026	11.8	2.0
Average	94.8	95.6	**96.2**	4.170	5.820	70.4	14.5

Table 3. The summary of results on authorship classification

	Accuracy			Running Time		Number of Patterns	
Author Pair	*DDPMine*	*SoftMine*	*NDPMine*	*SoftMine*	*NDPMine*	*SoftMine*	*NDPMine*
⟨J. Healy, E. Dash⟩	89.5	91.5	**93.5**	43.83	1.45	42.6	24.6
⟨J. Healy, D. Grady⟩	94.0	94.0	**96.5**	52.84	1.26	47.2	19.4
⟨J. Healy, G. Kolata⟩	93.0	95.0	**96.5**	46.48	0.86	40.0	8.8
⟨E. Dash, D. Grady⟩	91.0	89.5	**95.0**	35.43	1.77	32.0	28.2
⟨E. Dash, G. Kolata⟩	92.0	90.5	**98.0**	45.94	1.39	43.8	18.8
⟨D. Grady, G. Kolata⟩	78.0	84.0	**86.0**	71.01	6.89	62.0	53.4
Average	89.58	90.75	**94.25**	49.25	2.27	44.6	25.53

These results confirm our hypothesis that numerical features, like those mined by *NDPMine* and *SoftMine*, may be used to learn more accurate models than binary features like those mined by *DDPMine*. We also confirm that feature selection by LP results in a better feature set than feature selection by sequential coverage.

5.8 Expressiveness Evaluation

We also see from the results in Tables 2 and 3 that the numbers of patterns mined by *NDPMine* are typically smaller than those of *SoftMine*, yet the accuracy is similar or better. Because *NDPMine* and *SoftMine* both use SVM and mine numerical features in common, we can conclude that the feature set mined by *NDPMine* must be more expressive than the features mined by *SoftMine*.

Also, we observed that *NDPMine* mines more discriminative patterns for harder classification datasets and fewer for easier datasets under the same parameters θ, ν. We measured this by the correlation between the hardness of the classification task and the size of feature set mined by *NDPMine*. Among several hardness measures [8] we determine the separability of two classes in a given dataset as follows: (1) mine all frequent patterns, (2) build a SVM-classifier with linear kernel, and (3) measure the margin of the classifier. Note that SVM builds a classifier by searching the classification boundary with maximum margin. The margin can be interpreted as the separability of two classes.

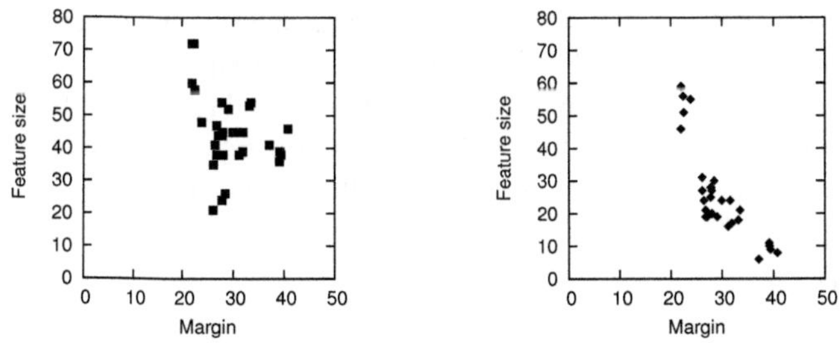

(a) *SoftMine* (b) *NDPMine*

Fig. 3. The correlation between the hardness of Classification tasks and feature sizes

If the margin is large, it implies that the classification task is easy. Next, we computed the correlation between the hardness of a classification task and the feature set size of *NDPMine* by using Pearson product-moment correlation coefficient (PMCC). A larger PMCC implies stronger correlation; conversely, a PMCC of 0 implies that there is no correlation between two variables. We investigated on the tree dataset, and drew the 30 points in Figure 3 (there are six pairs of authors and each pair has 5 testdata). The result in Figure 3 shows a correlation of -0.831 for *NDPMine* and -0.337 for *SoftMine*. For the sequence dataset, the correlations are -0.28 and -0.08 for *NDPMine* and *SoftMine*, respectively. Thus, we confirmed that *NDPMine* mines more patterns if the given classification task is more difficult. This is a very desired property for discriminative pattern mining algorithms in pattern-based classification.

6 Conclusions

Frequent pattern-based classification methods have shown their effectiveness at classifying large and complex datasets. Until recently, existing methods which mine a set of frequent patterns either use the two-step mining process which is computationally inefficient or can only operate on binary features. Due to the explosive number of potential features, the two-step process poses great computational challenges for feature mining. Conversely, those algorithms which use a direct pattern mining approach are not capable of mining numerical features. We showed that the number of occurrences of a pattern in an instance is more important than whether a pattern exists or not by extensive experiments on the software behavior classification and authorship classification datasets.

To our knowledge, there exists no discriminative pattern mining algorithm which can directly mine discriminative patterns as numerical features. In this study, we proposed a pattern-based classification approach which efficiently mines discriminative patterns as numerical features for classification *NDPMine*. A linear programming method is integrated into the pattern mining process, and a branch-and-bound search is employed to navigate the search space. A shrinking technique is applied to the search space storage procedure which reduces the search space significantly. Although *NDPMine* is a model-based algorithm, the final output from the algorithm is a set of features that can be used independently for other classification models.

Experimental results show that *NDPMine* achieves: (1) orders of magnitude speedup over two-step methods without degrading classification accuracy, (2) significantly higher accuracy than binary feature methods, and (3) better efficiency in space by using memory shrinking technique. In addition, we argue that the features mined by *NDPMine* can be more expressive than those mined by current techniques.

References

1. Chang, C.-C., Lin, C.-J.: LIBSVM: a Library for Support Vector Machines (2001), Software is available for download, at `http://www.csie.ntu.edu.tw/~cjlin/libsvm/`
2. Cheng, H., Yan, X., Han, J., Hsu, C.-W.: Discriminative frequent pattern analysis for effective classification. In: ICDE (2007)
3. Cheng, H., Yan, X., Han, J., Yu, P.S.: Direct discriminative pattern mining for effective classification. In: ICDE (2008)
4. Chi, Y., Xia, Y., Yang, Y., Muntz, R.R.: Mining closed and maximal frequent subtrees from databases of labeled rooted trees. IEEE Transactions on Knowledge and Data Engineering (TKDE) 17(2), 190–202 (2005)
5. Demiriz, A., Bennett, K.P., Shawe-Taylor, J.: Linear programming boosting via column generation. Machine Learning 46(1-3), 225–254 (2002)
6. Fan, W., Zhang, K., Cheng, H., Gao, J., Yan, X., Han, J., Yu, P.S., Verscheure, O.: Direct mining of discriminative and essential frequent patterns via model-based search tree. In: KDD (2008)
7. Freund, Y., Schapire, R.E.: A decision-theoretic generalization of on-line learning and an application to boosting. Journal of Computer and System Sciences 55(1), 119–139 (1997)
8. Ho, T.K., Basu, M.: Complexity measures of supervised classification problems. IEEE Trans. Pattern Anal. Mach. Intell. 24(3), 289–300 (2002)
9. Levy, S., Stormo, G.D.: Dna sequence classification using dawgs. In: Structures in Logic and Computer Science, A Selection of Essays in Honor of Andrzej Ehrenfeucht, London, UK, pp. 339–352. Springer, Heidelberg (1997)
10. Li, W., Han, J., Pei, J.: Cmar: Accurate and efficient classification based on multiple class-association rules. In: ICDM, pp. 369–376 (2001)
11. Liu, B., Hsu, W., Ma, Y.: Integrating classification and association rule mining. In: KDD, pp. 80–86 (1998)
12. Lo, D., Cheng, H., Han, J., Khoo, S.-C., Sun, C.: Classification of software behaviors for failure detection: A discriminative pattern mining approach. In: KDD (2009)
13. Nash, S.G., Sofer, A.: Linear and Nonlinear Programming. McGraw-Hill, New York (1996)
14. Nowozin, S., Gökhan Bakõr, K.T.: Discriminative subsequence mining for action classification. In: ICCV (2007)
15. Saigo, H., Kadowaki, T., Kudo, T., Tsuda, K.: A linear programming approach for molecular qsar analysis. In: MLG, pp. 85–96 (2006)
16. Saigo, H., Krämer, N., Tsuda, K.: Partial least squares regression for graph mining. In: KDD (2008)
17. Saigo, H., Nowozin, S., Kadowaki, T., Kudo, T., Tsuda, K.: gboost: a mathematical programming approach to graph classification and regression. Mach. Learn. 75(1), 69–89 (2009)
18. The Stanford Natural Language Processing Group. The Stanford Parser: A statistical parser, `http://www-nlp.stanford.edu/software/lex-parser.shtml`
19. Ye, L., Keogh, E.: Time series shapelets: a new primitive for data mining. In: KDD (2009)
20. Zaki, M.J., Aggarwal, C.C.: Xrules: an effective structural classifier for xml data. In: KDD (2003)

Hidden Conditional Ordinal Random Fields for Sequence Classification

Minyoung Kim and Vladimir Pavlovic

Rutgers University, Piscataway, NJ 08854, USA
{mikim,vladimir}@cs.rutgers.edu
http://seqam.rutgers.edu

Abstract. Conditional Random Fields and Hidden Conditional Random Fields are a staple of many sequence tagging and classification frameworks. An underlying assumption in those models is that the state sequences (tags), observed or latent, take their values from a set of nominal categories. These nominal categories typically indicate tag classes (e.g., part-of-speech tags) or clusters of similar measurements. However, in some sequence modeling settings it is more reasonable to assume that the tags indicate ordinal categories or ranks. Dynamic envelopes of sequences such as emotions or movements often exhibit intensities growing from neutral, through raising, to peak values. In this work we propose a new model family, Hidden Conditional Ordinal Random Fields (H-CORFs), that explicitly models sequençe dynamics as the dynamics of ordinal categories. We formulate those models as generalizations of ordinal regressions to structured (here sequence) settings. We show how classification of entire sequences can be formulated as an instance of learning and inference in H-CORFs. In modeling the ordinal-scale latent variables, we incorporate recent binning-based strategy used for static ranking approaches, which leads to a log-nonlinear model that can be optimized by efficient quasi-Newton or stochastic gradient type searches. We demonstrate improved prediction performance achieved by the proposed models in real video classification problems.

1 Introduction

In this paper we tackle the problem of time-series sequence classification, a task of assigning an entire measurement sequence a label from a finite set of categories. We are particularly interested in classifying videos of real human/animal activities, for example, facial expressions. In analyzing such video sequences, it is often observed that the sequences in nature undergo different phases or intensities of the displayed artifact. For example, facial emotion signals typically follow envelope-like shapes in time: `neutral`, `increase`, `peak`, and `decrease`, beginning with low intensity, reaching a maximum, then tapering off. (See Fig. 1 for the intensity envelope visually marked for an facial emotion video.) Modeling such an envelop is important for faithful representation of motion sequences and consequently for their accurate classification. A key challenge, however, is

J.L. Balcázar et al. (Eds.): ECML PKDD 2010, Part II, LNAI 6322, pp. 51–65, 2010.

that even though the action intensity follows the same qualitative envelope the rates of increase and decrease differ substantially across subjects (e.g., different subjects express the same emotion with substantially different intensities).

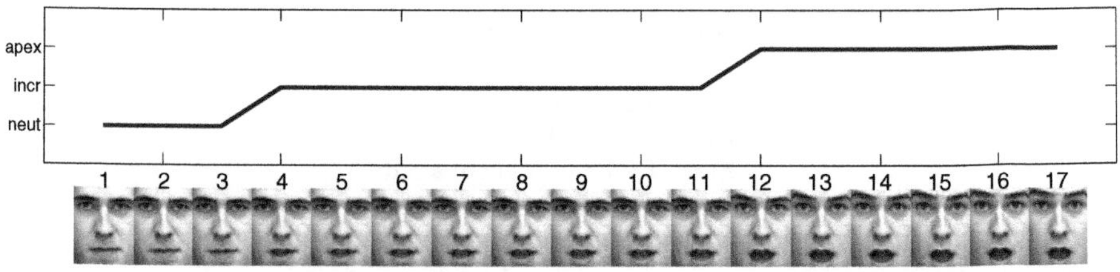

Fig. 1. An example of facial emotion video and corresponding intensity labels. The ordinal-scale labels over time form an intensity envelope (the first half shown here).

We propose a new modeling framework of Hidden Conditional Ordinal Random Fields (H-CORFs) to accomplish the task of sequence classification while imposing the qualitative intensity envelope constraint. H-CORF extends the framework of Hidden Conditional Random Fields (H-CRFs) [12,5] by replacing the hidden layer of H-CRFs category indicator variables with a layer of variables that represent the qualitative but latent intensity envelope. To model this envelope qualitatively yet accurately we require that the state space of each variable be *ordinal*, corresponding to the intensity rank of the modeled activity at any particular time. As a consequence, the hidden layer of H-CORF is a sequence of ordinal values whose differences model qualitative intensity dissimilarities between various stages of an activity. This is distinct from the way the latent dynamics are modeled in traditional H-CRFs, where states represent different categories without imposing their relative ordering. Modeling the dynamic envelope in a qualitative, ordinal manner is also critical for increased robustness. While the envelope could plausibly be modeled as a sequence of real-valued absolute intensity states, such models would inevitably introduce undesired dependencies. In such cases the differences in absolute intensities could be strongly tied to a subject or a manner in which the action is produced, making the models unnecessarily specific while obscuring the sought-after identity of the action.

To model the qualitative shape of the intensity envelope within H-CORF we extend the framework of ordinal regression to structured ordinal sequence spaces. The ordinal regression, often called the preference learning or ranking [6], has found applications in several traditional ranking problems, such as image classification and collaborative filtering [14,2], or image retrieval [7,8]. In the static setting, the goal is to predict the label of an item represented by feature vector $\mathbf{x} \in \mathbb{R}^p$ where the output label bears particular meaning of preference or order (e.g., low, medium or high). The ordinal regression is fundamentally different from the standard regression in that the actual absolute difference of output values is nearly meaningless, but only their relative order matters (e.g., low $<$ medium $<$ high). The ordinal regression problems may not be optimally handled by the standard multi-class classification either because of classifier's ignorance of the ordinal scale and symmetric treatment of different output categories (e.g., low would be equally different from high as it would be from medium).

Despite their success in static settings (i.e., a vectorial input associated with a singleton output label), ranking problems are rarely explored in structured problems, such as the segmentation of emotion signals into regions of neutral, increasing or peak emotion or actions into different intensity stages. In this case the ranks or ordinal labels at different time instances should vary smoothly, with temporally proximal instances likely to have similar ranks. For this purpose we propose an intuitive but principled Conditional Ordinal Random Field (CORF) model that can faithfully represent multiple ranking variables correlated in a combinatorial structure. The binning-based modeling strategy adopted by recent static ranking approaches (see (2) in Sec. 2.1) is incorporated into our structured models, CORF and H-CORF, through graph-based potential functions. While this formulation leads to a family of log-nonlinear models, we show that the models can still be estimated with high accuracy using general gradient-based search approaches.

We formally setup the problem and introduce basic notation below. We then propose a model for prediction of ordinal intensity envelopes in Sec. 2. Our classification model based on the ordinal modeling of the latent envelope is described in Sec. 3. In Sec. 4, the superior prediction performance of the proposed structured ranking model to the regular H-CRF model is demonstrated on two problems/datasets: emotion recognition from the CMU facial expression dataset [11] and behavior recognition from the UCSD mouse dataset [4].

1.1 Problem Setup and Notations

We consider a K-class classification problem, where we let $y \in \{1, ..., K\}$ be the class variable and $\mathbf{x}$ be the input covariate for predicting y. In the structured problems we assume that $\mathbf{x}$ is composed of individual input vectors $\mathbf{x}_r$ measured at the temporal and/or spatial positions r (i.e., $\mathbf{x} = \{\mathbf{x}_r\}$). Although our framework can be applied to arbitrary combinatorial structures for $\mathbf{x}$, in this paper we focus on the sequence data, written as $\mathbf{x} = \mathbf{x}_1 \dots \mathbf{x}_T$ where the sequence length T can vary from instance to instance. Throughout the paper, we assume a supervised setting: we are given a training set of n data pairs $\mathcal{D} = \{(y^i, \mathbf{x}^i)\}_{i=1}^n$, which are i.i.d. samples from an underlying but unknown distribution.

2 Structured Ordinal Modeling of Dynamical Envelope

In this section we develop the model which can be used to infer ordinal dynamical envelope from sequences of measurement. The model is reminiscent of a classical CRF model, where its graphical representation corresponds to the upper two layers in Fig. 2 with the variables $\mathbf{h} = h_1, \dots, h_T$ treated as observed outputs. But unlike the CRF it restricts the envelope (i.e., sequence of tags) to reside in a space of ordinal sequences. This requirement will impose ordinal, rank-like, similarities between different states instead of the nominal differences of

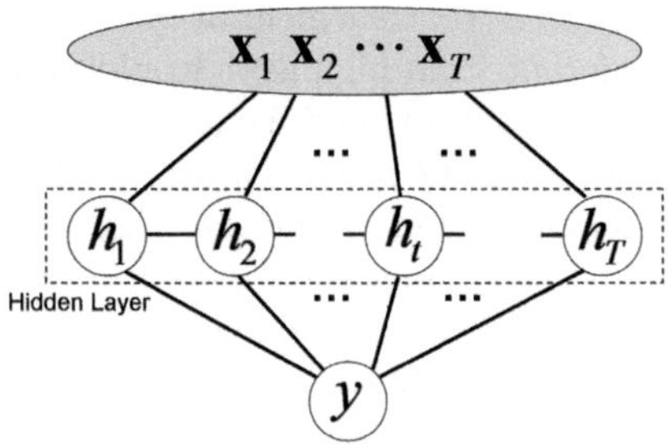

Fig. 2. Graphical representation of H-CRF. Our new model H-CORF (Sec. 3) shares the same structure. The upper two layers form CRF (and CORF in Sec. 2.3) when $\mathbf{h} = h_1, \ldots, h_T$ serves as observed outputs.

classical CRF states. We will refer to this model as the Conditional Ordinal Random Field (CORF). To develop the model we first introduce the framework of static ordinal regression and subsequently show how it can be extended into a structured, sequence setting.

2.1 Static Ordinal Regression

The goal of ordinal regression is to predict the label h of an item represented by a feature vector[1] $\mathbf{x} \in \mathbb{R}^p$ where the output indicates the preference or order of this item. Formally, we let $h \in \{1, \ldots, R\}$ for which R is the number of preference grades, and h takes an ordinal scale from the lowest preference $h = 1$ to the highest $h = R$, $h = 1 \prec h = 2 \prec \ldots \prec h = R$.

The most critical aspect that differentiates the ordinal regression approaches from the multi-class classification methods is the modeling strategy. Assuming a linear model (straightforwardly extendible to a nonlinear version by kernel tricks), the multi-class classification typically (c.f. [3]) takes the form of[2]

$$h = \arg\max_{c \in \{1,\ldots,R\}} \mathbf{w}_c^\top \mathbf{x} + b_c. \tag{1}$$

For each class c, the hyperplane $(\mathbf{w}_c \in \mathbb{R}^p, b_c \in \mathbb{R})$ defines the confidence toward the class c. The class decision is made by selecting the one with the largest confidence. The model parameters are $\{\{\mathbf{w}_c\}_{c=1}^R, \{b_c\}_{c=1}^R\}$. On the other hand, ordinal regression approaches adopt the following modeling strategy:

$$h = c \ \text{ iff } \mathbf{w}^\top \mathbf{x} \in (b_{c-1}, b_c], \quad \text{where } -\infty = b_0 \le b_1 \le \cdots \le b_R = +\infty. \tag{2}$$

The binning parameters $\{b_c\}_{c=0}^R$ form R different bins, where their adjacent placement and the output deciding protocol of (2) naturally enforce the ordinal scale criteria. The parameters of the model become $\{\mathbf{w}, \{b_c\}_{c=0}^R\}$, far fewer

[1] We use the notation $\mathbf{x}$ interchangeably for both a sequence observation $\mathbf{x} = \{\mathbf{x}_r\}$ and a vector, which is clearly distinguished by context.

[2] This can be seen as a general form of the popular one-vs-all or one-vs-one treatment for the multi-class problem.

in count than those of the classification models. The state-of-the-art Support Vector Ordinal Regression (SVOR) algorithms [14,2] conform to this representation while they aim to maximize margins at the nearby bins in the SVM-like formulation.

2.2 Conditional Random Field (CRF) for Sequence Segmentation

CRF [10,9] is a structured output model which represents the distribution of a set (sequence) of categorical tags $\mathbf{h} = \{h_r\}, h_r \in \{1, \ldots, R\}$, conditioned on input $\mathbf{x}$. More formally, the density $P(\mathbf{h}|\mathbf{x})$ has a Gibbs form clamped on the observation $\mathbf{x}$:

$$P(\mathbf{h}|\mathbf{x}, \boldsymbol{\theta}) = \frac{1}{Z(\mathbf{x}; \boldsymbol{\theta})} e^{s(\mathbf{x}, \mathbf{h}; \boldsymbol{\theta})}. \tag{3}$$

Here $Z(\mathbf{x}; \boldsymbol{\theta}) = \sum_{\mathbf{h} \in \mathcal{H}} e^{s(\mathbf{x}, \mathbf{h}; \boldsymbol{\theta})}$ is the partition function on the space of possible configurations $\mathcal{H}$, and $\boldsymbol{\theta}$ are the parameters[3] of the *score function* $s(\cdot)$.

The choice of the output graph $G = (V, E)$ on $\mathbf{h}$ critically affects model's representational capacity and the inference complexity. For convenience, we further assume that we have either *node* cliques ($r \in V$) or *edge* cliques ($e = (r, s) \in E$) with corresponding features, $\boldsymbol{\Psi}_r^{(V)}(\mathbf{x}, h_r)$ and $\boldsymbol{\Psi}_e^{(E)}(\mathbf{x}, h_r, h_s)$. By letting $\boldsymbol{\theta} = \{\mathbf{v}, \mathbf{u}\}$ be the parameters for node and edge features, respectively, the score function is typically defined as:

$$s(\mathbf{x}, \mathbf{h}; \boldsymbol{\theta}) = \sum_{r \in V} \mathbf{v}^\top \boldsymbol{\Psi}_r^{(V)}(\mathbf{x}, h_r) + \sum_{e=(r,s) \in E} \mathbf{u}^\top \boldsymbol{\Psi}_e^{(E)}(\mathbf{x}, h_r, h_s). \tag{4}$$

In conventional modeling practice, the node/edge features are often defined as products of measurement features confined to cliques and the output class indicators. For instance, in CRFs with sequence [10] and lattice outputs [9,17] we often have

$$\boldsymbol{\Psi}_r^{(V)}(\mathbf{x}, h_r) = \Big[I(h_r = 1), \cdots, I(h_r = R)\Big]^\top \otimes \boldsymbol{\phi}(\mathbf{x}_r), \tag{5}$$

where $I(\cdot)$ is the indicator function and $\otimes$ denotes the Kronecker product. Hence the k-th block ($k = 1, \ldots, R$) of $\boldsymbol{\Psi}_r^{(V)}(\mathbf{x}, h_r)$ is $\boldsymbol{\phi}(\mathbf{x}_r)$ if $h_r = k$, and the $\mathbf{0}$-vector otherwise. The edge feature may typically assess the absolute difference between the measurements at adjoining nodes,

$$\Big[I(h_r = k \ \wedge \ h_s = l)\Big]_{R \times R} \otimes \big|\boldsymbol{\phi}(\mathbf{x}_r) - \boldsymbol{\phi}(\mathbf{x}_s)\big|. \tag{6}$$

Learning and inference in CRFs has been studied extensively in the past decade, c.f. [10,9,17], with many efficient and scalable algorithms, particularly for sequential structures.

[3] For brevity, we often drop the dependency on $\boldsymbol{\theta}$ in our notation.

2.3 Conditional Ordinal Random Field (CORF)

A standard CRF model seeks to *classify*, treating each output category nominally and equally different from all other categories. The consequence is that the model's node potential has a direct analogy to the static multi-class classification model of (1): For $h_r = c$, the node potential equals $\mathbf{v}_c^\top \boldsymbol{\phi}(\mathbf{x}_r)$ where $\mathbf{v}_c$ is the c-th block of $\mathbf{v}$, or the c-th hyperplane $\mathbf{w}_c^\top \mathbf{x}_r + b_c$ in (1). The max can be replaced by the *softmax* function. To setup an exact equality, one can let $\boldsymbol{\phi}(\mathbf{x}_r) = [1, \mathbf{x}_r^\top]^\top$.

Conversely, the modeling strategy of the static ordinal regression methods such as (2) can be merged with the CRF through the node potentials to yield a structured output ranking model. However, the mechanism of doing so is not obvious because of the highly discontinuous nature of (2). Instead, we base our approach on the probabilistic model for ranking proposed by [1], which shares the notion of (2).

In [1], the noiseless probabilistic ranking likelihood is defined as

$$P_{ideal}(h = c|f(\mathbf{x})) = \begin{cases} 1 \text{ if } f(\mathbf{x}) \in (b_{c-1}, b_c] \\ 0 \text{ otherwise} \end{cases} \tag{7}$$

Here $f(\mathbf{x})$ is the model to be learned, which could be linear $f(\mathbf{x}) = \mathbf{w}^\top \mathbf{x}$. The effective ranking likelihood is constructed by contaminating the ideal model with noise. Under the Gaussian noise δ and after marginalization, one arrives at the ranking likelihood

$$P(h=c|f(\mathbf{x})) = \int_\delta P_{ideal}(h=c|f(\mathbf{x})+\delta) \cdot \mathcal{N}(\delta; 0, \sigma^2) d\delta = \Phi\left(\frac{b_c - f}{\sigma}\right) - \Phi\left(\frac{b_{c-1} - f}{\sigma}\right), \tag{8}$$

where $\Phi(\cdot)$ is the standard normal cdf, and σ is the parameter that controls the steepness of the likelihood function.

Now we set the node potential at node r of the CRF to be the log-likelihood of (8), that is,

$$\begin{aligned} &\mathbf{v}^\top \boldsymbol{\Psi}_r^{(V)}(\mathbf{x}, h_r) \longrightarrow \boldsymbol{\Gamma}_r^{(V)}(\mathbf{x}, h_r; \{\mathbf{a}, \mathbf{b}, \sigma\}), \text{ where} \\ &\boldsymbol{\Gamma}_r^{(V)}(\mathbf{x}, h_r) := \textstyle\sum_{c=1}^R I(h_r = c) \cdot \log \left(\Phi\left(\frac{b_c - \mathbf{a}^\top \boldsymbol{\phi}(\mathbf{x}_r)}{\sigma}\right) - \Phi\left(\frac{b_{c-1} - \mathbf{a}^\top \boldsymbol{\phi}(\mathbf{x}_r)}{\sigma}\right)\right). \end{aligned} \tag{9}$$

Here, $\mathbf{a}$ (having the same dimension as $\boldsymbol{\phi}(\mathbf{x}_r)$), $\mathbf{b} = [-\infty = b_0, \ldots, b_R = +\infty]^\top$, and σ are the new parameters, in contrast with the original CRF's node parameters $\mathbf{v}$. Substituting this expression into (4) leads to a new conditional model for structured ranking,

$$P(\mathbf{h}|\mathbf{x}, \boldsymbol{\omega}) \propto \exp\left(s(\mathbf{x}, \mathbf{h}; \boldsymbol{\omega})\right), \text{ where} \tag{10}$$

$$s(\mathbf{x}, \mathbf{h}; \boldsymbol{\omega}) = \sum_{r \in V} \boldsymbol{\Gamma}_r^{(V)}(\mathbf{x}, h_r; \{\mathbf{a}, \mathbf{b}, \sigma\}) + \sum_{e=(r,s) \in E} \mathbf{u}^\top \boldsymbol{\Psi}_e^{(E)}(\mathbf{x}, h_r, h_s). \tag{11}$$

We refer to this model as *CORF*, the Conditional Ordinal Random Field. The parameters of the CORF are denoted as $\boldsymbol{\omega} = \{\mathbf{a}, \mathbf{b}, \sigma, \mathbf{u}\}$, with the ordering constraint $b_i < b_{i+1}, \forall i$. Note that the number of parameters is significantly fewer than that of the regular CRF. Unlike CRF's log-linear form, the CORF becomes a *log-nonlinear* model, effectively imposing the ranking criteria via nonlinear binning-based modeling of the node potential $\boldsymbol{\Gamma}$.

Model Learning. We briefly discuss how the CORF model can be learned using gradient ascent. For the time being we assume that we are given labeled data pairs $(\mathbf{x}, \mathbf{h})$, a typical setting for CRF learning, although we treat $\mathbf{h}$ as latent variables for the H-CORF sequence classification model in Sec. 3.

First, it should be noted that CORF's log-nonlinear modeling does not impose any additional complexity on the inference task. Since the graph topology remains the same, once the potentials are evaluated, the inference follows exactly the same procedures as that of the standard log-linear CRFs. Second, it is not difficult to see that the node potential $\boldsymbol{\Gamma}_r^{(V)}(\mathbf{x}, h_r)$, although non-linear, remains concave.

Unfortunately, the overall learning of CORF is non-convex because of the log-partition function (*log-sum-exp* of nonlinear concave functions). However, the log-likelihood objective is bounded above by 0, and the quasi-Newton or the stochastic gradient ascent [17] can be used to estimate the model parameters. The gradient of the log-likelihood w.r.t. $\mathbf{u}$ is (the same as the regular CRF):

$$\frac{\partial \log P(\mathbf{h}|\mathbf{x}, \boldsymbol{\omega})}{\partial \mathbf{u}} = \sum_{e=(r,s)\in E} \left(\boldsymbol{\Psi}_e^{(E)}(\mathbf{x}, h_r, h_s) - \mathbb{E}_{P(h_r,h_s|\mathbf{x})}\left[\boldsymbol{\Psi}_e^{(E)}(\mathbf{x}, h_r, h_s) \right] \right). \tag{12}$$

The gradient of the log-likelihood w.r.t. $\mu = \{\mathbf{a}, \mathbf{b}, \sigma\}$ can be derived as:

$$\frac{\partial \log P(\mathbf{h}|\mathbf{x}, \boldsymbol{\omega})}{\partial \mu} = \sum_{r\in V} \left(\frac{\partial \boldsymbol{\Gamma}_r^{(V)}(\mathbf{x}, h_r)}{\partial \mu} - \mathbb{E}_{P(h_r|\mathbf{x})}\left[\frac{\partial \boldsymbol{\Gamma}_r^{(V)}(\mathbf{x}, h_r)}{\partial \mu} \right] \right), \tag{13}$$

where the gradient of the node potential can be computed analytically,

$$\frac{\partial \boldsymbol{\Gamma}_r^{(V)}(\mathbf{x}, h_r)}{\partial \mu} = \sum_{c=1}^{R} I(h_r = c) \frac{\mathcal{N}(z_0(r,c); 0, 1) \cdot \frac{\partial z_0(r,c)}{\partial \mu} - \mathcal{N}(z_1(r,c); 0, 1) \cdot \frac{\partial z_1(r,c)}{\partial \mu}}{\Phi(z_0(r,c)) - \Phi(z_1(r,c))},$$

$$\text{where } z_k(r,c) = \frac{b_{c-k} - \mathbf{a}^\top \phi(\mathbf{x}_r)}{\sigma} \text{ for } k = 0, 1. \tag{14}$$

Model Reparameterization for Unconstrained Optimization. The gradient-based learning proposed above has to be accomplished while respecting two sets of constraints: (i) the order constraints on $\mathbf{b}$: $\{b_{j-1} \leq b_j$ for $j = 1, \ldots, R\}$, and (ii) the positive scale constraint on σ: $\{\sigma > 0\}$. Instead of general constrained optimization, we introduce a reparameterization that effectively reduces the problem to an unconstrained optimization task.

To deal with the order constraints in the parameters $\mathbf{b}$, we introduce the displacement variables δ_k, where $b_j = b_1 + \sum_{k=1}^{j-1} \delta_k^2$ for $j = 2, \ldots, R-1$. So, $\mathbf{b}$

is replaced by the unconstrained parameters $\{b_1, \delta_1, \ldots, \delta_{R-2}\}$. The positiveness constraint for σ is simply handled by introducing the free parameter σ_0 where $\sigma = \sigma_0^2$. Hence, the unconstrained node parameters are: $\{\mathbf{a}, b_1, \delta_1, \ldots, \delta_{R-2}, \sigma_0\}$. Then the gradients for $\frac{\partial z_k(r,c)}{\partial \mu}$ in (14) then become:

$$\frac{\partial z_k(r,c)}{\partial \mathbf{a}} = -\frac{1}{\sigma_0^2}\boldsymbol{\phi}(\mathbf{x}_r), \quad \frac{\partial z_k(r,c)}{\partial \sigma_0} = -\frac{2\big(b_{c-k} - \mathbf{a}^\top \boldsymbol{\phi}(\mathbf{x}_r)\big)}{\sigma_0^3}, \quad \text{for } k = 0, 1. \quad (15)$$

$$\frac{\partial z_0(r,c)}{\partial b_1} = \begin{cases} 0 & \text{if } c = R \\ \frac{1}{\sigma_0^2} & \text{otherwise} \end{cases}, \quad \frac{\partial z_1(r,c)}{\partial b_1} = \begin{cases} 0 & \text{if } c = 1 \\ \frac{1}{\sigma_0^2} & \text{otherwise} \end{cases}. \quad (16)$$

$$\frac{\partial z_0(r,c)}{\partial \delta_j} = \begin{cases} 0 & \text{if } c \in \{1, \ldots, j, R\} \\ \frac{2\delta_j}{\sigma_0^2} & \text{otherwise} \end{cases}, \quad \frac{\partial z_1(r,c)}{\partial \delta_j} = \begin{cases} 0 & \text{if } c \in \{1, \ldots, j+1\} \\ \frac{2\delta_j}{\sigma_0^2} & \text{otherwise} \end{cases},$$
$$\text{for } j = 1, \ldots, R-2. \quad (17)$$

We additionally employ parameter regularization on the CORF model. For $\mathbf{a}$ and $\mathbf{u}$, we use the typical L2 regularizers $||\mathbf{a}||^2$ and $||\mathbf{u}||^2$. No specific regularization is necessary for the binning parameters b_1 and $\{\delta_j\}_{j=1}^{R-2}$ as they will be automatically adjusted according to the score $\mathbf{a}^\top \boldsymbol{\phi}(\mathbf{x}_r)$. For the scale parameter σ_0 we consider $(\log \sigma_0^2)^2$ as the regularizer, which essentially favors $\sigma_0 \approx 1$ and imposes quadratic penalty in log-scale.

3 Hidden Conditional Ordinal Random Field (H-CORF)

We now propose an extension of the CORF model to a sequence classification setting. The model builds upon the method for extending CRFs for classification, known as Hidden CRFs (H-CRF). H-CRF is a probabilistic classification model $P(y|\mathbf{x})$ that can be seen as a combination of K CRFs, one for each class. The CRF's output variables $\mathbf{h} = h_1, \ldots, h_T$ are now treated as latent variables (Fig. 2). H-CRF has been studied in the fields of computer vision [12,18] and speech recognition [5]. We use the same approach to combine individual CORF models as building blocks for sequence classification in the Hidden CORF setting, a structured ordinal regression model with latent variables.

To build a classification model from CORFs, we introduce a class variable $y \in \{1, \ldots, K\}$ and a new score function

$$s(y, \mathbf{x}, \mathbf{h}; \boldsymbol{\Omega}) = \sum_{k=1}^{K} I(y = k) \cdot s(\mathbf{x}, \mathbf{h}; \boldsymbol{\omega}_k)$$
$$= \sum_{k=1}^{K} I(y = k) \cdot \left[\sum_{r \in V} \boldsymbol{\Gamma}_r^{(V)}(\mathbf{x}, h_r; \{\mathbf{a}_k, \mathbf{b}_k, \sigma_k\}) + \sum_{e=(r,s) \in E} \mathbf{u}_k^\top \boldsymbol{\Psi}_e^{(E)}(\mathbf{x}, h_r, h_s) \right], \quad (18)$$

where $\boldsymbol{\Omega} = \{\boldsymbol{\omega}_k\}_{k=1}^{K}$ denotes the compound H-CORF parameters comprised of K CORFs $\boldsymbol{\omega}_k = \{\mathbf{a}_k, \mathbf{b}_k, \sigma_k, \mathbf{u}_k\}$ for $k = 1, \ldots, K$. The score function, in turn, defines the joint and class conditional distributions:

$$P(y, \mathbf{h}|\mathbf{x}) = \frac{\exp(s(y, \mathbf{x}, \mathbf{h}))}{Z(\mathbf{x})}, \quad P(y|\mathbf{x}) = \sum_{\mathbf{h}} P(y, \mathbf{h}|\mathbf{x}) = \frac{\sum_{\mathbf{h}} \exp(s(y, \mathbf{x}, \mathbf{h}))}{Z(\mathbf{x})}. \tag{19}$$

Evaluation of the class-conditional $P(y|\mathbf{x})$ depends on the partition function $Z(\mathbf{x}) = \sum_{y,\mathbf{h}} \exp(s(y, \mathbf{x}, \mathbf{h}))$ and the class-latent joint posteriors $P(y, h_r, h_s|\mathbf{x})$. Both can be computed from independent consideration of K individual CORFs. The compound partition function is the sum of individual partition functions, $Z(\mathbf{x}) = \sum_k Z(\mathbf{x}|y = k) = \sum_k \sum_{\mathbf{h}} \exp(s(k, \mathbf{x}, \mathbf{h}))$, computed in each CORF. Similarly, the joint posteriors can evaluated as $P(y, h_r, h_s|\mathbf{x}) = P(h_r, h_s|\mathbf{x}, y) \cdot P(y|\mathbf{x})$. Learning the H-CORF can be done by maximizing the class conditional log-likelihood $\log P(y|\mathbf{x})$, where its gradient can be derived as:

$$\frac{\partial \log P(y|\mathbf{x})}{\partial \Omega} = \mathbb{E}_{P(\mathbf{h}|\mathbf{x},y)}\left[\frac{\partial s(y, \mathbf{x}, \mathbf{h})}{\partial \Omega}\right] - \mathbb{E}_{P(y,\mathbf{h}|\mathbf{x})}\left[\frac{\partial s(y, \mathbf{x}, \mathbf{h})}{\partial \Omega}\right]. \tag{20}$$

Using the gradient derivation (12)-(14) for the CORF, it is straightforward to compute the expectations in (20). Finally, the assignment of a measurement sequence to a particular class, such as the action or emotion, is accomplished by the MAP rule $y^* = \arg\max_y P(y|\mathbf{x})$.

4 Evaluations

In this section we demonstrate the performance of our model with ordinal latent state dynamics, the H-CORF. We evaluate algorithms on two datasets/tasks: facial emotion recognition from the CMU facial expression video dataset and behavior recognition from the UCSD mouse dataset.

4.1 Recognizing Facial Emotions from Videos

We consider the task of the facial emotion recognition. We use the Cohn-Kanade facial expression database [11], which consists of six basic emotions (anger, disgust, fear, happiness, sadness, and surprise) performed by 100 students, 18 to 30 years old. In this experiment, we selected image sequences from 93 subjects, each of which enacts 2 to 6 emotions. Overall, the number of sequences is 352 where the class proportions are as follows: anger(36), disgust(42), fear(54), happiness(85), sadness(61), and surprise(74). For this 6-class problem, we randomly select 60%/40% of the sequences as training/testing, respectively. The training and the testing sets do not have sequences of the same subject. After detecting faces with the cascaded face detector [16], we normalize them into (64×64) images which are aligned based on the eye locations similar to [15].

Unlike the previous static emotion recognition approaches (e.g., [13]) where just the ending few peak frames are considered, we use the entire sequences that cover the onset state of the expression to the apex in order to conduct the task of dynamic emotion recognition. The sequence lengths are, on average, about 20 frames long. Fig. 3 shows some example sequences. We consider the qualitative

(a) Anger

(b) Disgust

(c) Fear

(d) Happiness

(e) Sadness

(f) Surprise

Fig. 3. Sample sequences for six emotions from the Cohn-Kanade dataset

intensity state of size $R = 3$, based on typical representation of three ordinal categories used to describe the emotion dynamics: `neutral` < `increasing` < `apex`. Note that we impose no actual prior knowledge of the category dynamics nor the correspondence of the three states to the qualitative categories described above. This correspondence can be established by interpreting the model learned in the estimation stage, as we demonstrate next. For the image features, we first extract the Haar-like features, following [20]. To reduce feature dimensionality, we apply PCA on the training frames for each emotion, which gives rise to 30-dimensional feature vectors corresponding to 90% of the total energy.

The recognition test errors are shown in Table 1. Here we also contrasted with the baseline generative approach based on a Gaussian Hidden Markov Model (GHMM). See also the confusion matrices of H-CRF and H-CORF in Fig. 4. Our model with ordinal dynamics leads to significant improvements in classification performance over both prior models.

To gain insight about the modeling ability of the new approach, we studied the latent intensity envelopes learned during the model estimation phase. Fig. 5 depicts a set of most likely latent envelopes estimated on a sample of test sequences. The decoded envelopes by our model correspond to typical visual changes in the emotion intensities, qualified by the three categories (neutral, increase, apex). On the other hand, the decoded states by the H-CRF model have weaker correlation with the three target intensity categories, typically exhibiting highly diverse scales and/or orders across the six emotions. The ability of the ordinal model to recover perceptually distinct dynamic categories from data may further explain the model's good classification performance.

Table 1. Recognition accuracy on CMU emotion video dataset

Methods	GHMM	H-CRF	H-CORF
Accuracy	72.99%	78.10%	89.05%

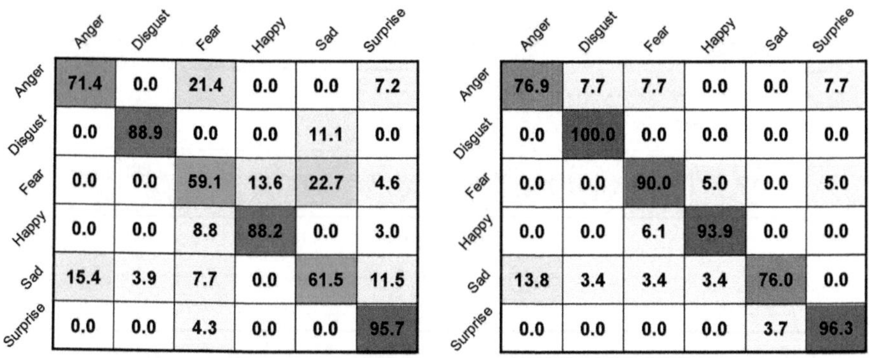

(a) H-CRF (b) (Proposed) H-CORF

Fig. 4. Confusion matrices for facial emotion recognition on CMU database

4.2 Behavior Recognition from UCSD Mouse Dataset

We next consider the task of behavior recognition from video, a very important problem in computer vision. We used the mouse dataset from the UCSD vision group[4]. The dataset contains videos of 5 different mouse behaviors (drink, eat, explore, groom, and sleep). See Fig. 6 for some sample frames. The video clips are taken at 7 different points in the day, separately kept as 7 different sets. The characteristics of each behavior vary substantially among each of the seven sets. From the original dataset, we select a subset comprised of 75 video clips (15 videos for each behavior) from 5 sets. Each video lasts between 1 and 10 seconds. For the recognition setting, we take one of the 5 sets having the largest number of instances (25 clips; 5 for each class) as the training set, while the remaining 50 videos from the other 4 sets are reserved for testing.

To obtain the measurement features from the raw videos, we extract dense spatio-temporal 3D *cuboid* features of [4]. Similar to [4], we construct a finite codebook of descriptors, and replace each cuboid descriptor by the corresponding codebook word. More specifically, after collecting the cuboid features from all videos, we cluster them into $C = 200$ centers using the k-means algorithm.

For the baseline performance comparison, we first run [4]'s static mixture approach where each video is represented as a static histogram of cuboid types contained in the video clip, essentially forming a bag-of-words representation. We then apply standard classification methods such as the nearest neighbor (NN)

[4] Available for download at http://vision.ucsd.edu

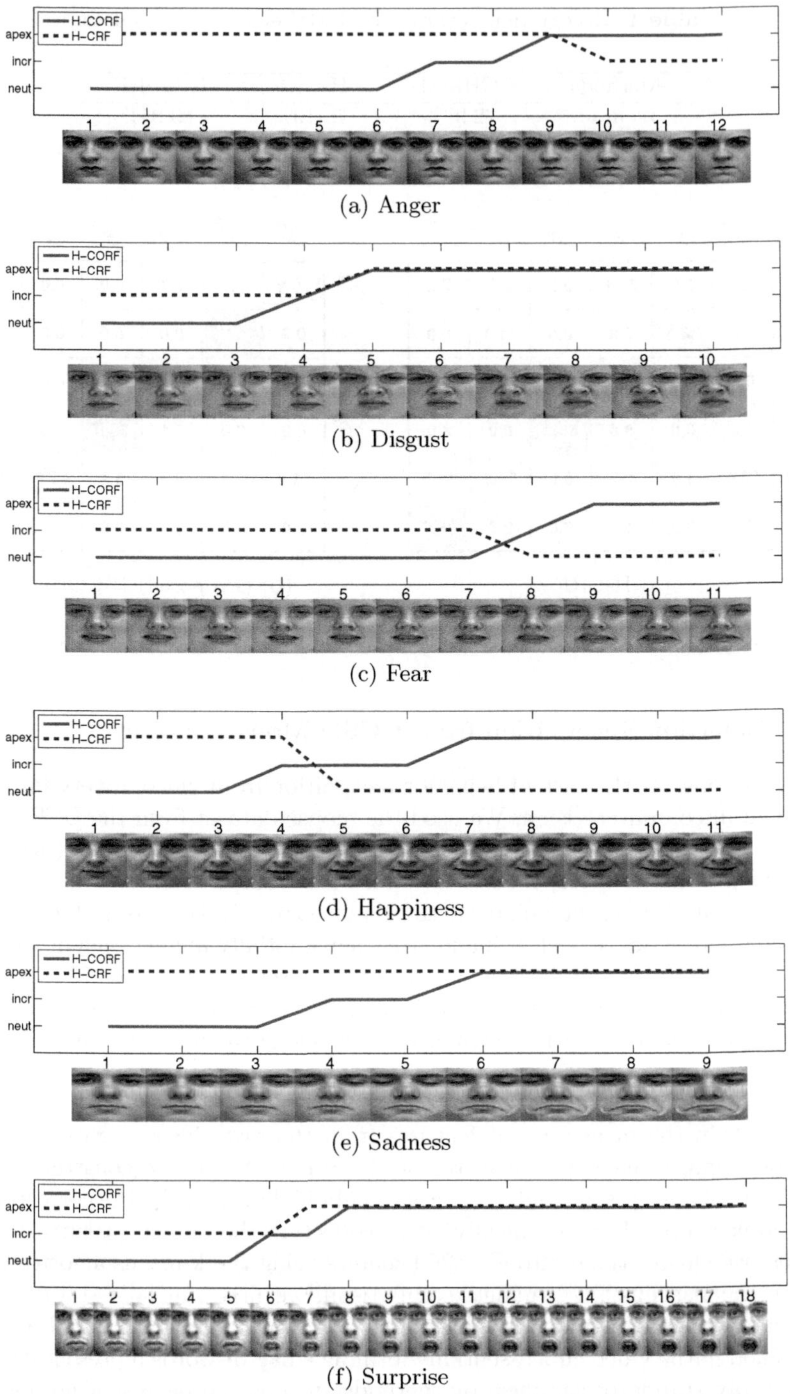

(a) Anger

(b) Disgust

(c) Fear

(d) Happiness

(e) Sadness

(f) Surprise

Fig. 5. Facial emotion intensity prediction for some test sequences. The decoded latent states by H-CORF are shown as red lines, contrasted with H-CRF's blue dotted lines.

Fig. 6. Sample frames from mouse dataset, representing each of the five classes (drink, eat, explore, groom, and sleep) from left to right

Table 2. Recognition accuracy on UCSD mouse dataset

Methods	NN Hist.-χ^2 [4]	GHMM	H-CRF	H-CORF
Accuracy	62.00%	64.00%	68.00%	78.00%

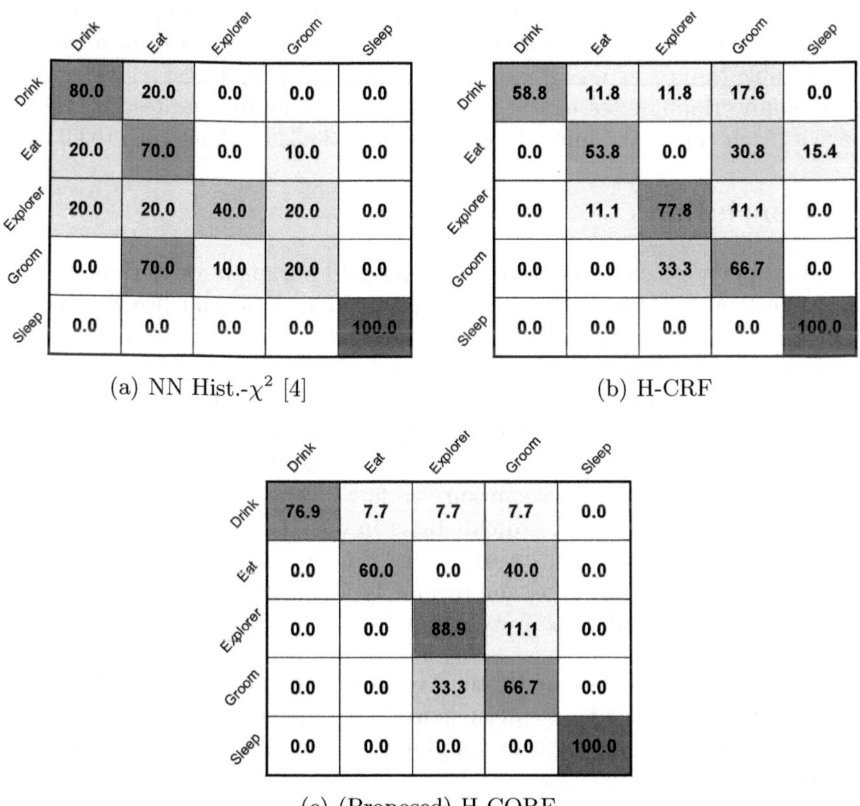

(a) NN Hist.-χ^2 [4]

(b) H-CRF

(c) (Proposed) H-CORF

Fig. 7. Confusion matrices for behavior recognition in UCSD mouse dataset

classifier based on the χ^2 distance measure on the histogram space. We obtain the test accuracy (Table 2) and the confusion matrix (Fig. 7) shown under the title "NN Hist.-χ^2". Note that the random guess would yield 20.00% accuracy.

Instead of representing the video as a single histogram, we consider a sequence representation for our H-CORF-based sequence models. For each time frame t, we set a time-window of size $W = 40$ centered at t. We then collect all detected cuboids with the window, and form a histogram of cuboid types as the node feature $\phi(\mathbf{x}_r)$. Note that some time slices may have no cuboids involved, in which case the feature vector is a zero-vector. To avoid a large number of parameters in the learning, we further reduce the dimensionality of features to 100-dim by PCA which corresponds to about 90% of the total energy.

The test errors and the confusion matrices of the H-CRF and our H-CORF are contrasted with the baseline approach in Table 2 and Fig. 7. Here the cardinality of the latent variables is set as $R = 3$ to account for different ordinal intensity levels of mouse motions, which is chosen among a set of values that produced highest prediction accuracy. Our H-CORF exhibits better performance than the H-CRF and [4]'s standard histogram-based approach. Results similar to ours have been reported in other works that use more complex models and are evaluated on the same dataset (c.f., [19]). However, they are not immediately comparable to ours as we have different experimental settings: a smaller subset with non-overlapping sessions (i.e., sets) between training and testing where we have a much smaller training data proportion (33.33%) than [19]'s (63.33%).

5 Conclusion

In this paper we have introduced a new modeling framework of Hidden Conditional Ordinal Random Fields to accomplish the task of sequence classification. The H-CORF, by introducing a set of ordinal-scale latent variables, aims at modeling the qualitative intensity envelope constraints often observed in real human/animal motions. The embedded sequence segmentation model, CORF, extends the regular CRF by incorporating the ranking-based potentials to model dynamically changing ordinal-scale signals. For the real datasets for facial emotion and mouse behavior recognition, we have demonstrated that the faithful representation of the linked ordinal states in our H-CORF is highly useful for accurate classification of entire sequences. In our future work, we will apply our method to more extensive and diverse types of sequence datasets including biological and financial data.

Acknowledgments. We are grateful to Peng Yang and Dimitris N. Metaxas for their help and discussions throughout the course of this work. This material is based upon work supported by the National Science Foundation under Grant No. IIS-0916812.

References

[1] Chu, W., Ghahramani, Z.: Gaussian processes for ordinal regression. Journal of Machine Learning Research 6, 1019–1041 (2005)

[2] Chu, W., Keerthi, S.S.: New approaches to support vector ordinal regression. In: International Conference on Machine Learning (2005)

[3] Crammer, K., Singer, Y.: On the algorithmic implementation of multiclass kernel-based vector machines. Journal of Machine Learning Research 2, 265–292 (2001)
[4] Dollar, P., Rabaud, V., Cottrell, G., Belongie, S.: Behavior recognition via sparse spatio-temporal features. In: IEEE International Workshop on Visual Surveillance and Performance Evaluation of Tracking and Surveillance (2005)
[5] Gunawardana, A., Mahajan, M., Acero, A., Platt, J.C.: Hidden conditional random fields for phone classification. In: International Conference on Speech Communication and Technology (2005)
[6] Herbrich, R., Graepel, T., Obermayer, K.: Large margin rank boundaries for ordinal regression. In: Advances in Large Margin Classifiers. MIT Press, Cambridge (2000)
[7] Hu, Y., Li, M., Yu, N.: Multiple-instance ranking: Learning to rank images for image retrieval. In: Computer Vision and Pattern Recognition (2008)
[8] Jing, Y., Baluja, S.: Pagerank for product image search. In: Proceeding of the 17th international conference on World Wide Web (2008)
[9] Kumar, S., Hebert, M.: Discriminative random fields. International Journal of Computer Vision 68, 179–201 (2006)
[10] Lafferty, J., McCallum, A., Pereira, F.: Conditional Random Fields: Probabilistic models for segmenting and labeling sequence data. In: International Conference on Machine Learning (2001)
[11] Lien, J., Kanade, T., Cohn, J., Li, C.: Detection, tracking, and classification of action units in facial expression. Journal of Robotics and Autonomous Systems (1999)
[12] Quattoni, A., Collins, M., Darrell, T.: Conditional random fields for object recognition. In: Neural Information Processing Systems (2004)
[13] Shan, C., Gong, S., McOwan, P.W.: Conditional mutual information based boosting for facial expression recognition. In: British Machine Vision Conference (2005)
[14] Shashua, A., Levin, A.: Ranking with large margin principle: Two approaches. In: Neural Information Processing Systems (2003)
[15] Tian, Y.: Evaluation of face resolution for expression analysis. In: Computer Vision and Pattern Recognition Workshop on Face Processing in Video (2004)
[16] Viola, P., Jones, M.: Robust real-time object detection. International Journal of Computer Vision 57(2), 137–154 (2001)
[17] Vishwanathan, S., Schraudolph, N., Schmidt, M., Murphy, K.: Accelerated training of conditional random fields with stochastic meta-descent. In: International Conference on Machine Learning (2006)
[18] Wang, S., Quattoni, A., Morency, L.P., Demirdjian, D., Darrell, T.: Hidden conditional random fields for gesture recognition. In: Computer Vision and Pattern Recognition (2006)
[19] Willems, G., Tuytelaars, T., Gool, L.: An efficient dense and scale-invariant spatio-temporal interest point detector. In: Forsyth, D., Torr, P., Zisserman, A. (eds.) ECCV 2008, Part II. LNCS, vol. 5303, pp. 650–663. Springer, Heidelberg (2008)
[20] Yang, P., Liu, Q., Metaxas, D.N.: Rankboost with l1 regularization for facial expression recognition and intensity estimation. In: International Conference on Computer Vision (2009)

A Unifying View of Multiple Kernel Learning

Marius Kloft*, Ulrich Rückert, and Peter L. Bartlett

University of California, Berkeley, USA
{mkloft,rueckert,bartlett}@cs.berkeley.edu

Abstract. Recent research on multiple kernel learning has lead to a number of approaches for combining kernels in regularized risk minimization. The proposed approaches include different formulations of objectives and varying regularization strategies. In this paper we present a unifying optimization criterion for multiple kernel learning and show how existing formulations are subsumed as special cases. We also derive the criterion's dual representation, which is suitable for general smooth optimization algorithms. Finally, we evaluate multiple kernel learning in this framework analytically using a Rademacher complexity bound on the generalization error and empirically in a set of experiments.

1 Introduction

Selecting a suitable kernel for a kernel-based [17] machine learning task can be a difficult task. From a statistical point of view, the problem of choosing a good kernel is a model selection task. To this end, recent research has come up with a number of *multiple kernel learning* (MKL) [11] approaches, which allow for an automated selection of kernels from a predefined family of potential candidates. Typically, MKL approaches come in one of these three different flavors:

(I) Instead of formulating an optimization criterion with a fixed kernel k, one leaves the choice of k as a variable and demands that k is taken from a linear span of base kernels $k := \sum_{i=1}^{M} \theta_i k_i$. The actual learning procedure then optimizes not only over the parameters of the kernel classifier, but also over $\boldsymbol{\theta}$ subject to the constraint that $\|\boldsymbol{\theta}\| \leq 1$ for some fixed norm. This approach is taken in [14,20] for 1-norm penalties and extended in [9] to ℓ_p-norms.

(II) A second approach takes k from a (non-)linear span of base kernels $k := \sum_{i=1}^{M} \theta_i^{-1} k_i$ subject to the constraint that $\|\boldsymbol{\theta}\| \leq 1$ for some fixed norm This approach was taken in [2] and [13] for p-norms and ℓ_∞ norm, respectively.

III) A third approach optimizes over all kernel classifiers for each of the M base kernels, but modifies the regularizer to a block norm, that is, a norm of the vector containing the individual kernel norms. This allows to trade-off the contributions of each kernel to the final classifier. This formulation was used, for example, in [4].

* Also at Machine Learning Group, Technische Universität Berlin, Franklinstr. 28/29, FR 6-9, 10587 Berlin, Germany.

J.L. Balcázar et al. (Eds.): ECML PKDD 2010, Part II, LNAI 6322, pp. 66–81, 2010.

(IV) Finally, since it appears to be sensible to have only the best kernels contribute to the final classifier, it makes sense to encourage sparse kernel weights. One way to do so is to extend the second setting with an *elastic net* regularizer, a linear combination of ℓ_1 and ℓ_2 regularizers. This approach was first considered in [4] as a numerical tool to approximate the ℓ_1-norm constraint and subsequently analyzed in [22] for its regularization properties.

While all of these formulations are based on similar considerations, the individual formulations and used techniques vary considerably. The particular formulations are tailored more towards a specific optimization approach rather than the inherent characteristics. Type (I) and (II) approaches, for instance, are generally solved using partially dualized wrapper approaches; (III) is directly optimized the dual; and (IV) solves MKL in the primal, extending the approach of [6]. This makes it hard to gain insights into the underpinnings and differences of the individual methods, to design general-purpose optimization procedures for the various criteria and to compare the different techniques empirically.

In this paper, we show that all the above approaches can be viewed under a common umbrella by extending the block norm framework (III) to more general norms; we thus formulate MKL as an optimization criterion with a block-norm regularizer. By using this specific form of regularization, we can incorporate all the previously mentioned formulations as special cases of a single criterion. We derive a modular dual representation of the criterion, which separates the contribution of the loss function and the regularizer. This allows practitioners to plug in specific (dual) loss functions and to adjust the regularizer in a flexible fashion.

We show how the dual optimization problem can be solved using standard smooth optimization techniques, report on experiments on real world data, and compare the various approaches according to their ability to recover sparse kernel weights. On the theoretical side, we give a concentration inequality that bounds the generalization ability of MKL classifiers obtained in the presented framework. The bound is the first known bound to apply to MKL with elastic net regularization; it matches the best previously known bound [8] for the special case of ℓ_1 and ℓ_2 regularization, and it is the first bound for ℓ_p block norm MKL with arbitrary p.

2 Multiple Kernel Learning—A Unifying View

In this section we cast multiple kernel learning in a unified framework. Before we go into the details, we need to introduce the general setting and notation.

2.1 MKL in the Primal

We begin with reviewing the classical supervised learning setup. Given a labeled sample $\mathcal{D} = \{(\boldsymbol{x}_i, y_i)\}_{i=1\dots,n}$, where the $\boldsymbol{x}_i$ lie in some input space $\mathcal{X}$ and

$y_i \in \mathcal{Y} \subset \mathbb{R}$, the goal is to find a hypothesis $f \in \mathcal{H}$, that generalizes well on new and unseen data. Regularized risk minimization returns a minimizer f^*,

$$f^* \in \operatorname{argmin}_f \mathrm{R_{emp}}(f) + \lambda\Omega(f),$$

where $\mathrm{R_{emp}}(f) = \frac{1}{n}\sum_{i=1}^{n} \ell\left(f(\boldsymbol{x}_i), y_i\right)$ is the empirical risk of hypothesis f w.r.t. a convex loss function $\ell : \mathbb{R} \times \mathcal{Y} \to \mathbb{R}$, $\Omega : \mathcal{H} \to \mathbb{R}$ is a regularizer, and $\lambda > 0$ is a trade-off parameter. We consider linear models of the form

$$f_{\boldsymbol{w}}(\boldsymbol{x}) = \langle \boldsymbol{w}, \Phi(\boldsymbol{x}) \rangle, \tag{1}$$

together with a (possibly non-linear) mapping $\Phi : \mathcal{X} \to \mathcal{H}$ to a Hilbert space $\mathcal{H}$ [18,12] and constrain the regularization to be of the form $\Omega(f) = \frac{1}{2}||\boldsymbol{w}||_2^2$ which allows to kernelize the resulting models and algorithms. We will later make use of kernel functions $k(\boldsymbol{x}, \boldsymbol{x}') = \langle \Phi(\boldsymbol{x}), \Phi(\boldsymbol{x}') \rangle_{\mathcal{H}}$ to compute inner products in $\mathcal{H}$. When learning with multiple kernels, we are given M different feature mappings $\Phi_m : \mathcal{X} \to \mathcal{H}_m$, $m = 1, \ldots M$, each giving rise to a reproducing kernel k_m of $\mathcal{H}_m$.

There are two main ways to formulate regularized risk minimization with MKL. The first approach, denoted by (I) in the introduction, introduces a linear kernel mixture $k_{\boldsymbol{\theta}} = \sum_{m=1}^{M} \theta_m k_m$, $\theta_m \geq 0$ and a blockwise weighted target vector $\boldsymbol{w}_{\boldsymbol{\theta}} := \left(\sqrt{\theta_1}\boldsymbol{w}_1^\top, ..., \sqrt{\theta_M}\boldsymbol{w}_M^\top\right)^\top$. With this, one solves

$$\begin{aligned} \inf_{\boldsymbol{w},\theta} \quad & C\sum_{i=1}^{n} \ell\left(\sum_{m=1}^{M} \langle \sqrt{\theta_m}\boldsymbol{w}_m, \Phi(\boldsymbol{x}_i)\rangle_{\mathcal{H}_m},\ y_i\right) + \|\boldsymbol{w}_{\boldsymbol{\theta}}\|_{\mathcal{H}}^2 \\ \text{s.t.} \quad & \|\boldsymbol{\theta}\|_q \leq 1. \end{aligned} \tag{2}$$

Alternatively, one can omit the explicit mixture vector $\boldsymbol{\theta}$ and use block-norm regularization instead (this approach was denoted by (III) in the introduction). In this case, denoting by $||\boldsymbol{w}||_{2,p} = \left(\sum_{m=1}^{M} ||\boldsymbol{w}_m||_{\mathcal{H}_m}^p\right)^{1/p}$ the ℓ_2/ℓ_p block norm, one optimizes

$$\inf_{\boldsymbol{w}} \quad C\sum_{i=1}^{n} \ell\left(\sum_{m=1}^{M} \langle \boldsymbol{w}_m, \Phi_m(\boldsymbol{x}_i)\rangle_{\mathcal{H}_m},\ y_i\right) + \|\boldsymbol{w}\|_{2,p}^2. \tag{3}$$

One can show that (2) is a special case of (3). In particular, one can show that setting the block-norm parameter to $p = \frac{2q}{q+1}$ is equivalent to having kernel mixture regularization with $\|\boldsymbol{\theta}\|_q \leq 1$ [10]. This also implies that the kernel mixture formulation is *strictly* less general, because it can not replace block norm regularization for $p > 2$.

Hence, we focus on the block norm criterion, and extend it to also include elastic net regularization. The resulting primal problem generalizes the approaches (I)–(IV); it is stated as follows:

Primal MKL Optimization Problem

$$\inf_{\boldsymbol{w}} \quad C\sum_{i=1}^{n} \ell\left(\langle \boldsymbol{w}, \Phi(\boldsymbol{x}_i)\rangle_{\mathcal{H}},\ y_i\right) + \frac{1}{2}||\boldsymbol{w}||_{2,p}^2 + \frac{\mu}{2}||\boldsymbol{w}||_2^2 \,, \tag{P}$$

where $\Phi = \Phi_1 \times \cdots \times \Phi_M$ denotes the Cartesian product of the Φ_m's. Using the above criterion it is possible to recover block norm regularization by setting $\mu = 0$ and the elastic net regularizer by setting $p = 1$.

Note that we use a slightly different—but equivalent—regularization than the one used in the original elastic net paper [25]: we square the term $||\boldsymbol{w}||_{2,p}$ while in the original criterion it appeared linearly. To see that the two formulations are equal, notice that the original regularizer can equivalently be encoded as a hard constraint $||\boldsymbol{w}||_{2,p} \leq \eta$ (this is similar to a well known result for SVMs; see [23]), which is equivalent to $||\boldsymbol{w}||_{2,p}^2 < \eta^2$ and subsequently can be incorporated into the objective, again. Hence, it is equivalent: regularizing with $||\boldsymbol{w}||_{2,p}$ and $||\boldsymbol{w}||_{2,p}^2$, respectively, leads to the same regularization path.

2.2 MKL in Dual Space

Optimization problems often have a considerably easier structure when studied in the dual space. In this section we derive the dual problem of the generalized MKL approach presented in the previous section. Let us begin with rewriting Optimization Problem (P) by expanding the decision values into slack variables as follows,

$$\begin{aligned} \inf_{\boldsymbol{w},\boldsymbol{t}} \quad & C\sum_{i=1}^{n} \ell\left(t_i,\ y_i\right) + \frac{1}{2}||\boldsymbol{w}||_{2,p}^2 + \frac{\mu}{2}||\boldsymbol{w}||_2^2 \\ \text{s.t.} \quad & \forall i: \ \langle \boldsymbol{w}, \Phi(\boldsymbol{x}_i)\rangle_{\mathcal{H}} = t_i. \end{aligned} \tag{4}$$

Applying Lagrange's theorem re-incorporates the constraints into the objective by introducing Lagrangian multipliers $\boldsymbol{\alpha} \in \mathbb{R}^n$. The Lagrangian saddle point problem is then given by

$$\begin{aligned} \sup_{\boldsymbol{\alpha}} \ \inf_{\boldsymbol{w},\boldsymbol{t}} \quad & C\sum_{i=1}^{n} \ell\left(t_i,\ y_i\right) + \frac{1}{2}||\boldsymbol{w}||_{2,p}^2 + \frac{\mu}{2}||\boldsymbol{w}||_2^2 \\ & -\sum_{i=1}^{n} \alpha_i \left(\langle \boldsymbol{w}, \Phi(\boldsymbol{x}_i)\rangle_{\mathcal{H}} - t_i\right). \end{aligned} \tag{5}$$

Setting the first partial derivatives of the above Lagrangian to zero w.r.t. $\boldsymbol{w}$ gives the following KKT optimality condition

$$\forall m: \quad \boldsymbol{w}_m \ = \left(||\boldsymbol{w}||_{2,p}^{2-p}||\boldsymbol{w}_m||^{p-2} + \mu\right)^{-1} \sum_i \alpha_i \Phi_m(\boldsymbol{x}_i)\,. \tag{KKT}$$

Inspecting the above equation reveals the representation $\boldsymbol{w}_m^* \in \text{span}(\Phi_m(\boldsymbol{x}_1), ..., \Phi_m(\boldsymbol{x}_n))$. Rearranging the order of terms in the Lagrangian,

$$\sup_{\boldsymbol{\alpha}} \quad -C\sum_{i=1}^{n} \sup_{t}\left(-\frac{\alpha_i t_i}{C} - \ell(t_i,\ y_i)\right) \\ -\sup_{\boldsymbol{w}}\left(\langle \boldsymbol{w}, \sum_{i=1}^{n} \alpha_i \Phi(\boldsymbol{x}_i)\rangle_{\mathcal{H}} - \frac{1}{2}||\boldsymbol{w}||_{2,p}^2 - \frac{\mu}{2}||\boldsymbol{w}||_2^2\right),$$

lets us express the Lagrangian in terms of Fenchel-Legendre conjugate functions $h^*(\boldsymbol{x}) = \sup_{\boldsymbol{u}} \boldsymbol{x}^\top \boldsymbol{u} - h(\boldsymbol{u})$ as follows,

$$\sup_{\boldsymbol{\alpha}} \quad -C\sum_{i=1}^{n} \ell^*\left(-\frac{\alpha_i}{C},\ y_i\right) - \left(\frac{1}{2}\left\|\sum_{i=1}^{n}\alpha_i\Phi(\boldsymbol{x}_i)\right\|_{2,p}^2 + \frac{\mu}{2}\left\|\sum_{i=1}^{n}\alpha_i\Phi(\boldsymbol{x}_i)\right\|_2^2\right)^*, \tag{6}$$

thereby removing the dependency of the Lagrangian on $\boldsymbol{w}$. The function ℓ^* is called *dual loss* in the following. Recall that the Inf-Convolution [16] of two functions f and g is defined by

$$(f \oplus g)(x) := \inf_{y} f(x-y) + g(y), \tag{7}$$

and that $(f^* \oplus g^*)(x) = (f+g)^*(x)$ and $(\eta f)^*(x) = \eta f^*(x/\eta)$ hold. Moreover, we have for the conjugate of the block norm $\left(\frac{1}{2}||\cdot||_{2,p}^2\right)^* = \frac{1}{2}||\cdot||_{2,p^*}^2$ [3] where p^* is the conjugate exponent, i.e., $\frac{1}{p} + \frac{1}{p^*} = 1$. As a consequence, we obtain the following *dual* optimization problem

Dual MKL Optimization Problem

$$\sup_{\boldsymbol{\alpha}} \quad -C\sum_{i=1}^{n} \ell^*\left(-\frac{\alpha_i}{C},\ y_i\right) - \left(\frac{1}{2}\,||\cdot||_{2,p*}^2 \oplus \frac{1}{2\mu}\,||\cdot||_2^2\right)\left(\sum_{i=1}^{n}\alpha_i\Phi(\boldsymbol{x}_i)\right). \tag{D}$$

Note that the supremum is also a maximum, if the loss function is continuous. The function $f \oplus \frac{1}{2\mu}||\cdot||^2$ is the so-called *Morea-Yosida Approximate* [19] and has been studied extensively both theoretically and algorithmically for its favorable regularization properties. It can "smoothen" an optimization problem—even if it is initially non-differentiable—and increase the condition number of the Hessian for twice differentiable problems.

The above dual generalizes multiple kernel learning to arbitrary convex loss functions and regularizers. Due to the mathematically clean separation of the loss and the regularization term—each loss term solely depends on a single real valued variable—we can immediately recover the corresponding dual for a specific choice of a loss/regularizer pair $(\ell, ||\cdot||_{2,p})$ by computing the pair of conjugates $(\ell^*, ||\cdot||_{2,p^*})$.

2.3 Obtaining Kernel Weights

While formalizing multiple kernel learning with block-norm regularization offers a number of conceptual and analytical advantages, it requires an additional step in practical applications. The reason for this is that the block-norm regularized dual optimization criterion does not include explicit kernel weights. Instead, this information is contained only implicitly in the optimal kernel classifier parameters, as output by the optimizer. This is a problem, for instance if one wishes to apply the induced classifier on new test instances. Here we need the kernel weights to form the final kernel used for the actual prediction. To recover the underlying kernel weights, one essentially needs to identify which kernel contributed to which degree for the selection of the optimal dual solution. Depending on the actual parameterization of the primal criterion, this can be done in various ways.

We start by reconsidering the KKT optimality condition given by Eq. (KKT) and observe that the first term on the right hand side,

$$\theta_m := \left(||\boldsymbol{w}||_{2,p}^{2-p} ||\boldsymbol{w}_m||^{p-2} + \mu \right)^{-1}. \tag{8}$$

introduces a scaling of the feature maps. With this notation, it is easy to see from Eq. (KKT) that our model given by Eq. (1) extends to

$$f_{\boldsymbol{w}}(\boldsymbol{x}) = \sum_{m=1}^{M} \sum_{i=1}^{n} \alpha_i \theta_m k_m(\boldsymbol{x}_i, \boldsymbol{x}).$$

In order to express the above model solely in terms of dual variables we have to compute $\boldsymbol{\theta}$ in terms of $\boldsymbol{\alpha}$.

In the following we focus on two cases. First, we consider ℓ_p block norm regularization for arbitrary $1 < p < \infty$ while switching the elastic net off by setting the parameter $\mu = 0$. Then, from Eq. (KKT) we obtain

$$||\boldsymbol{w}_m|| = ||\boldsymbol{w}||_{2,p}^{\frac{p-2}{p-1}} \left\| \sum_{i=1}^{n} \alpha_i \Phi_m(\boldsymbol{x}_i) \right\|_{\mathcal{H}_m}^{\frac{1}{p-1}} \quad \text{where} \quad \boldsymbol{w}_m = \theta_m \sum_i \alpha_i \Phi_m(\boldsymbol{x}_i).$$

Resubstitution into (8) leads to the proportionality

$$\exists\, c > 0 \;\; \forall\, m: \quad \theta_m = c \left(\left\| \sum_{i=1}^{n} \alpha_i \Phi_m(\boldsymbol{x}_i) \right\|_{\mathcal{H}_m} \right)^{\frac{2-p}{p-1}}. \tag{9}$$

Note that, in the case of classification, we only need to compute $\boldsymbol{\theta}$ up to a positive multiplicative constant.

For the second case, let us now consider the elastic net regularizer, i.e., $p = 1+\epsilon$ with $\epsilon \approx 0$ and $\mu > 0$. Then, the optimality condition given by Eq. (KKT) translates to

$$\boldsymbol{w}_m = \theta_m \sum_i \alpha_i \Phi_m(\boldsymbol{x}_i) \quad \text{where} \quad \theta_m = \left(\left(\sum_{m'=1}^{M} ||\boldsymbol{w}_{m'}||_{\mathcal{H}_{m'}}^{1+\epsilon} \right)^{1-\epsilon} ||\boldsymbol{w}_m||_{\mathcal{H}_m}^{\epsilon-1} + \mu \right)^{-1}.$$

Inserting the left hand side expression for $||\boldsymbol{w}_m||_{\mathcal{H}_m}$ into the right hand side leads to the non-linear system of equalities

$$\forall\, m: \quad \mu\theta_m||K_m||^{1-\epsilon} + \theta_m^{\epsilon}\left(\sum_{m'=1}^{M}\theta_{m'}^{1+\epsilon}||K_{m'}||^{1+\epsilon}\right)^{1-\epsilon} = ||K_m||^{1-\epsilon}, \qquad (10)$$

where we employ the notation $||K_m|| := ||\sum_{i=1}^{n}\alpha_i\Phi_m(\boldsymbol{x}_i)||_{\mathcal{H}_m}$. In our experiments we solve the above conditions numerically using $\epsilon \approx 0$. Notice, that this difficulty does not arise in [4] for $p = 1$ and in [22], which is an advantage of the latter approaches. The optimal mixing coefficients θ_m can now be computed solely from the dual $\boldsymbol{\alpha}$ variables by means of Eq. (9) and (10), and by the kernel matrices K_m using the identity

$$\forall m = 1, \cdots, M: \quad ||K_m|| = \sqrt{\boldsymbol{\alpha}K_m\boldsymbol{\alpha}}.$$

This enables optimization in the dual space as discussed in the next section.

3 Optimization Strategies

In this section we describe how one can simply solve the dual optimization problem by a common purpose quasi-Newton method. We do not claim that this is the fastest possible way to solve the problem; in the contrary, we conjecture that a SMO-type algorithm decomposition algorithm, as used in [4], might speed up the optimization. However, computational efficiency is not the focus of this paper; we focus on understanding and theoretically analyzing MKL and leave a more efficient implementation of our approach to future work.

For our experiments, we use the hinge loss $l(x) = \max(0, 1 - x)$ to obtain a support vector formulation, but the discussion also applies to most other convex loss functions. We first note that the dual loss of the hinge loss is $\ell^*(t, y) = \frac{t}{y}$ if $-1 \leq \frac{t}{y} \leq 0$ and ∞ elsewise [15]. Hence, for each i the term $\ell^*\left(-\frac{\alpha_i}{C},\, y_i\right)$ of the generalized dual, i.e., Optimization Problem (D), translates to $-\frac{\alpha_i}{Cy_i}$, provided that $0 \leq \frac{\alpha_i}{y_i} \leq C$. Employing a variable substitution of the form $\alpha_i^{\text{new}} = \frac{\alpha_i}{y_i}$, the dual problem (D) becomes

$$\sup_{\boldsymbol{\alpha}:\ \mathbf{0}\leq\boldsymbol{\alpha}\leq C\mathbf{1}} \quad \mathbf{1}^\top\boldsymbol{\alpha} - \left(\frac{1}{2}\left\|\cdot\right\|_{2,p*}^{2} \oplus \frac{1}{2\mu}\left\|\cdot\right\|_{2}^{2}\right)\left(\sum_{i=1}^{n}\alpha_i y_i\Phi(\boldsymbol{x}_i)\right),$$

and by definition of the Inf-convolution,

$$\sup_{\boldsymbol{\alpha},\boldsymbol{\beta}:\ \mathbf{0}\leq\boldsymbol{\alpha}\leq C\mathbf{1}} \quad \mathbf{1}^\top\boldsymbol{\alpha} - \frac{1}{2}\left\|\sum_{i=1}^{n}\alpha_i y_i\Phi(\boldsymbol{x}_i) - \boldsymbol{\beta}\right\|_{2,p*}^{2} - \frac{1}{2\mu}\left\|\boldsymbol{\beta}\right\|_2^2. \qquad (11)$$

We note that the representer theorem [17] is valid for the above problem, and hence the solution of (11) can be expressed in terms of kernel functions, i.e.,

$\beta_m = \sum_{i=1}^n \gamma_i k_m(x_i,\cdot)$ for certain real coefficients $\boldsymbol{\gamma} \in \mathbb{R}^n$ uniformly for all m, hence $\boldsymbol{\beta} = \sum_{i=1}^n \gamma_i \Phi(x_i)$. Thus, Eq. (11) has a representation of the form

$$\sup_{\boldsymbol{\alpha},\boldsymbol{\gamma}:\ \mathbf{0}\le\boldsymbol{\alpha}\le C\mathbf{1}} \quad \mathbf{1}^\top\boldsymbol{\alpha} - \frac{1}{2}\left\|\sum_{i=1}^n(\alpha_i y_i - \gamma_i)\Phi(\boldsymbol{x}_i)\right\|_{2,p*}^2 - \frac{1}{2\mu}\boldsymbol{\gamma}^\top K\boldsymbol{\gamma},$$

where we use the shorthand $K = \sum_{m=1}^M K_m$. The above expression can be written[1] in terms of kernel matrices as follows,

Support Vector MKL—The Hinge Loss Dual

$$\sup_{\boldsymbol{\alpha},\boldsymbol{\gamma}:\ \mathbf{0}\le\boldsymbol{\alpha}\le C\mathbf{1}} \quad \mathbf{1}^\top\boldsymbol{\alpha} - \frac{1}{2}\left\|\left((\boldsymbol{\alpha}\circ\boldsymbol{y}-\boldsymbol{\gamma})^\top K_m(\boldsymbol{\alpha}\circ\boldsymbol{y}-\boldsymbol{\gamma})\right)_{m=1}^M\right\|_{\frac{p^*}{2}} - \frac{1}{2\mu}\boldsymbol{\gamma}^\top K\boldsymbol{\gamma}. \qquad \text{(SV-MKL)}$$

In our experiments, we optimized the above criterion by using the limited memory quasi-Newton software L-BFGS-B [24]. L-BFGS-B is a common purpose solver that can simply be used out-of-the-box. It approximates the Hessian matrix based on the last t gradients, where t is a parameter to be chosen by the user. Note that L-BFGS-B can handle the box constraint induced by the hinge loss.

4 Theoretical Analysis

In this section we give two uniform convergence bounds for the generalization error of the multiple kernel learning formulation presented in Section 2. The results are based on the established theory on Rademacher complexities. Let $\sigma_1,\ldots,\sigma_n$ be a set of independent Rademacher variables, which obtain the values -1 or +1 with the same probability 0.5, and let $\mathcal{C}$ be some space of classifiers $c:\mathcal{X}\to\mathbb{R}$. Then, the *Rademacher complexity* of $\mathcal{C}$ is given by

$$\mathcal{R}_\mathcal{C} := \mathbf{E}\left[\sup_{c\in\mathcal{C}}\frac{1}{n}\sum_{i=1}^n \sigma_i c(x_i)\right].$$

If the Rademacher complexity of a class of classifiers is known, it can be used to bound the generalization error. We give one result here, which is an immediate corollary of Thm. 8 in [5] (using Thm. 12.4 in the same paper), and refer to the literature [5] for further results on Rademacher penalization.

Theorem 1. *Assume the loss* $\ell:\mathbb{R}\supseteq\mathcal{Y}\to[0,1]$ *is Lipschitz with constant* L. *Then, the following holds with probability larger than* $1-\delta$ *for all classifiers* $c\in\mathcal{C}$:

$$\mathbf{E}[\ell(yc(x))] \le \frac{1}{n}\sum_{i=1}^n \ell(y_i c(x_i)) + 2L\mathcal{R}_\mathcal{C} + \sqrt{\frac{8\ln\frac{2}{\delta}}{n}}. \qquad (12)$$

[1] We employ the notation $s = (s_1,\ldots,s_M)^\top = (s_m)_{m=1}^M$ for $s\in\mathbb{R}^M$ and denote by $\boldsymbol{x}\circ\boldsymbol{y}$ the elementwise multiplication of two vectors.

We will now give an upper bound for the Rademacher complexity of the block-norm regularized linear learning approach described above. More precisely, for $1 \leq i \leq M$ let $\|w\|_{\star i} := \sqrt{k_i(w,w)}$ denote the norm induced by kernel k_i and for $x \in \mathbb{R}^p$, $p, q \geq 1$ and $C_1, C_2 \geq 0$ with $C_1 + C_2 = 1$ define

$$\|x\|_O := C_1\|x\|_p + C_2\|x\|_q.$$

We now give a bound for the following class of linear classifiers:

$$\mathcal{C}_\star := \left\{ c : \begin{pmatrix} \Phi_1(x) \\ \vdots \\ \Phi_M(x) \end{pmatrix} \mapsto \begin{pmatrix} w_1 \\ \vdots \\ w_M \end{pmatrix}^T \begin{pmatrix} \Phi_1(x) \\ \vdots \\ \Phi_M(x) \end{pmatrix} \middle| \left\| \begin{pmatrix} \|w_1\|_{\star 1} \\ \vdots \\ \|w_M\|_{\star M} \end{pmatrix} \right\|_O \leq 1 \right\}.$$

Theorem 2. *Assume the kernels are normalized, i.e. $k_i(x,x) = \|x\|_{\star i}^2 \leq 1$ for all $x \in \mathcal{X}$ and all $1 \leq i \leq M$. Then, the Rademacher complexity of the class $\mathcal{C}_\star$ of linear classifiers with block norm regularization is upper-bounded as follows:*

$$\mathcal{R}_{\mathcal{C}_\star} \leq \frac{M}{C_1 M^{\frac{1}{p}} + C_2 M^{\frac{1}{q}}} \left(\sqrt{\frac{2\ln M}{n}} + \sqrt{\frac{1}{n}} \right). \tag{13}$$

For the special case with $p \geq 2$ and $q \geq 2$, the bound can be improved as follows:

$$\mathcal{R}_{\mathcal{C}_\star} \leq \frac{M}{C_1 M^{\frac{1}{p}} + C_2 M^{\frac{1}{q}}} \sqrt{\frac{1}{n}}. \tag{14}$$

Interpretation of Bounds. It is instructive to compare this result to some of the existing MKL bounds in the literature. For instance, the main result in [8] bounds the Rademacher complexity of the ℓ_1-norm regularizer with a $O(\sqrt{\ln M/n})$ term. We get the same result by setting $C_1 = 1, C_2 = 0$ and $p = 1$. For the ℓ_2-norm regularized setting, we can set $C_1 = 1, C_2 = 0$ and $p = \frac{4}{3}$ (because the kernel weight formulation with ℓ_2 norm corresponds to the block-norm representation with $p = \frac{4}{3}$) to recover their $O(M^{\frac{1}{4}}/\sqrt{n})$ bound. Finally, it is interesting to see how changing the C_1 parameter influences the generalization capacity of the elastic net regularizer ($p = 1, q = 2$). For $C_1 = 1$, we essentially recover the ℓ_1 regularization penalty, but as C_1 approaches 0, the bound includes an additional $O(\sqrt{M})$ term. This shows how the capacity of the elastic net regularizer increases towards the ℓ_2 setting with decreasing sparsity.

Proof (of Theorem 2). Using the notation $w := (w_1, \ldots, w_M)^T$ and $\|w\|_B := \|(\|w_1\|_{\star 1}, \ldots, \|w_M\|_{\star M})^T\|_O$ it is easy to see that

$$\mathbf{E}\left[\sup_{c \in \mathcal{C}_\star} \frac{1}{n} \sum_{i=1}^n \sigma_i y_i c(x_i)\right] = \mathbf{E}\left[\sup_{\|w\|_B < 1} \left\{ \begin{pmatrix} w_1 \\ \vdots \\ w_M \end{pmatrix}^T \begin{pmatrix} \frac{1}{n}\sum_{i=1}^n \sigma_i \Phi_1(x_i) \\ \vdots \\ \frac{1}{n}\sum_{i=1}^n \sigma_i \Phi_M(x_i) \end{pmatrix} \right\}\right]$$

$$= \mathbf{E}\left[\left\| \begin{pmatrix} \|\frac{1}{n}\sum_{i=1}^n \sigma_i \Phi_1(x_i)\|_{\star 1} \\ \vdots \\ \|\frac{1}{n}\sum_{i=1}^n \sigma_i \Phi_M(x_i)\|_{\star M} \end{pmatrix} \right\|_O^* \right],$$

where $\|x\|^* := \sup_z\{z^T x \| \|z\| \leq 1\}$ denotes the dual norm of $\|.\|$ and we use the fact that $\|w\|_B^* = \|(\|w_1\|_{\star 1}^*, \ldots, \|w_M\|_{\star M}^*)^T\|_O^*$ [3], and that $\|.\|_{\star i}^* = \|.\|_{\star i}$. We will show that this quantity is upper bounded by

$$\frac{M}{C_1 M^{\frac{1}{p}} + C_2 M^{\frac{1}{q}}} \left(\sqrt{\frac{2 \ln M}{n}} + \sqrt{\frac{1}{n}} \right). \tag{15}$$

As a first step we prove that for any $x \in \mathbb{R}^M$

$$\|x\|_O^* \leq \frac{M}{C_1 M^{\frac{1}{p}} + C_2 M^{\frac{1}{q}}} \|x\|_\infty \,. \tag{16}$$

For any $a \geq 1$ we can apply Hölder's inequality to the dot product of $x \in \mathbb{R}^M$ and $\mathbb{1}_M := (1, \ldots, 1)^T$ and obtain $\|x\|_1 \leq \|\mathbb{1}_M\|_{\frac{a}{a-1}} \cdot \|x\|_a = M^{\frac{a-1}{a}} \|x\|_a$. Since $C_1 + C_2 = 1$, we can apply this twice on the two components of $\|.\|_O$ to get a lower bound for $\|x\|_O$,

$$(C_1 M^{\frac{1-p}{p}} + C_2 M^{\frac{1-q}{q}}) \|x\|_1 \;\leq\; C_1 \|x\|_p + C_2 \|x\|_q \;=\; \|x\|_O.$$

In other words, for every $x \in \mathbb{R}^M$ with $\|x\|_O \leq 1$ it holds that

$$\|x\|_1 \leq 1 / \left(C_1 M^{\frac{1-p}{p}} + C_2 M^{\frac{1-q}{q}} \right) = M / \left(C_1 M^{\frac{1}{p}} + C_2 M^{\frac{1}{q}} \right).$$

Thus,

$$\{z^T x \| \|x\|_O \leq 1\} \subseteq \left\{ z^T x \middle| \|x\|_1 \leq \frac{M}{C_1 M^{\frac{1}{p}} + C_2 M^{\frac{1}{q}}} \right\}. \tag{17}$$

This means we can bound the dual norm $\|.\|_O^*$ of $\|.\|_O$ as follows:

$$\begin{aligned} \|x\|_O^* &= \sup_z \{z^T x \| \|z\|_O \leq 1\} \\ &\leq \sup_z \left\{ z^T x \middle| \|z\|_1 \leq \frac{M}{C_1 M^{\frac{1}{p}} + C_2 M^{\frac{1}{q}}} \right\} \\ &= \frac{M}{C_1 M^{\frac{1}{p}} + C_2 M^{\frac{1}{q}}} \|x\|_\infty \,. \end{aligned} \tag{18}$$

This accounts for the first factor in (15). For the second factor, we show that

$$\mathbf{E} \left[\left\| \begin{pmatrix} \|\frac{1}{n} \sum_{i=1}^n \sigma_i \Phi_1(x_i)\|_{\star 1} \\ \vdots \\ \|\frac{1}{n} \sum_{i=1}^n \sigma_i \Phi_M(x_i)\|_{\star M} \end{pmatrix} \right\|_\infty \right] \leq \sqrt{\frac{2 \ln M}{n}} + \sqrt{\frac{1}{n}} \,. \tag{19}$$

To do so, define

$$V_k := \left\| \frac{1}{n} \sum_{i=1}^n \sigma_i \Phi_k(x_i) \right\|_{\star k}^2 = \frac{1}{n^2} \sum_{i=1}^n \sum_{j=1}^n \sigma_i \sigma_j k_k(x_i, x_j) \,.$$

By the independence of the Rademacher variables it follows for all $k \leq M$,

$$\mathbf{E}\left[V_k\right] = \frac{1}{n^2}\sum_{i=1}^{n}\mathbf{E}\left[k_k(x_i,x_i)\right] \leq \frac{1}{n}. \tag{20}$$

In the next step we use a martingale argument to find an upper bound for $\sup_k[W_k]$ where $W_k := \sqrt{V_k} - \mathbf{E}[\sqrt{V_k}]$. For ease of notation, we write $\mathbf{E}_{(r)}[X]$ to denote the conditional expectation $\mathbf{E}[X|(x_1,\sigma_1),\ldots(x_r,\sigma_r)]$. We define the following martingale:

$$\begin{aligned} Z_k^{(r)} &:= \mathop{\mathbf{E}}_{(r)}[\sqrt{V_k}] - \mathop{\mathbf{E}}_{(r-1)}[\sqrt{V_k}] \\ &= \mathop{\mathbf{E}}_{(r)}\left[\left\|\frac{1}{n}\sum_{i=1}^{n}\sigma_i\Phi_k(x_i)\right\|_{\star k}\right] - \mathop{\mathbf{E}}_{(r-1)}\left[\left\|\frac{1}{n}\sum_{i=1}^{n}\sigma_i\Phi_k(x_i)\right\|_{\star k}\right]. \end{aligned} \tag{21}$$

The range of each random variable $Z_k^{(r)}$ is at most $\frac{2}{n}$. This is because switching the sign of σ_r changes only one summand in the sum from $-\Phi_k(x_r)$ to $+\Phi_k(x_r)$. Thus, the random variable changes by at most $\|\frac{2}{n}\Phi_k(x_r)\|_{\star k} \leq \frac{2}{n}k_k(x_r,x_r) \leq \frac{2}{n}$. Hence, we can apply Hoeffding's inequality, $\mathbf{E}_{(r-1)}\left[e^{sZ_k^{(r)}}\right] \leq e^{\frac{1}{2n^2}s^2}$. This allows us to bound the expectation of $\sup_k W_k$ as follows:

$$\begin{aligned} \mathbf{E}[\sup_k W_k] &= \mathbf{E}\left[\frac{1}{s}\ln\sup_k e^{sW_k}\right] \\ &\leq \mathbf{E}\left[\frac{1}{s}\ln\sum_{k=1}^{M}\exp\left[s\sum_{r=1}^{n}Z_k^{(r)}\right]\right] \\ &\leq \frac{1}{s}\ln\sum_{k=1}^{M}\prod_{r=1}^{n}\mathop{\mathbf{E}}_{(r)}\left[e^{sZ_k^{(n)}}\right] \\ &\leq \frac{1}{s}\ln\sum_{k=1}^{M}\left(e^{\frac{1}{2n^2}s^2}\right)^n \\ &= \frac{\ln M}{s} + \frac{s}{2n}, \end{aligned}$$

where we n times applied Hoeffding's inequality. Setting $s = \sqrt{2n\ln M}$ yields:

$$\mathbf{E}[\sup_k W_k] \leq \sqrt{\frac{2\ln M}{n}}. \tag{22}$$

Now, we can combine (20) and (22):

$$\mathbf{E}\left[\sup_k\sqrt{V_k}\right] \leq \mathbf{E}\left[\sup_k W_k + \sqrt{\mathbf{E}[V_k]}\right] \leq \sqrt{\frac{2\ln M}{n}} + \sqrt{\frac{1}{n}}.$$

This concludes the proof of (19) and therewith (13).

The special case (14) for $p, q \geq 2$ is similar. As a first step, we modify (16) to deal with the ℓ_2-norm rather than the ℓ_∞-norm:

$$\|x\|_O^* \leq \frac{\sqrt{M}}{C_1 M^{\frac{1}{p}} + C_2 M^{\frac{1}{q}}} \|x\|_2 \ . \tag{23}$$

To see this, observe that for any $x \in \mathbb{R}^M$ and any $a \geq 2$ Hölder's inequality gives $\|x\|_2 \leq M^{\frac{a-2}{2a}} \|x\|_a$. Applying this to the two components of $\|.\|_O$ we have:

$$(C_1 M^{\frac{2-p}{2p}} + C_2 M^{\frac{2-q}{2q}}) \|x\|_2 \leq C_1 \|x\|_p + C_2 \|x\|_q = \|x\|_O \ .$$

In other words, for every $x \in \mathbb{R}^M$ with $\|x\|_O \leq 1$ it holds that

$$\|x\|_2 \leq 1/\left(C_1 M^{\frac{2-p}{2p}} + C_2 M^{\frac{2-q}{2q}}\right) = \sqrt{M}/\left(C_1 M^{\frac{1}{p}} + C_2 M^{\frac{1}{q}}\right).$$

Following the same arguments as in (17) and (18) we obtain (23). To finish the proof it now suffices to show that

$$\mathbf{E}\left[\left\|\begin{pmatrix} \|\frac{1}{n}\sum_{i=1}^n \sigma_i \Phi_1(x_i)\|_{\star 1} \\ \vdots \\ \|\frac{1}{n}\sum_{i=1}^n \sigma_i \Phi_M(x_i)\|_{\star M} \end{pmatrix}\right\|_2\right] \leq \sqrt{\frac{M}{n}} \ .$$

This is can be seen by a straightforward application of (20):

$$\mathbf{E}\left[\sqrt{\sum_{k=1}^M \left\| \frac{1}{n}\sum_{i=1}^n \sigma_i \Phi_k(x_i) \right\|_{\star k}^2}\right] \leq \sqrt{\mathbf{E}\left[\sum_{k=1}^M V_k\right]} \leq \sqrt{\sum_{i=1}^M \frac{1}{n}} = \sqrt{\frac{M}{n}} \ .$$

5 Empirical Results

In this section we evaluate the proposed method on artificial and real data sets. To avoid validating over two regularization parameters simultaneously, we only only study elastic net MKL for the special case $p \approx 1$.

5.1 Experiments with Sparse and Non-sparse Kernel Sets

The goal of this section is to study the relationship of the level of sparsity of the true underlying function to the chosen block norm or elastic net MKL model. Apart from investigating which parameter choice leads to optimal results, we are also interested in the effects of suboptimal choices of p. To this aim we constructed several artificial data sets in which we vary the degree of sparsity in the true kernel mixture coefficients. We go from having all weight focused on a single kernel (the highest level of sparsity) to uniform weights (the least sparse scenario possible) in several steps. We then study the statistical performance of ℓ_p-block-norm MKL for different values of p that cover the entire range $[0, \infty]$.

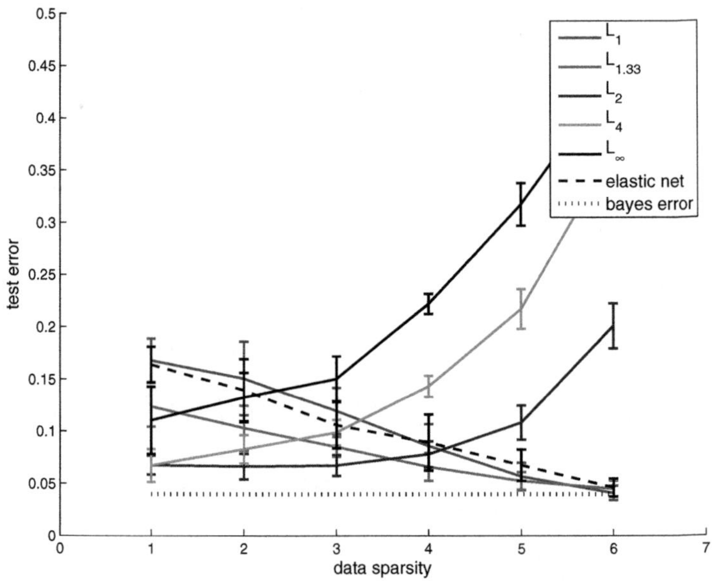

Fig. 1. Empirical results of the artificial experiment for varying true underlying data sparsity

We follow the experimental setup of [10] but compute classification models for $p = 1, 4/3, 2, 4, \infty$ block-norm MKL and $\mu = 10$ elastic net MKL. The results are shown in Fig. 1 and compared to the Bayes error that is computed analytically from the underlying probability model.

Unsurprisingly, ℓ_1 performs best in the sparse scenario, where only a single kernel carries the whole discriminative information of the learning problem. In contrast, the ℓ_∞-norm MKL performs best when all kernels are equally informative. Both MKL variants reach the Bayes error in their respective scenarios. The elastic net MKL performs comparable to ℓ_1-block-norm MKL. The non-sparse $\ell_{4/3}$-norm MKL and the unweighted-sum kernel SVM perform best in the balanced scenarios, i.e., when the noise level is ranging in the interval 60%-92%. The non-sparse ℓ_4-norm MKL of [2] performs only well in the most non-sparse scenarios. Intuitively, the non-sparse $\ell_{4/3}$-norm MKL of [7,9] is the most robust MKL variant, achieving an test error of less than 0.1% in all scenarios. The sparse ℓ_1-norm MKL performs worst when the noise level is less than 82%. It is worth mentioning that when considering the most challenging model/scenario combination, that is ℓ_∞-norm in the sparse and ℓ_1-norm in the uniformly non-sparse scenario, the ℓ_1-norm MKL performs much more robust than its ℓ_∞ counterpart. However, as witnessed in the following sections, this does not prevent ℓ_∞ norm MKL from performing very well in practice. In summary, we conclude that by tuning the sparsity parameter p for each experiment, block norm MKL achieves a low test error across all scenarios.

5.2 Gene Start Recognition

This experiment aims at detecting transcription start sites (TSS) of RNA Polymerase II binding genes in genomic DNA sequences. Many detectors rely on a combination of feature sets which makes the learning task appealing for MKL. For our experiments we use the data set from [21] and we employ five different kernels representing the TSS signal (weighted degree with shift), the promoter (spectrum), the 1st exon (spectrum), angles (linear), and energies (linear). The kernel matrices are normalized such that each feature vector has unit norm in Hilbert space. We reserve 500 and 500 randomly drawn instances for holdout and test sets, respectively, and use 250 elemental training sets. Table 1 shows the area under the ROC curve (AUC) averaged over 250 repetitions of the experiment. Thereby 1 and ∞ block norms are approximated by 64/63 and 64 norms, respectively. For the elastic net we use an $\ell_{1.05}$-block-norm penalty.

Table 1. Results for the bioinformatics experiment

	AUC $\pm$ stderr
$\mu = 0.01$ elastic net	85.91 ± 0.09
$\mu = 0.1$ elastic net	85.77 ± 0.10
$\mu = 1$ elastic net	87.73 ± 0.11
$\mu = 10$ elastic net	$\mathbf{88.24 \pm 0.10}$
$\mu = 100$ elastic net	87.57 ± 0.09
1-block-norm MKL	85.77 ± 0.10
4/3-block-norm MKL	87.93 ± 0.10
2-block-norm MKL	87.57 ± 0.10
4-block-norm MKL	86.33 ± 0.10
∞-block-norm MKL	87.67 ± 0.09

The results vary greatly between the chosen MKL models. The elastic net model gives the best prediction for $\mu = 10$. Out of the block norm MKLs the classical ℓ_1-norm MKL has the worst prediction accuracy and is even outperformed by an unweighted-sum kernel SVM (i.e., $p = 2$ norm MKL). In accordance with previous experiments in [9] the $p = 4/3$ block norm has the highest prediction accuracy of the models within the parameter range $p \in [1, 2]$. This performance can even be improved by the elastic net MKL with $\mu = 10$.

This is remarkable since elastic net MKL performs kernel *selection*, and hence the outputted kernel combination can be easily interpreted by domain experts. Note that the method using the unweighted sum of kernels [21] has recently been confirmed to be the leading in a comparison of 19 state-of-the-art promoter prediction programs [1]. It was recently shown to be outperformed by $\ell_{4/3}$-norm MKL [9], and our experiments suggest that its accuracy can be further improved by $\mu = 10$ elastic net MKL.

6 Conclusion

We presented a framework for multiple kernel learning, that unifies several recent lines of research in that area. We phrased the seemingly different MKL variants as a single generalized optimization criterion and derived its dual representation. By plugging in an arbitrary convex loss function many existing approaches can be recovered as instantiations of our model. We compared the different MKL variants in terms of their generalization performance by giving an concentration inequality for generalized MKL that matches the previous known bounds for ℓ_1 and $\ell_{4/3}$ block norm MKL.

Our empirical analysis shows that the performance of the MKL variants crucially depends on true underlying data sparsity. We compared several existing MKL variants on bioinformatics data. On the computational side, we derived derived a quasi Newton optimization method for unified MKL. It is up to future work to speed up optimization by a SMO-type decomposition algorithm.

Acknowledgments

The authors wish to thank Francis Bach and Ryota Tomioka for comments that helped improving the manuscript; we thank Klaus-Robert Müller for stimulating discussions. This work was supported in part by the Deutsche Forschungsgemeinschaft (DFG) through the grant RU 1589/1-1 and by the European Community under the PASCAL2 Network of Excellence (ICT-216886). We gratefully acknowledge the support of NSF through grant DMS-0707060. MK acknowledges a scholarship by the German Academic Exchange Service (DAAD).

References

1. Abeel, T., Van de Peer, Y., Saeys, Y.: Towards a gold standard for promoter prediction evaluation. Bioinformatics (2009)
2. Aflalo, J., Ben-Tal, A., Bhattacharyya, C., Saketha Nath, J., Raman, S.: Variable sparsity kernel learning — algorithms and applications. Journal of Machine Learning Research (submitted, 2010), `http://mllab.csa.iisc.ernet.in/vskl.html`
3. Agarwal, A., Rakhlin, A., Bartlett, P.: Matrix regularization techniques for online multitask learning. Technical Report UCB/EECS-2008-138, EECS Department, University of California, Berkeley (October 2008)
4. Bach, F.R., Lanckriet, G.R.G., Jordan, M.I.: Multiple kernel learning, conic duality, and the smo algorithm. In: Proc. 21st ICML. ACM, New York (2004)
5. Bartlett, P.L., Mendelson, S.: Rademacher and gaussian complexities: Risk bounds and structural results. Journal of Machine Learning Research 3, 463–482 (2002)
6. Chapelle, O.: Training a support vector machine in the primal. Neural Computation (2006)
7. Cortes, C., Mohri, M., Rostamizadeh, A.: L2 regularization for learning kernels. In: Proceedings, 26th ICML (2009)
8. Cortes, C., Mohri, M., Rostamizadeh, A.: Generalization bounds for learning kernels. In: Proceedings, 27th ICML (to appear, 2010), CoRR abs/0912.3309, `http://arxiv.org/abs/0912.3309`

9. Kloft, M., Brefeld, U., Sonnenburg, S., Laskov, P., Müller, K.-R., Zien, A.: Efficient and accurate lp-norm multiple kernel learning. In: Bengio, Y., Schuurmans, D., Lafferty, J., Williams, C.K.I., Culotta, A. (eds.) Advances in Neural Information Processing Systems, vol. 22, pp. 997–1005. MIT Press, Cambridge (2009)
10. Kloft, M., Brefeld, U., Sonnenburg, S., Zien, A.: Non-sparse regularization and efficient training with multiple kernels. Technical Report UCB/EECS-2010-21, EECS Department, University of California, Berkeley (February 2010), CoRR abs/1003.0079, `http://www.eecs.berkeley.edu/Pubs/TechRpts/2010/EECS-2010-21.html`
11. Lanckriet, G.R.G., Cristianini, N., Bartlett, P., El Ghaoui, L., Jordan, M.I.: Learning the kernel matrix with semidefinite programming. Journal of Machine Learning Research 5, 27–72 (2004)
12. Müller, K.-R., Mika, S., Rätsch, G., Tsuda, K., Schölkopf, B.: An introduction to kernel-based learning algorithms. IEEE Neural Networks 12(2), 181–201 (2001)
13. Nath, J.S., Dinesh, G., Ramanand, S., Bhattacharyya, C., Ben-Tal, A., Ramakrishnan, K.R.: On the algorithmics and applications of a mixed-norm based kernel learning formulation. In: Bengio, Y., Schuurmans, D., Lafferty, J., Williams, C.K.I., Culotta, A. (eds.) Advances in Neural Information Processing Systems, vol. 22, pp. 844–852 (2009)
14. Rakotomamonjy, A., Bach, F., Canu, S., Grandvalet, Y.: SimpleMKL. Journal of Machine Learning Research 9, 2491–2521 (2008)
15. Rifkin, R.M., Lippert, R.A.: Value regularization and fenchel duality. J. Mach. Learn. Res. 8, 441–479 (2007)
16. Rockafellar, R.T.: Convex Analysis. Princeton Landmarks in Mathemathics. Princeton University Press, New Jersey (1970)
17. Schölkopf, B., Smola, A.J.: Learning with Kernels. MIT Press, Cambridge (2002)
18. Schölkopf, B., Smola, A.J., Müller, K.-R.: Nonlinear component analysis as a kernel eigenvalue problem. Neural Computation 10, 1299–1319 (1998)
19. Showalter, R.E.: Monotone operators in banach space and nonlinear partial differential equations. Mathematical Surveys and Monographs 18 (1997)
20. Sonnenburg, S., Rätsch, G., Schäfer, C., Schölkopf, B.: Large Scale Multiple Kernel Learning. Journal of Machine Learning Research 7, 1531–1565 (2006)
21. Sonnenburg, S., Zien, A., Rätsch, G.: ARTS: Accurate Recognition of Transcription Starts in Human. Bioinformatics, 22(14), e472–e480 (2006)
22. Tomioka, R., Suzuki, T.: Sparsity-accuracy trade-off in mkl. In: arxiv (2010), CoRR abs/1001.2615
23. Vapnik, V.N.: Statistical Learning Theory. Wiley, Chichester (1998)
24. Zhu, C., Byrd, R.H., Lu, P., Nocedal, J.: Algorithm 778: L-bfgs-b: Fortran subroutines for large-scale bound-constrained optimization. ACM Trans. Math. Softw. 23(4), 550–560 (1997)
25. Zou, H., Hastie, T.: Regularization and variable selection via the elastic net. Journal of the Royal Statistical Society, Series B 67, 301–320 (2005)

Evolutionary Dynamics of Regret Minimization

Tomas Klos[1], Gerrit Jan van Ahee[2], and Karl Tuyls[3]

[1] Delft University of Technology, Delft, The Netherlands
[2] Yes2web, Rotterdam, The Netherlands
[3] Maastricht University, Maastricht, The Netherlands

Abstract. Learning in multi-agent systems (MAS) is a complex task. Current learning theory for single-agent systems does not extend to multi-agent problems. In a MAS the reinforcement an agent receives may depend on the actions taken by the other agents present in the system. Hence, the Markov property no longer holds and convergence guarantees are lost. Currently there does not exist a general formal theory describing and elucidating the conditions under which algorithms for multi-agent learning (MAL) are successful. Therefore it is important to fully understand the dynamics of multi-agent reinforcement learning, and to be able to analyze learning behavior in terms of stability and resilience of equilibria. Recent work has considered the replicator dynamics of evolutionary game theory for this purpose. In this paper we contribute to this framework. More precisely, we formally derive the evolutionary dynamics of the Regret Minimization polynomial weights learning algorithm, which will be described by a system of differential equations. Using these equations we can easily investigate parameter settings and analyze the dynamics of multiple concurrently learning agents using regret minimization. In this way it is clear why certain attractors are stable and potentially preferred over others, and what the basins of attraction look like. Furthermore, we experimentally show that the dynamics predict the real learning behavior and we test the dynamics also in non-self play, comparing the polynomial weights algorithm against the previously derived dynamics of Q-learning and various Linear Reward algorithms in a set of benchmark normal form games.

1 Introduction

Multi-agent systems (MAS) are a proven solution method for contemporary technological challenges of a distributed nature, such as e.g. load balancing and routing in networks [13,14]. Typical for these new challenges of today is that the environment in which those systems need to operate is dynamic, rather than static, and as such evolves over time, not only due to external environmental changes but also due to agents' interactions. The naive approach of providing all possible situations an agent can encounter along with the optimal behavior in each of them beforehand, is not feasible in this type of system. Therefore to successfully apply MAS, agents should be able to adapt themselves in response to actions of other agents and changes in the environment. For this purpose, researchers have investigated Reinforcement Learning (RL) [15,11].

J.L. Balcázar et al. (Eds.): ECML PKDD 2010, Part II, LNAI 6322, pp. 82–96, 2010.

RL is already an established and profound theoretical framework for learning in a single-agent framework. In this framework a single agent operates in an uncertain environment and must learn to act autonomously and achieve a certain goal. Under these circumstances it has been shown that as long as the environment an agent is experiencing is Markovian, and the agent can try out sufficiently many actions, RL guarantees convergence to the optimal strategy [22,16,12]. This task becomes more complex when multiple agents are concurrently learning and possibly interacting with one another. Furthermore, these agents have potentially different capabilities and goals. Consequently, learning in MAS does not guarantee the same theoretical grounding.

Recently an evolutionary game theoretic approach has been introduced to provide such a theoretical means to analyze the dynamics of multiple concurrently learning agents [19,17,18]. For a number of state of the art MAL algorithms, such as Q-learning and Learning Automata, the evolutionary dynamics have been derived. Using these derived dynamics one can visualize and thoroughly analyze the average learning behavior of the agents and stability of the attractors. For an important class of MAL algorithms, viz. Regret Minimization (RM), these dynamics are still unknown. The central idea of this type of algorithm is that after the agent has taken an action and received a reward in the learning process, he may look back at the history of actions and rewards taken so far, and regret not having played another action—namely the best action in hindsight. Based on this idea a loss function is calculated which is key to the update rule of an RM learning algorithm. To contribute to this EGT backbone for MAL, it is essential that we derive and examine the evolutionary dynamics of Regret Minimization as well, which we undertake in the present paper.

In this paper we follow this recent line of reasoning that captures the dynamics of MAL algorithms and formally derive the evolutionary dynamics of the Polynomial Weights Regret Minimization learning algorithm. Furthermore, we perform an extensive experimental study using these dynamics, illustrating how they predict the behavior of the associated learning algorithm. As such this allows for a quick and thorough analysis of the behavior of the learning agents in terms of learning traces, parameters, and stability and resilience of attractors. The derived dynamics provide theoretical insight in this class of algorithms and as such contribute to a theoretical backbone for MAL. Moreover, we do not only investigate the dynamics in self play but also compare the derived dynamics against the dynamics of Linear Reward-Inaction and Linear Reward-ϵPenalty Learning Automata. It is the first time that these MAL algorithms are compared using their derived dynamical systems instead of performing a time consuming experimental study with the learning algorithms themselves.

The remainder of this paper is structured as follows. In Sect. 2 we introduce the necessary background for the remainder of the paper. More precisely, we introduce Regret Minimization and the Replicator Dynamics of Evolutionary Game Theory. In Sect. 3 we formally derive the dynamics of RM, and we study them experimentally in Sect. 4. Section 5 summarizes related work and we conclude in Sect. 6.

2 Preliminaries

In this section we describe the necessary background for the remainder of the article. We start off by introducing Regret Minimization, the multi-agent learning algorithm of which we want to describe the evolutionary dynamics. Section 2.2 introduces the replicator dynamics of Evolutionary Game Theory.

2.1 Regret Minimization

Regret Minimizing algorithms are learning algorithms relating the history of an agents' play to his current choice of action. After acting, the agent looks back at the history of actions and corresponding rewards, and regrets not having played the best action in hindsight. Playing this action at all stages often results in a better total reward by removing the cost of exploration.

As keeping a history of actions and rewards is very expensive at best, most regret minimizing algorithms use the concept of *loss* l_i to aggregate the history per action i. Using the loss, the action selection probabilities are updated. Several algorithms have been constructed around computing loss.

In order to determine the best action in hindsight the agent needs to know what rewards he could have received, which could be provided by the system. Each action i played, results in a reward r_i and the best reward in hindsight r is determined, with the loss for playing i given by $l_i = r - r_i$: a measure for regret.

The Polynomial Weights algorithm [2] is a member of the Regret Minimization class. It assigns a weight w_i to each action i which is updated using the loss for not playing the best action in hindsight:

$$w_i^{(t+1)} = w_i^{(t)} \left(1 - \lambda l_i^{(t)}\right) , \tag{1}$$

where λ is a learning parameter to control the speed of the weight-change. The weights are now used to derive action selection probabilities by normalization:

$$x_i^{(t)} = \frac{w_i^{(t)}}{\sum_j w_j^{(t)}} . \tag{2}$$

2.2 Replicator Dynamics

The Replicator Dynamics (RD) are a system of differential equations describing how a population of strategies evolves through time [9]. Here we will consider an individual level of analogy between the related concepts of learning and evolution. Each agent has a set of possible strategies at hand. Which strategies are favored over others depends on the experience the agent has previously gathered by interacting with the environment and other agents. The collection of possible strategies can be interpreted as a population in an evolutionary game theory perspective [20]. The dynamical change of preferences within the set of strategies can be seen as the evolution of this population as described by the replicator dynamics. The continuous time two-population replicator dynamics are defined by the following system of ordinary differential equations:

$$\begin{aligned}\dot{x}_i &= x_i\left((A\mathbf{y})_i - \mathbf{x}^T A\mathbf{y}\right)\\ \dot{y}_i &= y_i\left((B\mathbf{x})_i - \mathbf{y}^T B\mathbf{x}\right)\end{aligned} \quad , \tag{3}$$

where A and B are the payoff matrices for player 1 (population x) and 2 (population y) respectively. For an example see Sect. 4. The probability vector $\mathbf{x}$ (resp. $\mathbf{y}$) describes the frequency of all pure strategies (also called replicators) for player 1 (resp. 2). Success of a replicator i in population x is measured by the difference between its current payoff $(A\mathbf{y})_i$ and the average payoff of the entire population x, i.e. $\mathbf{x}^T A\mathbf{y}$.

3 Modelling Regret Minimization

In this section we will derive a mathematical model, i.e. a system of differential equations, describing the dynamics of the polynomial no regret learning algorithm. Each learning agent will have his own system of differential equations describing the updates to his action selection probabilities. Just as in (3), our models will be using expected rewards to calculate the change in action selection probabilities. Here too, these rewards are determined by the other agents in the system.

The first step in finding a mathematical model for Polynomial Weights is to determine the update $\delta x_i^{(t)}$ at time t to the action selection probability for any action i:

$$\begin{aligned}\delta x_i^{(t)} &= x_i^{(t+1)} - x_i^{(t)}\\ &= \frac{w_i^{(t+1)}}{\sum_j w_j^{(t+1)}} - x_i^{(t)} \quad .\end{aligned}$$

This shows that the update δx_i to the action selection probability depends on the weight as well as the probabilties. If we want the model to consist of a coupled system of differential equations, we need to find an expression for these weights in terms of x_i and y_i. In other words we would like to find an expression of the weights in terms of their corresponding action selection probabilities. Therefore, using (2) we divide any two x_i and x_j:

$$\begin{aligned}\frac{x_i}{x_j} &= \frac{w_i}{\sum_k w_k}\frac{\sum_k w_k}{w_j}\\ w_i &= x_i\frac{w_j}{x_j} \quad .\end{aligned} \tag{4}$$

This allows to represent weights as the corresponding action selection probability multiplied by a common factor. Substituting (4) into (1) and subsequently (2) yields:

$$x_i^{(t+1)} = \frac{\frac{w_j^{(t)}}{x_j^{(t)}} x_i^{(t)} \left(1 - \lambda l_i^{(t)}\right)}{\sum_k \frac{w_j^{(t)}}{x_j^{(t)}} x_k^{(t)} \left(1 - \lambda l_k^{(t)}\right)} = \frac{x_i^{(t)} \left(1 - \lambda l_i^{(t)}\right)}{\sum_k x_k^{(t)} \left(1 - \lambda l_k^{(t)}\right)} \ . \tag{5}$$

The update $\delta x_i^{(t)}$ is found by subtracting $x_i^{(t)}$ from (5):

$$\begin{aligned}\delta x_i^{(t)} &= x_i^{(t+1)} - x_i^{(t)} \\ &= \frac{x_i^{(t)} \left(1 - \lambda l_i^{(t)}\right)}{\sum_j x_j^{(t)} \left(1 - \lambda l_j^{(t)}\right)} - x_i^{(t)} \\ &= \frac{x_i^{(t)} \left(\left(1 - \lambda l_i^{(t)}\right) - \sum_j x_j^{(t)} \left(1 - \lambda l_j^{(t)}\right)\right)}{\sum_j x_j^{(t)} \left(1 - \lambda l_j^{(t)}\right)} \ .\end{aligned} \tag{6}$$

In subsequent formulations, the reference to time will again be dropped, as all expressions reference the same time t.

The next step in the derivation requires the specification of the loss l_i. The best reward may be modeled as the maximum expected reward $r = \max_k(A\mathbf{y})_k$, the actual expected reward is given by $r_i = (A\mathbf{y})_i$. This yields the equation for the loss for action i:

$$l_i = \max_k (A\mathbf{y})_k - (A\mathbf{y})_i \ . \tag{7}$$

After substituting the loss l_i (7) into (6), the derivation of the model is nearly finished. Using the fact that $\sum_j x_j C = C$ for constant C, we may simplify the resulting equation by replacing these terms:

$$\begin{aligned}\mathbb{E}[\delta x_i](\mathbf{x}, \mathbf{y}) &= \frac{x_i \left(1 - \lambda(\max_k(A\mathbf{y})_k - (A\mathbf{y})_i) - \sum_j x_j \left(1 - \lambda(\max_k(A\mathbf{y})_k - (A\mathbf{y})_j)\right)\right)}{\sum_j x_j \left(1 - \lambda(\max_k(A\mathbf{y})_k - (A\mathbf{y})_j)\right)} \\ &= \frac{x_i \left(1 - \lambda \max_k(A\mathbf{y})_k + \lambda(A\mathbf{y})_i - \sum_j x_j + \lambda \sum_j x_j \max_k(A\mathbf{y})_k - \lambda \sum_j x_j (A\mathbf{y})_j\right)}{\sum_j x_j - \lambda \left(\sum_j x_j \max_k(A\mathbf{y})_k - \sum_j x_j (A\mathbf{y})_j\right)} \\ &= \frac{x_i \lambda \left((A\mathbf{y})_i - \sum_j x_j (A\mathbf{y})_j\right)}{1 - \lambda \left(\max_k(A\mathbf{y})_k - \sum_j x_j (A\mathbf{y})_j\right)} \ .\end{aligned} \tag{8}$$

Finally, we recognize $\sum_j x_j (A\mathbf{y})_j = \mathbf{x}^T A\mathbf{y}$ and we arrive at the general model:

$$\dot{x} = \frac{\lambda x_i \left((A\mathbf{y})_i - \mathbf{x}^T A\mathbf{y}\right)}{1 - \lambda \left(\max_k(A\mathbf{y})_k - \mathbf{x}^T A\mathbf{y}\right)} \ . \tag{9}$$

The derivation for $\dot{y}$ is completely analogous and yields:

$$\dot{y} = \frac{\lambda y_i \left((B\mathbf{x})_i - \mathbf{y}^T B\mathbf{x}\right)}{1 - \lambda\left(\max_k (B\mathbf{x})_k - \mathbf{y}^T B\mathbf{x}\right)} \ . \tag{10}$$

Equations 9 and 10 describe the dynamics of the Polynomial Weights learning algorithm. What is immediately interesting to note is that we recognize in this model the coupled replicator equations, described in (3), in the numerator. This value is then weighted based on the expected loss. At this point we can conclude that this learning algorithm can also be described based on the coupled RD from evolutionary game theory, just as has been shown before for Learning Automata and Q-learning (resulting in different equations that also contain the RD) [17].

4 Experiments

We performed numerical experiments to validate our model, by comparing its predictions with simulations of agents using the PW learning algorithm. In addition, we propose a method to investigate outcomes of interactions among agents using *different* learning algorithms, of which dynamics have been derived. First we present the games and algorithms we used in our experiments.

4.1 Sample Games

We limit ourselves to two-player, two-action, single state games. This class includes many interesting games, such as the Prisoner's Dilemma, and allows us to visualize the learning dynamics by plotting them in 2-dimensional trajectory fields. These plots show the direction of change for the two players' action selection probabilities.

Having two agents with two actions each yields games with action selection probabilities $\mathbf{x} = [x_1 x_2]^T$ and $\mathbf{y} = [y_1 y_2]^T$ and two 2-dimensional payoff matrices A and B for players 1 and 2 respectively:

$$A = \begin{bmatrix} a_{11} & a_{12} \\ a_{21} & a_{22} \end{bmatrix} \quad B = \begin{bmatrix} b_{11} & b_{12} \\ b_{21} & b_{22} \end{bmatrix} \ .$$

Note that these are payoff *matrices*, not payoff *tables* with row players and column players: put in these terms, each player is the *row* player in his own matrix.

This class of games can be partitioned into three subclasses [20]. We experimented with games from all three subclasses. The subclasses are the following.

1. At least one of the players has a dominant strategy when

$$(a_{11} - a_{21})(a_{12} - a_{22}) > 0 \text{ or}$$
$$(b_{11} - b_{21})(b_{12} - b_{22}) > 0 \ .$$

The Prisoner's Dilemma (PD) falls into this class. The reward matrices used for this class in the simulations and model are

$$A = B = \begin{bmatrix} 1 & 5 \\ 0 & 3 \end{bmatrix} .$$

This game has a single pure Nash equilibrium at $(\mathbf{x}, \mathbf{y}) = ([1,0]^T, [1,0]^T)$.

2. There are two pure equilibria and one mixed when

$$(a_{11} - a_{21})(a_{12} - a_{22}) < 0 \text{ and}$$
$$(b_{11} - b_{21})(b_{12} - b_{22}) < 0 \text{ and}$$
$$(a_{11} - a_{21})(b_{11} - b_{21}) > 0 \ .$$

The Battle of the Sexes (BoS) falls into this class. The reward matrices used for this class in the simulations and model are

$$A = \begin{bmatrix} 2 & 0 \\ 0 & 1 \end{bmatrix} \ B = \begin{bmatrix} 1 & 0 \\ 0 & 2 \end{bmatrix} .$$

This game has two pure Nash equilibria at $(\mathbf{x}, \mathbf{y}) = ([0,1]^T, [0,1]^T)$ and $([1,0]^T, [1,0]^T)$ and one mixed Nash equilibrium at $([2/3, 1/3]^T, [1/3, 2/3]^T)$.

3. There is just one mixed equilibrium when

$$(a_{11} - a_{21})(a_{12} - a_{22}) < 0 \text{ and}$$
$$(b_{11} - b_{21})(b_{12} - b_{22}) < 0 \text{ and}$$
$$(a_{11} - a_{21})(b_{11} - b_{21}) < 0.$$

This class contains Matching Pennies (MP). The reward matrices used for this class in the simulations and model are

$$A = B = \begin{bmatrix} 2 & 1 \\ 1 & 2 \end{bmatrix} .$$

This game has a single mixed Nash equilibrium at $\mathbf{x} = [1/2, 1/2]^T, \mathbf{y} = [1/2, 1/2]^T$.

4.2 Other Learning Algorithms

The dynamics we derived to model agents using the Polynomial Weights (PW) algorithm can be used to investigate the performance of the PW algorithm in selfplay. This provides an important test for the validity of a learning algorithm: in selfplay it should converge to a Nash equilibrium of the game [6]. Crucially, with our model, we are now also able to model and make predictions about interactions between agents using *different* learning algorithms.

In related work, analytical models for several other algorithms have been derived (see [19,17,7] and Table 1 for an overview). Here we use models for the Linear Reward-Inaction (L_{R-I}) and Linear Reward-ϵPenalty ($L_{R-\epsilon P}$) policy iteration algorithms.

L_{R-I} We study 2 algorithms from the class of linear reward algorithms. These are algorithms that update (increase or decrease) the action selection probability for the action selected by a fraction $0 < \lambda \leq 1$ of the payoff received. The parameter λ is called the *learning rate* of the algorithm. The L_{R-I} algorithm rewards, but does not punish actions for yielding low payoffs: the action selection probability of the selected action is increased whenever rewards $0 \leq r \leq 1$ are received.

$L_{R-\epsilon P}$ The $L_{R-\epsilon P}$ algorithm generalizes both the L_{R-I} algorithm and the L_{R-P} algorithm (which we therefore didn't include). In addition to rewarding high payoffs, the penalty algorithms also punish low payoffs. The $L_{R-\epsilon P}$ algorithm captures both other algorithms through the parameter $\epsilon > 0$ which specifies how severely low rewards are punished: as ϵ goes to 0 (1), there is no (full) punishment and the algorithm behaves like L_{R-I} (L_{R-P}).

The models for these algorithms are represented in Table 1, which also shows how they are all variations on the basic coupled replicator equations.

Table 1. Correspondence between Coupled Replicator Dynamics (CRD) and learning strategies. (Q refers to the Q-learning algorithm.)

Alg. name	Model	Reference
CRD	$x_i\left((A\mathbf{y})_i - \mathbf{x}^T A\mathbf{y}\right)$	[9]
L_{R-I}	$\lambda x_i\left((A\mathbf{y})_i - \mathbf{x}^T A\mathbf{y}\right)$	[17]
$L_{R\text{-}\epsilon P}$	$\lambda x_i\left((A\mathbf{y})_i - \mathbf{x}^T A\mathbf{y}\right) - \epsilon\lambda\left(-x^2(1-(A\mathbf{y})_i) + \frac{1-x_i}{r-1}\sum_{j\neq i} x_j(1-(A\mathbf{y})_j)\right)$	[1]
PW	$\lambda x_i\left((A\mathbf{y})_i - \mathbf{x}^T A\mathbf{y}\right) / \left(1 - \lambda\left(\max_k (A\mathbf{y})_k - \mathbf{x}^T A\mathbf{y}\right)\right)$	Sec. 3
Q	$\tau\lambda x_i\left((A\mathbf{y})_i - \mathbf{x}^T A\mathbf{y}\right) + \lambda x_i \sum_j x_j \ln\left(\frac{x_j}{x_i}\right)$	[19,7]

4.3 Results

To visualize learning, all of our plots show the action selection probability for the *first* actions x_1 and y_1 of the two agents on the two axes: the PW agent on the x-axis and the other agent on the y-axis (sometimes this other agent also uses the PW algoritm). Knowing these probabilities, we also know $x_2 = 1 - x_1$ and $y_2 = 1 - y_1$. When playing the repeated game, the learning strategy updates these probabilities after each iteration.

All learning algorithms have been simulated extensively in each of the above games. This has been done in order to validate the models we have derived or taken from the literature. The results show paths starting at the initial action selection probabilities for action 1 for both agents that, as learning progresses, move toward some equilibrium. The simulations depend on: (*i*) the algorithms and their parameters, (*ii*) the game played and (*iii*) the initial action selection probabilities. We simulate and model 3 different algorithms (see Sect 4.2: PW,

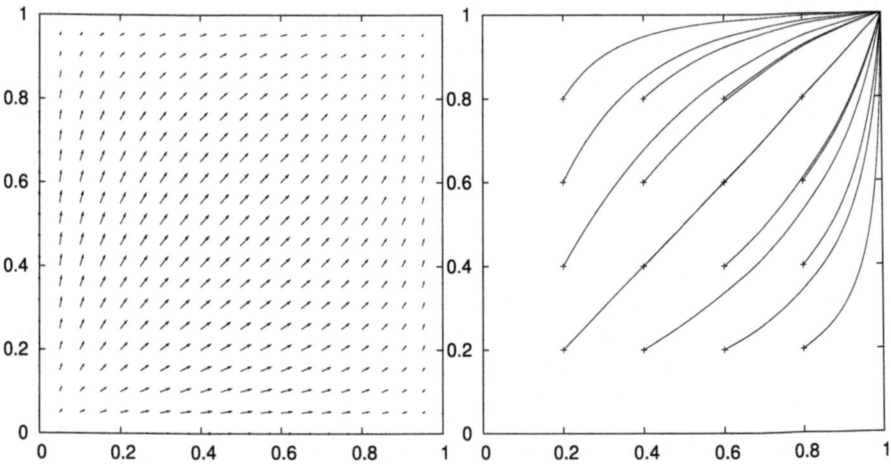

Fig. 1. PW selfplay, Prisoner's Dilemma

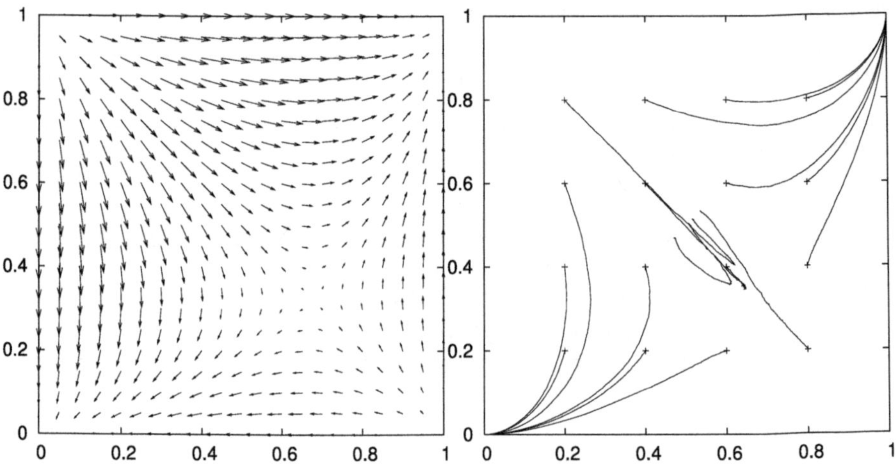

Fig. 2. PW selfplay, Battle of the Sexes

L_{R-I}, and $L_{R-\epsilon P}$), with $\lambda = \epsilon = 1$, in 3 different games (see Sect. 4.1: PD, BoS, and MP). The initial probabilities are taken from the grid $\{.2, .4, .6, .8\} \times \{.2, .4, .6, .8\}$; they are indicated by '+' in the plots. All figures show vector fields on the *left*, and average trajectories over 500 simulations of 1500 iterations of agents using the various learning algorithms on the *right*.

PW Selfplay. In Figures 1, 2, and 3 we show results for the PW algorithm in selfplay in the PD, the BoS, and MP, respectively. In all three figures, the models clearly show direction of motion towards each of the various Nash equilibria in

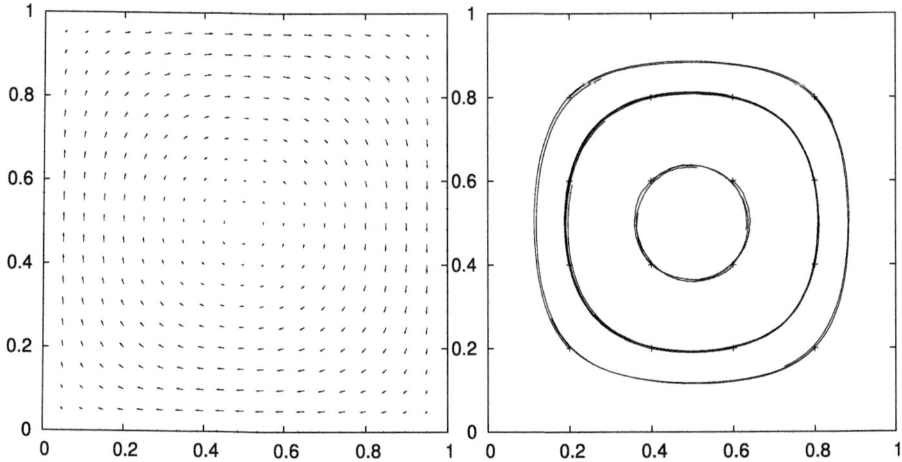

Fig. 3. PW selfplay, Matching Pennies

the respective games: the single pure strategy Nash equilibrium in the PD, the two pure strategy Nash equilibria in the BoS (not the unstable mixed one), and a circular oscillating pattern in the MP game.

The simulation paths (on the right) validate these models (on the left), in that the models are shown to accurately predict the simulation trajectories of interacting PW agents in all three games. The individual simulations in the BoS game (Fig. 2) that start from any one of the 4 initial positions on the diagonal (which is exactly the boundary between the two pure equilibria's basins of attraction) all end up in one of the two pure strategy Nash equilibria, but since we take the average of all 500 simulations, these plots end up in the center, somewhat spread out over the perpendicular (0,0)-(1,1) diagonal, because there is not a perfect 50%/50% division of the trajectories over the 2 Nash equilibria.

PW vs. L_{R-I}. Having established the external validity of our model of PW agents, we now turn to an analysis of interactions of PW agents and agents using other learning algorithms. To this end, in all subsequent figures, we propose to let movement in the x-direction be controlled by the PW dynamics and in the y-direction by the dynamics of one of the other models (see Table 1 and Sect. 4.2). We start with L_{R-I} agents, for which we only show interactions with PW agents in the PD (Fig. 4) and the BoS (Fig. 5). (The vector fields and trajectories in the MP game correspond closely again, and don't differ much from those in Fig. 3.)

This setting already shows that when agents use different learning algorithms, the interaction changes significantly. The direction of motion is still towards the single pure strategy Nash equilibrium in the PD and towards the two pure strategy Nash equilibria in the BoS, albeit along different lines. Again, the simulation paths follow the vector field closely, where the two seemingly anomalous average trajectories starting from $(0.2, 0.6)$ and from $(0.6, 0.4)$ in the BoS game (Fig. 5) can be explained in a similar manner as for Fig. 2. In this setting of PW vs.

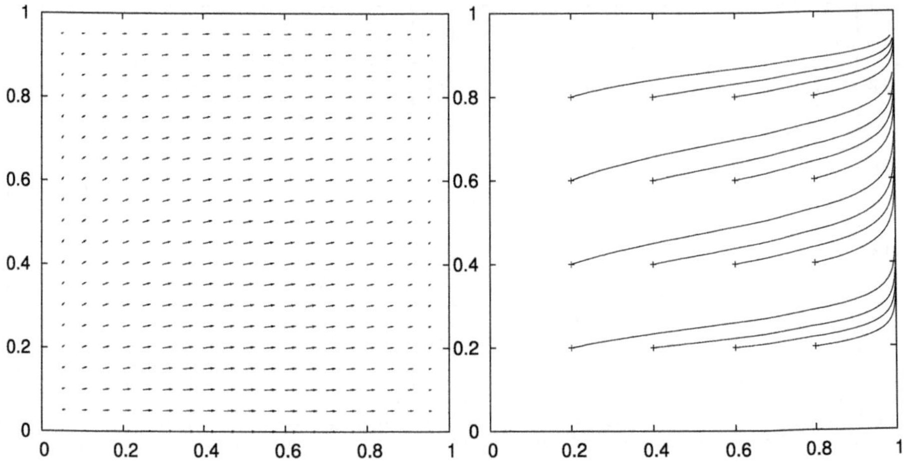

Fig. 4. PW vs. L_{R-I}, Prisoner's Dilemma

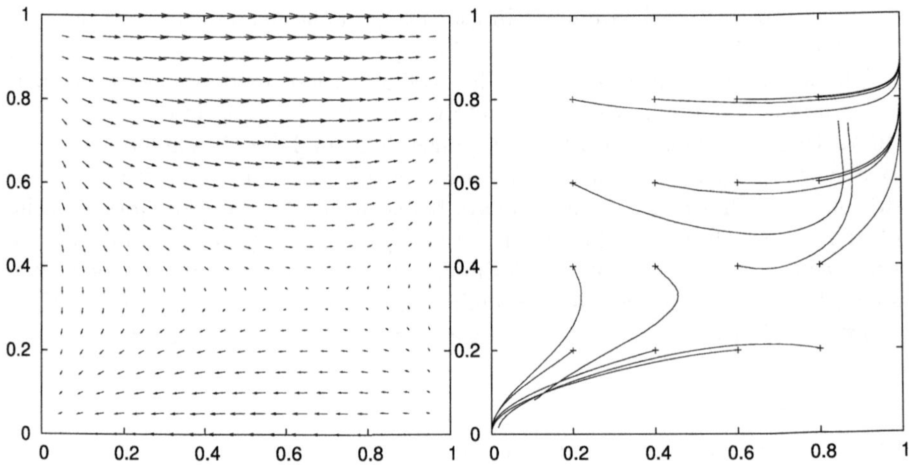

Fig. 5. PW vs. L_{R-I}, Battle of the Sexes

L_{R-I}, these points are now the initial points closest to the border between the basins of attraction of the two pure strategy equilibria, although they are not *on* the border, as the 4 points in Fig. 2 were, which is why these average trajectories end up closer to the equilibrium in whose basin of attraction they started. However, they are close enough to the border with the other basin to let stochasticity in the algorithm take some of the individual runs to the 'wrong' equilibrium.

An important observation we can make based on this novel kind of non-selfplay analysis, is that when agents use different learning algorithms, the outcomes of the game, or at least the trajectories agents may be expected to follow in reaching

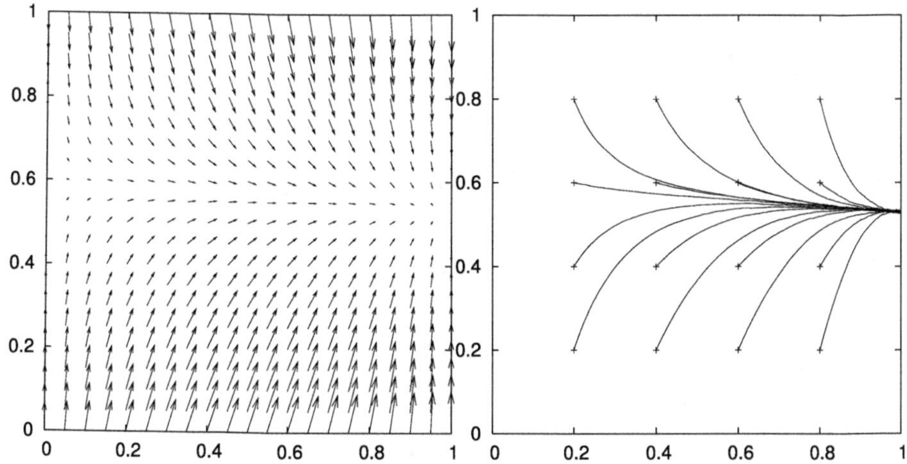

Fig. 6. PW vs. $L_{R-\epsilon P}$, Prisoner's Dilemma

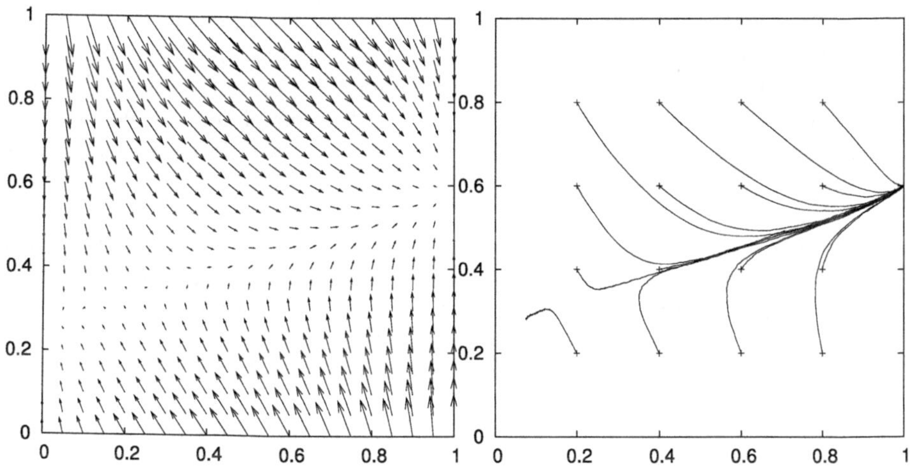

Fig. 7. PW vs. $L_{R-\epsilon P}$, Battle of the Sexes

those outcomes, as well as the basins of attraction of the various equilibria, change as a consequence. This gives insight into the outcomes we may expect from interactions among agents using different learning algorithms.

PW vs. $L_{R-\epsilon P}$. For this interaction, we again show plots for all games (Figures 6–8). The interaction is now changed not just quantitatively, but qualitatively as well. Clearly, the various Nash equilibria are not in reach of the interacting learners anymore: all these games, when played between one agent using the PW algorithm, and one agent using the $L_{R-\epsilon P}$ algorithm, have different equilibria than Nash.

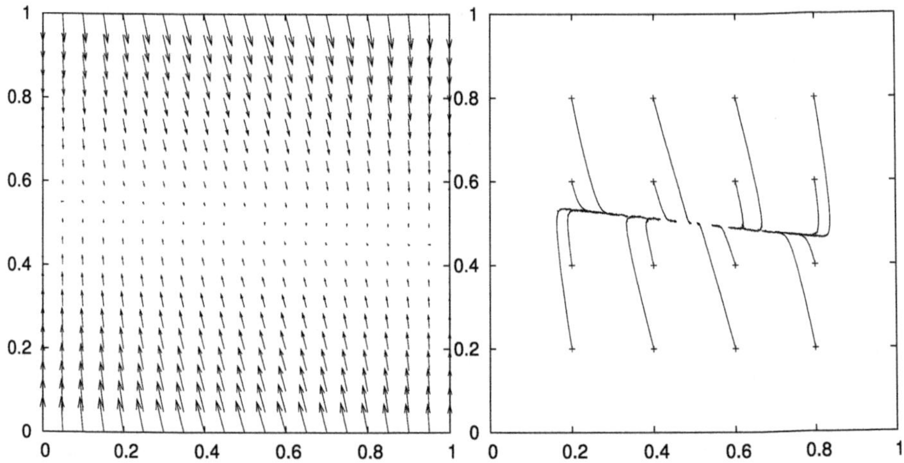

Fig. 8. PW vs. $L_{R-\epsilon P}$, Matching Pennies

What is particularly interesting to observe in Figures 6 and 7 is that the PW player is again playing the strategy prescribed by Nash while the $L_{R-\epsilon P}$ player randomizes over both strategies. (It is hard to tell just by visual inspection whether in Fig. 7 this leads to just a single equilibrium outcome on the right hand side, or whether there is another one on the left. This may be expected given the average trajectory starting at $(0.2, 0.2)$, but should be analyzed more thoroughly, for example using the Amoeba tool [21].) This implies that while the $L_{R-\epsilon P}$ player keeps on exploring its possible strategies the PW player already found the Nash strategy and consequently is regularly able to exploit the other player by always defecting when the $L_{R-\epsilon P}$ occasionally cooperates. Therefore the PW learner will receive, on average, more reward than when playing in self play for instance. While it can be seen from Fig. 4 that the L_{R-I} learner also evolves towards more or less the same point at the right vertical axis as the $L_{R-\epsilon P}$ against the PW learner, it can be observed that the L_{R-I} learner is still able to recover from his mixed strategy and is eventually able to find the Nash strategy as well. Consequently the L_{R-I} performs better against the PW learner than the $L_{R-\epsilon P}$ learner.

In the BoS game (Fig. 7), the equilibria don't just shift, but there even appears to be just a single equilibrium in the game played between a PW player and an $L_{R-\epsilon P}$ player, rather than 3 equilibria, as in the game played by rational agents. In the MP game (Fig. 8), the agents now converge to the mixed strategy equilibrium of the game: the PW player quickly, and the $L_{R-\epsilon P}$ player only after the PW agent has closely approached it's equilibrium strategy, much like in the case of the PW and the L_{R-I} players in the PD in Fig. 4.

5 Related Work

Modelling the learning dynamics of MAS using an evolutionary game theory approach has recently received quite some attention. In [3] Börgers and Sarin proved that the continuous time limit of Cross Learning converges to the most basic replicator dynamics model considering only selection. This work has been extended to Q-learning and Learning Automata in [19,17] and to multiple state problems in [8]. Based on these results the dynamcs of ϵ-greedy Q-learning have been derived in [7].

Other approaches investigating the dynamics of MAL have also been considered in [5,4,10]. For a survey on Multi-agent learning we refer to [13].

6 Conclusions

We have derived an analytical model describing the dynamics of a learning agent using the Polynomial Weights Regret Minimization algorithm. It is interesting to observe that the model for PW is connected to the Coupled Replicator Dynamics of evolutionary game theory, like other learning algorithms, e.g. Q-learning and L_{R-I}.

We use the newly derived model to describe agents in selfplay in the Prisoner's Dilemma, the Battle of the Sexes, and Matching Pennies. In extensive experiments, we have shown the validity of the model: the modeled behavior shows good resemblance with observations from simulation.

Moreover, this work has shown a way of modeling agent interactions when both agents use *different* learning algorithms. Combining two models in a single game provides much insight into the way the game may be played, as shown in Section 4.3. In this way it is not necessary to run time-consuming experiments to analyze the behavior of different algorithms against each other, but this can be analyzed directly by investigating the involved dynamical systems. We have analyzed the effect on the outcomes in several games when the agents use different combinations of learning algorithms, finding that the games change profoundly. This has significant implications for the analysis of Multi-agent systems, for which we believe our paper provides valuable tools.

In future work, we plan to extend our analysis to other games and learning algorithms and will perform an in-depth analysis of the differences in the mathematical models of the variety of learning algorithms connected to the replicator dynamics. We also plan to systematically analyze the outcomes of interactions among various learning algorithms in different games, by studying the equilibria that arise and their basins of attraction. Also, we need to investigate the sensitivity of the outcomes to changes in the algorithms' parameters.

References

1. Van Ahee, G.J.: Models for Multi-Agent Learning. Master's thesis, Delft University of Technology (2009)
2. Blum, A., Mansour, Y.: Learning, regret minimization and equilibria. In: Algorithmic Game Theory. Cambridge University Press, Cambridge (2007)
3. Börgers, T., Sarin, R.: Learning through reinforcement and replicator dynamics. J. Economic Theory 77 (1997)

4. Bowling, M.: Convergence problems of general-sum multiagent reinforcement learning. In: ICML (2000)
5. Claus, C., Boutilier, C.: The dynamics of reinforcement learning in cooperative multiagent systems. In: AAAI (1998)
6. Conitzer, V., Sandholm, T.: AWESOME: A general multiagent learning algorithm that converges in self-play and learns a best response against stationary opponents. Machine Learning 67 (2007)
7. Gomes, E.R., Kowalczyk, R.: Dynamic analysis of multiagent Q-learning with epsilon-greedy exploration. In: ICML (2009)
8. Hennes, D., Tuyls, K.: State-coupled replicator dynamics. In: AAMAS (2009)
9. Hofbauer, J., Sigmund, K.: Evolutionary Games and Population Dynamics. Cambridge University Press, Cambridge (1998)
10. Hu, J., Wellman, M.P.: Multiagent reinforcement learning: Theoretical framework and an algorithm. In: ICML (1998)
11. Kaelbling, L.P., Littman, M.L., Moore, A.W.: Reinforcement learning: A survey. J. Artificial Intelligence Research 4 (1996)
12. Narendra, K., Thathachar, M.: Learning Automata: An Introduction. Prentice-Hall, Englewood Cliffs (1989)
13. Panait, L., Luke, S.: Cooperative multi-agent learning: The state of the art. J. AAMAS 11 (2005)
14. Shoham, Y., Powers, R., Grenager, T.: If multi-agent learning is the answer, what is the question? Artificial Intelligence 171 (2007)
15. Sutton, R.S., Barto, A.G.: Reinforcement Learning: An Introduction. MIT Press, Cambridge (1998)
16. Tsitsiklis, J.: Asynchronous stochastic approximation and Q-learning. Tech. rep., LIDS Research Center, MIT (1993)
17. Tuyls, K., 't Hoen, P.J., Vanschoenwinkel, B.: An evolutionary dynamical analysis of multi-agent learning in iterated games. J. AAMAS 12 (2006)
18. Tuyls, K., Parsons, S.: What evolutionary game theory tells us about multiagent learning. Artificial Intelligence 171 (2007)
19. Tuyls, K., Verbeeck, K., Lenaerts, T.: A selection-mutation model for Q-learning in multi-agent systems. In: AAMAS (2003)
20. Vega-Redondo, F.: Game Theory and Economics. Cambridge University Press, Cambridge (2001)
21. Walsh, W.E., Das, R., Tesauro, G., Kephart, J.O.: Analyzing complex strategic interactions in multi-agent systems. In: Workshop on Game-Theoretic and Decision-Theoretic Agents (2002)
22. Watkins, C., Dayan, P.: Q-learning. Machine Learning 8 (1992)

Recognition of Instrument Timbres in Real Polytimbral Audio Recordings

Elżbieta Kubera[1,2], Alicja Wieczorkowska[2],
Zbigniew Raś[2,3], and Magdalena Skrzypiec[4]

[1] University of Life Sciences in Lublin, Akademicka 13, 20-950 Lublin, Poland
[2] Polish-Japanese Institute of Information Technology, Koszykowa 86, 02-008 Warsaw, Poland
[3] University of North Carolina, Dept. of Computer Science, Charlotte, NC 28223, USA
[4] Maria Curie-Skłodowska University in Lublin, Pl. Marii Curie-Skłodowskiej 5, 20-031 Lublin, Poland
elzbieta.kubera@up.lublin.pl,
alicja@poljap.edu.pl,
ras@uncc.edu,
mskrzypiec@hektor.umcs.lublin.pl

Abstract. Automatic recognition of multiple musical instruments in polyphonic and polytimbral music is a difficult task, but often attempted to perform by MIR researchers recently. In papers published so far, the proposed systems were validated mainly on audio data obtained through mixing of isolated sounds of musical instruments. This paper tests recognition of instruments in real recordings, using a recognition system which has multilabel and hierarchical structure. Random forest classifiers were applied to build the system. Evaluation of our model was performed on audio recordings of classical music. The obtained results are shown and discussed in the paper.

Keywords: Music Information Retrieval, Random Forest.

1 Introduction

Music Information Retrieval (MIR) gains increasing interest last years [24]. MIR is multi-disciplinary research on retrieving information from music, involving efforts of numerous researchers – scientists from traditional, music and digital libraries, information science, computer science, law, business, engineering, musicology, cognitive psychology and education [4], [33]. Topics covered in MIR research include [33]: auditory scene analysis, aiming at the recognition of e.g. outside and inside environments, like streets, restaurants, offices, homes, cars etc. [23]; music genre categorization – an automatic classification of music into various genres [7], [20]; rhythm and tempo extraction [5]; pitch tracking for query-by-humming systems that allows automatic searching of melodic databases using

J.L. Balcázar et al. (Eds.): ECML PKDD 2010, Part II, LNAI 6322, pp. 97–110, 2010.

sung queries [1]; and many other topics. Research groups design various intelligent MIR systems and frameworks for research, allowing extensive works on audio data, see e.g. [20], [29].

Huge repositories of audio recordings available from the Internet and private sets offer plethora of options for potential listeners. The listeners might be interested in finding particular titles, but they can also wish to find pieces they are unable to name. For example, the user might be in mood to listen to something joyful, romantic, or nostalgic; he or she may want to find a tune sung to the computer's microphone; also, the user might be in mood to listen to jazz with solo trumpet, or classic music with sweet violin sound. More advanced person (a musician) might need scores for the piece of music found in the Internet, to play it by himself or herself.

All these issues are of interest for researchers working in MIR domain, since meta-information enclosed in audio files lacks such data – usually recordings are labeled by title and performer, maybe category and playing time. However, automatic categorization of music pieces is still one of more often performed tasks, since the user may need more information than it is already provided, i.e. more detailed or different categorization. Automatic extraction of melody or possibly the full score is another aim of MIR. Pitch-tracking techniques yield quite good results for monophonic data, but extraction of polyphonic data is much more complicated. When multiple instruments play, information about timbre may help to separate melodic lines for automatic transcription of music [15] (spatial information might also be used here). Automatic recognition of timbre, i.e. of instrument, playing in polyphonic and polytimbral (multi-instrumental) audio recordings, is our goal in the investigations presented in this paper.

One of the main problems when working with audio recordings is labeling of the data, since without properly labeled data, testing is impossible. It is difficult to recognize all notes played by all instruments in each recording, and if numerous instruments are playing, this task is becoming infeasible. Even if a score is available for a given piece of music, still, the real performance actually differs from the score because of human interpretation, imperfections of tempo, minor mistakes, and so on. Soft and short notes pose further difficulties, since they might not be heard, and grace notes leave some freedom to the performer - therefore, consecutive onsets may not correspond to consecutive notes in the score. As a result, some notes can be omitted. The problem of score following is addressed in [28].

1.1 Automatic Identification of Musical Instruments in Sound Recordings

The research on automatic identification of instruments in audio data is not a new topic; it started years ago, at first on isolated monophonic (monotimbral) sounds. Classification techniques applied quite successfully for this purpose by many researchers include k-nearest neighbors, artificial neural networks, rough-set based classifiers, support vector machines (SVM) – a survey of this research is presented in [9]. Next, automatic recognition of instruments in audio data was performed on polyphonic polytimbral data, see e.g. [3], [12], [13], [14], [19], [30], [32], [35], also including investigations on separation of the sounds from the audio sources (see e.g. [8]).

The comparison of results of the research on automatic recognition of instruments in audio data is not so straightforward, because various scientists utilized different data sets: of different number of classes (instruments and/or articulation), different number of objects/sounds in each class, and basically different feature sets, so the results are quite difficult to compare. Obviously, the less classes (instruments) to recognize, the higher recognition rate was achieved, and identification in monophonic recordings, especially for isolated sounds, is easier than in polyphonic polytimbral environment. The recognition of instruments in monophonic recordings can reach 100% for a small number of classes, more than 90% if the instrument or articulation family is identified, or about 70% or less for recognition of an instrument when there are more classes to recognize. The identification of instruments in polytimbral environment is usually lower, especially for lower levels of the target sounds – even below 50% for same-pitch sounds and if more than one instrument is to be identified in a chord; more details can be found in the papers describing our previous work [16], [31]. However, this research was performed on sound mixes (created by automatic mixing of isolated sounds), mainly to make proper labeling of data easier.

2 Audio Data

In our previous research [17], we performed experiments using isolated sounds of musical instruments and mixes calculated from these sounds, with one of the sounds being of higher level than the others in the mix, so our goal was to recognize the dominating instrument in the mix. The obtained results for 14 instruments and one octave shown low classification error, depending on the level of sounds added to the main sound in the mix - the highest error was 10% for the level of accompanying sound equal to 50% of the level of the main sound. These results were obtained for random forest classifiers, thus proving usefulness of this methodology for the purpose of the recognition of the dominating instrument in polytimbral data, at least in case of mixes. Therefore, we applied the random forest technique for the recognition of plural (2–5) instruments in artificial mixes [16]. In this case we obtained lower accuracy, also depending of the level of the sounds used, and varying between 80% and 83% in total, and between 74% and 87% for individual instruments; some instruments were easier to recognize, and some were more difficult.

The ultimate goal of such work is to recognize instruments (as many as possible) in real audio recordings. This is why we decided to perform experiments on the recognition of instruments with tests on real polyphonic recordings as well.

2.1 Parameterization

Since audio data represent sequences of amplitude values of the recorded sound wave, such data are not really suitable for direct classification, and parameterization is performed as a preprocessing. An interesting example of a framework for modular sound parameterization and classification is given in [20], where collaborative scheme is used for feature extraction from distributed data sets, and further for audio data classification in a peer-to-peer setting.

The method of parameterization influences final classification results, and many parameterization techniques have been applied so far in research on automatic timbre classification. Parameterization is usually based on outcomes of sound analysis, such us Fourier transform, wavelet transform, or time-domain based description of sound amplitude or spectrum. There is no standard set of parameters, but low-level audio descriptors from the MPEG-7 standard of multimedia content description [11] are quite often used as a basis of musical instrument recognition. Since we have already performed similar research, we decided to use MPEG-7 based sound parameters, as well as additional ones.

In the experiments described in this paper, we used 2 sets of parameters: average values of sound parameters calculated through the entire sound (being a single sound or a chord), and temporal parameters, describing evolution of the same parameters in time. The following parameters were used for this purpose [35]:

- MPEG-7 audio descriptors [11], [31]:
  - *AudioSpectrumCentroid* - power weighted average of the frequency bins in the power spectrum of all the frames in a sound segment;
  - *AudioSpectrumSpread* - a RMS value of the deviation of the Log frequency power spectrum with respect to the gravity center in a frame;
  - *AudioSpectrumFlatness*, $flat_1, \ldots, flat_{25}$ - multidimensional parameter describing the flatness property of the power spectrum within a frequency bin for selected bins; 25 out of 32 frequency bands were used for a given frame;
  - *HarmonicSpectralCentroid* - the mean of the harmonic peaks of the spectrum, weighted by the amplitude in linear scale;
  - *HarmonicSpectralSpread* - represents the standard deviation of the harmonic peaks of the spectrum with respect to the harmonic spectral centroid, weighted by the amplitude;
  - *HarmonicSpectralVariation* - the normalized correlation between amplitudes of harmonic peaks of each 2 adjacent frames;
  - *HarmonicSpectralDeviation* - represents the spectral deviation of the log amplitude components from a global spectral envelope;
- other audio descriptors:
  - *Energy* - energy of spectrum in the parameterized sound;
  - MFCC - vector of 13 Mel frequency cepstral coefficients, describe the spectrum according to the human perception system in the mel scale [21];
  - *ZeroCrossingDensity* - zero-crossing rate, where zero-crossing is a point where the sign of time-domain representation of sound wave changes;

- $FundamentalFrequency$ - maximum likelihood algorithm was applied for pitch estimation [36];
- $NonMPEG7 - AudioSpectrumCentroid$ - a differently calculated version - in linear scale;
- $NonMPEG7 - AudioSpectrumSpread$ - different version;
- $RollOff$ - the frequency below which an experimentally chosen percentage equal to 85% of the accumulated magnitudes of the spectrum is concentrated. It is a measure of spectral shape, used in speech recognition to distinguish between voiced and unvoiced speech;
- $Flux$ - the difference between the magnitude of the DFT points in a given frame and its successive frame. This value was multiplied by 10^7 to comply with the requirements of the classifier applied in our research;
- $FundamentalFrequency'sAmplitude$ - the amplitude value for the predominant (in a chord or mix) fundamental frequency in a harmonic spectrum, over whole sound sample. Most frequent fundamental frequency over all frames is taken into consideration;
- *Ratio* $r_1, \ldots, r_{11}$ - parameters describing various ratios of harmonic partials in the spectrum;
  * r_1: energy of the fundamental to the total energy of all harmonic partials,
  * r_2: amplitude difference [dB] between 1st partial (i.e., the fundamental) and 2nd partial,
  * r_3: ratio of the sum of energy of 3rd and 4th partial to the total energy of harmonic partials,
  * r_4: ratio of the sum of partials no. 5-7 to all harmonic partials,
  * r_5: ratio of the sum of partials no. 8-10 to all harmonic partials,
  * r_6: ratio of the remaining partials to all harmonic partials,
  * r_7: brightness - gravity center of spectrum,
  * r_8: contents of even partials in spectrum,

$$r_8 = \frac{\sqrt{\sum_{k=1}^{M} A_{2k}^2}}{\sqrt{\sum_{n=1}^{N} A_n^2}}$$

    where A_n - amplitude of nth harmonic partial,
    N - number of harmonic partials in the spectrum,
    M - number of even harmonic partials in the spectrum,
  * r_9: contents of odd partials (without fundamental) in spectrum,

$$r_9 = \frac{\sqrt{\sum_{k=2}^{L} A_{2k-1}^2}}{\sqrt{\sum_{n=1}^{N} A_n^2}}$$

    where L – number of odd harmonic partials in the spectrum,
  * r_{10}: mean frequency deviation for partials 1-5 (when they exist),

$$r_{10} = \frac{\sum_{k=1}^{N} A_k \cdot |f_k - kf_1| / (kf_1)}{N}$$

where $N = 5$, or equals to the number of the last available harmonic partial in the spectrum, if it is less than 5,

* r_{11}: partial (i=1,...,5) of the highest frequency deviation.

Detailed description of popular features can be found in the literature; therefore, equations were given only for less commonly used features.

These parameters were calculated using fast Fourier transform, with 75 ms analyzing frame and Hamming window (hop size 15 ms). Such a frame is long enough to analyze the lowest pitch sounds of our instruments and yield quite good resolution of spectrum; since the frame should not be too long because the signal may then undergo changes, we believe that this length is good enough to capture spectral features and changes of these features in time, to be represented by temporal parameters. Our descriptors describe the entire sound, constituting one sound event, being a single note or a chord.

The sound timbre is believed to depend not only on the contents of sound spectrum (depending on the shape of the sound wave), but also on changes of spectrum (and the shape of the sound wave) over time. Therefore, the use of temporal sound descriptors was also investigated - we would like to check whether adding of such (even simple) descriptors will improve the accuracy of classification. The temporal parameters in our research were calculated in the following way. Temporal parameters describe temporal evolution of each original feature vector p, calculated as presented above. We were treating p as a function of time and searching for 3 maximal peaks. Maximum is described by k - the consecutive number of frame where the maximum appeared, and the value of this parameter in the frame k:

$M_i(p) = (k_i, p[k_i]), \quad i = 1,2,3 \quad k_1 < k_2 < k_3$

The temporal variation of each feature can be then presented by a vector T of new temporal parameters, built as follows:

$T_1 = k_2 - k_1$

$T_2 = k_3 - k_2$

$T_3 = k_3 - k_1$

$T_4 = p[k_2]/p[k_1]$

$T_5 = p[k_3]/p[k_2]$

$T_6 = p[k_3]/p[k_1]$

Altogether, we obtained a feature vector of 63 averaged descriptors, and another vector of $63 \cdot 6 = 378$ temporal descriptors for each sound object. We made a comparison of performance of classifiers built using only 63 averaged parameters and built using both averaged and temporal features.

2.2 Training and Testing Data

Our training and testing data were based on audio samples of the following 10 instruments: B-flat clarinet, cello, double bass, flute, French horn, oboe, piano, tenor trombone, viola, and violin. Full musical scale of these instruments was used for both training and testing purposes. Training data were taken from

Table 1. Number of pieces in RWC Classical Music Database with the selected instruments playing together

	clarinet	cello	dBass	flute	fHorn	piano	trbone	viola	violin	oboe
clarinet	0	8	7	5	6	1	3	8	8	5
cello	8	0	13	9	9	4	3	17	20	8
doublebass	7	13	0	9	9	2	3	13	13	8
flute	5	9	9	1	7	1	2	9	9	6
frenchhorn	6	9	9	7	3	4	4	9	11	8
piano	1	4	2	1	4	0	0	2	9	0
trombone	3	3	3	2	4	0	0	3	3	3
viola	8	17	13	9	9	2	3	0	17	8
violin	8	20	13	9	11	9	3	17	18	8
oboe	5	8	8	6	8	0	3	8	8	2

MUMS – McGill University Master Samples CDs [22] and The University of IOWA Musical Instrument Samples [26]. Both isolated single sounds and artificially generated mixes were used as training data. The mixes were generated using 3 sounds. Pitches of composing sounds were chosen in such a way that the mix constitutes a minor or major chord, or its part (2 different pitches), or even a unison. The probability of choosing instruments is based on statistics drawn from RWC Classical Music Database [6], describing in how many pieces these instruments play together in the recordings (see Table 1). The mixes were created in such a way that for a given sound, chosen as the first one, two other sounds were chosen. These two other sounds represent two different instruments, but one of them can also represent the instrument selected as the first sound. Therefore, the mixes of 3 sounds may represent only 2 instruments.

Since testing was already performed on mixes in our previous works, the results reported here describe tests on real recordings only, not based on sounds from the training set. Test data were taken from RWC Classical Music Database [6]. Sounds of length of at least 150 ms were used. For our tests we selected available sounds representing the 10 instruments used in training, playing in chords of at least 2 and no more than 6 instruments. The sound segments were manually selected and labeled (also comparing with available MIDI data) in order to prepare ground-truth information for testing.

3 Classification Methodology

So far, we applied various classifiers for the instrument identification purposes, including support vector machines (SVM, see e.g. [10]) and random forests (RF, [2]). The results obtained using RF for identification of instruments in mixes outperformed the results obtained via SVM by an order of magnitude. Therefore, the classification performed in the reported experiments was based on RF technique, using WEKA package [27].

Random forest is an ensemble of decision trees. The classifier is constructed using procedure minimizing bias and correlations between individual trees,

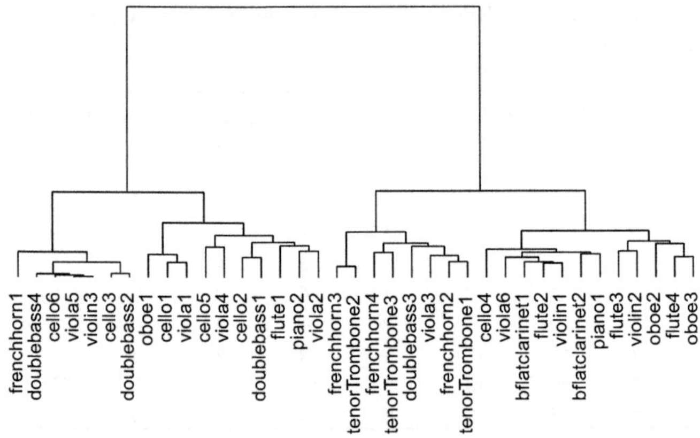

Fig. 1. Hierarchical classification of musical instrument sounds for the 10 investigated instruments

according to the following procedure [17]. Each tree is built using different N-element bootstrap sample of the training N-element set; the elements of the sample are drawn with replacement from the original set. At each stage of tree building, i.e. for each node of any particular tree in the random forest, p attributes out of all P attributes are randomly chosen ($p \ll P$, often $p = \sqrt{P}$). The best split on these p attributes is used to split the data in the node. Each tree is grown to the largest extent possible - no pruning is applied. By repeating this randomized procedure M times one obtains a collection of M trees – a random forest. Classification of each object is made by simple voting of all trees.

Because of similarities between timbres of musical instruments, both from psychoacoustic and sound-analysis point of view, hierarchical clustering of instrument sounds was performed using R – an environment for statistical computing [25]. Each cluster in the obtained tree represents sounds of one instrument (see Figure 1). More than one cluster may be obtained for each instrument; sounds representing similar pitch usually are placed in one cluster, so various pitch ranges are basically assigned to different clusters. To each leaf a classifier is assigned, trained to identify a given instrument. When the threshold of 50% is exceeded for this particular classifier alone, the corresponding instrument is identified.

We also performed node-based classification in additional experiments, i.e. when any node exceeded the threshold, but no its children did, then the instruments represented in this node were returned as a result. The instruments from this node can be considered similar, and they give a general idea on what sort of timbre was recognized in the investigated chord.

Data cleaning. When this tree was built, pruning was performed and the leaves representing less than 5% of sounds of a given instruments were removed, and these sounds were removed from the training set. As a result, the training data in case of 63-element feature vector consisted of 1570 isolated single sounds, and the same number of mixes. For the extended feature vector (with temporal parameters added), 1551 isolated sounds and the same number of mixes was used. The difference in number is caused by different pruning for the different hierarchical classification tree, built for the extended feature vector. Testing data set included 100 chords.

Since we are recognizing instruments in chords, we are dealing with multi-label data. The use of multi-label data makes reporting of results more complicated, and the results depend on the way of counting the number of correctly identified instruments, omissions and false recognitions [18], [34]. We are aware of influence of these factors on the precision and recall of the performed classification. Therefore, we think the best way to present the results is to show average values of precision and recall for all chords in the test set, and f-measures calculated from these average results.

4 Experiments and Results

General results of our experiments are shown in Table 2, for various experimental settings regarding training data, classification methodology, and feature vector applied. As we can see, the classification quality is not as good as in case of our previous research, thus showing the increased level of difficulty in case of our current research.

The presented experiments were performed for various sets of training data, i.e. for isolated musical instrumental sounds only, and for mixes added to the training set. Classification was basically performed aiming at identification of each instrument (i.e. down to the leaves of hierarchical classification), but we also performed classification using information from nodes of the hierarchical tree, as described in Section 3. Experiments was performed for 2 versions of feature vector, including 63 parameters describing average values of sound features

Table 2. General results of recognition of 10 selected musical instruments playing in chords taken from real audio recording from RWC Classical Music Database [6]

Training data	Classification	Feature vector	Precision	Recall	F-measure
Isolated sounds + mixes	Leaves + nodes	Averages only	63.06%	49.52%	0.5547
Isolated sounds + mixes	Leaves only	Averages only	62.73%	45.02%	0.5242
Isolated sounds only	Leaves + nodes	Averages only	**74.10%**	32.12%	0.4481
Isolated sounds only	Leaves only	Averages only	71.26%	18.20%	0.2899
Isolated sounds + mixes	Leaves + nodes	Averages + temporal	57.00%	**59.22%**	**0.5808**
Isolated sounds + mixes	Leaves only	Averages + temporal	57.45%	53.07%	0.5517
Isolated sounds only	Leaves + nodes	Averages + temporal	51.65%	25.87%	0.3447
Isolated sounds only	Leaves only	Averages + temporal	54.65%	18.00%	0.2708

Table 3. Results of recognition of 10 selected musical instruments playing in chords taken from real audio recording from RWC Classical Music Database [6] - the results for best settings for each instruments are shown

	precision	recall	f-measure
bflatclarinet	50.00%	16.22%	0.2449
cello	69.23%	77.59%	0.7317
doublebass	40.00%	61.54%	0.4848
flute	31.58%	33.33%	0.3243
frenchhorn	20.00%	47.37%	0.2813
oboe	16.67%	11.11%	0.1333
piano	14.29%	16.67%	0.1538
tenorTrombone	25.00%	25.00%	0.2500
viola	63.24%	72.88%	0.6772
violin	89.29%	86.21%	0.8772

calculated through the entire sound in the first version of the feature vector, and additionally temporal parameters describing the evolution of these features in time in the second version. Precision and recall for these settings, as well as F-measure, are shown in Table 2.

As we can see, when training is performed on isolated sound only, the obtained recall is rather low, and it is increased when mixes are added to the training set. On the other hand, when training is performed on isolated sound only, the highest precision is obtained. This is not surprising, as illustrating a usual trade-off between precision and recall. The highest recall is obtained when information from nodes of hierarchical classification is taken into account. This was also expected; when the user is more interested in high recall than in high precision, then such a way of classification should be followed. Adding temporal descriptors to the feature vector does not make such a clear influence on the obtained precision and recall, but it increases recall when mixes are present in the training set.

One might be also interested in inspecting the results for each instrument. These results are shown in Table 3, for best settings of the classifiers used. As we can see, some string instruments (violin, viola and cello) are relatively easy to recognize, both in terms of precision and recall. Oboe, piano and trombone are difficult to be identified, both in terms of precision and recall. For double bass recall is much better than precision, whereas for clarinet the obtained precision is better than recall. Some results are not very good, but we must remember that correct identification of all instruments playing in a chord is generally a difficult task, even for humans.

It might be interesting to see which instruments are confused with which ones, and this is illustrated in confusion matrices. As we mentioned before, omissions and false positives can be considered in various ways, thus we can present different confusion matrices, depending on how the errors are counted. In Table 4 we presents the results when $1/n$ is added in each cell when identification happens (n represents the number of instruments actually playing in the mix).

Table 4. Confusion matrix for the recognition of 10 selected musical instruments playing in chords taken from real audio recording from RWC Classical Music Database [6]. When n instruments are actually playing in the recording, $1/n$ is added in case of each identification.

Instrument \ Classified as	clarinet	cello	dBass	flute	fHorn	oboe	piano	trombone	viola	violin
clarinet	6	2	1	3.08	4.42	1.75	2.42	0.75	4.92	0.58
cello	2	45	4.67	0.75	8.15	1.95	3.2	1.08	1.5	0.58
dBass	0	0.25	16	0.5	2.23	0.45	1.12	0	0.5	0.25
flute	0.67	0.58	1.17	6	1.78	1.37	0.95	0	0.58	0.5
fHorn	0	4.33	1.83	0.17	9	0	0.33	0	4.83	3
oboe	0	0.67	0.33	1.33	1.67	2	1.5	0.33	0	0.5
piano	0	4.83	2.83	0	0	0	3	0	4.83	3
trombone	0	0	0	0.17	0.53	0	0.92	2	0.58	0.58
viola	1.33	1.75	4.5	2.25	7.32	1.03	3.28	1.92	43	0
violin	2	5.58	7.67	4.75	9.9	3.45	4.28	1.92	7.25	75

Table 5. Confusion matrix for the recognition of 10 selected musical instruments playing in chords taken from real audio recording from RWC Classical Music Database [6]. In case of each identification, 1 is added in a given cell.

Instrument \ Classified as	clarinet	cello	dBass	flute	fHorn	oboe	piano	trombone	viola	violin
clarinet	6	4	2	8	17	4	8	3	11	2
cello	6	45	14	4	31	7	13	4	5	2
dBass	0	1	16	3	12	2	6	0	2	1
flute	2	2	4	6	7	5	3	0	2	1
fHorn	0	10	4	1	9	0	2	0	12	6
oboe	0	2	1	5	9	2	5	1	0	1
piano	0	11	6	0	0	0	3	0	12	6
trombone	0	0	0	1	2	0	4	2	2	2
viola	4	5	14	8	29	4	13	6	43	0
violin	6	14	21	13	35	10	15	0	18	75

To compare with, the confusion matrix is also shown when each identification is counted as 1 instead (Table 5). We believe that Table 4 more properly describes the classification results than Table 5, although the latter is more clear to look at. We can observe from both tables which instruments are confused with which ones, but we must remember that we are aiming at identifying actually a group of instruments, and our output also represents a group. Therefore, concluding about confusion between particular instruments is not so simple and straightforward, because we do not know exactly which instrument caused which confusion.

5 Summary and Conclusions

The investigations presented in this paper aimed at identification of instruments in real audio polytimbral (multi-instrumental) recordings. The parameterization included temporal descriptors, which improved recall when training was performed on both single isolated sounds and mixes. The use of real recordings not included in training set posed high level of difficulties for the classifiers; not only the sounds of instruments originated from different audio sets, but also the recording conditions were different. Taking this into account, we can conclude that the results were not bad, especially that some sounds were soft, and still several instruments were quite well recognized (certainly higher than random choice). In order to improve classification, we can take into account usual settings of instrumentation and the probability of use of particular instruments and instrument groups playing together. The classifiers adjusted specifically to given genres and sub-genres may yield much higher results, further improved by taking into account cleaning of results (removal of spurious single indications in the context of neighboring recognized sounds). Basing on the results of other research [20], we also believe that adjusting the feature set and performing feature selection in each node should improve our results. Finally, adjusting thresholds of firing of the classifiers may improve the results.

Acknowledgments. This project was partially supported by the Research Center of PJIIT, supported by the Polish National Committee for Scientific Research (KBN) and also by the National Science Foundation under Grant Number *IIS 0968647*. Any opinions, findings, and conclusions or recommendations expressed in this material are those of the author(s) and do not necessarily reflect the views of the National Science Foundation.

References

1. Birmingham, W.P., Dannenberg, R.D., Wakefield, G.H., Bartsch, M.A., Bykowski, D., Mazzoni, D., Meek, C., Mellody, M., Rand, B.: MUSART: Music retrieval via aural queries. In: Proceedings of ISMIR 2001, 2nd Annual International Symposium on Music Information Retrieval, Bloomington, Indiana, pp. 73–81 (2001)
2. Breiman, L., Cutler, A.: Random Forests, `http://stat-www.berkeley.edu/users/breiman/RandomForests/cc_home.htm`
3. Dziubinski, M., Dalka, P., Kostek, B.: Estimation of musical sound separation algorithm effectiveness employing neural networks. J. Intel. Inf. Syst. 24(2-3), 133–157 (2005)
4. Downie, J.S.: Wither music information retrieval: ten suggestions to strengthen the MIR research community. In: Downie, J.S., Bainbridge, D. (eds.) Proceedings of the Second Annual International Symposium on Music Information Retrieval: ISMIR 2001, pp. 219–222. Bloomington, Indiana (2001)
5. Foote, J., Uchihashi, S.: The Beat Spectrum: A New Approach to Rhythm Analysis. In: Proceedings of the International Conference on Multimedia and Expo ICME 2001, Tokyo, Japan, pp. 1088–1091 (2001)

6. Goto, M., Hashiguchi, H., Nishimura, T., Oka, R.: RWC Music Database: Popular, Classical, and Jazz Music Databases. In: Proceedings of the 3rd International Conference on Music Information Retrieval (ISMIR 2002), pp. 287–288 (2002)
7. Guaus, E., Herrera, P.: Music Genre Categorization in Humans and Machines, AES 121st Convention, San Francisco (2006)
8. Heittola, T., Klapuri, A., Virtanen, T.: Musical instrument recognition in polyphonic audio using source-filter model for sound separation. In: 10th ISMIR, pp. 327–332 (2009)
9. Herrera, P., Amatriain, X., Batlle, E., Serra, X.: Towards instrument segmentation for music content description: a critical review of instrument classification techniques. In: International Symposium on Music Information Retrieval ISMIR (2000)
10. Hsu, C.-W., Chang, C.-C., Lin, C.-J.: A Practical Guide to Support Vector Classification, `http://www.csie.ntu.edu.tw/~cjlin/papers/guide/guide.pdf`
11. ISO: MPEG-7 Overview, `http://www.chiariglione.org/mpeg/`
12. Itoyama, K., Goto, M., Komatani, K., Ogata, T., Okuno, H.G.: Instrument Equalizer for Query-By-Example Retrieval: Improving Sound Source Separation Based on Integrated Harmonic and Inharmonic Models. In: 9th ISMIR (2008)
13. Jiang, W.: Polyphonic Music Information Retrieval Based on Multi-Label Cascade Classification System. Ph.D thesis, Univ. North Carolina, Charlotte (2009)
14. Kitahara, T., Goto, M., Komatani, K., Ogata, T., Okuno, H.: Instrogram: Probablilistic Representation of Instrument Existence for Polyphonic Music. IPSJ Journal 48(1), 214–226 (2007)
15. Klapuri, A.: Signal processing methods for the automatic transcription of music. Ph.D. thesis, Tampere University of Technology, Finland (2004)
16. Kursa, M.B., Kubera, E., Rudnicki, W.R., Wieczorkowska, A.A.: Random Musical Bands Playing in Random Forests. In: Szczuka, M., Kryszkiewicz, M., Ramanna, S., Jensen, R., Hu, Q. (eds.) RSCTC 2010. LNCS (LNAI), vol. 6086, pp. 580–589. Springer, Heidelberg (2010)
17. Kursa, M., Rudnicki, W., Wieczorkowska, A., Kubera, E., Kubik-Komar, A.: Musical Instruments in Random Forest. In: Rauch, J., Raś, Z.W., Berka, P., Elomaa, T. (eds.) Foundations of Intelligent Systems. LNCS, vol. 5722, pp. 281–290. Springer, Heidelberg (2009)
18. Lauser, B., Hotho, A.: Automatic multi-label subject indexing in a multilingual environment. FAO, Agricultural Information and Knowledge Management Papers (2003)
19. Little, D., Pardo, B.: Learning Musical Instruments from Mixtures of Audio with Weak Labels. In: 9th ISMIR (2008)
20. Mierswa, I., Morik, K., Wurst, M.: Collaborative Use of Features in a Distributed System for the Organization of Music Collections. In: Shen, J., Shephard, J., Cui, B., Liu, L. (eds.) Intelligent Music Information Systems: Tools and Methodologies, pp. 147–176. IGI Global (2008)
21. Niewiadomy, D., Pelikant, A.: Implementation of MFCC vector generation in classification context. Journal of Applied Computer Science 16(2), 55–65 (2008)
22. Opolko, F., Wapnick, J.: MUMS – McGill University Master Samples. CD's (1987)
23. Peltonen, V., Tuomi, J., Klapuri, A., Huopaniemi, J., Sorsa, T.: Computational Auditory Scene Recognition. In: International Conference on Acoustics Speech and Signal Processing, Orlando, Florida (2002)
24. Raś, Z.W., Wieczorkowska, A.A. (eds.): Advances in Music Information Retrieval. Studies in Computational Intelligence, vol. 274. Springer, Heidelberg (2010)

25. R Development Core Team: R: A language and environment for statistical computing. R Foundation for Statistical Computing, Vienna, Austria (2009)
26. The University of IOWA Electronic Music Studios: Musical Instrument Samples, http://theremin.music.uiowa.edu/MIS.html
27. The University of Waikato: Weka Machine Learning Project, http://www.cs.waikato.ac.nz/~ml/
28. Miotto, R., Montecchio, N., Orio, N.: Statistical Music Modeling Aimed at Identification and Alignment. In: Raś, Z.W., Wieczorkowska, A.A. (eds.) Advances in Music Information Retrieval. SCI, vol. 274, pp. 187–212. Springer, Heidelberg (2010)
29. Tzanetakis, G., Cook, P.: Marsyas: A framework for audio analysis. Organized Sound 4(3), 169–175 (2000)
30. Viste, H., Evangelista, G.: Separation of Harmonic Instruments with Overlapping Partials in Multi-Channel Mixtures. In: IEEE Workshop on Applications of Signal Processing to Audio and Acoustics WASPAA 2003, New Paltz, NY (2003)
31. Wieczorkowska, A.A., Kubera, E.: Identification of a dominating instrument in polytimbral same-pitch mixes using SVM classifiers with non-linear kernel. J. Intell. Inf. Syst. (2009), doi: 10.1007/s10844-009-0098-3
32. Wieczorkowska, A., Kubera, E., Kubik-Komar, A.: Analysis of Recognition of a Musical Instrument in Sound Mixes Using Support Vector Machines. In: Nguyen, H.S. (ed.) SCKT 2008 Hanoi, Vietnam (PRICAI), pp. 110–121 (2008)
33. Wieczorkowska, A.A.: Music Information Retrieval. In: Wang, J. (ed.) Encyclopedia of Data Warehousing and Mining, 2nd edn., pp. 1396–1402. IGI Global (2009)
34. Wieczorkowska, A., Synak, P.: Quality Assessment of k-NN Multi-Label Classification for Music Data. In: Esposito, F., Raś, Z.W., Malerba, D., Semeraro, G. (eds.) ISMIS 2006. LNCS (LNAI), vol. 4203, pp. 389–398. Springer, Heidelberg (2006)
35. Zhang, X.: Cooperative Music Retrieval Based on Automatic Indexing of Music by Instruments and Their Types. Ph.D thesis, Univ. North Carolina, Charlotte (2007)
36. Zhang, X., Marasek, K., Raś, Z.W.: Maximum Likelihood Study for Sound Pattern Separation and Recognition. In: 2007 International Conference on Multimedia and Ubiquitous Engineering MUE 2007, pp. 807–812. IEEE, Los Alamitos (2007)

Finding Critical Nodes for Inhibiting Diffusion of Complex Contagions in Social Networks

Chris J. Kuhlman[1], V.S. Anil Kumar[1], Madhav V. Marathe[1], S.S. Ravi[2], and Daniel J. Rosenkrantz[2]

[1] Virginia Bioinformatics Institute, Virginia Tech, Blacksburg, VA 24061, USA
{ckuhlman,akumar,mmarathe}@vbi.vt.edu
[2] Computer Science Department, University at Albany – SUNY, Albany, NY 12222, USA
{ravi,djr}@cs.albany.edu

Abstract. We study the problem of inhibiting diffusion of complex contagions such as rumors, undesirable fads and mob behavior in social networks by removing a small number of nodes (called **critical nodes**) from the network. We show that, in general, for any $\rho \geq 1$, even obtaining a ρ-approximate solution to these problems is **NP**-hard. We develop efficient heuristics for these problems and carry out an empirical study of their performance on three well known social networks, namely `epinions`, `wikipedia` and `slashdot`. Our results show that the heuristics perform well on the three social networks.

1 Introduction and Motivation

Analyzing social networks has become an important research topic in data mining (e.g. [31, 9, 20, 21, 7, 32]). With respect to diffusion in social networks, researchers have studied the propagation of favorite photographs in a Flickr network [6], the spread of information [16, 23] via Internet communication, and the effects of online purchase recommendations [26], to name a few. In some instances, models of diffusion are combined with data mining to predict social phenomena (e.g., product marketing [9, 31] and trust propagation [17]).

Here we are interested in the diffusion of a particular class of contagions, namely **complex contagions**. As stated by Centola and Macy [5], "Complex contagions require social affirmation from multiple sources." That is, a person acquires a social contagion through interaction with $t > 1$ other individuals, as opposed to a single individual (i.e., $t = 1$); the latter is called a **simple contagion**.

As described by Granovetter [15], the idea of complex contagions dates back to the 1960's, and more current studies are referenced in [5, 11]. Such phenomena include diffusion of innovations, rumors, worker strikes, educational attainment, fashion, and social movements. For example, in strikes, mob violence, and political upheavals, individuals can be reluctant to participate for fear of reprisals to themselves and their families. It is safer to wait for a critical mass of people to commit before committing oneself. Researchers have used data mining

J.L. Balcázar et al. (Eds.): ECML PKDD 2010, Part II, LNAI 6322, pp. 111–127, 2010.

techniques to study the propagation of complex contagions such as online DVD purchases [26] and teenage smoking initiation [19]. As discussed by Easley and Kleinberg [11], complex contagion is also closely related to *Coordination Games*. Motivation for our work came partially from recent quantitative work [5] showing that simple contagions and complex contagions can differ significantly in behavior. Further, it is well known [14] that weak edges play a dominant role in spreading a simple contagion between clusters within a population, thereby dictating whether or not a contagion will reach a large segment of a population. However, for complex contagions, this effect is greatly diminished [5] because chances are remote that multiple members, who are themselves connected within a group, are each linked to multiple members of another group.

The focus of our work is inhibiting the diffusion of complex contagions such as rumors, undesirable fads, and mob behavior in social networks. In our formulation, the goal is to minimize the spread of a contagion by removing a small number of nodes, called **critical nodes**, from the network. Other formulations of this problem have been considered in the literature for simple contagions (e.g. [18]). We will discuss the differences between our work and that reported in other references in Section 3. Applications of finding critical nodes in a network include thwarting the spread of sensitive information that has been leaked [7], disrupting communication among adversaries [1], marketing to counteract the advertising of a competing product [31, 9], calming a mob [15], and changing people's opinions [10].

We present both theoretical and empirical results. (A more technical summary of our results is given in Section 3.) On the theoretical side, we show that for two versions of the problem, even obtaining efficient approximations is **NP**-hard. These results motivate the development and evaluation of heuristics that work well in practice. We develop two efficient heuristics for finding critical sets and empirically evaluate their performance on three well known social networks, namely `epinions`, `wikipedia` and `slashdot`.

This paper is organized as follows. Section 2 describes the model employed in this work and presents the formal problem statement. Section 3 contains related work and a summary of results. Theoretical results are provided in Section 4. Two heuristics are described in Section 5 and are evaluated against three social networks in Section 6. Directions for future work are provided in Section 7.

2 Dynamical System Model and Problem Formulation

2.1 System Model and Associated Definitions

We model the propagation of complex contagions over a social network using discrete dynamical systems [2, 24]. We begin with the necessary definitions.

Let $\mathbb{B}$ denote the Boolean domain {0,1}. A **Synchronous Dynamical System** (SyDS) $\mathcal{S}$ over $\mathbb{B}$ is specified as a pair $\mathcal{S} = (G, \mathcal{F})$, where
(a) $G(V, E)$, an undirected graph with n nodes, represents the underlying social network over which the contagion propagates, and

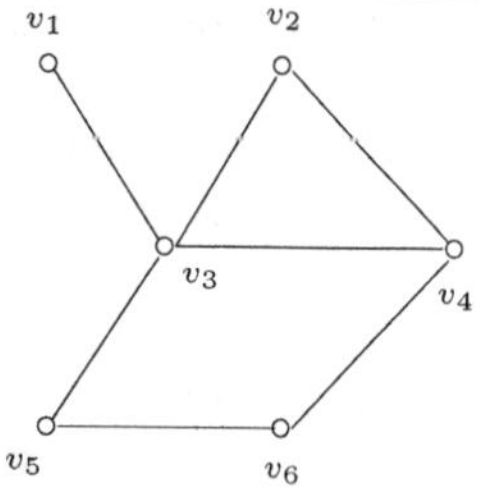

Initial Configuration:	(1, 1, 0, 0, 0, 0)
Configuration at time 1:	(1, 1, 1, 0, 0, 0)
Configuration at time 2:	(1, 1, 1, 1, 0, 0)

Note: Each configuration has the form $(s_1, s_2, s_3, s_4, s_5, s_6)$, where s_i is the state of node v_i, $1 \leq i \leq 6$. The configuration at time 2 is a fixed point.

Fig. 1. An example of a synchronous dynamical system

(b) $\mathcal{F} = \{f_1, f_2, \ldots, f_n\}$ is a collection of functions in the system, with f_i denoting the **local transition function** associated with node v_i, $1 \leq i \leq n$.

Each function f_i specifies the local interaction between node v_i and its neighbors in G. We note that each node of G has a state value from $\mathbb{B}$. To encompass various types of social contagions as described in Section 1, nodes in state 0 (1) are said to be **unaffected** (**affected**). In the case of information flow, an affected node could be one that has received the information. It is assumed that once a node reaches state 1, it cannot return to state 0. A discrete dynamical system with this property is referred to as a **ratcheted dynamical system** [24].

We can now formally describe the local interaction functions. The inputs to function f_i are the state of v_i and those of the neighbors of v_i in G; function f_i maps each combination of inputs to a value in $\mathbb{B}$. For the propagation of contagions in social networks, it is appropriate to model each function f_i $(1 \leq i \leq n)$ as a t_i-**threshold function** [12,7,10,4,5,20,22] for an appropriate nonnegative integer t_i. Such a threshold function (taking into account the ratcheted nature of the dynamical system) is defined as follows: (a) If the state of v_i is 1, then f_i is 1, regardless of the values of the other inputs to f_i, and (b) If the state of v_i is 0, then f_i is 1 if at least t_i of the inputs are 1; otherwise, f_i is 0.

A **configuration** $\mathcal{C}$ of a SyDS at any time is an n-vector $(s_1, s_2, \ldots, s_n)$, where $s_i \in \mathbb{B}$ is the value of the state of node v_i $(1 \leq i \leq n)$. A single SyDS transition from one configuration to another is implemented by using all states s_i at time j for the computation of the next states at time $j+1$. Thus, in a SyDS, nodes update their states *synchronously*. Other update disciplines (e.g. sequential updates) for discrete dynamical systems have also been studied [2]. A configuration $\mathcal{C}$ is called a **fixed point** if the successor of $\mathcal{C}$ is $\mathcal{C}$ itself.

Example: Consider the graph shown in Figure 1. Suppose the local interaction function at each node is the 2-threshold function. Initially, v_1 and v_2 are in state 1 and all other nodes are in state 0. During the first time step, the state of node v_3 changes to 1 since two of its neighbors (namely v_1 and v_2) are in state 1; the states of other nodes remain the same. In the second time step, the state of node v_4 changes to 1 since two of its neighbors (namely v_2 and v_3) are in state 1;

again the states of the other nodes remain the same. The resulting configuration $(1, 1, 1, 1, 0, 0)$ is a fixed point for this system.

The SyDS in the above example reached a fixed point. This is not a coincidence. The following general result (which holds for any ratcheted dynamical system over $\mathbb{B}$) is shown in [24].

Theorem 1. *Every ratcheted SyDS over $\mathbb{B}$ reaches a fixed point in at most n transitions, where n is the number of nodes in the underlying graph.* ■

2.2 Problem Formulation

For simplicity, statements of problems and results in this paper use terminology from the context of information propagation in social networks, such as that for social unrest in a group; these can be easily extended to other contagions.

Suppose we have a social network in which some nodes are initially affected. In the absence of any action to contain the unrest, it may spread to a large part of the population. Decision-makers must decide on suitable actions to inhibit information spread, such as quarantining a subset of people, subject to resource constraints and societal pressures (e.g., quarantining too many people may fuel unrest or it may be cost prohibitive to apprehend particular individuals).

We assume that people who are as yet unaffected can be quarantined or isolated. Under the dynamical system model, quarantining a person is represented by removing the corresponding node (and all the edges incident on that node) from the graph. Equivalently, removing a node v corresponds to changing the local transition function at v so that v's state remains 0 for all combinations of input values. The goal of isolation is to minimize the number of new affected nodes that occur over time until the system reaches a fixed point (when no additional nodes can be affected). We use the term **critical set** to refer to the set of nodes removed from the graph to reduce the number of newly affected nodes. Recall that resource constraints impose a budget constraint on the size of the critical set. We can now provide a precise statement of the problem of finding critical sets. (This problem was first formulated in [12] for the case where each node computes a 1-threshold function.)

Small Critical Set to Minimize New Affected Nodes (SCS-MNA)

Given: A social network represented by the SyDS $\mathcal{S} = (G(V, E), \mathcal{F})$ over $\mathbb{B}$, with each function $f \in \mathcal{F}$ being a threshold function; the set I ($n_s = |I|$) of nodes which are initially in state 1; an upper bound β on the size of the critical set.

Requirement: A critical set C (i.e., $C \subseteq V - I$) such that $|C| \leq \beta$ and among all subsets of $V - I$ of size at most β, the removal of C from G leads to the smallest number of new affected nodes.

An alternative formulation, where the objective is to maximize the number of people who are not affected, can also be considered. We use the name "Small Critical Set to Maximize Unaffected Nodes" for this problem and abbreviate it as SCS-MUN. Clearly, any optimal solution for SCS-MUN is also an optimal solution for SCS-MNA. Our results in Section 4 provide an indication of the difficulties in obtaining provably good approximation algorithms for either version of the problem. So, our focus is on devising heuristics that work well in practice.

2.3 Additional Terminology

Here, we present some terminology used in the later sections of this paper. The term "t-threshold system" is used to denote a SyDS in which each local transition function is the t-threshold function for some integer $t \geq 0$. (The value of t is the same for all nodes of the system.)

Let $\mathcal{S} = (G(V, E), \mathcal{F})$ be a SyDS and let $I \subseteq V$ denote the set of nodes whose initial state is 1. We say that a node $v \in V - I$ is **salvageable** if there is a critical set $C \subseteq V - I$ whose removal ensures that v remains in state 0 when the modified SyDS (i.e., the SyDS obtained by removing C) reaches a fixed point. Otherwise, v is called an **unsalvageable** node. Thus, in any SyDS, only salvageable nodes can possibly be saved from becoming affected.

We also need some terminology with respect to approximation algorithms for optimization problems [13]. For any $\rho \geq 1$, a **ρ-approximation** for an optimization problem is an efficient algorithm that produces a solution which is within a factor of ρ of the optimal value for all instances of the problem. Such an approximation algorithm is also said to provide a **performance guarantee** of ρ. Clearly, the smaller the value of ρ, the better is the performance of the approximation algorithm.

The following terms are used in describing empirical results of Section 6. A **cascade** occurs when diffusion starts from a set of **seed** nodes (set I) and 95% or more of nodes that can be affected are affected. **Halt** means that a set of critical nodes will stop the diffusion process, thus preventing a cascade. A **delay** means that the set of critical nodes will increase the time at which the peak number of newly affected nodes occurs, but will not necessarily halt diffusion.

3 Summary of Results and Related Work

Our main results can be summarized as follows.

(a) We show that for any $t \geq 2$ and any $\rho \geq 1$, it is **NP**-hard to obtain a ρ-approximation for either the SCS-MNA or the SCS-MUN problem for t-threshold systems. (The result holds even when ρ is a function of the form n^{δ}, where $\delta < 1$ is a constant and n is the number of nodes in the network.)

(b) We show that the problem of saving all salvageable nodes (SCS-SASN) can be solved in linear time for 1-threshold systems and that the required critical set is unique. In contrast, we show that the problem is **NP**-hard for t-threshold systems for any $t \geq 2$. We also develop an $O(\log n)$-approximation algorithm for this problem, where n is the number of nodes in the network.

(c) We develop two intuitively appealing heuristics for the SCS-MNA problem and carry out an empirical study of their performance on three social networks, namely `epinions`, `wikipedia` and `slashdot`. Our experimental results show that in many cases, the two heuristics are similar in their ability to delay and halt the diffusion process. In general, one of the heuristics runs faster but there are cases where the other heuristic is more effective in inhibiting diffusion.

Related work on finding critical sets has been confined to threshold $t = 1$. Further, the focus is on selecting critical nodes to inhibit diffusion starting from a small random set I of initially infected (or **seed**) nodes. Our approach, in contrast, is focused on $t \geq 2$ and our heuristics compute a critical set for any specified set of seed nodes. Critical nodes are called "blockers" in [18]. They examine dynamic networks and use a probabilistic diffusion model with threshold $= 1$. They rely on graph metrics such as degree, diameter, and betweenness to identify critical nodes. In [7], the largest eigenvalue of the adjacency matrix of a graph is used to identify a node that causes the maximum decrease in the epidemic threshold. Vaccinating such a node reduces the likelihood of a large outbreak. A variety of network-based candidate measures for identifying critical nodes under threshold 1 conditions are described in [3]; however, the applications are confined to small networks. Hubs, or high degree nodes in scale free networks, have also been investigated as critical nodes, using mean field theory, in [8]. Reference [12] presents an approximation algorithm for the problem of minimizing the number of new affected nodes for 1-threshold systems. Reference [28] considers the problem of detecting cascades in networks and develops submodularity-based algorithms to determine the size of the affected population before a cascade is detected.

4 Theoretical Results for the Critical Set Problem

In this section, we first present complexity results for finding critical sets. We also present results that show a significant difference between 1-threshold systems and t-threshold systems where $t \geq 2$. Owing to space limitations, we have omitted proofs of these results; they can be found in [25].

4.1 Complexity Results

As mentioned earlier, the SCS-MNA problem was shown to be NP-complete in [12] for the case when each node has a 1-threshold function. We now extend that result, and include a result for the SCS-MUN problem, to show that even obtaining ρ-approximate solutions is **NP**-hard for systems in which each node computes the t-threshold function for any $t \geq 2$.

Theorem 2. *Assuming that the bound β on the size of the critical set cannot be violated, for any $\rho \geq 1$ and any $t \geq 2$, there is no polynomial time ρ-approximation algorithm for either the SCS-MNA problem or the SCS-MUN problem for t-threshold systems, unless $\mathbf{P} = \mathbf{NP}$.*

Proof: See [25]. ∎

4.2 Critical Sets for Saving All Salvageable Nodes

Recall from Section 2.3 that a node v of a SyDS is salvageable if there is a critical set whose removal ensures that v will not be affected. We now consider the following problem which deals with saving all salvageable nodes.

Small Critical Set to Save All Salvageable Nodes (SCS-SASN):

Given: A social network represented by the SyDS $\mathcal{S} = (G(V, E), \mathcal{F})$ over $\mathbb{B}$, with each function $f \in \mathcal{F}$ being a threshold function; the set I of nodes which are initially in state 1.

Requirement: A critical set C (i.e., $C \subseteq V - I$) of minimum cardinality whose removal ensures that all salvageable nodes are saved from being affected.

For the above problem, we present results that show a significant difference between 1-threshold systems and t-threshold systems where $t \geq 2$.

Theorem 3. *Let $\mathcal{S} = (G(V, E), \mathcal{F})$ be a 1-threshold SyDS. The SCS-SASN problem for $\mathcal{S}$ can be solved in $O(|V| + |E|)$ time. Moreover, the solution is unique.*

Proof: See [25]. ∎

The next result concerns the SCS-SASN problem for t-threshold systems, where $t \geq 2$.

Theorem 4. *The SCS-SASN problem is **NP**-hard for t-threshold systems, where $t \geq 2$. However, there is an $O(\log n)$-approximation algorithm for this problem, where n is the number of nodes in the network.*

Proof: See [25]. ∎

5 Heuristics for Finding Small Critical Sets

5.1 Overview

As can be seen from the complexity results presented in Section 4, it is difficult to develop heuristics with provably good performance guarantees for the SCS-MNA and SCS-MUN problems. So, we focus on the development of heuristics that work well in practice for one of these problems, namely SCS-MNA. In this section, we present two such heuristics that are evaluated in Section 6. The first heuristic uses a set cover computation. The second heuristic relies on a potential function, which provides an indication of a node's ability to affect other nodes.

5.2 Covering-Based Heuristic

Given a SyDS $\mathcal{S} = (G(V, E), \mathcal{F})$ and the set $I \subseteq V$ of nodes whose initial state is 1, one can compute the set $S_j \subseteq V$ of nodes that change to state 1 at the jth time step, $1 \leq j \leq \ell$, for some suitable $\ell \leq |V|$. The covering-based heuristic (CBH) chooses a critical set C as a subset of S_j for some suitable j. The intuitive reason for doing this is that each node w in S_{j+1} has at least one neighbor v in

Input: A SyDS $\mathcal{S} = (G(V, E), \mathcal{F})$, the set $I \subseteq V$ of nodes whose initial state is 1, the upper bound β on the size of the critical set and the number of initial simulation steps $\ell \leq |V|$.

Output: A critical set $C \subseteq V - I$ whose removal leads to a small number of new affected nodes.

Steps:

1. Simulate the system for ℓ time steps and determine sets $S_1, S_2, \ldots, S_\ell$, where S_j is the set of newly affected nodes at time j, $1 \leq j \leq \ell$.
2. **if** any set S_j has at most β nodes, **then** output such a set as the critical set and **stop**. (When there are ties, choose the set S_j with the smallest value of j.)
3. **Comment:** Here, all the S_j's have $\beta + 1$ or more nodes.
   (i) **for** $j = 1$ **to** $\ell - 1$ **do**
      (a) For each node $v_j \in S_j$, construct the set Γ_j which consists of all the neighbors of v_j in S_{j+1} that can be prevented from becoming affected by removing v_j. Let Γ denote the collection of all the sets constructed.
      (b) Use a greedy approach to find a subcollection Γ' of Γ containing at most β sets so as to cover as many elements of S_{j+1} as possible.
      (c) Let the critical set C consist of the nodes of S_j corresponding to the elements of Γ'.
   (ii) Among all the critical sets C considered in Step 3(i)(c), output the one C that occurs earliest in time that covers all nodes of Γ, and if no such C exists, output the earliest C such that $|S_j| - |C|$ is minimum.

Fig. 2. Details of the covering-based heuristic

S_j. (Otherwise, w would have changed to 1 in an earlier time step.) Therefore, if a suitable subset of S_j can be chosen so that none of the nodes in S_{j+1} changes to 1 during the $(j+1)^{\text{st}}$ time step, the contagion cannot spread beyond S_j. In general, when nodes have thresholds ≥ 2, the problem of choosing at most β nodes from S_j to prevent a maximum number of nodes in S_{j+1} from changing to 1 is also **NP**-hard. (This result can be proven in a manner similar to that of Theorem 2.) Therefore, we use a greedy approach for this step. In each iteration, this approach chooses a node from S_j that saves the largest number of nodes in S_{j+1} from becoming affected. The greedy approach is repeated for each j, $1 \leq j \leq \ell - 1$. The steps of the covering-based heuristic are shown in Figure 2. In Step 2, when two or more sets have β or fewer nodes, we choose the one that corresponds to an earlier time step since such a choice can save more nodes from becoming affected.

5.3 Potential-Based Heuristic

The idea of the potential-based heuristic (PBH) is to assign a potential to each node v depending on how early v is affected and how many nodes it can affect

Input: A SyDS $\mathcal{S} = (G(V,E), \mathcal{F})$, the set $I \subseteq V$ of nodes whose initial state is 1, the upper bound β on the size of the critical set.

Output: A critical set $C \subseteq V - I$ whose removal leads to a small number of new affected nodes.

Steps:

1. Simulate the system $\mathcal{S}$ and determine sets $S_1, S_2, \ldots, S_T$, where T is the time step at which $\mathcal{S}$ reaches a fixed point and S_j is the set of newly affected nodes at time j, $1 \leq j \leq T$.
2. **for** each node $x \in S_T$ **do**
   $P[x] = 0$.
3. **for** $j = T - 1$ **downto** 1 **do**
   for each node $x \in S_j$ **do**
   (a) Find $N_{j+1}[x]$ and let $P[x] = |N_{j+1}[x]|$.
   (b) **for** each node $y \in N_{j+1}[x]$ **do**
   $P[x] \;=\; P[x] + P[y]$
   (d) Set $P[x] \;=\; (T-j)^2 P[x]$.
4. Let the critical set C contain β nodes with the highest potential among all the nodes. (Break ties arbitrarily.) Output C.

Fig. 3. Details of the potential-based heuristic

later. Nodes with larger potential values are more desirable for inclusion in the critical set. While CBH chooses a critical set from one of the S_j sets, the potential based approach may select nodes in a more global fashion from the whole graph. One can obtain different versions of PBH by choosing different potential functions. We have chosen one that is easy to compute. Details of PBH are shown in Figure 3.

We assume that set S_j of newly affected nodes at time j has been computed for each j, $1 \leq j \leq T$, where T is the time at which the system reaches a fixed point. For any node $x \in S_j$, let $N_{j+1}[x]$ denote the set of nodes in S_{j+1} which are adjacent to x in G. The potential $P[x]$ of a node x is computed as follows: (a) For each node x in S_T, $P[x] = 0$. (**Justification:** There is no diffusion beyond level T. So, it is not useful to include nodes from S_T in the critical set.) (b) For each node x in level j, $1 \leq j \leq T-1$,

$$P[x] = (T-j)^2 \left[|N_{j+1}[x]| + \sum_{y \in N_{j+1}[x]} P[y] \right]$$

(**Justification:** The term $(T-j)^2$ decreases as j increases. Thus, higher potentials are assigned to nodes that are affected earlier. The term $|N_{j+1}[x]|$ gives more weight to nodes that have a large number of neighbors in the next level.)

6 Empirical Evaluation of Heuristics

6.1 Networks, Study Parameters and Test Procedures

Table 1 provides selected features of three social networks used in this study. We assume all edges are undirected to foster greater diffusion and thereby test more stringently the heuristics. The degree and clustering coefficient[1] distributions for the three networks are given elsewhere [25].

Table 1. Three networks [30, 29, 27] and selected characteristics

Network	Number of Nodes	Number of Edges	Average Degree	Average Clustering Coefficient
`epinions`	75879	405740	10.7	0.138
`wikipedia`	7115	100762	28.3	0.141
`slashdot`	77360	469180	12.1	0.0555

Table 2 lists the parameters and values used in the parametric study with the networks to evaluate the two heuristics. For a given value of number of seeds n_s, 100 sets of size n_s were determined from each network to provide a range of cases for testing the heuristics. Each seed node was taken from a 20-core, a subgraph in which each node has a degree of at least 20. The 20-core was a good compromise between selecting high-degree nodes, and having a sufficiently large pool of nodes to choose from so that sets of seeds overlapped little. Moreover, every seed node in a set is adjacent to at least one other seed node, so the seeds were "clumped," in order to foster diffusion. Thus, the test cases utilized two means, namely seeding of high-degree nodes and clumping the seed nodes, to foster diffusion and hence tax the heuristics.

Table 2. Parameters and values of parametric study

Thresholds, t	Numbers of Seeds, n_s	Budgets of Critical Nodes, β	Number of Replicates
2, 3, 5	2, 3, 5, 10, 20	5, 10, 20, 50, 100, 500	100

The test plan consists of running simulations of 100 iterations each (1 iteration for each seed node set) on the three networks for all combinations of t, n_s, and β. All nodes except seed nodes are initially in the unaffected state. Our simulator outputs for each node the time at which it is affected. The heuristics use this as input data and calculate one set of β critical nodes for each iteration. The simulations are then repeated, but now they include the critical nodes, so that

[1] For a node v in a graph G, the **clustering coefficient** c_v is defined as follows. Let $N(v)$ denote the set of nodes adjacent to v. Then, c_v is the ratio of the number of edges in the subgraph induced on $N(v)$ to the number of edges in a complete graph on $N(v)$.

the decrease in the total number of affected nodes caused by a critical set can be quantified. Heuristic computations and simulations were performed on a 96-node cluster (2 processors/node; 4 cores/processor), with 3 GHz Intel Xeon cores and 2 MB memory per core.

6.2 Results

A summary of our main experimental findings is as follows. The discussion uses some of the terminology (namely, cascade, halt and delay) from Section 2.3.

Structural results

(a) Critical node sets either halt diffusion with very small affected set sizes or do not prevent a cascade; thus, critical nodes generate phase transitions (unless all iterations halt the diffusion).
(b) The fraction of iterations cascading behaves as $(1/\beta)$ for $n_s \leq 5$, so to halt diffusion over all iterations can require $\beta \geq 500 = 100n_s$. This is in part attributable to the stochastic nature of the seeding process. While a heuristic may be successful on average, there will be combinations of seed nodes that are particularly difficult to halt.
(c) In some cases, if diffusion is not halted, a delay in the time to reach the peak number of newly affected nodes can be achieved, thus providing a retarding effect. This is a consequence of computed critical nodes impeding initial diffusion near the start time, but being insufficient to halt the spread. For the deterministic diffusion of this study, it is virtually impossible to impede diffusion after time step 2 or 3 because by this time, too many nodes have been affected.

Quality of solution

(d) The heuristics perform far better than setting high-degree nodes critical, and setting random nodes critical ("null" condition).
(e) For $n_s \leq 5$ and $\beta \leq 50$, the two heuristics often give similar results, and do not always halt diffusion. For small numbers of seeds, PBH, which is purposely biased toward selecting nodes affected early in the diffusion process, selects nodes at early times. CBH also seeks to halt at early times. Hence, both heuristics are trying to accomplish the same thing.
(f) However, when $\beta \geq 100$ nodes are required to stop diffusion because of a larger number of seeds, CBH is more effective in halting diffusion because it focuses critical nodes at one time step as explained below; hence there can be a tradeoff between speed of computation and effectiveness of the heuristics since PBH executes faster.

Figure 4 depicts the execution times for each heuristic for $\beta = 5$. For the `epinions` network, Figure 4(a), these times translate into a maximum of roughly 1.5 hours for CBH to determine 100 sets of critical nodes, versus less than 5 minutes for PBH. For the `wikipedia` network, Figure 4(b), comparable execution

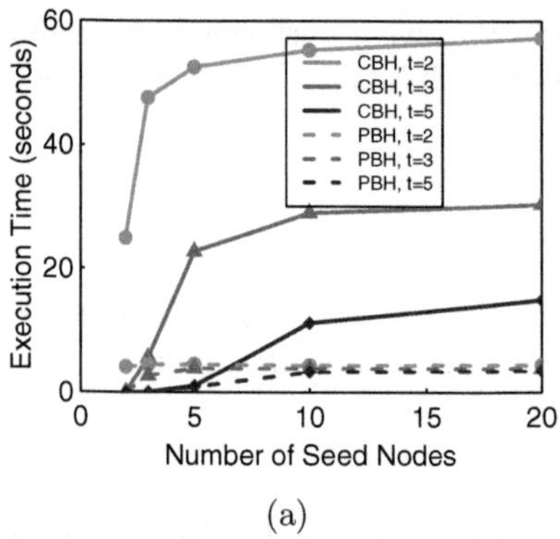

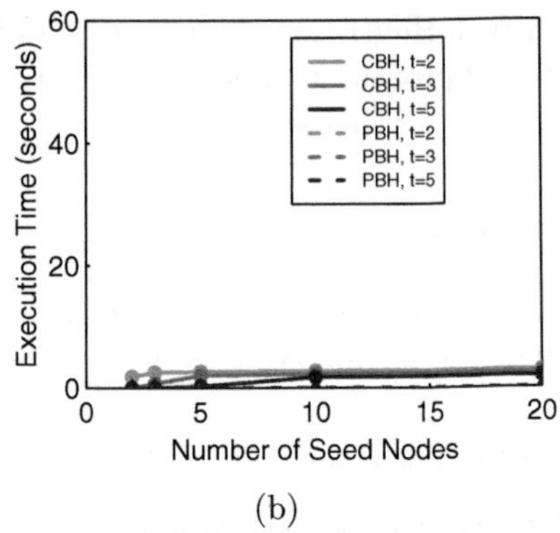

Fig. 4. Times for CBH and PBH to compute one set of critical nodes as a function of threshold and number of seeds for the (a) `epinions` network; (b) `wikipedia` network. Times are averages over 100 iterations.

times are observed when the number of nodes decreases by an order of magnitude. As described in Section 5, PBH evaluates every node once, whereas a node in CBH is often analyzed at many different time steps.

We now turn to evaluating the heuristics in halting and delaying diffusion by first comparing the heuristics with the heuristics of (1) randomly setting nodes critical (RCH), and (2) setting high-degree nodes critical (HCH). Table 3 summarizes selected results where we have a high ratio of β/n_s to give RCH and HCH the best chances for success (i.e., for minimizing the fraction of cascades). While CBH and PBH halt almost all 100 iterations, RCH and HCH allow cascades in 38% to 100% of iterations. To obtain the same fraction of cascades as for random and high-degree critical nodes, CBH would require only about $\beta = 5$ critical nodes. Neither RCH nor HCH focus on specific seed sets, and RCH can select nodes of degree 1 as critical (of which there are many in the three networks); these nodes do not propagate complex contagions, so specifying them as critical is wasteful.

Figure 5(a) shows cumulative number of affected nodes as a function of time for the `slashdot` network with CBH. Results from the 40 iterations that cascade

Table 3. Comparison of CBH and PBH against random critical nodes and high-degree critical nodes, with respect to the fraction of iterations in which cascades occur, for $t = 2$ and $\beta = 500$. Each cell has two entries: one value for $n_s = 2$ and one for $n_s = 3$.

Network	Numbers of Seeds	Fraction of Cascades Random	Fraction of Cascades High-Degree	Fraction of Cascades CBH	Fraction of Cascades PBH
`epinions`	2/3	0.94/1.00	0.75/0.99	0.00/0.00	0.00/0.01
`wikipedia`	2/3	0.96/1.00	0.65/0.99	0.00/0.00	0.00/0.01
`slashdot`	2/3	0.60/0.95	0.38/0.80	0.00/0.00	0.00/0.00

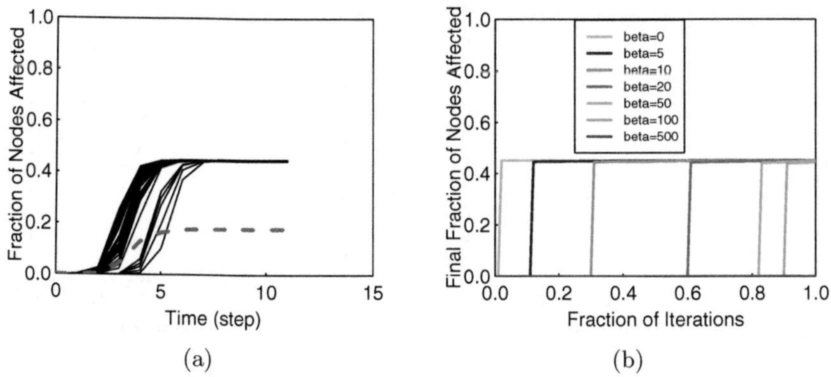

Fig. 5. (a) Cumulative number of affected nodes for each iteration (solid lines) and average over all 100 iterations (dashed line) for heuristic CBH, for the case $t = 3$, $n_s = 10$, and $\beta = 20$ with the `slashdot` network. (b) Final number of affected nodes for the `slashdot` network and CBH heuristic for $t = 3$ and $n_s = 10$.

are plotted as solid lines; all iterations plateau at 44% of the network nodes. (To be precise, the number of nodes affected varies by a very small amount—about 2% or less—for different sets of seed nodes. Also, in a very few instances, an iteration with no critical nodes also halts the diffusion. We ignore these minor effects throughout for clarity of presentation.) These features are observed in *all* simulation results and are dictated by the deterministic state transition model: if the diffusion is not halted by the critical nodes, then the size of the outbreak is the same for all iterations, although the progression may vary.

The final fractions of nodes affected for each of the 100 iterations, arranged in increasing numerical order, are plotted as the $\beta = 20$ curve in Figure 5(b). The other curves correspond to different budget values, and all exhibit a sharp phase transition, except for $\beta = 500$, which halts all iterations. Over all the simulations conducted in this study, both heuristics produce this type of phase transition.

Figure 6 examines the regime of small numbers of seed nodes, and depicts the fraction of iterations that cascade as a function of β for two networks. Note the larger discrepancy between heuristics in Figure 6(a) for $\beta = 10$; this is explained below. In both plots a $(1/\beta)$ behavior is observed, so that the number of cascades drops off sharply with increasing budget, but to completely eliminate all cascades in the `wikipedia` network in Figure 6(a) for example, $\beta = 500$ is required for both heuristics when $n_s = 5$.

Figure 7 provides results that show the greatest differences in the fraction of iterations to cascade for the two heuristics, which generally occur for the largest sizes of seed sets. The results are for the same conditions as in Figure 6(b). For example, in Figure 7, only 17% of iterations result in a cascade with CBH, while PBH permits 63% for $n_s = 10$.

In all cases, CBH is at least as effective as PBH. This is because CBH focuses on conditions at one time step that are required to halt diffusion. PBH, in contrast, can span multiple time steps in that a parent of a high-potential

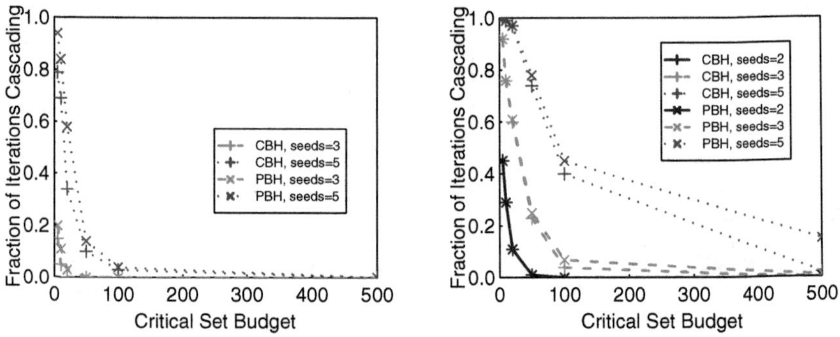

(a) (b)

Fig. 6. Comparisons of CBH and PBH in inhibiting diffusion in (a) the `wikipedia` network for $t = 3$; (b) the `epinions` network for $t = 2$.

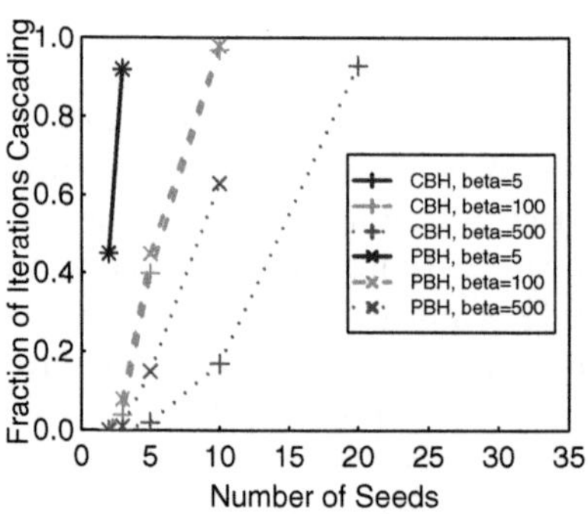

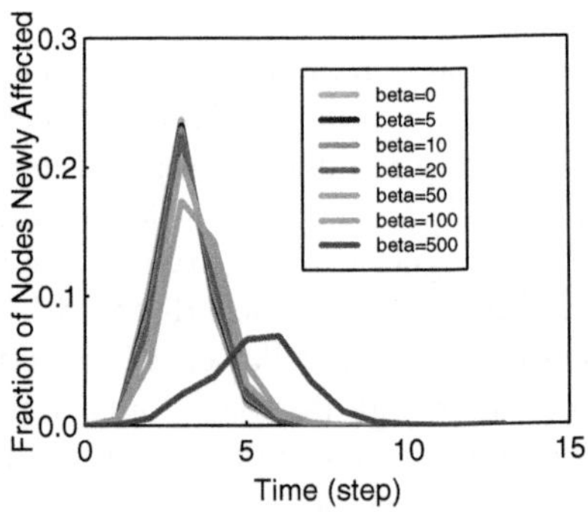

Fig. 7. Comparisons of CBH and PBH in inhibiting diffusion in the `epinions` network for $t = 2$

Fig. 8. Average curves of newly affected nodes for PBH for the case $t = 2$, $n_s = 10$, and different values of β with the `epinions` network

node will itself be a high potential node, and hence both may be determined critical. Consequently, there is greater chance for critical nodes to redundantly save salvageable nodes at the expense of others, rendering the critical set less effective. This behavior is the cause of PBH allowing more cascades for $\beta = 10$ and $n_s = 5$ in Figure 6(a).

In Figure 8, the *average* number of newly affected nodes in each time step over 100 iterations is given for simulations with different numbers of critical nodes. While a budget of $\beta = 500$ does not halt the diffusion process, it does slow the diffusion, moving the time of peak number of newly affected nodes from 3 to 6. This may be useful in providing decision-makers more time for suitable interventions.

7 Future Work

There are several directions for future work. Among these are: (a) development of practical heuristics for the critical set problem for complex contagions when there are weights on edges (to model the degree to which a node is influenced by a neighbor); (b) investigation of the critical set problem for complex contagions when the diffusion process is probabilistic; and (c) formulation and study of the problem for time-varying networks in which nodes and edges may appear and disappear over time.

Acknowledgment. We thank the referees from ECML PKDD 2010. We also thank our external collaborators and members of the Network Dynamics and Simulation Science Laboratory (NDSSL) for their suggestions and comments. This work has been partially supported by NSF Nets Grant CNS- 0626964, NSF HSD Grant SES-0729441, NIH MIDAS project 2U01GM070694-7, NSF PetaApps Grant OCI-0904844, DTRA R&D Grant HDTRA1-0901-0017, DTRA CNIMS Grant HDTRA1-07-C-0113, NSF NETS CNS-0831633, DHS 4112-31805, NSF CNS-0845700 and DOE DE-SC003957.

References

1. Arulselvan, A., Commander, C.W., Elefteriadou, L., Pardalos, P.M.: Detecting Critical Nodes in Sparse Graphs. Comput. Oper. Res. 36(7), 2193–2200 (2009)
2. Barrett, C.L., Hunt III, H.B., Marathe, M.V., Ravi, S.S., Rosenkrantz, D.J., Stearns, R.E.: Complexity of Reachability Problems for Finite Discrete Dynamical Systems. J. Comput. Syst. Sci. 72(8), 1317–1345 (2006)
3. Borgatti, S.: Identifying sets of key players in a social network. Comput. Math. Organiz. Theor. 12, 21–34 (2006)
4. Centola, D., Eguiluz, V., Macy, M.: Cascade Dynamics of Complex Propagation. Physica A 374, 449–456 (2006)
5. Centola, D., Macy, M.: Complex Contagions and the Weakness of Long Ties. American Journal of Sociology 113(3), 702–734 (2007)
6. Cha, M., Mislove, A., Adams, B., Gummadi, K.: Characterizing Social Cascades in Flickr. In: Proc. of 1st First Workshop on Online Social Networks, pp. 13–18 (2008)
7. Chakrabarti, D., Wang, Y., Wang, C., Leskovec, J., Faloutsos, C.: Epidemic Thresholds in Real Networks. ACM Trans. Inf. Syst. Secur. 10(4), 13-1–13-26 (2008)
8. Dezso, Z., Barabasi, A.: Halting Viruse. In: Scale-Free Networks. Physical Review E 65, 055103-1–055103-4 (2002)
9. Domingos, P., Richardson, M.: Mining the Network Value of Customers. In: Proc. ACM Intl. Conf. on Data Mining and Knowledge Discovery (KDD 2001), pp. 57–61 (2001)
10. Dreyer, P., Roberts, F.: Irreversible k-Threshold Processes: Graph-Theoretical Threshold Models of the Spread of Disease and Opinion. Discrete Applied Mathematics 157, 1615–1627 (2009)
11. Easley, D., Kleinberg, J.: Networks, Crowds, and Markets. Cambridge University Press, Cambridge (2010)

12. Eubank, S., Kumar, V.S.A., Marathe, M.V., Srinivasan, A., Wang, N.: Structure of Social Contact Networks and Their Impact on Epidemics. In: Abello, J., Cormode, G. (eds.) Discrete Methods in Epidemiology. DIMACS Series in Discrete Mathematics and Theoretical Computer Science, pp. 179–200. American Mathematical Society, Providence (2006)
13. Garey, M.R., Johnson, D.S.: Computers and Intractability: A Guide to the Theory of NP-completeness. W. H. Freeman and Co., San Francisco (1979)
14. Granovetter, M.: The Strength of Weak Ties. American Journal of Sociology 78(6), 1360–1380 (1973)
15. Granovetter, M.: Threshold Models of Collective Behavior. American Journal of Sociology 83(6), 1420–1443 (1978)
16. Gruhl, D., Guha, R., Liben-Nowell, D., Tomkins, A.: Information Diffusion Through Blogspace. In: Proc. of the 13th International World Wide Web Conference (WWW 2004), pp. 491–501 (2004)
17. Guha, R., Kumar, R., Raghavan, P., Tomkins, A.: Propagation of Trust and Distrust. In: Proc. of the 13th International World Wide Web Conference (WWW 2004), pp. 403–412 (2004)
18. Habiba, Yu, Y., Berger-Wolf, T., Saia, J.: Finding Spread Blockers in Dynamic Networks. In: The 2nd SNA-KDD Workshop 2008, SNA-KDD 2008 (2008)
19. Harris, K.: The National Longitudinal Study of Adolescent Health (Add Health), Waves I and II, 1994-1996; Wave III, 2001-2002 [machine-readable data file and documentation]. arolina Population Center, University of North Carolina at Chapel Hill, Chapel Hill, NC (2008)
20. Kempe, D., Kleinberg, J., Tardos, E.: Maximizing the Spread of Influence Through a Social Network. In: Proc. ACM Intl. Conf. on Data Mining and Knowledge Discovery (KDD 2003), pp. 137–146 (2003)
21. Kempe, D., Kleinberg, J., Tardos, E.: Influential Nodes in a Diffusion Model for Social Networks. In: Caires, L., Italiano, G.F., Monteiro, L., Palamidessi, C., Yung, M. (eds.) ICALP 2005. LNCS, vol. 3580, pp. 1127–1138. Springer, Heidelberg (2005)
22. Kleinberg, J.: Cascading Behavior in Networks: Algorithmic and Economic Issues. In: Nissan, N., Roughgarden, T., Tardos, E., Vazirani, V. (eds.) Algorithmic Game Theory, ch. 24, pp. 613–632. Cambridge University Press, New York (2007)
23. Kossinets, G., Kleinberg, J., Watts, D.: The Structure of Information Pathways in a Social Communication Network. In: Proc. ACM Intl. Conf. on Data Mining and Knowledge Discovery, KDD 2008 (2008)
24. Kuhlman, C.J., Anil Kumar, V.S., Marathe, M.V., Ravi, S.S., Rosenkrantz, D.J.: Computational Aspects of Ratcheted Discrete Dynamical Systems (April 2010) (under preparation)
25. Kuhlman, C.J., Anil Kumar, V.S., Marathe, M.V., Ravi, S.S., Rosenkrantz, D.J.: NDSSL Technical Report No. 10-060 (2010), `http://ndssl.vbi.vt.edu/download/kuhlman/tr-10-60.pdf`
26. Leskovec, J., Adamic, L., Huberman, B.: The Dynamics of Viral Marketing. ACM Transactions on the Web, 1(1) (2007)
27. Leskovec, J., Huttenlocher, D., Kleinberg, J.: Predicting Positive and Negative Links in Online Social Networks. In: WWW 2010 (2010)
28. Leskovec, J., Krause, A., Guestrin, C., Faloutsos, C., VanBriesen, J., Glance, N.: Cost-Effective Outbreak Detection in Networks. In: Proc. ACM Intl. Conf. on Data Mining and Knowledge Discovery, KDD 2007 (2007)
29. Leskovec, J., Lang, K., Dasgupta, A., Mahoney, M.: Community Structure in Large Networks: Natural Cluster Sizes and the Absence of Large Well-Defined Clusters (2008), Appears as arXiv.org:0810.1355

30. Richardson, M., Agrawal, R., Domingos, P.: Trust Management for the Semantic Web. In: Fensel, D., Sycara, K., Mylopoulos, J. (eds.) ISWC 2003. LNCS, vol. 2870, pp. 351–368. Springer, Heidelberg (2003)
31. Richardson, M., Domingos, P.: Mining Knowledge-Sharing Sites for Viral Marketing. In: Proc. ACM Intl. Conf. on Data Mining and Knowledge Discovery (KDD 2002), pp. 61–70 (2002)
32. Tantipathananandh, C., Berger-Wolf, T.Y., Kempe, D.: A Framework for Community Identification in Dynamic Social Networks. In: Proc. ACM Intl. Conf. on Data Mining and Knowledge Discovery (KDD 2007), pp. 717–726 (2007)

Semi-supervised Abstraction-Augmented String Kernel for Multi-level Bio-Relation Extraction

Pavel Kuksa[1], Yanjun Qi[2], Bing Bai[2], Ronan Collobert[2], Jason Weston[3], Vladimir Pavlovic[1], and Xia Ning[4]

[1] Department of Computer Science, Rutgers University, USA
[2] NEC Labs America, Princeton, USA
[3] Google Research, New York City, USA
[4] Computer Science Department, University of Minnesota, USA

Abstract. Bio-relation extraction (bRE), an important goal in bio-text mining, involves subtasks identifying relationships between bio-entities in text at multiple levels, e.g., at the article, sentence or relation level. A key limitation of current bRE systems is that they are restricted by the availability of annotated corpora. In this work we introduce a semi-supervised approach that can tackle multi-level bRE via string comparisons with mismatches in the string kernel framework. Our string kernel implements an abstraction step, which groups similar words to generate more abstract entities, which can be learnt with unlabeled data. Specifically, two unsupervised models are proposed to capture contextual (local or global) semantic similarities between words from a large unannotated corpus. This Abstraction-augmented String Kernel (ASK) allows for better generalization of patterns learned from annotated data and provides a unified framework for solving bRE with multiple degrees of detail. ASK shows effective improvements over classic string kernels on four datasets and achieves state-of-the-art bRE performance without the need for complex linguistic features.

Keywords: Semi-supervised string kernel, Relation extraction, Sequence classification, Learning with auxiliary information.

1 Introduction

The task of relation extraction from text is important in biomedical domains, since most scientific discoveries describe biological relationships between bio-entities and are communicated through publications or reports. A range of text mining and NLP strategies have been proposed to convert natural language in the biomedical literature into formal computer representations to facilitate sophisticated biomedical literature access [14]. However, the lack of annotated data and the complex nature of biomedical discoveries have limited automatic literature mining from having large impact.

In this paper, we consider "bio-relation extraction" tasks, i.e. tasks that aim to discover biomedical relationships of interest reported in the literature through

J.L. Balcázar et al. (Eds.): ECML PKDD 2010, Part II, LNAI 6322, pp. 128–144, 2010.

Table 1. Examples of sentence-level task and relation-level task

Task 2: Sentence-level PPI extraction	
Negative	TH, AADC and GCH were effectively co-expressed in transduced cells with three separate AAV vectors.
Positive	This study demonstrates that IL - 8 recognizes and activates CXCR1, CXCR2, and the Duffy antigen by distinct mechanisms .
Task 3: Relation-level PPI extraction	
Input Sentence	The protein product of c-cbl proto-oncogene is known to interact with several proteins, including Grb2, Crk, and PI3 kinase, and is known to regulate signaling...
Output Interacting Pairs	(c-cbl, Grb2), (c-cbl, Crk), (c-cbl, PI3).

identifying the textual triggers with different levels of detail in the text [14]. Specifically we cover three tasks in our experiments associated with one important biological relation: protein-protein-interaction (PPI). In order to identify PPI events, the tasks aim to: (1) retrieve PubMed abstracts describing PPIs; (2) classify text sentences as PPI relevant or not relevant; (3) when protein entities have been recognized in the sentence, extract which protein-protein pairs having interaction relationship, i.e. pairwise PPI relations from the sentence. Table 2 gives examples of the second and third tasks. Examples of the first task are long text paragraphs and are omitted due to space limitations.

There exist very few annotated training datasets for all three tasks above. For bRE tasks at article-level, researchers [14] handled them as text categorization problems and support vector machines were shown to give good results with careful pre-processing, stemming, POS and named-entity tagging, and voting. For bRE tasks at the relation level, most systems in the literature are rule-based, cooccurrence-based or hybrid approaches (survey in [29]). Recently several researchers proposed the all-paths graph kernel [1], or an ensemble of multiple kernels and parsers [21], which were reported to yield good results.

Generally speaking, these tasks are all important instances of information extraction problems where entities are protein names and relationships are protein-protein interactions. Early approaches for the general "relation extraction" problem in natural languages are based on patterns [23], usually expressed as regular expressions for words with wildcards. Later researchers proposed kernels for dependency trees [7] or extended the kernel with richer structural features [23]. Considering the complexity of generating dependency trees from parsers, we try to avoid this step in our approach. Also bRE systems at article/long-text levels need to handle very long word sequences, which are problematic for previous tree/graph kernels to handle.

Here we propose to detect and extract relations from biomedical literature using string kernels with semi-supervised extensions, named Abstraction-augmented String Kernels (ASK). A novel semi-supervised "abstraction" augmentation strategy is applied on a string kernel to leverage supervised event extraction with unlabeled data. The "abstraction" approach includes two stages: (1) Two unsupervised auxiliary tasks are proposed to learn accurate word representations from contextual semantic similarity of words in biomedical literature, with one task focusing on short local neighborhoods (local ASK), and the other using long paragraphs as word context (global ASK). (2) Words are grouped to generate more abstract entities according to their learned representations. On benchmark PPI extraction data sets targeting three text levels, the proposed kernel achieves state-of-the-art performance and improves over classic string kernels.

Furthermore, we want to point out that ASK is a general sequence modeling approach and not tied to the multi-level bRE applications. We show this generality by extending ASK to a benchmark protein sequence classification task (the 4th dataset), and get improved performances over all tested supervised and semi-supervised string kernel baselines.

2 String Kernels

All of our targeted bRE tasks can be treated as problems of classifying sequences of words into certain types related to the relation of interest (i.e., PPI). For example, in bRE tasks at the article-level, we classify input articles or long paragraphs as PPI-relevant (positive) or not (negative). For the bRE task at the sentence-level, we classify sentence into PPI-related or not, which again is a string classification problem.

Various methods have been proposed to solve the string classification problem, including generative (e.g., HMMs) or discriminative approaches. Among the discriminative approaches, string kernel-based machine learning methods provide some of the most accurate results [27,19,16,28].

The key idea of basic string kernels is to apply a mapping $\phi(\cdot)$ to map text strings of variable length into a vectorial feature space of fixed length. In this space a standard classifier such as a support vector machine (SVM) can then be applied. As SVMs require only inner products between examples in the feature space, rather than the feature vectors themselves, one can define a *string kernel* which *implicitly* computes an inner product in the feature space:

$$K(x, y) = \langle \phi(x), \phi(y) \rangle, \tag{1}$$

where $x, y \in S$, S is the set of all sequences composed of elements which take on a finite set of possible values, e.g., sequences of words in our case, and $\phi : S \rightarrow \mathbb{R}^m$ is a feature mapping from a word sequence (text) to a m-dim. feature vector.

Feature extraction and feature representation play key roles in the effectiveness of sequence analysis since text sequences cannot be readily described as feature vectors. Traditional text categorization methods use feature vectors indexed by all possible words (e.g., bag of words [25]) in a certain dictionary (vocabulary $\mathcal{D}$) to represent text documents, which can be seen as a simple form of string kernel. This "bag of words" strategy treats documents as an unordered set of features (words), where critical word ordering information is not preserved.

Table 2. Subsequences considered for string matching in different kernels

Type	Parameters	Subsequences to Consider
Spectrum Kernel	$k = 3$	(SM binds RNA), (binds RNA in), (RNA in vitro), ...
Mismatch Kernel	$k = 3, m = 1$	(X binds RNA), (SM X RNA), (SM binds X), (X RNA in), (binds X in), (binds RNA X), ...
Gapped Kernel	$k = 3, m = 1$	(SM [ ] RNA in), (binds RNA in [ ]), (binds [ ] in vitro), ...

To take word ordering into account, documents can be considered as bags of short sequences of words with feature vectors corresponding to all possible word n-grams (n adjacent words from vocabulary $\mathcal{D}$). With this representation, the high similarity between two text documents means they have many n-grams in common. One can then define a corresponding string kernel as follows,

$$K(x, y) = \sum_{\gamma \in \Gamma} c_x(\gamma) \cdot c_y(\gamma), \tag{2}$$

where γ is a n-gram, Γ is the set of all possible n-grams, and $c_x(\gamma)$ is the number of occurrences (with normalization) of n-gram γ in a text string x. This is also called the spectrum kernel in the literature [18]. More general, the so-called substring kernels [27] measure similarity between sequences based on common co-occurrence of exact sub-patterns (e.g., substrings).

Inexact comparison, which is critical for effective matching (similarity evaluation) between text documents due to naturally occurring word substitutions, insertions, or deletions, is typically achieved by using different families of mismatch [19]. The mismatch kernel considers word (or character) n-gram counts with inexact matching of word (or character) n-grams. The gapped kernel calculates dot-product of (non-contiguous) word (or character) n-gram counts with gaps allowed between words. That is we revise $c_x(\gamma)$ as the number of subsequences matching the n-gram γ with up to k gaps. For example, as shown in Table 2, when calculating counts of trigram in a given sentence "SM binds RNA in vitro ..." , three string kernels we tried in our experiments consider different subsequences into the counts. As can be seen from examples, string kernels can capture relationship patterns using mixtures of words (n-grams with gaps or mismatch) as features.

String kernel implementations in practice typically require efficient methods for dot-product computation without explicitly constructing potentially very high-dimensional feature vectors. A number of algorithmic approaches have been proposed [27,24,17] for efficient string kernel computation and we adopt a sufficient statistic strategy from [16] for fast calculation of mismatch and gapped kernels. It provides a new family of *linear* time string kernel computation that scale well with large alphabet size and input length, e.g., word vocabulary in our context.

3 ASK: Abstraction-Augmented String Kernel

Currently there exist very few annotated training data for the tasks of bio-relation extractions. For example, the largest (to the best of our knowledge) publicly available training set for identifying "PPI relations" from PubMed abstracts includes only about four thousands annotated examples. This small set of training data could hardly cover most of the words in the vocabulary (about 2 million words in PubMed, which is the central collection of biomedical papers). On the other hand, PubMed stores more than 17 million citations (papers/reports), and provides free downloads of all abstracts (with over ~1.3G tokens after preprocessing). Thus our goal is to use a large unlabeled corpus to boost the performance of string kernels where only a small number of labeled examples are provided for sequence classification.

We describe a new semi-supervised string kernel, called "Abstraction-augmented String Kernel" (ASK). The key term "abstraction" describes an operation of grouping similar words to generate more abstract entities. We also refer to the resulting abstract entities as "abstraction". ASK is accomplished in two steps: (i) learning word abstractions with unsupervised embedding and clustering (Figure 2); (ii) constructing a string kernel on both words and word abstractions (Figure 1).

3.1 Word Abstraction with Embedding

ASK relies on the key observation that individual words carry significant semantic information in natural language text. We learn a mapping of each word to a vector of real values (called an "embedding" in the following) which describes this word's semantic meaning. Figure 2 illustrates this mapping step with an exemplar sentence. Two types of unsupervised auxiliary tasks are exploited to learn embedded feature representations from unlabeled text, which aim to capture:

- Local semantic patterns: an unsupervised model is trained to capture words' semantic meanings in short text segments (e.g. text windows of 7 words).
- Global semantic distribution: an unsupervised model is trained to capture words' semantic patterns in long text sequences (e.g. long paragraphs or full documents).

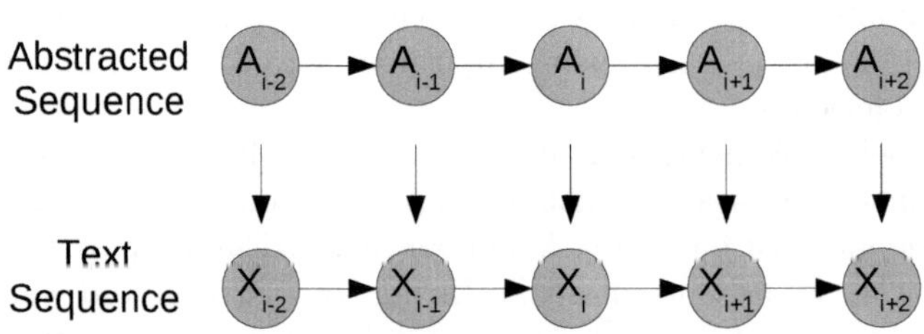

Fig. 1. Semi-supervised Abstraction-Augmented String Kernel. Both text sequence X and learned abstracted sequence A are used jointly.

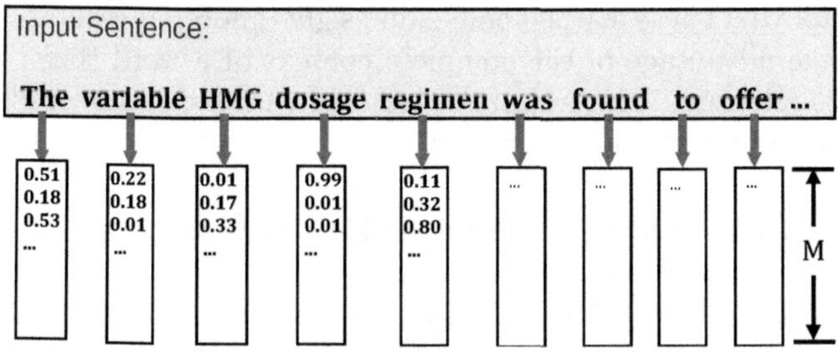

Fig. 2. The word embedding step maps each word in an input sentence to a vector of real values (with dimension M) by learning from a large unlabeled corpus

Local Word Embedding (Local ASK). It can be observed that in most natural language text, semantically similar words can usually be exchanged with no impact on the sentence's basic meaning. For example, in a sentence like "EGFR interacts with an inhibitor" one can replace "interacts" with "binds" with no change in the sentence labeling. With this motivation, traditional language models estimate the probability of the next word being w in a language sequence. In a related task, [6] proposed a different type of "language modeling" (LM) which learns to embed normal English words into a M dimensional feature space by utilizing unlabeled sentences with an unsupervised auxiliary task. We adapt this approach to bio-literature texts and train the language model on unlabeled sentences in PUBMED abstracts.

We construct an auxiliary task which learns to predict whether a given text sequence (short word window) exists naturally in biomedical literature, or not. The real text fragments are labeled as positive examples, and negative text fragments are generated by random word substitution (in this paper we substitute the middle word by a random word). That is, LM tries to recognize if the word in the middle of the input window is related to its context or not. Note, the end goal is not the solution to the classification task itself, but the embedding of words into an M-dimensional space that are the parameters of the model. These will be used to effectively *learn the abstraction* for ASK.

Following [6], a Neural Network (NN) architecture is used for this LM embedding learning. With a sliding window approach, values of words in the current window are concatenated and fed into subsequent layers which are classical neural network (NN) layers (with one hidden layer and another output layer, using sliding text windows of size 11). The word embeddings and parameters of the subsequent NN layers are all automatically trained by backpropagation. The model is trained with a ranking-type cost (with margin):

$$\sum_{s \in \mathcal{S}} \sum_{w \in \mathcal{D}} \max\left(0,\, 1 - f(s) + f(s^w)\right) , \tag{3}$$

where $\mathcal{S}$ is the set of possible local windows of text, $\mathcal{D}$ is the vocabulary of words, and $f(\cdot)$ represents the output of NN architecture and s^w is a text window where the middle word has been replaced by a random word w (negative window as mentioned above). These learned embeddings give good representations of words where we take advantage of the complete context of a word (before and after) to predict its relevance. The training is handled with stochastic gradient descent which samples the cost *online* w.r.t. (s, w).

Global Word Embedding (Global ASK). Since the local word embedding learns from very short text segments, it cannot capture similar words having long range relationships. Thus we propose a novel auxiliary task which aims to catch word semantics within longer text sequences, e.g., full documents.

We still represent each word as a vector in an M dimensional feature space as in Figure 2. To capture semantic patterns in longer texts, we try to model real articles in an unlabeled language corpus. Considering that words happen multiple times in documents, we represent each document as a weighted sum of its included words' embeddings,

$$g(d) = \sum_{w \in d} c_d(w) E(w) \tag{4}$$

where scalar $c_d(w)$ means the normalized tf-idf weight of word w on document d, and vector $E(w)$ is the M-dim embedded representation of word w which would be learned automatically through backpropagation. The M-dimensional feature vector $g(d)$ thus represents the semantic embedding of the current document d.

Similar to the LM, we try to force $g(\cdot)$ of two documents with similar meanings to have closer representations, and force two documents with different meanings to have dissimilar representations. For an unlabeled document set, we adopt the following procedure to generate a pseudo-supervised signals for training of this model. We split a document a into two sections: a_0 and a_1, and assume that (in natural language) the similarity between two sections a_0 and a_1 is larger than the similarity between a_i $(i \in \{0,1\})$ and one section b_j $(j \in \{0,1\})$ from another random document b: that is

$$f(g(a_0), g(a_1)) > f(g(a_i), g(b_j)) \tag{5}$$

where $f(\cdot)$ represents a similarity measure on the document representation $g(\cdot)$. $f(\cdot)$ is chosen as the cosine similarity in our experiments. Naturally the above assumption comes to minimize a margin ranking loss:

$$\sum_{(a,b) \in \mathcal{A}} \sum_{i,j=0,1} \max(0, 1 - f(g(a_i), g(a_{1-i})) + f(g(a_i), g(b_j))) \tag{6}$$

where $i \in \{0,1\}$, $j \in \{0,1\}$ and $\mathcal{A}$ represents all documents in the unlabeled set. We train $E(w)$ using stochastic gradient descent, where iteratively, one picks a random tuple from (a_i and b_j) and makes a gradient step for that tuple. The stochastic method scales well to our large unlabeled corpus and is easy to implement.

Abstraction using Vector Quantization. As we mentioned, "abstraction" means grouping similar words to generate more abstract entities. Here we try to

Table 3. Example words mapped to the same "abstraction" as the query word (first column) according to two different embeddings. We can see that "local" embedding captures part-of-speech and "local" semantics, while "global" embedding found words semantically close in their long range topics across a document.

Query	Local ASK	Global ASK
protein	ligand, subunit, receptor, molecule	proteins, cosNUM, phosphoprotein, isoform
medical	surgical, dental, preventive, reconstructive	hospital, investigated, research, urology
interact	cooperate, compete, interfere, react	interacting, interacts, associate, member
immunoprecipitation	co-immunoprecipitation, EMSA, autoradiography, RT-PCR	coexpression, two-hybrid, phosphorylated, tbp

group words according to their embedded feature representations from either of the two embedding tasks described above. For a given word w, the auxiliary tasks learn to define a feature vector $E(w) \in \mathbb{R}^M$. Similar feature vectors $E(w)$ can indicate semantic closeness of the words. Grouping similar $E(w)$ into compact entities might give stronger indications of the target patterns. Simultaneously, this will also make the resulting kernel tractable to compute[1].

As a classical lossy data compression method in the field of signal processing, Vector quantization (VQ) [10] is utilized here to achieve the abstraction operation. The input vectors are quantized (clustered) into different groups via "prototype vectors". VQ summarizes the distribution of input vectors with their matched prototype vectors. The set of all *prototype vectors* is called the *codebook*. We use $\mathcal{C}$ to represent the codebook set which includes N prototype vectors, $\mathcal{C} = \{C_1, C_2, ..., C_N\}$.

Formally speaking, VQ tries to optimize (minimize) the following objective function, in order to find the codebook $\mathcal{C}$ and in order to best quantize each input vector into its matched prototype vector,

$$\sum_{i=1...|\mathcal{D}|} ||E(w_i) - C_n||^2, \; n \in \{1...N\} \tag{7}$$

where $E(w_i) \in \mathbb{R}^M$ is the embedding of word w_i. Hence, our basic VQ is essentially a k-means clustering approach.

For a given word w we call the index of the prototype vector C_j that is closest to $E(w)$ its *abstraction*.

According to the two different embeddings, Table 3 gives the lists of example words mapped to the same "abstraction" as the query word (first column). We can see that "local" embedding captures part-of-speech and "local" semantics, while "global" embedding found words semantically close in their long range topics across a document.

[1] One could avoid the VQ step by considering the direct kernel $k(x, y) = \sum_{i,j} \exp(-\gamma||E(x_i) - E(y_j))||)$ that measures the similarity of embeddings between all pairs of words between two documents, but this would be slow to compute.

3.2 Semi-supervised String Kernel

Unlike standard string kernels which use words directly from the input text, semi-supervised ASK combines word sequences with word abstractions (Figure 1). The word abstractions are learned to capture local and global semantic patterns of words (described in previous sections). As Table 3 shows, using learned embeddings to group words into abstractions could give stronger indications of the target pattern. For example, in local ASK, the word "protein" is grouped with terms like "ligand", "receptor", or "molecule". Clearly, this abstraction could improve the string kernel matching since it provides a good summarization of the involved parties related to target event patterns.

We define the semi-supervised abstraction-augmented string kernel as follows

$$K(x,y) = \Big\langle \big(\phi(x), \phi'(a(x))\big), \big(\phi(y), \phi'(a(y))\big) \Big\rangle \tag{8}$$

where $(\phi(x), \phi'(a(x)))$ extends the basic n-gram representation $\phi(x)$ with the representation $\phi'(a(x))$. $\phi'(a(x))$ is a n-gram representation of the abstraction sequence, where

$$a(x) = (a(x_1), \ldots, a(x_{|x|})) = (A_1, \ldots, A_{|x|}) \tag{9}$$

$|x|$ means the length of the sequence and its i_{th} item is $A_i \in \{1...N\}$. The abstraction sequence $a(x)$ is learned through the embedding and abstraction steps.

The abstraction kernel exhibits a number of properties:

- It is a wrapper approach and can be used to extend *both supervised* and *semi-supervised* string kernels.
- It is very efficient as it has linear cost in the input length.
- It provides two unsupervised models for word-feature learning from unlabeled text.
- The baseline supervised or semi-supervised models can learn if the learned abstractions are relevant or not.
- It provides a unified framework for bRE at multiple levels where tasks have small training sets.
- It is quite general and not restricted to the biomedical text domain, since no domain specific knowledge is necessary for the training.
- It can incorporate other types of word similarities (e.g., obtained from classical latent semantic indexing [8]).

4 Related Work

4.1 Semi-supervised Learning

Supervised NLP techniques are restricted by the availability of labeled examples. Semi-supervised learning has become popular, since unlabeled language data is abundant. Many semi-supervised learning algorithms exist, including self-training, co-training, Transductive SVMs, graph-based regularization [30], entropy regularization [11] and EM with generative mixture models [22], see [5] for a review. Except self-training and co-training, most of these semi-supervised methods have scalability problems for large scale tasks.

Some other methods utilized auxiliary information from large unlabeled corpora for training sequence models (e.g., through multi-task learning). Ando and Zhang [2] proposed a method based on defining multiple tasks using unlabeled data that are multi-tasked with the task of interest, which they showed to perform very well on POS and NER tasks. Similarly, the language model strategy proposed in [6] is another type of auxiliary task. Both our local and global embedding methods belong to this semi-supervised category.

4.2 Semi-supervised String Kernel

For text categorization, the word sequence kernel proposed in [4] utilizes soft matching of words based on a certain similarity matrix used within the strin kernels. This similarity matrix could be derived from cooccurrence of words in unlabled text, i.e. adding semi-supervision to string kernel. Adding soft matching in the string kernel results qudratic complexity, though ASK does not add to complexity more than a linear cost to the input length (in practice we observed at most a factor of 1.5-2x slowdown compared to classic string kernels), while improving predictive performance significantly (Section "Results").

In terms of semi-supervised extensions of string kernels, another very simple method, called the "sequence neighborhood" kernel or "cluster" kernel has been employed [28] previously. This method replaces every example with a new representation obtained by averaging representations of the example's neighbors found in the unlabeled data using some standard sequence similarity measure. This kernel applies well in biological sequence analysis since relatively accurate measures exist (e.g., PSI-BLAST). Formally speaking, the *sequence neighborhood* kernels take advantage of the unlabeled data using the process of neighborhood induced regularization. But its application in most other domains (like text) is not straightforward since no accurate and standard measure of similarity exists.

4.3 Word Abstraction Based Models

Several previous works ([20]) tried to solve information extraction tasks with word clustering (abstraction). For example, Miller et al. [20] proposed to augment annotated training data with hierarchical word clusters that are automatically derived from a large unannotated corpus according to occurrence. Another group of closely related methods treat word clusters as hidden variables in their models. For instance, [12] proposed a conditional log-linear model, with hidden variables representing the assignment of atomic items to word clusters or word senses. The model learns to automatically make the cluster assignments based on a discriminative training criterion. Furthermore, researchers proposed to augment probabilistic models with abstractions in a hierarchical structure [26]. Our proposed ASK differs by building words similarity from two unsupervised models to capture auxiliary information implicit in large text corpus and employs VQ to build discrete word groups for string kernels.

5 Experimental Results

We now present experimental results for comparing ASk to classic string kernels and the state-of-art bRE results at multiple levels. Moreover to show generality, we extend ASK and apply it to a benchmark protein sequence classification dataset as the fourth experiment.

5.1 Three Benchmark bRE Data Sets

In our experiments, we explore three benchmark data sets related to PPI relation extractions. (1) The first one was provided from BioCreative II [13], a competition in 2006 for the extraction of protein-protein interaction (PPI) annotations from the literature. The competition evaluated multiple teams' submissions against a manually curated "gold standard" carried out by expert database annotators. Multiple subtasks were tested and we choose one specific task called "IAS" which aims to classify PubMed abstracts, based on whether they are relevant to protein interaction annotation or not. (2) The second data set is the "AIMED PPI sentence classification" data set. Extraction of relevant text segments (sentences) containing reference to important biomedical relationships is one of the first steps in annotation pipelines of biomedical database curation. Focusing on PPI, this step could be accomplished through classification of text fragments (sentences) as either relevant (i.e. containing PPI relation) or not relevant (non-PPI sentences). Sentences with PPI relations in the AIMED dataset [3] are treated as positive examples, while all other sentences (without PPI) are negative examples. In this data set, protein names are not annoated. (3) The third data set is called "AIMED PPI Relation Extraction", which uses a benchmark set aiming to extract binary protein-protein interaction (PPI) pairs from bio-literature sentences [3]. An example of such extraction is listed in Table 2. In this set, the sentences have been annotated with protein names if any. To ensure generalization of the learned extraction model, protein names are replaced with PROT1, PROT2 or PROT, where PROT1 and PROT2 are the pair of interests. The PPI relation extraction task is treated as a binary classification, where protein pairs that are stated to interact are positive examples and other co-occurring pairs negative. This means, for each sentence, $\binom{n}{2}$ relation examples are generated, with n as the number of protein names in the sentence. We downloaded this corpus from [9].

We use over 4.5M PubMed abstracts from 1994 to 2009 as our unlabeled corpus for learning word abstractions. The size of the training/test/unlabeled sets is given in Table 4.

Baselines. As each of these datasets has been used extensively, we will also compare our methods with the best reported results in the literature (see Table 5

Table 4. Size of datasets used in three "relation extraction" tasks

Dataset	Labeled	Unlabeled
BioCreativeII IAS Train	5495 (abstracts)1142559(tokens)	4.5M (abstracts)~1.3G (tokens)
BioCreativeII IAS Test	677 (abstracts)143420 (tokens)	
AIMED Relation	4026 (sentences) 143774 (tokens)	4.5M (abstracts)~1.3G (tokens)
AIMED Sentence	1730 (sentences)50675 (tokens)	4.5M (abstracts)~1.3G (tokens)

Table 5. Comparison with previous results and baselines on IAS task

Method	Precision	Recall	F1	ROC	Accuracy
Baseline 1: BioCreativeII compet. (best)	70.31	**87.57**	78.00	81.94	75.33
Baseline 2: BioCreativeII compet. (rank-2)	75.07	81.07	77.95	84.71	77.10
Baseline 3: TF-IDF	66.83	82.84	73.98	79.22	70.90
Spectrum (n-gram) kernel	69.29	80.77	74.59	81.49	72.53
Mismatch kernel	69.02	83.73	75.67	81.70	73.12
Gapped kernel	67.84	85.50	75.65	82.01	72.53
Global ASK	73.59	84.91	78.85	84.96	77.25
Local ASK	**76.06**	84.62	**80.11**	**85.67**	**79.03**

and 7). In the following, we also compare global and local ASK with various other baselines string kernels, including fully-supervised and semi-supervised approaches.

Method. We used the word n-grams as base features with ASK. Note we did not use any syntactic or linguistic features (e.g., no POS, chunk types, parse tree attributes, etc).

For global ASK, we use PubMed abstracts to learn word embedding vectors using a vocabulary of the top 40K most frequent words in PubMed. These word representations are clustered to obtain word abstractions (1K prototypes). Similarly, local ASK learns word embeddings on text windows (11 words, with 50-dim. embedding) extracted from the PubMed abstracts. Word embeddings are again clustered to obtain 1K abstraction entities. We set parameters of the string kernels to typical values, with spectrum n-gram using $k = 1$ *to* 5, the maximum number of mismatches is set to $m = 1$ and the maximum number of gaps uses up to $g = 6$).

Metric. The methods are evaluated using F1 score (including precision and recall) as well as ROC score. (1) For BioCreativeII IAS, evaluation is performed at the document level. (2) For two "AIMED" tasks, PPI extraction performance is measured at the sentence level for predicted/extracted interacting protein pairs using 10-fold cross-validation.

5.2 Task 1: PPI Extract at Article-Level: IAS

The lower part of Table 5 summarizes results for the IAS task from Global and Local ASK to baseline methods (spectrum n-gram kernel, n-gram kernel with

Table 6. AIMED PPI sentence classification task (F1 score). Both local ASK and global ASK improve over string kernel baselines.

Method	Baseline	+Global ASK	+Local ASK
Words	61.49	67.83	**69.46**
Words+Stems	65.94	67.99	**70.49**

mismatches, and gapped n-gram kernel using different base feature sets (words only, stems, characters)). Both Local and Global ASK provide improvements over baseline n-gram based string kernels.

Using word and character n-gram features, the best performance obtained with global ASK (F1 78.85), and the best performance by local ASK (F1 80.11) are superior to the best performance reported in the BioCreativeII competition (F1 78.00), as well as baseline bag-of-words with TF-IDF weighting (F1 73.98) and the best supervised string kernel result in the competition (F1 77.17). Observed improvements are significant, e.g., local ASK (F1 80.11) performs better than the best string kernel (F1 77.17), with p-values 5.8e-3 (calculating with standard z-test).

Note that all the top systems in the competition used more extensive feature sets than ours, including protein names, interaction keywords, part of speech tags and/or parse trees, etc. Thus, in summary, ASK effectively improves interaction article retrieval and achieves state-of-the-art performance with only plain words as features. We also note that using both local and global ASK together (multiple kernel) provides further improvements in performance compared to individual kernel results (e.g., we observe an increase in F1 score to 80.22).

5.3 Task 2: PPI Extraction Sentence Level: AIMED PPI Sentence

For the third benchmark task, "Classification of Protein Interaction Sentences", we summarize comparison results of both local and global ASK in Table 6.

The task here is to classify sentences as containing PPI relations or not. Both ASK models effectively improve over the traditional spectrum n-gram string kernels. For example, F1 70.49% from local ASK is significantly better than F1 65.94% from the best string kernel.

5.4 Task 3: PPI Extraction Relation-Level: AIMED

Table 7 summarizes the comparison results between ASK to baseline bag-of-words and supervised string kernel baselines. Both local and global ASK show effective improvements over the word n-gram based string kernels. We find that the observed improvements are statistically significant with $p < 0.05$ for the case with the best performance (F1 64.54) achieved by global ASK. One state-of-the-art relation-level bRE system (as far as we know) is listed as "baseline 2" in Table 7, which was tested on the same AIMED dataset as we used. Clearly our approach (with 64.54 F-score) performs better than this baseline (59.96 F-score) while using only basic words. Moreover, this baseline system utilized

Table 7. Comparison with previous results and baselines on AIMED relation-leve data

Method	Precision	Recall	F1	ROC	Accuracy
Baseline 1: Bag of words	41.39	62.46	49.75	74.58	70.22
Baseline 2: Transductive SVM [9]	59.59	60.68	59.96	-	-
Spectrum n-gram	58.35	62.77	60.42	83.06	80.57
Mismatch kernel	52.88	59.83	56.10	77.88	71.89
Gapped kernel	57.33	64.35	60.59	82.47	80.53
Global ASK	60.68	**69.08**	**64.54**	84.94	82.07
Local ASK	**61.18**	67.92	64.33	**85.27**	82.24

many complex, expensive techniques such as, dependency parsers, to achieve good performance.

Furthermore as pointed out by [1], though the AIMED corpus has been applied in numerous evaluations for PPI relation extraction, the datasets used in different papers varied largely due to diverse postprocessing rules used to create the relation-level examples. For instance, the corpus used to test our ASK in Table 7 was downloaded from [9] which contains 4026 examples with 951 as positive and 3075 as negatives. However, the AIMED corpus used in [1] includes more relation examples, i.e. 1000 positive relations and 4834 negative examples. The difference between the two reference sets make it impossible to compare our results in Table 7 to this state-of-the-art bRE system as claimed by [1] (with 56.4 F-score). Therefore we re-experiment ASK on this new AIMED relation corpus with both local and global ASK using the mismatch or spectrum kernel. Under the same (abstract-based) cross-validation splits from [1], our best performing case could achieve 54.7 F-score from local ASK on spectrum n-gram kernel with k from 1 to 5. We conclude that using only basic words ASK is comparable (slightly lower) to the bRE system from [1] where complex POS tree structures were used.

5.5 Task 4: Comparison on Biological Sequence Task

As mentioned in the introduction, the proposed ASK method is general to any sequence modeling problem, and good for cases with few labeled examples and a large unlabeled corpus. In the following, we extend ASK to biological domain and compare it with semi-supervised and supervised string kernels . The related work Section pointed out that the "Cluster kernel" is the only realistic semi-supervised competitor we know so far proposed for string kernels. However it needs a similarity measure specific to "protein sequences", which is not applicable to most sequence mining tasks. Three benchmark datasets evaluated above are all within the scope of text mining, where the cluster kernel is not applicable. In this experiment, we compare ASK with the cluster kernel and other string kernels in the biological domain on the problem of structural classification from protein sequences.

Measuring the degree of structural homology between protein sequences (also known as remote protein homology prediction) is a fundamental and difficult

Table 8. Mean ROC50 score on remote protein homology problem. Local ASK improves over string kernel baselines, both supervised and semi-supervised.

Method	Baseline	+Local ASK
Spectrum (n-gram)[18]	27.91	**33.06**
Mismatch [19]	41.92	**46.68**
Spatial sample kernel [15]	50.12	**52.75**
Semi-supervised Cluster kernel [28]	67.91	**70.14**

problem in biomedical research. For this problem, we use a popular benchmark dataset for structural homology prediction (SCOP) that corresponds to 54 remote homology detection experiments [28,17]. We test local ASK (with local embedding trained on a UNIPROT dataset, a collection of about 400,000 protein sequences) and compare with the supervised string kernels commonly used for the remote homology detection [19,28,15,17]. Each amino acid is treated as a word in this case. As shown in Table 8, local ASK effectively improves the performance of the traditional string kernels. For example, the mean ROC50 score (commonly used metric for this task) improves from 41.92 to 46.68 in the case of the mismatch kernel. One reason for this may be the use of the abstracted alphabet (rather than using standard amino-acid letters) which effectively captures similarity between otherwise symbolically different amino-acids. We also observe that adding ASK on the semi-supervised cluster kernel approach [28] improves over the standard mismatch string kernel-based cluster kernel. For example, for the cluster kernel computed on the unlabeled subset ($\sim$ 4000 protein sequences) of the SCOP dataset, the cluster kernel with ASK achieves mean ROC50 70.14 compared to ROC50 67.91 using the cluster kernel alone.

Furthermore the cluster kernel introduces new examples (sequences) and requires semi-supervision at testing time, while our unsupervised auxiliary tasks are feature learning methods, i.e. the learned features could be directly added to the existing feature set. From the experiments, it appears that the learned features from embedding models provide an orthogonal method for improving accuracy, e.g., these features could be combined with the cluster kernel to further improve its performance.

6 Conclusion

In this paper we propose to extract PPI relationships from sequences of biomedical text using a novel semi-supervised string kernel. The abstraction-augmented string kernel tries to improve supervised extractions with word abstractions learned from unlabeled data. Semi-supervision relies on two unsupervised auxiliary tasks that learn accurate word representations from contextual semantic similarity of words. On three bRE data sets, the proposed kernel matches state-of-the-art performance and improves over all string kernel baselines we tried without the need to get complex linguistic features. Moreover, we extend ASK to protein sequence analysis and on a classic benchmark dataset we found improved performance compared to all existing string kernels we tried.

Future work includes extension of ASK to more complex data types that have richer structures, such as graphs.

References

1. Airola, A., Pyysalo, S., Bjorne, J., Pahikkala, T., Ginter, F., Salakoski, T.: All-paths graph kernel for protein-protein interaction extraction with evaluation of cross-corpus learning. BMC Bioinformatics 9(S11), S2 (2008)
2. Ando, R.K., Zhang, T.: A framework for learning predictive structures from multiple tasks and unlabeled data. J. of Machine Learning Research 6, 1817–1853 (2005)
3. Bunescu, R., Mooney, R.: Subsequence kernels for relation extraction. In: Weiss, Y., Schölkopf, B., Platt, J. (eds.) NIPS 2006, pp. 171–178 (2006)
4. Cancedda, N., Gaussier, E., Goutte, C., Renders, J.M.: Word sequence kernels. J. Mach. Learn. Res. 3, 1059–1082 (2003)
5. Chapelle, O., Schölkopf, B., Zien, A. (eds.): Semi-Supervised Learning (Adaptive Computation and Machine Learning). MIT Press, Cambridge (2006)
6. Collobert, R., Weston, J.: A unified architecture for nlp: deep neural networks with multitask learning. In: ICML 2008, pp. 160–167 (2008)
7. Culotta, A., Sorensen, J.: Dependency tree kernels for relation extraction. In: ACL 2004, p. 423 (2004)
8. Deerwester, S., Dumais, S.T., Furnas, G.W., Landauer, T.K., Harshman, R.: Indexing by latent semantic analysis. JASIS 41(6), 391–407 (1990)
9. Erkan, G., Ozgur, A., Radev, D.R.: Semi-supervised classification for extracting protein interaction sentences using dependency parsing. In: EMNLP-CoNLL 2007, pp. 228–237 (2007)
10. Gersho, A., Gray, R.M.: Vector quantization and signal compression, Norwell, MA, USA (1991)
11. Grandvalet, Y., Bengio, Y.: Semi-supervised learning by entropy minimization. In: NIPS 2005, pp. 529–536 (2005)
12. Koo, T., Collins, M.: Hidden-variable models for discriminative reranking. In: HLT 2005, pp. 507–514 (2005)
13. Krallinger, M., Morgan, A., Smith, L., Hirschman, L., Valencia, A., et al.: Evaluation of text-mining systems for biology: overview of the second biocreative community challenge. Genome Biol. 9(S2), S1 (2008)
14. Krallinger, M., Valencia, A., Hirschman, L.: Linking genes to literature: text mining, information extraction, and retrieval applications for biology. Genome Biol. 9(S2), S8 (2008)
15. Kuksa, P., Huang, P.H., Pavlovic, V.: Fast protein homology and fold detection with sparse spatial sample kernels. In: ICPR 2008 (2008)
16. Kuksa, P., Huang, P.H., Pavlovic, V.: Scalable algorithms for string kernels with inexact matching. In: NIPS, pp. 881–888 (2008)
17. Leslie, C., Kuang, R.: Fast string kernels using inexact matching for protein sequences. J. Mach. Learn. Res. 5, 1435–1455 (2004)
18. Leslie, C., Eskin, E., Noble, W.S.: The spectrum kernel: A string kernel for SVM protein classification. In: PSB, pp. 566–575 (2002)
19. Leslie, C., Eskin, E., Weston, J., Noble, W.S.: Mismatch string kernels for SVM protein classification. In: NIPS, pp. 1417–1424 (2002)
20. Miller, S., Guinness, J., Zamanian, A.: Name tagging with word clusters and discriminative training. In: HLT-NAACL 2004, pp. 337–342 (2004)

21. Miwa, M., Sætre, R., Miyao, Y., Tsujii, J.: A rich feature vector for protein-protein interaction extraction from multiple corpora. In: EMNLP, pp. 121–130 (2009)
22. Nigam, K., McCallum, A.K., Thrun, S., Mitchell, T.: Text classification from labeled and unlabeled documents using EM. Mach. Learn. 39(2-3), 103–134 (2000)
23. Reichartz, F., Korte, H., Paass, G.: Dependency tree kernels for relation extraction from natural language text. In: ECML, pp. 270–285 (2009)
24. Rousu, J., Shawe-Taylor, J.: Efficient computation of gapped substring kernels on large alphabets. J. Mach. Learn. Res. 6, 1323–1344 (2005)
25. Salton, G., McGill, M.: Introduction to Modern Information Retrieval. McGraw-Hill Inc., New York (1986)
26. Segal, E., Koller, D., Ormoneit, D.: Probabilistic abstraction hierarchies. In: NIPS 2001 (2001)
27. Vishwanathan, S., Smola, A.: Fast kernels for string and tree matching, vol. 15, pp. 569–576. MIT Press, Cambridge (2002)
28. Weston, J., Leslie, C., Ie, E., Zhou, D., Elisseeff, A., Noble, W.S.: Semi-supervised protein classification using cluster kernels. Bioinformatics 21(15), 3241–3247 (2005)
29. Zhou, D., He, Y.: Extracting interactions between proteins from the literature. J. Biomed. Inform. 41(2), 393–407 (2008)
30. Zhu, X., Ghahramani, Z., Lafferty, J.: Semi-supervised learning using gaussian fields and harmonic functions. In: ICML 2003, pp. 912–919 (2003)

Online Knowledge-Based Support Vector Machines

Gautam Kunapuli[1], Kristin P. Bennett[2], Amina Shabbeer[2], Richard Maclin[3], and Jude Shavlik[1]

[1] University of Wisconsin-Madison
[2] Rensselaer Polytechnic Insitute
[3] University of Minnesota, Duluth

Abstract. Prior knowledge, in the form of simple advice rules, can greatly speed up convergence in learning algorithms. Online learning methods predict the label of the current point and then receive the correct label (and learn from that information). The goal of this work is to update the hypothesis taking into account not just the label feedback, but also the prior knowledge, in the form of soft polyhedral advice, so as to make increasingly accurate predictions on subsequent examples. Advice helps speed up and bias learning so that generalization can be obtained with less data. Our passive-aggressive approach updates the hypothesis using a hybrid loss that takes into account the margins of both the hypothesis and the advice on the current point. Encouraging computational results and loss bounds are provided.

1 Introduction

We propose a novel online learning method that incorporates advice into passive-aggressive algorithms, which we call the *Adviceptron*. Learning with advice and other forms of inductive transfer have been shown to improve machine learning by introducing bias and reducing the number of samples required. Prior work has shown that advice is an important and easy way to introduce domain knowledge into learning; this includes work on knowledge-based neural networks [15] and prior knowledge via kernels [12]. More specifically, for SVMs [16], knowledge can be incorporated in three ways [13]: by modifying the data, the kernel or the underlying optimization problem. While we focus on the last approach, we direct readers to a recent survey [9] on prior knowledge in SVMs.

Despite advances to date, research has not addressed how to incorporate advice into incremental SVM algorithms from either a theoretical or computational perspective. In this work, we leverage the strengths of Knowledge-Based Support Vector Machines (KBSVMs) [6] to effectively incorporate advice into the passive-aggressive framework introduced by Crammer et al., [4]. Our work explores the various difficulties and challenges in incorporating prior knowledge into online approaches and serves as a template to extending these techniques to other online algorithms. Consequently, we present an appealing framework

J.L. Balcázar et al. (Eds.): ECML PKDD 2010, Part II, LNAI 6322, pp. 145–161, 2010.

for generalizing KBSVM-type formulations to online algorithms with simple, closed-form weight-update formulas and known convergence properties.

We focus on the binary classification problem and demonstrate the incorporation of advice that leads to a new algorithm called the passive-aggressive *Adviceptron*. In the Adviceptron, as in KBSVMs, advice is specified for convex, polyhedral regions in the input space of data. As shown in Fung et al., [6], advice takes the form of (a set of) simple, possibly conjunctive, implicative rules. Advice can be specified about *every* potential data point in the input space which satisfies certain advice constraints, such as the rule

$$(\texttt{feature}_7 \geq 5) \wedge (\texttt{feature}_{12} \geq 4) \Rightarrow (\texttt{class} = +1),$$

which states that the class should be +1 when $\texttt{feature}_7$ is at least 5 *and* $\texttt{feature}_{12}$ is at least 4. Advice can be specified for individual features as above and for linear combinations of features, while the conjunction of multiple rules allows more complex advice sets. However, just as label information of data can be noisy, the advice specification can be noisy as well. The purpose of advice is twofold: first, it should help the learner reach a good solution with *fewer training data points*, and second, advice should help the learner reach a potentially *better solution* (in terms of generalization to future examples) than might have been possible learning from data alone.

We wish to study the generalization of KBSVMs to the online case within the well-known framework of passive-aggressive algorithms (PAAs, [4]). Given a loss function, the algorithm is *passive* whenever the loss is zero, i.e., the data point at the current round t is correctly classified. If misclassified, the algorithm updates the weight vector ($\mathbf{w}^t$) *aggressively*, such that the loss is minimized over the new weights ($\mathbf{w}^{t+1}$). The update rule that achieves this is derived as the optimal solution to a constrained optimization problem comprising two terms: a loss function, and a *proximal term* that requires $\mathbf{w}^{t+1}$ to be as close as possible to $\mathbf{w}^t$. There are several advantages of PAAs: first, they readily apply to standard SVM loss functions used for batch learning. Second, it is possible to derive closed-form solutions and consequently, simple update rules. Third, it is possible to formally derive relative loss bounds where the loss suffered by the algorithm is compared to the loss suffered by some arbitrary, fixed hypothesis.

We evaluate the performance of the Adviceptron on two real-world tasks: diabetes diagnosis, and *Mycobacterium tuberculosis* complex (MTBC) isolate classification into major genetic lineages based on DNA fingerprints. The latter task is an essential part of tuberculosis (TB) tracking, control, and research by health care organizations worldwide [7]. MTBC is the causative agent of tuberculosis, which remains one of the leading causes of disease and morbidity worldwide. Strains of MTBC have been shown to vary in their infectivity, transmission characteristics, immunogenicity, virulence, and host associations depending on their phylogeographic lineage [7]. MTBC biomarkers or DNA fingerprints are routinely collected as part of molecular epidemiological surveillance of TB.

Classification of strains of MTBC into genetic lineages can help implement suitable control measures. Currently, the United States Centers for Disease Control and Prevention (CDC) routinely collect DNA fingerprints for all culture positive TB patients in the United States. Dr. Lauren Cowan at the CDC has developed expert rules for classification which synthesize "visual rules" widely used in the tuberculosis research and control community [1,3]. These rules form the basis of the expert advice employed by us for the TB task. Our numerical experiments demonstrate the Adviceptron can speed up learning better solutions by exploiting this advice. In addition to experimental validation, we also derive regret bounds for the Adviceptron.

We introduce some notation before we begin. Scalars are denoted by lowercase letters (e.g., y, τ), all vectors by lowercase bold letters (e.g., $\mathbf{x}$, $\boldsymbol{\eta}$) and matrices by uppercase letters (e.g., D). Inner products between two vectors are denoted $\mathbf{x}'\mathbf{z}$. For a vector $\mathbf{p}$, the notation $\mathbf{p}_+$ denotes the componentwise plus-function, $\max(p_j, 0)$ and $\mathbf{p}_\star$ denotes the componentwise step function. The step function is defined for a scalar component p_j as $(p_j)_\star = 1$ if $p_j > 0$ and $(p_j)_\star = 0$ otherwise.

2 Knowledge-Based SVMs

We now review knowledge-based SVMs [6]. Like classical SVMs, they learn a linear classifier ($\mathbf{w}'\mathbf{x} = b$) given data $(\mathbf{x}^t, y_t)_{t=1}^T$ with $\mathbf{x}^t \in \mathbb{R}^n$ and labels $y_t \in \{\pm 1\}$. In addition, they are also given *prior knowledge* specified as follows: *all points* that satisfy constraints of the polyhedral set $D_1\mathbf{x} \leq \mathbf{d}^1$ belong to class $+1$. That is, the advice specifies that $\forall \mathbf{x},\ D_1\mathbf{x} \leq \mathbf{d}^1 \Rightarrow \mathbf{w}'\mathbf{x} - b \geq 0$. Advice can also be given about the other class using a second set of constraints: $\forall \mathbf{x},\ D_2\mathbf{x} \leq \mathbf{d}^2 \Rightarrow \mathbf{w}'\mathbf{x} - b \leq 0$. Combining both cases using *advice labels*, $z = \pm 1$, advice is given by specifying $(D, \mathbf{d}, z)$, which denotes the implication

$$D\mathbf{x} \leq \mathbf{d} \Rightarrow z(\mathbf{w}'\mathbf{x} - b) \geq 0. \tag{1}$$

We assume that m advice sets $(D_i, \mathbf{d}^i, z_i)_{i=1}^m$ are given in addition to the data, and if the i-th advice set has k_i constraints, we have $D_i \in \mathbb{R}^{k_i \times n}$, $\mathbf{d}^i \in \mathbb{R}^{k_i}$ and $z_i = \{\pm 1\}$. Figure 1 provides an example of a simple two-dimensional learning problem with both data and polyhedral advice.

Note that due to the implicative nature of the advice, it says *nothing* about points that do not satisfy $D\mathbf{x} \leq \mathbf{d}$. Also note that the notion of margin can easily be introduced by requiring that $D\mathbf{x} \leq \mathbf{d} \Rightarrow z(\mathbf{w}'\mathbf{x} - b) \geq \gamma$, i.e., that the advice sets (and all the points contained in them) be separated by a margin of γ analogous to the notion of margin for individual data points. Advice in implication form cannot be incorporated into an SVM directly; this is done by exploiting theorems of the alternative [11]. Observing that $p \Rightarrow q$ is equivalent to $\neg p \vee q$, we require that the latter be true; this is same as requiring that the negation $(p \wedge \neg q)$ be false or that the system of equations

$$\{D\mathbf{x} - \mathbf{d}\,\tau \leq 0,\ z\mathbf{w}'\mathbf{x} - zb\,\tau < 0,\ -\tau < 0\} \text{ has no solution } (\mathbf{x}, \tau). \tag{2}$$

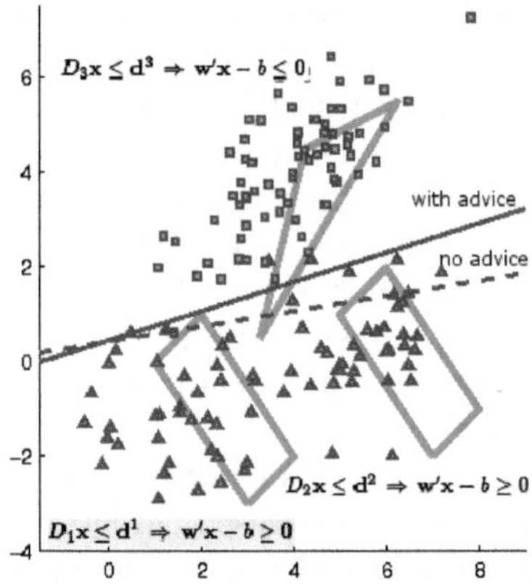

Fig. 1. Knowledge-based classifier which separates data and advice sets. If the advice sets were perfectly separable we have hard advice, (3). If subregions of advice sets are misclassified (analogous to subsets of training data being misclassified), we soften the advice as in (4). We revisit this data set in our experiments.

The variable τ is introduced to bring the system to nonhomogeneous form. Using the nonhomogeneous Farkas theorem of the alternative [11] it can be shown that (2) is equivalent to

$$\{D'\mathbf{u} + z\mathbf{w} = 0, \ -\mathbf{d}'\mathbf{u} - zb \geq 0, \ \mathbf{u} \geq 0\} \text{ has a solution } \mathbf{u}. \tag{3}$$

The set of (hard) constraints above incorporates the advice specified by *a single rule/advice set.* As there are m advice sets, each of the m rules is added as the equivalent set of constraints of the form (3). When these are incorporated into a standard SVM, the formulation becomes a hard-KBSVM; the formulation is hard because the advice is assumed to be linearly separable, that is, always feasible. Just as in the case of data, linear separability is a very limiting assumption and can be relaxed by introducing slack variables ($\boldsymbol{\eta}^i$ and ζ_i) to soften the constraints (3). If $\mathcal{P}$ and $\mathcal{L}$ are some convex regularization and loss functions respectively, the full soft-advice KBSVM is

$$\begin{array}{rl} \underset{(\boldsymbol{\xi},\mathbf{u}^i,\boldsymbol{\eta}^i,\zeta_i)\geq 0,\mathbf{w},b}{\text{minimize}} & \mathcal{P}(\mathbf{w}) + \lambda\,\mathcal{L}_{data}(\boldsymbol{\xi}) + \mu \sum_{i=1}^{m} \mathcal{L}_{advice}(\boldsymbol{\eta}^i, \zeta_i) \\ \text{subject to} & Y(X\mathbf{w} - b\mathbf{e}) + \boldsymbol{\xi} \geq \mathbf{e}, \\ & D_i'\mathbf{u}^i + z_i\mathbf{w} + \boldsymbol{\eta}^i = 0, \\ & -\mathbf{d}^{i\prime}\mathbf{u}^i - z_i b + \zeta_i \geq 1, \; i = 1, \ldots, m, \end{array} \tag{4}$$

where X is the $T \times n$ set of data points to be classified with labels $\mathbf{y} \in \{\pm 1\}^T$, $Y = diag(\mathbf{y})$ and $\mathbf{e}$ is a vector of ones of the appropriate dimension. The variables $\boldsymbol{\xi}$ are the standard slack variables that allow for soft-margin classification of the data. There are two regularization parameters λ, $\mu \geq 0$, which tradeoff the data and advice errors with the regularization.

While converting the advice from implication to constraints, we introduced new variables for each advice set: the *advice vectors* $\mathbf{u}^i \geq 0$. The advice vectors perform the same role as the dual multipliers α in the classical SVM. Recall that points with non-zero α's are the *support vectors* which additively contribute to $\mathbf{w}$. Here, for *each* advice set, the constraints of the set which have non-zero $\mathbf{u}^i$s are called *support constraints.*

3 Passive-Aggressive Algorithms with Advice

We are interested in an online version of (4) where the algorithm is given T labeled points $(\mathbf{x}^t, y_t)_{t=1}^T$ sequentially and required to update the model hypothesis, $\mathbf{w}^t$, as well as the advice vectors, $\mathbf{u}^{i,t}$, at every iteration. The batch formulation (4) can be extended to an online passive-aggressive formulation by introducing proximal terms for the advice variables, $\mathbf{u}^i$:

$$\begin{aligned}
\underset{\xi,\mathbf{u}^i,\boldsymbol{\eta}^i,\zeta_i,\mathbf{w}}{\arg\min}\ & \frac{1}{2}\|\mathbf{w}-\mathbf{w}^t\|^2 + \frac{1}{2}\sum_{i=1}^m \|\mathbf{u}^i - \mathbf{u}^{i,t}\|^2 + \frac{\lambda}{2}\xi^2 + \frac{\mu}{2}\sum_{i=1}^m \left(\|\boldsymbol{\eta}^i\|^2 + \zeta_i^2\right) \\
\text{subject to } & y_t \mathbf{w}'\mathbf{x}^t - 1 + \xi \geq 0, \\
& \left.\begin{aligned} D_i'\mathbf{u}^i + z_i\mathbf{w} + \boldsymbol{\eta}^i &= 0 \\ -\mathbf{d}^{i\prime}\mathbf{u}^i - 1 + \zeta_i &\geq 0 \\ \mathbf{u}^i &\geq 0 \end{aligned}\right\}\ i = 1,\ldots,m.
\end{aligned} \tag{5}$$

Notice that while L_1 regularization and losses were used in the batch version [6], we use the corresponding L_2 counterparts in (5). This allows us to derive passive-aggressive closed-form solutions. We address this illustrative and effective special case, and leave the general case of dynamic online learning of advice and weight vectors for general losses as future work.

Directly deriving the closed-form solutions for (5) is impossible owing to the fact that satisfying the many inequality constraints at optimality is a combinatorial problem which can only be solved iteratively. To circumvent this, we adopt a two-step strategy when the algorithm receives a new data point $(\mathbf{x}^t, y_t)$: first, fix the advice vectors $\mathbf{u}^{i,t}$ in (5) and use these to update the weight vector $\mathbf{w}^{t+1}$, and second, fix the newly updated weight vector in (5) to update the advice vectors and obtain $\mathbf{u}^{i,t+1}$, $i = 1, \ldots, m$. While many decompositions of this problem are possible, the one considered above is arguably the most intuitive and leads to an interpretable solution and also has good regret minimizing properties. In the following subsections, we derive each step of this approach and in the section following, analyze the regret behavior of this algorithm.

3.1 Updating $\mathbf{w}$ Using Fixed Advice Vectors $\mathbf{u}^{i,t}$

At step t $(= 1, \ldots, T)$, the algorithm receives a new data point $(\mathbf{x}^t, y_t)$. The hypothesis from the previous step is $\mathbf{w}^t$, with corresponding advice vectors $\mathbf{u}^{i,t}$, $i = 1, \ldots, m$, one for each of the m advice sets. In order to update $\mathbf{w}^t$ based on the advice, we can simplify the formulation (5) by fixing the advice variables $\mathbf{u}^i = \mathbf{u}^{i,t}$. This gives a fixed-advice online passive-aggressive step, where the variables ζ_i drop out of the formulation (5), as do the constraints that involve those variables. We can now solve the following problem (the corresponding Lagrange multipliers for each constraint are indicated in parentheses):

$$\begin{aligned}\mathbf{w}^{t+1} &= \underset{\mathbf{w},\xi,\boldsymbol{\eta}^i}{\text{minimize}} \ \frac{1}{2}\|\mathbf{w}-\mathbf{w}^t\|_2^2 + \frac{\lambda}{2}\xi^2 + \frac{\mu}{2}\sum_{i=1}^m \|\boldsymbol{\eta}^i\|_2^2 \\ &\quad \text{subject to } y_t\mathbf{w}'\mathbf{x}^t - 1 + \xi \geq 0, \qquad (\alpha) \\ &\qquad\qquad D_i'\mathbf{u}^i + z_i\mathbf{w} + \boldsymbol{\eta}^i = 0, \ i = 1,\ldots,m. \ (\boldsymbol{\beta}^i)\end{aligned} \tag{6}$$

In (6), $D_i'\mathbf{u}^i$ is the classification hypothesis according to the i-th knowledge set. Multiplying $D_i'\mathbf{u}^i$ by the label z_i, the labeled i-th hypothesis is denoted $\mathbf{r}^i = -z_i D_i'\mathbf{u}^i$. We refer to the $\mathbf{r}^i$s as the *advice-estimates of the hypothesis* because they represent each advice set as a point in hypothesis space. We will see later that the next step when we update the advice using the fixed hypothesis can be viewed as representing the hypothesis-estimate of the advice as a point in that advice set. The effect of the advice on $\mathbf{w}$ is clearly through the equality constraints of (6) which force $\mathbf{w}$ at each round to be as close to *each of* the advice-estimates as possible by aggressively minimizing the error, $\boldsymbol{\eta}^i$. Moreover, Theorem 1 proves that the optimal solution to (6) can be computed in closed-form and that this solution requires only the centroid of the advice estimates, $\mathbf{r} = (1/m)\sum_{i=1}^m \mathbf{r}^i$. For fixed advice, the centroid or *average advice* vector $\mathbf{r}$, provides a compact and sufficient summary of the advice.

Update Rule 1 (Computing $\mathbf{w}^{t+1}$ from $\mathbf{u}^{i,t}$). *For λ, $\mu > 0$, and given advice vectors $\mathbf{u}^{i,t} \geq 0$, let $\mathbf{r}^t = 1/m\sum_{i=1}^m \mathbf{r}^{i,t} = -1/m\sum_{i=1}^m z_i D_i'\mathbf{u}^{i,t}$, with $\nu = 1/(1+m\mu)$. Then, the optimal solution of (6) which also gives the closed-form update rule is given by*

$$\mathbf{w}^{t+1} = \mathbf{w}^t + \alpha_t y_t \mathbf{x}^t + \sum_{i=1}^m z_i\boldsymbol{\beta}^{i,t} = \nu\left(\mathbf{w}^t + \alpha_t y_t \mathbf{x}^t\right) + (1-\nu)\,\mathbf{r}^t,$$

$$\alpha_t = \frac{\left(1 - \nu\, y_t {\mathbf{w}^t}'\mathbf{x}^t - (1-\nu)\, y_t {\mathbf{r}^t}'\mathbf{x}^t\right)_+}{\frac{1}{\lambda} + \nu\|\mathbf{x}^t\|^2}, \quad \frac{z_i\boldsymbol{\beta}^{i,t}}{\mu} = \mathbf{r}^{i,t} - \frac{\mathbf{w}^t + \alpha_{t\star}\,\lambda\, y_t\mathbf{x}^t + m\mu\,\mathbf{r}^t}{\frac{1}{\nu} + \alpha_{t\star}\,\lambda\|\mathbf{x}^t\|^2}. \tag{7}$$

The numerator of α_t is the *combined loss function,*

$$\ell_t = \max\left(1 - \nu\, y_t {\mathbf{w}^t}'\mathbf{x}^t - (1-\nu)\, y_t {\mathbf{r}^t}'\mathbf{x}^t,\ 0\right), \tag{8}$$

which gives us the condition upon which the update is implemented. This is exactly the hinge loss function where the margin is computed by a convex combination of the current hypothesis $\mathbf{w}^t$ and the current advice-estimate of the hypothesis $\mathbf{r}^t$. Note that for any choice of $\mu > 0$, the value of $\nu \in (0, 1]$ with $\nu \to 0$ as $\mu \to \infty$. Thus, ℓ_t is simply the hinge loss function applied to a *convex combination* of the margin of the hypothesis, $\mathbf{w}^t$ from the *current iteration* and the margin of the *average advice-estimate*, $\mathbf{r}^t$. Furthermore, if there is no advice, $m = 0$ and $\nu = 1$, and the updates above become exactly identical to online passive-aggressive algorithms for support vector classification [4]. Also, it is possible to eliminate the variables $\boldsymbol{\beta}^i$ from the expressions (7) to give a very simple update rule that depends only on α_t and $\mathbf{r}^t$:

$$\mathbf{w}^{t+1} = \nu(\mathbf{w}^t + \alpha_t y_t \mathbf{x}^t) + (1 - \nu)\mathbf{r}^t. \tag{9}$$

This update rule is a convex combination of the current iterate updated by the data, $\mathbf{x}^t$ and the advice, $\mathbf{r}^t$.

3.2 Updating $\mathbf{u}^{i,t}$ Using the Fixed Hypothesis $\mathbf{w}^{t+1}$

When $\mathbf{w}$ is fixed to $\mathbf{w}^{t+1}$, the master problem breaks up into m smaller subproblems, the solution of each one yielding updates to each of the $\mathbf{u}^i$ for the i-th advice set. The i-th subproblem (with the corresponding Lagrange multipliers) is shown below:

$$\begin{aligned} \mathbf{u}^{i,t+1} = \arg\min_{\mathbf{u}^i, \boldsymbol{\eta}, \zeta} \; & \frac{1}{2}\|\mathbf{u}^i - \mathbf{u}^{i,t}\|^2 + \frac{\mu}{2}\left(\|\boldsymbol{\eta}^i\|_2^2 + \zeta_i^2\right) & \\ \text{subject to } & D_i'\mathbf{u}^i + z_i\mathbf{w}^t + \boldsymbol{\eta}^i = 0, & (\boldsymbol{\beta}^i) \\ & -\mathbf{d}^{i\prime}\mathbf{u}^i - 1 + \zeta_i \geq 0, & (\gamma_i) \\ & \mathbf{u}^i \geq 0. & (\boldsymbol{\tau}^i) \end{aligned} \tag{10}$$

The first-order gradient conditions can be obtained from the Lagrangian:

$$\mathbf{u}^i = \mathbf{u}^{i,t} + D_i\boldsymbol{\beta}^i - \mathbf{d}^i\gamma_i + \boldsymbol{\tau}^i, \quad \boldsymbol{\eta}^i = \frac{\boldsymbol{\beta}^i}{\mu}, \quad \zeta_i = \frac{\gamma_i}{\mu}. \tag{11}$$

The complicating constraints in the above formulation are the cone constraints $\mathbf{u}^i \geq 0$. If these constraints are dropped, it is possible to derive a closed-form intermediate solution, $\tilde{\mathbf{u}}^i \in \mathbb{R}^{k_i}$. Then, observing that $\boldsymbol{\tau}^i \geq 0$, we can compute the final update by projecting the intermediate solution onto $\mathbf{u}^i \geq 0$.

$$\mathbf{u}^{i,t+1} = \left(\mathbf{u}^{i,t} + D_i\boldsymbol{\beta}^i - \mathbf{d}^i\gamma_i\right)_+. \tag{12}$$

When the constraints $\mathbf{u}^i \geq 0$ are dropped from (10), the resulting problem can be solved (analogous to the derivation of the update step for $\mathbf{w}^{t+1}$) to give a closed-form solution which depends on the dual variables: $\tilde{\mathbf{u}}^{i,t+1} = \mathbf{u}^{i,t} + D_i\boldsymbol{\beta}^i - \mathbf{d}^i\zeta_i$. This solution is then projected into the positive orthant by applying the plus function: $\mathbf{u}^{i,t+1} = \tilde{\mathbf{u}}_+^{i,t+1}$. This leads to the advice updates, which need to applied to each advice vector $\mathbf{u}^{i,t}$, $i = 1, \ldots, m$ individually.

Update Rule 2 (Computing $\mathbf{u}^{i,t+1}$ from $\mathbf{w}^{t+1}$). *For $\mu > 0$, and given the current hypothesis $\mathbf{w}^{t+1}$, for each advice set, $i = 1, \ldots, m$, the update rule is given by*

$$\mathbf{u}^{i,t+1} = \left(\mathbf{u}^{i,t} + D_i\boldsymbol{\beta}^i - \mathbf{d}^i\gamma_i\right)_+, \quad \begin{bmatrix} \boldsymbol{\beta}^i \\ \gamma_i \end{bmatrix} = H_i^{-1}\mathbf{g}^i,$$

Algorithm 1. The Passive-Aggressive Adviceptron Algorithm

input: data $(\mathbf{x}^t, y_t)_{t=1}^T$, advice sets $(D_i, \mathbf{d}^i, z_i)_{i=1}^m$, parameters λ, $\mu > 0$
initialize: $\mathbf{u}^{i,1} = 0$, $\mathbf{w}^1 = 0$
let $\nu = 1/(1+m\mu)$
for $(\mathbf{x}^t, y_t)$ **do**
    predict label $\hat{y}_t = \mathrm{sign}(\mathbf{w}^{t\prime}\mathbf{x}^t)$
    receive correct label y_t
    suffer loss $\ell_t = 1 - \nu y_t \mathbf{w}^{t\prime}\mathbf{x}^t - (1-\nu) y_t \mathbf{r}^{t\prime}\mathbf{x}^t$ where $\mathbf{r}^t = -\frac{1}{m}\sum_{i=1}^m z_i D_i' \mathbf{u}^{i,t}$
    update hypothesis using $\mathbf{u}^{i,t}$, as defined in Update 1

$$\alpha = \ell_t / (\frac{1}{\lambda} + \nu\|\mathbf{x}^t\|^2), \quad \mathbf{w}^{t+1} = \nu\,(\,\mathbf{w}^t + \alpha\, y_t \mathbf{x}^t\,) + (1-\nu)\,\mathbf{r}^t$$

    update advice using $\mathbf{w}^{t+1}$, $(H_i, \mathbf{g}^i)$ as defined in Update 2

$$(\boldsymbol{\beta}^i, \gamma_i) = H_i^{-1}\mathbf{g}^i, \quad \mathbf{u}^{i,t+1} = \left(\mathbf{u}^{i,t} + D_i\boldsymbol{\beta}^i - \mathbf{d}^i\gamma_i\right)_+$$

end for

$$H_{i,t} = \begin{bmatrix} -(D_i'D_i + \frac{1}{\mu}I_n) & D_i'\mathbf{d}^i \\ \mathbf{d}^{i\prime}D_i & -(\mathbf{d}^{i\prime}\mathbf{d}^i + \frac{1}{\mu}) \end{bmatrix}, \quad \mathbf{g}^{i,t} = \begin{bmatrix} D_i'\mathbf{u}^{i,t} + z_i\mathbf{w}^t \\ -\mathbf{d}^{i\prime}\mathbf{u}^{i,t} - 1 \end{bmatrix}, \tag{13}$$

with the untruncated solution being the optimal solution to (10) without the cone constraints $\mathbf{u}^i \geq 0$.

Recall that, when updating the hypothesis $\mathbf{w}^t$ using new data points $\mathbf{x}^t$ and the fixed advice (i.e., $\mathbf{u}^{i,t}$ is fixed), each advice set contributes an estimate of the hypothesis ($\mathbf{r}^{ti} = -z_i D_i'\mathbf{u}^{i,t}$) to the update. We termed the latter the *advice-estimate* of the hypothesis. Here, given that when there is an update, $\beta \neq 0$, $\gamma_i > 0$, we denote $\mathbf{s}^i = \boldsymbol{\beta}^i/\gamma_i$ as the *hypothesis-estimate of the advice.* Since $\boldsymbol{\beta}^i$ and γ_i depend on $\mathbf{w}^t$, we can reinterpret the update rule (12) as

$$\mathbf{u}^{i,t+1} = \left(\mathbf{u}^{i,t} + \gamma_i(D_i\mathbf{s}^i - \mathbf{d}^i)\right)_+ . \tag{14}$$

Thus, the advice variables are refined using the hypothesis-estimate of that advice set according to the current $\mathbf{w}^t$; here the update is the *error or the amount of violation of the constraint* $D_i\mathbf{x} \leq \mathbf{d}^i$ *by an ideal data point,* $\mathbf{s}^i$ *estimated by the current hypothesis,* $\mathbf{w}^t$. Note that the error is scaled by a factor γ_i.

Now, update Rules 1 and 2 can be combined together to yield the full passive-aggressive *Adviceptron* (Algorithm 1).

4 Analysis

In this section, we analyze the behavior of the passive-aggressive adviceptron by studying its regret behavior and loss-minimizing properties. Returning to (4) for a moment, we note that there are three loss functions in the objective, each one penalizing a slack variable in each of the three constraints. We formalize the definition of the three loss functions here. The loss function $L_\xi(\mathbf{w};\, \mathbf{x}^t,\, y_t)$ measures the error of the labeled data point $(\mathbf{x}^t,\, y_t)$ from the hyperplane $\mathbf{w}$; $L_\eta(\mathbf{w}^t, \mathbf{u}^i;\, D_i,\, z_i)$ and $L_\zeta(\mathbf{u}^i;\, \mathbf{d}^i,\, z_i)$ cumulatively measure how well $\mathbf{w}$ satisfies the advice constraints $(D_i,\, \mathbf{d}^i,\, z_i)$. In deriving Updates 1 and 2, we used the following loss functions:

$$L_\xi(\mathbf{w};\, \mathbf{x}^t,\, y_t) = (1 - y_t \mathbf{w}'\mathbf{x}^t)_+, \tag{15}$$

$$L_\eta(\mathbf{w}, \mathbf{u};\, D_i,\, z_i) = \|D_i'\mathbf{u} + z_i \mathbf{w}\|^2, \tag{16}$$

$$L_\zeta(\mathbf{u};\, \mathbf{d}^i,\, z_i) = (1 + \mathbf{d}^{i\prime}\mathbf{u})_+. \tag{17}$$

Also, in the context of (4), $\mathcal{L}_{data} = \frac{1}{2}L_\xi^2$ and $\mathcal{L}_{advice} = \frac{1}{2}(L_\eta + L_\zeta^2)$. Note that in the definitions of the loss functions, the arguments after the semi-colon are the data and advice, which are fixed.

Lemma 1. *At round t, if we define the updated advice vector* before projection *for the i-th advice set as* $\tilde{\mathbf{u}}^i$, *the following hold for all* $\mathbf{w} \in \mathbb{R}^n$*:*

1. $\tilde{\mathbf{u}}^i = \mathbf{u}^{i,t} - \mu \nabla_{\mathbf{u}^i} \mathcal{L}_{advice}(\mathbf{u}^{i,t})$,
2. $\|\nabla_{\mathbf{u}^i} \mathcal{L}_{advice}(\mathbf{u}^{i,t})\|^2 \leq \left(\|D_i\|^2 L_\eta(\mathbf{u}^{i,t}, \mathbf{w}) \;+\; \|\mathbf{d}^i\|^2 L_\zeta^2(\mathbf{u}^{i,t})\right).$

The first inequality above can be derived from the definition of the loss functions and the first-order conditions (11). The second inequality follows from the first condition using convexity: $\|\nabla_{\mathbf{u}^i} \mathcal{L}_{advice}(\mathbf{u}^{i,t})\|^2 = \|D_i \boldsymbol{\eta}^i - \mathbf{d}^i \gamma_i\|^2 = \|D_i(D_i'\mathbf{u}^{i,t} + z_i \mathbf{w}^{t+1}) + \mathbf{d}^i(\mathbf{d}^{i\prime}\mathbf{u}^{i,t} + 1)\|^2 \leq \|D_i(D_i'\mathbf{u}^{i,t} + z_i \mathbf{w}^{t+1})\|^2 + \|\mathbf{d}^i(\mathbf{d}^{i\prime}\mathbf{u}^{i,t} + 1)\|^2$. The inequality follows by applying $\|A\mathbf{x}\| \leq \|A\|\|\mathbf{x}\|$. We now state additional lemmas that can be used to derive the final regret bound. The proofs are in the appendix.

Lemma 2. *Consider the rules given in Update 1, with* $\mathbf{w}^1 = 0$ *and* $\lambda,\, \mu > 0$. *For all* $\mathbf{w}^* \in \mathbb{R}^n$ *we have*

$$\|\mathbf{w}^* - \mathbf{w}^{t+1}\|^2 - \|\mathbf{w}^* - \mathbf{w}^t\|^2 \leq$$
$$\nu\lambda L_\xi(\mathbf{w}^*)^2 - \frac{\nu\lambda}{1 + \nu\lambda X^2} L_\xi(\widehat{\mathbf{w}}^t)^2 + (1-\nu)\|\mathbf{w}^* - \mathbf{r}^t\|^2.$$

where $\widehat{\mathbf{w}}^t = \nu \mathbf{w}^t + (1-\nu)\mathbf{r}^t$, *the combined hypothesis that determines if there is an update,* $\nu = 1/(1 + m\mu)$, *and we assume that* $\|\mathbf{x}^t\|^2 \leq X^2, \forall t = 1, \ldots, T$.

Lemma 3. *Consider the rules given in Update 2, for the i-th advice set with* $\mathbf{u}^{i,1} = 0$, *and* $\mu > 0$. *For all* $\mathbf{u}^* \in \mathbb{R}_+^{k_i}$, *we have*

$$\begin{aligned}\|\mathbf{u}^* - \mathbf{u}^{i,t+1}\|^2 - \|\mathbf{u}^* - \mathbf{u}^{i,t}\|^2 \;&\leq\; \mu L_\eta(\mathbf{u}^*, \mathbf{w}^t) + \mu L_\zeta(\mathbf{u}^*)^2 \\ &-\mu\left((1 - \mu\Delta^2) L_\eta(\mathbf{u}^{i,t}, \mathbf{w}^t) + (1 - \mu\delta^2) L_\zeta(\mathbf{u}^{i,t})^2\right).\end{aligned}$$

where we assume that $\|D_i\|^2 \leq \Delta^2$ *and* $\|\mathbf{d}^i\|^2 \leq \delta^2$.

Lemma 4. *At round t, given the current hypothesis and advice vectors $\mathbf{w}^t$ and $\mathbf{u}^{i,t}$, for any $\mathbf{w}^* \in \mathbb{R}^n$ and $\mathbf{u}^{i,*} \in \mathbb{R}^{k_i}_+$, $i = 1, \ldots, m$, we have*

$$\|\mathbf{w}^* - \mathbf{r}^t\|^2 \leq \frac{1}{m}\sum_{i=1}^{m} L_\eta(\mathbf{w}^*, \mathbf{u}^{i,t}) = \frac{1}{m}\sum_{i=1}^{m} \|\mathbf{w}^* - \mathbf{r}^{i,t}\|^2$$

The overall loss suffered over one round $t = 1, \ldots, T$ is defined as follows:

$$R(\mathbf{w}, \mathbf{u}; c_1, c_2, c_3) = \left(c_1 L_\xi(\mathbf{w})^2 + \sum_{i=1}^{m} (c_2 L_\eta(\mathbf{w}, \mathbf{u}) + c_3 L_\zeta(\mathbf{u})^2) \right).$$

This is identical to the loss functions defined in the batch version of KBSVMs (4) and its online counterpart (10). The Adviceptron was derived such that it minimizes the latter. The lemmas are used to prove the following regret bound for the Adviceptron[1].

Theorem 1. *Let $\mathfrak{S} = \{(\mathbf{x}^t, y_t)\}_{t=1}^T$ be a sequence of examples with $(\mathbf{x}^t, y_t) \in \mathbb{R}^n \times \{\pm 1\}$, and $\|\mathbf{x}^t\|_2 \leq X \quad \forall t$. Let $\mathfrak{A} = \{(D_i, \mathbf{d}^i, z_i)\}_{i=1}^m$ be m advice sets with $\|D_i\|_2 \leq \Delta$ and $\|\mathbf{d}^i\|_2 \leq \delta$. Then the following holds for all $\mathbf{w}^* \in \mathbb{R}^n$ and $\mathbf{u}^i \in \mathbb{R}^{k_i}_+$:*

$$\begin{aligned}
&\frac{1}{T}\sum_{t=1}^{T} R\left(\mathbf{w}^t, \mathbf{u}^t; \frac{\lambda}{1+\nu\lambda X^2},\ \mu(1-\mu\Delta^2),\ \mu(1-\mu\delta^2)\right) \\
&\leq \frac{1}{T}\sum_{t=1}^{T} R(\mathbf{w}^*, \mathbf{u}^*;\ \lambda, 0, \mu) + R(\mathbf{w}^*, \mathbf{u}^t;\ 0, \mu, 0) + R(\mathbf{w}^{t+1}, \mathbf{u}^*;\ 0, \mu, 0) \qquad (18) \\
&\qquad + \frac{1}{\nu T}\|\mathbf{w}^*\|^2 + \frac{1}{T}\sum_{i=1}^{M} \|\mathbf{u}^{i,*}\|^2.
\end{aligned}$$

If the last two R terms in the right hand side are bounded by $2R(\mathbf{w}^*, \mathbf{u}^*; 0, \mu, 0)$, then the regret behavior becomes similar to truncated-gradient algorithms [8].

5 Experiments

We performed experiments on three data sets: one artificial (see Figure 1) and two real world. Our real world data sets are Pima Indians Diabetes data set from the UCI repository [2] and *M. tuberculosis* spoligotype data set (both are described below). We also created a synthetic data set where one class of the data corresponded to a mixture of two small σ Gaussians and the other (overlapping) class was represented by a flatter (large σ) Gaussian. For this set, the learner is provided with three hand-made advice sets (see Figure 1).

[1] The complete derivation can be found at `http://ftp.cs.wisc.edu/machine-learning/shavlik-group/kunapuli.ecml10.proof.pdf`

Table 1. The number of isolates for each MTBC class and the number of positive and negative pieces of advice for each classification task. Each task consisted of 50 training examples drawn randomly from the isolates with the rest becoming test examples.

Class	#isolates	#pieces of Positive Advice	#pieces of Negative Advice
East-Asian	4924	1	1
East-African-Indian	1469	2	4
Euro-American	25161	1	2
Indo-Oceanic	5309	5	5
M. africanum	154	1	3
M. bovis	693	1	3

5.1 `Diabetes` Data Set

The `diabetes` data set consists of 768 points with 8 attributes. For domain advice, we constructed two rules based on statements from the NIH web site on risks for Type-2 Diabetes[2]. A person who is obese, characterized by high body mass index ($\texttt{BMI} \geq 30$) and high `bloodglucose` level (≥ 126) is at strong risk for diabetes, while a person who is at normal weight ($\texttt{BMI} \leq 25$) and low `bloodglucose` level (≤ 100) is unlikely to have diabetes. As `BMI` and `bloodglucose` are features of the data set, we can give advice by combining these conditions into conjunctive rules, one for each class. For instance, the rule predicting that diabetes is false is $(\texttt{BMI} \leq 25) \wedge (\texttt{bloodglucose} \leq 100) \Rightarrow \neg\texttt{diabetes}$.

5.2 `Tuberculosis` Data Set

These data sets consist of two types of DNA fingerprints of *M. tuberculosis* complex (MTBC): the spacer oglionucleotide types (spoligotypes) and Mycobacterial Interspersed Repetitive Units (MIRU) types of 37942 clinical isolates collected by the US Centers for Disease Control and Prevention (CDC) during 2004–2008 as part of routine TB surveillance and control. The spoligotype captures the variability in the direct repeat (DR) region of the genome of a strain of MTBC and is represented by a 43 bit long binary string constructed on the basis of presence or absence of spacers (non-repeating sequences interspersed between short direct repeats) in the DR. In addition, the number of repeats present at the 24th locus of the MIRU type (`MIRU24`) is used as an attribute.

Six major lineages of strains of the MTBC have been previously identified: the "modern" lineages: Euro-American, East-Asian and East-African-Indian and the "ancestral" lineages: *M. bovis*, *M. africanum* and Indo-Oceanic. Prior studies report high classification accuracy of the major genetic lineages using Bayesian Networks on spoligotypes and up to 24 loci of MIRU [1] on this dataset. Expert-defined rules for the classification of MTBC strains into these lineages have been previously documented [3,14]. The rules are based on observed patterns in the presence or absence of spacers in the spoligotypes, and in the number of tandem

[2] `http://diabetes.niddk.nih.gov/DM/pubs/riskfortype2`

repeats at MIRU of a single MIRU locus – `MIRU24`, associated with each lineage. The `MIRU24` locus is known to distinguish ancestral versus modern lineages with high accuracy for most isolates with a few exceptions.

The six TB classification tasks are to distinguish each lineage from the rest. The advice consists of positive advice to identify each lineage, as well as negative advice that rules out specific lineages. We found that incorporation of negative advice for some classes like *M. africanum* significantly improved performance. The number of isolates for each class and the number of positive and negative pieces of advice for each classification task are given in Table 1. Examples of advice are provided below[3].

```
Spacers(1-34) absent ⇒ East-Asian
At least one of Spacers(1-34) present ⇒ ¬East-Asian
Spacers(4-7, 23-24, 29-32) absent ∧ MIRU24≤1 ⇒ East-African-Indian
Spacers(4-7, 23-24) absent ∧ MIRU24≤1 ∧ at least one spacer of (29-32)
present ∧ at least one spacer of (33-36) present⇒ East-African-Indian
Spacers(3, 9, 16, 39-43) absent ∧ spacer 38 present ⇒ M. bovis
Spacers(8, 9, 39) absent ∧ MIRU24>1 ⇒ M. africanum
Spacers(3, 9, 16, 39-43) absent ∧ spacer 38 present ⇒ ¬ M. africanum
```

For each lineage, both negative and positive advice can be naturally expressed. For example, the positive advice for *M. africanum* closely corresponds to a known rule: `if spacers(8, 9, 13) are absent` $\wedge$ `MIRU24` $\leq 1 \Rightarrow$ *M. africanum*. However, this rule is overly broad and is further refined by exploiting the fact that *M. africanum* is an ancestral strain. Thus, the following rules out all modern strains: `if MIRU24` $\leq 1 \Rightarrow \neg$ *M. africanum*. The negative advice captures the fact that spoligotypes do not regain spacers once lost. For example, `if at least one of Spacers(8, 9, 39) is present` $\Rightarrow \neg$ *M. africanum*. The final negative rule rules out *M. bovis*, a close ancestral strain easily confused with *M. africanum*.

5.3 Methodology

The results for each data set are averaged over multiple randomized iterations (20 iterations for synthetic and `diabetes`, and 200 for the `tuberculosis` tasks). For each iteration of the synthetic and `diabetes` data sets, we selected 200 points at random as the training set and used the rest as the test set. For each iteration of the `tuberculosis` data sets, we selected 50 examples at random from the data set to use as a training set and tested on the rest. Each time, the training data was presented in a random order, one example at a time, to the learner to generate the learning curves shown in Figures 2(a)–2(h). We compare the results to well-studied incremental algorithms: standard passive-aggressive algorithms [4], margin-perceptron [5] and ROMMA [10]. We also compare it to the standard batch KBSVM [6], where the learner was given all of the examples used in training the online learners (e.g., for the synthetic data we had 200 data points to create the learning curve, so the KBSVM used those 200 points).

[3] The full rules can be found in `http://ftp.cs.wisc.edu/machine-learning/shavlik-group/kunapuli.ecml10.rules.pdf`

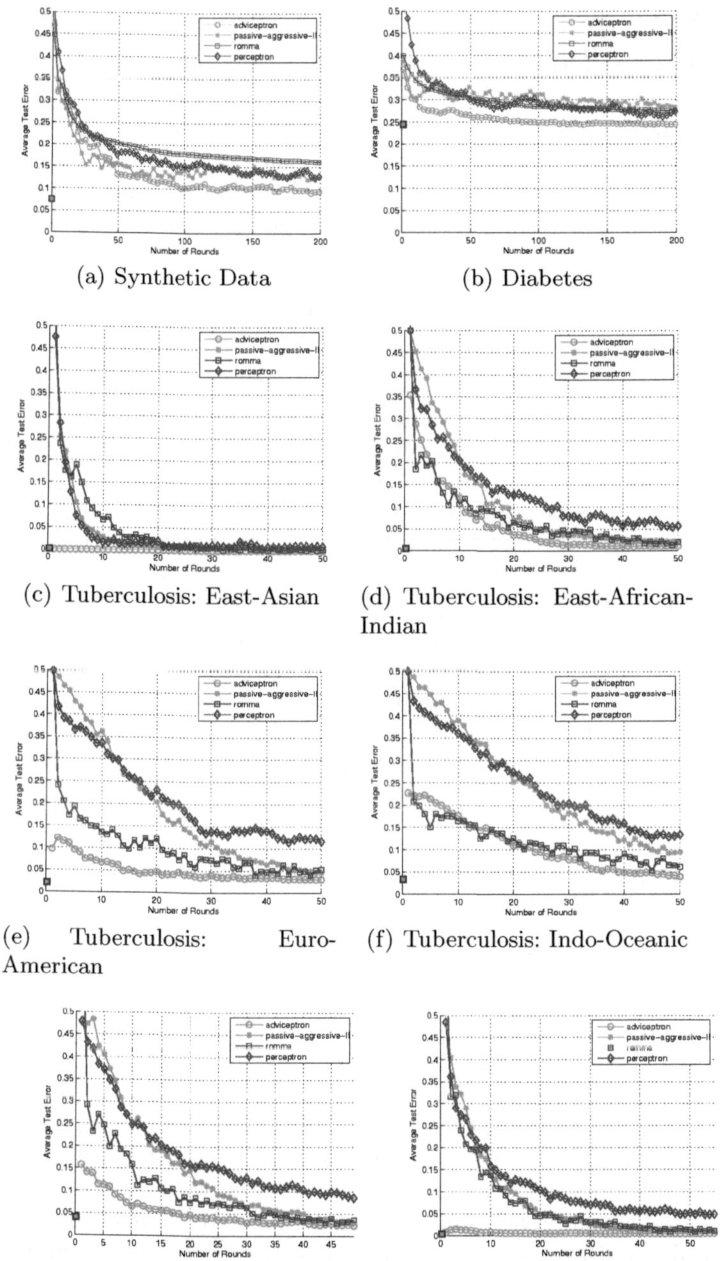

(a) Synthetic Data

(b) Diabetes

(c) Tuberculosis: East-Asian

(d) Tuberculosis: East-African-Indian

(e) Tuberculosis: Euro-American

(f) Tuberculosis: Indo-Oceanic

(g) Tuberculosis: *M. africanum*

(h) Tuberculosis: *M. bovis*

Fig. 2. Results comparing the Adviceptron to standard passive-aggressive, ROMMA and perceptron, where one example is presented at each round. The baseline KBSVM results are shown as a square on the y-axis for clarity; in each case, batch-KBSVM uses the entire training set available to the online learners.

5.4 Analysis of Results

For both artificial and real world data sets, the advice leads to significantly faster convergence of accuracy over the no-advice approaches. This reflects the intuitive idea that a learner, when given prior knowledge that is useful, will be able to more quickly find a good solution. In each case, note also, that the learner is able to use the learning process to improve on the starting accuracy (which would be produced by advice only). Thus, the Adviceptron is able to learn effectively from both data *and* advice.

A second point to note is that, in some cases, prior knowledge allows the learner to converge on a level of accuracy that is not achieved by the other methods, which do not benefit from advice. While the results demonstrate that advice can make a significant difference when learning with small data sets, in many cases, large amounts of data may be needed by the advice-free algorithms to eventually achieve performance similar to the Adviceptron. This shows that advice can provide large improvements over just learning with data.

Finally, it can be seen that, in most cases, the generalization performance of the Adviceptron converges rapidly to that of the batch-KBSVM. However, the batch-KBSVMs take, on average, 15–20 seconds to compute an optimal solution as they have to solve a quadratic program. In contrast, owing to the simple, closed-form update rules, the Adviceptron is able to obtain identical test-set performance in under 5 seconds on average. Further scalability experiments represent one of the more immediate directions of future work. One minor point to note is regarding the results on East-Asian and *M. bovis* (Figures 2(e) and 2(h)): the advice (provided by a tuberculosis domain expert) was so effective that these problems were almost immediately learned (with few to no examples).

6 Conclusions and Related Work

We have presented a new online learning method, the Adviceptron, that is a novel approach that makes use of prior knowledge in the form of polyhedral advice. This approach is an online extension to KBSVMs [6] and differs from previous polyhedral advice-taking approaches and the neural-network-based KBANN [15] in two significant ways: it is an online method with closed-form solutions and it provides a theoretical mistake bound.

The advice-taking approach was incorporated into the passive-aggressive framework because of its many appealing properties including efficient update rules and simplicity. Advice updates in the adviceptron are computed using a projected-gradient approach similar to the truncated-gradient approaches by Langford et al., [8]. However, the advice updates are truncated far more aggressively. The regret bound shows that as long as the projection being considered is non-expansive, it is still possible to minimize regret.

We have presented a bound on the effectiveness of this method and a proof of that bound. In addition, we performed several experiments on artificial and real world data sets that demonstrate that a learner with reasonable advice can significantly outperform a learner without advice. We believe our approach can serve as a template for other methods to incorporate advice into online learning methods.

One drawback of our approach is the restriction to certain types of loss functions. More direct projected-gradient approach or other related online convex programming [17] approaches can be used to develop algorithms with similar properties. This also allows for the derivation of general algorithms for different loss functions. KBSVMs can also be extended to kernels as shown in [6], and is yet another direction of future work.

Acknowledgements

The authors would like to thank Dr. Lauren Cowan of the CDC for providing the TB dataset and the expert-defined rules for lineage classification. The authors gratefully acknowledge support of the Defense Advanced Research Projects Agency under DARPA grant FA8650-06-C-7606 and the National Institute of Health under NIH grant 1-R01-LM009731-01. Views and conclusions contained in this document are those of the authors and do not necessarily represent the official opinion or policies, either expressed or implied of the US government or of DARPA.

References

1. Aminian, M., Shabbeer, A., Bennett, K.P.: A conformal Bayesian network for classification of Mycobacterium tuberculosis complex lineages. BMC Bioinformatics, 11(suppl. 3), S4 (2010)
2. Asuncion, A., Newman, D.J.: UCI machine learning repository (2007)
3. Brudey, K., Driscoll, J.R., Rigouts, L., Prodinger, W.M., Gori, A., Al-Hajoj, S.A., Allix, C., Aristimuño, L., Arora, J., Baumanis, V., et al.: Mycobacterium tuberculosis complex genetic diversity: Mining the fourth international spoligotyping database (spoldb 4) for classification, population genetics and epidemiology. BMC Microbiology 6(1), 23 (2006)
4. Crammer, K., Dekel, O., Keshet, J., Shalev-Shwartz, S., Singer, Y.: Online passive-aggressive algorithms. J. of Mach. Learn. Res. 7, 551–585 (2006)
5. Freund, Y., Schapire, R.E.: Large margin classification using the perceptron algorithm. Mach. Learn. 37(3), 277–296 (1999)
6. Fung, G., Mangasarian, O.L., Shavlik, J.W.: Knowledge-based support vector classifiers. In: Becker, S., Thrun, S., Obermayer, K. (eds.) NIPS, vol. 15, pp. 521–528 (2003)
7. Gagneux, S., Small, P.M.: Global phylogeography of Mycobacterium tuberculosis and implications for tuberculosis product development. The Lancet Infectious Diseases 7(5), 328–337 (2007)
8. Langford, J., Li, L., Zhang, T.: Sparse online learning via truncated gradient. J. Mach. Learn. Res. 10, 777–801 (2009)
9. Lauer, F., Bloch, G.: Incorporating prior knowledge in support vector machines for classification: A review. Neurocomp. 71(7-9), 1578–1594 (2008)
10. Li, Y., Long, P.M.: The relaxed online maximum margin algorithm. Mach. Learn. 46(1/3), 361–387 (2002)

11. Mangasarian, O.L.: Nonlinear Programming. McGraw-Hill, New York (1969)
12. Schölkopf, B., Simard, P., Smola, A., Vapnik, V.: Prior knowledge in support vector kernels. In: NIPS, vol. 10, pp. 640–646 (1998)
13. Schölkopf, B., Smola, A.: Learning with Kernels: Support Vector Machines, Regularization Optimization and Beyond. MIT Press, Cambridge (2001)
14. Shabbeer, A., Cowan, L., Driscoll, J.R., Ozcaglar, C., Vandenberg, S.L., Yener, B., Bennett, K.P.: TB-Lineage: An online tool for classification and analysis of strains of Mycobacterium tuberculosis Complex (2010) (unpublished manuscript)
15. Towell, G.G., Shavlik, J.W.: Knowledge-based artificial neural networks. AIJ 70(1-2), 119–165 (1994)
16. Vapnik, V.: The Nature of Statistical Learning Theory. Springer-Verlag (2000)
17. Zinkevich, M.: Online convex programming and generalized infinitesimal gradient ascent. In: Proc. 20th Int. Conf. on Mach. Learn, ICML 2003 (2003)

Appendix

Proof of Lemma 2

The progress at trial t is $\Delta_t = \frac{1}{2}\|\mathbf{w}^* - \mathbf{w}^{t+1}\|^2 - \frac{1}{2}\|\mathbf{w}^* - \mathbf{w}^t\|^2 = \dfrac{1}{2}\|\mathbf{w}^{t+1} - \mathbf{w}^t\|^2 + (\mathbf{w}^t - \mathbf{w}^*)'(\mathbf{w}^{t+1} - \mathbf{w}^t)$. Substituting $\mathbf{w}^{t+1} - \mathbf{w}^t = \nu\,\alpha_t y_t \mathbf{x}^t + (1-\nu)(\mathbf{r}^t - \mathbf{w}^t)$, from the update rules, we have

$$\Delta_t \leq \frac{1}{2}\nu^2\alpha_t^2\|\mathbf{x}^t\|^2 + \nu\alpha_t\left(\nu\, y_t {\mathbf{w}^t}'\mathbf{x}^t + (1-\nu)\, y_t {\mathbf{r}^t}'\mathbf{x}^t - y_t {\mathbf{w}^*}'\mathbf{x}^t\right) + \frac{1}{2}(1-\nu)\,\|\mathbf{r}^t - \mathbf{w}^*\|^2.$$

The loss suffered by the adviceptron is defined in (8). We focus only on the case when the loss is > 0. Define $\widehat{\mathbf{w}}^t = \nu\mathbf{w}^t + (1-\nu)\mathbf{r}^t$. Then, we have $1 - L_\xi(\widehat{\mathbf{w}}^t) = \nu\, y_t {\mathbf{w}^t}'\mathbf{x}^t + (1-\nu)\, y_t {\mathbf{r}^t}'\mathbf{x}^t$. Furthermore, by definition, $L_\xi(\mathbf{w}^*) \geq 1 - y_t {\mathbf{w}^*}'\mathbf{x}^t$. Using these two results,

$$\Delta_t \;\leq\; \frac{1}{2}\nu^2\alpha_t^2\|\mathbf{x}^t\|^2 + \nu\,\alpha_t(L_\xi(\mathbf{w}^*) - L_\xi(\mathbf{w}^t)) + \frac{1}{2}(1-\nu)\,\|\mathbf{r}^t - \mathbf{w}^*\|^2. \quad (19)$$

Adding $\frac{1}{2}\nu\,(\frac{\alpha_t}{\sqrt{\lambda}} - \sqrt{\lambda}\ell_t^*)^2$ to the left-hand side of the and simplifying, using Update 1:

$$\Delta_t \leq \frac{1}{2}\,\frac{\nu\, L_\xi(\mathbf{w}^t)^2}{\left(\dfrac{1}{\lambda} + \nu\,\|\mathbf{x}^t\|^2\right)} - \frac{1}{2}\nu\,\lambda L_\xi(\mathbf{w}^*)^2 + \frac{1}{2}(1-\nu)\,\|\mathbf{r}^t - \mathbf{w}^*\|^2.$$

Rearranging the terms above and using $\|\mathbf{x}^t\|^2 \leq X^2$ gives the bound. □

Proof of Lemma 3

Let $\widetilde{\mathbf{u}}^{i,t} = \mathbf{u}^{i,t} + D_i\boldsymbol{\beta}^i - \mathbf{d}^i\gamma_i$ be the update before the projection onto $\mathbf{u} \geq 0$. Then, $\mathbf{u}^{i,t+1} = \widetilde{\mathbf{u}}^{i,t}_+$. We also write $\mathcal{L}_{advice}(\mathbf{u}^{i,t})$ compactly as $\mathcal{L}(\mathbf{u}^{i,t})$. Then,

$$\begin{aligned}
\frac{1}{2}\|\mathbf{u}^* - \mathbf{u}^{i,t+1}\|^2 &\leq \frac{1}{2}\|\mathbf{u}^* - \widetilde{\mathbf{u}}^{i,t}\|^2 \\
&= \frac{1}{2}\|\mathbf{u}^* - \mathbf{u}^{i,t}\|^2 + \frac{1}{2}\|\mathbf{u}^{i,t} - \widetilde{\mathbf{u}}^{i,t}\|^2 + (\mathbf{u}^* - \mathbf{u}^{i,t})'(\mathbf{u}^{i,t} - \widetilde{\mathbf{u}}^{i,t}) \\
&= \frac{1}{2}\|\mathbf{u}^* - \mathbf{u}^{i,t}\|^2 + \frac{\mu^2}{2}\|\nabla_{\mathbf{u}^i}\mathcal{L}(\mathbf{u}^{i,t})\|^2 + \mu(\mathbf{u}^* - \mathbf{u}^{i,t})'\nabla_{\mathbf{u}^i}\mathcal{L}(\mathbf{u}^{i,t})
\end{aligned}$$

The first inequality is due to the non-expansiveness of projection and the next steps follow from Lemma 1.1. Let $\Delta_t = \frac{1}{2}\|\mathbf{u}^* - \mathbf{u}^{i,t+1}\|^2 - \frac{1}{2}\|\mathbf{u}^* - \mathbf{u}^{i,t}\|^2$. Using Lemma 1.2, we have

$$\begin{aligned}\Delta_t \le &\frac{\mu^2}{2}\left(\|D_i\|^2 L_\eta(\mathbf{u}^{i,t}, \mathbf{w}^t) + \|\mathbf{d}^i\|^2 L_\varsigma(\mathbf{u}^{i,t})^2\right)\\ &\quad + \mu(\mathbf{u}^* - \mathbf{u}^{i,t})'\left(\frac{1}{2}\nabla_{\mathbf{u}^i} L_\eta(\mathbf{u}^{i,t}, \mathbf{w}^t) + L_\varsigma(\mathbf{u}^{i,t})\nabla_{\mathbf{u}^i} L_\varsigma(\mathbf{u}^{i,t})\right)\\ \le &\frac{\mu^2}{2}\left(\|D_i\|^2 L_\eta(\mathbf{u}^{i,t}, \mathbf{w}^t) + \|\mathbf{d}^i\|^2 L_\varsigma(\mathbf{u}^{i,t})^2\right)\\ &\quad + \frac{\mu}{2}\left(L_\eta(\mathbf{u}^*, \mathbf{w}^t) - L_\eta(\mathbf{u}^{i,t}, \mathbf{w}^t)\right) + \frac{\mu}{2}\left(L_\varsigma(\mathbf{u}^*)^2 - L_\varsigma(\mathbf{u}^{i,t})^2\right)\end{aligned}$$

where the last step follows from the convexity of the loss function L_η and the fact that

$$\begin{aligned}&L_\varsigma(\mathbf{u}^{i,t})(\mathbf{u}^* - \mathbf{u}^{i,t})'\nabla_{\mathbf{u}^i} L_\varsigma(\mathbf{u}^{i,t})\\ &\qquad \le L_\varsigma(\mathbf{u}^{i,t})\left(L_\varsigma(\mathbf{u}^*) - L_\varsigma(\mathbf{u}^{i,t})\right) \quad \text{(convexity of } L_\varsigma)\\ &\qquad \le L_\varsigma(\mathbf{u}^{i,t})\left(L_\varsigma(\mathbf{u}^*) - L_\varsigma(\mathbf{u}^{i,t})\right) + \frac{1}{2}\left(L_\varsigma(\mathbf{u}^*) - L_\varsigma(\mathbf{u}^{i,t})\right)^2.\end{aligned}$$

Rearranging the terms and bounding $\|D_i\|^2$ and $\|\mathbf{d}^i\|^2$ proves the lemma. □

Learning with Randomized Majority Votes

Alexandre Lacasse, François Laviolette,
Mario Marchand, and Francis Turgeon-Boutin

Department of Computer Science and Software Engineering,
Laval University, Québec (QC), Canada

Abstract. We propose algorithms for producing weighted majority votes that learn by probing the empirical risk of a randomized (uniformly weighted) majority vote—instead of probing the zero-one loss, at some margin level, of the deterministic weighted majority vote as it is often proposed. The learning algorithms minimize a risk bound which is convex in the weights. Our numerical results indicate that learners producing a weighted majority vote based on the empirical risk of the randomized majority vote at some finite margin have no significant advantage over learners that achieve this same task based on the empirical risk at zero margin. We also find that it is sufficient for learners to minimize only the empirical risk of the randomized majority vote at a fixed number of voters without considering explicitly the entropy of the distribution of voters. Finally, our extensive numerical results indicate that the proposed learning algorithms are producing weighted majority votes that generally compare favorably to those produced by AdaBoost.

1 Introduction

Randomized majority votes (RMVs) were proposed by [9] as a theoretical tool to provide a margin-based risk bound for weighted majority votes such as those produced by AdaBoost [2]. Given a distribution Q over a (possibly continuous) space $\mathcal{H}$ of classifiers, a RMV is a uniformly weighted majority vote of N classifiers where each classifier is drawn independently at random according to Q. For infinitely large N, the RMV becomes identical to the Q-weighted majority vote over $\mathcal{H}$. The RMV is an example of a stochastic classifier having a risk (*i.e.*, a generalization error) that can be tightly upper bounded by a PAC-Bayes bound. Consequently, [6] have used the PAC-Bayes risk bound of [8] (see also [7]) to obtain a tighter margin-based risk bound than the one proposed by [9]. Both of these bounds depend on the empirical risk, at some margin level θ, made by the Q-weighted majority vote and on some regularizer. In the case of [9], the regularizer depends on the cardinality of $\mathcal{H}$ (or its VC dimension in the case of a continuous set) and, consequently, the only learning principle that can be inferred from this bound is to choose Q in order to maximize the margin. In the case of [6], the regularizer depends on the Kullback-Leibler divergence $\mathrm{KL}(Q\|P)$ between a prior distribution P and the posterior Q. Consequently, when P is uniform, the design principle inferred from this bound is to choose Q to maximize

J.L. Balcázar et al. (Eds.): ECML PKDD 2010, Part II, LNAI 6322, pp. 162–177, 2010.

both the margin and the entropy. We emphasize that both of these risk bounds are NP-hard to minimize because they depend on the empirical risk at margin θ of the Q-weighted majority vote as measured by the zero-one loss. Maximum entropy discrimination [4] is a computationally feasible method to maximize the entropy while maintaining a large margin by using classification constraints on each training examples. This is done by incorporating a prior on margin values for each training example which yield slack variables similar as those used for the SVM. These classification constraints are introduced, however, in an ad-hoc way which does not follow from a risk bound.

In this paper, we propose to use PAC-Bayes bounds for the risk of the Q-weighted majority vote, tighter than the one proposed by [6], which depend on the empirical risk of the randomized (uniformly weighted) majority vote—instead of depending on the empirical risk of the deterministic Q-weighted majority vote. As we shall see, the risk of the randomized majority vote (RMV) on a single example is non convex. But it can be tightly upper-bounded by a convex surrogate—thus giving risk bounds which are convex in Q and computationally cheap to minimize. We therefore propose learning algorithms that minimize these convex risk bounds. These are algorithms that basically learn by finding a distribution Q over $\mathcal{H}$ having large entropy and small empirical risk for the associated randomized majority vote. Hence, instead of learning by probing the empirical risk of the deterministic majority vote (as suggested by [9] and [6]), we propose to learn by probing the empirical risk of the randomized (uniformly weighted) majority vote. Our approach thus differs substantially from maximum entropy discrimination [4] where the empirical risk of the RMV is not considered.

Recently, [3] have also proposed learning algorithms that construct a weighted majority vote by minimizing a PAC-Bayes risk bound. However, their approach was restricted to the case of isotropic Gaussian posteriors over the set of linear classifiers. In this paper, both the posterior Q and the set $\mathcal{H}$ of basis functions are completely arbitrary. However, the algorithms that we present here apply only to the case where $\mathcal{H}$ is finite.

Our numerical results indicate that learners producing a weighted majority vote based on the empirical risk of the randomized majority vote at some finite margin θ have no significant advantage over learners that achieve this same task based on the empirical risk at zero margin. Perhaps surprisingly, we also find that it is sufficient for learners to minimize only the empirical risk of the randomized majority vote at a fixed number of voters without considering explicitly the entropy of the distribution of voters. Finally, our extensive numerical results indicate that the proposed learning algorithms are producing weighted majority votes that generally compare favorably to those produced by AdaBoost.

2 Definitions and PAC-Bayes Theory

We consider the binary classification problem where each example is a pair $(\mathbf{x}, y)$ such that $y \in \{-1, +1\}$ and $\mathbf{x}$ belongs to an arbitrary set $\mathcal{X}$. As usual, we assume that each example $(\mathbf{x}, y)$ is drawn independently according to a fixed, but unknown, distribution D on $\mathcal{X} \times \{-1, +1\}$.

The *risk* $R(h)$ of any classifier h is defined as the probability that h misclassifies an example drawn according to D. Given a training set $S = \{(\mathbf{x}_i, y_i), \ldots, (\mathbf{x}_m, y_m)\}$ of m examples, the *empirical risk* $R_S(h)$ of h is defined by its frequency of training errors on S. Hence,

$$R(h) \stackrel{\text{def}}{=} \mathop{\mathbf{E}}_{(\mathbf{x},y)\sim D} I(h(\mathbf{x}) \neq y) \quad ; \quad R_S(h) \stackrel{\text{def}}{=} \frac{1}{m}\sum_{i=1}^{m} I(h(\mathbf{x}_i) \neq y_i),$$

where $I(a) = 1$ if predicate a is true and 0 otherwise.

The so-called PAC-Bayes theorems [10,7,5,3] provide guarantees for a stochastic classifier called the *Gibbs classifier*. Given a distribution Q over a (possibly continuous) space $\mathcal{H}$ of classifiers, the Gibbs classifier G_Q is defined in the following way. Given an input example $\mathbf{x}$ to classify, G_Q chooses randomly a (deterministic) classifier h according to Q and then classifies $\mathbf{x}$ according to $h(\mathbf{x})$. The risk $R(G_Q)$ and the empirical risk $R_S(G_Q)$ of the Gibbs classifier are thus given by

$$R(G_Q) = \mathop{\mathbf{E}}_{h\sim Q} R(h) \quad ; \quad R_S(G_Q) = \mathop{\mathbf{E}}_{h\sim Q} R_S(h).$$

To upper bound $R(G_Q)$, we will make use of the following PAC-Bayes theorem due to [1]; see also [3]. In contrast, [6] used the looser bound of [8].

Theorem 1. *For any distribution D, any set $\mathcal{H}$ of classifiers, any distribution P of support $\mathcal{H}$, any $\delta \in (0,1]$, and any positive real number C, we have*

$$\Pr_{S\sim D^m}\left(\begin{array}{l} \forall\, Q \text{ on } \mathcal{H}: \\ R(G_Q) \leq \frac{1}{1-e^{-C}}\Big\{1 - \exp\Big[-\Big(C\cdot R_S(G_Q) \\ \qquad\qquad + \frac{1}{m}\big[\mathrm{KL}(Q\|P) + \ln\frac{1}{\delta}\big]\Big)\Big]\Big\} \\ \qquad \leq \frac{1}{1-e^{-C}}\Big\{C\cdot R_S(G_Q) + \frac{1}{m}\big[\mathrm{KL}(Q\|P) + \ln\frac{1}{\delta}\big]\Big\} \end{array}\right) \geq 1-\delta,$$

where $\mathrm{KL}(Q\|P) \stackrel{\text{def}}{=} \mathop{\mathbf{E}}_{h\sim Q} \ln\frac{Q(h)}{P(h)}$ *is the Kullback-Leibler divergence between Q and P.*

The second inequality, obtained by using $1-e^{-x} \leq x$, gives a looser bound which is, however, easier to interpret.

In Theorem 1, the prior distribution P must be defined *a priori* without reference to the training data S. Hence, P cannot depend on S whereas arbitrary dependence on S is allowed for the posterior Q. Finally note that the bound of Theorem 1 holds for any constant C. Thanks to the standard union bound argument, the bound can be made valid uniformly for k different values of C by replacing δ with δ/k.

3 Specialization to Randomized Majority Votes

Given a distribution Q over a space $\mathcal{H}$ of classifiers, we are often more interested in predicting according to the deterministic *weighted majority vote* B_Q instead of the stochastic Gibbs classifier G_Q. On any input example $\mathbf{x}$, the output $B_Q(\mathbf{x})$ of B_Q is given by

$$B_Q(\mathbf{x}) \stackrel{\text{def}}{=} \text{sgn}\left[\mathop{\mathbf{E}}_{h\sim Q} h(\mathbf{x})\right],$$

where $\text{sgn}(s) = +1$ if $s > 0$ and -1 otherwise.

Theorem 1, however, provides a guarantee for $R(G_Q)$ and not for the risk $R(B_Q)$ of the weighted majority vote B_Q. As an attempt to characterize the quality of weighted majority votes, let us analyze a special type of Gibbs classifier, closely related to B_Q, that we call the *randomized majority vote* (RMV).

Given a distribution Q over $\mathcal{H}$ and a natural number N, the randomized majority vote G_{Q^N} is a *uniformly weighted* majority vote of N classifiers chosen independently at random according to Q. Hence, to classify $\mathbf{x}$, G_{Q^N} draws N classifiers $\{h_{k(1)}, \ldots, h_{k(N)}\}$ from $\mathcal{H}$ independently according to Q and classifies $\mathbf{x}$ according to $\text{sgn}\,(g\,(\mathbf{x}))$, where

$$g\,(\mathbf{x}) \stackrel{\text{def}}{=} \frac{1}{N}\sum_{i=1}^{N} h_{k(i)}\,(\mathbf{x})\ .$$

We denote by $g \sim Q^N$, the above-described process of choosing N classifiers according to Q^N to form g.

Let us denote by $W_Q(\mathbf{x}, y)$ the fraction of classifiers, under measure Q, that misclassify example $(\mathbf{x}, y)$:

$$W_Q\,(\mathbf{x}, y) \stackrel{\text{def}}{=} \mathop{\mathbf{E}}_{h\sim Q} I\,(h\,(\mathbf{x}) \neq y)\ .$$

For simplicity, let us limit ourselves to the case where N is odd. In that case, $g\,(\mathbf{x}) \neq 0\ \forall \mathbf{x} \in \mathcal{X}$.

Similarly, denote by $W_{Q^N}(\mathbf{x}, y)$ the fraction of uniformly weighted majority votes of N classifiers, under measure Q^N, that err on $(\mathbf{x}, y)$:

$$\begin{aligned} W_{Q^N}(\mathbf{x}, y) &\stackrel{\text{def}}{=} \mathop{\mathbf{E}}_{g\sim Q^N} I\,(\text{sgn}\,[g(\mathbf{x})] \neq y) \\ &= \mathop{\mathbf{E}}_{g\sim Q^N} I\,(yg\,(\mathbf{x}) < 0) \\ &= \mathop{\Pr}_{g\sim Q^N} (yg(\mathbf{x}) < 0)\ . \end{aligned}$$

Recall that $W_Q\,(\mathbf{x}, y)$ is the probability that a classifier $h \in \mathcal{H}$, drawn according to Q, err on $\mathbf{x}$. Since $yg\,(\mathbf{x}) < 0$ iff more than half of the classifiers drawn according to Q err on $\mathbf{x}$, we have

$$W_{Q^N}(\mathbf{x}, y) = \sum_{k=\lceil \frac{N}{2} \rceil}^{N} \binom{N}{k} W_Q^k\,(\mathbf{x}, y)\,[1 - W_Q\,(\mathbf{x}, y)]^{N-k}\ .$$

Note that, with these definitions, the risk $R(G_{Q^N})$ of the randomized majority vote G_{Q^N} and its empirical estimate $R_S(G_{Q^N})$ on a training set S of m examples are respectively given by

$$R\left(G_{Q^N}\right) = \mathop{\mathbf{E}}_{(\mathbf{x},y)\sim D} W_{Q^N}(\mathbf{x},y) \quad ; \quad R_S\left(G_{Q^N}\right) = \frac{1}{m}\sum_{i=1}^{m} W_{Q^N}(\mathbf{x}_i,y_i)\,.$$

Since the randomized majority vote G_{Q^N} is a Gibbs classifier with a distribution Q^N over the set of all uniformly weighted majority votes that can be realized with N base classifiers chosen from $\mathcal{H}$, we can apply to G_{Q^N} the PAC-Bayes bound given by Theorem 1. To achieve this specialization, we only need to replace Q and P by Q^N and P^N respectively and use the fact that

$$\mathrm{KL}\left(Q^N \| P^N\right) = N \cdot \mathrm{KL}\left(Q \| P\right) .$$

Consequently, given this definition for G_{Q^N}, Theorem 1 admits the following corollary.

Corollary 1. *For any distribution D, any set $\mathcal{H}$ of base classifiers, any distribution P of support $\mathcal{H}$, any $\delta \in (0,1]$, any positive real number C, and any non-zero positive integer N, we have*

$$\Pr_{S\sim D^m}\left(\begin{array}{l}\forall\, Q \text{ on } \mathcal{H}: \\ R(G_{Q^N}) \leq \frac{1}{1-e^{-C}}\Big\{1-\exp\Big[-\Big(C\cdot R_S(G_{Q^N}) \\ \qquad\qquad +\frac{1}{m}\big[N\cdot \mathrm{KL}(Q\|P)+\ln\frac{1}{\delta}\big]\Big)\Big]\Big\}\end{array}\right) \geq 1-\delta.$$

By the standard union bound argument, the above corollary will hold uniformly for k values of C and all $N > 1$ if we replace δ by $\frac{6\delta}{k(N\pi)^2}$ (in view of the fact that $\sum_{i=1}^{\infty} i^{-2} = \pi^2/6$). Figure 1 shows the behavior of $W_{Q^N}(\mathbf{x},y)$ as a function of $W_Q(\mathbf{x},y)$ for different values of N. We can see that $W_{Q^N}(\mathbf{x},y)$ tends to the

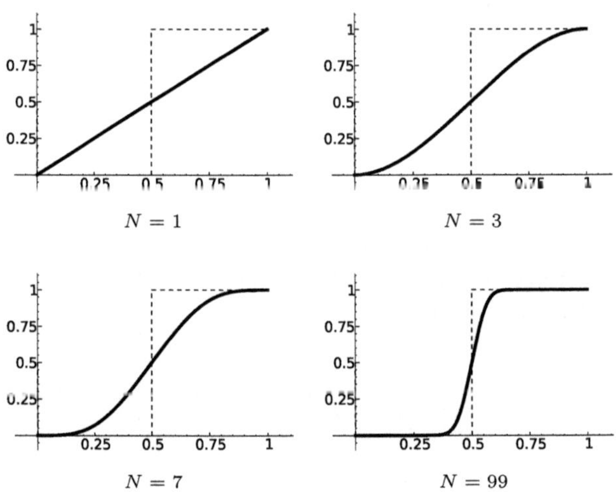

Fig. 1. Plots of $W_{Q^N}(\mathbf{x},y)$ as a function of $W_Q(\mathbf{x},y)$ for different values of N

zero-one loss $I(W_Q(\mathbf{x}, y) > 1/2)$ of the weighted majority vote B_Q as N is increased.

Since $W_{Q^N}(\mathbf{x}, y)$ is monotone increasing in $W_Q(\mathbf{x}, y)$ and $W_{Q^N}(\mathbf{x}, y) = 1/2$ when $W_Q(\mathbf{x}, y) = 1/2$, it immediately follows that $I(W_Q(\mathbf{x}, y) > 1/2) \leq 2W_{Q^N}(\mathbf{x}, y)$ for all N and $(\mathbf{x}, y)$. Consequently $R(B_Q) \leq 2R(G_{Q^N})$ and Corollary 1 provides an upper bound to $R(B_Q)$ via this "factor of two" rule.

4 Margin Bound for the Weighted Majority Vote

Since the risk of the weighted majority vote can be substantially smaller that Gibbs' risk, it may seem too crude to upper bound $R(B_Q)$ by $2R(G_{Q^N})$. One way to get rid of this factor of two is to consider the relation between $R(B_Q)$ and Gibbs' risk $R^\theta(G_{Q^N})$ at some positive margin θ. [9] have shown that

$$R(B_Q) \leq R^\theta(G_{Q^N}) + e^{-N\theta^2/2}, \tag{1}$$

where

$$R^\theta(G_{Q^N}) \stackrel{\text{def}}{=} \mathop{\mathbf{E}}_{(\mathbf{x},y)\sim D} \Pr_{g\sim Q^N} \left(yg(\mathbf{x}) \leq \theta \right).$$

Hence, for sufficiently large $N\theta^2$, Equation 1 provides an improvement over the "factor of two rule" as long as $R^\theta(G_{Q^N})$ is less than $2R(G_{Q^N})$.

Following this definition of $R^\theta(G_{Q^N})$, let us denote by $W^\theta_{Q^N}(\mathbf{x}, y)$ the fraction of uniformly weighted majority votes of N classifiers, under measure Q^N, that err on $(\mathbf{x}, y)$ at some margin θ, *i.e.*,

$$W^\theta_{Q^N}(\mathbf{x}, y) \stackrel{\text{def}}{=} \Pr_{g\sim Q^N} \left(yg(\mathbf{x}) \leq \theta \right).$$

Consequently,

$$R^\theta\left(G_{Q^N}\right) = \mathop{\mathbf{E}}_{(\mathbf{x},y)\sim D} W^\theta_{Q^N}(\mathbf{x}, y) \quad ; \quad R^\theta_S\left(G_{Q^N}\right) = \frac{1}{m}\sum_{i=1}^{m} W^\theta_{Q^N}(\mathbf{x}_i, y_i).$$

For N odd, $Nyg(\mathbf{x})$ can take values only in $\{-N, -N-2, \ldots, -1, +1, \ldots, +N\}$. We can thus assume, without loss of generality (w.l.o.g.), that θ can only take $(N+1)/2+1$ values.

To establish the relation between $W^\theta_{Q^N}(\mathbf{x}, y)$ and $W_Q(\mathbf{x}, y)$, note that $yg(\mathbf{x}) \leq \theta$ iff

$$1 - \frac{2}{N}\sum_{i=1}^{N} I(h_{k(i)}(\mathbf{x}) \neq y) \leq \theta.$$

The randomized majority vote G_{Q^N} thus misclassifies $(\mathbf{x}, y)$ at margin θ iff at least $\lceil \frac{N}{2}(1-\theta) \rceil$ of its voters err on $(\mathbf{x}, y)$. Consequently,

$$W^\theta_{Q^N}(\mathbf{x}, y) = \sum_{k=\zeta^\theta_N}^{N} \binom{N}{k} W^k_Q(\mathbf{x}, y)\left[1 - W_Q(\mathbf{x}, y)\right]^{N-k},$$

where, for positive θ,

$$\zeta_N^\theta \stackrel{\text{def}}{=} \max\left(\left\lceil \frac{N}{2}(1-\theta)\right\rceil, 0\right).$$

Figure 2 shows the behavior of $W^\theta_{Q^N}$ as a function of W_Q. The inflexion point of $W^\theta_{Q^N}$, when $N > 1$ and $\zeta_N^\theta > 1$ occurs[1] at $W_Q = \xi_N^\theta$ where

$$\xi_N^\theta \stackrel{\text{def}}{=} \frac{\zeta_N^\theta - 1}{N-1}.$$

Since N is a odd number, $\xi_N^\theta = 1/2$ when $\theta = 0$.

Equation 1 was the key starting point of [9] and [6] to obtain a margin bound for the weighted majority vote B_Q. The next important step is to upper bound $R^\theta(G_{Q^N})$. [9] achieved this task by upper bounding

$$\Pr_{(\mathbf{x},y)\sim D}\left(yg(\mathbf{x}) \le \theta\right)$$

uniformly for all $g \in \mathcal{H}^N$ in terms of their empirical risk at margin θ. Unfortunately, in the case of a finite set $\mathcal{H}$ of base classifiers, this step introduces a term in $O\left(\sqrt{(N/m)\log|\mathcal{H}|}\right)$ in their risk bound by the application of the union bound over the set of at most $|\mathcal{H}|^N$ uniformly weighted majority votes of N classifiers taken from $\mathcal{H}$.

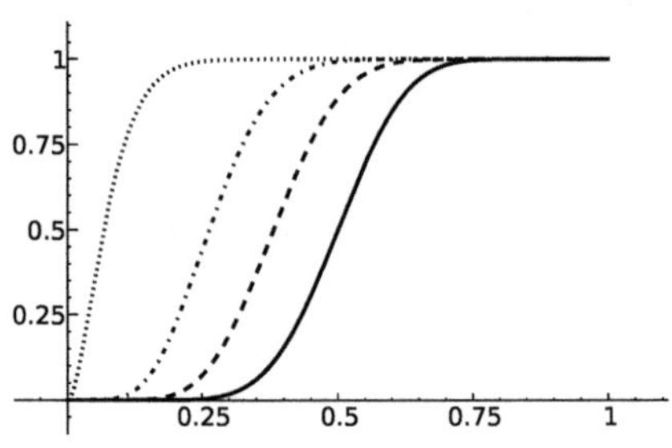

Fig. 2. Plots of $W^\theta_{Q^N}(\mathbf{x}, y)$ as a function of $W_Q(\mathbf{x}, y)$ for $N = 25$ and $\theta = 0, 0.2, 0.5$ and 0.9 (for curves from right to left respectively)

In contrast, [6] used the PAC-Bayes bound of [8] to upper bound $R^\theta(G_{Q^N})$. This introduces a term in $O\left(\sqrt{(N/m)\mathrm{KL}(Q\|P)}\right)$ in the risk bound and thus provides a significant improvement over the bound of [9] whenever $\mathrm{KL}(Q\|P) \ll \ln|\mathcal{H}|$. Here we propose to obtain an even tighter bound by making use of Theorem 1 to upper bound $R^\theta(G_{Q^N})$. This gives the following corollary.

[1] $W^\theta_{Q^N}$ has no inflexion point when $N = 1$ or $\zeta_N^\theta \le 1$.

Corollary 2. *For any distribution D, any set $\mathcal{H}$ of base classifiers, any distribution P of support $\mathcal{H}$, any $\delta \in (0,1]$, any $C > 0$ and $\theta \geq 0$, and any integer $N > 0$, we have*

$$\Pr_{S\sim D^m}\left(\begin{array}{l}\forall\, Q \text{ on } \mathcal{H}: \\ R(B_Q) \leq \frac{1}{1-e^{-C}}\Big\{1-\exp\Big[-\Big(C\cdot R_S^{\theta}(G_{Q^N}) \\ \qquad\qquad\qquad +\frac{1}{m}\big[N\cdot \mathrm{KL}(Q\|P)+\ln\frac{1}{\delta}\big]\Big)\Big]\Big\} + e^{-N\theta^2/2}\end{array}\right) \geq 1-\delta .$$

To make this bound valid uniformly for any odd number N and for any of the $(N+1)/2+1$ values of θ mentioned before, the standard union bound argument tells us that it is sufficient to replace δ by $\frac{12}{\pi^2}\frac{1}{N^2(N+3)}\delta$ (in view of the fact that $\sum_{i=1}^{\infty} i^{-2} = \pi^2/6$). Moreover, we should chose the value of N to keep $e^{-N\theta^2/2}$ comparable to the corresponding regularizer. This can be achieved by choosing

$$N = \left\lceil \frac{2}{\theta^2}\ln\Big(m[1-e^{-C}]\Big)\right\rceil . \tag{2}$$

Finally, it is important to mention that both [9] and [6] used

$$R_S^{\theta}(G_{Q^N}) \leq R_S^{2\theta}(B_Q) + e^{-N\theta^2/2}$$

to write their upper bound only in terms of the empirical risk $R_S^{2\theta}(B_Q)$ at margin 2θ of the weighted majority vote B_Q, where

$$R_S^{\theta}(B_Q) \stackrel{\text{def}}{=} \mathop{\mathbf{E}}_{(\mathbf{x},y)\sim D} I\left(y \mathop{\mathbf{E}}_{h\sim Q} h(\mathbf{x}) \leq \theta\right) .$$

This operation, however, contributes to an additional deterioration of the bound which, because of the presence of $R_S^{2\theta}(B_Q)$, is now NP-hard to minimize. Consequently, for the purpose of bound minimization, it is preferable to work with a bound like Corollary 2 which depends on $R_S^{\theta}(G_{Q^N})$ (and not on $R_S^{2\theta}(B_Q)$).

5 Proposed Learning Algorithms

The task of the learning algorithm is to find the posterior Q that minimizes the upper bound of Corollary 1 or Corollary 2 for fixed parameters C, N and θ. Note that for both of these cases, this is equivalent to find Q that minimizes

$$F(Q) \stackrel{\text{def}}{=} C\cdot R_S^{\theta}\left(G_{Q^N}\right) + \frac{1}{m}\cdot N\cdot \mathrm{KL}\left(Q\|P\right) . \tag{3}$$

Indeed, minimizing $F(Q)$, when $\theta = 0$, gives the posterior Q minimizing the upper bound on $R(G_{Q^N})$ given by Corollary 1. Whereas minimizing $F(Q)$, when $\theta > 0$, gives the posterior Q minimizing the upper bound on $R(B_Q)$ given by Corollary 2.

Note that, for any fixed example $(\mathbf{x}, y)$, $W^{\theta}_{Q^N}(\mathbf{x}, y)$ is a quasiconvex function of Q. However, a sum of quasiconvex functions is generally not quasiconvex and thus, not convex. Consequently, $R^{\theta}_S\left(G_{Q^N}\right)$, which is a sum of quasiconvex functions, is generally not convex with respect to Q. To obtain a convex optimization problem, we replace $R^{\theta}_S\left(G_{Q^N}\right)$ by the convex function $\mathcal{R}^{\theta}_S\left(G_{Q^N}\right)$ defined as

$$\mathcal{R}^{\theta}_S\left(G_{Q^N}\right) \stackrel{\text{def}}{=} \frac{1}{m}\sum_{i=1}^{m} \mathcal{W}^{\theta}_{Q^N}\left(\mathbf{x}_i, y_i\right) ,$$

where $\mathcal{W}^{\theta}_{Q^N}$ is the convex function of W_Q which is the closest to $W^{\theta}_{Q^N}$ with the property that $\mathcal{W}^{\theta}_{Q^N} = W^{\theta}_{Q^N}$ when $W_Q \leq \xi^{\theta}_N$ (the inflexion point of $W^{\theta}_{Q^N}$). Hence,

$$\mathcal{W}^{\theta}_{Q^N}\left(\mathbf{x}, y\right) \stackrel{\text{def}}{=} \begin{cases} W^{\theta}_{Q^N}\left(\mathbf{x}, y\right) & \text{if } W^{\theta}_{Q}\left(\mathbf{x}, y\right) \leq \xi^{\theta}_N \\ W^{\theta}_{Q^N}\Big|_{W_Q=\xi^{\theta}_N} + \Delta^{\theta}_N \cdot \left(W_Q\left(\mathbf{x}, y\right) - \xi^{\theta}_N\right) & \text{otherwise}, \end{cases}$$

where Δ^{θ}_N is the first derivative of $W^{\theta}_{Q^N}$ evaluated at its inflexion point ξ^{θ}_N, *i.e.*,

$$\Delta^{\theta}_N \stackrel{\text{def}}{=} \left.\frac{\partial W^{\theta}_{Q^N}}{\partial W_Q}\right|_{W_Q=\xi^{\theta}_N} .$$

We thus propose to find Q that minimizes[2]

$$\mathcal{F}\left(Q\right) \stackrel{\text{def}}{=} C \cdot \mathcal{R}^{\theta}_S\left(G_{Q^N}\right) + \frac{1}{m} \cdot N \cdot \mathrm{KL}\left(Q \| P\right). \tag{4}$$

We now restrict ourselves to the case where the set $\mathcal{H}$ of basis classifiers is finite. Let $\mathcal{H} \stackrel{\text{def}}{=} \{h_1, h_2, \ldots, h_n\}$. Given a distribution Q, let Q_i be the weight assigned by Q to classifier h_i.

For fixed N, $\mathcal{F}$ is a convex function of Q with continuous first derivatives. Moreover, $\mathcal{F}$ is defined on a bounded convex domain which is the n-dimensional probability simplex for Q. Consequently, any local minimum of $\mathcal{F}$ is also a global minimum. Under these circumstances, coordinate descent minimization is guaranteed to converge to the global minimum of $\mathcal{F}$. To deal with the constraint that Q is a distribution, we propose to perform the descent of $\mathcal{F}$ by using pairs of coordinates. For that purpose, let $Q^{j,k}_{\lambda}$ be the distribution obtained from Q by transferring a weight λ from classifier h_k to classifier h_j while keeping all other weights unchanged. Thus, for all $i \in \{1, \ldots, n\}$, we have

$$(Q^{j,k}_{\lambda})_i \stackrel{\text{def}}{=} \begin{cases} Q_i + \lambda \text{ if } i = j \\ Q_i - \lambda \text{ if } i = k \\ Q_i \quad\quad \text{otherwise}. \end{cases}$$

[2] Since $R^{\theta}_S\left(G_{Q^N}\right) \leq \mathcal{R}^{\theta}_S\left(G_{Q^N}\right)$, the bound of Corollary 2 holds whenever we replace $R^{\theta}_S\left(G_{Q^N}\right)$ by $\mathcal{R}^{\theta}_S\left(G_{Q^N}\right)$.

Algorithm 1 : $\mathcal{F}$ minimization

Input: $S = \{(\mathbf{x}_1, y_1), \ldots, (\mathbf{x}_m, y_m)\}$,
$\mathcal{H} = \{h_1, h_2, \ldots, h_n\}$
Initialization: $Q_j = \frac{1}{n}$ for $j = 1, \ldots, n$
repeat
  Choose j and k randomly from $\{1, 2, \ldots, n\}$.
  $\lambda_{min} \leftarrow -\min(Q_j, 1 - Q_k)$
  $\lambda_{max} \leftarrow \min(Q_k, 1 - Q_j)$
  $\lambda_{opt} \leftarrow \underset{\lambda \in [\lambda_{min}, \lambda_{max}]}{\operatorname{argmin}} \mathcal{F}\left(Q_\lambda^{j,k}\right)$
  $Q_j \leftarrow Q_j + \lambda_{opt}$
  $Q_k \leftarrow Q_k - \lambda_{opt}$
until Stopping criteria attained

Note that in order for $Q_\lambda^{j,k}$ to remain a valid distribution, we need to choose λ in the range $[-\min(Q_j, 1 - Q_k), \min(Q_k, 1 - Q_j)]$.

As described in Algorithm 1, each iteration of $\mathcal{F}$ minimization consists of the following two steps. We first choose j and k from $\{1, 2, \ldots, n\}$ and then find λ that minimizes $\mathcal{F}\left(Q_\lambda^{j,k}\right)$.

Since $\mathcal{F}$ is convex, the optimal value of λ at each iteration is given by

$$\frac{\partial \mathcal{F}\left(Q_\lambda^{j,k}\right)}{\partial \lambda} = C \frac{\partial \mathcal{R}_S^\theta\left(G_{(Q_\lambda^{j,k})^N}\right)}{\partial \lambda} + \frac{1}{m} \cdot N \cdot \frac{\partial \mathrm{KL}\left(Q_\lambda^{j,k} \| P\right)}{\partial \lambda} = 0 . \qquad (5)$$

For the uniform prior ($P_i = 1/n \; \forall i$), we have

$$\frac{\partial \mathrm{KL}\left(Q_\lambda^{j,k} \| P\right)}{\partial \lambda} = \ln\left(\frac{Q_j + \lambda}{Q_k - \lambda}\right) .$$

Now, let $\mathcal{V}_N^\theta(W_Q) \stackrel{\text{def}}{=} \frac{\partial \mathcal{W}_{Q^N}^\theta}{\partial W_Q}$. We have

$$\mathcal{V}_N^\theta(W_Q) = \begin{cases} \Delta_N^\theta & \text{if } W_Q \geq \xi_N^\theta \\ \dfrac{N! \, W_Q^{\zeta_N^\theta - 1} (1 - W_Q)^{N - \zeta_N^\theta}}{(\zeta_N^\theta - 1)! \, (N - \zeta_N^\theta)!} & \text{otherwise} . \end{cases}$$

Then we have

$$\frac{\partial \mathcal{R}_S^\theta\left(G_{(Q_\lambda^{j,k})^N}\right)}{\partial \lambda} = \frac{1}{m} \sum_{i=1}^m \mathcal{V}_N^\theta\left(W_{Q_\lambda^{j,k}}(\mathbf{x}_i, y_i)\right) \frac{\partial W_{Q_\lambda^{j,k}}(\mathbf{x}_i, y_i)}{\partial \lambda} .$$

From the definition of $W_{Q_\lambda^{j,k}}(\mathbf{x}_i, y_i)$, we find

$$W_{Q_\lambda^{j,k}}(\mathbf{x}_i, y_i) = W_Q(\mathbf{x}_i, y_i) + \lambda \cdot D_i^{j,k} ,$$

where

$$D_i^{j,k} \stackrel{\text{def}}{=} I\big(h_j(\mathbf{x}_i) \neq h_k(\mathbf{x}_i)\big)\, y_i h_k(\mathbf{x}_i)\,.$$

Hence, $\frac{\partial W_{Q_\lambda^{j,k}}(\mathbf{x}_i,y_i)}{\partial\lambda} = D_i^{j,k}$. Equation 5 therefore becomes

$$\ln\left(\frac{Q_j+\lambda}{Q_k-\lambda}\right) + \frac{C}{N}\sum_{i=1}^{m} D_i^{j,k}\, \mathcal{V}_N^\theta\left(W_{Q_\lambda^{j,k}}(\mathbf{x}_i,y_i)\right) = 0\,.$$

Since $\mathcal{V}_N^\theta$ is multiplied by $D_i^{j,k}$, we can replace in the above equation $W_{Q_\lambda^{j,k}}(\mathbf{x}_i,y_i)$ by $W_Q(\mathbf{x}_i,y_i) + \lambda y_i h_k(\mathbf{x}_i)$. If we now use $W_Q(i)$ as a shorthand notation for $W_Q(\mathbf{x}_i,y_i)$, Equation 5 finally becomes

$$\ln\left(\frac{Q_j+\lambda}{Q_k-\lambda}\right) + \frac{C}{N}\sum_{i=1}^{m} D_i^{j,k}\, \mathcal{V}_N^\theta\Big(W_Q(i) + \lambda y_i h_k(\mathbf{x}_i)\Big) = 0\,. \tag{6}$$

An iterative root-finding method, such as Newton's, can be used to solve Equation 6. Since we cannot factor out λ from the summation in Equation 6 (as it can be done for AdaBoost), each iteration step of the root-finding method costs $\Theta(m)$ time. Therefore, Equation 6 is solved in $O(mk(\epsilon))$ time, where $k(\epsilon)$ denotes the number k of iterations needed by the root-finding method to find λ_{opt} within precision ε. Once we have found λ_{opt}, we update Q with the new weights for Q_j and Q_k and update[3] each $W_Q(i)$ according to

$$W_Q(i) \;\leftarrow\; W_Q(i) + \lambda D_i^{j,k} \quad \text{for } i \in \{1,\ldots,m\}\,,$$

in $\Theta(m)$ time. We repeat this process until all the weight modifications are within a desired precision ε .

Finally, if we go back to Equation 4 and consider the fact that $\mathrm{KL}(Q\|P) \leq \ln|\mathcal{H}|$, we note $\mathcal{R}_S^\theta(G_{Q^N})$ can dominate $N\cdot\mathrm{KL}(Q\|P)$. This is especially true whenever S has the property that for any Q there exist some training examples having $W_Q(\mathbf{x},y) > 1/2$. Indeed, in that case, the convexity of $\mathcal{W}_Q(\mathbf{x},y)$ can force $\mathcal{R}_S^\theta(G_{Q^N})$ to be always much larger than $N\cdot\mathrm{KL}(Q\|P)$ for any Q. In these circumstances, the posterior Q that minimizes $\mathcal{R}_S^\theta(G_{Q^N})$ should be similar to the one that minimizes $\mathcal{F}(Q)$. Consequently, we have also decided to minimize $\mathcal{R}_S^\theta(G_{Q^N})$ at fixed N. In this case, we can drop the C parameter and each iteration of the algorithm consists of solving Equation 6 without the presence of the logarithm term. Parameter N then becomes the regularizer of the learning algorithm.

6 Experimental Results

We have tested our algorithms on more than 20 data sets. Except for MNIST, all data sets come from the UCI repository. Each data set was randomly split

[3] Initially we have $W_Q(i) = \frac{1}{n}\sum_{j=1}^{n} I(h_j(\mathbf{x}_i) \neq y_i)$ for $i \in \{1,\ldots,m\}$.

Table 1. Results for Algorithm 1, $\mathcal{F}$ minimization, at zero margin ($\theta = 0$)

Dataset	Bound					CV - $\mathbf{R(B_Q)}$				
Name	$R(B_Q)$	$R(G_{Q^N})$	N	C	Bnd	$R(B_Q)$	$R(G_{Q^N})$	N	C	Bnd
Adult	0.206	0.206	1	0.2	0.245	0.152	0.171	499	20	0.958
Letter:AB	0.093	0.092	1	0.5	0.152	0.009	0.043	49	2	0.165
Letter:DO	0.141	0.143	1	0.5	0.199	0.027	0.040	999	50	0.808
Letter:OQ	0.257	0.257	1	0.5	0.313	0.041	0.052	4999	200	0.994
MNIST:0vs8	0.046	0.054	1	1	0.102	0.007	0.015	49	50	0.415
MNIST:1vs7	0.045	0.058	1	1	0.115	0.011	0.017	49999	100	0.506
MNIST:1vs8	0.042	0.108	25	1	0.233	0.021	0.030	499	500	0.835
MNIST:2vs3	0.138	0.159	1	0.5	0.215	0.045	0.066	75	20	0.600
Mushroom	0.019	0.035	49	1	0.097	0.000	0.001	999	100	0.317
Ringnorm	0.046	0.117	999999	1	0.252	0.026	0.034	9999	200	0.998
Waveform	0.083	0.117	25	0.5	0.172	0.081	0.114	49	0.5	0.172

into a training set S of $|S|$ examples and a testing set T of $|T|$ examples. The number d of attributes for each data set is also specified in Table 2.

For all tested algorithms, we have used *decision stumps* for the set $\mathcal{H}$ of base classifiers. Each decision stump $h_{\ell,t,b}$ is a threshold classifier that outputs $+b$ if the ℓth attribute of the input example exceeds a threshold value t, and $-b$ otherwise, where $b \in \{-1, +1\}$. For each attribute, at most ten equally spaced possible values for t were determined *a priori*.

The results for the first set of experiments are summarized in Table 1. For these experiments, we have minimized the objective function $\mathcal{F}$ at zero margin as described by Algorithm 1 and have compared two different ways of choosing the hyperparameters N and C of $\mathcal{F}$. For the **Bound** method, the values chosen for N and C were those minimizing the risk bound given by Corollary 1 on S whereas, for the **CV - R($\mathbf{B_Q}$)** method, the values of these hyperparameters were those minimizing the 10-fold cross-validation score (on S) of the weighted majority vote B_Q. For all cases, $R(B_Q)$ and $R(G_{Q^N})$ refer, respectively, to the empirical risk of the weighted majority vote B_Q and of the randomized majority vote G_{Q^N} computed on the testing set T. Also indicated, are the values found for N and C. In all cases, N was chosen among a set[4] of 17 values between 1 and $10^6 - 1$ and C was chosen among a set[5] of 15 values between 0.02 and 1000.

As we can see in Table 1, the bound values are indeed much smaller when N and C are chosen such as to minimize the risk bound. However, both the weighted majority vote B_Q and the randomized majority vote G_{Q^N} obtained in this manner performed much worse than those obtained when C and N were

[4] Values for N :
$\{1, 3, 5, 7, 9, 25, 49, 75, 99, 499, 999, 4999, 9999, 49999, 99999, 499999, 999999\}$.

[5] Values for C :
$\{0.02, 0.05, 0.1, 0.2, 0.5, 1, 2, 5, 10, 20, 50, 100, 200, 500, 1000\}$.

Table 2. Results for Algorithm 1, $\mathcal{F}$ minimization, compared with AdaBoost (AB)

Dataset				AB	Algo 1, $\theta = 0$			Algo 1, $\theta > 0$			
Name	$\|S\|$	$\|T\|$	d	$R(B_Q)$	$R(B_Q)$	N	C	$R(B_Q)$	N	C	θ
Adult	1809	10000	14	**0.149**	0.152	499	20	0.153	49999	1000	0.017
BreastCancer	343	340	9	0.053	0.041	7	1	**0.038**	499	1000	0.153
Credit-A	353	300	15	0.170	**0.150**	9999	2	**0.150**	49999	5	0.015
Glass	107	107	9	0.178	**0.131**	49	500	**0.131**	499	200	0.137
Haberman	144	150	3	**0.260**	0.273	1	0.001	0.273	5	0.02	0.647
Heart	150	147	13	0.252	0.177	75	1	**0.170**	4999	5	0.045
Ionosphere	176	175	34	0.120	**0.103**	499	200	0.114	4999	200	0.045
Letter:AB	500	1055	16	0.010	0.009	49	2	**0.006**	4999	10	0.050
Letter:DO	500	1058	16	0.036	**0.027**	999	50	0.032	999	50	0.112
Letter:OQ	500	1036	16	**0.038**	0.041	4999	200	0.044	49999	20	0.016
Liver	170	175	6	0.320	0.349	25	2	**0.314**	999	20	0.101
MNIST:0vs8	500	1916	784	0.008	**0.007**	49	50	**0.007**	499	50	0.158
MNIST:1vs7	500	1922	784	0.013	**0.011**	49999	100	0.013	9999	50	0.035
MNIST:1vs8	500	1936	784	0.025	0.021	499	500	**0.020**	999	50	0.112
MNIST:2vs3	500	1905	784	0.047	0.045	75	20	**0.034**	4999	20	0.050
Mushroom	4062	4062	22	**0.000**	**0.000**	999	100	**0.000**	4999	200	0.058
Ringnorm	3700	3700	20	0.043	**0.026**	9999	200	0.028	49999	500	0.018
Sonar	104	104	60	0.231	**0.192**	25	20	0.231	999	500	0.096
Usvotes	235	200	16	**0.055**	**0.055**	1	0.2	**0.055**	25	1	0.633
Waveform	4000	4000	21	0.085	**0.081**	49	0.5	**0.081**	999	100	0.129
Wdbc	285	284	30	0.049	**0.035**	499	20	0.039	9999	100	0.034

selected by cross-validation. The difference is statistically significant[6] in every cases except on the Waveform data set. We can also observe that the values of N chosen by cross-validation are much larger than those selected by the risk bound. When N is large, the stochastic predictor G_{Q^N} becomes close to the deterministic weighted majority vote B_Q but we can still observe an overall superiority for the B_Q predictor.

The results for the second set of experiments are summarized in Table 2 where we also provide a comparison to AdaBoost[7]. The hyperparameters C and N for these experiments were selected based on the 10-fold cross-validation score (on S) of B_Q. We have also compared the results for Algorithm 1 when θ is fixed to zero and when θ can take non-zero values. The results presented here for Algorithm 1 at non-zero margin are those when θ is fixed to the value given by Equation 2. Interestingly, as indicated in Table 3, we have found that fixing θ in this way gave, overall, equivalent performance as choosing it by cross-validation (the difference is never statistically significant).

[6] To determine whether or not a difference of empirical risk measured on the testing set T is statistically significant, we have used the test set bound method of [5] (based on the binomial tail inversion) with a confidence level of 95%.

[7] For these experiments, the number of boosting rounds was fixed to 200.

Table 3. Comparison of results when θ is chosen by 10-fold cross-validation to those when θ is fixed to the value given by Equation 2

Dataset	**AB**	**Algo 1, CV-θ**				**Algo 1, θ^***			
Name	$R(B_Q)$	$R(B_Q)$	N	C	θ	$R(B_Q)$	N	C	θ
Adult	**0.149**	0.152	499	20	0.005	0.153	49999	1000	0.017
BreastCancer	0.053	0.041	25	10	0.05	**0.038**	499	1000	0.153
Credit-A	0.170	**0.150**	9999	2	0	**0.150**	49999	5	0.015
Glass	0.178	**0.131**	9999	200	0.05	**0.131**	499	200	0.137
Haberman	0.260	**0.253**	3	500	0.5	0.273	5	0.02	0.647
Heart	0.252	0.177	49999	5	0.025	**0.170**	4999	5	0.045
Ionosphere	0.120	**0.114**	49999	1000	0.1	**0.114**	4999	200	0.045
Letter:AB	0.010	**0.006**	4999	10	0.025	**0.006**	4999	10	0.050
Letter:DO	0.036	**0.029**	499	10	0.025	0.032	999	50	0.112
Letter:OQ	**0.038**	0.041	999	100	0.005	0.044	49999	20	0.016
Liver	0.320	0.349	25	2	0	**0.314**	999	20	0.101
MNIST:0vs8	0.008	**0.007**	99	1000	0.1	**0.007**	499	50	0.158
MNIST:1vs7	0.013	**0.012**	4999	50	0.025	0.013	9999	50	0.035
MNIST:1vs8	0.025	0.021	499	500	0	**0.020**	999	50	0.112
MNIST:2vs3	0.047	0.049	99	100	0.1	**0.034**	4999	20	0.050
Mushroom	**0.000**	**0.000**	999	100	0	**0.000**	4999	200	0.058
Ringnorm	0.043	**0.027**	9999	200	0.005	0.028	49999	500	0.018
Sonar	0.231	**0.144**	49	500	0.025	0.231	999	500	0.096
Usvotes	**0.055**	**0.055**	1	0.2	0	**0.055**	25	1	0.633
Waveform	0.085	**0.081**	49	0.5	0	**0.081**	999	100	0.129
Wdbc	0.049	**0.035**	499	20	0	0.039	9999	100	0.034

Going back to Table 2, we see that the results for Algorithm 1 when $\theta > 0$ are competitive (but different) with those obtained at $\theta = 0$. There is thus no competitive advantage at choosing a non-zero margin value (but there is no disadvantage either and no computational disadvantage since the value of θ is not chosen by cross-validation). Finally, the results indicate that both of these algorithms perform generally better than AdaBoost but the results are significant only on the Ringnorm data set.

As described in the previous section, we have also minimized $\mathcal{R}_S^\theta(G_{Q^N})$ at a fixed number of voters N, which now becomes the regularizer of the learning algorithm. This algorithm has the significant practical advantage of not having an hyperparameter C to tune. Three versions of this algorithm are compared in Table 4. In the first version, $\mathcal{R}_S^\theta(G_{Q^N})$**-min**, the value of θ was selected based on the 10-fold cross-validation score (on S) of B_Q. In the second version, the value of θ was fixed to $\sqrt{(1/N)\ln(2m)}$. In the third version, the value of θ was fixed to zero. We see, in Table 4, that all three versions are competitive to one another. The difference in the results was never statistically significant. Hence, again, there is no competitive advantage at choosing a non-zero margin value for the empirical risk of the randomized majority vote. We also find that results for all three versions are competitive with AdaBoost. The difference was significant

Table 4. Results for the Algorithm that minimizes $\mathcal{R}_S^\theta(G_{Q^N})$

Dataset	AB	$\mathcal{R}_\mathbf{S}^\theta(\mathbf{G_{Q^N}})$-min.			$\mathcal{R}_\mathbf{S}^\theta(\mathbf{G_{Q^N}})$, θ^*			$\mathcal{R}_\mathbf{S}^0(\mathbf{G_{Q^N}})$	
Name	$R(B_Q)$	$R(B_Q)$	N	θ	$R(B_Q)$	N	θ	$R(B_Q)$	N
Adult	**0.149**	0.153	999	0.091	0.153	999	0.091	0.151	75
BreastCancer	0.053	**0.044**	499	0.114	**0.044**	499	0.114	**0.044**	25
Credit-A	0.170	**0.133**	25	0.512	**0.133**	25	0.512	0.137	9
Glass	0.178	**0.131**	499	0.104	**0.131**	499	0.104	**0.131**	49
Haberman	**0.260**	0.273	7	0.899	0.273	7	0.899	0.273	1
Heart	0.252	0.190	499	0.107	0.190	499	0.107	**0.177**	49
Ionosphere	**0.120**	0.131	4999	0.034	0.131	4999	0.034	0.143	999
Letter:AB	0.010	**0.001**	99999	0.008	**0.001**	99999	0.008	0.006	99999
Letter:DO	0.036	**0.026**	49999	0.012	**0.026**	49999	0.012	0.028	49999
Letter:OQ	**0.038**	0.043	4999	0.037	0.043	4999	0.037	0.048	999
Liver	**0.320**	0.343	999	0.076	0.343	999	0.076	0.349	49999
MNIST:0vs8	0.008	0.008	4999	0.037	0.008	4999	0.037	**0.007**	75
MNIST:1vs7	0.013	0.011	99999	0.008	0.011	99999	0.008	**0.010**	49999
MNIST:1vs8	0.025	0.020	4999	0.037	0.020	4999	0.037	**0.018**	4999
MNIST:2vs3	0.047	0.041	4999	0.037	0.041	4999	0.037	**0.035**	49999
Mushroom	**0.000**	**0.000**	4999	0.042	**0.000**	4999	0.042	**0.000**	999
Ringnorm	0.043	**0.028**	49999	0.013	**0.028**	49999	0.013	0.029	9999
Sonar	0.231	0.212	4999	0.033	0.212	4999	0.033	**0.192**	99
Usvotes	**0.055**	**0.055**	25	0.496	**0.055**	25	0.496	**0.055**	1
Waveform	0.085	**0.080**	499	0.134	**0.080**	499	0.134	0.081	99
Wdbc	0.049	0.039	9999	0.025	0.039	9999	0.025	**0.035**	75

on the Ringnorm and Letter:AB data sets (in favor of $\mathcal{R}_S^\theta(G_{Q^N})$ minimization). Hence, $\mathcal{R}_S^\theta(G_{Q^N})$ minimization at a fixed number N of voters appears to be a good substitute to regularized variants of boosting.

7 Conclusion

In comparison with other state-of-the-art learning strategies such as boosting, our numerical experiments indicate that learning by probing the empirical risk of the randomized majority vote is an excellent strategy for producing weighted majority votes that generalize well. We have shown that this learning strategy is strongly supported by PAC-Bayes theory because the proposed risk bound immediately gives the objective function to minimize. However, the precise weighting of the KL regularizer versus the empirical risk that appears in the bound is not the one giving the best generalization. In practice, substantially less weighting should be given to the regularizer. In fact, we have seen that minimizing the empirical risk of the randomized majority vote at a fixed number of voters, without considering explicitly the KL regularizer, gives equally good results. Among the different algorithms that we have proposed, the latter appears to be the best substitute to regularized variants of boosting because the number of voters is the only hyperparameter to tune.

We have also found that probing the empirical risk of the randomized majority vote at zero margin gives equally good weighted majority votes as those produced by probing the empirical risk at finite margin.

Acknowledgments

Work supported by NSERC discovery grants 122405 and 262067.

References

1. Catoni, O.: PAC-Bayesian supervised classification: the thermodynamics of statistical learning. Monograph series of the Institute of Mathematical Statistics (December 2007), `http://arxiv.org/abs/0712.0248`
2. Freund, Y., Schapire, R.E.: A decision-theoretic generalization of on-line learning and an application to boosting. Journal of Computer and System Sciences 55, 119–139 (1997)
3. Germain, P., Lacasse, A., Laviolette, F., Marchand, M.: PAC-Bayesian Learning of Linear Classifiers. In: Bottou, L., Littman, M. (eds.) Proceedings of the 26th International Conference on Machine Learning (ICML 2009), pp. 353–360. Omnipress, Montreal (June 2009)
4. Jaakkola, T., Meila, M., Jebara, T.: Maximum entropy discrimination. In: Advances in neural information processing systems, vol. 12. MIT Press, Cambridge (2000)
5. Langford, J.: Tutorial on practical prediction theory for classification. Journal of Machine Learning Research 6, 273–306 (2005)
6. Langford, J., Seeger, M., Megiddo, N.: An improved predictive accuracy bound for averaging classifiers. In: Brodley, C.E., Danyluk, A.P. (eds.) Proceedings of the Eighteenth International Conference on Machine Learning (ICML 2001), June 28-July 1, pp. 290–297. Morgan Kaufmann, San Francisco (2001)
7. McAllester, D.: PAC-Bayesian stochastic model selection. Machine Learning 51, 5–21 (2003)
8. McAllester, D.A.: PAC-Bayesian model averaging. In: COLT, pp. 164–170 (1999)
9. Schapire, R.E., Freund, Y., Bartlett, P., Lee, W.S.: Boosting the margin: A new explanation for the effectiveness of voting methods. The Annals of Statistics 26, 1651–1686 (1998)
10. Seeger, M.: PAC-Bayesian generalization bounds for gaussian processes. Journal of Machine Learning Research 3, 233–269 (2002)

Exploration in Relational Worlds

Tobias Lang[1], Marc Toussaint[1], and Kristian Kersting[2]

[1] Machine Learning and Robotics Group, Technische Universität Berlin, Germany
tobias.lang@tu-berlin.de, mtoussai@cs.tu-berlin.de
[2] Fraunhofer Institute IAIS, Sankt Augustin, Germany
kristian.kersting@iais.fraunhofer.de

Abstract. One of the key problems in model-based reinforcement learning is balancing exploration and exploitation. Another is learning and acting in large relational domains, in which there is a varying number of objects and relations between them. We provide one of the first solutions to exploring large relational Markov decision processes by developing relational extensions of the concepts of the Explicit Explore or Exploit (E^3) algorithm. A key insight is that the inherent generalization of learnt knowledge in the relational representation has profound implications also on the exploration strategy: what in a propositional setting would be considered a novel situation and worth exploration may in the relational setting be an instance of a well-known context in which exploitation is promising. Our experimental evaluation shows the effectiveness and benefit of relational exploration over several propositional benchmark approaches on noisy 3D simulated robot manipulation problems.

1 Introduction

Acting optimally under uncertainty is a central problem of artificial intelligence. In reinforcement learning, an agent's learning task is to find a policy for action selection that maximizes its reward over the long run. Model-based approaches learn models of the underlying Markov decision process from the agent's interactions with the environment, which can then be analyzed to compute optimal plans. One of the key problems in reinforcement learning is the exploration-exploitation tradeoff, which strives to balance two competing types of behavior of an autonomous agent in an unknown environment: the agent can either make use of its current knowledge about the environment to maximize its cumulative reward (i.e., to exploit), or sacrifice short-term rewards to gather information about the environment (i.e., to explore) in the hope of increasing future long-term return, for instance by improving its current world model. This exploration/exploitation tradeoff has received a lot of attention in propositional and continuous domains. Several powerful technique have been developed such as E^3 [14], $R^{\max}$ [3] and Bayesian reinforcement learning [19].

Another key problem in reinforcement learning is learning and acting in large relational domains, in which there is a varying number of objects and relations among them. Nowadays, relational approaches become more and more important [9]: information about one object can help the agent to reach conclusions

J.L. Balcázar et al. (Eds.): ECML PKDD 2010, Part II, LNAI 6322, pp. 178–194, 2010.

about other, related objects. Such relational domains are hard – or even impossible – to represent meaningfully using an enumerated state space. For instance, consider a hypothetical household robot which just needs to be taken out of the shipping box, turned on, and which then explores the environment to become able to attend its cleaning chores. Without a compact knowledge representation that supports abstraction and generalization of previous experiences to the current state and potential future states, it seems to be difficult – if not hopeless – for such a "robot-out-of-the-box" to explore one's home in reasonable time. There are too many objects such as doors, plates and water-taps. For instance, after having opened one or two water-taps in bathrooms, the priority for exploring further water-taps in bathrooms, and also in other rooms such as the kitchen, should be reduced. This is impossible to express in a propositional setting where we would simply encounter a new and therefore non-modelled situation.

So far, however, the important problem of exploration in stochastic relational worlds has received surprisingly little attention. This is exactly the problem we address in the current paper. Simply applying existing, propositional exploration techniques is likely to fail: what in a propositional setting would be considered a novel situation and worth exploration may in the relational setting be an instance of a well-known context in which exploitation is promising. This is the key insight of the current paper: *the inherent generalization of learnt knowledge in the relational representation has profound implications also on the exploration strategy*. Consequently, we develop relational exploration strategies in this paper. More specifically, our work is inspired by Kearns and Singh's seminal exploration technique E^3 (Explicit Explore or Exploit, discussed in detail below). By developing a similar family of strategies for the relational case and integrating it into the state-of-the-art model-based relational reinforcement learner PRADA [16], we provide a practical solution to the exploration problem in relational worlds. Based on actively generated training trajectories, the exploration strategy and the relational planner together produce in each round a learned world model and in turn a policy that either reduces uncertainty about the environment, i.e., improves the current model, or exploits the current knowledge to maximize utility of the agent. Our extensive experimental evaluation in a 3D simulated complex desktop environment with an articulated manipulator and realistic physics shows that our approaches can solve tasks in complex worlds where non-relational methods face severe efficiency problems.

We proceed as follows. After touching upon related work, we review background work. Then, we develop our relational exploration strategies. Before concluding, we present the results of our extensive experimental evaluation.

2 Related Work

Several exploration approaches such as E^3 [14], $R^{\max}$ [3] and extensions [13,10] have been developed for propositional and continuous domains, i.e., assuming the environment to be representable as an enumerated or vector space. In recent years, there has been a growing interest in using rich representations such as relational languages for reinforcement learning (RL). While traditional RL requires (in principle) explicit state and action enumeration, these symbolic approaches seek to avoid explicit state and action enumeration through a symbolic representation of states and actions. Most work in this context has focused on model-free approaches estimating a value function and has not developed relational exploration strategies. Essentially, a number of relational regression algorithms have been developed for use in these relational RL systems such as relational regression trees [8] or graph kernels and Gaussian processes [7]. Kersting and Driessens [15] have proposed a relational policy gradient approach. These approaches use some form of ϵ-greedy strategy to handle explorations; no special attention has been paid to the exploration-exploitation problem as done in the current paper. Driessens and Džeroski [6] have proposed the use of "reasonable policies" to provide guidance, i.e., to increase the chance to discover sparse rewards in large relational state spaces. This is orthogonal to exploration. Ramon *et al.* [20] presented an incremental relational regression tree algorithm that is capable of dealing with concept drift and showed that it enables a relational Q-learner to transfer knowledge from one task to another. They, however, do not learn a model of the domain and, again, relational exploration strategies were not developed. Croonenborghs *et al.* [5] learn a relational world model online and additionally use lookahead trees to give the agent more informed Q-values by looking some steps into the future when selecting an action. Exploration is based on sampling random actions instead of informed exploration. Walsh [23] provides the first principled investigation into the exploration-exploitation tradeoff in relational domains and establishes sample complexity bounds for specific relational MDP learning problems. In contrast, we learn more expressive domain models and propose a variety of different relational exploration strategies.

There is also an increasing number of (approximate) dynamic programming approaches for solving relational MDPs, see e.g. [2,21]. In contrast to the current paper, however, they assume a given model of the world. Recently, Lang and Toussaint [17] and Joshi *et al.* [12] have shown that successful planning typically involves only a small subset of relevant objects respectively states and how to make use of this fact to speed up symbolic dynamic programming significantly. A principled approach to exploration, however, has not been developed.

3 Background on MDPs, Exploration, and Relational Worlds

A Markov decision process (MDP) is a discrete time stochastic control process used to model the interaction of an agent with its environment. At each time-step, the process is in one of a fixed set of discrete states S and the agent can choose an action from a set A. The conditional transition probabilities $P(s'|a, s)$ specify the distribution over successor states when executing an action in a given state. The agent receives rewards in states according to a function $R : S \to \mathbb{R}$. The goal is to find a policy $\pi : S \to A$ specifying which action to take in a given state in order to maximize the future rewards. For a discount factor $0 < \gamma < 1$, the value of a policy π for a state s is defined as the sum of discounted rewards $V^{\pi}(s) = E[\sum_t \gamma^t R(s_t) \mid s_0 = s, \pi]$. In our context, we do not know the transition probabilities $P(s'|a, s)$ so that we face the problem of reinforcement learning (RL). We pursue a model-based approach: we estimate $P(s'|a,s)$ from our experiences and compute (approximately) optimal policies based on the estimated model. The quality of these policies depends on the accuracy of this estimation. We need to ensure that we learn enough about the environment in order to be able to plan for high-value states (explore). At the same time, we have to ensure not to spend too much time in low-value parts of the state space (exploit). This is known as the exploitation/exploration-tradeoff.

Kearns and Singh's E^3 (Explicit Explore or Exploit) algorithm [14] provides a near-optimal model-based solution to the exploitation/exploration problem. It distinguishes explicitly between exploitation and exploration phases. The central concept are *known states* where all actions have been observed sufficiently often. If E^3 enters an *unknown* state, it takes the action it has tried the fewest times there ("**direct exploration**"). If it enters a *known* state, it tries to calculate a high-value policy within an MDP built from all known states (where its model estimates are sufficiently accurate). If it finds such a policy which stays with high probability in the set of known states, this policy is executed ("**exploitation**"). Otherwise, E^3 plans in a different MDP in which the unknown states are assumed to have very high value ("optimism in the face of uncertainty"), ensuring that the agent explores unknown states efficiently ("**planned exploration**").

One can prove that with high probability E^3 performs near optimally for all but a polynomial number of time-steps. The theoretical guarantees of E^3 and similar algorithms such as $R^{\max}$ are strong. In practice, however, the number of exploratory actions becomes huge so that in case of large state spaces, such as in relational worlds, it is unrealistic to meet the theoretical thresholds of state visits. To address this drawback, variants of E^3 for factored but propositional MDP representations have been explored [13,10]. Our evaluations will include variants of factored exploration strategies (PEX and OPT-PEX) where the factorization is based on the grounded relational formulas. However, such factored MDPs still do not generalize over objects. Relational worlds can be represented more compactly using relational MDPs. The state space S of a relational MDP (RMDP) has a relational structure defined by predicates $\mathcal{P}$ and functions $\mathcal{F}$, which yield the set of ground atoms with arguments taken from the set of domain objects $\mathcal{O}$. The action space A is defined by atoms $\mathcal{A}$ with arguments from $\mathcal{O}$. In contrast to ground atoms, abstract atoms contain logical variables as arguments. We will speak of grounding an abstract formula ψ if we apply a substitution σ that maps all of the variables appearing in ψ to objects in $\mathcal{O}$.

A compact relational transition model $P(s'|a,s)$ uses formulas to abstract from concrete situations and object identities. The principle ideas of relational exploration we develop in this paper work with any type of relational model. In this paper, however, we employ *noisy indeterministic deictic (NID) rules* [18] to illustrate and empirically evaluate our ideas. A NID rule r is given as

$$a_r(\mathcal{X}) : \ \phi_r(\mathcal{X}) \quad \rightarrow \quad \begin{cases} p_{r,1} & : \Omega_{r,1}(\mathcal{X}) \\ & \vdots \\ p_{r,m_r} & : \Omega_{r,m_r}(\mathcal{X}) \\ p_{r,0} & : \Omega_{r,0} \end{cases} , \tag{1}$$

where $\mathcal{X}$ is a set of logic variables in the rule (which represent a (sub-)set of abstract objects). The rule r consists of preconditions, namely that action a_r is applied on $\mathcal{X}$ and that the abstract state context ϕ_r is fulfilled, and m_r+1 different abstract outcomes with associated probabilities $p_{r,i} > 0$, $\sum_{i=0} p_{r,i} = 1$. Each outcome $\Omega_{r,i}(\mathcal{X})$ describes which atoms "change" when the rule is applied. The context $\phi_r(\mathcal{X})$ and outcomes $\Omega_{r,i}(\mathcal{X})$ are conjunctions of literals constructed from the literals in $\mathcal{P}$ as well as equality statements comparing functions from $\mathcal{F}$ to constant values. The so-called *noise outcome* $\Omega_{r,0}$ subsumes all possible action outcomes which are not explicitly specified by one of the other $\Omega_{r,i}$. The arguments of the action $a(\mathcal{X}_a)$ may be a true subset $\mathcal{X}_a \subset \mathcal{X}$ of the variables $\mathcal{X}$ of the rule. The remaining variables are called deictic references $DR = \mathcal{X} \setminus \mathcal{X}_a$ and denote objects relative to the agent or action being performed.

So, how do we apply NID rules? Let σ denote a substitution that maps variables to constant objects, $\sigma : \mathcal{X} \to \mathcal{O}$. Applying σ to an abstract rule $r(\mathcal{X})$ yields a *grounded rule* $r(\sigma(\mathcal{X}))$. We say a grounded rule r *covers* a state s and a ground action a if $s \models \phi_r$ and $a = a_r$. Let Γ be our set of rules and $\Gamma(s,a) \subset \Gamma$ the set of rules covering (s,a). If there is a unique covering rule $r_{(s,a)} \in \Gamma(s,a)$, we use it to model the effects of action a in state s. If no such rule exists (including the case that more one rule covers the state-action pair), we use a noisy default rule r_ν which predicts all effects as noise. The semantics of NID rules allow one to efficiently plan in relational domains, i.e. to find a "satisficing" action sequence that will lead with high probability to states with large rewards. In this paper, we use the PRADA algorithm [16] for planning in grounded relational domains. PRADA converts NID rules into dynamic Bayesian networks, predicts the effects of action sequences on states and rewards by means of approximate inference and samples action sequences in an informed way. PRADA copes with different types of reward structures, such as partially abstract formulas or maximizing derived functions. We learn NID rules from the experiences $\mathcal{E} = \{(s_t, a_t, s_{t+1})_{t=0}^{T-1}\}$ of an actively exploring agent, using a batch algorithm that trades off the likelihood of these triples with the complexity of the learned rule-set. $\mathcal{E}(r) = \{(s,a,s') \in \mathcal{E} \,|\, r = r_{(s,a)}\}$ are the experiences which are uniquely covered by a learned rule r. For more details, we refer the reader to Pasula et al. [18].

4 Exploration in Relational Domains

We first discuss the implications of a relational knowledge representation for exploration on a conceptual level. We adopt a density estimation view to pinpoint the differences between propositional and relational exploration (Sec. 4.1). This conceptual discussion opens the door to a large variety of possible exploration strategies – we cannot test all such approaches within this paper. Thus, we focus on specific choices to estimate novelty and hence of the respective exploration strategies (Sec. 4.2), which we found effective as a first proof of concept.

4.1 A Density Estimation View on Known States and Actions

The theoretical derivations of the non-relational near-optimal exploration algorithms E^3 and $R^{\max}$ show that the concept of known states is crucial. On the one hand, the confidence in estimates in known states drives exploitation. On the other hand, exploration is guided by seeking for novel (yet unknown) states and actions. For instance, the direct exploration phase in E^3 chooses novel actions, which have been tried the fewest; the planned exploration phase seeks to visit novel states, which are labeled as yet unknown.

In the case of the original E^3 algorithm (and $R^{\max}$ and similar methods) operating in an enumerated state space, states and actions are considered known based directly on the number of times they have been visited. In relational domains, there are two reasons for why we should go beyond simply counting state-action visits to estimate the novelty of states and actions:

1. The size of the state space is exponential in the number of objects. If we base our notion of known states directly on visitation counts, then the overwhelming majority of all states will be labeled yet-unknown and the exploration time required to meet the criteria for known states of E^3 even for a small relevant fraction of the state space becomes exponential in large domains.
2. The key benefit of relational learning is the ability to generalize over yet unobserved instances of the world based on relational abstractions. This implies a fundamentally different perspective on what is novel and what is known and permits qualitatively different exploration strategies compared to the propositional view.

A constructive approach to pinpoint the differences between propositional and relational notions of exploration, novelty and known states is to focus on a density estimation view. This is also inspired by the work on active learning which typically selects points that, according to some density model of previously seen points, are novel (see, e.g., [4] where the density model is an implicit mixture of Gaussians). In the following we first discuss different approaches to model a distribution of known states and actions in a relational setting. These methods estimate which relational states are considered known with some useful confidence measures according to our experiences $\mathcal{E}$ and world model $\mathcal{M}$.

Propositional: Let us first consider briefly the propositional setting from a density estimation point of view. We have a finite enumerated state space S and action space A. Assume our agent has so far observed the set of state transitions $\mathcal{E} = \{(s_t, a_t, s_{t+1})\}_{t=1}^{T-1}$. This translates directly to a density estimate

$$P(s) \propto c_{\mathcal{E}}(s)\,, \quad \text{with } c_{\mathcal{E}}(s) = \sum\nolimits_{(s_e,a_e,s'_e)\in\mathcal{E}} I(s_e = s)\,, \tag{2}$$

where $c_{\mathcal{E}}(s)$ counts the number of occasions state s has been visited in $\mathcal{E}$ (in the spirit of [22]) and $I(\cdot)$ is the indicator function which is 1 if the argument evaluates to true and 0 otherwise. This density implies that all states with low $P(s)$ are considered novel and should be explored, as in E^3. There is no generalization in this notion of known states. Similar arguments can be applied on the level of state-action counts and the joint density $P(s, a)$.

Predicate-based: Given a relational structure with the set of logical predicates $\mathcal{P}$, an alternative approach to describe what are known states is based on counting how often a ground or abstract predicate has been observed true or false in the experiences $\mathcal{E}$ (all statements equally apply to functions $\mathcal{F}$, but we neglect this case here). First, we consider grounded predicates $p \in \mathcal{P}^G$ with arguments taken from the domain objects $\mathcal{O}$. This leads to a density estimate

$$P_p(s) \propto c_p(s)\, I(s \models p) + c_{\neg p}(s)\, I(s \models \neg p) \quad \text{with} \quad c_p(s) := \sum_{(s_e,a_e,s'_e)\in\mathcal{E}} I(s_e \models p). \tag{3}$$

Each p implies a density $P_p(s)$ which counts how often p has the same truth values in s and in experienced states. We take the product to combine all $P_p(s)$. This implies that a state is considered familiar (with non-zero $P(s)$) if each predicate that is true (false) in this state has been observed true (false) before. We will use this approach for our planned exploration strategy (Sec. 4.2). We can follow the same approach for partially grounded predicates $\mathcal{P}^{PG}$. For $p \in \mathcal{P}^{PG}$ and a state s, we examine whether there are groundings of the logical variables in p such that s covers p. More formally, we replace $s \models p$ by $\exists\sigma : s \models \sigma(p)$. E.g., we may count how often the blue ball was on top of *some* other object. If this was rarely the case this implies a notion of novelty which guides exploration.

Context-based: Assume that we are given a finite set Φ of contexts, which are formulas of abstract predicates and functions. While many relational knowledge representations have some notion of context or rule precondition, in our case these may correspond to the set of NID rule contexts $\{\phi_r\}$. These are learnt from the experiences $\mathcal{E}$, which have specifically been optimized to be a compact context representation that covers the experiences and allows for the prediction of action effects (cf. Sec. 3). Analogous to the above, given a set of such formulas we may consider the density

$$P_\phi(s) \propto \sum_{\phi\in\Phi} c_{\mathcal{E}}(\phi)\, I(\exists\sigma : s \models \sigma(\phi)) \quad \text{with} \quad c_{\mathcal{E}}(\phi) = \sum_{(s_e,a_e,s'_e)\in\mathcal{E}} I(\exists\sigma : s_e \models \sigma(\phi)). \tag{4}$$

$c_{\mathcal{E}}(\phi)$ counts in how many experiences $\mathcal{E}$ the context ϕ was covered with arbitrary groundings. Intuitively, the context of the NID rules may be understood as describing situation classes based on whether the same predictive rules can be applied. Taking this approach, states are considered novel if they are not covered by any existing context ($P_\phi(s) = 0$) or covered by a context that has rarely occurred in $\mathcal{E}$ ($P_\phi(s)$ is low). That is, the description of novelty which drives exploration is lifted to the level of abstraction of these relational contexts. Similarly, we formulate a density estimation over states and actions based on the set of NID rules, where each rule defines a state-action context,

$$P_r(s,a) \propto \sum_{r\in\Gamma} c_{\mathcal{E}}(r)\, I(r = r_{s,a}), \quad \text{with} \quad c_{\mathcal{E}}(r) := |\mathcal{E}(r)|, \tag{5}$$

which is based on counting how many experiences are covered by the unique covering rule $r_{s,a}$ for a in s. Recall that $\mathcal{E}(r)$ are the experiences which are covered by r. Thus, the more experiences the corresponding unique covering rule $r_{(s,a)}$ covers the larger is $P_r(s,a)$ and it can be seen as a measure of confidence in r. We will use $P_r(s,a)$ to guide direct exploration below.

Distance-based: As mentioned in the related work discussion, different methods to estimate the similarity of relational states exist. These can be used for relational density estimation (in the sense of 1-class SVMs) which, when applied in our context, would readily imply alternative notions of novelty and thereby exploration strategies. To give an example, [7] and [11] present relational reinforcement learning approaches which use relational graph kernels to estimate the similarity of relational states. Applying such a method to model $P(s)$ from $\mathcal{E}$ would imply that states are considered novel (with low $P(s)$) if they have a low kernel value (high "distance") to previous explored states. For a given state s, we directly define a measure of distance to all observed data, $d(s) = \min_{(s_e,a_e,s'_e)\in\mathcal{E}} d(s, s_e)$, and set

$$P_d(s) \propto \frac{1}{d(s)+1} \,. \tag{6}$$

Here, $d(s, s')$ can be any distance measure, for instance based on relational graph kernels. We will use a similar but simplified approach as part of a specific direct exploration strategy on the level of $P_d(s, a)$, as described in detail in Sec. 4.2. In our experiments, we use a simple distance based on least general unifiers.

All three relational density estimation techniques emphasize different aspects and we combine them in our algorithms.

4.2 Relational Exploration Algorithms

The density estimation approaches discussed above open a large variety of possibilities for concrete exploration strategies. In the following, we derive model-based relational reinforcement learning algorithms which explicitly distinguish between exploration and exploitation phases in the sense of E^3. Our methods are based on simple, but empirically effective relational density estimators. We are certain that more elaborate and efficient exploration strategies can be derived from the above principles in the future. Our algorithms perform the following general steps: *(i)* In each step they first adapt the relational model $\mathcal{M}$ with the set of experiences $\mathcal{E}$. *(ii)* Based on $\mathcal{M}$, s and $\mathcal{E}$, they select an action a – we focus on this below. *(iii)* The action a is executed, the resulting state s' observed and added to the experiences $\mathcal{E}$, and the process repeated.

Our first algorithm transfers the general E^3 approach (distinguishing between exploration and exploitation based on whether the current state is fully known) to the relational domain to compute actions. The second tries to exploit more optimistically, even when the state is not known or only partially known. Both algorithms are based on a set of subroutines which instantiate the ideas mentioned above and which we describe first:

plan(world model $\mathcal{M}$, reward function τ, state s_0): Returns the first action of a plan of actions that maximizes τ. Typically, τ is expressed in terms of logical formulas, describing goal situations to which a reward of 1 is associated. If the planner estimates a maximum expected reward close to zero (i.e., no good plan is found), it returns a 0 instead of the first action. In this paper, we employ NID rules as $\mathcal{M}$ and use the PRADA algorithm for planning.

isKnown(world model $\mathcal{M}$, state s): s is known if the estimated probabilities $P(s, a)$ of all actions a are larger than some threshold. We employ the rule-context based density estimate $P_r(s, a)$ (Eq. 5).

Algorithm 1. REX – Action Computation

Input: World model $\mathcal{M}$, Reward function τ, State s_0, Experiences $\mathcal{E}$
Output: Action a

if isKnown($\mathcal{M}$, s_0) **then**
  a = plan($\mathcal{M}$, τ, s_0) ▷ Try to exploit
  if $a \neq 0$ **then**
    return a ▷ Exploit succeeded
  end if
  $\tau_{explore}$ = getPlannedExplorationReward($\mathcal{M}$, $\mathcal{E}$)
  a = plan($\mathcal{M}$, $\tau_{explore}$, s_0) ▷ Try planned exploration
  if $a \neq 0$ **then**
    return a ▷ Planned exploration succeeded
  end if
end if
$\mathbf{w}$ = getDirectExplorationWeights($\mathcal{M}$, $\mathcal{E}$, s_0) ▷ Sampling weights for actions
a = sample($\mathbf{w}$) ▷ Direct exploration (without planning)
return a

isPartiallyKnown(world model $\mathcal{M}$, reward function τ, state s): In contrast to before, we only consider relevant actions. These refer to objects which appear explicitly in the reward description or are related to them in s by some binary predicate.

getPlannedExplorationReward(world model $\mathcal{M}$, experiences $\mathcal{E}$): Returns a reward function for planned exploration, expressed in terms of logical formulas as for plan, describing goal situations worth for exploration. We follow the predicate-based density estimation view (Eq. (3)) and set the reward function to $\frac{1}{P_p(s)}$.

getDirectExplorationWeights(world model $\mathcal{M}$, experiences $\mathcal{E}$, state s): Returns weights according to which an action is sampled for direct exploration. Here, the two algorithms use different heuristics:
(i) REX sets the weights for actions a with minimum value $|\mathcal{E}(r_{s,a})|$ to 1 and for all others to 0, thereby employing $P_r(s,a)$. This combines E^3 (choosing the action with the fewest "visits") with relational generalization (defining "visits" by means of confidence in abstract rules).
(ii) OPT-REX combines three scores to decide on direct exploration weights. The first score is inverse proportional to $P_r(s,a)$. The second is inverse proportional to the distance-based density estimation $P_d(s,a)$ (Eq. 6). The third score is an additional heuristic to increase the probability of relevant actions (with the same idea as in partially known states, that we care more about the supposedly relevant parts of the action space).

These subroutines are the basic building blocks for the two relational exploration algorithms REX and OPT-REX that we discuss now in turn.

REX **(Relational Explicit Explore or Exploit).** (Algorithm 1) REX lifts the E^3 planner to relational exploration and uses the same phase order as E^3. If the current state is known, it tries to exploit $\mathcal{M}$. In contrast to E^3, REX also plans through unknown states as it is unclear how to efficiently build and

Algorithm 2. OPT-REX – Action Computation

Input: World model $\mathcal{M}$, Reward function τ, State s_0, Experiences $\mathcal{E}$
Output: Action a

$a = \mathsf{plan}(\mathcal{M}, \tau, s_0)$ ▷ Try to exploit
if $a \neq 0$ **then**
  return a ▷ Exploit succeeded
end if
if $\mathsf{isPartiallyKnown}(\mathcal{M}, \tau, s_0)$ **then**
  $\tau_{explore} = \mathsf{getPlannedExplorationReward}(\mathcal{M}, \mathcal{E})$
  $a = \mathsf{plan}(\mathcal{M}, \tau_{explore}, s_0)$ ▷ Try planned exploration
  if $a \neq 0$ **then**
    return a ▷ Planned exploration succeeded
  end if
end if
$\mathbf{w} = \mathsf{getDirectExplorationWeights}(\mathcal{M}, \mathcal{E}, s_0)$ ▷ Sampling weights for actions
$a = \mathrm{sample}(\mathbf{w})$ ▷ Direct exploration (without planning)
return a

exclusively use an MDP of known relational states. However, in every state only sufficiently known actions are taken into account. In our experiments, for instance, our planner PRADA achieves this by only considering actions with unique covering rules in a given state. If exploitation fails, an exploration goal is set up for planned exploration. In case planned exploration fails as well or the current state is unknown, the action with the lowest confidence is carried out (similarly, as E^3 chooses the action which was performed the least often in the current state).

OPT-REX **(Optimistic** REX). (Algorithm 2) OPT-REX modifies REX according to the intuition that there is no need to understand the world dynamics to full extent: rather it makes sense to focus on the relevant parts of the state and action space. OPT-REX exploits the current knowledge optimistically to plan for the goal. For a given state s_0, it tries immediately to come up with an exploitation plan. If this fails, it checks whether s_0 is partially known, i.e., whether the world model $\mathcal{M}$ can predict the actions which are relevant for the reward τ. If the state s_0 is partially known, planned exploration is tried. If this fails or s_0 is partially unknown, direct exploration is undertaken, with action sampling weights as described above.

5 Evaluation

Our intention here is to compare propositional and relational techniques for exploring relational worlds. More precisely, we investigate the following questions:

- $Q1$: Can relational knowledge improve exploration performance?
- $Q2$: How do propositional and relational explorers scale with the number of domain objects?
- $Q3$: Can relational explorers transfer knowledge to new situations, objects and tasks?

Fig. 1. In our experiments, a robot has to explore a 3D simulated desktop environment with cubes, balls and boxes of different sizes and colors to master various tasks

To do so, we compare five different methods inspired by E^3 based on propositional or abstract symbolic world models. In particular, we learn (propositional or abstract) NID rules after each new observation from scratch using the algorithm of Pasula et al. [18] and employ PRADA [16] for exploitation or planned exploration. All methods deem an action to be known in a state if the confidence in its covering rule is above a threshold ς. Instead of deriving ς from the E^3 equations which is not straightforward and will lead to overly large thresholds (see [10]), we set it heuristically such that the confidence is high while still being able to explore the environments of our experiments within a reasonable number of actions (< 100).

PEX (propositional E^3) is a variant of E^3 based on propositional NID rules (with ground predicates and functions). While it abstracts over states using the factorization of rules, it cannot transfer knowledge to unseen objects. OPT-PEX (optimistic PEX) is similar, but always tries to exploit first, independently of whether the current state is known or not. REX and OPT-REX (cf. Sec. 4.2) use abstract relational NID rules for exploration and exploitation. In addition, we investigate a relational baseline method RAND-REX (Relational exploit or random) which tries to exploit first (being as optimistic as OPT-REX) and if this is impossible produces a random action.

Our test domain is a simulated complex desktop environment where a robot manipulates cubes, balls and boxes scattered on a table (Fig. 1). We use a 3D rigid-body dynamics simulator (ODE) that enables a realistic behavior of the objects. For instance, piles of objects may topple over or objects may even fall off the table (in which case they become out of reach for the robot). Depending on their type, objects show different characteristics. For example, it is almost impossible to successfully put an object on top of a ball, and building piles with small objects is more difficult. The robot can grab objects, try to put them on top of other objects, in a box or on the table. Boxes have a lid; special actions may open or close the lid; taking an object out of a box or putting it into it is possible only when the box is opened. The actions of the robot are affected by noise so that resulting object piles are not straight-aligned. We assume full observability of triples (s, a, s') that specify how the world changed when an action was executed in a certain state. We represent the data with

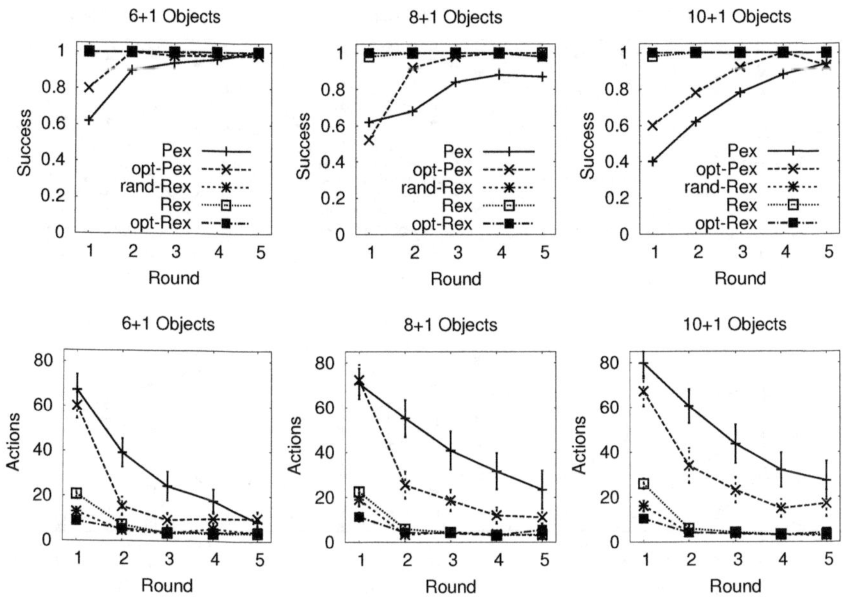

Fig. 2. *Experiment 1: Unchanging Worlds of Cubes and Balls.* A run consists of 5 subsequent rounds with the same start situations and goal objects. The robot starts with no knowledge in the first round. The success rate and the mean estimators of the action numbers with standard deviations over 50 runs are shown (5 start situations, 10 seeds).

predicates $cube(X)$, $ball(X)$, $box(X)$, $table(X)$, $on(X,Y)$, $contains(X,Y)$, $out(X)$, $inhand(X)$, $upright(X)$, $closed(X)$, $clear(X) \equiv \forall Y.\neg on(Y,X)$, $inhandNil() \equiv \neg\exists X.inhand(X)$ and functions $size(X)$, $color(X)$ for state descriptions and $grab(X)$, $puton(X)$, $openBox(X)$, $closeBox(X)$ and $doNothing()$ for actions. If there are o objects and f different object sizes and colors in a world, the state space is huge with $f^{2o}2^{2o^2+7o}$ different states (not excluding states one would classify as "impossible" given some intuition about real world physics). This points at the potential of using abstract relational knowledge for exploration.

We perform four increasingly complex series of experiments[1] where we pursue the same or similar tasks over multiple rounds. In all experiments *the robot starts from zero knowledge* ($\mathcal{E} = \emptyset$) *in the first round* and carries over experiences to the next rounds. In each round, we execute a maximum of 100 actions. If the task is still not solved by then, the round fails. We report the success rates and the action numbers to which failed trials contribute with the maximum number.

Unchanging Worlds of Cubes and Balls: The goal in each round is to pile two specific objects, $on(obj1, obj2)$. To collect statistics we investigate worlds of

[1] The website `http://www.user.tu-berlin.de/lang/explore/` provides videos of exemplary rounds as well as pointers to the code of our simulator, the learning algorithm of NID rules and PRADA.

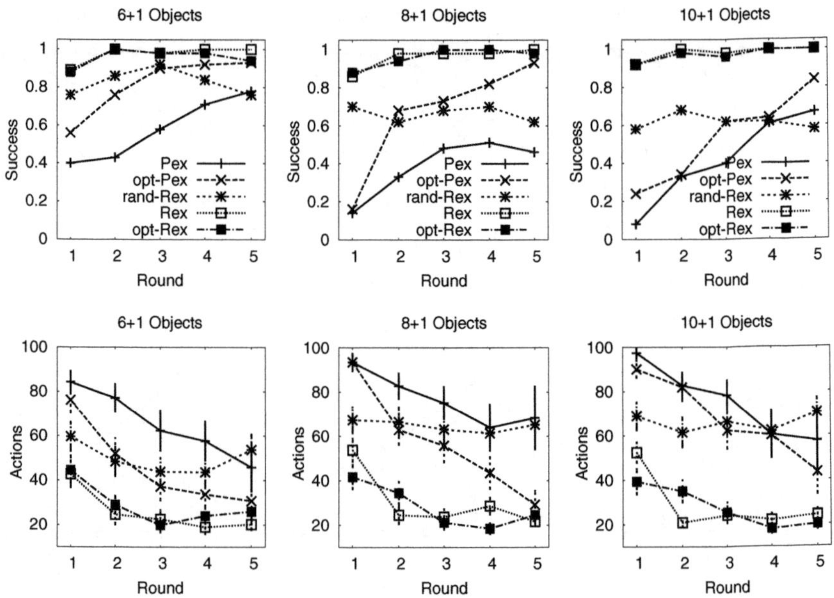

Fig. 3. *Experiment 2: Unchanging Worlds of Boxes.* A run consists of 5 subsequent rounds with the same start situations and goal objects. The robot starts with no knowledge in the first round. The success rate and the mean estimators of the action numbers with standard deviations over 50 runs are shown (5 start situations, 10 seeds).

varying object numbers and for each object number, we create five worlds with different objects. For each such world, we perform 10 independent runs with different random seeds. Each run consists of 5 rounds with the same goal instance and the same start situation. The results presented in Fig. 2 show that already in the first round the relational explorers solve the task with significantly higher success rates and require up to 8 times fewer actions than the propositional explorers. OPT-REX is the fastest approach which we attribute to its optimistic exploitation bias. In subsequent rounds, the relational methods use previous experiences much better, solving those in almost minimal time. In contrast, the action numbers of the propositional explorers fall only slowly.

Unchanging Worlds with Boxes: We keep the task and the experimental setup as before, but in addition the worlds contain boxes, resulting in more complex action dynamics. In particular, some goal objects are put in boxes in the beginning, necessitating more intense exploration to learn how to deal with boxes. Fig. 3 shows that again the relational explorers have superior success rates, require significantly fewer actions and reuse their knowledge effectively in subsequent rounds. While the performance of the propositional planners deteriorates with increasing numbers of objects, OPT-REX and REX scale well. In worlds with many objects, the cautious exploration of REX has the effect that it requires about one third more actions than OPT-REX in the first round, but performs better in subsequent rounds due to the previous thorough exploration.

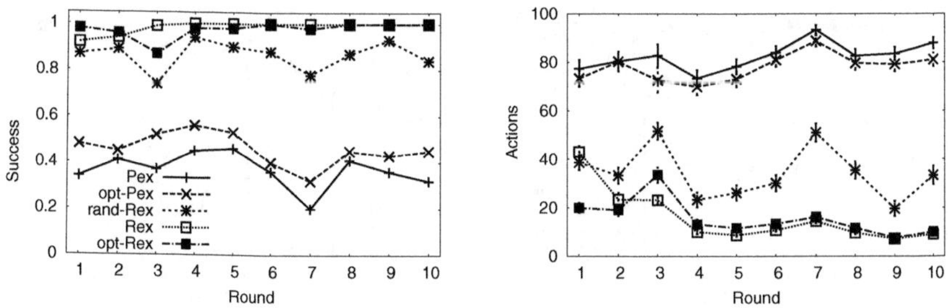

Fig. 4. *Experiment 3: Generalization to New Worlds.* A run consists of a problem sequence of 10 subsequent rounds with different objects, numbers of objects (6 - 10 cubes/balls/boxes + table) and start situations in each round. The robot starts with no knowledge in the first round. The success rate and the mean estimators of the action numbers with standard deviations over 100 runs are shown (10 sequences, 10 seeds).

After the first two experiments we conclude that the usage of relational knowledge improves exploration (question $Q1$) and relational explorers scale better with the number of objects than propositional explorers (question $Q2$).

Generalization to New Worlds: In this series of experiments, the objects, their total numbers and the specific goal instances are different in each round (worlds of 7, 9 and 11 objects). We create 10 problem sequences (each with 10 rounds) and perform 10 trials for each sequence with different random seeds. As Fig.4 shows the performance of the relational explorers is good from the beginning and becomes stable at a near-optimal level after 3 rounds. This answers the first part of question $Q3$: relational explorers can transfer their knowledge to new situations and objects. In contrast, the propositional explorers cannot transfer the knowledge to different worlds and thus neither their success rates nor their action numbers improve in subsequent rounds. Similarly as before, OPT-REX requires less than half of the actions of REX in the first round due to its optimistic exploitation strategy; in subsequent rounds, REX is on par as it has sufficiently explored the system dynamics before.

Generalization to New Tasks: In our final series of experiments, we perform in succession three tasks of increasing difficulty: piling two specific objects in simple worlds with cubes and balls (as in Exp. 1), in worlds extended by boxes (as in Exp. 2 and 3) and building a tower on top of a box where the required objects are partially contained in boxes in the beginning. Each task is performed for three rounds in different worlds with different goal objects. The results presented in Fig. 5 confirm the previous results: the relational explorers are able to generalize over different worlds for a fixed task, while the propositional explorers fail. Beyond that, again in contrast to the propositional explorers, the relational explorers are able to transfer the learned knowledge from simple to difficult tasks in the sense of curriculum learning [1], answering the second part of question $Q3$. To see that, one has to compare the results of round 4 (where the second task of piling two objects in worlds of boxes is given the first time) with the results of round 1 in Experiments 2 and 3. In the latter, no experience from previous tasks

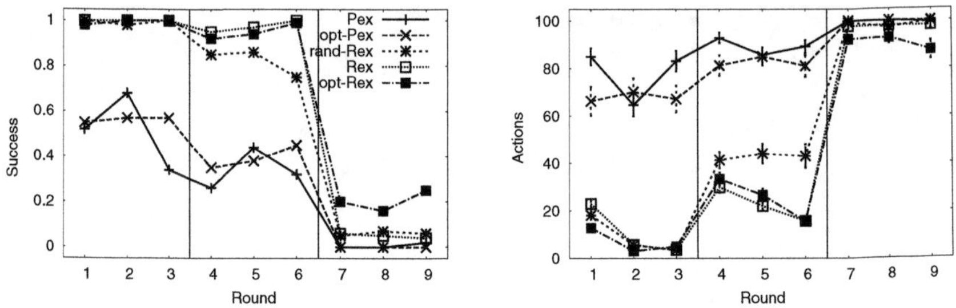

Fig. 5. *Experiment 4: Generalization to New Tasks.* A run consists of a problem sequence of 9 subsequent rounds with different objects, numbers of objects (6 - 10 cubes/balls/boxes + table) and start situations in each round. The tasks are changed between round 3 and 4 and round 6 and 7 to more difficult tasks. The robot starts with no knowledge in the first round. The success rate and the mean estimators of the action numbers with standard deviations over 100 runs are shown (10 sequences, 10 seeds).

is available and REX requires $43.0 - 53.8 \pm 2.5$ actions. In contrast, here it can reuse the knowledge of the simple task (rounds 1-3) and needs about 29.9 ± 2.3 actions. It is instructive to compare this with OPT-REX which performs about the same or even slightly better in the first rounds of Exp. 2 and 3: here, it can fall victim to its optimistic bias which is not appropriate given the changed world dynamics due to the boxes. As a final remark, the third task (rounds 7-9) was deliberately chosen to be very difficult to test the limits of the different approaches. While the propositional planners almost always fail to solve it, the relational planners achieve 5 to 25 times higher success rates.

6 Conclusions

Efficient exploration in relational worlds is an interesting problem that is fundamental to many real-life decision-theoretic planning problems, but has only received little attention so far. We have approached this problem by proposing relational exploration strategies that borrow ideas from efficient techniques for propositional and continuous MDPs. A few principled and practical issues of relational exploration have been discussed, and insights are drawn by relating it to its propositional counterpart. The experimental results show a significant improvement over established results for solving difficult, highly stochastic planning tasks in a 3D simulated complex desktop environment, even in a curriculum learning setting where different problems have to be solved one after the other.

There are several interesting avenues for future work. One is to investigate incremental learning of rule-sets. Another is to explore the connection between relational exploration and transfer learning. Finally, one should start to explore statistical relational reasoning and learning techniques for the relational density estimation problem implicit in exploring relational worlds.

Acknowledgements. TL and MT were supported by the German Research Foundation (DFG), Emmy Noether fellowship TO 409/1-3. KK was supported by the European Commission under contract number FP7-248258-First-MM and the Fraunhofer ATTRACT Fellowship STREAM.

References

1. Bengio, Y., Louradour, J., Collobert, R., Weston, J.: Curriculum learning. In: Proc. of the Int. Conf. on Machine Learning (ICML), pp. 41–48 (2009)
2. Boutilier, C., Reiter, R., Price, B.: Symbolic dynamic programming for first-order MDPs. In: Proc. of the Int. Conf. on Artificial Intelligence (IJCAI), pp. 690–700 (2001)
3. Brafman, R.I., Tennenholtz, M.: R-max - a general polynomial time algorithm for near-optimal reinforcement learning. Journal of Machine Learning Research 3, 213–231 (2002)
4. Cohn, D.A., Ghahramani, Z., Jordan, M.I.: Active learning with statistical models. Journal of Artificial Intelligence Research 4(1), 129–145 (1996)
5. Croonenborghs, T., Ramon, J., Blockeel, H., Bruynooghe, M.: Online learning and exploiting relational models in reinforcement learning. In: Proc. of the Int. Conf. on Artificial Intelligence (IJCAI), pp. 726–731 (2007)
6. Driessens, K., Džeroski, S.: Integrating guidance into relational reinforcement learning. Machine Learning 57(3), 271–304 (2004)
7. Driessens, K., Ramon, J., Gärtner, T.: Graph kernels and Gaussian processes for relational reinforcement learning. In: Machine Learning (2006)
8. Džeroski, S., de Raedt, L., Driessens, K.: Relational reinforcement learning. Machine Learning 43, 7–52 (2001)
9. Getoor, L., Taskar, B. (eds.): A Introduction to Statistical Relational Learning. MIT Press, Cambridge (2007)
10. Guestrin, C., Patrascu, R., Schuurmans, D.: Algorithm-directed exploration for model-based reinforcement learning in factored MDPs. In: Proc. of the Int. Conf. on Machine Learning (ICML), pp. 235–242 (2002)
11. Halbritter, F., Geibel, P.: Learning models of relational MDPs using graph kernels. In: Proc. of the Mexican Conf. on A.I (MICAI), pp. 409–419 (2007)
12. Joshi, S., Kersting, K., Khardon, R.: Self-taught decision theoretic planning with first order decision diagrams. In: Proceedings of ICAPS 2010 (2010)
13. Kearns, M., Koller, D.: Efficient reinforcement learning in factored MDPs. In: Proc. of the Int. Conf. on Artificial Intelligence (IJCAI), pp. 740–747 (1999)
14. Kearns, M., Singh, S.: Near-optimal reinforcement learning in polynomial time. Machine Learning 49(2-3), 209–232 (2002)
15. Kersting, K., Driessens, K.: Non–parametric policy gradients: A unified treatment of propositional and relational domains. In: Proceedings of the 25th International Conference on Machine Learning (ICML 2008), July 5-9 (2008)
16. Lang, T., Toussaint, M.: Approximate inference for planning in stochastic relational worlds. In: Proc. of the Int. Conf. on Machine Learning, ICML (2009)
17. Lang, T., Toussaint, M.: Relevance grounding for planning in relational domains. In: Proc. of the European Conf. on Machine Learning (ECML) (September 2009)
18. Pasula, H.M., Zettlemoyer, L.S., Kaelbling, L.P.: Learning symbolic models of stochastic domains. Artificial Intelligence Research 29, 309–352 (2007)
19. Poupart, P., Vlassis, N., Hoey, J., Regan, K.: An analytic solution to discrete bayesian reinforcement learning. In: Proc. of the Int. Conf. on Machine Learning (ICML), pp. 697–704 (2006)

20. Ramon, J., Driessens, K., Croonenborghs, T.: Transfer learning in reinforcement learning problems through partial policy recycling. In: Kok, J.N., Koronacki, J., Lopez de Mantaras, R., Matwin, S., Mladenič, D., Skowron, A. (eds.) ECML 2007. LNCS (LNAI), vol. 4701, pp. 699–707. Springer, Heidelberg (2007)
21. Sanner, S., Boutilier, C.: Practical solution techniques for first order MDPs. Artificial Intelligence Journal 173, 748–788 (2009)
22. Thrun, S.: The role of exploration in learning control. In: White, D., Sofge, D. (eds.) Handbook for Intelligent Control: Neural, Fuzzy and Adaptive Approaches, Van Nostrand Reinhold, Florence (1992)
23. Walsh, T.J.: Efficient learning of relational models for sequential decision making. PhD thesis, Rutgers, The State University of New Jersey, New Brunswick, NJ (2010)

Efficient Confident Search in Large Review Corpora

Theodoros Lappas[1] and Dimitrios Gunopulos[2]

[1] UC Riverside
[2] University of Athens

Abstract. Given an extensive corpus of reviews on an item, a potential customer goes through the expressed opinions and collects information, in order to form an educated opinion and, ultimately, make a purchase decision. This task is often hindered by false reviews, that fail to capture the true quality of the item's attributes. These reviews may be based on insufficient information or may even be fraudulent, submitted to manipulate the item's reputation. In this paper, we formalize the *Confident Search* paradigm for review corpora. We then present a complete search framework which, given a set of item attributes, is able to efficiently search through a large corpus and select a compact set of high-quality reviews that accurately captures the overall *consensus* of the reviewers on the specified attributes. We also introduce **CREST** (Confident REview Search Tool), a user-friendly implementation of our framework and a valuable tool for any person dealing with large review corpora. The efficacy of our framework is demonstrated through a rigorous experimental evaluation.

1 Introduction

Item reviews are a vital part of the modern e-commerce model, due to their large impact on the opinions and, ultimately, the purchase decisions of Web users. The nature of the reviewed items is extremely diverse, spanning everything from commercial products to restaurants and holiday destinations. As review-hosting websites become more popular, the number of available reviews per item increases dramatically. Even though this can be viewed as a healthy symptom of online information sharing, it can also be problematic for the interested user: as of February of 2010, Amazon.com hosted over 11,480 reviews on the popular "Kindle" reading device. Clearly, it is impractical for a user to read through such an overwhelming review corpus, in order to make a purchase decision. In addition, this massive volume of reviews on a single item inevitably leads to redundancy: many reviews are often repetitious, exhaustively expressing the same (or similar) opinions and contributing little additional knowledge. Further, reviews may also be misleading, reporting false information that does not accurately represent the attributes of an item. Possible causes of such reviews include:

- **Insufficient information:** The reviewer proceeds to an evaluation without having enough information on the item. Instead, opinions are based on partial or irrelevant information.
- **Fraud:** The reviewer maliciously submits false information on an item, in order to harm or boost its reputation.

J.L. Balcázar et al. (Eds.): ECML PKDD 2010, Part II, LNAI 6322, pp. 195–210, 2010.

The main motivation of our work is that a user should not have to manually go through massive volumes of redundant and ambiguous data in order to obtain the required information. The search engines that are currently employed by major review-hosting sites do not consider the particular nature of opinionated text. Instead, reviews are evaluated as typical text segments, while focused queries that ask for reviews with opinions on *specific* attributes are not supported. In addition, reviews are ranked based on very basic methods (e.g. by date) and information redundancy is not considered.

Ideally, false or redundant reviews could be filtered before they become available to users. However, simply labeling a review as "true" or "false" is over-simplifying, since a review may only be partially false. Instead, we propose a framework that evaluates the validity of the opinions expressed in a review and assigns an appropriate *confidence score*. High confidence scores are assigned to reviews expressing opinions that respect the *consensus* formed by the entire review corpus. For example, if 90% of the reviews compliment the battery-life of a new laptop, there is a strong positive consensus on the specific attribute. Therefore, any review that criticizes the battery-life will suffer a reduction in its confidence score, *proportional to the strength of the positive consensus.* At this point, it is important to distinguish between the two types of rare opinions: 1) those that are expressed on attributes that are rarely reviewed and 2) those that contradict the opinion of the majority of the reviewers on a specific attribute. Our approach only penalizes the latter, since the rare opinions in the first group can still be valid (e.g. expert opinions, commenting on attributes that are often overlooked by most users). Further, we employ a simple and efficient method to deal with ambiguous attributes, for which the numbers of positive and negative opinions differ marginally.

Confidence evaluation is merely the first phase of our framework; high-confidence reviews may still be redundant, if they express identical opinions on the same attributes. To address this, we propose an efficient redundancy filter, based on the skyline operator [2]. As shown in the experiments section, the filter achieves a significant reduction of the size of the corpus.

The final component of our framework deals with the evaluation of focused queries: given a set of attributes that the user is interested in, we want to identify *a minimal* set of high-confidence reviews that covers all the specified attributes. To address this, we formalize the *Review Selection* problem for large review corpora and propose a customized search engine for its solution. A complete diagram of our framework can be seen in Figure (1). Figure (2) shows a screenshot of **CREST** (Confident REview Search Tool), a user-friendly tool that implements the full functionality of our framework. In the shown example, **CREST** is applied on a corpus of reviews on a popular Las Vegas hotel. As soon as a review corpus is loaded, **CREST** evaluates the confidence of the available reviews and filters out redundant artifacts. The user can then select a set of features from a list extracted automatically from the corpus. The chosen set is submitted as a query to the search engine, which returns a compact and informative set of reviews. It is important to stress that our engine has no bias against attributes that appear sparsely in the corpus: as long as the user includes an attribute in the query, an appropriate review will be identified and included in the solution.

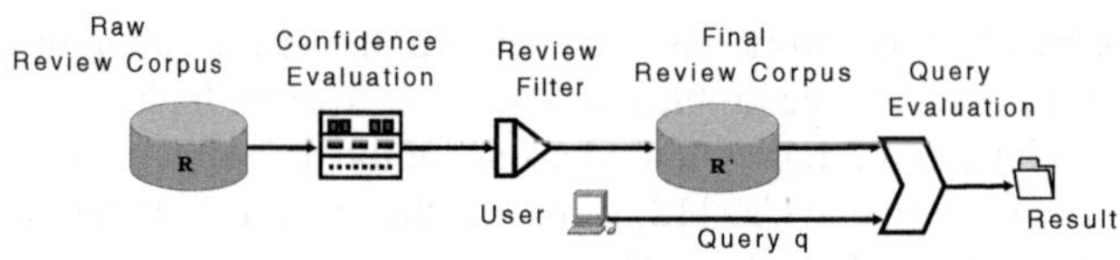

Fig. 1. Given a review corpus $\mathcal{R}$, we first evaluate the *confidence* of each review $r \in \mathcal{R}$. Then, the corpus is filtered, in order to eliminate redundant reviews. Finally, given a query of attributes, the search engine goes through the processed corpus to evaluate the query and select an appropriate set of reviews.

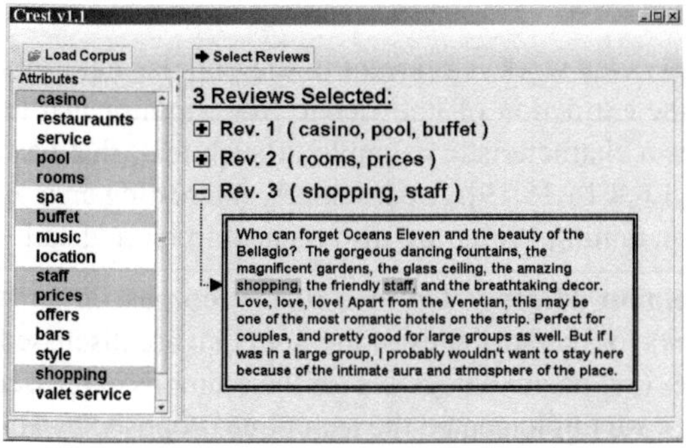

Fig. 2. A user loads a corpus of reviews and then chooses a query of attributes from the automatically-extracted list on the left. The "Select Reviews" button prompts the system to return an appropriate minimal set of reviews.

Contribution: Our primary contribution is an efficient search engine that is customized for large review corpora. The proposed framework can respond to any attribute-based query by returning an appropriate minimal subset of high-quality reviews.

Roadmap: We begin in Section 2 with a discussion on related work. In section 3 we introduce the Confident Search paradigm for large review corpora. In Section 4 we describe how we measure the quality of a review through evaluating the confidence in the opinions it expresses. In Section 5 we discuss how we can effectively reduce the size of the corpus by filtering-out redundant reviews. In Section 6 we propose a review-selection mechanism for the evaluation of attribute-based queries. Then, in Section 7, we conduct a thorough experimental evaluation of the methods proposed in our paper. Finally, we conclude in Section 8 with a brief discussion of the paper.

2 Background

Our work is the first to formalize and address the Confident Search paradigm for review corpora. Even though there has been progress in relevant areas individually, ours is the first work to synthesize elements from all of them toward a customized search engine for review corpora. Next, we review the relevant work from various fields.

Review Assessment: Some work has been devoted on the evaluation of review *helpfulness* [21,13], formalizing the problem as one of regression. Jindal and Liu [10] also adopt an approach based on regression, focusing on the detection of spam (e.g. duplicate reviews). Finally, Liu and Cao [12] formulate the problem as binary classification, assigning a quality rating of "high" or "low" to reviews. Our concept of review assessment differs dramatically from the above-mentioned approaches: first, our framework has no requirement of tagged training data (e.g. spam/not spam, helpful/not helpful). Second, our work is the first to address redundant reviews in a principled and effective manner (Section 5). In any case, we consider prior work on review assessment complementary to ours, since it can be used to filter spam before the application of our framework.

Sentiment Analysis: Our work is relevant to the popular field of sentiment analysis, which deals with the extraction of knowledge from opinionated text. The domain of customer reviews is a characteristic example of such text, that has attracted much attention in the past [1,6,8,14,15,19]. A particularly interesting area of this field is that of attribute and opinion mining, which we discuss next in more detail.

Attribute and Opinion Mining: Given a review corpus on an item, opinion mining [9,17,7,18], looks for the attributes of the item that are discussed in each review, as well as the polarities (i.e. positive/negative) of the opinions expressed on each attribute. For our experiments, we implemented the technique proposed by Hu and Liu [9]: given a review corpus $\mathcal{R}$ on an item, the technique extracts the set of the item's attributes $\mathcal{A}$, and also identifies opinions of the form $(a \rightarrow p), p \in \{-1+1\}, \alpha \in \mathcal{A}$ in each review. We refer the reader to the original paper for further details. Even though this method worked superbly in practice, it is important to note that our framework is compatible with *any* method for attribute and opinion extraction.

Opinion Summarization: In the field of opinion summarization [12,22,11], the given review corpus is processed to produce a cumulative summary of the expressed opinions. The produced summaries are statistical in nature, offering information on the distribution of positive and negative opinions on the attributes of the reviewed item. We consider this work complementary to our own: we present an efficient search engine, able to select a minimal set of actual reviews in response to a specific query of attributes. This provides the user with actual comments written by humans, instead of a less user-friendly and intuitive statistical sheet.

3 Efficient Confident Search

Next, we formalize the *Confident Search* paradigm for large review corpora. We begin with an example, shown in Figure (3). The figure shows the attribute-set and the available review corpus $\mathcal{R}$ for a laptop computer. Out of the 9 available attributes, a user selects only those that interest him. In this case: {"Hard Drive", "Price", "Processor", "Memory"}. Given this query, our search engine goes through the corpus and selects a set of reviews $\mathcal{R}^* = \{r_1, r_7, r_9, r_{10}\}$ that accurately evaluates the specified attributes. Taking this example into consideration, we can now define the three requirements that motivate our concept of *Confident Search*:

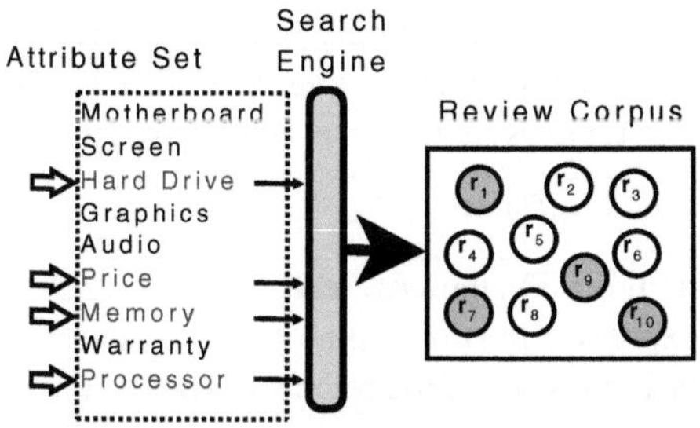

Fig. 3. A use case of our search engine: The user submits a query of 4 attributes, selected from the attribute-set of a computer. Then, the engine goes through a corpus of reviews and locates those that best cover the query (highlighted circles).

1. **Quality:** Given a query of attributes, a user should be presented with a set of high-quality reviews that accurately evaluates the attributes in the query.
2. **Efficiency:** The search engine should minimize the time required to evaluate a query, by appropriately pre-processing the corpus and eliminating redundancy.
3. **Compactness:** The set of retrieved reviews should be informative but also compact, so that a user can read through it in a reasonable amount of time.

Next, we will go over each of the three requirements, and discuss how they are addressed in our framework.

4 Quality through Confidence

We address the requirement for quality by introducing the concept of *confidence* in the opinions expressed within a review. Intuitively, a high-confidence review is one that provides accurate information on the item's attributes. Formally:

[Review Confidence Problem]: Given a review corpus $\mathcal{R}$ on an item, we want to define a function $conf(r, \mathcal{R})$ that maps each review $r \in \mathcal{R}$ to a score, representing the overall confidence in the *opinions* expressed within r.

Let $\mathcal{A}$ be the set of attributes of the reviewed item. Then, an *opinion* refers to one of the attributes in $\mathcal{A}$, and can be either positive or negative. Formally, we define an opinion as a mapping $(\alpha \rightarrow p)$ of an attribute $\alpha \in \mathcal{A}$ to a polarity $p \in \{-1, +1\}$. In our experiments, we extract the set of attributes $\mathcal{A}$ and the respective opinions using the method proposed in [9]. Further, let $\mathcal{O}^-_{r,a}$ and $\mathcal{O}^+_{r,a}$ represent the sets of negative and positive opinions expressed on an attribute α in review r, respectively. Then, we define $pol(\alpha, r)$ to return the *polarity* of α in r. Formally:

$$pol(\alpha, r) = \left\{ \begin{array}{l} +1, \text{ if } |\mathcal{O}^+_{r,\alpha}| > |\mathcal{O}^-_{r,\alpha}| \\ \\ -1, \text{ if } |\mathcal{O}^+_{r,\alpha}| < |\mathcal{O}^-_{r,\alpha}| \end{array} \right\} \quad (1)$$

Note that, for $|\mathcal{O}^+_{r,\alpha}| = |\mathcal{O}^-_{r,\alpha}|$, we simply ignore α, since the expressed opinion is clearly ambiguous. Now, given a review corpus $\mathcal{R}$ and an attribute α, let $n(\alpha \rightarrow p, \mathcal{R})$ be equal to the number of reviews in $\mathcal{R}$, for which $pol(\alpha, r) = p$. Formally:

$$n(\alpha \rightarrow p, \mathcal{R}) = |\{r : pol(\alpha, r) = p,\ r \in \mathcal{R}\}| \tag{2}$$

For example, if the item is a TV, then $n(\text{"screen"} \rightarrow +1, \mathcal{R})$ would return the number of reviews in $\mathcal{R}$ that express a positive opinion on its screen. Given Eq. (2), we can define the concept of the *consensus* of the review-corpus $\mathcal{R}$ on an attribute α as follows:

Definition 1 [Consensus]: Given a set of reviews $\mathcal{R}$ and an attribute α, we define the *consensus of* $\mathcal{R}$ *on* α as:

$$C_{\mathcal{R}}(a) = \underset{p \in \{-1,+1\}}{\operatorname{argmax}}\ n(\alpha \rightarrow p, \mathcal{R}) \tag{3}$$

Conceptually, the consensus expresses the polarity $\in \{-1, +1\}$ that was assigned to the attribute by the majority of the reviews. Formally, given a review corpus $\mathcal{R}$ and an opinion $\alpha \rightarrow p$, we define the *strength* $d(\alpha \rightarrow p, \mathcal{R})$ of the opinion as follows:

$$d(\alpha \rightarrow p, \mathcal{R}) = n(\alpha \rightarrow p) - n(\alpha \rightarrow -p) \tag{4}$$

Since the consensus expresses the majority, we know that $d(\alpha \rightarrow C_{\mathcal{R}}(\alpha), \mathcal{R}) \geq 0$. Further, the higher the value of $d(\alpha \rightarrow C_{\mathcal{R}}(\alpha))$, the higher is our confidence in the consensus. Given Eq. (4), we can now define the overall confidence in the opinions expressed within a given review. Formally:

Definition 2 [Review Confidence]: Given a review corpus $\mathcal{R}$ on an item and the set of the item's attributes $\mathcal{A}$, let $\mathcal{A}_r \subseteq \mathcal{A}$ be the subset of attributes that are actually evaluated within a review $r \in \mathcal{R}$. Then, we define the overall *confidence* of r as follows:

$$conf(r, \mathcal{R}) = \frac{\sum_{\alpha \in \mathcal{A}_r} d(\alpha \rightarrow pol(\alpha, r), \mathcal{R})}{\sum_{\alpha \in \mathcal{A}_r} d(\alpha \rightarrow C_{\mathcal{R}}(\alpha), \mathcal{R})} \tag{5}$$

The confidence in a review takes values in $[-1, 1]$, and is maximized when all the opinions expressed in the review agree with the consensus (i.e. $pol(\alpha, r) = C_{\mathcal{R}}(\alpha), \forall \alpha \in \mathcal{A}_r$). By dividing by the sum of the confidence values in the consensus on each $\alpha \in \mathcal{A}_r$, we ensure that the effect of an opinion $(\alpha \rightarrow p)$ on the confidence of r is proportional to the strength of the consensus on attribute α.

High-confidence reviews are more trustworthy and preferable sources of information, while those with low confidence values contradict the majority of the corpus. The confidence scores are calculated offline and are then stored and readily available for the search engine to use on demand.

5 Efficiency through Filtering

In this Section, we formalize the concept of *redundancy* within a set of reviews and propose a filter for its elimination. As we show with experiments on real datasets, the filter can drastically reduce the size of the corpus. The method is based on the following observation:

Observation 1. Given two reviews r_1 and r_2 in a corpus $\mathcal{R}$, let $\mathcal{A}_{r1} \subseteq \mathcal{A}_{r2}$ and $pol(\alpha, r_1) = pol(\alpha, r_2), \forall \alpha \in \mathcal{A}_{r1}$. Further, let $conf(r_1, \mathcal{R}) \leq conf(r_2, \mathcal{R})$. Then r_1 is redundant, since r_2 expresses the same opinions on the same attributes, while having a higher confidence score.

According to Observation 1, some of the reviews in the corpus can be safely pruned, since they are *dominated* by another review. This formulation matches the definition of the well-known *Skyline* operator [2][16][4], formally defined as follows:

Definition 3 [Skyline]: Given a set of multi-dimensional points $\mathcal{K}$, $Skyline(\mathcal{K})$ is a subset of $\mathcal{K}$ such that, for every point $k \in Skyline(\mathcal{K})$, there exists no point $k' \in \mathcal{K}$ that *dominates* k. We say that k' *dominates* k, if k' is no worse than k in all dimensions.

The computation of the skyline is a highly-studied problem, that comes up in different domains [16]. In the context of our problem, the set of dimensions is represented by the set of possible opinions $\mathcal{O}_{\mathcal{R}}$ that can be expressed within a review corpus $\mathcal{R}$. In the general skyline scenario, a point can assume any value in any of its multiple dimensions. In our case, however, the value of a review $r \in \mathcal{R}$ with respect to an opinion $op \in \mathcal{O}_{\mathcal{R}}$ can only assume one of two distinct values: if the opinion is actually expressed in r, then the value on the respective dimension is equal to $conf(r, \mathcal{R})$. Otherwise, we assign a value of -1, which is the minimum possible confidence score for a review. This ensures that a review r_1 can never be dominated by another review r_2, as long as it expresses at least one opinion that is not expressed in r_2 (since the value of r_2 for the respective dimension will be the lowest possible, i.e. -1).

Most skyline algorithms employ multi-dimensional indexes and techniques for high-dimensional search. However, in a constrained space such as ours, such methods lose their advantage. Instead, we propose a simple and efficient approach that is customized for our problem. The proposed method, which we refer to as `ReviewSkyline`, is shown in Algorithm (2).

Analysis of Algorithm (2): The input consists of a review corpus $\mathcal{R}$, along with the confidence score of each review $r \in \mathcal{R}$ and the set of possible opinions $O_{\mathcal{R}}$. The output is the skyline of $\mathcal{R}$.

Lines [1-2]: The algorithm first sorts the reviews in descending order by confidence. This requires $O(|\mathcal{R}| \log |\mathcal{R}|)$ time. It then builds an inverted index, mapping each opinion to the list of reviews that express it, sorted by confidence. Since we already have a sorted list of all the review from the previous step, this can be done in $O(|\mathcal{R}| \times M)$ time, where M is the size of the review with the most opinions in $\mathcal{R}$.

Lines [3-15]: The algorithm iterates over the reviews in $\mathcal{R}$ in sorted order, eliminating reviews that are dominated by the current Skyline. In order to efficiently check for this, we keep he reviews in the Skyline sorted by confidence. Therefore, since a review can only be dominated by one of higher or equal confidence, a binary search probe is used to check if a review r is dominated.

In line (6), we define a collection of lists $\mathcal{L} = \{L[op] | \forall op \in \mathcal{O}_r\}$, where $L[op]$ is the sorted list of reviews that express the opinion op (from the inverted index created in line (2)). The lists in $\mathcal{L}$ are searched in a round-robin fashion: the first $|\mathcal{L}|$ reviews to be

Algorithm 2. `ReviewSkyline`

Input: review corpus $\mathcal{R}$, $conf(r,\mathcal{R})\forall r \in \mathcal{R}$, set of possible opinions $\mathcal{O}_{\mathcal{R}}$
Output: $Skyline$ of $\mathcal{R}$

Sort all reviews in $\mathcal{R}$ in descending order by $conf(r,\mathcal{R})$
Create an Inverted Index, mapping each opinion $op \in \mathcal{O}_{\mathcal{R}}$ to a list $L[op]$ of the reviews that express it, sorted by confidence.
for every review $r \in \mathcal{R}$ **do**
  if (r is dominated by some set in $Skyline$) **then**
    GOTO 3: // skip r
  $\mathcal{L} = \{L[op] \mid \forall o \in \mathcal{O}_r\}$
  while (NOT all Lists in $\mathcal{L}$ are exhausted) **do**
    for every opinion $op \in \mathcal{O}_r$ **do**
      $r' = getNext(L[op])$
      if $(conf(r,\mathcal{R}) < conf(r',\mathcal{R}))$ **then**
        Consider $L[op]$ to be exhausted
        GOTO 8:
      if (r' dominates r) **then**
        GOTO 3: // skip r
  $Skyline \leftarrow Skyline \cup \{r\}$
return $Skyline$

checked are those that are ranked first in each of the lists. We then check the reviews ranked 2nd and continue until all the lists have been exhausted.

The $getNext(L[op])$ routine returns the next review r' to be checked from the given list. If r' has a lower confidence than r, then we can safely stop checking $L[op]$, since any sets ranked lower will have an even lower score. Therefore, $L[op]$ is considered exhausted and we go back to check the list of the next opinion. If r' dominates r, we eliminate r and go back to examine the next review. If all the lists in $\mathcal{L}$ are exhausted without finding any review that dominates r, then we add it to the skyline.

Performance: In the worst case, all the reviews represent skyline points. Then, the complexity of the algorithm is quadratic in the number of reviews. In practice, however, the skyline includes only a small subset of the corpus. We demonstrate this on real datasets in the experiments section. We also show that `ReviewSkyline` is several times faster and more scalable than the state-of-the art for the general skyline computation problem. In addition, by using an inverted index instead of the multi-dimensional index typically employed by skyline algorithms, `ReviewSkyline` saves both memory and computational time.

6 Compactness through Selection

The requirement for compactness implies that simply evaluating the quality of the available reviews is not enough: top-ranked reviews may still express identical opinions on the same attributes and, thus, a user may have to read through a large number of reviews in order to obtain all the required information. Instead, given a query of attributes, a review should be included in the result, *only if it evaluates at least one attribute that is not evaluated in any of the other included reviews.* Note that our problem differs significantly from conventional document retrieval tasks: instead of independently evaluating documents with respect to a given query, we want a *set* of reviews that collectively cover a subset of item-features. In addition, we want the returned set to contain opinions that respect the consensus reached by the reviewers on the specified features. Taking this into consideration, we define the *Review Selection Problem* as follows:

Problem 1 [Review Selection Problem]: Given the review corpus $\mathcal{R}$ on an item and a subset of the item's attributes $\mathcal{A}^* \subseteq \mathcal{A}$, find a subset $\mathcal{R}^*$ of $\mathcal{R}$, such that:

1. All the attributes in $\mathcal{A}^*$ are covered in $\mathcal{R}^*$
2. $pol(\alpha, r) = C_{\mathcal{R}}(\alpha), \forall \alpha \in \mathcal{A}^*, r \in \mathcal{R}^*$.
3. Let $\mathcal{X} \subseteq 2^{\mathcal{R}}$ be the collection of review-subsets that satisfy the first 2 conditions. Then:
$\mathcal{R}^* = \underset{\mathcal{R}' \in \mathcal{X}}{\text{argmax}} \sum_{r \in \mathcal{R}'} conf(r, \mathcal{R}')$

The 1st condition is straightforward. The 2nd condition ensures that the selected reviews contain no opinions that contradict the consensus on the specified attributes, in order to avoid selecting reviews with contradictory opinions. Finally, the 3rd condition asks for the set with the maximum overall confidence, among those that satisfy the first 2 conditions.

Ambiguous attributes: For certain attributes, the number of negative opinions may be only marginally higher than the number of positive ones (or vice versa), leading to a weak consensus. In order to identify such attributes, we define the $weight$ of an attribute α to be proportional to the $strength$ of its respective consensus (defined in Eq. (4)). Formally, given a review corpus $\mathcal{R}$ and an attribute α, we define $w(\alpha, \mathcal{R})$ as follows:

$$w(\alpha, \mathcal{R}) = \frac{d(\alpha \to C_{\mathcal{R}}(\alpha), \mathcal{R})}{|\mathcal{R}|} \tag{6}$$

Observe that, since $0 \leq d(\alpha \to C_{\mathcal{R}}(\alpha) \leq |\mathcal{R}|$, we know that $w(\alpha, \mathcal{R})$ takes values in $[0, 1]$. Conceptually, a low weight shows that the reviews on the specific attribute are mixed. Therefore, a set of reviews that contains only positive (or negative) opinions will not deliver a complete picture to the user. To address this, we relax the 2nd condition as follows: if the weight of an attribute α is less than some pre-defined lower bound b (i.e. $w(\alpha, \mathcal{R}) < b$), then the reported set $\mathcal{R}^*$ will be allowed to include reviews that contradict the (weak) consensus on α. In addition, $\mathcal{R}^*$ will be $required$ to contain at least one positive and one negative review with respect to α. The value of b depends on our concept of a weak consensus. For our experiments, we used $b = 0.5$.

6.1 A Combinatorial Solution

Next, we propose a combinatorial solution for the Review Selection problem. We show that the problem can be mapped to the popular Weighted Set Cover problem [3,5] (`WSC`), from which we can leverage solution techniques. Formally, the `WSC` problem is defined as follows:

Routine 3. `Transformation Routine`

Input: Set of attributes $\mathcal{A}$, Set of reviews $\mathcal{R}$
Output: Collection of subsets $\mathcal{S}$, $cost[s] \forall s \in \mathcal{S}$
for (every review $r \in \mathcal{R}$) **do**
  $s \leftarrow \emptyset$ // New empty set
  for (every attribute $\alpha \in \mathcal{A}$) **do**
    if $pol(\alpha, r) = +1$ **then** $s \leftarrow s \cup \{\alpha^+\}$
    else if $pol(\alpha, r) = -1$ **then** $s \leftarrow s \cup \{\alpha^-\}$
  $cost[s] \leftarrow (1 - conf(r, \mathcal{R}))/2$
  $\mathcal{S}.add(s)$
return $\mathcal{S}, cost[\,]$

[Weighted Set Cover Problem]: We are given a universe of elements $\mathcal{U} = \{e_1, e_2, \ldots, e_n\}$ and a collection $\mathcal{S}$ of subsets of $\mathcal{U}$, where each subset $s \in \mathcal{S}$ has a positive cost $cost[s]$. The problem asks for a collection of subsets $\mathcal{S}^* \subseteq \mathcal{S}$, such that $\bigcup_{s \in \mathcal{S}^*} \{s\} = \mathcal{U}$ and the cost $\sum_{s \in \mathcal{S}^*} cost[s]$ is minimized.

Given a review corpus $\mathcal{R}$, Routine (3) is used to generate a collection of sets $\mathcal{S}$, including a set s for every review $r \in \mathcal{R}$. The produced sets consist of elements from the same universe and have their respective costs, as required by the `WSC` problem.

Algorithm 1. `Greedy-Reviewer`

Input: $\mathcal{S}, \mathcal{A}^* \subseteq \mathcal{A}$, lower bound b
Output: weighted set-cover $\mathcal{S}^*$
$\mathcal{U} \leftarrow \emptyset$
for every attribute $\alpha \in \mathcal{A}^*$ **do**
  if $w(\alpha, \mathcal{R}) < b$ **then** $\mathcal{U} \leftarrow \mathcal{U} \cup \{\alpha^+\} \cup \{\alpha^-\}$
  else if $C_{\mathcal{R}}(\alpha) = +1$ **then** $\mathcal{U} \leftarrow \mathcal{U} \cup \{\alpha^+\}$
  else $\mathcal{U} \leftarrow \mathcal{U} \cup \{\alpha^-\}$
$\mathcal{S}^* \leftarrow \emptyset$ `// The set-cover`
$\mathcal{Z} \leftarrow \emptyset$ `// The still-uncovered part of` $\mathcal{U}$
while ($\mathcal{S}^*$ is not a cover of $\mathcal{U}$) **do**
  o ($\underset{s' \in \mathcal{S},\ s' \cap \mathcal{U} = \emptyset}{\operatorname{argmin}} \left(\frac{cost[s']}{|s' \cap \mathcal{Z}|} \right)$
  $\mathcal{S}^*.add(s)$
return $\mathcal{S}^*$

The `Greedy-Reviewer` Algorithm: Next, we present an algorithm that can efficiently solve the Review Selection problem. The input consists of the collection of sets $\mathcal{S}$ returned by the transformation routine, a query of attributes $\mathcal{A}^* \subseteq \mathcal{A}$, and a number $b \in [0, 1]$, used to determine if the consensus on an attribute is weak (as described earlier in this section). The algorithm returns a subset $\mathcal{S}^*$ of $\mathcal{S}$. The pseudocode is given in Algorithm (1).

The Algorithm begins by populating the universe $\mathcal{U}$ of elements to be covered (lines 2-6). For each attribute $\alpha \in \mathcal{A}^*$, if the consensus on the attribute is weak ($w(\alpha, \mathcal{R}) < b$), two elements α^+ and α^- are added to $\mathcal{U}$. Otherwise, if the consensus is strong and positive (negative), an element α^+ (α^-) is added.

The universe of elements $\mathcal{U}$, together with the collection of sets $\mathcal{S}$, constitute an instance of the WSC Problem. The problem is known to be NP-Hard, but can be approximated by a well-known Greedy algorithm, with an $\ln n$ approximation ratio [5]. First, we define 2 variables $\mathcal{S}^*$ and $\mathcal{Z}$ to maintain the final solution and the still-uncovered subset of $\mathcal{U}$, respectively. The greedy-choice is conducted in lines 9-11: the algorithm selects the set that minimizes the quotient of the cost, over the still-uncovered part of $\mathcal{U}$ that is covered by the set. Since there is a 1-to-1 correspondence between sets and reviews, we can trivially obtain the set of selected reviews $\mathcal{R}^*$ from the reported set-cover $\mathcal{S}^*$ and return it to the user.

7 Experiments

In this section, we present the experiments we conducted toward the evaluation of our search framework. We begin with a description of the used datasets. We then proceed to discuss the motivation and setup of each experiment, followed by a discussion of the results. All experiments were run on a desktop with a Dual-Core 2.53GHz Processor and 2G of RAM.

7.1 Datasets

• `GPS`: For this dataset, we collected the complete review corpora for 20 popular GPS Systems from Amazon.com. The average number of reviews per item was 203.5. For each review, we extracted the stars rating, the date the review was submitted and the review content.

• `TVs`: For this dataset, we collected the complete review corpora for 20 popular TV Sets from Amazon.com. The average number of reviews per item was 145. For each review, we extracted the same information as in the `GPS` dataset.

• `Vegas-Hotels`: For this dataset, we collected the review corpora for 20 popular Las Vegas Hotels from yelp.com. Yelp is a popular review-hosting website, where users can evaluate business and service providers from different parts of the United States. The average number of reviews per item was 266. For each review, we extracted the content, the stars rating and the date of submission.

• `SF-Restaurants`: For this dataset, we collected the reviews for 20 popular San Francisco restaurants from yelp.com. The average number of reviews per item was 968. For each review, we extracted the same information as in the `Vegas-Hotels` dataset. **The data is available upon request.**

7.2 Qualitative Evidence

We begin with some qualitative results, obtained by using the proposed search framework on real data. For lack of space, we cannot present the sets of reviews reported for numerous queries. Instead, we focus on 2 indicative queries, 1 from

`SF-Restaurants` and 1 from `Vegas-Hotels`. For reasons of discretion, we omit the names of the specific items. For each item, we present the query, as well as the relevant parts of the retrieved reviews.

`SF-Restaurants`

Item 1, Query: {*food, service, atmosphere, restrooms*} 3 Reviews:

• *"...The dishes were creative and delicious ... The only drawback was the single unisex restroom."*

• *"Excellent food, excellent service. Only taking one star for the size and cramp seating. The wait can get long, and i mean long..."*

• *"... Every single dish is amazing. Solid food, nice cozy atmosphere, extremely helpful waitstaff, and close proximity to MY house..."*

Item 2, Query: {*location, price, music*}, 2 Reviews:

• *"...Great location, its across from 111 Minna. Considering the decor, the prices are really reasonable...."*

• *"..Another annoying thing is the noise level. The music is so loud that it's really difficult to have a conversation..."*

`Vegas-Hotels`

Item 3, Query: {*pool, location, rooms*}, 1 Review:

• *"...It was also a fantastic location, right in the heart of things...The pool was a blast with the eiffel tower overlooking it with great frozen drinks and pool side snacks. The room itself was perfectly fine, no complaints."*

Item 4, Query: {*pool, location, buffet, staff*}, 2 Reviews:

• *"This is one of my favorite casinos on the strip; good location; good buffet; nice rooms; nice pool(s); huge casino..."*

• *"...The casino is huge and there is an indoor nightclub on the ground floor. All staff are professional and courteous..."*

As can be seen from the results, our engine returns a compact set of reviews that accurately captures the consensus on the query-attributes and, thus, serves as a valuable tool for the interested user.

7.3 Skyline Pruning for Redundant Reviews

In this section, we present a series of experiments for the evaluation of the redundancy filter described in Section 5.

Number of Pruned Reviews: First, we examine the percentage of reviews that are discarded by our filter: for every item in each of the 4 datasets, we find the set of reviews that represents the skyline of the item's review corpus. We then calculate the average percentage of pruned reviews (i.e. reviews not included in the skyline), taken over all the items in each dataset. The computed values for `TVs`, `GPS`, `Vegas-Hotels` and `SF-Restaurants` were 0.4, 0.47, 0.54 and 0.79, respectively. The percentage of pruned reviews reaches up to 79%. This illustrates the redundancy in the corpora, with numerous reviewers expressing identical opinions on the same attributes. By focusing on the skyline, we can drastically reduce the number of reviews and effectively reduce the query response time.

Evolution of the Skyline: Next, we explore the correlation between the size of the skyline and the size of the review corpus, as the latter grows over time. First, we sort the reviews for each item in ascending order, by date of submission. Then, we calculate the cardinality of the skyline of the first K reviews. We repeat the process for $K \in \{50, 100, 200, 400\}$. For each value of K, we report the average percentage of the reviews that is covered by the skyline, taken over all the items in each dataset. The results are shown in Table 1.

Table 1. Skyline Cardinality Vs. Total #Reviews

	Avg #Reviews in the Skyline (Per Item)			
#Reviews	TVs	GPS	Vegas-Hotels	SF-Restaurants
50	0.64	0.53	0.47	0.35
100	0.56	0.47	0.44	0.28
200	0.55	0.43	0.4	0.24
400	0.55	0.43	0.39	0.19

The table shows that the introduction of more reviews has a decreasing effect on the percentage of the corpus that is covered by the skyline, which converges after a certain point. This is an encouraging finding, indicating that a compact skyline can be extracted regardless of the size of the corpus.

Running Time: Next, we evaluate the performance of the `ReviewSkyline` algorithm (Section 5). We compare the required computational time against that of the state-of-the-art Branch-and-Bound Algorithm (`BnB`) by Papadias et al. [16]. Our motivation is to show how our specialized algorithm compares to one made for the general problem.

The results, shown in Table 2, show that `ReviewSkyline` achieved superior performance in all 4 datasets. `BnB` treats each corpus as a very-high dimensional dataset, assuming a new dimension for every distinct opinion. As a result, the computational time is dominated by the construction of the required R-tree structure, which is known to deteriorate for very high dimensions [20]. `ReviewSkyline` avoids these shortcomings by taking into consideration the constrained nature of the review space.

Table 2. Avg Running Time Skyline Computation (in seconds)

	TVs	GPS	Vegas-Hotels	SF-Restaurants
ReviewSkyline	0.2	0.072	0.3	0.11
BnB	24.8	39.4	28.9	116.2

Scalability: In order to demonstrate the scalability of `ReviewSkyline`, we created a benchmark with very large batches of artificial reviews. As a seed, we used the reviews corpus for the "slanted door" restaurant from the `SF-Restaurants` dataset, since it had the largest corpus across all datasets (about 1400 reviews). The data was generated as follows: first, we extracted the set $\mathcal{Y}$ of distinct opinions (i.e. attribute-to-polarity mappings) from the corpus, along with their respective frequencies. A total of 25 distinct attributes were extracted from the corpus, giving us a set of 50 distinct opinions. In the context of the skyline problem, this number represents the dimensionality of the data.

Each artificial review was then generated as follows: first, we flip an unbiased coin. If the coin comes up heads, we choose an opinion from $\mathcal{Y}$ and add it to the review. The probability of choosing an opinion from $\mathcal{Y}$ is proportional to its frequency in the original corpus. We flip the coin 10 times. Since the coin is unbiased, the expected average number of opinions per review of is 5, which is equal to the actual average observed in the corpus. We created 6 artificial corpora, where each corpus had a population of p reviews, $p \in \{10^4, 2 \times 10^4, 4 \times 10^4, 8 \times 10^4, 16 \times 10^4\}$. We compare `ReviewSkyline` with the `BnB` Algorithm, as we did in the previous experiment. The Results are shown in Figure (4). The entries on the x-axis represent the 5 artificial corpora, while the values on the y-axis represent the computational time *(in logarithmic scale)*. The results show that `ReviewSkyline` achieves superior performance for all 5 corpora. The algorithm exhibited great scalability, achieving a low computational time even for the largest corpus (less than 3 minutes). In contrast to `ReviewSkyline`, `BnB` is burdened by the construction and poor performance of the R-tree in very high-dimensional datasets.

7.4 Query Evaluation

In this section, we evaluate the search engine described in Section 6. Given the set of attributes $\mathcal{A}$ of an item, we choose 100 subsets of $\mathcal{A}$, where each subset contains exactly k elements. The probability of including an attribute to a query is proportional to the attribute's frequency in the corpus. The motivation is to generate more realistic queries,

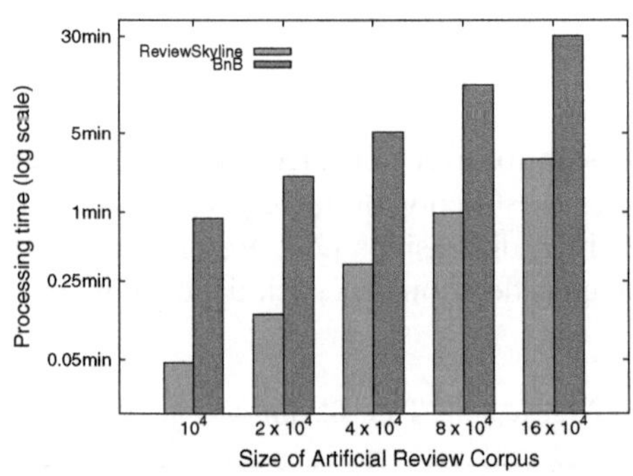

Fig. 4. Scalability of `ReviewSkyline` and `BnB`

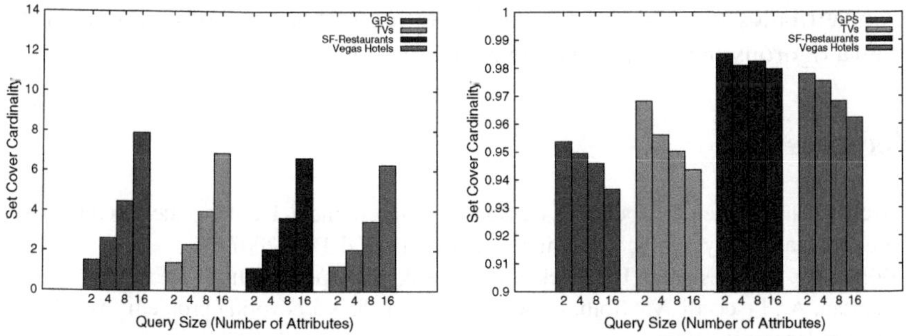

(a) Avg. Result size

(b) Avg. Confidence per review

Fig. 5. Figures (a) and (b) show the average number of reviews included in the result and the average confidence per reported review, respectively

since users tend to focus on the primary and more popular attributes of an item. We repeat the process for $k \in \{2, 4, 8, 16\}$, for a total of $100 \times 4 = 400$ queries per item.

Query size Vs. Result size: First, we evaluate how the size of the query affects the cardinality of the returned sets. Ideally, we would like to retrieve a small number of reviews, so that a user can read them promptly and obtain the required information. Given a specific item I and a query size k, let $Avg[I, k]$ be the average number of reviews included in the result, taken over the 100 queries of size k for the item. We then report the mean of the $Avg[I, k]$ values, taken over all 20 items in each dataset. The results are shown in Figure 5(a): The reported sets were consistently small, with less than 8 reviews were enough to cover queries containing up to 16 different attributes. Such compact sets are desirable since they can promptly be read by the user.

Query Size Vs. Confidence: Next, we evaluate how the size of the query affects the average confidence of the selected reviews. The experimental setup is similar to that of the previous experiment. However, instead of the average result cardinality, we report the average confidence per selected review. Figure 5(b) shows the very promising results. An average confidence of 0.93 or higher was consistently reported for all query sizes, and for all 4 datasets. Combined with the findings of the previous experiment, we conclude that our framework produces compact sets of high-quality reviews.

8 Conclusion

In this paper, we formalized the Confident Search paradigm for large review corpora. Taking into consideration the requirements of the paradigm, we presented a complete search framework, able to efficiently handle large sets of reviews. Our framework employs a principled method for evaluating the confidence in the opinions expressed in reviews. In addition, it is equipped with an efficient method for filtering redundancy. The filtered corpus maintains all the useful information and is considerably smaller, which makes it easier to store and to search. Finally, we formalized and addressed the problem of selecting a minimal set of high-quality reviews that can effectively cover any query of attributes submitted by the user. The efficacy of our methods was demonstrated through a rigorous and diverse experimental evaluation.

References

1. Archak, N., Ghose, A., Ipeirotis, P.: Show me the money! Deriving the pricing power of product features by mining consumer reviews. In: SIGKDD (2007)
2. Börzsönyi, S., Kossmann, D., Stocker, K.: The skyline operator. In: ICDE (2001)
3. Caprara, A., Fischetti, M., Toth, P.: Algorithms for the set covering problem. Annals of Operations Research (1996)
4. Chomicki, J., Godfrey, P., Gryz, J., Liang, D.: Skyline with presorting. In: ICDE (2003)
5. Chvatal, V.: A greedy heuristic for the set-covering problem. Mathematics of Operations Research (1979)
6. Dave, K., Lawrence, S., Pennock, D.M.: Mining the peanut gallery: opinion extraction and semantic classification of product reviews. In: WWW 2003 (2003)
7. Ghani, R., Probst, K., Liu, Y., Krema, M., Fano, A.: Text mining for product attribute extraction. SIGKDD Explorations Newsletter (2006)
8. Hu, M., Liu, B.: Mining and summarizing customer reviews. In: SIGKDD (2004)
9. Hu, M., Liu, B.: Mining opinion features in customer reviews. In: AAAI (2004)
10. Jindal, N., Liu, B.: Opinion spam and analysis. In: WSDM 2008 (2008)
11. Ku, L.-W., Liang, Y.-T., Chen, H.-H.: Opinion extraction, summarization and tracking in news and blog corpora. In: AAAI Symposium on Computational Approaches to Analysing Weblogs, AAAI-CAAW (2006)
12. Liu, J., Cao, Y., Lin, C.-Y., Huang, Y., Zhou, M.: Low-quality product review detection in opinion summarization. In: EMNLP-CoNLL (2007)
13. Min Kim, S., Pantel, P., Chklovski, T., Pennacchiotti, M.: Automatically assessing review helpfulness. In: EMNLP 2006 (2006)
14. Pang, B., Lee, L.: Seeing stars: Exploiting class relationships for sentiment categorization with respect to rating scales. In: ACL (2005)
15. Pang, B., Lee, L., Vaithyanathan, S.: Thumbs up? Sentiment classification using machine learning techniques. In: EMNLP (2002)
16. Papadias, D., Tao, Y., Fu, G., Seeger, B.: Progressive skyline computation in database systems. ACM Trans. Database Syst. (2005)
17. Popescu, A.-M., Etzioni, O.: Extracting product features and opinions from reviews. In: HLT 2005 (2005)
18. Riloff, E., Patwardhan, S., Wiebe, J.: Feature subsumption for opinion analysis. In: EMNLP (2006)
19. Turney, P.D.: Thumbs up or thumbs down? semantic orientation applied to unsupervised classification of reviews. In: ACL (2002)
20. Weber, R., Schek, H.-J., Blott, S.: A quantitative analysis and performance study for similarity-search methods in high-dimensional spaces. In: VLDB 1998 (1998)
21. Zhang, Z., Varadarajan, B.: Utility scoring of product reviews. In: CIKM (2006)
22. Zhuang, L., Jing, F., Zhu, X., Zhang, L.: Movie review mining and summarization. In: CIKM (2006)

Learning to Tag from Open Vocabulary Labels

Edith Law, Burr Settles, and Tom Mitchell

Machine Learning Department
Carnegie Mellon University
{elaw,bsettles,tom.mitchell}@cs.cmu.edu

Abstract. Most approaches to classifying media content assume a fixed, closed vocabulary of labels. In contrast, we advocate machine learning approaches which take advantage of the millions of free-form tags obtainable via online crowd-sourcing platforms and social tagging websites. The use of such open vocabularies presents learning challenges due to typographical errors, synonymy, and a potentially unbounded set of tag labels. In this work, we present a new approach that organizes these noisy tags into well-behaved semantic classes using topic modeling, and learn to predict tags accurately using a mixture of topic classes. This method can utilize an arbitrary open vocabulary of tags, reduces training time by 94% compared to learning from these tags directly, and achieves comparable performance for classification and superior performance for retrieval. We also demonstrate that on open vocabulary tasks, human evaluations are essential for measuring the true performance of tag classifiers, which traditional evaluation methods will consistently underestimate. We focus on the domain of tagging music clips, and demonstrate our results using data collected with a human computation game called TagATune.

Keywords: Human Computation, Music Information Retrieval, Tagging Algorithms, Topic Modeling.

1 Introduction

Over the years, the Internet has become a vast repository of multimedia objects, organized in a rich and complex way through tagging activities. Consider music as a prime example of this phenomenon. Many applications have been developed to collect tags for music over the Web. For example, Last.fm is collaborative *social tagging* network which collects users' listening habits and roughly 2 million tags (e.g., "acoustic," "reggae," "sad," "violin") per month [12]. Consider also the proliferation of *human computation* systems, where people contribute tags as a by-product of doing a task they are naturally motivated to perform, such as playing causal web games. TagATune [14] is a prime example of this, collecting tags for music by asking two players to describe their given music clip to each other with tags, and then guess whether the music clips given to them are the same or different. Since deployment, TagATune has collected over a million annotations from tens of thousands of players.

J.L. Balcázar et al. (Eds.): ECML PKDD 2010, Part II, LNAI 6322, pp. 211–226, 2010.

In order to effectively organize and retrieve the ever-growing collection of music over the Web, many so-called *music taggers* have been developed [2,10,24] to automatically annotate music. Most previous work has assumed that the labels used to train music taggers come from a small fixed vocabulary and are devoid of errors, which greatly simplifies the learning task. In contrast, we advocate using tags collected by collaborative tagging websites and human computation games, since they leverage the effort and detailed domain knowledge of many enthusiastic individuals. However, such tags are noisy, i.e., they can be misspelled, overly specific, irrelevant to content (e.g., "albums I own"), and virtually unlimited in scope. This creates three main learning challenges: (1) *over-fragmentation*, since many of the enormous number of tags are synonymous or semantically equivalent, (2) *sparsity*, since most tags are only associated with a few examples, and (3) *scalability issues*, since it is computationally inefficient to train a classifier for each of thousands (or millions) of tags.

In this work, we present a new technique for classifying multimedia objects by tags that is scalable (i.e., makes full use of noisy, open-vocabulary labels that are freely available on the Web) and efficient (i.e., the training time remains reasonably short as the tag vocabulary grows). The main idea behind our approach is to organize these noisy tags into well-behaved semantic classes using a topic model [4], and learn to predict tags accurately using a mixture of topic classes. Using the TagATune [14] dataset as a case study, we compare the tags generated by our topic-based approach against a traditional baseline of predicting each tag independently with a binary classifier. These methods are evaluated in terms of both tag annotation and music retrieval performance. We also highlight a key limitation of traditional evaluation methods—comparing against a ground truth label set—which is especially severe for open-vocabulary tasks. Specifically, using the results from several Mechanical Turk studies, we show that human evaluations are essential for measuring the *true* performance of music taggers, which traditional evaluation methods will consistently underestimate.

2 Background

The ultimate goal of music tagging is to enable the automatic annotation of large collections of music, such that users can then browse, organize, and retrieve music in an semantic way. Although tag-based search querying is arguably one of the most intuitive methods for retrieving music, until very recently [2,10,24], most retrieval methods have focused on querying metadata such as artist or album title [28], similarity to an audio input query [6,7,8], or a small fixed set of category labels based on genre [26], mood [23], or instrument [9]. The lack of focus on music retrieval by rich and diverse semantic tags is partly due to a historical lack of labeled data for training music tagging systems.

A variety of machine learning methods have been applied to music classification, such as logistic regression [1], support vector machines [17,18], boosting [2], and other probabilistic models [10,24]. All of these approaches employ binary classifiers—one per label—to map audio features directly to a limited number (tens to few hundreds) of tag labels independently. This is in contrast to the TagATune data set used in this paper, which has over 30,000 clips, over 10,000 unique tags collected from tens of thousands of users.

The drawback of learning to tag music from open-vocabulary training data is that it is *noisy*[1], by which we mean the over-fragmentation of the label space due to synonyms ("serene" vs. "mellow"), misspellings ("chello") and compound phrases ("guitar plucking"). Synonyms and misspellings cause music that belongs to the same class to be labeled differently, and compound phrases are often overly descriptive. All of these phenomena can lead to label sparsity, i.e., very few training examples for a given tag label.

It is possible to design data collection mechanisms to minimize such label noise in the first place. One obvious approach is to impose a controlled vocabulary, as in the Listen Game [25] which limits the set of tags to 159 labels pre-defined by experts. A second approach is to collect tags by allowing players to enter free-form text, but filter out the ones that have not been verified by multiple users, or that are associated with too few examples. For example, of the 73,000 tags acquired through the music tagging game MajorMiner [20], only 43 were used in the 2009 MIREX benchmark competition to train music taggers [15]. Similarly, the Magnatagatune data set [14] retains only tags that are associated with more than 2 annotators and 50 examples. Some recent work has attempted to mitigate these problems by distinguishing between content relevant and irrelevant tags [11], or by discovering higher-level concepts using tag co-occurrence statistics [13,16]. However, none of these works explore the use of these higher-level concepts in training music annotation or retrieval systems.

3 Problem Formulation

Assume we are given as training data a set of N music clips $\mathcal{C} = \{c_1, \ldots, c_N\}$ each of which has been annotated by humans using tags $\mathcal{T} = \{t_1, \ldots, t_V\}$ from a vocabulary of size V. Each music clip $c_i = (\boldsymbol{a_i}, \boldsymbol{x_i})$ is represented as a tuple, where $\boldsymbol{a_i} \in \mathbb{Z}^V$ is a the *ground truth tag* vector containing the frequency of each tag in $\mathcal{T}$ that has been used to annotate the music clip by humans, and $\boldsymbol{x_i} \in \mathbb{R}^M$ is a vector of M real-valued acoustic features, which describes the characteristics of the audio signal itself.

The goal of music annotation is to learn a function $\hat{f} : X \times T \to \mathbb{R}$, which maps the acoustic features of each music clip to a set of scores that indicate the relevance of each tag for that clip. Having learned this function, music clips can be retrieved for a search query q by rank ordering the distances between the query vector (which has value 1 at position j if the tag t_j is present in the search query, 0 otherwise) and the tag probability vector for each clip. Following [24], we measure these "distances" using KL divergence, which is a common information-theoretic measure of the difference between two distributions.

[1] We use *noise* to refer to the challenging side-effects of open tagging described here, which differs slightly from the common interpretation of mislabeled training data.

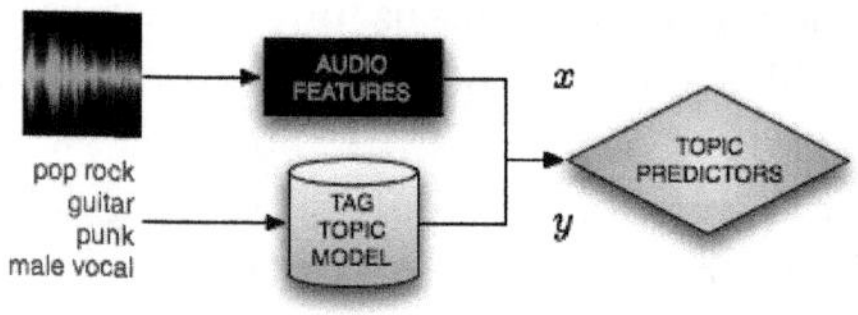

(a) Training Phase

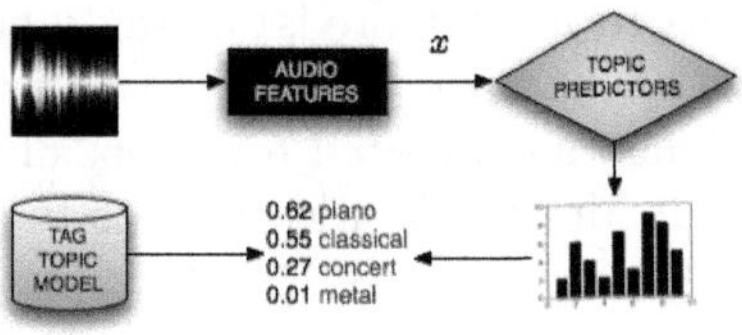

(b) Inference phase

Fig. 1. The training and inference phases of the proposed approach

3.1 Topic Method (Proposed Approach)

We propose a new method for automatically tagging music clips, by first mapping from the music clip's audio features to a small number of semantic classes (which account for all tags in the vocabulary), and then generating output tags based on these classes. Training involves learning classes, or "topics," with their associated tag distributions, and the mapping from audio features to a topic class distribution. An overview of the approach is presented in Figure 1.

Training Phase. As depicted in Figure 1(a), training is a two-stage process. First, we induce a *topic model* [4,22] using the ground truth tags associated with each music clip in the training set. The topic model allows us to infer distribution over topics for each music clip in the training set, which we use to replace the tags as training labels. Second, we train a classifier that can predict topic class distributions directly from audio features.

In the first stage of training, we use Latent Dirichlet Allocation (LDA) [4], a common topic modeling approach. LDA is a hierarchical probabilistic model that describes a process for generating constituents of an entity (e.g., words of a document, musical notes in a score, or pixels in an image) from a set of latent class variables called topics. In our case, constituents are tags and an entity is the semantic description of a music clip (i.e., set of tags). Figure 2(a) shows an example model of 10 topics induced from music annotations collected by TagATune. Figure 2(b) and Figure 2(c) show the topic distributions for two very distinct music clips and their ground truth annotations (in the caption; note synonyms and typos among the tags entered by users). The music clip from Figure 2(b) is associated with both topic 4 (classical violin) and topic 10 (female opera singer). The music clip from Figure 2(c) is associated with both topic 7 (flute) and topic 8 (quiet ambient music).

In the second stage of training, we learn a function that maps the audio features for a given music clip to its topic distribution. For this we use a maximum entropy (MaxEnt) classifier [5], which is a multinomial generalization of logistic regression. We use the LDA and MaxEnt implementations in the MALLET toolkit[2], with a slight modification of the optimization procedure [29] which enables us to train a MaxEnt model from class distributions rather than a single class label. We refer to this as the *Topic Method.*

[2] http://mallet.cs.umass.edu

1	electronic beat fast drums synth dance beats jazz
2	male choir man vocal male_vocal vocals choral singing
3	indian drums sitar eastern drum tribal oriental middle_eastern
4	classical violin strings cello violins classic slow orchestra
5	guitar slow strings classical country harp solo soft
6	classical harpsichord fast solo strings harpsicord classic harp
7	flute classical flutes slow oboe classic clarinet wind
8	ambient slow quiet synth new_age soft electronic weird
9	rock guitar loud metal drums hard_rock male fast
10	opera female woman vocal female_vocal singing female_voice vocals

(a) Topic Model

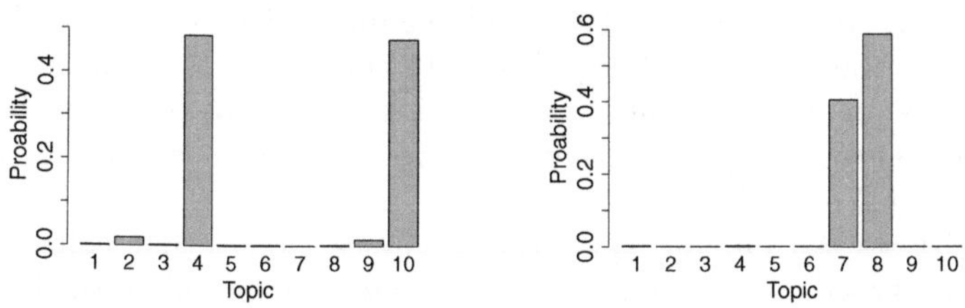

(b) woman, classical, classsical, opera, male, violen, violin, voice, singing, strings, italian

(c) chimes, new age, spooky, flute, quiet, whistle, fluety, ambient, soft, high pitch, bells

Fig. 2. An example LDA model of 10 topic classes learned over music tags, and the representation of two sample music clips annotations by topic distribution

Our approach tells an interesting generative story about how players of TagATune might decide on tags for the music they are listening to. According to the model, each listener has a latent topic structure in mind when thinking of how to describe the music. Given a music clip, the player first selects a topic according to the topic distribution for that clip (as determined by audio features), and then selects a tag according to the posterior distribution of the chosen topics. Under this interpretation, our goal in learning a topic model over tags is to discover the topic structure that the players use to generate tags for music, so that we can leverage a similar topic structure to automatically tag new music.

Inference Phase. Figure 1(b) depicts the process of generating tags for novel music clips. Given the audio features $\boldsymbol{x_i}$ for a test clip c_i, the trained MaxEnt classifier is used to predict a topic distribution for that clip. Based on this predicted topic distribution, each tag t_j is then given a relevance score $P(t_j|\boldsymbol{x_i})$ which is its expected probability over all topics:

$$P(t_j|\boldsymbol{x_i}) = \sum_{k=1}^{K} P(t_j|y_k)P(y_k|\boldsymbol{x_i}),$$

where $j = 1, \ldots, V$ ranges over the tag vocabulary, and $k = 1, \ldots, K$ ranges over all topic classes in the model.

3.2 Tag Method (Baseline)

To evaluate the efficiency and accuracy of our method, we compare it against an approach that predicts $P(t_j|\boldsymbol{x}_i)$ directly using a set of binary logistic regression classifiers (one per tag). This second approach is consistent with previous approaches to music tagging with closed vocabularies [1,17,18,2,10,24]. We refer to it as the *Tag Method.* In some experiments we also compare against a method that assigns tags randomly.

4 Data Set

The data is collected via a two-player online game called called TagATune [14]. Figure 3 shows the interface of TagATune. In this game, two players are given either the same or different music clips, and are asked to describe their given music clip. Upon reviewing each other's description, they must guess if the music clips are the same or different.

There exist several human computation games [20,25] that collect tags for music that are based on the *output-agreement mechanism* (a.k.a. the ESP Game [27] mechanism), where two players must match on a tag in order for that tag to become a valid label for a music clip. In our previous work [14], we have showed that output-agreement games, although effective for image annotation, are restrictive for music data: there are so many ways to describe music and sounds that players often have a difficult time agreeing on any tags. In TagATune, the problem of agreement is alleviated by allowing players to communicate with each other. Furthermore, by requiring that the players guess whether the music are the same or different based on each other's tags, the quality and validity of the tags are ensured. The downside of opening up the communication between players is that the tags entered are more noisy.

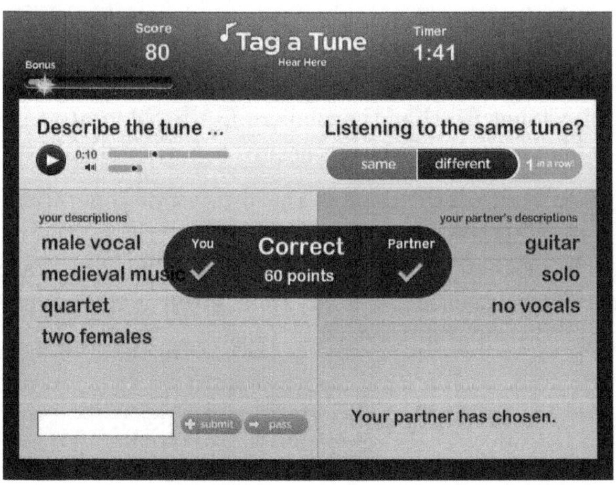

Fig. 3. A screen shot of the TagATune user interface

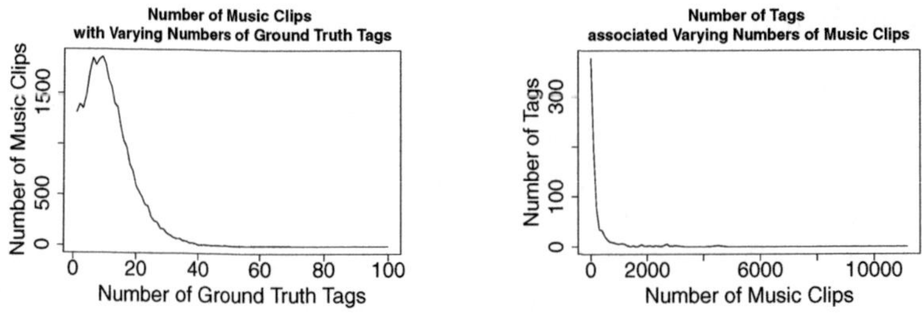

(a) Number of music clips with X number of ground truth tags

(b) Number of tags associated X number of music clips

Fig. 4. Characteristics of the TagATune data set

Figure 4 shows the characteristics of the TagATune dataset. Figure 4(a) is a rank frequency plot showing the number of music clips that have a certain number of ground truth tags. The plot reveals a disparity in the number of ground truth tags each music clip has – a majority of the clips (1,500+) have under 10, approximately 1,300 music clips have only 1 or 2, and very few have a large set (100+). This creates a problem in our evaluation – many of the generated tags that are relevant for the clip may be missing from the ground truth tags, and therefore will be considered incorrect. Figure 4(b) is a rank frequency plot showing the number of tags that have a certain number of music clips available to them as training examples. The plot shows that the vast majority of the tags have few music clips to use as training examples, while a small number of tags are endowed with a large number of examples. This highlights the aforementioned sparsity problem that emerges when tags are used directly as labels, a problem that is addressed by our proposed method.

We did a small amount of pre-processing on a subset of the data set, tokenizing tags, removing punctuation and four extremely common tags that are not related to the content of the music, i.e. "yes," "no," "same," "diff". In order to accommodate the baseline Tag Method, which requires a sufficient number of training examples for each binary classification task, we also eliminated tags that have fewer than 20 training music clips. This reduces the number of music clips from 31,867 to 31,251, the total number of ground truth tags from 949,138 to 699,440, and the number of *unique* ground truth tags from 14,506 to 854. Note that we are throwing away a substantial amount of tag data in order to accommodate the baseline Tag Method. A key motivation for using our Topic Method is that we do not need to throw away any tags at all. Rare tags, i.e. tags that are associated with only one or two music clips, can still be grouped into a topic, and used in the annotation and retrieval process.

Each of the 31,251 music clips is 29 seconds in duration, and is represented by a set of ground truth tags collected via the TagATune game, as well as a set of content-based (spectral and temporal) audio features extracted using the technique described in [19].

5 Experiments

We conducted several experiments guided by five central questions about our proposed approach. (1) *Feasibility*: given a set of noisy music tags, is it possible to learn a low-dimensional representation of the tag space that is both semantically meaningful and predictable by music features? (2) *Efficiency*: how does training time compare against the baseline method? (3) *Annotation Performance*: how accurate are the generated tags? (4) *Retrieval Performance*: how well do the generated tags facilitate music retrieval? (5) *Human Evaluation*: to what extent are the performance evaluations a reflection of the true performance of the music taggers? All results are averaged over five folds using cross-validation.

5.1 Feasibility

Table 1 (on the next page) shows the top 10 words for each topic learned by LDA with the number of topics fixed at 10, 20 and 30. In general, the topics are able to capture meaningful groupings of tags, e.g., synonyms (e.g., "choir/choral/chorus" or "male/man/male_vocal"), misspellings (e.g., "harpsichord/harpsicord" or "cello/chello"), and associations (e.g., "indian/drums/sitar/eastern/oriental" or "rock/guitar/loud/metal"). As we increase the number of topics, new semantic grouping appear that were not captured by models which use a fewer number of topics. For example, in 20-topic model, topic 3 (which describes soft classical music), topic 13 (which describes jazz), and topic 17 (which describes rap, hiphop and reggae) are new topics that are not evident in the model with only 10 topics. We also observe some repetition or refinement of topics as the number of topic increases (e.g., topics 8, 25 and 27 in the 30-topic model all describe slightly different variations on female vocal music).

It was difficult to know exactly how many topics can succinctly capture the concepts underlying the music in our data set. Therefore, in all our experiments we empirically tested how well the topic distribution and the best topic can be predicted using audio features, fixing the number of topics at 10, 20, 30, 40, and 50. Figure 5 summarizes the results. We evaluated performance using several

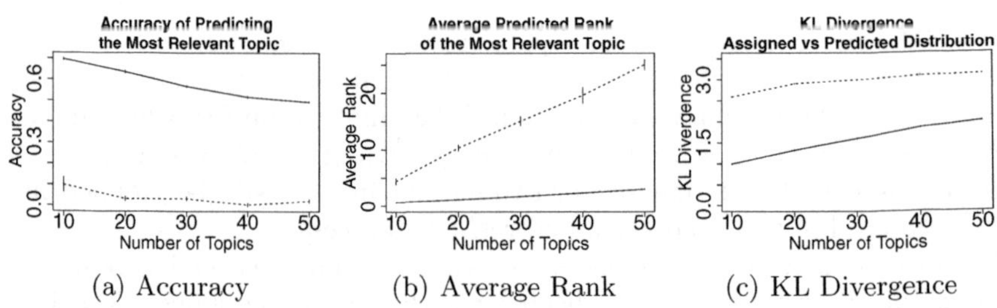

(a) Accuracy (b) Average Rank (c) KL Divergence

Fig. 5. Results showing how well topic distributions or the best topic can be predicted from audio features. The metrics include accuracy and average rank of the most relevant topic, and KL divergence between the assigned and predicted topic distribution.

Table 1. Topic Model with 10, 20, and 30 topics. The topics in bold in the 20-topic model are examples of new topics that emerge when the number of topics is increased from 10 to 20. The topics marked by * in the 30-topic model are examples of topics that start to repeat as the number of topics is increased.

	10 Topics
1	electronic beat fast drums synth dance beats jazz electro modern
2	male choir man vocal male_vocal vocals choral singing male_voice pop
3	indian drums sitar eastern drum tribal oriental middle_eastern foreign fast
4	classical violin strings cello violins classic slow orchestra string solo
5	guitar slow strings classical country harp solo soft quiet acoustic
6	classical harpsichord fast solo strings harpsicord classic harp baroque organ
7	flute classical flutes slow oboe classic clarinet wind pipe soft
8	ambient slow quiet synth new_age soft electronic weird dark low
9	rock guitar loud metal drums hard_rock male fast heavy male_vocal
10	opera female woman vocal female_vocal singing female_voice vocals female_vocals voice

	20 Topics
1	indian sitar eastern oriental strings middle_eastern foreign guitar arabic india
2	flute classical flutes oboe slow classic pipe wind woodwind horn
3	**slow quiet soft classical solo silence low calm silent very_quiet**
4	male male_vocal man vocal male_voice pop vocals singing male_vocals guitar
5	cello violin classical strings solo slow classic string violins viola
6	opera female woman classical vocal singing female_opera female_vocal female_voice operatic
7	female woman vocal female_vocal singing female_voice vocals female_vocals pop voice
8	guitar country blues folk irish banjo fiddle celtic harmonica fast
9	guitar slow classical strings harp solo classical_guitar soft acoustic spanish
10	electronic synth beat electro ambient weird new_age drums electric slow
11	drums drum beat beats tribal percussion indian fast jungle bongos
12	fast beat electronic dance drums beats synth electro trance loud
13	**jazz jazzy drums sax bass funky guitar funk trumpet clapping**
14	ambient slow synth new_age electronic weird quiet soft dark drone
15	classical violin strings violins classic orchestra slow string fast cello
16	harpsichord classical harpsicord strings baroque harp classic fast medieval harps
17	**rap talking hip_hop voice reggae male male_voice man speaking voices**
18	classical fast solo organ classic slow soft quick upbeat light
19	choir choral opera chant chorus vocal vocals singing voices chanting
20	**rock guitar loud metal hard_rock drums fast heavy electric_guitar heavy_metal**

	30 Topics
1	choir choral opera chant chorus vocal male chanting vocals singing
2	classical solo classic oboe fast slow clarinet horns soft flute
3	rap organ talking hip_hop voice speaking man male_voice male man_talking
4	rock metal loud guitar hard_rock heavy fast heavy_metal male punk
5	guitar classical slow strings solo classical_guitar acoustic soft harp spanish
6	cello violin classical strings solo slow classic string violins chello
7	violin classical strings violins classic slow cello string orchestra baroque
*8	female woman female_vocal vocal female_voice pop singing female_vocals vocals voice
9	bells chimes bell whistling xylophone whistle chime weird high_pitch gong
10	ambient slow synth new_age electronic soft spacey instrumental quiet airy
11	rock guitar drums loud electric_guitar fast pop guitars electric bass
12	slow soft quiet solo classical sad calm mellow very_slow low
13	water birds ambient rain nature ocean waves new_age wind slow
14	irish violin fiddle celtic folk strings clapping medieval country violins
15	electronic synth beat electro weird electric drums ambient modern fast
16	indian sitar eastern middle_eastern oriental strings arabic guitar india foreign
17	drums drum beat beats tribal percussion indian fast jungle bongos
18	classical strings violin orchestra violins classic orchestral string baroque fast
19	quiet slow soft classical silence low very_quiet silent calm solo
20	flute classical flutes slow wind woodwind classic soft wind_instrument violin
21	guitar country blues banjo folk harmonica bluegrass acoustic twangy fast
22	male man male_vocal vocal male_voice pop singing vocals male_vocals voice
23	jazz jazzy drums sax funky funk bass guitar trumpet reggae
24	harp strings guitar dulcimer classical sitar slow string oriental plucking
*25	vocal vocals singing foreign female voices women woman voice choir
26	fast loud upbeat quick fast_paced very_fast happy fast_tempo fast_beat faster
*27	opera female woman vocal classical singing female_opera female_voice female_vocal operatic
28	ambient slow dark weird drone low quiet synth electronic eerie
29	harpsichord classical harpsicord baroque strings classic harp medieval harps guitar
30	beat fast electronic dance drums beats synth electro trance upbeat

metrics, including accuracy and average rank of the most probable topic, as well as the KL divergence between the ground truth topic distribution and the predicted distribution.

Although we see a slight degradation of performance as the number of topics increases, all models significantly outperform the random baseline, which uses random distributions as labels for training. Moreover, even with 50 topics, the average rank of the top topic is still around 3, which suggests that the classifier is capable of predicting the most relevant topic, an important pre-requisite for the generation of accurate tags.

5.2 Efficiency

A second hypothesis is that the Topic Method would be more computationally efficient to train, since it learns to predict a joint topic distribution in a reduced-dimensionality tag space (rather than a potentially limitless number of independent classifiers). Training the Topic Method (i.e., inducing the topic model *and* the training the classifier for mapping audio features to a topic distribution) took anywhere from 18.3 minutes (10 topics) to 48 minutes (50 topics) per fold, but quickly plateaus after 30 topics: . The baseline Tag Method, by contrast, took 845.5 minutes (over 14 hours) per fold. Thus, the topic approach can reduce training time by 94% compared to the Tag Method baseline, which confirms our belief that the proposed method will be significantly more scalable as the size of the tag vocabulary grows, while eliminating the need to filter low-frequency tags.

5.3 Annotation Performance

Following [10], we evaluate the accuracy of the 10 tags with the highest probabilities for each music clip, using three different metrics: per-clip metric, per-tag metric, and omission-penalizing per-tag metric.

Per-Clip Metrics. The per-clip precision@N metric measures the proportion of correct tags (according to agreement with the ground truth set) amongst the N most probable tags for each clip according to the tagger, averaged over all the clips in the test set. The results are presented in Figure 6. The Topic Method and baseline Tag Method both significantly outperform the random baseline, and the Topic Method with 50 topics is indistinguishable from the Tag Method.

Per-Tag Metric. Alternatively, we can evaluate the annotation performance by computing the precision, recall, and F-1 scores for each tag, averaged over all the tags that are output by the algorithm (i.e. if the music tagger does not output a tag, it is ignored). Specifically, given a tag t, we calculate its precision $P_t = \frac{c_t}{a_t}$, recall $R_t = \frac{c_t}{g_t}$, and and F-1 measure $F_t = \frac{2 \times P_t \times R_t}{P_t + R_t}$, where g_t is the number of test music clips that have t in their ground truth sets, a_t is the number of clips that are annotated with t by the tagger, and c_t is the number of clips that have been *correctly* annotated with the tag t by the tagger (i.e., t is found in the ground truth set). The overall per-tag precision, recall and F-1

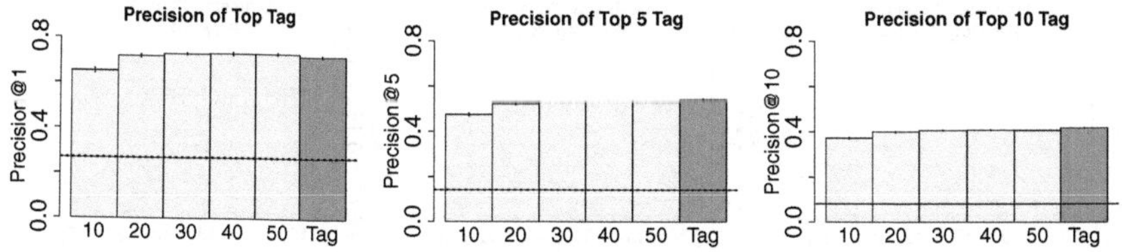

(a) Precision@1 (b) Precision@5 (c) Precision@10

Fig. 6. Per-clip Metrics. The light-colored bars represent Topic Method with 10, 20, 30, 40 and 50 topics. The dark-colored bar represents the Tag Method. The horizontal line represent the random baseline, and the dotted lines represent its standard deviation.

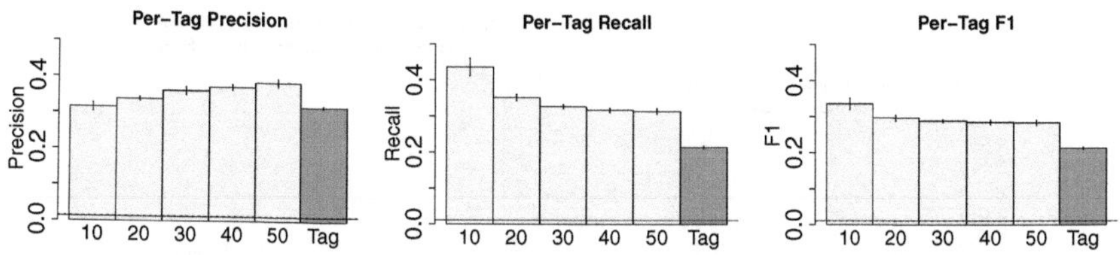

(a) Precision (b) Recall (c) F-1

Fig. 7. Per-tag Metrics. The light-colored bars represent Topic Method with 10, 20, 30, 40 and 50 topics. The dark-colored bar represents the Tag Method. The horizontal line represent the random baseline, and the dotted lines represent its standard deviation.

scores for a test set are P_t, R_t and F_t for each tag t, averaged over all tags in the vocabulary. Figure 7 presents these results, showing that the Topic Method significantly outperforms the baseline Tag Method under this set of metrics.

Omission-Penalizing Per-Tag Metrics. A criticism of some of the previous metrics, in particular the *per-clip* and *per-tag precision* metrics, is that a tagger that simply outputs the most common tags (omitting rare ones) can still perform reasonably well. Some previous work [2,10,24] has adopted a set of per-tag metrics that penalize omissions of tags that could have been used to annotate music clips in the test set. Following [10,24], we alter tag precision P_t to be the empirical frequency E_t of the tag t in the test set *if* the tagger failed to predict t for any instances at all (otherwise, $P_t = \frac{c_t}{a_t}$ as before). Similarly, the tag recall $R_t = 0$ if the tagger failed to predict t for any music clips (and $R_t = \frac{c_t}{g_t}$ otherwise). This specification penalizes classifiers that leave out tags, especially rare ones. Note these metrics are upper-bounded by a quantity that depends on the number of tags output by the algorithm. This quantity can be computed empirically by setting the precision and recall to 1 when a tag is present, and to E_t and 0 (respectively) when a tag is omitted.

Results (Figure 8) show that for the Topic Method, performance increases with more topics, but reaches a plateau as the number of topics approaches 50. One possible explanation is revealed by Figure 9(a), which shows that the number

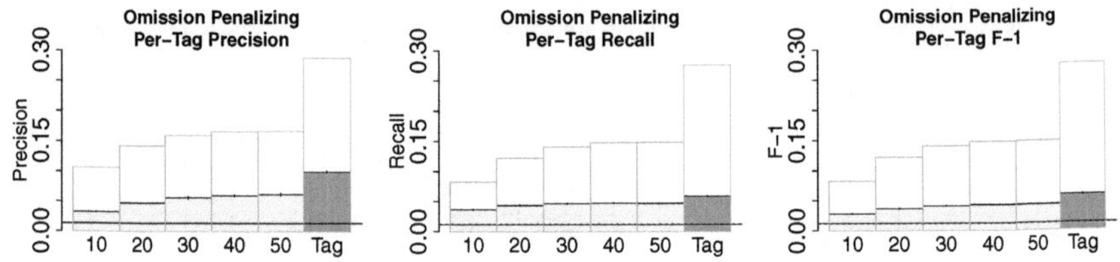

(a) Precision (b) Recall (c) F-1

Fig. 8. Omission-Penalizing Per-tag Metrics. Light-colored bars represent the Topic Method with 10, 20, 30, 40 and 50 topics. Dark-colored bars represent the Tag Method. Horizontal lines represent the random baseline. Grey outlines indicate upper bounds.

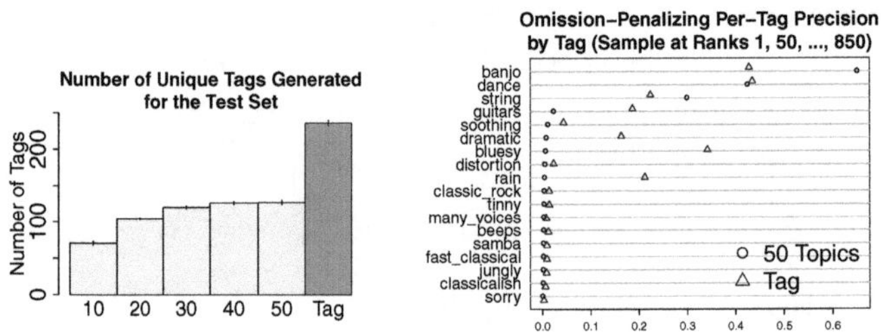

(a) Tag Coverage (b) Precision by Tag

Fig. 9. Tag coverage and loss of precision due to omissions

of unique tags generated by the Topic Method reaches a plateau at around this point. In additional experiments using 60 to 100 topics, we found that this plateau persists. This might explain why the Tag Method outperforms the Topic Method under this metric—it generates many more unique tags.

Figure 9(b), which shows precision scores for sample tags achieved by each method, confirms this hypothesis. For the most common tags (e.g., "banjo," "dance," "string"), the Topic Method achieves superior or comparable precision, while for rarer tags (e.g., "dramatic," "rain" etc.), the Tag Method is better and the Topic Method receives lower scores due to omissions. Note that these low-frequency tags contain more noise (e.g., "jungly," "sorry"), so it could be that the Tag Method is superior simply on its ability to output noisy tags.

5.4 Retrieval Performance

The tags generated by a music tagger can be used to facilitate retrieval. Given a search query, music clips can be ranked by the KL divergence between the query tag distribution and the tag probability distribution for each clip. We measure the quality of the top 10 music clips retrieved using the mean average precision [24] metric, $M_{10} = \frac{1}{10}\sum_{r=1}^{10} \frac{s_r}{r}$, where s_r is the number of "relevant" (i.e., the search query can be found in the ground truth set) songs at rank r.

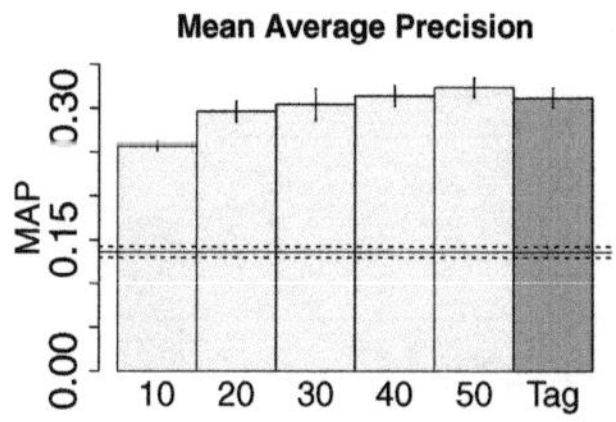

Fig. 10. Retrieval performance in terms of mean average precision

Figure 10 shows the performance of the three methods under this metric. The retrieval performance of the Topic Method with 50 topics is slightly better than the Tag Method, but otherwise indistinguishable. Both methods perform significantly better than random (the horizontal line).

5.5 Human Evaluation

We argue that the performance metrics used so far can only *approximate* the quality of the generated tags. The reason is that generated tags that cannot be found amongst ground truth tags (due to missing tags or vocabulary mismatch) are counted as wrong, when they might in fact be relevant but missing due to the subtleties of using an open tag vocabulary.

In order to compare the true merit of the tag classifiers, we conducted several Mechanical Turk experiments asking humans to evaluate the annotation and retrieval capabilities of the Topic Method (with 50 topics), Tag Method and Random Method. For the annotation task, we randomly selected a set of 100 music clips, and solicited evaluations from 10 unique evaluators per music clip. For each clip, the user is given three lists of tags generated by each of the three methods. The order of the lists is randomized each time to eliminate presentation bias. The users are asked to (1) click the checkbox beside a tag if it describes the music clip well, and (2) rank order their overall preference for each list.

Figure 11 shows the per-tag precision, recall and F-1 scores as well as the per-clip precision scores for the three methods, using both ground truth set evaluation and using human evaluators. Results show that when tags are judged based on whether they are present in the ground truth set, performance of the tagger is grossly underestimated for all metrics. In fact, of the predicted tags that the users considered "appropriate" for a music clip (generated by either the Topic Method or the Tag Method method), on average, approximately half of them are missing from the ground truth set.

While the human-evaluated performance of the Tag Method and Topic Method are virtually identical, when asked to rank the tag lists evaluators preferred the the Tag Method (62.0% of votes) over the Topic Method (33.4%) or Random (4.6%). Our hypothesis is that people prefer the Tag Method because its has better coverage (Section 5.3). Since evaluation is based on 10 tags generated by the tagger, we conjecture that a new way of generating this *set* of

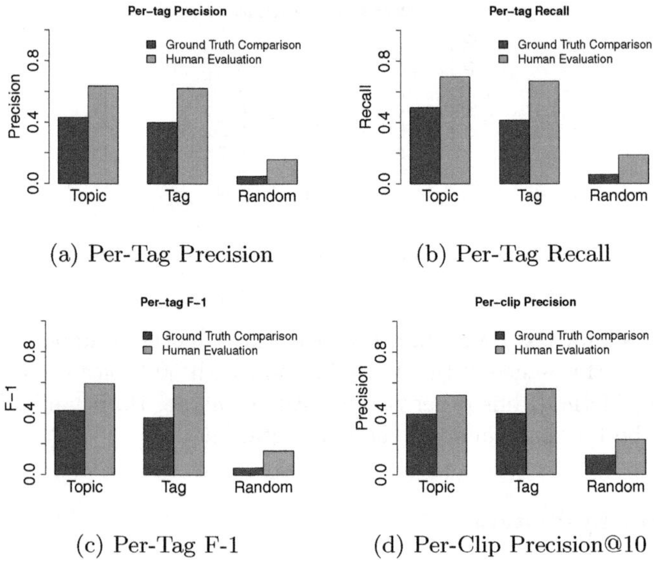

(a) Per-Tag Precision (b) Per-Tag Recall

(c) Per-Tag F-1 (d) Per-Clip Precision@10

Fig. 11. Mechanical Turk results for annotation performance

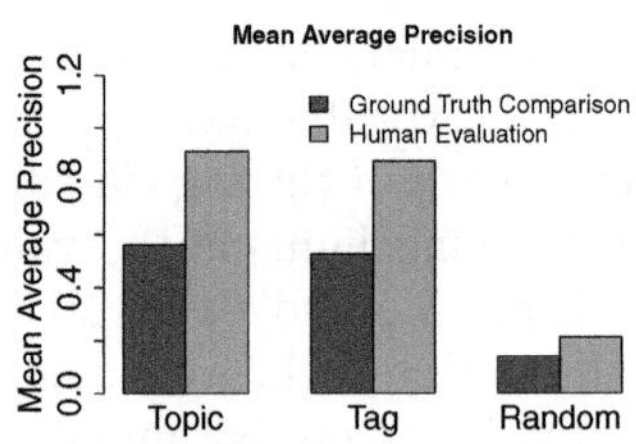

Fig. 12. Mechanical Turk results for music retrieval performance

output tags from topic posteriors (e.g., to improve diversity) may improve in this regard.

We also conducted an experiment to evaluating retrieval performance, where we provided each human evaluator a single-word search query and three lists of music clips retrieved by each method. We used 100 queries and 3 evaluators per query. Users were asked to check each music clip that they considered to be "relevant" for the query. In addition, they are asked to rank order the three lists in terms of their overall relevance to the query.

Figure 12 shows the mean average precision, when the ground truth tags versus human judgment is used to evaluate the relevance of each music clip in the retrieved set. As with annotation performance, the performance of all methods is significantly lower when evaluated using the ground truth set than when using human evaluations. Finally, when asked to rank music lists, users strongly preferred our Topic Method (59.3% of votes) over the Tag Method (39.0%) or Random (1.7%).

6 Conclusion and Future Work

The purpose of this work is to show how tagging algorithms can be trained, in an efficient way, to generate labels for objects (e.g., music clips) when the training data consists of a huge vocabulary of noisy labels. Focusing on music tagging as the domain of interest, we showed that our proposed method is both time and data efficient, while capable of achieving comparable (or superior, in the case of retrieval) performance to the traditional method of using tags as labels directly. This work opens up the opportunity to leverage the huge number of tags freely available on the Web for training annotation and retrieval systems.

Our work also exposes the problem of evaluating tags when the ground truth sets are noisy or incomplete. Following the lines of [15], an interesting direction would be to build a human computation game that is suited specifically for evaluating tags, and which can become a service for evaluating any music tagger.

There have been recent advances on topic modeling [3,21] that induce topics not only text, but also from other metadata (e.g., audio features in our setting). These methods may be good alternatives for training the topic distribution classifier in a one-step process as opposed to two, although our preliminary work in this direction has so far yielded mixed results.

Finally, another potential domain for our Topic Method is birdsong classification. To date, there are not many (if any) databases that allow a birdsong search by arbitrary tags. Given the myriad ways of describing birdsongs, it would be difficult to train a tagger that maps from audio features to tags directly, as most tags are likely to be associated with only a few examples. In collaboration with Cornell's Lab of Ornithology, we plan to use TagATune to collect birdsong tags from the tens of thousands of "citizen scientists" and apply our techniques to train an effective birdsong tagger and semantic search engine.

Acknowledgments. We gratefully acknowledge support for this work from a Microsoft Graduate Fellowship and DARPA under contract AF8750-09-C-0179.

References

1. Bergstra, J., Lacoste, A., Eck, D.: Predicting genre labels for artists using freedb. In: ISMIR, pp. 85–88 (2006)
2. Bertin-Mahieux, T., Eck, D., Maillet, F., Lamere, P.: Autotagger: a model for predicting social tags from acoustic features on large music databases. TASLP 37(2), 115–135 (2008)
3. Blei, D., McAuliffe, J.D.: Supervised topic models. In: NIPS (2007)
4. Blei, D., Ng, A., Jordan, M.: Latent dirichlet allocation. Journal of Machine Learning Research 3, 993–1022 (2003)
5. Csiszar, I.: Maxent, mathematics, and information theory. In: Hanson, K., Silver, R. (eds.) Maximum Entropy and Bayesian Methods. Kluwer Academic Publishers, Dordrecht (1996)
6. Dannenberg, R.B., Hu, N.: Understanding search performance in query-by-humming systems. In: ISMIR, pp. 41–50 (2004)

7. Eisenberg, G., Batke, J.M., Sikora, T.: Beatbank – an mpeg-7 compliant query by tapping system. Audio Engineering Society Convention, 6136 (2004)
8. Goto, M., Hirata, K.: Recent studies on music information processing. Acoustic Science and Technology, 419–425 (2004)
9. Herrera, P., Peeters, G., Dubnov, S.: Automatic classification of music instrument sounds. Journal of New Music Research, 3–21 (2003)
10. Hoffman, M., Blei, D., Cook, P.: Easy as CBA: A simple probabilistic model for tagging music. In: ISMIR, pp. 369–374 (2009)
11. Iwata, T., Yamada, T., Ueda, N.: Modeling social annotation data with content relevance using a topic model. In: NIPS (2009)
12. Lamere, P.: Social tagging and music information retrieval. Journal of New Music Research 37(2), 101–114 (2008)
13. Laurier, C., Sordo, M., Serra, J., Herrera, P.: Music mood representations from social tags. In: ISMIR, pp. 381–386 (2009)
14. Law, E., von Ahn, L.: Input-agreement: A new mechanism for collecting data using human computation games. In: CHI, pp. 1197–1206 (2009)
15. Law, E., West, K., Mandel, M., Bay, M., Downie, S.: Evaluation of algorithms using games: The case of music tagging. In: ISMIR, pp. 387–392 (2009)
16. Levy, M., Sandler, M.: A semantic space for music derived from social tags. In: ISMIR (2007)
17. Li, T., Ogihara, M., Li, Q.: A comparative study on content-based music genre classification. In: SIGIR, pp. 282–289 (2003)
18. Mandel, M., Ellis, D.: Song-level features and support vector machines for music classification. In: ISMIR (2005)
19. Mandel, M., Ellis, D.: Labrosa's audio classification submissions (2009)
20. Mandel, M., Ellis, D.: A web-based game for collecting music metadata. Journal of New Music Research 37(2), 151–165 (2009)
21. Mimno, D., McCallum, A.: Topic models conditioned on arbitrary features with dirichlet-multinomial regression. In: UAI (2008)
22. Steyvers, M., Griffiths, T.: Probabilistic topic models. In: Landauer, T., McNamara, D.S., Dennis, S., Kintsch, W. (eds.) Handbook of Latent Semantic Analysis. Erlbaum, Hillsdale (2007)
23. Trohidis, K., Tsoumakas, G., Kalliris, G., Vlahavas, I.: Multi-label classification of music emotions. In: ISMIR, pp. 325–330 (2008)
24. Turnbull, D., Barrington, L., Torres, D., Lanckriet, G.: Semantic annotation and retrieval of music and sound effects. TASLP 16(2), 467–476 (2008)
25. Turnbull, D., Liu, R., Barrington, L., Lanckriet, G.: A game-based approach for collecting semantic annotations of music. In: ISMIR, pp. 535–538 (2007)
26. Tzanetakis, G., Cook, P.: Musical genre classification of audio signals. IEEE Transactions on Speech and Audio Processing 10(5), 293–302 (2002)
27. von Ahn, L., Dabbish, L.: Labeling images with a computer game. In: CHI, pp. 319–326 (2004)
28. Whitman, B., Smaragdis, P.: Combining musical and cultural features for intelligent style detection. In: ISMIR (2002)
29. Yao, L., Mimno, D., McCallum, A.: Efficient methods for topic model inference on streaming document collections. In: KDD, pp. 937–946 (2009)

A Robustness Measure of Association Rules

Yannick Le Bras[1,3], Patrick Meyer[1,3], Philippe Lenca[1,3], and Stéphane Lallich[2]

[1] Institut Télécom, Télécom Bretagne,
UMR CNRS 3192 Lab-STICC,
Technopôle Brest Iroise CS 83818, 29238 Brest Cedex 3
{yannick.lebras,patrick.meyer,philippe.lenca}@telecom-bretagne.eu
[2] Université de Lyon
Laboratoire ERIC, Lyon 2, France
stephane.lallich@univ-lyon2.fr
[3] Université européenne de Bretagne, France

Abstract. We propose a formal definition of the robustness of association rules for interestingness measures. It is a central concept in the evaluation of the rules and has only been studied unsatisfactorily up to now. It is crucial because a good rule (according to a given quality measure) might turn out as a very fragile rule with respect to small variations in the data. The robustness measure that we propose here is based on a model we proposed in a previous work. It depends on the selected quality measure, the value taken by the rule and the minimal acceptance threshold chosen by the user. We present a few properties of this robustness, detail its use in practice and show the outcomes of various experiments. Furthermore, we compare our results to classical tools of statistical analysis of association rules. All in all, we present a new perspective on the evaluation of association rules.

Keywords: association rules, robustness, measure, interest.

1 Introduction

Since their seminal definition [1] and the APRIORI algorithm [2], association rules have generated a lot of research activities around algorithmic issues. Unfortunately, the numerous deterministic and efficient algorithms inspired by APRIORI tend to produce a huge number of rules. A widespread method to evaluate the interestingness of association rules consists of the quantification of this interest through objective quality measures on the basis of the contingency table of the rules. However, the provided rankings may strongly differ with respect to the chosen measure [3]. The large number of measures and their several properties have given rise to many research activities. We suggest that the interested reader refers to the following surveys: [4], [5], [6], [7] and [8].

Let us recall that an association rule $\mathtt{A} \rightarrow \mathtt{B}$, extracted from a database $\mathcal{B}$, is considered as an interesting rule according to the measure m and the user-specified threshold $m_{\min}$, if $m(\mathtt{A} \rightarrow \mathtt{B}) \geq m_{\min}$. This qualification of the rules raises some legitimate questions: to what extent is a good rule the result of

J.L. Balcázar et al. (Eds.): ECML PKDD 2010, Part II, LNAI 6322, pp. 227–242, 2010.

chance; is its evaluation significantly above the threshold; would it still be valid if the data had been different to some extent (noise) or if the acceptance threshold had been slightly raised; are there interesting rules which have been filtered out because of a threshold which is somewhat too high.

These questions lead very naturally to the intuitive notion of robustness of an association rule, i.e., the sensibility of the evaluation of its interestingness with respect to modifications of $\mathcal{B}$ and/or $m_{\min}$. Besides, it is already obvious here and now that this concept is closely related to the addition of counterexamples and/or the loss of examples of the rule. In this perspective, the study of the measures according to the number of such counterexamples becomes crucial: their decrease according to the number of counterexamples is a necessary condition for their eligibility, whereas their more or less high decrease rate when the first counterexamples appear is a property depending on the user's goals. We recommend that the interested reader has a look at [7] for a detailed study of 20 measures on these two characteristics.

To our knowledge, only very few works concentrate on the robustness of association rules, and can roughly be divided into three approaches: the first one is experimental and is mainly based on simulations [9,10,11], the second one uses statistical tests [5,12], whereas the third one is more formal as it studies the derivative of the measures [13,14,15].

Our proposal, which develops the ideas presented in [13] and [14], gives on the one hand a precise definition of the notion of robustness, and on the other hand presents a formal and coherent measure of the robustness of association rules. In Section 2 we briefly recall some general notions on association rules before presenting the definition of the measure of robustness and its use in practice. Then, in Section 3, we detail some experiments on classical databases with this notion of robustness. We then compare this concept to that of statistical significance in Section 4 and conclude in Section 5.

2 Robustness

2.1 Association Rules and Quality Measures

In a previous work [16], we have focused on a formal framework to study association rules and quality measures, which was initiated by [17]. Our main result in that article is the combination of an association rule with a projection in the unit cube of $\mathbb{R}^3$. As the approach detailed in this article is based on this framework, we briefly recall it here. Let us note $r : \mathtt{A} \rightarrow \mathtt{B}$ an association rule in a database $\mathcal{B}$. A quality measure is a function which associates a rule with a real number characterizing its interest. In this article, we focus exclusively on objective measures, whose value on r is determined solely by the contingency table of the rule. Figure 1 presents such a contingency table, in which we write p_x for the frequency of the pattern $\mathtt{X}$.

Once the three degrees of freedom of the contingency table are chosen, it is possible to consider a measure as a function from $\mathbb{R}^3$ to $\mathbb{R}$ and to use the classical results and techniques from mathematical analysis. In our previous work [16], we

	B	$\overline{\text{B}}$	
A	p_{ab}	$p_{a\bar{b}}$	p_a
$\overline{\text{A}}$	$p_{\bar{a}b}$	$p_{\bar{a}\bar{b}}$	$p_{\bar{a}}$
	p_b	$p_{\bar{b}}$	1

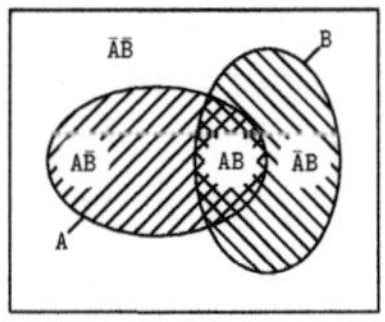

Fig. 1. Contingency table of $r : \mathtt{A} \rightarrow \mathtt{B}$

have shown that it is possible to make a link between algorithmic and analytical properties of certain measures, in particular those related to their variations.

In order to study the measures as functions of three variables, it is necessary to thoroughly define their domain of definition. This domain depends on the chosen parametrization: via the examples, the counterexamples or even the confidence. [14], [7] have stressed out the importance of the number of counterexamples in the evaluation of the interestingness of an association rule. As a consequence, in this work we analyze the behavior of the measures according to the variations of the counterexamples, i.e., an association rule $r : \mathtt{A} \rightarrow \mathtt{B}$ is characterized via the triplet $(p_{a\bar{b}}, p_a, p_b)$. In this configuration, the interestingness measures are functions on a subset $\mathcal{D}$ of the unit cube of $\mathbb{R}^3$ whose definition is given hereafter [16]:

$$\mathcal{D} = \left\{ (x, y, z) \left| \begin{array}{ccc} 0 & < y < & 1 \\ 0 & < z < & 1 \\ \max(0, y - z) & < x < & \min(y, 1 - z) \end{array} \right. \right\}$$

where x (resp. y, z) represent $p_{a\bar{b}}$ (resp. p_a, p_b).

In $\mathbb{R}^3$ a rule can now be considered as a vector, and it is possible to study its neighborhood and to observe its behavior in this neighborhood. This represents the starting point for the new characterization of the robustness of association rules, which we introduce in the following section.

2.2 A Definition of the Robustness

Let us suppose that a user wishes to evaluate association rules extracted from a database $\mathcal{B}$ via an objective interestingness measure m. In such a case, he has fixed a threshold $m_{\min}$ above which the rules are considered as interesting. These selected rules depend on many parameters, among which:

- the threshold $m_{\min}$: the user can modify it at any time and let appear or disappear a large number of rules;
- the noise: a given selected rule might not resist variations of the data, as, e.g., the addition of new transactions or the presence of erroneous recordings.

In this article we propose a contribution to the study of this latter point, namely the weakness of a rule according to variations in the data. [14] suggest different approaches for the study of the variations of the measures according to counterexamples of the rules. They develop various models to study the variations in the data that a rule can withstand in order to remain interesting. However, the authors do not give a general model which aggregates their multiple proposals, which does not allow to obtain a general measure of the robustness.

Our vision of the robustness is quite different and is based on the concept of *limit rule*. Note right beforehand that such a rule can be abstract, as it is not necessarily a rule which is achieved in the database $\mathcal{B}$. We define a distance between two rules r and r', $d_2(r, r')$, which is the euclidian distance between the projection of r and r' in $\mathcal{D}$.

Definition 1 (Limit rule). *A limit rule is an association rule $r_{\min}$, possibly abstract, such that $m(r_{\min}) = m_{\min}$. Let r be an association rule. We write r^* for a limit rule which minimizes $d(r, r_{\min})$ in $\mathbb{R}^3$. Formally,*

$$r^* \in argmin\{d_2(r, r_{\min}) | r_{\min} \text{ limit rule}\}$$

The limit rules which are actually realized in the database are those rules which have been barely selected according to the threshold $m_{\min}$. For a given rule r, r^* is not necessarily unique. However, its choice is not crucial for the notion of robustness that we are introducing in the sequel.

As a limit rule is an association rule, associated with $(x_{\min}, y_{\min}, z_{\min})$, it is necessarily an element of $\mathcal{D}$. Therefore, $d(r, r^*)$ is not simply the distance between r and the surface $\mathcal{S}$ of equation $m = m_{\min}$, but rather the distance to $\mathcal{S} \cap \mathcal{D}$.

Definition 2 (Robustness of an association rule). *Let m be an interestingness measure and $m_{\min}$ a threshold fixed by the user. Let r be an association rule on a database $\mathcal{B}$ such that $m(r) \geq m_{\min}$. The robustness of r according to m and $m_{\min}$ is defined by:*

$$rob_m(r, m_{\min}) = \frac{d(r, r^*)}{\sqrt{3}}$$

Figure 2 shows our concept of robustness for two rules. The important factor in this formula is the numerator $d(r, r^*)$, the division by $\sqrt{3}$ is a normalization factor which allows to fit the quantity in the interval $[0, 1]$. Other normalizations are indeed possible. If there is no ambiguity, we will write this robustness rob(r). In the following section we discuss this definition to show why it represents a notion of robustness, and present some of its properties.

2.3 Properties of the Robustness

Let us start by justifying the designation of robustness. Consider a database $\mathcal{B}$ and an association rule r : $\mathtt{A} \rightarrow \mathtt{B}$ in $\mathcal{B}$ such that $m(r) > m_{\min}$. We note $(p_{a\bar{b}}, p_a, p_b)$ the corresponding supports. Let us now add some noise in the database $\mathcal{B}$ in order to obtain a database $\mathcal{B}'$ in which the rule r' : $\mathtt{A} \rightarrow \mathtt{B}$ is characterized by $(p'_{a\bar{b}}, p'_a, p'_b)$. For short, after the noise introduction the patterns remain the same, but their supports change. Let us now suppose that the noise which is added respects:

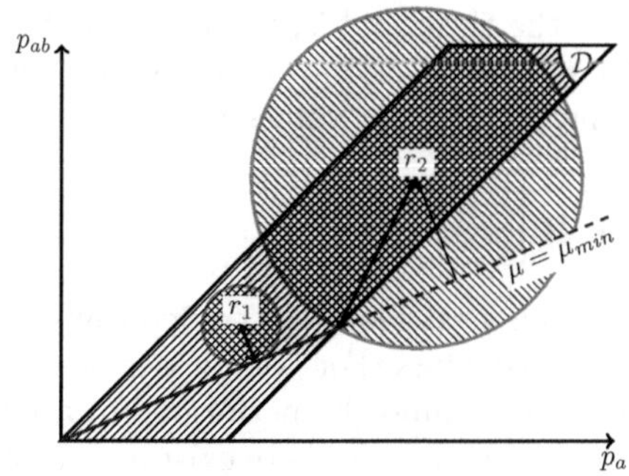

Fig. 2. Visualization of robustness for two different rules r_1 and r_2. Here, p_b is fixed, and the measure is the measure of confidence.

$$|p'_{a\bar{b}} - p_{a\bar{b}}| \leq \frac{d(r, r^*)}{\sqrt{3}} \ ; \ |p'_a - p_a| \leq \frac{d(r, r^*)}{\sqrt{3}} \ ; \ |p'_b - p_b| \leq \frac{d(r, r^*)}{\sqrt{3}}$$

In such a case, $d(r, r') = \sqrt{|p'_{a\bar{b}} - p_{a\bar{b}}|^2 + |p'_a - p_a|^2 + |p'_b - p_b|^2} \leq d(r, r^*)$, and thus by the definition of r^*, $m(r') \geq m_{\min}$. Thus, rob(r) clearly expresses the quantity of noise that the rule can withstand and still stay interesting. We can see that our definition of the robustness is closely linked to a notion of safety: if the noise is sufficiently controlled, then an interesting rule will stay interesting. The inverse is however not true, as a poorly robust rule can evolve to become more interesting and more robust.

This notion of robustness can be easily understood if the noise is inserted *by transaction*. Indeed, if one inserts the noise into less than rob(r)% of the transactions, the rule r will stay interesting according to $m_{\min}$. However, if the noise is inserted *by attribute* [9], it is harder to control it accurately.

Inversely, if the percentage of noise in a database is known, then the interesting robust rules (for this amount of noise) extracted from the noisy database will also be interesting in the *ideal* noiseless one.

Property 1. The robustness measure *rob*(r) has the following interesting analytical characteristics :

- the robustness of a rule is a real number of $[0, 1]$;
- $\text{rob}_m(r, m_{\min}) = 0$ if r is a limit rule, i.e., if $m(r) = m_{\min}$;[1]
- if the measure m, seen as a function of 3 variables, is continuous from $\mathcal{D} \subset \mathbb{R}^3$ to $\mathbb{R}$, then the robustness is decreasing with respect to $m_{\min}$;
- the robustness is continuous according to r.

[1] Note that the value $\text{rob}_m(r, m_{\min}) = 1$ is a theoretical value which corresponds to a very special configuration of r, $m_{\min}$ and m. In practice, in our experiments, we have not encountered this value.

These properties allow us to confirm certain expected behaviors of the robustness notion. First, the higher the threshold is, the less robust are the rules, and the more important is the reliability of the data. Second, two rules having close projections in $\mathbb{R}^3$ will have equivalent values for the robustness.

2.4 Calculating the Robustness

The calculation of the robustness requires the determination of the distance to a surface under certain constraints. For complex measures (Klosgen, collective strength, ...), this calculation cannot be performed in a formal way, and necessitates numerical techniques. However, there exist a certain number of measures based on frequencies for which the calculation is quite simple. In this paper we concentrate exclusively on these measures, which we call *planar measures*.

Definition 3 (Planar measure). *An interestingness measure m is called* planar *if the surface defined by $m(r) = m_{\min}$ is a plane.*

In particular, this is the case for measures like Sebag-Shoenauer, example-counter-example rate, Jaccard, contramin, precision, recall, specificity. In this case, the distance between a rule r_1 with coordinates (x_1, y_1, z_1) and the plane $\mathcal{P} : ax + by + cz + d = 0$ is given by:

$$d(r_1, \mathcal{P}) = \frac{|ax_1 + by_1 + cz_1 + d|}{\sqrt{a^2 + b^2 + c^2}}$$

However, to obtain the robustness measure, r^* must belong to the domain $\mathcal{D}$. Therefore, if it is not the case for the orthogonal projection of the rule on the plane, the distance of interest is the one between the rule and the intersection polygon $\mathcal{P} \cap \mathcal{D}$. We therefore determine the corners of this convex polygon to obtain the distance between the rule and the perimeter of the polygon as the minimal distance between the rule and the edges of the polygon (as segments).

Consequently, the calculation algorithm of the robustness measure for planar measures is given hereafter:

- Determine $r^{\perp}$, the orthogonal projection of r on $\mathcal{P}$;
- If $r^{\perp} \in \mathcal{D}$, $r^* = r^{\perp}$ and return $d(r, r^*)$;
- Else, return the distance between the rule and the perimeter of the intersection polygon.

Example 1. The following measures are planar. Their level lines $m = m_0$ define the following planes:

- confidence: $x - (1 - m_0)y = 0$;
- Sebag-Shoenauer: $(1 + m_0)x - y = 0$;
- example-counterexample rate: $(2 - m_0)x - (1 - m_0)y = 0$;
- Jaccard: $(1 + m_0)x - y + m_0 z = 0$.

Let us now study in further details the case of the confidence measure. In a parametrization via the counterexamples, the plane defined by the confidence threshold $m_{\min}$ is $\mathcal{P} : x - (1 - m_{\min})y = 0$. The distance between a rule r_1 (with coordinates (x_1, y_1, z_1) and confidence $m(r_1) > m_{\min}$) and the plane is given by

$$d = y_1 \frac{m(r_1) - m_{\min}}{\sqrt{1 + (1 - m_{\min})^2}}. \tag{1}$$

Thereby, for a given value of $m_{\min}$, the robustness depends on two parameters:

- y_1, the support of the antecedent;
- $m(r_1)$, the value taken by the interestingness measure of the rule.

Thus, two rules having the same confidence, can have very different robustness values. Similarly, two rules having the same robustness, can have various confidences. Therefore, it will not be surprising to observe rules with a low value for the interestingness measure and a high robustness, as well as rules with a high interestingness and a low robustness. Indeed, it is possible to discover rules which are simultaneously very interesting and very fragile.

Example 2. Consider a fictive database of 100000 transactions. We write n_x for the number of occurrences of the pattern X. In this database, we can find a first rule r_1 : A $\rightarrow$ B such that $n_a = 100$ and $n_{a\bar{b}} = 1$. Its confidence equals 99%. However, its robustness, at the level of confidence of 0.8 equals $\text{rob}(r_1) = 0.0002$. A second rule r_2 : C $\rightarrow$ D has the following characteristics: $n_c = 50000$ and $n_{c\bar{d}} = 5000$. Its confidence only equals 90%, whereas its robustness measure is 0.05. As r_2 has proportionally to its antecedent more counterexamples than r_1, at first sight it could be mistakenly considered as less reliable.

In the first case, the closest limit rule can be described by $n_a^* = 96$ et $n_{a\bar{b}}^* = 19$. The original rule therefore only resists very few variations on the entries. The second rule however has a closest limit rule with parameters $n_c = 49020$ et $n_{c\bar{d}} = 9902$, which shows that r_2 can bear about a thousand changes in the database.

As a conclusion, r_2 is much less sensitive to noise as r_1, even if r_1 appears to be more interesting according to the confidence measure.

These observations show that the determination of the *real interestingness* of a rule is more difficult than it seems: how should we arbitrate between a rule which is interesting according to a quality measure but poorly robust, and one which is less interesting but which is more reliable with respect to noise.

2.5 Use of the Robustness in Practice

The robustness, as defined earlier, can have two immediate applications. First, the robustness measure allows to compare any two rules and to compute a weak order on the set of selected rules (a ranking with ties). Second, the robustness measure can be used to filter the rules if the user fixes a limit threshold.

However, similarly as for the interestingness measures, the determination of this robustness threshold might be a difficult task. In practice, it should therefore be avoided to impose the determination of another threshold on a user. This notion can nevertheless be a further parameter in the comparison of two rules.

When considering the interestingness measure of a rule according to its robustness measure, it is possible to distinguish between two situations. When comparing rules which are fragile and uninteresting to rules which are robust and interesting, it is obvious that a user will prefer the second ones. However, this choice is more demanding for a fragile but interesting rule compared to a robust but uninteresting one. Is it better to have an interesting rule which depends a lot on the noise in the data or a very robust one, which will resist changes in the data? The answers to this question depend of course on the practical situation and the confidence of the user in the quality of his data.

In the sequel we will observe that the interestingness vs. robustness plots show a lot of robust rules which are dominated in terms of quality measures by less robust ones.

3 Experiments

In this section we present the results obtained on 4 databases for 5 planar measures. First we present the experimental protocol, then we study the plots that we generated in order to stress out the link between the interestingness measures and the robustness. Finally, we analyze the influence of noise on association rules.

3.1 Experimental Protocol

Extraction of the rules. Recall that we focus here on planar measures. For this experiment, we have selected 5 of them: confidence, Jaccard, Sebag-Shoenauer, example-counterexample rate, and specificity. Table 1 summarizes their definition in terms of the counterexamples and the plane they define in $\mathbb{R}^3$.

For our experiments we have chosen 4 of the usual databases [18]. We have extracted class rules, i.e. rules for which the consequent is constrained, both from Mushroom and a discretized version of Census. The databases Chess and Connect have been binarized in order to extract unconstrained rules. All the rules

Table 1. The planar measures, their definition, the plane defined by m_0 and the selected threshold value

name	formula	plane	threshold
confidence	$\frac{p_a - p_{a\bar{b}}}{p_a}$	$x - (1 - m_0)y = 0$	0.984
Jaccard	$\frac{p_a - p_{a\bar{b}}}{p_b + p_{a\bar{b}}}$	$(1 + m_0)x - y + m_0 z = 0$	0.05
Sebag-Shoenauer	$\frac{p_a - p_{a\bar{b}}}{p_{a\bar{b}}}$	$(1 + m_0)x - y = 0$	10
specificity	$\frac{1 - p_b - p_{a\bar{b}}}{1 - p_a}$	$x - m_0 y + z = 1 - m_0$	0.5
example-counterexample rate	$1 - \frac{p_{a\bar{b}}}{p_a - p_{a\bar{b}}}$	$(2 - m_0)x - (1 - m_0)y = 0$	0.95

Table 2. Databases used in our experiments. The fifth column indicates the maximal size of the extracted rules.

database	attributes	transactions	type	size	# rules
census	137	48842	class	5	244487
chess	75	3196	unconstrained	3	56636
connect	129	67557	unconstrained	3	207703
mushroom	119	8124	class	4	42057

have been generated via the APRIORI algorithm of [19], in order to obtain rules with a positive support, a confidence above 0.8 and of variable size according to the database. These information are summarized in Table 2. Note that the generated rules are interesting, the nuggets of knowledge have not been left out, and the number of rules is fairly high.

Calculation of the robustness. For each set of rules and each measure we have applied the same calculation method for the robustness of the association rules. In a first step, we have selected only the rules with an interestingness measure above a predefined threshold. We have chosen to fix this threshold according to the values of Table 1. These thresholds have been determined by observing the behavior of the measures on the rules extracted from the Mushroom database, in order to obtain interesting and uninteresting rules in similar proportions.

Then we have implemented an algorithm, based on the description of Section 2.4 for the specific case of planar measures, which determines the robustness of a rule according to the value it takes for the interestingness measure and the threshold. As an output we obtain a list of rules with their corresponding support, robustness and interestingness measure values. The complexity of this algorithm depends mostly on the number of rules which have to be analyzed. These results, presented in Section 3.2, allow us to generate the interestingness vs. robustness plots mentioned earlier.

Noise insertion. As indicated earlier, we analyze the influence of noise in the data on the rules, according to their robustness. This noise is introduced transaction-wise, for the reasons mentioned in Section 2.3, as follows: in 5% of randomly selected rows of each database, the values of the attributes are modified randomly (equally likely and without replacement). Once the noise is inserted, we recalculate the supports of the initially generated rules. We then extract the interesting rules according to the given measures and evaluate their robustness. The study of the noise is discussed in Section 3.3.

3.2 Robustness Analysis

For each database and each interestingness measure, we plot the value taken by the rule for the measure according to its robustness. Figure 3 shows a representative sample of these results (for a given interestingness measure, the plots are, in general, quite similar for all the databases).

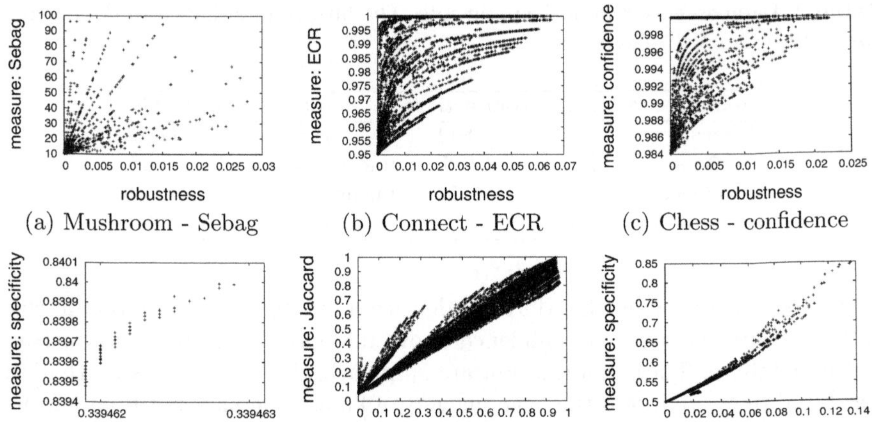

(a) Mushroom - Sebag (b) Connect - ECR (c) Chess - confidence

(d) Census - specificity (e) Connect - Jaccard (f) Mushroom - specificity

Fig. 3. Value of the interestingness measure according to the robustness for a sample of databases and measures

Various observations can be deduced from these plots. First, the interestingness measure is in general increasing with the robustness. A closer analysis shows that a large number of rules are dominated in terms of their interestingness by less robust rules. This is specially the case for the Sebag measure (Figure 3(a)), for which we observe that a very interesting rule r_1 ($\text{Sebag}(r_1) = 100$) can be significantly less robust ($\text{rob}(r_1) = 10^{-4}$) than a less interesting rule r_2 ($\text{Sebag}(r_2) = 20$ and $\text{rob}(r_2) = 2 \cdot 10^{-3}$). The second rule resists twenty times more changes than the first one.

Second, in most of the cases, we observe quite distinct level lines. Sebag and Jaccard bring forward straight lines, confidence and example-counterexample rate generate concave curves, and the specificity seems to produce convex ones.

Let us analyze the case of the level curves for the confidence. Note that similar calculations can be done for the other interestingness measures. Equation (1) presents the robustness according to the measure, where y represents p_a. As $p_a = \frac{p_{a\bar{b}}}{1-m(r)}$, we can write the measure $m(r)$ according to the distance d:

$$m(r) = \frac{m_{\min} + \sqrt{1 + (1 - m_{\min})^2} \cdot \frac{d}{x}}{1 + \sqrt{1 + (1 - m_{\min})^2} \cdot \frac{d}{x}} \tag{2}$$

Thus, for a given x (i.e. for a constant number of counterexamples), the rules are situated on a well defined concave and increasing curve. This shows that the level lines in the case of the confidence are made of rules which have the same number of counterexamples.

Another behavior seems common for most of the studied measures: there exists no rule which is close to the threshold and very robust. Sebag is the only measure which does not completely fit to this observation. We think that this might be strongly linked to the restriction of the study to planar measures.

3.3 Study of the Influence of the Noise

In this section we are studying the links between the addition of noise to the database and the evolution of the rules sets, with respect to robustness. To do so, we create 5 noisy databases from the original ones (see 3.1) and for each of them, analyze the robustness of the rules resisting these changes and of the ones disappearing. In order to validate our notion of robustness, we expect that the robustness of the rules which have vanished is lower on average than the robustness of the rules which stay in the noisy databases.

Table 3 presents the results of this experiment, by showing the average of the robustness values for the two sets of rules, for the 5 noisy databases. In most of the cases, the rules which resisted the noise are approximatively 10 times more robust than those which vanished. The only exception is the Census database for the measures example-counterexample rate, Sebag and confidence, which do not confirm this result. However, this is not negating our theory. Indeed, the initial robustness values for the rules of the Census database are around 10^{-6}, which makes them vulnerable to 5% of noise. It is therefore not surprising that all the rules can potentially become uninteresting.

On the opposite, the measure of specificity underlines a common behavior of the Census and the Connect databases. For both of them, no rule vanishes after the insertion of 5% of noise. The average value of the robustness of the rules which resisted the noise is significantly higher than these 5%, which means that all the rules are well protected. In the case of the Census base, the lowest specificity value equals 0.839, which is well above the threshold which has been fixed beforehand. This explains why the rules originating from the Census database

Table 3. Comparison between the average robustness values for the vanished rules and those which resisted the noise, for each of the studied measures

(a) example-counter-example rate

base	vanished	stayed
census	0.83e-6	0.79e-6
chess	1.16e-3	0.96e-2
connect	5.26e-4	7.72e-3
mushroom	9.4e-5	6.6e-4

(b) Sebag

base	vanished	stayed
census	1.53e-6	1.53e-6
chess	1.63e-3	1.72e-2
connect	8.38e-4	1.42e-2
mushroom	1.28e-4	1.22e-3

(c) specificity

base	vanished	stayed
census	0	0.19
chess	7.23e-5	8.76e-2
connect	0	1.2e-1
mushroom	2.85e-4	1.37e-2

(d) confidence

base	vanished	stayed
census	2.61e-7	2.61e-7
chess	5.59e-4	3.77e-3
connect	2.16e-4	2.73e-3
mushroom	5.51e-5	2.34e-4

(e) Jaccard

base	vanished	stayed
census	0	0
chess	3.2e-4	1.69e-1
connect	1.94e-3	1.43e-1
mushroom	3.20e-4	1.90e-2

all resist the noise. In the case of the Connect database, the average value of the specificity measure equals 0.73 with a standard deviation of 0.02. The minimal value equals 0.50013 and corresponds to a robustness of $2.31e-5$. However, this rule has been saved in the 5 noise additions. This underlines the fact that our definition of the robustness corresponds to the definition of a security zone around a rule. If the rule changes and leaves this area, it can evolve freely in the space, without ever getting to the threshold surface. Nevertheless, the risk still prevails.

In the following section we compare the approach via the robustness measure to a more classical one to determine if a rule is considered as statistically significant.

4 Robustness vs. Statistical Significance

In the previous sections, we have defined the robustness of a rule as its capacity to overcome variations in the data, like a loss of examples and / or a gain of counter-examples, so that its evaluation $m(r)$ remains above the given threshold $m_{\min}$. This definition looks quite similar to the notion of statistical significance. In this section we explore the links between both approaches.

4.1 Significant Rule

From a statistical point of view, we have to distinguish between the following notions: $m(r)$ is the empirical value of the rules computed over a given data sample, that is the observed value of the random variable $M(r)$, and $\mu(r)$ is the theoretical value of the interestingness measure. A *statistically significant rule* r for a threshold $m_{\min}$ and the chosen measure is a rule for which we can consider that $\mu(r) > m_{\min}$.

Usually, for each rule, the null-hypothesis $H_0 : \mu(r) = m_{\min}$ is tested against the alternative hypothesis $H_1 : \mu(r) > m_{\min}$. A rule r is considered as significant at the significance level α_0 (type I error, false positive) if its p-value is at most α_0. Recall that the p-value of a rule r whose empirical value is $m(r)$ is defined as $P(M(r) \geq m(r)|H_0)$.

However, due to the high number of tests which need to be performed, and the resulting multitude of false discoveries, the p-values need to be adapted (see [20] for a general presentation, and [5] for the specific case of association rules with respect to independency).

The algebraic form of the p-value can be determined only if the law of M under H_0 is (at least approximately) known. This is the case for the measure of confidence, for which $M = N_{ab}/N_a$ where N_x is the number of instances of the itemset x. The distribution of M under H_0 is established via the models proposed by [21] and generalized by [22], provided that the margins N_a and N_b are fixed. However, this is somewhat simplistic, like for the χ^2 test. Furthermore, in many cases, as e.g. for the planar measure of Jaccard, it is impossible to establish the law of M under H_0.

Therefore, we here prefer to estimate the risk that the interestingness measure of the rule falls below the threshold $m_{\min}$ via a bootstrapping technique which allows to approximate the variations of the rule in the real population. In our case we draw with replacement 400 samples of size n from the original population of size n. The risk is then estimated via the proportion of samples in which the evaluation of the rule fell under the threshold. Note that this value is smoothed by using the normal law. Only the rules with a risk less or equal to α_0 are considered as significant.

However, even if no rule is significant, $n\alpha_0$ rules will be selected. In the case where $n = 10000$ and $\alpha_0 = 0.05$, this would lead to 500 false discoveries. Among all the false discoveries control methods, one is of particular interest. In [23], Benjamini and Liu proposed a sequential method: the risk values are sorted in increasing order and named $p_{(i)}$. A rule is selected if its corresponding $p_{(i)} \leq i\frac{\alpha_0}{n}$. This procedure allows to control the expected proportion of wrongfully selected rules in the set of selected rules (False Discovery Rate) conditionally to the independence of the data. This is compatible with positively dependent data.

4.2 Comparison of the Two Approaches on an Example

In order to get a better understanding of the difference between these rules stability approaches, we compare the results of the robustness calculation and the complementary risk resulting from the bootstrapping. Our experiments are based on the SolarFlare database [18]. We detail here the case of the two measures mentioned above, Confidence and Jaccard, for which an algebraic approach of the p-value is either problematic (fixed margins), or impossible.

We first extract the rules with the classical Apriori algorithm with support and confidence thresholds set to 0.13 and 0.85 respectively. This produces 9890 rules with their confidence and robustness. A bootstrapping technique with 400 iterations allows to compute the risk of each rule to fall below the threshold. From the 9890 rules, one should note that even if 8481 have a bootstrapping risk of less than 5%, only 8373 of them are kept when applying the procedure of Benjamini and Liu.

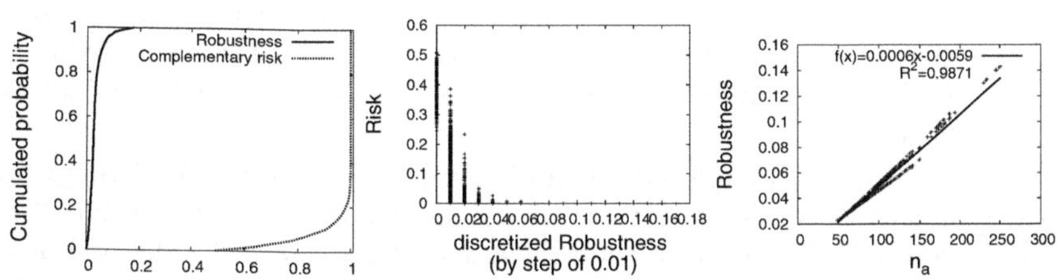

(a) Empirical cumulative distribution functions (b) Risk and robustness (c) Robustness and n_a

Fig. 4. Confidence Case

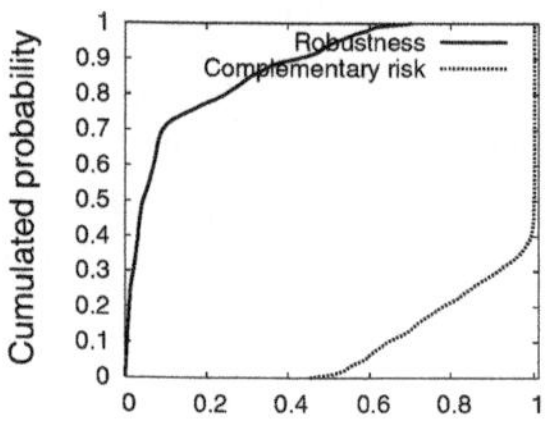

(a) Empirical cumulative distribution functions

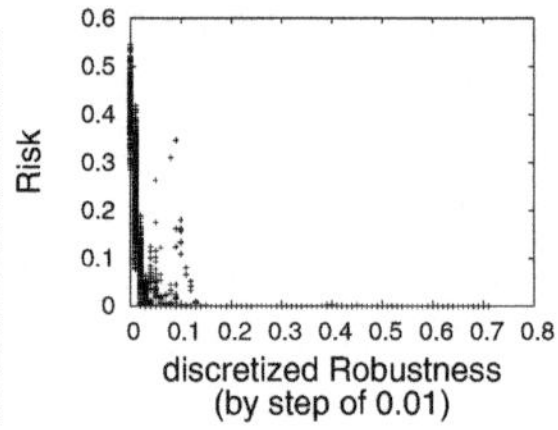

(b) Risk and robustness

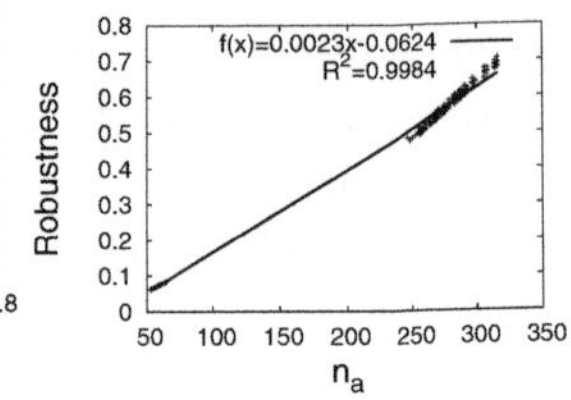

(c) Robustness and n_a

Fig. 5. Jaccard index case

Figure 4(a) shows the empirical cumulative distribution function of the robustness and the complementary risk resulting from the bootstrapping. It shows that the robustness is clearly more discriminatory than the complementary risk, especially for interesting rules. Figure 4(b) represents the risk with regard to the class of robustness (discretized by steps of 0.01). It shows that the risk is globally correlated with robustness.

However, the outputs of two approaches are clearly different. On the one side, the process of Benjamini returns 1573 unsignificant rules having a robustness less than 0.025 (except for 3 of them). On the other side, 3616 rules of the significant ones have a robustness less than 0.05. Besides, it is worth noticing that the robustness of the 2773 logical rules takes many different values between 0.023 and 0.143. Finally, as shown in Figure 4(c), the robustness of a rule is linearly correlated with its coverage.

The results obtained with the Jaccard measure are of the same kind. The support threshold is set to 0.0835, whereas the Jaccard index is fixed to 0.3. We obtain 6066 rules, from which 4059 are declared significant at the 5% level by the bootstrapping technique (400 iterations), and 3933 by the process of Benjamini (that is 2133 unsignificant rules). Once again, the study of the empirical cumulative distribution functions (see Figure 5(a)) shows that the robustness is more discriminatory than the complementary risk of the bootstrapping for the more interesting rules. Similarly, Figure 5(b) shows that the risk for the Jaccard measure is globally correlated with the robustness, but again, there are significant differences between the two approaches. The rules determined as significant for the process of Benjamini have a robustness less than 0.118 when significant rules at the 5% level have robustness spread from 0.018 and 0.705, which is a quite big range.

There are 533 rules with a Jaccard index greater than 0.8. All of them have a zero complementary risk, and their robustness value vary between 0.062 and 0.705. As shown by Figure 5(c), the robustness of the Jaccard index is linearly correlated to the coverage of the rule for high values of the index (> 0.80).

As a conclusion of this comparison, the statistical approach of bootstrapping to estimate the type I error has the major drawback that it is not very discriminatory, especially for high values of n, which is the case in datamining.

In addition, the statistical analysis assume that the actual data are a random subset of the whole population, which is not really the case in datamining. All in all, the robustness study for a given measure gives a more precise idea of the stability of interesting rules.

5 Conclusion

The robustness of association rules is a crucial topic, which has only been poorly studied by formal approaches. The robustness of a rule with respect to variations in the database adds a further argument for its interestingness and increases the validity of the information which is given to the user.

In this article, we have presented a new operational notion of robustness which depends on the chosen interestingness measure and the corresponding acceptability threshold. As we have shown, our definition of this notion is consistent with the natural intuition linked to the concept of robustness. We have analyzed the case of a subset of measures, called planar measures, for which we are able to give a formal characterization of the robustness. Our experiments on 5 measures and 4 classical databases illustrate and corroborate the theoretical discourse. The proposed robustness measure is also compared to a more classical statistical analysis of the significance of a rule, which turns out to be less discriminatory in the context of data mining.

In practice, the robustness measure allows to rank rules according to their ability to withstand changes in the data. However, the determination of a robustness threshold by a user remains an issue. In the future, we plan to propose a generic protocol to calculate the robustness of association rules with respect to any interestingness measure via the use of numerical methods.

References

1. Agrawal, R., Imieliski, T., Swami, A.: Mining association rules between sets of items in large databases. In: ACM SIGMOD International Conference on Management of Data, Washington, D.C., United States, pp. 207–216 (1993)
2. Agrawal, R., Srikant, R.: Fast algorithms for mining association rules in large databases. In: 20th International Conference on Very Large Data Bases, Santiago de Chile, Chile, pp. 478–499 (1994)
3. Vaillant, B., Lenca, P., Lallich, S.: A clustering of interestingness measures. In: 7th International Conference on Discovery Science, Padova, Italy, pp. 290–297 (2004)
4. Geng, L., Hamilton, H.J.: Interestingness measures for data mining: A survey. ACM Computing Surveys 38(3, Article 9) (2006)
5. Lallich, S., Teytaud, O., Prudhomme, E.: Association rule interestingness: Measure and statistical validation. In: Quality Measures in Data Mining, pp. 251–275 (2007)
6. Geng, L., Hamilton, H.J.: Choosing the right lens: Finding what is interesting in data mining. Quality Measures in Data Mining, 3–24 (2007)
7. Lenca, P., Meyer, P., Vaillant, B., Lallich, S.: On selecting interestingness measures for association rules: user oriented description and multiple criteria decision aid. European Journal of Operational Research 184, 610–626 (2008)

8. Suzuki, E.: Pitfalls for categorizations of objective interestingness measures for rule discovery. In: Statistical Implicative Analysis, Theory and Applications, pp. 383–395 (2008)
9. Azé, J., Kodratoff, Y.: Evaluation de la résistance au bruit de quelques mesures d'extraction de règles d'association. In: 2nd Extraction et Gestion des Connaissances conference, Montpellier, France, pp. 143–154 (2002)
10. Cadot, M.: A simulation technique for extracting robust association rules. In: Computational Statistics & Data Analysis, Limassol, Chypre (2005)
11. Azé, J., Lenca, P., Lallich, S., Vaillant, B.: A study of the robustness of association rules. In: The 2007 Intl. Conf. on Data Mining, Las Vegas, Nevada, USA, pp. 163–169 (2007)
12. Rakotomalala, R., Morineau, A.: The TVpercent principle for the counterexamples statistic. In: Statistical Implicative Analysis, Theory and Applications, pp. 449–462. Springer, Heidelberg (2008)
13. Lenca, P., Lallich, S., Vaillant, B.: On the robustness of association rules. In: 2nd IEEE International Conference on Cybernetics and Intelligent Systems and Robotics, Automation and Mechatronics, Bangkok, Thailand, pp. 596–601 (2006)
14. Vaillant, B., Lallich, S., Lenca, P.: Modeling of the counter-examples and association rules interestingness measures behavior. In: The 2006 Intl. Conf. on Data Mining, Las Vegas, Nevada, USA, pp. 132–137 (2006)
15. Gras, R., David, J., Guillet, F., Briand, H.: Stabilité en A.S.I. de l'intensité d'implication et comparaisons avec d'autres indices de qualité de règles d'association. In: 3rd Workshop on Qualite des Donnees et des Connaissances, Namur Belgium, pp. 35–43 (2007)
16. Le Bras, Y., Lenca, P., Lallich, S.: On optimal rules discovery: a framework and a necessary and sufficient condition of antimonotonicity. In: Theeramunkong, T., Kijsirikul, B., Cercone, N., Ho, T.-B. (eds.) PAKDD 2009. LNCS, vol. 5476, pp. 705–712. Springer, Heidelberg (2009)
17. Hébert, C., Crémilleux, B.: A unified view of objective interestingness measures. In: 5th Intl. Conf. on Machine Learning and Data Mining, Leipzig, Germany, pp. 533–547 (2007)
18. Asuncion, A., Newman, D.: UCI machine learning repository (2007)
19. Borgelt, C., Kruse, R.: Induction of association rules: Apriori implementation. In: 15th Conference on Computational Statistics, Berlin, Germany, pp. 395–400 (2002)
20. Dudoit, S., van der Laan, M.J.: Multiple Testing Procedures with Applications to Genomics (2007)
21. Lerman, I.C., Gras, R., Rostam, H.: Elaboration d'un indice d'implication pour les données binaires, I et II. Mathématiques et Sciences Humaines, 5–35, 5–47 (1981)
22. Lallich, S., Vaillant, B., Lenca, P.: A probabilistic framework towards the parameterization of association rule interestingness measures. Methodology and Computing in Applied Probability, 447–463 (2007)
23. Benjamini, Y., Liu, W.: A step-down multiple hypotheses testing procedure that controls the false discovery rate under independence. Journal of Statistical Planning and Inference 82(1-2), 163–170 (1999)

Automatic Model Adaptation for Complex Structured Domains

Geoffrey Levine, Gerald DeJong, Li-Lun Wang, Rajhans Samdani, Shankar Vembu, and Dan Roth

Department of Computer Science
University of Illinois at Champaign-Urbana
Urbana, IL 61801
{levine,dejong,lwang4,rsamdan2,svembu,danr}@cs.illinois.edu

Abstract. Traditional model selection techniques involve training all candidate models in order to select the one that best balances training performance and expected generalization to new cases. When the number of candidate models is very large, though, training all of them is prohibitive. We present a method to automatically explore a large space of models of varying complexities, organized based on the structure of the example space. In our approach, one model is trained by minimizing a minimum description length objective function, and then derivatives of the objective with respect to model parameters over distinct classes of the training data are analyzed in order to suggest what model specifications and generalizations are likely to improve performance. This directs a search through the space of candidates, capable of finding a high performance model despite evaluating a small fraction of the total number of models. We apply our approach in a complex fantasy (American) football prediction domain and demonstrate that it finds high quality model structures, tailored to the amount of training data available.

1 Motivation

We consider a *model* to be a parametrically related family of hypotheses. Having a good model can be crucial to the success of a machine learning endeavor. A model that is too flexible for the amount of data available or a model whose flexibility is poorly positioned for the information in the data will perform badly on new inputs. But crafting an appropriate model by hand is both difficult and, in a sense, self-defeating. The learning algorithm (the focus of ML research) is then only partially responsible for any success; the designer's ingenuity becomes an integral component. This has given rise to a long-term trend in machine learning toward weaker models which in turn demand a great deal of world data in the form of labeled or unlabeled examples. Techniques such as structural risk minimization which *a priori* specify a nested family of increasingly complex models are an important direction. The level of flexibility is then variable and can be adjusted automatically based on the data itself.

This research reports our first steps in a new direction for automatically adapting model flexibility to the distinction that seem most important given the data. This allows adaptation to the *kind* of flexibility in addition to the level of complexity. We are interested in automatically constructing generative models to be used as a computational

J.L. Balcázar et al. (Eds.): ECML PKDD 2010, Part II, LNAI 6322, pp. 243–258, 2010.

proxy for the real world. Once constructed and calibrated, such a model guides decisions within some prescribed domain of interest. Importantly, its utility is judged only within this limited domain of application. It can (and will likely) be wildly inaccurate elsewhere as the best model will concentrate its expressiveness where it will do the most good.

We contrast model learning with classification learning, which attempts to characterize a pattern given some appropriate representation of the data. Learning a model focuses on how best to represent the data. Real world data can be very complex. Given a myriad of distinctions that are possible, which are worth noticing? Which interactions should be emphasized and which ignored? Should certain combinations of distinctions be merged so as to pool the data and allow more accurate parameter estimation? In short, what is the most useful model from a space of related models? The answer depends on 1) the purpose to which the model will be applied: distinctions crucial for one purpose will be insignificant in others, 2) the amount of training data available: more data will generally support a more complex model, and 3) the emergent patterns in the training data: a successful model will carve the world "at its joints" respecting the natural equivalences and significant similarities within the domain.

The conventional tools for model selection include the minimum description length principle [1], the Akaike information criterion [2], and the Bayesian information criterion [3], as well as cross-validation. The disadvantage of these techniques is that in order to evaluate a model, it must be trained to the data. For many models, this is time consuming, and so the techniques do not scale well to cases where we would like to consider a large number of candidate models.

In adapting models, it is paramount to reduce the danger of overfitting. Just as an overfit hypothesis will perform poorly, so will an overfit model. Such a model would make distinctions relevant to the particular data set but not to the underlying domain. The flexibility that it exposes will not match the needs of future examples, and even the best from such a space will perform poorly. To drive our adaption process we employ available prior domain knowledge. For us, this consists of information about distinctions that many experts through many years, or even many generations, have discovered about the domain. This sort of prior knowledge has the potential to provide far more information than can reasonably be extracted from any training set.

For example, in evaluating the future earnings of businesses, experts introduce distinctions such as the Sector: {*Manufacturing, Service, Financial, Technology*} which is a categorical multiset and Market Capitalization: {*Micro, Small, Medium, Large*} which is ordinal. Distinctions may overlap, a company's numeric Beta and its Cyclicality: {*Cyclical, Non-cyclical, Counter-cyclical*} represent different views of the same underlying property. Such distinctions are often latent, in the sense that they are derived or emergent properties; the company is perfectly well-formed and well-defined without them. Rather they represent conceptualizations that experts have invented. When available such prior knowledge should be taken as potentially useful. Ignoring it when relevant may greatly increase the required amount of data to essentially re-derive the expert knowledge. But blindly adopting it can also lead to degraded performance. If the distinction is unnecessary for the task or if there is insufficient data to confidently make use, performance will also suffer. We explore how the space of distinctions interacts with training data. Our algorithm conducts a directed search through model structures, and performs much better than simply trying every possibility. In our approach, one model is trained and analyzed to suggest alternative model formulations that are likely to result in better general performance. These suggestions guide a general search through the full space of alternative model formulations, allowing us to find a high quality model despite evaluating only a small fraction of the total number.

2 Preliminaries

To introduce our notation, we use as a running example the business earnings domain. Assume we predict future earnings with a linear function of N numerical features: f_1 to f_N. Thus, each prediction takes the form $\Phi \cdot F = \phi_1 f_1 + \phi_2 f_2 + ... + \phi_N f_N$. One possibility is to learn a single vector Φ. Of course, the individual ϕ_i's must still be estimated from training data, but once learned this single linear function will apply to all future examples. Another possibility is to treat companies in different sectors differently. Then we might learn one Φ for Manufacturing, a different one for Service companies, another for Financial companies, and another for Technology companies. A new example company is then treated according to its (primary) sector. On the other hand, perhaps Service companies and Financial companies seem to behave similarly in the training data. Then we might choose to lump those together, but to treat Manufacturing and Technology separately. Furthermore, we need not make the same distinctions for each feature. Consider f_i = unemployment rate. If there is strong evidence that the unemployment rate will have a different effect on companies based on their sector *and* size, we would estimate (and later apply) a specialized ϕ_i (for unemployment) based on sector and size together. Let D_i be the finest grain distinction for feature f_i (here, $Sector \times Size \times Cyclicity$). We refer to this set as the *domain of applicability* for parameter ϕ_i. Depending on the evidence, though, we may choose not to distinguish all elements D_i. The space of distinctions we can consider are the partitions of the D_i's. These form the alternative models that we must choose among.

Generally, the partitions of D_i form a lattice (as shown in Figure 1). This lattice, which we will refer to as Λ_{D_i}, is ordered by the *finer-than* operator (a partition P is said to be finer than partition P', if every elements of P is a subset of some element of P'). In turn, we can construct the Cartesian product lattice $\Lambda = \Lambda_{D_1} \times \Lambda_{D_2} \times ... \times \Lambda_{D_N}$. Note that Λ can have a very large number of elements. For example, if $|N| = 4$, and $|D_i| = 4$ for all i, then each lattice Λ_{D_i} has 15 elements and the joint lattice Λ has 15^4 = 50625 elements.

We formally characterize a model by (M, Θ_M), where $M = (P_1, P_2, ..., P_N)$ and $P_i = (S_{i,1}, S_{i,2}, ..., S_{i,|P_i|})$ is a partition of D_i, the domain of applicability for parameter type i. $\Theta_M = (\phi_{1,1}, \phi_{1,2}, ..., \phi_{1,|P_i|}, \phi_{2,1}, ..., \phi_{N,|P_N||})$, where each $\phi_{i,j}$ is the value of parameter i applicable to data points corresponding to $S_{i,j} \subseteq D_i$. We denote this, $c_i(x) \in S_{i,j}$, where c_i is a "characteristic" function. We refer to M as the *model structure* and Θ_M as the *model parameterization*.

Our goal is to find the trained model (M, Θ_M) that best balances simplicity with goodness of fit to the training data. For this paper, we choose to use the minimum description length principle [1]. Consider training data $X = \{x_1, x_2, ..., x_m\}$. In order

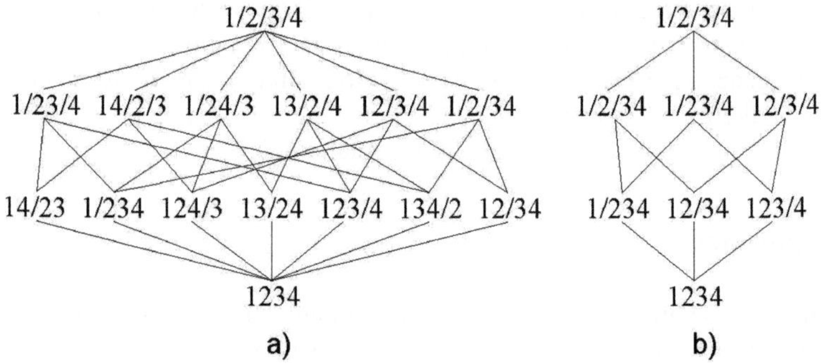

Fig. 1. Lattices of distinctions for four-class sets. If the classes are unstructured, for example the set of business sectors {Manufacturing, Service, Financial, Technology}, then we entertain the distinctions in lattice a). If the classes are ordinal, for example business sizes {Micro, Small, Medium, Large}, then we entertain only the partitions that are consistent with the class ordering, b). For example, we would not consider grouping small and large businesses while distinguishing them from medium businesses.

to evaluate the description length of the data, we use a two-part code combining the description length of the model and the description length of the data given the model:

$$L = DataL(X|(M, \Theta_M)) + ModelL((M, \Theta_M)) \tag{1}$$

for our approach we assume that $ModelL((M, \Theta_M))$ is a function only of the model structure M. Thus, adding or removing parameters affects the value of $ModelL$, but solely changing their values does not. We also assume that $DataL(X|(M, \Theta_M))$ can be decomposed into a sum of individual description lengths for each x_k. That is:

$$DataL(X|(M, \Theta_M)) = \sum_{x_k \in X} ExampleL(x_k|(M, \Theta_M)) \tag{2}$$

We assume that the function $ExampleL(x_k|(M, \Theta_M))$ is twice differentiable with respect to the model parameters, $\phi_{i,j}$, $1 \leq i \leq N$,$1 \leq j \leq |P_i|$.

Consider a model structure $M = (P_1, P_2, ..., P_N)$, $P_i = (S_{i,1}, S_{i,2}, ..., S_{i,|P_i|})$. We refer to a neighboring (in Λ) model $M' = (P'_1, P'_2, ..., P'_N)$ as a refinement of M if there exists a value j such that $P'_j = (S_{j,1}, ..., S_{j,k-1}, Y, Z, S_{j,k+1}, ..., S_{j,|P_j|})$, (Y, Z) is a partition of $S_{i,j}$, and $P'_i = P_i$ for all i $\neq$ j. That is, a refinement of M is a model which makes one additional distinction that M does not make.

Likewise, M' is a generalization of M if there exists a value j such that $P'_j = (S_{j,1}, ..., S_{j,k-1}, S_{j,k} \cup S_{j,k+1}, S_{j,k+2}, ..., S_{j,|P_j|})$, and $P'_i = P_i$ for all i $\neq$ j. That is, a generalization of M is a model that makes one less distinction than M.

3 Model Exploration

The number of candidate model structures in Λ explodes very quickly as the number of potential distinctions increases. Thus, for all but the simplest spaces, it is computationally infeasible to train and evaluate all such model structures.

Instead, we offer an efficient exploration method to explore lattice Λ in order to find a (locally) optimal value of (M, Θ_M). The general idea is to train a model structure M, arriving at parameter values Θ_M, and then leverage the differentiability of the description length function to estimate the value for other model structures, in order to direct the search through Λ.

3.1 Objective Estimation

Note that at convergence, $\frac{\partial L}{\partial \phi_{i,j}} = 0$ for all $\phi_{i,j}$. As $ModelL$ is fixed for a fixed M, and data description length is the sum of description lengths for each training example we have that

$$\sum_{x_k \in X} \left(\frac{\partial ExampleL(x_k|(M, \Theta_M))}{\partial \phi_{i,j}} \right) = 0 \tag{3}$$

Recalling that $\phi_{i,j}$ is applicable only over $S_{i,j} \subseteq D_i$, this can be rewritten:

$$\sum_{w \in S_{i,j}} \quad \sum_{x_k \ s.t. \ c_i(x_k)=w} \left(\frac{\partial ExampleL(x_k|(M, \Theta_M))}{\partial \phi_{i,j}} \right) = 0 \tag{4}$$

Note that the inner summation (over each w) need not equal zero. That is, the training data may suggest that for class $w \in D_i$, parameter ϕ_i should be different than the value $\phi_{i,j}$. However, because the current model structure does not distinguish w from the other elements of $S_{i,j}$, $\phi_{i,j}$ is the best value across the entire domain of $S_{i,j}$.

In order to determine what distinctions we might want to add or remove, we consider the effect that each parameter has on each fine-grained class of data. Let $w \in S_{i,j} \subseteq D_i$ and $v \in S_{g,h} \subseteq D_g$. Let $w \wedge v$ denote the set $\{x_k | c_i(x_k) = w \wedge c_g(x_k) = v\}$. We define the following quantities:

$$d_{\phi_{i,j},w} = \sum_{x_k \ s.t. \ c_i(x_k)=w} \left(\frac{\partial ExampleL(x_k|(M, \Theta_M))}{\partial \phi_{i,j}} \right) \tag{5}$$

$$d_{\phi_{g,h},v} = \sum_{x_k \ s.t. \ c_g(x_k)=v} \left(\frac{\partial ExampleL(x_k|(M, \Theta_M))}{\partial \phi_{g,h}} \right) \tag{6}$$

$$dd_{\phi_{i,j},\phi_{g,h},w,v} = \sum_{x_k \in w \wedge v} \left(\frac{\partial^2 ExampleL(x_k|(M, \Theta_M))}{\partial \phi_{i,j} \partial \phi_{g,h}} \right) \tag{7}$$

The first two values are the first derivatives of the objective with respect to $\phi_{i,j}$ and $\phi_{g,h}$ for the examples corresponding to $w \in D_i$ and $v \in D_g$ respectively. The third equation is the second derivative taken once with respect to each parameter. Note that the value is zero for all examples other than those in $w \wedge v$. Consider the model M^* that makes every possible distinction (the greatest element of lattice Λ). Computed over all $1 \leq i, g \leq N$, $w \in D_i$, $v \in D_j$, these values allow us to construct a second order Taylor expansion polynomial estimation for the value of $L((M^*, \Theta_{M^*}))$ for all values of Θ_{M^*}.

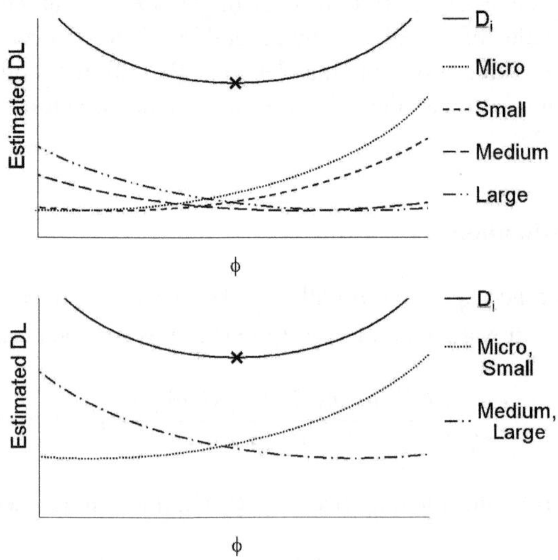

Fig. 2. Example polynomial estimation of description length considering distinctions based on business size. For no distinctions, description length is minimized at point x. However, Taylor expansion estimates that the behavior of micro and small businesses is substantially different than medium or large businesses. This suggest that the distinction {{Micro, Small}, {Medium, Large}} should be entertained if the expected reduction in description length of the data is greater than the cost associated with the additional parameter.

$$
\begin{aligned}
\widehat{DataL}(X|(M^*,\Theta_{M^*})) &= DataL(X|(M,\Theta_M)) \\
&+ \sum_{\substack{1\leq i\leq N \\ w\in D_i}} (\phi^*_{i,w} - \hat{\phi}_{i,j_w}) \times d_{\phi_{i,j_w},w} \\
+ \sum_{\substack{1\leq i\leq N \\ w\in D_i}} \sum_{\substack{1\leq g\leq N \\ v\in D_g}} & (\phi^*_{i,w} - \hat{\phi}_{i,j_w})(\phi^*_{g,v} - \hat{\phi}_{g,h_v}) \times \frac{dd_{\phi_{i,j_w},\phi_{g,h_v},w\wedge v}}{2}
\end{aligned} \tag{8}
$$

where $\hat{\phi}_{i,j_w}$ is the value of $\phi_{i,j}$, where $w \in S_{i,j}$. Note that this polynomial is the same polynomial that would be constructed from the gradient and Hessian matrix in Newton's method. By minimizing this polynomial, we can estimate the minimum L for M^*. More generally, we can use the polynomial to estimate the minimum L for any model stucture M' in Λ. Suppose we wish to consider a model that does not distinguish between classes w and $w' \in D_i$ with respect to parameter i. To do this, we enforce the constraint $\phi_{i,w} = \phi_{i,w'}$, which results in a polynomial with one fewer parameters. Minimizing this polynomial gives us an estimate for the minimum value of $DataL$ of the more general model structure. In this manner, any model structure can be estimated by placing equality constraints over parameters corresponding to classes not distinguished. A simple one dimensional example is detailed in Figure 2.

We can then estimate the complete minimum description length M':

$$\min_{\Theta_{M'}} \widehat{L}((M', \Theta_{M'}) = ModelL(M') + \min_{\Theta_{M'}} \widehat{DataL}((M', \Theta_{M'})) \tag{9}$$

3.2 Theoretical Guarantees

When considering alternative model structures, we are guided by estimates of their minimum description length. However, if the domain satisfies certain criteria, we can compute a lower bound for this value, which may result in greater efficiency.

Consider model formulation (M, Θ_M); let values $d_{\phi_{i,j},w}$ be computed as described above.

Theorem 1 *Consider maximal model (M^*, Θ_{M^*}). Assume $DataL(X|(M^*, \Theta_{M^*}))$ is twice continuously differentiable with respect to elements of Θ_{M^*} Let $H(\Theta_{M^*})$ be the Hessian matrix of the $DataL(X|(M^*, \Theta_{M^*}))$ with respect to Θ_{M^*}. If $y^T H(\Theta_{M^*})y \geq b > 0 \; \forall \; \Theta_{M^*}, y$ st. $||y||_2 = 1$, then*

$$\overline{DataL}(X|(M^*, \Theta_{M^*})) = DataL(X|(M, \Theta_M)) + \sum_{\substack{1 \leq i \leq N \\ w \in D_i}} \left((\phi^*_{i,w} - \hat{\phi}_{i,j_w}) \times d_{\phi_{i,j_w},w} + (\phi^*_{i,w} - \hat{\phi}_{i,j_w})^2 \times \frac{b}{2} \right) \tag{10}$$

is a lower bound polynomial on the value of $DataL(X|(\Phi_{M^*}, \Theta_{M^*}))$.

Proof. The Hessian of $\overline{DataL}(X|M^*, \Theta_{M^*}))$ with respect to Θ, $\overline{H}(\Theta_{M^*})$, is equal to b times the identity matrix. Thus $\forall \; \Theta_{M^*}, y$ st. $||y||_2 = 1$, $y^T \overline{H}(\Theta_{M^*})y = b$. Let $z = ((\phi^*_{i,w_1} - \hat{\phi}_{i,j_{w_1}}), \ldots, (\phi^*_{N,w_{|D_N|}} - \hat{\phi}_{i,j_{w_{|D_N|}}}))$, and let $y = \frac{z}{|z|}$. By Taylor's Theorem,

$$DataL(X|(M^*, \Theta_{M^*})) = DataL(X|(M, \Theta_M)) + \sum_{\substack{1 \leq i \leq N \\ w \in D_i}} \left((\phi^*_{i,w} - \hat{\phi}_{i,j_w}) \times d_{\phi_{i,j_w},w} + \frac{(\phi^*_{i,w} - \hat{\phi}_{i,j_w})^2 y^T H(\Theta'_{M^*})y}{2} \right) \tag{11}$$

for some Θ'_{M^*} on the line connected Θ_M and Θ_M*. Thus, by our assumptions on the Hessian matrix we know that,

$$DataL(X|(M^*, \Theta_{M^*})) \geq DataL(X|(M, \Theta_M)) + \sum_{\substack{1 \leq i \leq N \\ w \in D_i}} \left((\phi^*_{i,w} - \hat{\phi}_{i,j_w}) \times d_{\phi_{i,j_w},w} + \frac{(\phi^*_{i,w} - \hat{\phi}_{i,j_w})^2 b}{2} \right) \tag{12}$$

□

When the condition holds, this derivation allows us not only to lower bound the data description length of M^*, but the length for any model structure in Λ. In the same manner as above, placing equality constraints on sets of $\phi^*_{i,j}$'s results in a lower order polynomial estimation for $DataL$. In the same format as Equation 9, we can compute an optimistic lower bound for any model's value of L. $b > 0$ is satisfied for all cases where the objective function is strongly convex, however, the value is data and model format sensitive, so we can not offer a general solution to compute it.

Note that the gap between the estimated lower bound on $\min_{\Theta'_M} L(M', \Theta'_M)$ and its actual value will generally grow as the first derivative of $DataL(X|(M', \Theta_{M'}))$ increases. That is, we will compute more meaningful lower bounds of $\min_{\Theta'_M} L(M', \Theta'_M)$ for models whose optimal parameter values are "close" to our current values.

3.3 Model Search

Given a trained model, using the techniques described above, we can estimate and lower bound the value of L and estimate the optimal parameter settings for any alternative model, M'. We offer two general techniques to search through the space of model structures. In the first approach, we maintain an optimistic lower bound on the value $min_{\Theta_{M'}} L((M', \Theta_{M'}))$ for all M'. At each step, we select for training the model structure M' with the lowest optimistic bound for L. After training, we learn it's optimal parameter values, Θ'_M, and associated description length $L((M', \Theta'_M))$. We then use Equation 10 to generate the lower bounding Taylor expansion polynomial around $\Theta_{M'}$. This polynomial is then used to update the optimistic description lengths for all alternative models (increasing but never decreasing each bound). We proceed until a model M' has been evaluated whose description length is within ϵ of the minimum optimistic bound across all unevaluated models. At this point we adopt model M'.

Of course, the number of such models grows exponentially with the number of example classes. Thus, even maintaining optimistic bounds for all such models may be prohibitive. Thus, we present an alternative model exploration technique that hill-climbs in the lattice of models. In this approach, we iterate by training a model M, and then estimate $\widehat{L}((M', \Theta_{M'}))$ only for model structures M' that are neighbors (immediate generalizations and specializations) of M in lattice Λ. These alternative model structures are limited in number, making estimation computationally feasible, and similar to the current trained model. Thus, we expect the optimal parameter settings for these models will be "close" to our current parameter values, so that the objective estimates will be reasonably accurate. At this point, we can transition to and evaluate model M' with the lowest estimated value of $\widehat{L}((M', \Theta_{M'})$, from the neighbors of M. This cycle repeats until no neighboring models are estimated to decrease the description length, at which point the evaluated model with minimum L is adopted. For the complex fantasy football domain presented in the following section, the number of models in Λ is computationally infeasible, and so we use the alternative greedy exploration approach.

4 Fantasy Football

Fantasy football [4] is a popular game that millions of people participate in each fall during the American National Football League season. The NFL season extends 17 weeks, in which each of the 32 "real" teams plays 16 games, with one bye (off) week. In fantasy football, participants manage virtual (fantasy) teams composed of real players, and compete in virtual games against other managers. In these games, managers must choose which players on their roster to make active for the upcoming week's games, while taking into account constraints on the maximum number of active players in each position. A fantasy team's score is then derived from the active players' performances in their real-world games. While these calculations vary somewhat from league to league, a typical formula is:

$$\begin{aligned} FantasyPoints = &+ \frac{RushingYards}{10} + 6 \times RushingTouchDowns \\ &+ \frac{ReceivingYards}{10} + 6 \times ReceivingTouchDowns \\ &+ \frac{PassingYards}{25} + 4 \times PassingTouchDowns - 1 \times PassingInterceptions \\ &+ 3 \times FieldGoalsMade + ExtraPointsMade \end{aligned} \tag{13}$$

The sum of points earned by the active players during the week is the fantasy team's score, and the team wins if its score is greater than its opponent's. Thus, being successful in fantasy football necessitates predicting as accurately as possible the number of points players will earn in future real games.

Many factors affect how much and how effectively a player will play. For one, the player will be faced with a different opponent each week, and the quality of these opponents can vary significantly. Second, American football is a very physical sport and injuries, both minor and serious, are common. While we expect an injury to decrease the injured player's performance, it may increase the productivity of teammates who may then accrue more playing time.

American football players all play a primary position on the field. The positions that are relevant to fantasy football are quarterbacks (QB), running backs (RB), wide receivers (WR), tight ends (TE), and kickers (K). Players at each of these positions perform different roles on the team, and players at the same position on the same NFL team act somewhat like interchangeable units. In a sense, these players are in competition with each other to earn playing time during the games, and the team exhibits a preference over the players, in which high priority players (starters) are on the field most of the game and other players (reserves) are used sparingly.

4.1 Modeling

Our task is to predict the number of points each fantasy football player will earn in the upcoming week's games. Suppose the current week is week w (let week 1 refer to the first week for which historical data exists, not the first week of the current season). In order to make these predictions, we have access to the following data:

- The roster of each team for weeks 1 to w
- For each player, for each week 1 to $w-1$ we have:
  - The number of fantasy points that the player earned, and
  - The number of plays in which the player actively participated (gained possession of or kicked the ball)

– For each player, for each week 1 to w we have the players pregame injury status

If we normalize the number of plays in which a player participated by the total number across all players at the same position on the same team, we get a fractional number which we will refer to as *playing time*. For example, if a receiver catches 6 passes in a game, and amongst his receiver teammates a total of 20 passes are caught, we say the player's playing time = .3. Injury statuses are reported by each team several days before each game and classify each player into one of five categories:

1. Healthy (H): Will play
2. Probable (P): Likely to play
3. Questionable (Q): Roughly 50% likely to play
4. Doubtful (D): Unlikely to play
5. Out (O): Will not play

In what follows we define a space of generative model structures to predict fantasy football performance. The construction is based on the following ideas. We assume that each player has two inherent latent features: *priority* and *skill*. Priority indicates how much they are favored in terms of playing time compared to the other players at the same position on the same team. Skill is the number of points a player earns, on average, per unit of playing time. Likewise, each team has a latent skill value, indicating how many points better or worse than average the team gives up to average players. Our generative model assumes that these values are generated from Gaussian prior distributions $N(\mu_{pp}, \sigma^2_{pp})$, $N(\mu_{ps}, \sigma^2_{ps})$, and $N(\mu_{ts}, \sigma^2_{ts})$ respectively.

Consider the performance of player i on team t in week w. We model the playing time and number of points earned by player i as random variables with the following means:

$$\overline{PlayingTime}_{i,w} = \frac{e^{pp_i + injury(i,w)}}{\sum_{j \in R_t, pos(j)=pos(i)} e^{pp_j + injury(j,w)}} \tag{14}$$

$$\overline{Points}_{i,w} = PlayingTime_{i,w} \times ps_i \times ts(opp(t,w), pos(i)) \tag{15}$$

where R_t is the set of players on team t's roster, $pos(i)$ is the position of player i, $injury(i,w)$ is a function mapping to the real numbers that corresponds to the player's injury's effect on his playing time. We assume, then, that the actual values are distributed as follows:

$$PlayingTime_{i,w} \sim N(\overline{PlayingTime}_{i,w}, \sigma^2_{time}) \tag{16}$$

$$Points_{i,w} \sim N(\overline{Points}_{i,w}, PlayingTime_{i,w}\sigma^2_{points}) \tag{17}$$

We do not know *a priori* what distinctions are worth noticing, in terms of variances, prior distributions, and injury effects. For example, do high priority players have significantly higher skill values than medium priority players? Does a particular injury status have different implications for tight ends than for kickers? Of course, the answer to these questions depends on the amount of training data we have to calibrate our model. We utilize the greedy model structure exploration procedure defined in Section 3.3 to answer these questions. For this domain, we entertain alternative models based on the following parameters and domains of applicability:

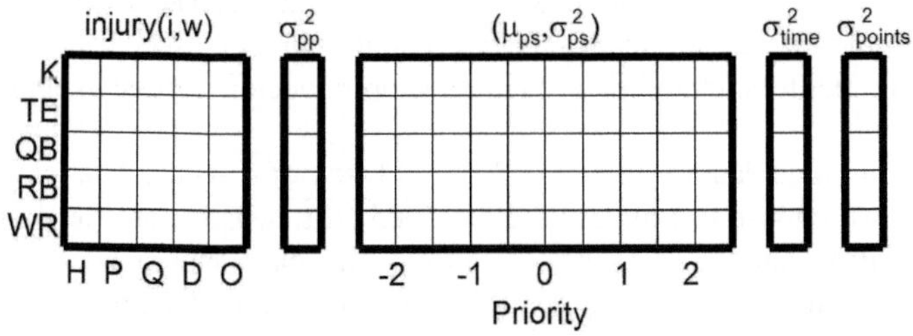

Fig. 3. The space of model distinctions. For each parameter, the domain of applicability is carved up into one or more regions along the grid lines, and each region is associated with a distinct parameter value.

1. $injury(i, w) : D_1 = Position \times InjuryStatus$
2. $\sigma^2_{pp} : D_2 = Position$ (μ_{pp} is arbitrarily set to zero)
3. $(\mu_{ps}, \sigma^2_{ps}) : D_3 = Position \times Priority$
4. $\sigma^2_{time} : D_4 = Position$
5. $\sigma^2_{points} : D_5 = Position$

Figure 3 illustrates the space of distinctions. We initialize the greedy model structure search with the simplest model, that makes no distinctions for any of the five parameters.

Given a fixed model structure M, we utilize the expectation maximization [5] procedure to minimize $DataL(X|(M, \Theta_M)) = -log_2 P(X|(M, \Theta_M))$. This procedure alternates between computing posterior distributions for the latent player priorities, skills, and team skills for fixed Θ_M, and then re-estimating Θ_M based on these distributions and the observed data. In learning these values, we limit the contributing data to a one year sliding window preceding the week in question. Additionally, because players' priorities change with time, we apply an exponential discount factor for earlier weeks and seasons. This allows the model to bias the player priority estimates to reflect the players' current standings on their team. We found that player and team skill features change little within the time frame of a year, and so discounting for these values was not necessary.

$ModelL((M, \Theta_M))$, the description length of the model, has two components, the representation of the model structure M, and the representation of Θ_M. We choose to make the description length of M constant (equivalent to a uniform prior over all model structures). The description length of Θ_M scales linearly with the number of parameters. Although in our implementation, these values are represented as 32-bit floating point values, 32 bits is not necessarily the correct description length for each parameter as it fails to capture the useful range and grain-size. Therefore, this parameter penalty, along with the week and year discount factors, are learned via cross validation.

5 Experiments

A suite of experiments demonstrates the following: First, given an amount of training data, the greedy model structure exploration procedure suitably selects a model (of the appropriate complexity), to generalize to withheld data. Second, when trained on the full set of training data, the model selected by our approach exceeds the performance of suitable competitors, including a standard support vector regression approach and a human expert.

We have compiled data for the 2004-2008 NFL seasons. As the data must be treated sequentially, we choose to utilize the 2004-2005 NFL season data for training, 2006 data for validation, and the 2007-2008 data for testing each approach.

First we demonstrate that, for a given amount of training data, our model structure search selects an appropriate model structure. We do this by using our validation data to selecting model structures based on various amounts of training data, and then evaluate them in alternative scenarios where different amounts of data are available.

Due to the interactions of players and teams in the fantasy football domain, we cannot simply throw out some fraction of the players to learn a limited-data model. Instead, we impose the following schema to learn different models corresponding to different amount of data. We randomly assign the players into G artificial groups. That is, for G = 10, each group contains (on average) one tenth of the total number of players. Then, we learn different model structures and parameter values for each group, although all players still interact in terms of predicted playing time and points are as describe in Equation 14.

For example, consider the value μ_{ps}, the mean player skill for some class of players. Even if no other distinctions are made (those that could be made based on position or priority), we learn G values for μ_{ps}, one for each group, and each parameter value is based only on the players in one group. As G increases, these parameters are estimated based on fewer players. As making additional distinctions carries a greater risk of over-fitting, in general, we expect the complexity of the best model to decrease as G increases.

In order to evaluate how well our approach selects a model tailored to an amount of training data, we utilize the 2006 validation data to learn models for each of G_{train} = 1, 4, and 16. In each case we learn G_{train} different models (one for each group). Then for each week w in 2007-2008, we again randomly partition the players, but into a different number of groups, G_{test}. For each of the G_{test} groups, we sample at random a model structure uniformly from those learned. Then, model parameters and player/team latent variables are re-estimated using EM with data for the one year data window leading up to week w, for each of the G_{test} models. Finally, predictions are made for week w and compared to the players' actual performances. We repeat this process three times for each (G_{train}, G_{test}) pair and report the average results. We also report results for each value of G_{test} when the model structure is selected uniformly at random from the entire lattice, Λ.

We expect that if our model structure selection technique behaves appropriately, for each value of G_{test}, performance should peak when $G_{train} = G_{test}$. For cases where $G_{train} < G_{test}$ the model structures will be too flexible for the more limited parameter estimation data available during testing, and performance will suffer due to overfitting. On the other hand, when $G_{train} > G_{test}$, the model structures cannot appreciate all the patterns in the calibration data. The root mean squared error of each model for each test grouping is shown in Figure 4. In fact, for each value of G_{test} we see that

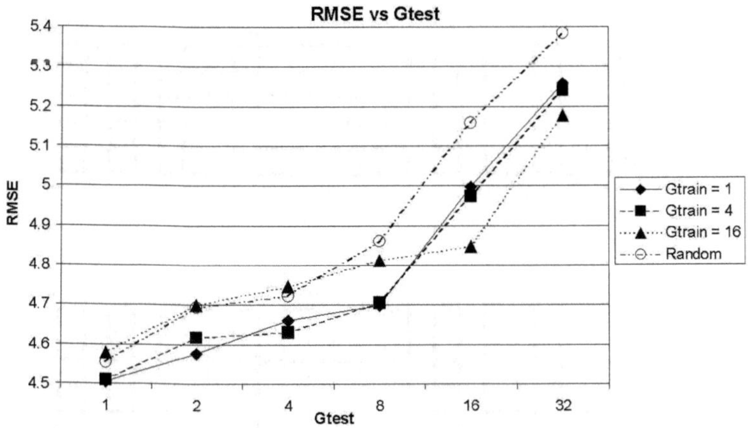

Fig. 4. Root mean squared errors for values of G_{test}. Model structures are learned from the training data for different values of G_{train} or sampled randomly from Λ.

performance is maximized when $G_{train} = G_{test}$, suggesting that our model structure selection procedure is appropriately balancing flexibility with generalization, for each amount of training data.

Figure 5 shows the model structure learned when $G_{train} = 1$, as well as a lower-complexity model learned when $G_{train} = 16$. For $G_{train} = 1$, the model structure selection procedure observes sufficient evidence to distinguish σ^2_{time} and σ^2_{points} with respect to each position. The model makes far more distinctions for high priority players than their lower priority counterparts. This is likely due to two reasons. First, the elite players' skills are further spaced out than the reserve level players, whose skills are closer to average and thus more common across all players. Second, because the high-priority players play more often than the reserves, there is more statistical evidence to justify these distinctions. The positions of quarterback, kicker and tight end all have the characteristic that playing time tends to be dominated by one player, and the learned model structure makes no distinction for the variance of priorities across these positions. Finally, the model does not distinguish the injury statuses *healthy* and *probable*, nor does it distinguish *doubtful* and *out*. Thus, *probable* appears to suggest that the player will almost certainly participate at close to his normal level, and *doubtful* means the player is quite unlikely to play at all.

In general, models learned for $G_{train} = 16$ contain fewer overall distinctions. In this case the model is similar to its $G_{train} = 1$ counterpart, except that it makes far fewer distinctions with regard to the priority skill prior.

Finally, we compare the prediction accuracy of our approach to those of a standard support vector regression technique and a human expert. For the support vector regression approach we use the LIBSVM [6] implementation of ϵ-SVR with a RBF kernel.

Consider the prediction for the performance of player i on team t in week w. We train four SVR's with different feature sets, starting with a small set of the most informative features and enlarging it to include less relevant teammate and opponent features. The

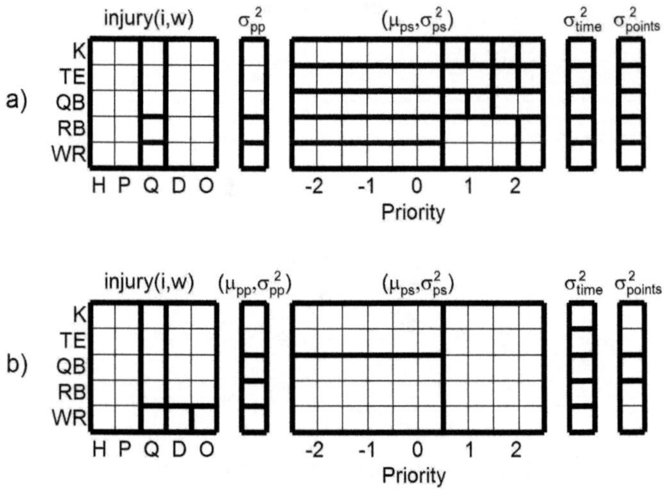

Fig. 5. Model structure learned for a) G_{train} = 1, and b) G_{train} = 16. Distinctions made with respect to 1) injury weight, 2) priority prior variance, 3) skill prior mean/variance, 4) playing time variance, and 5) points variance are shown in bold.

first SVR (SVR_1) includes only the points earned by player i in each of his games in the past year. Bye weeks are ignored, so f_1 is the points earned by player i in his most recent game, f_2 corresponds to his second most recent game, etc. For SVR_2, we also include in the feature set player i's playing time for each game, as well as his injury status each game (including the upcoming game). SVR_3 adds the points, playing times, and injury statuses for each teammate of player i at the same position each game. Finally, SVR_4 adds for teams that player i has played against in the last year, as well as his upcoming opponent, the total number of fantasy points given up by the team for each of their games in the data window. At each week w, we train one SVR for each position, using one example for each player at each week y, $w-h \leq y \leq w-1$ (an example for week y has features based on weeks $y-h$ to y). All features are scaled to have absolute range [0,1] within the the training examples. We utilize a grid search on the validation data to choose values for ϵ, γ, and C.

We also compare our accuracy against statistical projections made by the moderator of the fantasy football website (www.fftoday.com) [7]. These projections, made before each week's games, include predictions on each of the point earning statistical categories for many of the league's top players. From these values, we compute a projected number of fantasy points according to Equation 13. There are two caveats, the expert does not make projections for all players, and the projected statistical values are always integral, whereas our approach can predict any continuous number of fantasy points. To have a fair comparison, we compare results based only on the players for which the expert has made a prediction using the normalized Kendall tau distance. For this comparison, we construct two orderings each week, one based on projected points, the other based on actual points. The distance is then the number of disagreements between the

Table 1. Performance of our approach versus human expert and support vector regressors with various feature sets

	All Data		Expert Predicted Data	
	RMSE	Normalized Kenall Tau	RMSE	Normalized Kenall Tau
Our Approach	4.498	.2505	6.125	.3150
Expert	N/A	N/A	6.447	.3187
SVR_1	4.827	.2733	6.681	.3311
SVR_2	4.720	.2674	6.449	.3248
SVR_3	4.712	.2731	6.410	.3259
SVR_4	4.773	.2818	6.436	.3323

two orderings, normalized to the range [0,1] (0 if the orderings are the same, 1 for complete disagreement). By considering only the predicted ordering of players and not their absolute projected number of points, the expert is not handicapped by his limited prediction vocabulary. We compute the Kendall tau distances for each method each week, and present the average value across all weeks 2007-2008.

Table 1 shows that our approach compares favorably with both the SVR and the expert. Again, note that because of the constrained vocabulary in which the expert predicts points, the final column is the only completely fair comparison with the expert. Of the candidate SVR feature sets, SVR_2 (with player i's points, playing times, and injury statuses) and SVR_3 (adding teammates' points, playing times, and injury statuses) perform the best.

6 Related Work

Our work on learning model structure is related to previous work on graphical-model structure learning, including Bayesian networks. In cases where a Bayes net is generating the data, a greedy procedure to explore the space of networks is guaranteed to converge to the correct structure as the number of training cases increases [8]. Friedman and Yakhini [9] suggest exploring the space of Bayes nets structures using simulated annealing and a BIC scoring function. The general task of learning the best Bayesian Network according to a scoring function that favors simple networks is NP-hard [10]. For undirected graphical models such as Markov Random Fields, application of typical model selection criteria is hindered by the necessary calculation of a probability normalization constant, although progress has been made on constrained graphical structures, such as trees [11,12]. Our approach differs most notably from these in that we not only consider the relevancy of each feature, but the possible grouping of that feature's value. We also present a global search strategy for selecting model structure, and our approach applies when variables are continuous and interactions are more complex than a Bayesian network can capture.

Another technique, reversible jump Markov chain Monte Carlo [13], generalizes Markov chain Monte Carlo to entertain jumps between alternative spaces of differing dimensions. Using this approach, it is possible to perform model selection based on the posterior probability of models with different parameter spaces. The approach requires that significant care be taken in defining the MCMC proposal distributions in order to avoid exorbitant mixing times. This difficulty is magnified when the models are organized in a high-degree fashion, as is the case for our lattice.

7 Conclusion

In this paper, we present an approach to select a model structure from a large space by only evaluating a small number of candidates. We present two search strategies, one global strategy guaranteed to find a model within ϵ of the best scoring candidate in terms of MDL. The second approach hill climbs in the space of model structures. We demonstrate our approach on a difficult fantasy football prediction task, showing that the model selection technique appropriately selects structures for various amounts of training data, and that the overall performance of the system compares favorably with the performance of a support vector regressor as well as a human expert.

Acknowledgments

This work is supported by an ONR Award on "Guiding Learning and Decision Making in the Presence of Multiple Forms of Information."

References

1. Grunwald, P.: The Minimum Description Length Principle. MIT Press, Cambridge (2007)
2. Akaike, H.: A new look at the statistical model identification. IEEE Transactions on Automatic Control 19(6), 716–723 (1974)
3. Schwarz, G.E.: Estimating the dimension of a model. Annals of Statistics 6(2), 461–464 (1978)
4. ESPN: Fantasy football, `http://games.espn.go.com/frontpage/football` (Online; accessed 15-April-2008)
5. Hogg, R., McKean, J., Craig, A.: Introduction to Mathematical Statistics. Pearson Prentice Hall, London (2005)
6. Chang, C.C., Lin, C.J.: LIBSVM: A Library for Support Vector Machines (2001), `http://www.csie.ntu.edu.tw/~cjlin/libsvm`
7. Krueger, M.: Player rankings and projections - ff today, `http://www.fftoday.com/rankings/index.html` (Online; accessed 8-April-2008)
8. Chickering, D.: Optimal structure identification with greedy search. Journal of Machine Learning Research 3, 507–554 (2002)
9. Friedman, N., Yakhini, Z.: On the sample complexity of learning bayesian networks. In: The 12th Conference on Uncertainty in Artificial Intelligence (1996)
10. Chickering, D.: Large-sample learning of bayesian networks is np-hard. Journal of Machine Learning Research 5, 1287–1330 (2004)
11. Chow, C., Liu, C.: Approximating discrete probability distributions with dependence trees. IEEE Transactions on Information Theory 14(3), 462–467 (1968)
12. Srebro, N.: Maximum likelihood bounded tree-width markov networks. Artificial Intelligence 143, 123–138 (2003)
13. Brooks, S., Giudici, P., Roberts, G.: Efficient construction of reversible jump markov chain monte carlo proposal distributions. Journal of the Royal Statistical Society (65), 3–55 (2003)

Collective Traffic Forecasting

Marco Lippi, Matteo Bertini, and Paolo Frasconi

Dipartimento Sistemi e Informatica, Università degli Studi di Firenze
{lippi,bertinim,p-f}@dsi.unifi.it

Abstract. Traffic forecasting has recently become a crucial task in the area of intelligent transportation systems, and in particular in the development of traffic management and control. We focus on the simultaneous prediction of the congestion state at multiple lead times and at multiple nodes of a transport network, given historical and recent information. This is a highly relational task along the spatial and the temporal dimensions and we advocate the application of statistical relational learning techniques. We formulate the task in the supervised learning from interpretations setting and use Markov logic networks with grounding-specific weights to perform collective classification. Experimental results on data obtained from the California Freeway Performance Measurement System (PeMS) show the advantages of the proposed solution, with respect to propositional classifiers. In particular, we obtained significant performance improvement at larger time leads.

1 Introduction

Intelligent Transportation Systems (ITSs) are widespread in many densely urbanized areas, as they give the opportunity to better analyze and manage the growing amount of traffic flows, due to increased motorization, urbanization, population growth, and changes in population density.

One of the main targets of an ITS is to reduce congestion times, as they seriously affect the efficiency of a transportation infrastructure, usually measured as a multi-objective function taking into account several aspects of a traffic control system, like travel time, air pollution, and fuel consumption. As for travel time, for example, it is often important to minimize both the mean value and its variability [13], which represents an added cost for a traveler making a given journey.

This management effort is supported by the growing amount of data gathered by ITSs, coming from a variety of different sources. Loop detectors are the most commonly used vehicle detectors for freeway traffic monitoring, which can typically register the number of vehicles passed in a certain time interval (flow), and the percentage of time the sensor is occupied per interval (occupancy). In recent years, there has been also a spread of employment of wireless sensors, like GPS and floating car data (FCD) [11], which will eventually reveal in real-time the position of almost every vehicle, by collecting information from mobile phones in vehicles that are being driven. These different kinds of data are heterogeneous,

J.L. Balcázar et al. (Eds.): ECML PKDD 2010, Part II, LNAI 6322, pp. 259–273, 2010.

and therefore would need a pre-processing phase in order to be integrated and used as support to the decision processes. A large sensor network corresponds to a large number of potentially noisy or faulty components. In particular, in the case of traffic detectors, several different fault typologies might affect the system: communication problems on the line, intermittent faults resulting in insufficient or incomplete data transmitted by the sensors, broken controllers, bad wiring, etc.

In Urban Traffic Control (UTC) systems, such as the Split Cycle Offset Optimization Technique (SCOOT) system [15] and the Sydney Coordinated Adaptive Traffic (SCAT) system [16], short-term forecasting modules are used to adapt system variables and maintain optimal performances. Systems without a forecasting module can only operate in a reactive manner, after some event has occurred. Classic short-term forecasting approaches usually focus on 10-15 minutes ahead predictions [19,24,20]. Effective proactive transportation management (e.g. car navigation systems), arguably needs forecasts extending on longer horizons in order to be effective.

Most of the predictors employed in these traffic control systems are based on a time series forecasting technology. Time series forecasting is a vast area of statistics, with a wide range of application domains [3]. Given the history of past events sampled at certain time intervals, the goal is to predict the continuation of the series. Formally, given a time series $\mathcal{X} = \{x_1, \ldots, x_t\}$ describing the dynamic behavior of some observed physical quantity x_j, the task is to predict x_{t+1}. In the traffic management domain, common physical quantities of interest are (i) the traffic *flow* of cars passing at a given location in a fixed time interval, (ii) the *average speed* observed at a certain location, (iii) the *average time* needed to travel between two locations.

Historically, many statistical methods have been developed to address the problem of traffic forecasting: these include methods based on auto-regression and moving average, such as ARMA, ARIMA, SARIMA and other variants, or non-parametric regression. See [19] and references therein for an overview of these statistical methodologies. Also from a machine learning perspective, the problem of traffic forecasting has been addressed using a wide number of different algorithms, like support vector regression (SVR) [24], Bayesian networks [20] or time-delay neural networks (TDNNs) [1]. Most of these methods address the problem as single-point forecasting, intended as the ability to predict future values of a certain physical quantity at a certain location, given only past measurements of the same quantity at the same location. Yet, given a graph representing a transportation network, predicting the traffic conditions at multiple nodes and at multiple temporal steps ahead is an inherently relational task, both in the spatial and in the temporal dimension: for example, at time t, the predictions for two measurement sites s_1 and s_2, which are spatially close in the network, can be strongly interrelated, as well as predictions at t and $t+1$ for the same site s. Inter-dependencies between different time series are usually referred to as Granger's causality [8], a concept initially introduced in the domain of economy and marketing: time series A is said to Granger-cause time series B if A can be used to enhance the forecasts on B. Few methods until now have taken into account the relational structure of the data: multiple Kalman filters [23], the STARIMA model (space-time ARIMA) [10] and structural time series models [7] are the first attempts in this direction.

The use of a statistical relational learning (SRL) framework for this kind of task might be crucial in order to improve predictive accuracy. First of all, SRL allows to represent the domain in terms of logical predicates and rules, and therefore to easily include background knowledge in the model, and to describe relations and dependencies, such as the topological characteristics of a transportation network. Within this setting, the capability of SRL models to integrate multiple sources and levels of information might become a key feature for future transportation control systems. Moreover, the SRL framework allows to perform collective classification or regression, by jointly predicting traffic conditions in the whole network in a single inference process: in this way, a single model can represent a wide set of locations, while propositional methods should typically train a different predictor for each node in the graph.

Dealing with large data sets within SRL is a problem which still has to receive adequate attention, but it is one of the key challenges of the whole research area [5]. Traffic forecasting is a very interesting benchmark from this point of view: for example, just considering highways in California, over 30,000 detectors continuously generate flow and occupancy data, producing a huge amount of information. Testing the scalability of inference algorithms on such a large model is a crucial point for SRL methodologies.

Moreover, many of the classic time series approaches like ARIMA, SARIMA and most of their variants, are basically linear models. Non-linearity, on the other hand, is a crucial issue in many application domains in order to build a competitive predictor: for this reason, some attempts to extend statistical approaches towards non-linear models have been proposed, as in the KARIMA or VARMA models [22,4].

Among the many SRL methodologies that have been proposed in recent years, we employ Markov logic [6], extended with grounding-specific weights (GS-MLNs) [12]. The first-order logic formalism allows to incorporate background knowledge of the domain in a straightforward way. The use of probabilities within such a model allow us to handle noise to take into account statistical interdependencies. The grounding-specific weights extension enables the use of vectors of continuous features and non-linear classifiers (like neural networks) within the model.

2 Grounding-Specific Markov Logic Networks

Markov logic [6] integrates first-order logic with probabilistic graphical models, providing a formalism which allows us to describe a domain in terms of logic predicates and probabilistic formulae. While a first-order knowledge base can be seen as a set of *hard* constraints over possible worlds (or Herbrand interpretations), where a world violating even a single formula has zero probability, in Markov logic such a world would be *less probable*, but not impossible. Formally, a Markov logic network (MLN) is defined by a set of first-order logic formulae $\mathcal{F} = \{F_1, \ldots, F_n\}$ and a set of constants $\mathcal{C} = \{C_1, \ldots, C_k\}$. A Markov random field is then created by introducing a binary node for each possible ground atom and an edge between two nodes if the corresponding atoms appear together in a ground formula. Uncertainty is handled by attaching a real-valued weight w_j to each formula F_j: the higher the weight, the lower the probability of a world violating that formula, others things being equal. In the discriminative setting, MLNs essentially define a template for arbitrary (non linear-chain) conditional random fields that would be hard to specify and maintain if hand-coded. The language of first-order logic, in fact, allows to describe relations and inter-dependencies between the different domain objects in a straightforward way. In this paper, we are interested in the supervised learning setting. In Markov logic, the usual distinction between the input and output portions of the data is reflected in the distinction between evidence and query atoms. In this setting, an MLN defines a conditional probability distribution of query atoms Y given evidence atoms X, expressed as a log-linear model in a feature space described by all possible groundings of each formula:

$$P(Y = y|X = x) = \frac{\exp\left(\sum_{F_i \in F_Y} w_i n_i(x, y)\right)}{Z_x} \tag{1}$$

where F_Y is the set of clauses involving query atoms and $n_i(x, y)$ is the number of groundings of formula F_i satisfied in world (x, y). Note that the feature space jointly involves X and Y as in other approached to structured output learning. MAP inference in this setting allows us to collectively predict the truth value of all query ground atoms: $f(x) = y^* = \arg\max_y P(Y = y|X = x)$. Solving the MAP inference problem is known to be intractable but even if we could solve it exactly, the prediction function f is still linear in the feature space induced by the logic formulae. Hence, a crucial ingredient for obtaining an expressive model (which often means an accurate model) is the ability of tailoring the feature space to the problem at hand. For some problems, this space needs to be high-dimensional. For example, it is well known that linear chain conditional random fields (which we can see as a special case of discriminative MLNs), often work better in practice when using high-dimensional feature spaces. However, the logic language behind MLNs only offers a limited ability for controlling the size of the feature space. We will explain this using the following example. Suppose we have a certain query predicate of interest, $\texttt{Query}(\texttt{t}, \texttt{s})$ (where, e.g., the variable $\texttt{t}$ and $\texttt{s}$ represent time and space) that we know to be predictable from a certain set of attributes, one for each $(\texttt{t}, \texttt{s})$ pair, represented by the evidence predicate $\texttt{Attributes}(\texttt{t}, \texttt{s}, \texttt{a}_1, \texttt{a}_2, \ldots, \texttt{a}_\texttt{n})$. Also, suppose that performance for this hypothetical problem crucially depends, for each t and s, on our ability of defining a nonlinear mapping between the attributes and the query. To fix our ideas, imagine that an SVM with RBF kernel taking $\texttt{a}_1, \texttt{a}_2, \ldots, \texttt{a}_\texttt{n}$ as inputs (treating each (s, t) pair as an independent example) already produces a good classifier, while a linear classifier fails. Finally, suppose we have some available background knowledge, which might help us to write formulae introducing statistical interdependencies between different query ground atoms (at different $\texttt{t}$

and s), thus giving us a potential advantage in using a non-iid classifier for this problem. An MLN would be a good candidate for solving such a problem, but emulating the already good feature space induced by the RBF kernel may be tricky. One possibility for producing a very high dimensional feature space is to define a feature for each possible configuration of the attributes. This can be achieved by writing several ground formulae with different associated weights. For this purpose, in the Alchemy system[1], one might write an expression like

$$\texttt{Attributes}(\texttt{t}, \texttt{s}, +\texttt{a}_1, +\texttt{a}_2, \ldots, +\texttt{a}_n) \Rightarrow \texttt{Query}(\texttt{t}, \texttt{s})$$

where the + symbol preceding some of the variables expands the expression into separate formulae resulting from the possible combination of constants from those variables. Different weights are attached to each formula in the resulting expansion. Yet, this solution presents two main limitations: first, the number of parameters of the MLN grows exponentially with the number of variables in the formula; second, if some of the attributes ai are continuous, they need to be discretized in order to be used within the model.

GS-MLNs [12] allow us to use weights that depend on the specific grounding of a formula, even if the number of possible groundings can in principle grow exponentially or can be unbound in the case of real-valued constants. Under this model, we can write formulae of the kind:

$$\texttt{Attributes}(\texttt{t}, \texttt{s}, \$\texttt{v}) \Rightarrow \texttt{Query}(\texttt{t}, \texttt{s})$$

where v has the type of an n-dimensional real vector, and the \$ symbol indicates that the weight of the formula is a parameterized function of the specific constant substituted for the variable v. In our approach, the function is realized by a discriminative classifier, such as a neural network with adjustable parameters θ. The idea of integrating non-linear classifiers like neural networks within conditional random fields has been also recently proposed in conditional neural fields [14].

In MLN with grounding-specific weights, the conditional probability of query atoms given evidence can therefore be rewritten as follows:

$$P(Y = y|X = x) = \frac{\exp\left(\sum_{F_i \in F_Y} \sum_j w_i(c_{ij}, \theta_i) n_{ij}(x, y)\right)}{Z_x} \quad (2)$$

where $w_i(c_{ij}, \theta_i)$ is a function of some constants depending on the specific grounding, indicated by c_{ij}, and of a set of parameters θ_i.

Any inference algorithm for standard MLNs can be applied with no changes. During the parameter learning phase, on the other hand, MLN and neural network weights need to be adjusted jointly. The resulting algorithm can implement gradient ascent, exploiting the chain rule:

$$\frac{\partial P(y|x)}{\partial \theta_k} = \frac{\partial P(y|x)}{\partial w_i} \frac{\partial w_i}{\partial \theta_k}$$

[1] http://alchemy.cs.washington.edu

where the first term is computed by MLN inference and the second term is computed by backpropagation. As in standard MLNs, the computation of the first term requires to compute the expected counts $E_w[n_i(x,y)]$:

$$\frac{\partial P(y|x)}{\partial w_i} = n_i(x,y) - \sum_{y'} P(y'|x) n_i(x,y^*) = n_i(x,y) - E_w[n_i(x,y)]$$

which are usually approximated with the counts in the MAP state y^*:

$$\frac{\partial P(y|x)}{\partial w_i} \simeq n_i(x,y) - n_i(x,y^*)$$

From the above equation, we see that if all the groundings of formula F_j are correctly assigned their truth values in the MAP state y^*, then that formula gives a zero contribution to the gradient, because $n_j(x,y) = n_j(x,y^*)$. For grounding-specific formulae, each grounding corresponds to a different example for the neural network: therefore, there will be no backpropagation term for a given example if the truth value of the corresponding atom has been correctly assigned by the collective inference.

When learning from many independent interpretations, it is possible to split the data set into minibatches and apply stochastic gradient descent [2]. Basically this means that gradients of the likelihood are only computed for small batches of interpretations and weights (both for the MLN and for the neural networks) are updated immediately, before working with the subsequent interpretations. Stochastic gradient descent can be more generally applied to minibatches consisting of the connected components of the Markov random field generated by the MLN. This trick is inspired by a common practice when training neural networks and can very significantly speedup training time.

3 Data Preparation and Experimental Setting

3.1 The Data Set

We performed our experiments on the California Freeway Performance Measurement System (PeMS) data set [21], which is a wide collection of measurements obtained by over 30,000 sensors and detectors placed around nine districts in California. The system covers 164 Freeways, including a total number of 6,328 mainline Vehicle Detector Stations and 3,470 Ramp Detectors.

The loop detectors used within the PeMS are frequently deployed as single detectors, one loop per lane per detector station. The raw single loop signal is noisy and can be used directly to obtain only the raw count (traffic flow) and the occupancy (lapse of time the loop detector is active) but cannot measure the speed of the vehicles. The PeMS infrastructure collects filtered and aggregated flow and occupancy from single loop detectors, and provides an estimate of the speed [9] and other derived quantities. In some locations, a double loop detector is used to directly measure the instantaneous speed of the vehicles. All traffic detectors report measurements every 30 seconds.

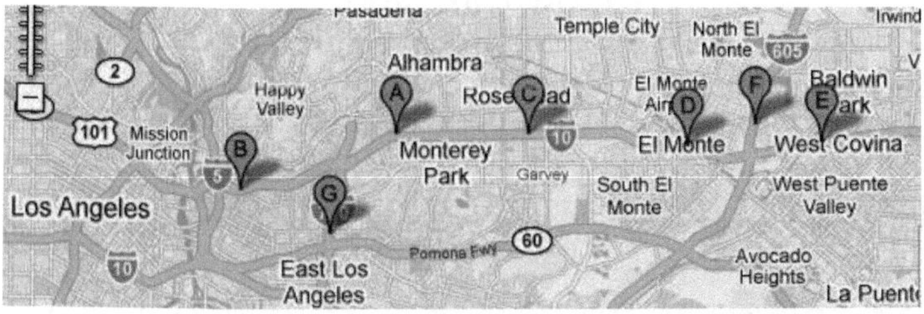

Fig. 1. The case study used in the experiments: 7 measurement stations placed on three different Highways in the area of East Los Angeles

In our experiments, the goal is to predict whether the average speed at a certain time in the future falls under a certain threshold. This is the measurement employed by Google^{TM} Maps[2] for the coloring scheme encoding the different levels of traffic congestions: the yellow code, for example, means that the average speed is below 50 mph, which is the threshold adopted in all our experiments.

Table 1. Summary of stations used in experiments. VDS stays for Vehicle Detector Station and identifies each station in the PeMS data set.

Station	VDS	Highway	# Lanes
A	716091	I10-W	4
B	717055	I10-W	4
C	717119	I10-W	4
D	717154	I10-W	5
E	717169	I10-W	4
F	717951	I605-S	4
G	718018	I710-S	3

In our case study, we focused on seven locations in the area of East Los Angeles (see Figure 1), five of which are placed on the I10 Highway (direction West), one on the I5 (direction South) and one on the I710 (direction South) (see Table 1). We aggregated the available raw data into 15-minutes samples, averaging the measurements taken on the different lanes. In all our experiments we used the previous three hours of measurements as the input portion of the data. For all considered locations we predict traffic congestions at the next four lead times (i.e., 15, 30, 45 and 60 minutes ahead). Thus each interpretation spans a time interval of four hours. We used two months of data (Jan-Feb 2008) as training set, one month (Mar 2008) as tuning set, and two months (Apr-May 2008) for test. Time intervals of four hours containing missing measurements due to temporary faults in the sensors were discarded from the data set. The tuning

[2] http://maps.google.com

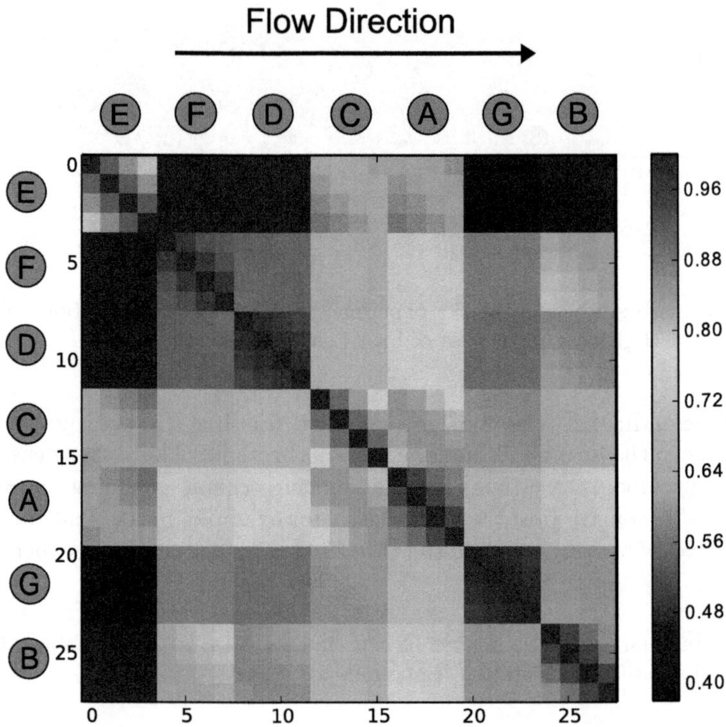

Fig. 2. Spatiotemporal correlations in the training set data. There are 28 boolean congestion variables corresponding to 7 measurement stations and 4 lead times. Rows and columns are lexicographically sorted on the station-lead time pair. With the exception of station E, spatial correlations among nearby stations are very strong and we can observe the spatiotemporal propagation of the congestion state along the direction of flow (traffic is westbound).

set was used to choose the C and γ parameters for the SVM predictor, and to perform early stopping for the GS-MLNs.

The inter-dependencies between nodes which are close in the transportation network are evident from the simple correlation diagram shown in Figure 2.

3.2 Experimental Setup

The GS-MLN model was trained under the learning from interpretations setting. An interpretation in this case corresponds to a typical forecasting session, where at time t we want to forecast the congestion state of the network at future lead times, given previous measurements. Hence interpretations are indexed by their time stamp t, which is therefore be omitted in all formulae (the temporal index h in the formulae below refers to the time lead of the prediction, i.e. 1,2,3, and 4 for 15,30,45, and 60 minutes ahead). Interpretations are assumed to be independent, and this essentially follows the setting of other supervised learning approaches such as [24,18,17]. However, in our approach congestion states at different lead times and at different sites are predicted collectively. Dependencies are introduced by spatiotemporal neighborhood rules, such as

$$\begin{aligned}&\texttt{Congestion}(+\texttt{s},\texttt{h}) \wedge \texttt{Weekday}(+\texttt{wd}) \wedge \texttt{TimeSlot}(+\texttt{ts})\\ &\Rightarrow \texttt{Congestion}(+\texttt{s},\texttt{h}+1)\end{aligned} \tag{3}$$

$$\texttt{Congestion}(+\texttt{s1},\texttt{h}) \wedge \texttt{Next}(\texttt{s1},\texttt{s2}) \Rightarrow \texttt{Congestion}(+\texttt{s2},\texttt{h}+1) \tag{4}$$

where $\texttt{Congestion}(\texttt{S},\texttt{H})$ is true of the velocity at site $\texttt{S}$ and lead time $\texttt{H}$ falls below the 50mph threshold, and the + symbol before a site variable assigns a different weight to each site or site pair. The predicate $\texttt{Next}(\texttt{s1},\texttt{s2})$ is true if site $\texttt{s2}$ follows site $\texttt{s1}$ in the flow direction. The predicate $\texttt{Weekday}(\texttt{wd})$ distinguishes between workdays and holidays, while $\texttt{TimeSlot}(\texttt{ts})$ encodes the part of the day (morning, afternoon, etc.) of the current timestamp.

Of course the road congestion state also depends on previously observed velocity or flow. Indeed, literature results [24,18,17] suggest that good local forecasts can be obtained as a nonlinear function of the recent sequence of observed traffic flow or speed. Using GS-MLNs, continuous attributes describing the observed time series can be introduced within the model, using a set of grounding-specific formulae, e.g.:

$$\texttt{SpeedSeries}(\texttt{SD},\$\texttt{SeriesD}) \Rightarrow \texttt{Congestion}(\texttt{SD},1) \tag{5}$$

where the grounding-specific weights are computed by a neural network taking as input a real vector associated with constant $\texttt{Series\_SD}$ (being $\texttt{SD}$ the station identifier), containing past speed measurements during the previous 12 time steps. Note that a separate formula (and a separate neural network) is employed for each site and for each lead time.

Seasonality was encoded by the predicate $\texttt{SeasonalCongestion}(\texttt{s})$, which is true if, on average, station $\texttt{s}$ presents a congestion at the time of the day referred to by the current interpretation (this information was extracted from averages on the training set). Other pieces of background knowledge were encoded in the MLN. For example, the number of lanes at a given site can be influence bottleneck behaviors:

$$\begin{aligned}&\texttt{Congestion}(\texttt{s1},\texttt{h}) \wedge \texttt{NodeClose}(\texttt{s1},\texttt{s2}) \wedge \texttt{NLanes}(\texttt{s1},\texttt{l1}) \wedge \texttt{NLanes}(\texttt{s2},\texttt{l2}) \wedge\\ &\texttt{l2} < \texttt{l1} \Rightarrow \texttt{Congestion}(\texttt{s2},\texttt{h}+1)\end{aligned}$$

The MLN contained 14 formulae in the background knowledge and 125 parameters after grounding variables prefixed by a +. The 28 neural networks had 12 continuous inputs and 5 hidden units each, yielding about 2000 parameters in total. Our software implementation is a modified version of the Alchemy system to incorporate neural network as pluggable components. Inference was performed by MaxWalkSat algorithm. Twenty epochs of stochastic gradient ascent were performed, with a learning rate $\epsilon = 0.03$ for the MLN weights, and $\mu = \frac{0.00003}{n}$ for the neural networks, being n the number of misclassifications in the current minibatch. In order to further speed up the training procedure, all neural networks were pre-trained for a few epochs (using the congestion state as the target) before plugging them into the GS-MLN jointly and tuning the whole set of parameters.

We compared the obtained results against three competitors:

Trivial predictor. The seasonal average classifier predicts, for any time of the day, the congestion state observed on average in the training set at that time. Although it is a baseline predictor, it is widely used in literature as a competitor.

SVM. We used SVM as a representative of state-of-the-art propositional classifiers. A different SVM with RBF kernel was trained for each station and for each lead time, performing a separated model selection for the C and γ values to be adopted for each measurement station. The measurements used by the SVM predictor consist in the speed time series observed in the past 180 minutes, aggregated at 15 minutes intervals, hence producing 12 inputs, plus an additional one representing the seasonal average at current time. A gaussian standardization was applied to all these inputs.

Standard MLN. When implementing the classifier based on standard MLNs, the speed time series had to be discretized in order to be used within the model. Five different speed classes were used, and the quantization thresholds were chosen by following a maximum entropy strategy. The trend of the speed time series was modeled by the following set of formulae that were used in place of formula 5:

$$\texttt{Speed\_Past\_1}(\texttt{n}, +\texttt{v}) \Rightarrow \texttt{Congestion}(\texttt{n}, 1)$$
$$\ldots$$
$$\texttt{Speed\_Past\_k}(\texttt{n}, +\texttt{v}) \Rightarrow \texttt{Congestion}(\texttt{n}, 1)$$

where predicate $\texttt{Speed\_Past\_j}(\texttt{node}, \texttt{speed\_value})$ encodes the discrete values of the speed at the j-th time step before the current time. Note that an MLN containing only the above formulae essentially represents a logistic regression classifier taking the discretized features as inputs. All remaining formulae were identical to those used in conjunction with the GS-MLN.

As for the predictor based on GS-MLNs, there is no need to use discretized features, but the same vectors of features used by the SVM classifier can be adopted.

4 Results and Discussion

4.1 Performance Analysis

The congestion state in the analyzed highway segment is a very unbalanced task even at the 50mph threshold. Table 2 shows the percentage of positive query atoms in the training set and in the test set, for each station. The last two columns report the percentage of days containing at least one congestion. The

Table 2. Percentage of true ground atoms, for each measurement station. The percentage of days in the train/test set containing at least one congestion is reported in the last two columns.

Station	% pos train	% pos test	% pos days train	% pos days test
A	11.8	9.2	78.3	70.7
B	5.8	4.9	60.0	53.4
C	16.8	13.7	66.6	86.9
D	3.4	2.3	45.0	31.0
E	28.2	22.9	86.7	72.4
F	3.9	1.8	51.6	31.0
G	1.9	1.7	30.0	22.4

Table 3. Comparison between the tested predictors. Results show the F_1 on the positive class, averaged on the seven nodes. The symbol $\ominus$ indicates a significant loss of the method with respect to GS-MLN, according to a Wilcoxon paired test (p-value<0.05).

	15 m		30 m		45 m		60 m	
Seasonal Avg	38.3	$\ominus$	38.3	$\ominus$	38.3	$\ominus$	38.3	$\ominus$
SVM	81.7		68.6		56.4	$\ominus$	51.8	$\ominus$
MLN	59.5	$\ominus$	56.5	$\ominus$	53.6	$\ominus$	50.4	$\ominus$
GS-MLN	80.9		69.2		61.6		56.9	

data distribution shows that the stations present different behaviors, corroborating the choice of using different neural networks for each station.

Given the unbalanced data set, we compare the predictors on the F_1 measure, as the harmonic mean between precision $P = \frac{TP}{TP+FP}$ and recall $R = \frac{TP}{TP+FN}$: $F_1 = \frac{2PR}{P+R}$. Table 3 shows the F_1 measure, averaged per station. The advantages of the relational approach are much more evident when increasing the prediction horizon: at 45 and 60 minutes ahead, the improvement of the GS-MLN model is statistically significant, according to a Wilcoxon paired test, with p-value< 0.05. Detailed comparisons for each sensor station at 15, 30, 45, and 60 minutes ahead are reported in Tables 4 , Tables 5 , Tables 6 and 7, respectively. These tables show that congestion at some of the sites are clearly "easier" to predict than at other sites. Comparing Tables 4-7 to Table 2 we see that the difficulty strongly correlates with the data set imbalance, an effect which is hardly surprising. It is also often the case that GS-MLN significantly outperforms the SVM classifier for "difficult" sites. The comparison between the standard MLN and the GS-MLN shows that input quantization can significantly deteriorate performance, all other things being equal. This supports the proposed strategy of embedding neural networks as a key component of the model.

An interesting performance measure considers only those test cases in which traffic conditions are anomalous with respect to the typical seasonal behavior. To this aim, we restricted the test set, by collecting only those interpretations for which the baseline seasonal average classifier would miss the prediction of

Table 4. Details on the predictions per station, at 15 minutes ahead

	SVM	MLN	GS-MLN
A	82.9	64.0	80.4
B	78.0	50.8	74.5
C	91.2	66.5	89.1
D	77.5	51.9	79.5
E	92.0	69.4	92.9
F	70.6	51.9	66.7
G	80.0	61.7	83.4

Table 5. Details on the predictions per station, at 30 minutes ahead.

	SVM	MLN	GS-MLN
A	76.2	50.6	74.2
B	60.9	46.5	60.5
C	85.6	81.5	86.0
D	64.4	57.0	65.5
E	85.7	74.3	86.0
F	36.0	30.4	45.6
G	71.6	55.5	66.7

Table 6. Details on the predictions per station, at 45 minutes ahead.

	SVM	MLN	GS-MLN
A	74.3	71.1	73.5
B	41.6	29.3	44.5
C	82.7	75.1	83.9
D	46.2	49.9	59.4
E	80.7	78.2	82.9
F	33.8	28.7	37.3
G	35.5	43.2	50.0

the current congestion state. Table 8 shows that the advantage of the relational approach is still evident for long prediction horizons.

The experiments were performed on a 3GHz processor with 4Mb cache. The total training time for SVM is 40 minutes, and 7-8 hours for GS-MLNs. As for testing times, both systems perform in real-time.

4.2 Dealing with Missing Data

The problem of missing or incomplete data is crucial in all time series forecasting applications [3,4]: in the case of punctual missing information, a reconstruction algorithm might be employed in order to interpolate the signal, so that prediction methods might be applied unchanged. Occasionally, sensor faults can last several

Table 7. Details on the predictions per station, at 60 minutes ahead

	SVM	MLN	GS-MLN
A	72.5	71.6	72.1
B	29.9	27.9	37.0
C	83.5	80.9	84.7
D	38.0	41.0	52.4
E	79.7	75.4	79.9
F	26.0	21.0	29.9
G	33.3	32.6	42.4

Table 8. Comparison between the tested predictors, only on those cases where the seasonal average predictor fails. Results show the F_1 on the positive class, averaged on the seven nodes.

	15 m	30 m	45 m	60 m
SVM	81.4	69.1	59.1	59.2
MLN	39.9	47.6	48.4	41.6
GS-MLN	78.4	68.2	68.4	65.5

Table 9. Comparison between the tested predictors, using a test set containing missing values, reconstructed using the seasonal average. Results show the F_1 on the positive class.

	15 m	30 m	45 m	60 m
SVM	79.0	63.2	53.6	48.8
GS-MLN	80.5	70.4	62.6	58.1

time steps, and when this happens, a large part of the input can be unavailable to a standard propositional predictor until the sensor recovers from the failure state. Of course, cases containing missing data can be filtered from the training set as we did for our previous experiments. However, in order to deploy a predictor on a real-time task, it is necessary also to handle the case of missing values at prediction time. A relational model can be in principle more robust than its propositional counterpart by exploiting information from nearby sites.

In this section we report results obtained by simulating the absence of several values within the observed time series, using the trivial seasonal average predictor (Section 3.2) as reconstruction algorithm for these unobserved data . Producing an accurate model of sensor faults is clearly beyond the scope of this paper and we built a naive observation model based on a two states first-order Markov chain with $P(\text{observed} \mapsto \text{observed}) = 0.99$ and $P(\text{reconstructed} \mapsto \text{reconstructed}) = 0.9$. The performance of the predictors on this task are shown in Table 9.

5 Conclusions

We have proposed a statistical relational learning approach to traffic forecasting, in order to collectively classify the congestion state at several nodes of a transportation network, and at multiple lead times in the future, exploiting the relational structure of the domain. Our method is based on grounding-specific Markov logic networks, which extend the framework of Markov logic in order to include discriminative classifiers and generic vectors of features within the model. Experimental results performed on a case study extracted from the Californian PeMS data set show that the relational approach outperforms the propositional one, in particular when the prediction horizon grows.

Although we performed experiments on a binary classification task, we plan to extend the framework also to the case of multiclass classification or ordinal regression. As a further direction of research, the use of Markov logic gives the possibility to extend the model by applying structure learning algorithms to learn relations and dependencies directly from data in an automatic way.

The proposed methodology is not restricted to traffic management, but it can be applied to several different time series application domains, such as ecologic time series, for air pollution monitoring, or economic time series, for marketing analysis.

Acknowledgments

This research is partially supported by grant SSAMM-2009 from the Foundation for Research and Innovation of the University of Florence.

References

1. Abdulhai, B., Porwal, H., Recker, W.: Short-term freeway traffic flow prediction using genetically optimized time-delay-based neural networks. In: Transportation Research Board, 78th Annual Meeting, Washington D.C (1999)
2. Bottou, L.: Stochastic learning. In: Bousquet, O., von Luxburg, U., Rätsch, G. (eds.) Machine Learning 2003. LNCS (LNAI), vol. 3176, pp. 146–168. Springer, Heidelberg (2004)
3. Box, G., Jenkins, G.M., Reinsel, G.: Time Series Analysis: Forecasting & Control, 3rd edn. Prentice-Hall, Englewood Cliffs (1994)
4. Chatfield, C.: The Analysis of Time Series: An Introduction, 6th edn. Chapman & Hall/CRC, Boca Raton (2003)
5. Dietterich, T.G., Domingos, P., Getoor, L., Muggleton, S., Tadepalli, P.: Structured machine learning: the next ten years. Machine Learning 73(1), 3–23 (2008)
6. Domingos, P., Kok, S., Lowd, D., Poon, H., Richardson, M., Singla, P.: Markov logic. In: De Raedt, L., Frasconi, P., Kersting, K., Muggleton, S.H. (eds.) Probabilistic Inductive Logic Programming. LNCS (LNAI), vol. 4911, pp. 92–117. Springer, Heidelberg (2008)
7. Ghosh, B., Basu, B., O'Mahony, M.: Multivariate short-term traffic flow forecasting using time-series analysis. Trans. Intell. Transport. Sys. 10(2), 246–254 (2009)

8. Granger, C.W.J., Newbold, P.: Forecasting Economic Time Series (Economic Theory and Mathematical Economics). Academic Press, London (1977)
9. Jia, Z., Chen, C., Coifman, B., Varaiya, P.: The pems algorithms for accurate, real-time estimates of g-factors and speeds from single-loop detectors. pp. 536 – 541 (2001)
10. Kamarianakis, Y., Prastacos, P.: Space-time modeling of traffic flow. Comput. Geosci. 31, 119–133 (2005)
11. Kerner, B.S., Demir, C., Herrtwich, R.G., Klenov, S.L., Rehborn, H., Aleksic, M., Haug, A.: Traffic state detection with floating car data in road networks. In: Proceedings of Intelligent Transportation Systems, pp. 44–49. IEEE, Los Alamitos (2005)
12. Lippi, M., Frasconi, P.: Prediction of protein beta-residue contacts by markov logic networks with grounding-specific weights. Bioinformatics 25(18), 2326–2333 (2009)
13. Noland, R.B., Polak, J.W.: Travel time variability: a review of theoretical and empirical issues. Transport Reviews: A Transnational Transdisciplinary Journal 22, 39–54 (2002)
14. Peng, J., Bo, L., Xu, J.: Conditional neural fields. In: Bengio, Y., Schuurmans, D., Lafferty, J., Williams, C.K.I., Culotta, A. (eds.) Advances in Neural Information Processing Systems, vol. 22, pp. 1419–1427 (2009)
15. Selby, D.L., Powell, R.: Urban traffic control system incorporating scoot: design and implementation. In: Proceedings of Institution of Civil Engineers, vol. 82, pp. 903–920 (1987)
16. Sims, A.: S.C.A.T. The Sydney Co-ordinated Adaptive Traffic System. In: Symposium on Computer Control of Transport 1981: Preprints of Papers, pp. 22–26 (1981)
17. Smith, B.L., Demetsky, M.J.: Short-term traffic flow prediction: neural network approach. Transportation Research Record 1453, 98–104 (1997)
18. Smith, B.L., Demetsky, M.J.: Traffic flow forecasting: Comparison of modeling approaches. Journal of Transportation Engineering-Asce 123(4), 261–266 (1997)
19. Smith, B.L., Williams, B.M., Keith Oswald, R.: Comparison of parametric and nonparametric models for traffic flow forecasting. Transportation Research Part C 10(4), 303–321 (2002)
20. Sun, S., Zhang, C., Yu, G.: A bayesian network approach to traffic flow forecasting. IEEE Transactions on Intelligent Transportation Systems 7(1), 124–132 (2006)
21. Varaiya, P.: Freeway Performance Measurement System: Final Report. PATH Working Paper UCB-ITS-PWP-2001-1, University of California Berkley (2001)
22. Watson, S.: Combining kohonen maps with arima time series models to forecast traffic flow. Transportation Research Part C: Emerging Technologies 4(12), 307–318 (1996)
23. Whittaker, J., Garside, S., Lindveld, K.: Tracking and predicting a network traffic process. International Journal of Forecasting 13(1), 51–61 (1997)
24. Wu, C.H., Ho, J.M., Lee, D.T.: Travel-time prediction with support vector regression. IEEE Transactions On Intelligent Transportation Systems 5(4), 276–281 (2004)

On Detecting Clustered Anomalies Using SCiForest

Fei Tony Liu[1], Kai Ming Ting[1], and Zhi-Hua Zhou[2,*]

[1] Gippsland School of Information Technology
Monash University, Victoria, Australia
{tony.liu,kaiming.ting}@infotech.monash.edu.au
[2] National Key Laboratory for Novel Software Technology
Nanjing University, Nanjing 210093, China
zhouzh@lamda.nju.edu.cn

Abstract. Detecting local clustered anomalies is an intricate problem for many existing anomaly detection methods. Distance-based and density-based methods are inherently restricted by their basic assumptions—anomalies are either far from normal points or being sparse. Clustered anomalies are able to avoid detection since they defy these assumptions by being dense and, in many cases, in close proximity to normal instances. In this paper, without using any density or distance measure, we propose a new method called SCiForest to detect clustered anomalies. SCiForest separates clustered anomalies from normal points effectively even when clustered anomalies are very close to normal points. It maintains the ability of existing methods to detect scattered anomalies, and it has superior time and space complexities against existing distance-based and density-based methods.

1 Introduction

"*The identification of clusters of outliers can lead to important types of knowledge discovery.*" Edwin M. Knorr [12]

Anomaly detection identifies unusual data patterns that are different from the majority of data. In this paper, we use the terms anomalies and outliers interchangeably. In general, anomalies can be divided into four different types using two dimensions. The first distinguishes anomalies by their proximity to normal instances — *local* versus *global*. The second divides anomalies based on their data distribution — *clustered* versus *scattered*. For example, global clustered anomalies refer to anomalies that are far from normal points, and very close to each others forming a cluster.

A number of existing anomaly detection methods, including distance-based [22,20] and density-based methods [6], carry the assumption that anomalies are distant or sparse with respect to normal instances. Therefore, these methods solely target scattered anomalies, often only global scattered anomalies. However, this assumption does *not* always hold. When anomalies gathered to form clusters, they become very difficult to detect [23], due to their proximity and density, which is also known as the 'masking' effect [18].

* Z.-H. Zhou was partially supported by the National Science Foundation of China (60635030, 60721002), the National Fundamental Research Program of China (2010CB327903) and the Jiangsu Science Foundation (BK2008018).

J.L. Balcázar et al. (Eds.): ECML PKDD 2010, Part II, LNAI 6322, pp. 274–290, 2010.

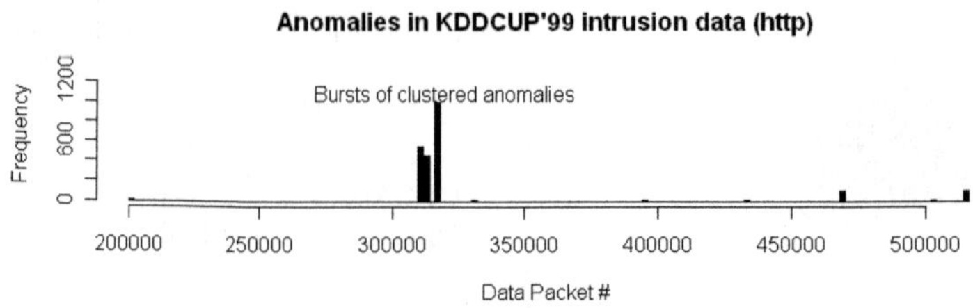

Fig. 1. Burst of clustered anomalies can be observed through out the $http$ data set

Identifying clustered anomalies is important since they may carry critical information in circumstances such as disease outbreaks [27], burst of intrusions and fraudulent activities [10]. In particular, detecting clustered anomalies are usually more rewarding as such discovery often lead to greater benefits as compared to scattered anomalies. For example, the detection of frequent fraudsters potentially prevents higher financial loss as compared to occasional fraudsters.

A publicly available example of clustered anomalies can be found in KDDCUP 1999 data set [1], where bursts of attacks (clustered anomalies) can be observed in a subset known as $http$ [28] as shown in Figure 1. Three bursts of attacks are clustered, first in the middle of the data stream; and two smaller ones appeared at the end of the stream. These attacks are characterized by their arrival in a short period of time, and having the same values in three attributes, i.e., 2091 out of 2211 anomalies in $http$ have the same values in attributes: `duration, src_bytes and dst_bytes`. It shows that the problem of clustered anomalies exist and it is worthy for further investigation.

The detection of clustered anomalies is identified as a challenging future working by Knorr [12] in 2002. Knorr motivates that occasional anomalies may be tolerated or ignored in some applications, however when similar anomalies appear many times; it is unwise to ignored them. Knorr defines that clustered anomalies are points which are close to each other and far from normal points. When anomalies come very close to normal points, the problem of detecting clustered anomalies becomes even more challenging.

The challenges to detect the four types of anomalies are illustrated in Figure 2, where clustered anomalies c_g, c_l, c_n and scattered anomalies x_g, x_l are shown together with two clusters of normal points. Subscript g denotes global anomalies, and l, n local anomalies. Each anomaly cluster has twelve data points. Using popular anomaly detectors, `LOF` [6], `ORCA` [5], `iForest` [16] and `SCiForest` – our proposed method in this paper, the ranking result for each method is provided in Figure 2. There are a total of 38 anomalies and SCiForest is the only method that correctly ranks all these anomalies at the top of the list. The local clustered anomalies are very challenging to the other three detectors for two reasons:

– *Plurality and density* — when the number of clustered anomalies is more than a certain threshold, e.g., the k parameter of k-nn based methods, then clustered

[1] http://kdd.ics.uci.edu/databases/kddcup99/kddcup99.html

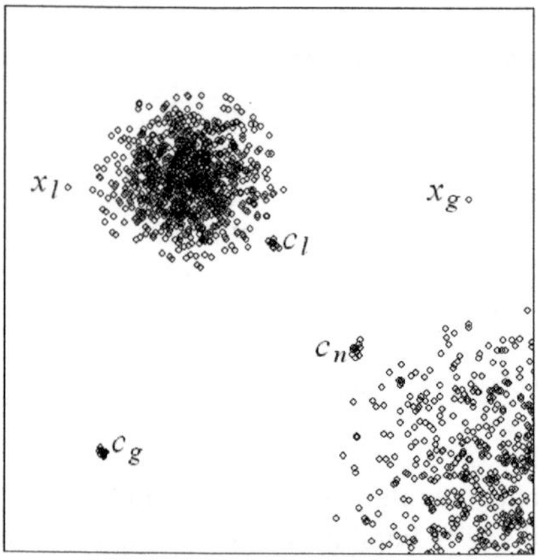

(a) **Data Distribution**

Ranking	x_g	x_l	c_g	c_n, c_l
SCiForest	**7**	**38**	**1-6,8-13**	**14-37**
iForest	**12**	28	**1-11,13**	-
LOF($k = 15$)	**1**	**14**	**2-13**	-
LOF($k = 10$)	**1**	**2**	-	-
ORCA($k = 15$)	**1**	27	**2-13**	-
ORCA($k = 10$)	**1**	10	-	-

(b) **Rankings of Anomalies**
Local clustered anomalies c_n, c_l are difficult to detect. '-' means ranking > 38. Consecutive rankings are bold-faced. Non-consecutive rankings mean false-positives are ranked higher than anomalies.

Fig. 2. SCiForest is the only detector that is able to detect all the anomalies in the data set above. (a) illustrates the data distribution. (b) reports the anomaly rankings provided by different anomaly detectors.

anomalies become undetectable by these methods; both LOF and ORCA miss detecting C_g when $k < 15$; and
- *Proximity* — when anomalies are located close to normal instances, they are easily mistaken as normal instances. All except SCiForest miss detecting local clustered anomalies, C_n and C_l.

We propose SCiForest—an anomaly detector that is specialised in detecting local clustered anomalies, in an efficient manner. Our contributions are four-fold:

- **we tackle the problem of clustered anomalies**, in particular local clustered anomalies. We employ a split selection criterion to choose a split that separates clustered anomalies from normal points. To the best of our knowledge, no existing methods use the same technique to detect clustered anomalies;
- **we analyse the properties of this split selection criterion** and show that it is effective even when anomalies are very close to normal instances, which is the most challenging scenario presented in Figure 2;
- **we introduce the use of randomly generated hyper-planes** in order to provide suitable projections that separate anomalies from normal points. The use of multiple hyper-planes avoids costly computation to search for the optimal hyper-plane as in SVM [25]; and
- **the proposed method is able to separate anomalies without a significant increase in processing time.** In contrast to SVM, distance-based and density-based methods, our method is superior in processing time especially in large data sets.

This paper is organised as follows: Section 2 defines key terms used in this paper. Section 3 reviews existing methods in detecting clustered anomalies, especially local clustered anomalies. Section 4 describes the construction of SCiForest, including the proposed split-selection criterion, randomly generated hyper-planes and SCiForest's computational time complexity. Section 5 empirically evaluates the proposed method with real-life data sets. We also evaluate the robustness of the proposed method using different scenarios with (i) high number of anomalies, (ii) clustered, and (iii) close proximity to normal instances. Section 6 concludes this paper.

2 Definition

In this paper, we use the term 'Isolation' to refer to *"separating each instance from the rest"*. Anomalies are data points that are more susceptible to isolation.

Definition 1. ***Anomalies*** *are points that are few and different as compared with normal points.*

We define two different types of anomalies as follows:

Definition 2. ***Scattered anomalies*** *are anomalies scattered outside the range of normal points.*

Definition 3. ***Clustered anomalies*** *are anomalies which form clusters outside the range of normal points.*

3 Literature Review

Distance-based methods can be implemented in three ways: anomalies have (1) very few neighbours within a certain distance [13] or (2) a distant kth nearest neighbour or (3) distant k nearest neighbours [22,3]. If anomalies have short pair-wise distances among themselves, then k is required to be larger than the size of the largest anomaly cluster in order to detect them successfully. Note that increases k also increases processing time. It is also known that distance-based methods break down when data contain varying densities since distance is measured uniformly across a data set. Most distance-based methods have a time complexity of $O(n^2)$. Many recent implementations improve performance in terms of speed, e.g., ORCA [5], DOLPHIN [2]. However, very little work is done to detect clustered anomalies.

Density-based methods assume that normal instances have higher density than anomalies. Under this assumption, density-based methods also have problem with varying densities. In order to cater for this problem, Local Outlier Factor (LOF) [6] was proposed which measures relative density rather than absolute density. This improves the ability to detect local scattered anomalies. However, the ability to detect clustered anomalies is still limited by LOF's underlying algorithm—k nearest neighbours, in which k has to be larger than the size of the largest anomaly cluster. The time complexity for LOF is also $O(n^2)$.

Clustering-based methods. Some methods use clustering methods to detect anomalies. The three assumptions in clustering-based methods are: a) anomalies are points that do not belong to any cluster, b) anomalies are points that are far away from their closest cluster centroid, and c) anomalies belong to small or sparse clusters [8]. Since many clustering methods are based on distance and density measures, clustering-based methods suffer similar problems as distance or density based methods in which anomalies can evade detection by being very dense or by being very close to normal clusters. The time complexity of clustering algorithms is often $O(n^2d)$.

Other methods. In order for density-based methods to address the problem of clustered anomalies, LOCI [21] utilizes multi-granularity deviation factor (MDEF) which captures the discrepancy between a point and its neighbours at different granularities. Anomalies are detected by comparing, for a point, the number of neighbours with the average number of neighbours' neighbours. For each point, difference between the two counts at a coarse granularity indicates clustered anomaly. LOCI requires to have a working radius larger than the radius of an anomaly cluster in order to achieve successful detection. A grid-based variant aLOCI has a time complexity of $O(nLdg)$ for building a quad tree, and $O(nL(dg+2^d))$ for scoring and flagging, where L is the total numbers of levels and $10 \leq g \leq 30$. LOCI is able to detect clustered anomalies, however, detecting anomalies is not a straight-forward exercise, it requires an interpretation of LOCI curve for each point.

OutRank [17] is another method which can handle clustered anomalies, OutRank maps a data set to a weighted undirected graph. Each node represents a data point and each edge represents the similarity between instances. The edge weights are transformed to transition probabilities so that the dominant eigenvector can be found. The eigenvector is then used to determine anomalies. The weighted graph requires a significant amount of computing resources which is a bottleneck for real life applications.

At the time of writing, none of LOCI and OutRank implementations is available for comparison and none of them are to handle local clustered anomalies.

Collective anomalies are different from clustered anomalies. Collective anomalies are anomalous due to their unusual temporal or sequential relationship among themselves [9]. In comparison, cluster anomalies are anomalous because they are clustered and different from normal points.

A recently proposed method *Isolation Forest* (iForest) [16], adopts a fundamentally different approach that takes advantage of anomalies' intrinsic properties of being '*few and different*'. In many methods, these two properties are measured individually by different measurements, e.g., density and distance. By applying the concept of *isolation* expressed as path length of isolation tree, iForest simplifies the fundamental mechanism to detect anomalies which avoids many costly computations, e.g., distance calculation. The time complexity of iForest is $O(t\psi\log\psi + nt\log\psi)$, where ψ and t are small constants. SCiForest and iForest share the use of path length to formulate anomaly scores; they are different in terms of how they construct their models.

4 Constructing SCiForest

The proposed method consists of two stages. In the first stage (**training stage**), t number of trees are generated, and the process of building a tree is illustrated in Algorithm 1 which trains a tree in SCiForest from a randomly selected sub-sample. Let $X = \{\mathbf{x}^1, ..., \mathbf{x}^n\}$ be a data set with d-variate distribution and an instance $\mathbf{x} = [x_1, ..., x_d]$. An isolation tree is constructed by (a) selecting a random sub-sample of data (without replacement for each tree), $X' \subset X$, $|X'| = \psi$, and (b) selecting a separating hyperplane f using Sd_{gain} criterion in every recursive subdivision of X'. We call our method

Algorithm 1. Building a single tree in SCiForest(X', q, τ)

Input: X' - input data, q - number of attributes used in a hyperplane, τ - number of hyperplanes considered in a node

Output: an iTree T

if $|X'| \leq 2$ **then**
  return *exNode*$\{Size \leftarrow |X'|\}$
else
  $f \leftarrow$ a hyper-plane with the best split point p that yields the highest Sd_{gain} among τ hyper-planes of q randomly selected attributes.
  $X^l \leftarrow \{\mathbf{x} \in X' | f(\mathbf{x}) < 0\}$
  $X^r \leftarrow \{\mathbf{x} \in X' | f(\mathbf{x}) \geq 0\}$
  $v \leftarrow \max_{\mathbf{x} \in X'}(f(\mathbf{x})) - \min_{\mathbf{x} \in X'}(f(\mathbf{x}))$
  return inNode$\{Left \leftarrow$ *iTree*(X^l, q, τ),
    $Right \leftarrow$ *iTree*(X^r, q, τ),
    $SplitPlane \leftarrow f$,
    $UpperLimit \leftarrow +v$,
    $LowerLimit \leftarrow -v\}$
end if

SCiForest, which stands for ***I**solation **Forest** with **S**plit-selection **C**riterion*. The formulation of hyperplane will be explained in Section 4.1 and Sd_{gain} criterion in Section 4.2.

The second stage (**evaluation stage**) is illustrated in Algorithm 2 to evaluate path length $h(\mathbf{x})$ for each data point $\mathbf{x}$. The path length $h(\mathbf{x})$ of a data point $\mathbf{x}$ of a tree is measured by counting the number of edges $\mathbf{x}$ traverses from the root node to a leaf node. The expected path length $E(h(\mathbf{x}))$ over t trees is used as an anomaly measure which encapsulates the two properties of anomalies: long expected path length implies normal instances and short expected path length implies anomalies which are few and different as compared with normal points.

The $PathLength$ function in Algorithm 2 basically counts the number of edges e $\mathbf{x}$ traverses from the root node to an external node in T. A acceptable range is defined at each node to omit the counting of path length for unseen anomalies; this facility will be explained in details in Section 4.3. When $\mathbf{x}$ reaches an external node, the value of $c(T.Size)$ is used as a path length estimation for an unbuilt sub-tree; $c(m)$ the average tree height of binary tree is defined as :

$$c(m) = 2H(m-1) - 2(m-1)/n \text{ for } m > 2, \tag{1}$$

$c(m) = 1$ for $m = 2$ and $c(m) = 0$ otherwise; $H(i)$ is the harmonic number which can be estimated by $ln(i) + 0.5772156649$ (Euler's constant).

The time complexity to construct SCiForest consists of three major components: a) computing hyper-plane values, b) sorting hyper-plane values and c) computing the criterion. They are repeated τ times in a node and there are maximum $\psi - 1$ internal nodes in a tree. Using the three major components mentioned above, the time complexity of training a SCiForest of t trees is $O(t\tau\psi(q\psi + \log\psi + \psi))$. In the evaluation stage, the time complexity of SCiForest is $O(qnt\psi)$, where n is the number of instances to

Algorithm 2. PathLength$(\mathbf{x}, T, e)$

Inputs : $\mathbf{x}$ - an instance, T - an iTree, e - number of edges from the root node; it is to be initialised to zero when the function is first called

Output: path length of $\mathbf{x}$

if T is an exNode **then**
  return $e + c(T.size)$ $\{c(.)$ is defined in Equation 1$\}$
end if
$y \leftarrow T.SplitPlane(\mathbf{x})$
if $0 \leq y$ **then**
  return PathLength$(\mathbf{x}, T.right, e + (y < T.UpperLimit\ ?\ 1 : 0))$
else if $y < 0$ **then**
  return PathLength$(\mathbf{x}, T.left, e + (T.LowerLimit \leq y\ ?\ 1 : 0))$
end if

be evaluated. The time complexity of SCiForest is low since t, τ, ψ and q are small constants and only the evaluation stage grows linear with n.

4.1 Random Hyper-Planes

When anomalies can only be detected by considering multiple attributes at the same time, individual attributes are not effective to separate anomalies from normal points. Hence, we introduce random hyper-planes which are non-axis-parallel to the original attributes. SCiForest is a tree ensemble model; it is not necessary to have the optimal hyper-plane in every node. In each node, given sufficient trials of randomly generated hyper-planes, a good enough hyper-plane will emerge, guided by Sd_{gain}. Although individual hyper-planes may be less than optimal, the resulting model is still highly effective as a whole, due to the aggregating power of ensemble learner.

The idea of hyper-plane is similar to Oblique Decision Tree [19]; but we generate hyper-planes with randomly chosen attributes and coefficients, and we use them in the context of isolation trees rather than decision trees.

At each division in constructing a tree, a separating hyper-plane f is constructed using the best split point p and the best hyperplane that yields the highest Sd_{gain} among τ randomly generated hyper-planes. f is formulated as follows:

$$f(\mathbf{x}) = \sum_{j \in Q} c_j \frac{x_j}{\sigma(X'_j)} - p, \quad (2)$$

where Q has q attribute indices, randomly selected without replacement from $\{1, 2, ..., d\}$; c_j is a coefficient, randomly selected between $[-1, 1]$; X'_j are j^{th} attribute values of X'. After f is constructed, steps 5 and 6 in Algorithm 1 return subsets X^l and X^r, $X^l \cup X^r = X'$, according to f. This tree building process continues recursively with the filtered subsets X^l and X^r until the size of a subset is less than or equal to two.

4.2 Detecting Clustered Anomalies Using Sd_{gain} Criterion

Hawkins defines, "anomalies are suspicious of being generated by a different mechanism" [11], this infers that clustered anomalies are likely to have their own distribution

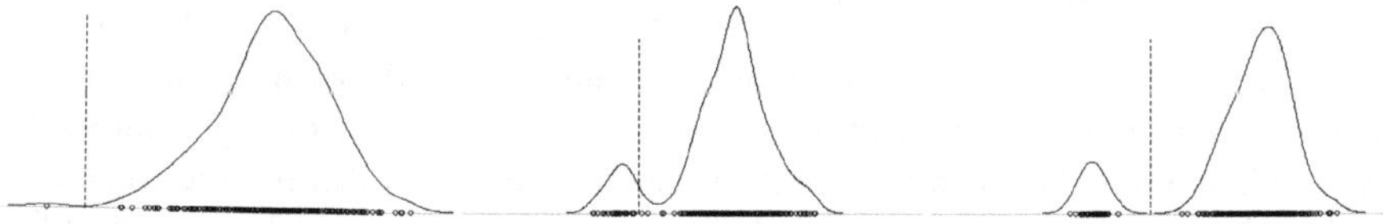

(a) Separate an anomaly from the main distribution (b) Isolate an anomaly cluster close to the main distribution (c) Separate an anomaly cluster from the main distribution

Fig. 3. Examples of Sd_{gain} selected split points in three projected distributions

under certain projections. For this reason, we introduce a split-selection criterion that isolates clustered anomalies from normal points based on their distinct distributions.

When a split clearly separates two different distributions, their dispersions are minimized. Using this simple but effective mechanism, our proposed split-selection criterion (Sd_{gain}) is defined as:

$$Sd_{gain}(Y) = \frac{\sigma(Y) - avg(\sigma(Y^l), \sigma(Y^r))}{\sigma(Y)}, \tag{3}$$

where $Y^l \cup Y^r = Y$; Y is a set of real values obtained by projecting X' onto a hyperplane f. $\sigma(.)$ is the standard deviation function and $avg(a, b)$ simply returns $\frac{a+b}{2}$. A split point p is required to separate Y into Y^l and Y^r such that $y^l < p \leq y^r, y^l \in Y^l, y^r \in Y^r$. The criterion is normalised using $\sigma(Y)$, which allows a comparison of different scales from different attributes. To find the best split p from a given sample Y, we pass the data twice. The first pass computes the base standard deviation $\sigma(Y)$. The second pass finds the best split p which gives the maximum Sd_{gain} across all possible combinations of Y^l and Y^r, using Equation 3. Standard deviation measures the dispersion of a data distribution; when an anomaly cluster is presented in Y, it is separated first as this reduces the average dispersion of Y^l and Y^r the most. To calculate standard deviation, a reliable one-pass solution can be found in [14, p. 232, vol. 2, 3rd ed.]. This solution is not subjected to cancellation error[2] and allows us to keep the computational cost to a minimum.

We illustrate the effectiveness of Sd_{gain} in Figure 3. This criterion is shown to be able to (a) separate a normal cluster from an anomaly, (b) separate an anomaly cluster which is very close to the main distribution, and (c) separate an anomaly cluster from the main distribution.

Sd_{gain} is able to separate two overlapping distributions. Using the analysis in [24], we can see that as long as the combined distribution for any two distributions is bimodal, Sd_{gain} is able to separate the two distributions early in the tree construction process. Using two distributions of the same variance i.e. $\sigma_1^2 = \sigma_2^2$, with their respective means μ_1 and μ_2, it is shown that the combined distribution can only be bimodal when $|\mu_2 - \mu_1| > 2\sigma$ [24]. In the case when $\sigma_1^2 \neq \sigma_2^2$, the condition of bi-modality is $|\mu_2 - \mu_1| > S(r)(\sigma_1 + \sigma_2)$, where the ratio $r = \sigma_1^2/\sigma_2^2$ and separation factor

[2] Cancellation error refers to the inaccuracy in computing very large or very small numbers, which are out of the precision of ordinary computational representation.

$S(r) = \frac{\sqrt{-2+3r+3r^2-2r^3+2(1-r+r^2)^{\frac{3}{2}}}}{\sqrt{r}(1+\sqrt{r})}$ [24]. $S(r)$ equals to 1 when $r = 1$, and S decreases slowly when r increases. That means bi-modality holds when one-standard deviation regions of the two distributions do not overlap. This condition is generalised for any population ratio between the two distributions and it is further relaxed when their standard derivations are different. Based on this condition of bi-modality, it is clear that Sd_{gain} is able to separate any two distributions that are indeed very close to each other.

In SciForest, Sd_{gain} has two purposes: (a) to select the best split point among all possible split points and (b) to select the best hyper-plane among randomly generated hyper-planes.

4.3 Acceptable Range

In the training stage, SCiForest always focuses on separating clustered anomalies. For this reason, setting up a acceptable range at the evaluation stage is helpful to fence off any unseen anomalies that are out-of-range. An illustration of acceptable range is shown in Figure 4. In steps 6 and 8 of Algorithm 2, any instance $\mathbf{x}$ that falls outside of the acceptable range of a node, i.e. $f(\mathbf{x}) > UpperLimit$ or $f(\mathbf{x}) < LowerLimit$, is penalized without a path length increment for that node. The effect of acceptable range is to reduce the path length measures of unseen data points which are more suspicious of being anomalies.

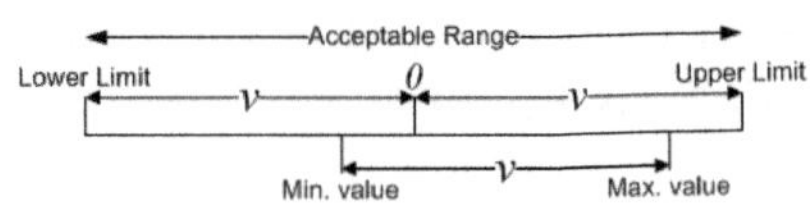

Fig. 4. An example of acceptable range with reference to hyper-plane f $(SplitPlane)$

5 Empirical Evaluation

Our empirical evaluation consists of five subsections. Section 5.1 provides a comparison in detecting clustered anomalies in real-life data sets. Section 5.2 contrasts the detection behaviour between SCiForest and iForest, and explores the utility of hyper-plane. Section 5.3 examines the robustness of the four anomaly detectors against dense anomaly clusters in terms of density and plurality of anomalies. Section 5.4 examines the breakdown behaviours of the four detectors in terms of the proximity of both clustered and scattered anomalies. Section 5.5 provides a comparison with other real-life data sets, which contain different scattered anomalies.

Performance measures include Area Under receiver operating characteristic Curve (AUC) and processing time (training time plus evaluation time). Ten runs averages are reported. Significance tests are conducted using paired t-test at 5% significance level. Experiments are conducted as single-threaded jobs processed at 2.3GHz in a Linux cluster (www.vpac.org).

In our empirical evaluation, the panel of anomaly detectors includes SCiForest, iForest [16], ORCA [5], LOF [6] (from R's package `dprep`) and one-class SVM [26]. As for SCiForest and iForest, the common default settings are $\psi = 256$ and $t = 100$, as used in [16]. For SCiForest, the default settings for hyper-plane are $q = 2$ and $\tau = 10$.

Table 1. Performance comparison of five anomalies detectors on selected data sets containing only clustered anomalies. Boldfaced are best performance. Mulcross' setting is ($D = 1, d = 4, n = 262144, cl = 2, a = 0.1$).

		AUC					Time (seconds)				
	size	SCiF	iF	ORCA	LOF	SVM	SCiF	iF	ORCA	LOF	SVM
Http	567,497	**1.00**	**1.00**	0.36	NA	0.90	39.22	**14.13**	9487.47	NA	34979.76
Mulcross	262,144	**1.00**	0.93	0.83	0.90	0.59	61.64	**8.37**	2521.55	156,044.13	7366.09
Annthyroid	6,832	**0.91**	0.84	0.69	0.72	0.63	5.91	**0.39**	2.39	121.58	4.17
Dermatology	366	**0.89**	0.78	0.77	0.41	0.74	1.04	0.27	**0.04**	0.91	**0.04**

The use of parameter q depends on the characteristic of anomalies; an analysis can be found in Section 5.2. Setting $q = 2$ is suitable for most data. Parameter τ produces similar result when $\tau > 5$ in most data sets, the average variance of AUC for the eight data sets used is 0.00087 for $30 \geq \tau \geq 5$. Setting $\tau = 10$ is adequate for most data sets.

In this paper, ORCA's parameter settings[3] are $k = 10$ and $N = \frac{n}{8}$, where N the number of anomalies detected. LOF's default parameter is the commonly used $k = 10$. One-class SVM is using the Radial Basis Function kernel and its inverse width parameter is estimated by the method suggested in [7].

5.1 Performance on Data Sets Containing Only Clustered Anomalies

In our first experiment, we compare five detectors with data sets containing known clustered anomalies. Using data visualization, we find that the following four data sets contains only clustered anomalies. Data sets included are: a data generator Mulcross[4] [23] which is designed to evaluate anomaly detectors, and three other anomaly detection data sets from UCI repository [4]: *http*, *Annthyroid* and *Dermatology*. Previous usage can be found in [28,23,15]. *Http* is the largest subset from KDD CUP 99 network intrusion data [28]; attack instances are treated as anomalies. *Annthyroid* and *Dermatology* are selected as they have known clustered anomalies. In Dermatology, the smallest class is defined as anomalies; in Annthyroid classes 1 and 2. All nominal and binary attributes are removed.

Mulcross has five parameters, which control the number of dimensions d, the number of anomaly clusters cl, the distance between normal instance and anomalies D, the percentage of anomalies a (contamination level) and the number of generated data points n. Settings for Mulcross will be provided for different experiments.

Their detection performance and processing time are reported in Table 1. SCiForest (SCiF) has the best detection performance, attributed by its ability to detect clustered anomalies in data. SCiForest is significant better than iForest, ORCA and SVM using paired t-test. iForest (iF) has slightly lower AUC in Mulcross, Annthyroid and Dermatology as compared with SCiForest. In terms of processing time, iForest and SCiForest are very competitive, especially in large data sets, including http and Mulcross. LOF result on http is not reported as the process runs for more than two weeks.

[3] ORCA's default setting of $k = 5$, $N = 30$ returns $AUC = 0.5$ for most data sets.

[4] http://lib.stat.cmu.edu/jasasoftware/rocke

Table 2. SCiForest targets clustered anomalies while iForest targets scattered anomalies. SCiForest has a higher hit rate in Annthyroid data. Instances with similar high z-scores implies clustered anomalies, i.e., attribute t3 under SCiForest. Top ten identified anomalies are presented with their z-scores which measure their deviation from the mean values. Z-scores > 3 are bold-faced meaning outlying values. $*$ denotes ground truth anomaly.

SCiForest							iForest						
id	tsh	t3	tt4	t4u	tfi	tbg	id	tsh	t3	tt4	t4u	tfi	tbg
*3287	-1.7	**21.5**	-2.0	-2.9	1.1	-3.0	1645	-1.5	-0.2	**21.2**	**8.9**	-1.6	**14.6**
*5638	-1.8	**20.6**	-1.4	-1.8	1.7	-2.2	2114	1.3	-0.2	**15.0**	**8.4**	-1.0	11.2
*1640	1.5	**21.3**	-2.0	-2.7	2.2	-2.9	*3287	-1.7	**21.5**	-2.0	-2.9	1.1	-3.0
*2602	-1.4	**19.8**	-2.0	-2.4	2.1	-2.7	*1640	1.5	**21.3**	-2.0	-2.7	2.2	-2.9
*4953	-2.6	**20.3**	-0.4	-2.1	1.0	-2.3	3323	1.7	0.4	**6.2**	**4.7**	-0.7	**6.0**
*5311	-1.4	**20.2**	-1.7	-2.5	0.6	-2.6	*6203	-1.8	**18.9**	-2.0	-2.4	1.8	-2.6
*5932	0.4	**22.9**	0.0	-2.8	0.7	-2.9	*2602	-1.4	**19.8**	-2.0	-2.4	2.1	-2.7
*6203	-1.8	**18.9**	-2.0	-2.4	1.8	-2.6	2744	-1.2	0.4	**4.8**	**4.7**	-1.0	6.7
*1353	0.1	**18.8**	-1.4	-2.7	0.2	-2.8	*4953	-2.6	**20.3**	-0.4	-2.1	**1.0**	-2.3
*6360	0.4	**17.2**	-2.0	-2.7	1.1	-2.9	4171	-0.6	-0.2	**7.0**	**8.9**	**0.6**	**7.8**

Top 10 anomalies' z-scores on Annthyroid data set.

5.2 SCiForest's Detection Behaviour and the Utility of Hyper-Plane

By examining attributes' z-scores in top anomalies, we can contrast the behavioural differences between SCiForest and iForest in terms of their ranking preferences. In Table 2, SCiForest (on the left hand side) prefers to rank an anomaly cluster first, which has distinct values in attribute 't3', as shown by similar high z-scores in 't3'. However, iForest (on the right hand side of Table 2) prefers to rank scattered anomalies first the same anomaly cluster. SCiForest's preference allows it to focus on clustered anomalies, while iForest focuses on scattered anomalies in general.

When anomalies are depended on multiple attributes, SCiForest's detection performance increases when q the number of attributes used in hyper-planes increases. In Figure 5, Dermatology data set has an increasing AUC as q increases due to the dependence of its anomalies on multiple attributes. On the other hand, Annthyroid data set has a decrease in detection performance since its anomalies are depended on only a single attribute "t3" as shown above. Both data sets are presented with AUC of SCiForest with various q values in comparison with iForest, LOF and ORCA in their default settings. In both cases, their maximum AUC are above 0.95, which show that room for further improvement is minimal. From these examples, we can see that the parameter q allows further tuning of hyperplanes in order to obtain a better detection performance in SCiForest.

5.3 Global Clustered Anomalies

To demonstrate the robustness of SCiForest, we analyse performance of four anomaly detectors using data generated by Mulcross with various contamination levels. This provides us with an opportunity to examine the robustness of detectors in detecting

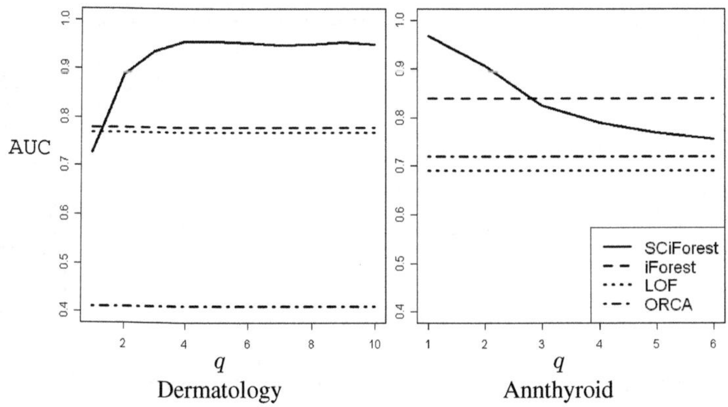

Fig. 5. Performance analysis on the utility of Hyper-plane. AUC (y-axis) increases with q the number of attributes used in the hyper-plane (x-axis) when anomalies are depends on multiple attributes as in Dermatology

global clustered anomalies under increasing density and plurality of anomalies. Mulcross is designed to generate dense anomaly clusters when the contamination level increases, in which case the density and the number of anomalies also increase, making the problem of detecting global clustered anomalies gradually harder. When the contamination level increases, the number of normal points remains at 4096, which provides the basis for comparison. When AUC drops to 0.5 or below, the performance is equal to random ranking. Figure 6(c) illustrates an example of Mulcross's data with one anomaly cluster.

In Figure 6(a) where there is only one anomaly cluster, SCiForest clearly performs better than iForest. SCiForest is able to stay above $AUC = 0.8$ even when the contamination level reaches $a = 0.3$; whereas iForest drops below $AUC = 0.6$ at around $a = 0.15$. The other two detectors; ORCA and LOF, have sharper drop rates as compared to SCiForest and iForest between $a = \frac{1}{2^{12}}$ to 0.05. In Figure 6(b) where there are ten anomaly clusters, it is actually an easier problem because the size of anomaly clusters becomes smaller and the density of anomaly clusters is reduced for the same contamination level as compared to Figure 6(a). In this case, SCiForest is still the most robust detector, having AUC stay above 0.95 for the entire range. iForest is a close second with a sharper drop between $a = 0.02$ to $a = 0.3$. The other two detectors have a marginal improvement from the case with one anomaly cluster. This analysis confirms that SCiForest is robust in detecting dense global anomaly clusters even when they are large and dense. SVM's result is omitted for clarity.

5.4 Local Clustered Anomalies and Local Scattered Anomalies

When clustered anomalies become too close to normal instances, anomaly detectors based on density and distance measures breakdown due to the proximity of anomalies. To examine the robustness of different detectors against local clustered anomalies, we generate a cluster of twelve anomalies with various distances from a normal cluster in

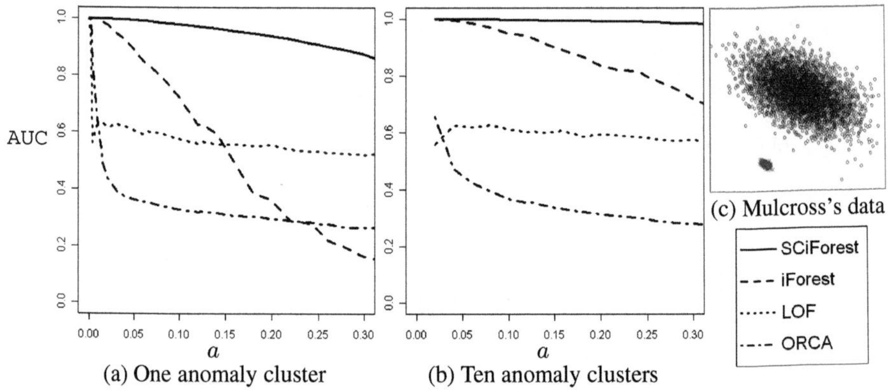

Fig. 6. SCiForest is robust against dense clustered anomalies at various contamination levels Presented is the AUC performance (y-axis) of the four detectors on Mulcross $(D = 1, d = 2, n = 4096/(1 - a))$ data with contamination level $a = \{\frac{1}{2^{12}}, ..., 0.3\}$ (x-axis).

the context of two normal clusters. We use a distance factor $= \frac{h}{r}$, where h is the distance between anomaly cluster and the center of a normal cluster and r is the radius of the normal cluster. When the distance factor is equal to one, the anomaly cluster is located right at the edge of the dense normal cluster. In this evaluation, LOF and ORCA are given $k = 15$ so that k is larger than the size of anomaly groups.

As shown in Figure 7(a), the result confirms that SCiForest has the best performance in detecting local clustered anomalies, followed by iForest, LOF and ORCA. Figure 7(b) shows the scenario of distance factor = 1.5. When distance factor is equal to or slightly less than one in Figure 7(a), SCiForest's AUC remains high despite the fact that local anomalies have come into contact with normal instances. By inspecting the actual model, we find that many hyper-planes close to the root node are still separating anomalies from normal instances, resulting in a high detection performance.

A similar evaluation is also conducted for scattered anomalies. In Figure 7(c), SCiForest also has the best performance in detecting local scattered anomalies, then followed by LOF, iForest and ORCA. Note that LOF is slightly better than iForest from distance factor > 0.7 onwards. Figure 7(d) illustrates the data distribution when distance factor is equal to 1.5.

5.5 Performance on Data Sets Containing Scattered Anomalies

As for data sets which contain scattered anomalies, we find that SCiForest has a similar and comparable performance as compared with other detectors. In Table 3, four data sets from UCI repository [4] including Satellite, Pima, Breastw and Ionosphere are used for a comparison. They are selected as they are previously used in literature, e.g., [15] and [1]. In terms of anomaly class definition, the three smallest classes in Satellite are defined as anomalies, class positive in Pima, class malignant in Breastw and class bad in Ionosphere.

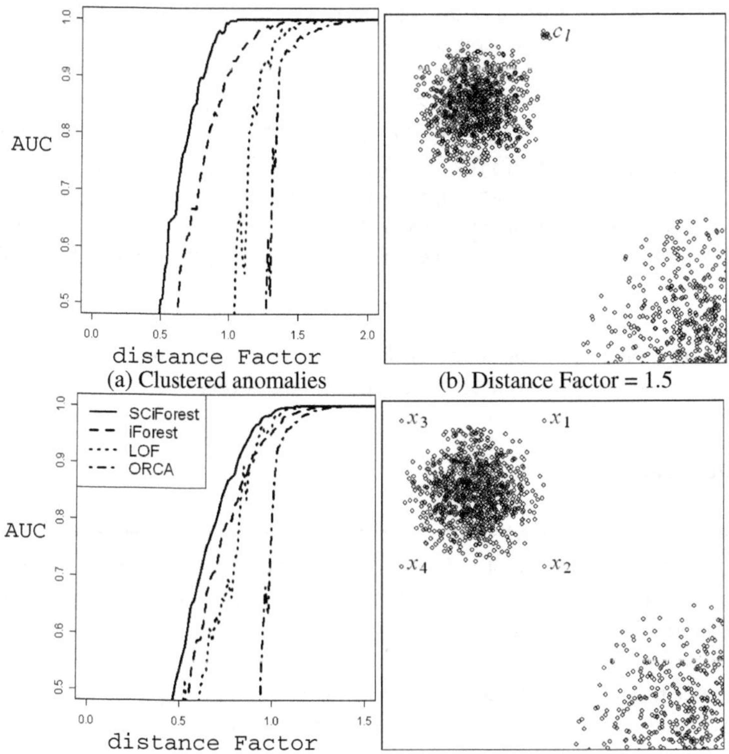

(a) Clustered anomalies (b) Distance Factor = 1.5

(c) Scattered anomalies (d) Distance Factor = 1.5

Fig. 7. Performance in detecting Local Anomalies. Results are shown in (a) and (c) with AUC (y-axis) versus distance factor (x-axis). (b) and (d) illustrate the data distributions in both scattered and clustered cases when distance factor = 1.5.

SCiForest's detection performance is significantly better than LOF and SVM, and SCiForest is not significantly different from iForest and ORCA. This result shows that SCiForest maintains the ability to detect scattered anomalies as compared with other detectors. In terms of processing time, although SCiForest is not the fastest detector among thc fives in these small data sets, its processing time is in the same order as compared with other detectors.

One may ask how SCiForest can detect anomalies if none of the anomalies is seen by the model due to a small sampling size ψ. To answer this question, we provide a short discussion below. Let a be the number of clustered anomalies over n the number of data instances in a data set and ψ the sampling size for each tree used in SCiForest. The probability P for selecting anomalies in a sub-sample is $P = a\psi$. Once a member of the anomalies is considered, appropriate hyper-planes will be formed in order to detect anomalies from the same cluster. ψ can be increased to increase P. The higher the P, the higher the number of trees in SCiForest's model that are catered to detect this kind of anomalies. In cases where P is small, the facility of acceptable range would reduce the path lengths for unseen anomalies, hence exposes them for detection, as long as

Table 3. Performance comparison of five anomalies detectors on data sets containing scattered anomalies. Boldfaced are best performance

		AUC					Time (seconds)				
	size	SCiF	iF	ORCA	LOF	SVM	SCiF	iF	ORCA	LOF	SVM
Satellite	6,435	**0.74**	0.72	0.65	0.52	0.61	5.38	**0.74**	8.97	528.58	9.13
Pima	768	0.65	0.67	**0.71**	0.49	0.55	1.10	0.21	2.08	1.50	**0.06**
Breastw	683	**0.98**	**0.98**	**0.98**	0.37	0.66	1.16	0.21	**0.04**	2.14	0.08
Ionosphere	351	0.91	0.84	**0.92**	0.90	0.71	4.43	0.28	**0.04**	0.96	**0.04**

they are located outside of the range of normal instances. In either cases, SCiForest is equipped with the facilities to detect anomalies, seen or unseen.

6 Conclusions

In this study, we find that when local clustered anomalies are present, the proposed method — SCiForest consistently delivers better detection performance than other detectors and the additional time cost of this performance is small. The ability to detect clustered anomalies is brought about by a simple and effective mechanism, which minimizes the post-split dispersion of the data in the tree growing process. We introduce random hyper-planes for anomalies that are undetectable by single attributes. When the detection of anomalies depends on multiple attributes, using higher number of attributes in hyper-planes yields better detection performance.

Our analysis shows that SCiForest is able to separate clustered anomalies from normal points even when clustered anomalies are very close to or at the edge of normal cluster. In our experiments, SCiForest is shown to have better detection performance than iForest, ORCA, SVM and LOF in detecting clustered anomalies, global or local. Our empirical evaluation shows that SCiForest maintains a fast processing time in the same order of magnitude as iForest's.

References

1. Aggarwal, C.C., Yu, P.S.: Outlier detection for high dimensional data. In: SIGMOD 2001: Proceedings of the 2001 ACM SIGMOD International Conference on Management of Data, pp. 37–46. ACM Press, New York (2001)
2. Angiulli, F., Fassetti, F.: Dolphin: An efficient algorithm for mining distance-based outliers in very large datasets. ACM Trans. Knowl. Discov. Data 3(1), 1–57 (2009)
3. Angiulli, F., Pizzuti, C.: Outlier mining in large high-dimensional data sets. IEEE Transactions on Knowledge and Data Engineering 17(2), 203–215 (2005)
4. Asuncion, A., Newman, D.J.: UCI machine learning repository (2007)

5. Bay, S.D., Schwabacher, M.: Mining distance-based outliers in near linear time with randomization and a simple pruning rule. In: Proceedings of the Ninth ACM SIGKDD International Conference on Knowledge Discovery and Data Mining, pp. 29–38. ACM Press, New York (2003)
6. Breunig, M.M., Kriegel, H.-P., Ng, R.T., Sander, J.: LOF: identifying density-based local outliers. ACM SIGMOD Record 29(2), 93–104 (2000)
7. Caputo, B., Sim, K., Furesjo, F., Smola, A.: Appearance-based object recognition using svms: which kernel should i use? In: Proc. of NIPS Workshop on Statitsical Methods for Computational Experiments in Visual Processing and Computer Vision, Whistler (2002)
8. Chandola, V., Banerjee, A., Kumar, V.: Outlier detection - a survey. Technical Report TR 07-017, Univeristy of Minnesota, Minneapolis (2007)
9. Chandola, V., Banerjee, A., Kumar, V.: Anomaly detection: A survey. ACM Comput. Surv. 41(3), 1–58 (2009)
10. Fawcett, T., Provost, F.: Adaptive fraud detection. Data Mining and Knowledge Discovery 1(3), 291–316 (1997)
11. Hawkins, D.M.: Identification of Outliers. Chapman and Hall, London (1980)
12. Knorr, E.M.: Outliers and data mining: Finding exceptions in data. PhD thesis, University of British Columbia (2002)
13. Knorr, E.M., Ng, R.T.: Algorithms for mining distance-based outliers in large datasets. In: VLDB 1998: Proceedings of the 24rd International Conference on Very Large Data Bases, pp. 392–403. Morgan Kaufmann, San Francisco (1998)
14. Knuth, D.E.: The art of computer programming. Addison-Wiley (1968)
15. Lazarevic, A., Kumar, V.: Feature bagging for outlier detection. In: KDD 2005: Proceedings of the eleventh ACM SIGKDD International Conference on Knowledge Discovery in Data Mining, pp. 157–166. ACM Press, New York (2005)
16. Liu, F.T., Ting, K.M., Zhou, Z.-H.: Isolation forest. In: Proceedings of the 8th IEEE International Conference on Data Mining (ICDM 2008), pp. 413–422 (2008)
17. Moonesignhe, H.D.K., Tan, P.-N.: Outlier detection using random walks. In: ICTAI 2006: Proceedings of the 18th IEEE International Conference on Tools with Artificial Intelligence, Washington, DC, USA, pp. 532–539. IEEE Computer Society Press, Los Alamitos (2006)
18. Murphy, R.B.: On Tests for Outlying Observations. PhD thesis, Princeton University (1951)
19. Murthy, S.K., Kasif, S., Salzberg, S.: A system for induction of oblique decision trees. Journal of Artificial Intelligence Research 2, 1–32 (1994)
20. Otey, M.E., Ghoting, A., Parthasarathy, S.: Fast distributed outlier detection in mixed-attribute data sets. Data Mining and Knowledge Discovery 12(2-3), 203–228 (2006)
21. Papadimitriou, S., Kitagawa, H., Gibbons, P.B., Faloutsos, C.: Loci: Fast outlier detection using the local correlation integral. In: Proceedings of the 19th International Conference on Data Engineering (ICDE 2003), pp. 315–326 (2003)
22. Ramaswamy, S., Rastogi, R., Shim, K.: Efficient algorithms for mining outliers from large data sets. In: SIGMOD 2000: Proceedings of the 2000 ACM SIGMOD International Conference on Management of Data, pp. 427–438. ACM Press, New York (2000)
23. Rocke, D.M., Woodruff, D.L.: Identification of outliers in multivariate data. Journal of the American Statistical Association 91(435), 1047–1061 (1996)
24. Schilling, M.F., Watkins, A.E., Watkins, W.: Is human height bimodal? The American Statistician 56, 223–229 (2002)
25. Schölkopf, B., Platt, J.C., Shawe-Taylor, J., Smola, A.J., Williamson, R.C.: Estimating the support of a high-dimensional distribution. Technical Report MSR-TR-99-87, Microsoft Research (1999)

26. Schölkopf, B., Platt, J.C., Shawe-Taylor, J.C., Smola, A.J., Williamson, R.C.: Estimating the support of a high-dimensional distribution. Neural Computation 13(7), 1443–1471 (2001)
27. Wong, W.-K., Moore, A., Cooper, G., Wagner, M.: Rule-based anomaly pattern detection for detecting disease outbreaks. In: Eighteenth national conference on Artificial intelligence, pp. 217–223. AAAI, Menlo Park (2002)
28. Yamanishi, K., Takeuchi, J.-I., Williams, G., Milne, P.: On-line unsupervised outlier detection using finite mixtures with discounting learning algorithms. In: Proceedings of the Sixth ACM SIGKDD International Conference on Knowledge Discovery and Data Mining, pp. 320–324. ACM Press, New York (2000)

Constrained Parameter Estimation for Semi-supervised Learning: The Case of the Nearest Mean Classifier

Marco Loog*

Pattern Recognition Laboratory
Delft University of Technology
Delft, The Netherlands
m.loog@tudelft.nl,
prlab.tudelft.nl**

Abstract. A rather simple semi-supervised version of the equally simple nearest mean classifier is presented. However simple, the proposed approach is of practical interest as the nearest mean classifier remains a relevant tool in biomedical applications or other areas dealing with relatively high-dimensional feature spaces or small sample sizes. More importantly, the performance of our semi-supervised nearest mean classifier is typically expected to improve over that of its standard supervised counterpart and typically does not deteriorate with increasing numbers of unlabeled data. This behavior is achieved by constraining the parameters that are estimated to comply with relevant information in the unlabeled data, which leads, in expectation, to a more rapid convergence to the large-sample solution because the variance of the estimate is reduced. In a sense, our proposal demonstrates that it may be possible to properly train a known classification scheme such that it can benefit from unlabeled data, while avoiding the additional assumptions typically made in semi-supervised learning.

1 Introduction

Many, if not all, research works that discuss semi-supervised learning techniques stress the need for additional assumptions on the available data in order to be able to extract relevant information not only from the labeled, but especially from the unlabeled examples. Known presuppositions include the cluster assumption, the smoothness assumption, the assumption of low density separation, the manifold assumption, and the like [6,23,30].

While it is undeniably true that having more precise knowledge on the distribution of data could, or even should, help in training a better classifier, the question to what extent such data assumptions are at all necessary has not

* Partly supported by the Innovational Research Incentives Scheme of the Netherlands Research Organization [NWO, VENI Grant 639.021.611].
** Secondary affiliation with the Image Group, University of Copenhagen, Denmark.

J.L. Balcázar et al. (Eds.): ECML PKDD 2010, Part II, LNAI 6322, pp. 291–304, 2010.

been studied to a great extent. Theoretical contributions have both discussed the benefits and the absence of it of the inclusion of unlabeled data in training [4,13,24,25]. With few exceptions, however, these results rely on assumptions being made with respect to the underlying data. Reference [25] aims to make the case that, in fact, it may be so that no extra requirements on the data are needed to obtain improved performance using unlabeled data in addition to labeled data.

A second, related issue is that in many cases, the proposed semi-supervised learning technique has little in common with any of the classical decision rules that many of us know and use; it seems as if semi-supervised learning problems call for a completely different approach to classification. Nonetheless, one still may wonder to what extent substantial gains in classification performance are possible when properly training a known type of classifier, e.g. LDA, QDA, 1NN, in the presence of unlabeled data.

There certainly are exceptions to the above. There even exist methods that are able to extend the use of any known classifier to the semi-supervised setting. In particular, we would like to mention the iterative approaches that rely on expectation maximization or self-learning (or self-training), as can for instance be found in [16,18,19,26,29,27] or the discussion of [10]. The similarity between self-learning and expectation maximization (in some cases equivalence even) has been noted in various papers, e.g. [1,3], and it is to no surprise that such approaches suffer from the same drawback: As soon as the underlying model assumptions do not fit the data, there is the real risk that adding too much unlabeled data leads to a substantial decrease of classification performance [8,9,19]. This is in contrast with the supervised setting, where most classifiers, generative or not, are capable of handling mismatched data assumptions rather well and adding more data generally improves performance.

We aim to convince the reader that, in a way, it may actually also be possible to guarantee a certain improvement with increased numbers of unlabeled data. This possibility is illustrated using the nearest mean classifier (NMC) [11,17], which is adapted to learn from unlabeled data in such a way that some of the parameters become better estimated with increasing amounts of data. The principal idea is to exploit known constraints on the these parameters in the training of the NMC, which results in faster convergence to their real values. The main caveat is that this reduction of variance does not necessarily translate into a reduction of classification error. Section 4 shows, however, that the possible increase in error is limited.

Regarding the NMC, it is needless to say that it is a rather simple classifier, which nonetheless can provide state-of-the-art performance, especially in relatively high-dimensional problems, and which is still, for instance, used in novel application areas [15,14,21,28] (see also Subsection 4.1). Neither the simplicity of the classifier nor the caveat indicated above should distract one from the point we like to illustrate, i.e., it may be feasibility to perform semi-supervised learning without making the assumptions typically made in the current literature.

1.1 Outline

The next section introduces, through a simple, fabricated illustration, the core technical idea that we like to put forward. Subsequently, Section 3 provides a particular implementation of this idea for the nearest mean classifier in a more realistic setting and briefly analyzes convergence properties for some of its key variables. Section 4 shows, by means of some controlled experiments on artificial data, some additional properties of our semi-supervised classifier and compares it to the supervised and the self-learned solutions. Results on six real-world data sets are given as well. Section 5 completes the paper and provides a discussion and conclusions.

2 A Cooked-Up Example of Exponentially Fast Learning

Even though the classification problem considered in this section may be unrealistically simple, it does capture very well the essence of the general proposal to improve semi-supervised learners that we have in mind.

Let us assume that we are dealing with a two-class problem in a one-dimensional feature space where both classes have equal prior probabilities, i.e., $\pi_1 = \pi_2$. Suppose in addition, the NMC is our classifier of choice to tackle this problem with. NMC simply estimates the mean of every class and assigns new feature vectors to the class corresponding to the nearest class mean. Finally, assume that an arbitrarily large set of unlabeled data points is at our disposal. The obvious question to ask is: Can the unlabeled data be exploited to our benefit? The maybe surprising answer is a frank: Yes.

To see this, one should first of all realize that in general, when employing an NMC, the two class means, m_1 and m_2, and the overall mean of the data, μ, fulfill the constraint

$$\mu = \pi_1 m_1 + \pi_2 m_2 . \tag{1}$$

In our particular example based on equal priors, this mean that the total mean should be right in between the two class means. Moreover, again in the current case, the total mean is exactly on the decision boundary. In fact, in our one-dimensional setting, the mean equals the actual decision boundary. Now, if there is anything one can estimate rather accurate from an unlimited amount of data for which labels are not necessarily provided, it would be this overall mean. In other words, provided our training set contains a large number of labeled or unlabeled data points, the zero-dimensional decision boundary can be located to arbitrary precision. That is, it is identifiable, cf. [5].

The only thing we do not know yet is which class is located on what side of the decision boundary. In order to decide this, we obviously do need labeled data. As the decision boundary is already fixed, however, the situation compares directly to the one described in Castelli and Cover [5] and, in a similar spirit, training can be done exponentially fast in the number of labeled samples.

The key point in this example is that the actual distribution of the two classes does in fact not matter. The rapid convergence takes place without making any assumptions on the underlying data, except for the equal class priors. What really leads to the improvement is proper use of the constraint in Equation (1). In the following, we demonstrate how such convergence behavior can generally be obtained for the NMC.

3 Semi-supervised NMC and Its (Co)variance

One of the major lacuna in the example above, is that one rarely has an unlimited amount of samples at ones disposal. We therefore propose a simple adaptation of the NMC in case one has a limited amount of labeled and unlabeled data. Subsequently, a general convergence property of this NMC solution is considered in some detail, together with two special situations.

3.1 Semi-supervised NMC

The semi-supervised version of NMC proposed in this work is rather straightforward and it might only be adequate to a moderate extent in the finite sample setting. The solution suggested simply shifts all K sample class means m_i ($i \in 1, \ldots, K$) by a similar amount such that the overall sample mean $m' = \sum_{i=1}^{K} p_i m'_i$ of the shifted class means m'_i coincides with the total sample mean m_{t}. The latter has been obtained using all data, both labeled and unlabeled. In the foregoing p_i is the estimated posterior corresponding to class i.

More precisely, we take

$$m'_i = m_i - \sum_{i=1}^{K} p_i m_i + m_{\mathrm{t}} \tag{2}$$

for which one can easily check that $\sum_{i=1}^{K} p_i m'_i$ indeed equals m_{t}.

Merely considering the two-class case from now on, there are two vectors that play a role in building the actual NMC [20]. The first one, $\Delta = m_1 - m_2$, determines the direction perpendicular to the linear decision boundary. The second one, $m_1 + m_2$, determines—after taking the inner product with Δ and dividing it by two—the position of the threshold or the bias. Because $\Delta = m_1 - m_2 = m'_1 - m'_2$, the orientations of the two hyperplanes correspond and therefore the only estimates we are interested in are $m_1 + m_2$ and $m'_1 + m'_2$.

3.2 Covariance of the Estimates

To compare the standard supervised NMC and its semi-supervised version, the squared error that measures the deviation of these estimated to their true values is considered. Or rather, as both estimates are unbiased, we consider their covariance matrices.

The first covariance matrix, for the supervised case, is easy to obtain:

$$\operatorname{cov}(m_1 + m_2) = \frac{C_1}{N_1} + \frac{C_2}{N_2}, \tag{3}$$

where C_i is the true covariance matrix of class i and N_i is the number of samples from that class.

To get to the covariance matrix related to the semi-supervised approach, we first express $m_1' + m_2'$ in terms of the variables defined earlier plus m_u, the mean of the unlabeled data, and N_u, the number of unlabeled data points:

$$\begin{aligned} m_1' + m_2' &= m_1 + m_2 - 2\frac{N_1 m_1 + N_2 m_2}{N_1 + N_2} + 2\frac{N_1 m_1 + N_2 m_2 + N_\mathrm{u} m_\mathrm{u}}{N_1 + N_2 + N_\mathrm{u}} \\ &= \left(1 - \tfrac{2N_1}{N_1+N_2} + \tfrac{2N_1}{N_1+N_2+N_\mathrm{u}}\right) m_1 \\ &+ \left(1 - \tfrac{2N_2}{N_1+N_2} + \tfrac{2N_2}{N_1+N_2+N_\mathrm{u}}\right) m_2 + \tfrac{2N_\mathrm{u}}{N_1+N_2+N_\mathrm{u}} m_\mathrm{u} \,. \end{aligned} \quad (4)$$

Realizing that the covariance matrix of the unlabeled samples equals the total covariance T, it now is easy to see that

$$\begin{aligned} \mathrm{cov}(m_1' + m_2') = &\left(1 - \frac{2N_1}{N_1 + N_2} + \frac{2N_1}{N_1 + N_2 + N_\mathrm{u}}\right)^2 \frac{C_1}{N_1} \\ + &\left(1 - \frac{2N_2}{N_1 + N_2} + \frac{2N_2}{N_1 + N_2 + N_\mathrm{u}}\right)^2 \frac{C_2}{N_2} \\ + &\left(\frac{2N_\mathrm{u}}{N_1 + N_2 + N_\mathrm{u}}\right)^2 \frac{T}{N_\mathrm{u}} \,. \end{aligned} \quad (5)$$

3.3 Some Further Considerations

Equations (3) and (5) basically allow us to compare the variability in the two NMC solutions. To get a feel for how these indeed compare, let as consider the situation similar to the one from Section 2 in which the amount of unlabeled data is (virtually) unlimited. It holds that

$$\lim_{N_\mathrm{u}\to\infty} \mathrm{cov}(m_1' + m_2') = \left(1 - \frac{2N_1}{N_1 + N_2}\right)^2 \frac{C_1}{N_1} + \left(1 - \frac{2N_2}{N_1 + N_2}\right)^2 \frac{C_2}{N_2} \,. \quad (6)$$

The quantity $(1 - \frac{2N_i}{N_1+N_2})^2$ is smaller or equal to one and we can readily see that $\mathrm{cov}(m_1' + m_2') \preceq \mathrm{cov}(m_1 + m_2)$, i.e., the variance of the semi-supervised estimate is smaller or equal to the supervised variance for every direction in the feature space and, generally, the former will be a better estimate than the latter. Again as an example, when the true class priors are equal, $1 - \frac{2N_i}{N_1+N_2}$ tends to be nearer zero with increasing number of labeled samples, which implies a dramatic decrease of variance in case of semi-supervision.

Another situation that provides some insight in Equations (3) and (5) is the one in which we consider $C = C_1 = C_2$ and $N = N_1 = N_2$ (for the general case the expression becomes somewhat unwieldy). For this situation we can derive that the two covariance matrices of the sum of means become equal when

$$T = \frac{(4N + N_\mathrm{u})C}{2N} \,. \quad (7)$$

What we might be more interested in is, for example, the situation in which $2NT \preceq (4N + N_u)C$ as this would mean that the expected deviation from the true NMC solution is smaller for the semi-supervised approach, in which case this would be the preferred solution. Note also that from Equation (7) it can be observed that if the covariance C is very small, the semi-supervised method is not expected to give any improvement over the standard approach unless N_u is large.

In a real-world setting, the decisions of which approach to use, necessarily has to rely on the finite number of observations in the training set and sample estimates have to be employed. Moreover, the equations above merely capture the estimates' covariance, which explains only part of the actual variance in the classification error. For the remainder, we leave this issue untouched and turn to the experiments using the suggested approach, which is compared to supervised NMC and a self-learned version.

4 Experimental Results

We carried out several experiments to substantiate some of the earlier findings and claims and to potentially further our understanding of the novel semi-supervised approach. We are interested to what extent NMC can be improved by semi-supervision and a comparison is made to the standard, supervised setting and an NMC trained by means of self-learning [16,18,29].

The latter is a technique in which a classifier of choice is iteratively updated. It starts by the supervised classifier, labels all unlabeled data and retrains the classifier given the newly labeled data. Using this classifier, the initially unlabeled data is reclassified, based on which the next classifier is learned. This is iterated until convergence.

As the focus is on the semi-supervised training of NMC, other semi-supervised learning algorithms are indeed not of interest in the comparisons presented here.

4.1 Initial Experimental Setup and Experiments

As it is not directly of interest to this work, we do not consider learning curves for the number of labeled observations. Obviously, NMC might not need too many labeled examples to perform reasonably and strongly limit the number of labeled examples. We experimented mainly with two, the bare minimum, and ten labeled training objects. In all cases we made sure every class has at least one training sample.

We do, however, consider learning curves as a function of the number of unlabeled instances. This setting easily disclosed both the sensitivity of the self-learning to an abundance of unlabeled data and the improvements that may generally be obtained given various quantities of unlabeled data. The number of unlabeled objects considered in the main experiments are 2, 8, 32, 128, 512, 2048, and 8192.

The tests carried out involve three artificial and eight real-world data set all having two classes. Six of the latter are taken from the UCI Machine Learning

Table 1. Error rates on the two benchmark data sets from [7]

data set	Text		SecStr		
number of labeled objects	10	100	100	1000	10000
error NMC	0.4498	0.2568	0.4309	0.3481	0.3018
error constrained NMC	0.4423	0.2563	0.4272	0.3487	0.3013

Repository [2]. On these, extensive experimentation has been implemented in which for every combination of number of unlabeled objects and labeled objects 1,000 repetitions were executed. In order to be able to do so on the limited amount of samples in the UCI data sets, we allowed to draw instances with replacement, basically assuming that the empirical distribution of every data set is its true distributions. This approach enabled us to properly study the influence of the constraint estimation on real-world data without having to deal with the extra variation due to cross validation or the like. The artificial sets do not suffer from limited amounts of data.

The two other data sets, `Text` and `SecStr`, are benchmarks from [7], which were chosen for their feature dimensionality and for which we followed the protocol as prescribed in [7]. We consider the results, however, of limited interest as the semi-supervised constrained approach gave results only minimally different from those obtained by regular, supervised NMC (after this we did not try the self-learner). Nevertheless, we do not want to withhold these results from the reader, which can be found in Table 1. In fact, we can make at least two interesting observations from them. To start with, the constrained NMC does not perform worse than the regular NMC, for none of the experiments. Compared to the results in [7] both the supervised and the semi-supervised perform acceptable on the `Text` data set when 100 labeled samples are available and both obtain competitive error rates on `SecStr` for all numbers of labeled training data, again confirming the validity of the NMC.

4.2 The Artificial Data

The first artificial data set, `1D`, consists of a one-dimensional data set with two normally distributed classes with unit variance for which the class means are 2 units apart. This setting reflects the situation considered in Section 2. The two top subfigures in Figures 1 and 2 plot the error rates against different numbers of unlabeled data points for the supervised, semi-supervised, and self-learned classifier. All graphs are based on 1,000 repetitions of every experiment. In every round, the classification error is estimated by a new sample of size 10,000. Figure 1 displays the results with two labeled samples, while Figure 2 gives error rates in case of ten labeled samples. Note that adding more unlabeled data indeed further improves the performance.

As second artificial data set, `2D correlated`, we again consider two normally distributed classes, but now in two dimensions. The covariance matrix has the form $\left(\begin{smallmatrix}4 & 3\\ 3 & 4\end{smallmatrix}\right)$, meaning the features are correlated, which, in some sense, does not

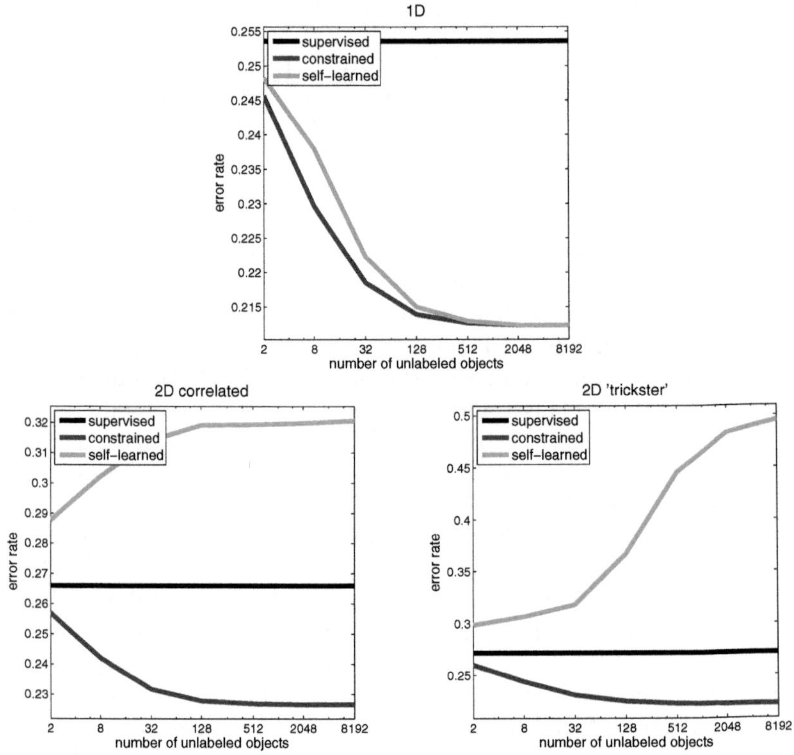

Fig. 1. Error rates on the artificial data sets for various unlabeled sample sizes and a single labeled sample per class. Top subfigure: `1D` data set. Left subfigure: `2D correlated`. Right: `2D 'trickster'`.

fit the underlying assumptions of NMC. Class means in one dimension are 4 apart and the optimal error rate is about 0.159. Further results, like those for the first artificial data set, are again presented in the two figures.

The last artificial data set, `2D 'trickster'`, has been constructed to trick the self-learner. The total data distribution consists of two two-dimensional normal distributions with unit covariance matrices whose means differ in the first feature dimension by 1 unit. The classes, however, are completely determined by the second feature dimension: If this value is larger than zero we assign to class 1, if smaller we assign to class 2. This means that the optimal decision boundary is perpendicular to the boundary that would keep the two normal distributions apart. By construction, the optimal error rate is 0.

Both Figures 1 and 2 illustrate the deteriorating effect adding too much unlabeled data can have on the self-learner, while the constrained semi-supervised approach does not seem to suffer from such behavior and in most cases clearly improves upon the supervised NMC, even though absolute gains can be moderate.

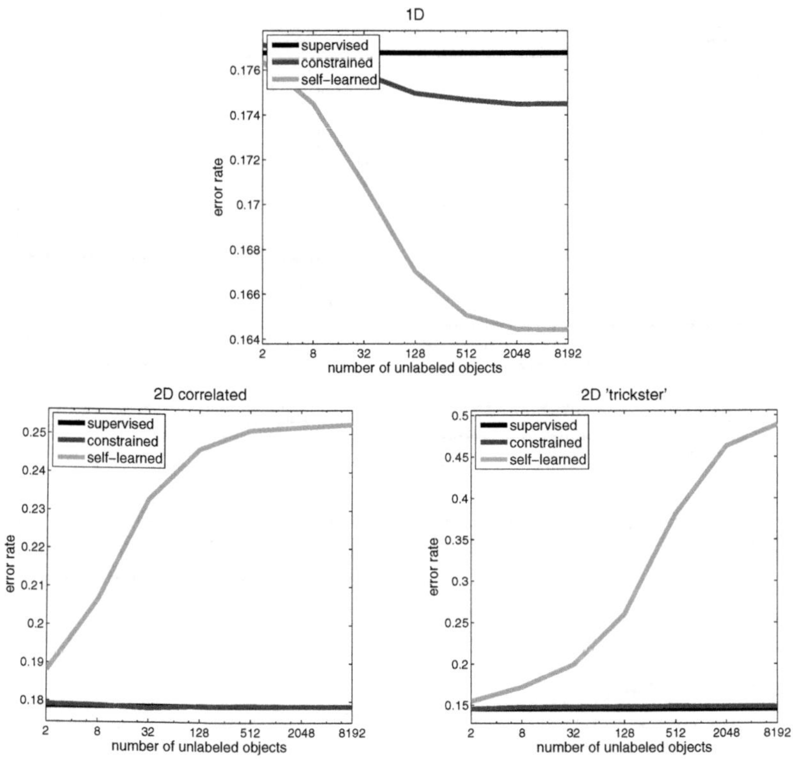

Fig. 2. Error rates on the artificial data sets for various unlabeled sample sizes and a total of ten labeled samples. Top subfigure: `1D` data set. Left subfigure: `2D correlated`. Right: `2D 'trickster'`.

4.3 Six UCI Data Sets

The UCI data sets used are `parkinsons`, `sonar`, `spect`, `spectf`, `transfusion`, and `wdbc'` for which some specifications can be found in Table 2. The classification performance of supervision, semi-supervision, and self-learning are displayed in Figures 3 and 4, for two and ten labeled training objects, respectively.

Table 2. Basic properties of the six real-world data sets

data set	number of objects	dimensionality	smallest class prior
`parkinsons`	195	22	0.25
`sonar`	208	60	0.47
`spect`	267	22	0.21
`spectf`	267	44	0.21
`transfusion`	748	3	0.24
`wdbc`	569	30	0.37

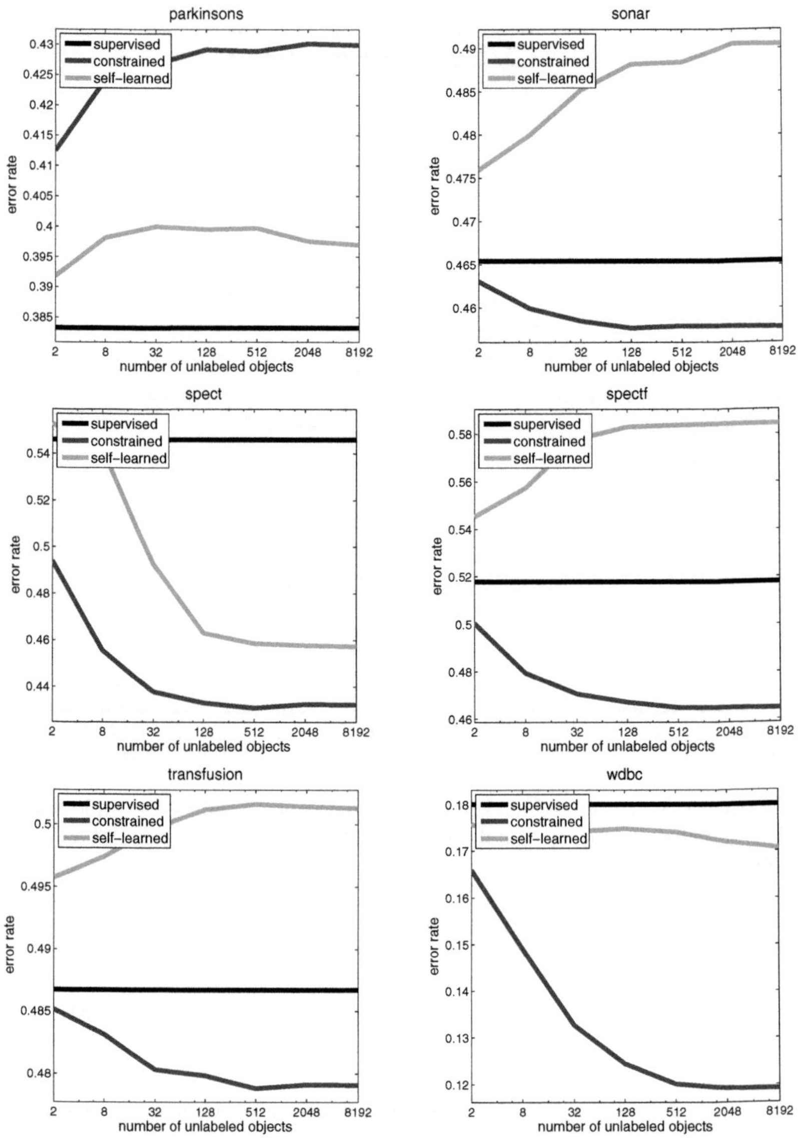

Fig. 3. Error rates for the supervised, semi-supervised, and self-learned classifiers on the six real-world data sets for various unlabeled sample sizes and a single labeled sample per class

In the first place, one should notice that in most of the experiments the constrained NMC performs best of the three schemes employed and that the self-learner in many cases leads to deteriorated performance with increasing unlabeled data sizes. There are various instances in which our semi-supervised approach starts off at an error rate similar to the one obtained by regular supervision, but

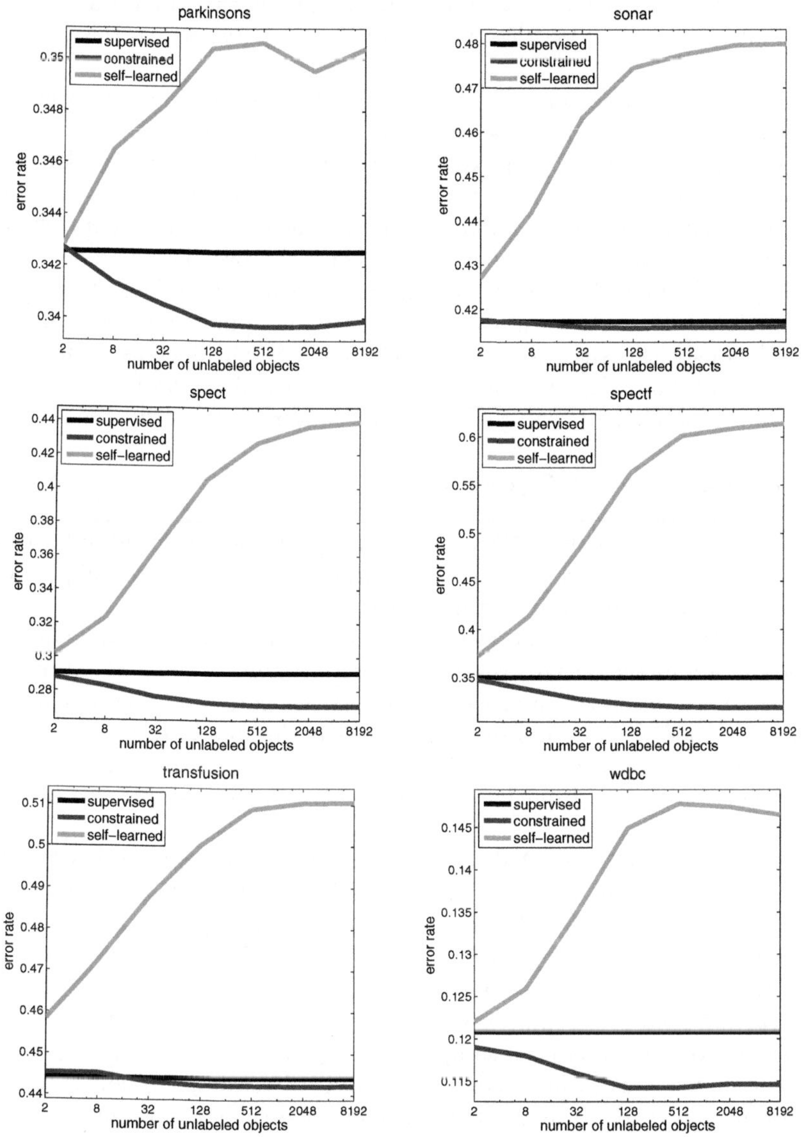

Fig. 4. Error rates for the supervised, semi-supervised, and self-learned classifiers on the six real-world data sets for various unlabeled sample sizes and a total of ten labeled training samples

adding a moderate amount of additional unlabeled objects already ensures that the improvement in performance becomes significant.

The notable outlier is the very first plot in Figure 3 in which constrained NMC performs worse than the other two approaches and even deteriorates with increasing amounts of unlabeled data. How come? We checked the estimates for the covariance matrices in Equations 3 and 3 and saw that the variability of the sum of the means is indeed less in case of semi-supervision, so this is not the problem.

What comes to the fore here, however, is that a reduction in variance for these parameters does not necessarily directly translate into a gain in classification performance. Not even in expectation. The main problem we identified is basically the following (consider the example from Section 2): The more accurately a classifier manages to approximate the true decision boundary, the more errors it will typically make if the side on which the two classes are located are mixed up in the first place. Such a configuration would indeed lead to worse and worse performance for the semi-supervised NMC with more and more unlabeled data. Obviously, this situation is less likely to occur with increasing numbers of labeled samples and Figure 4 shows that the constrained NMC is expected to attain improved classification results on `parkinsons` for as few as ten labels.

5 Discussion and Conclusion

The nearest mean classifier (NMC) and some of its properties have been studied in the semi-supervised setting. In addition to the known technique of self-learning, we introduced a constrained-based approach that typically does not suffer from the major drawback of the former for which adding more and more unlabeled data might actually result in a deterioration. As pointed out, however, this non-deterioration concerns the parameter estimates and does not necessarily reflect immediately in improved classifier's performance. In the experiments, we identified an instance where a deterioration indeed occurs, but the negative effect seems limited and quickly vanishes with a moderate increase of labeled training data.

Recapitulating our general idea, we suggest that particular constraints, which relate estimates coming from both labeled and unlabeled data, should be met by the parameters that have to be estimated in the training phase of the classifier. For the nearest mean we rely on Equation (1) that connects the two class means to the overall mean of the data. Experiments show that enforcing this constraint in a straightforward way improves the classification performance in the case of moderately to large unlabeled sample sizes. Qualitatively, this partly confirms the theory in Section 3, which shows that adding increasing numbers of unlabeled data, eventually leads to reduced variance in the estimates and, in a way, faster convergence to the true solution.

A shortcoming of the general idea of constrained estimation is that it is not directly clear which constraints to apply to most of the other classical decision rules, if at all applicable. The main question obviously being if there is a more general principle of constructing and applying constraints that is more broadly applicable. On the other hand, one should realize that the NMC may act as a basis for LDA and its penalized and flexible variations, as described in [12] for instance. Moreover, kernelization by means of a Gaussian kernel, reveals similarities to the classical Parzen classifier, cf. [22]. Our findings may be directly applicable in these situations.

In any case, the important point we did convey is that, in a way, it is possible to perform semi-supervised learning without making additional assumptions on the characteristics of the data distribution, but by exploiting some characteristics of the classifier. We consider it also important that it is possible to do this based on a known classifier and in such a way that adding more and more data does not lead to its deterioration. A final advantage is that our semi-supervised NMC is as easy to train as the regular NMC with no need for complex regularization schemes or iterative procedures.

References

1. Abney, S.: Understanding the Yarowsky algorithm. Computational Linguistics 30(3), 365–395 (2004)
2. Asuncion, A., Newman, D.: UCI machine learning repository (2007), `http://www.ics.uci.edu/~mlearn/MLRepository.html`
3. Basu, S., Banerjee, A., Mooney, R.: Semi-supervised clustering by seeding. In: Proceedings of the Nineteenth International Conference on Machine Learning, pp. 19–26 (2002)
4. Ben-David, S., Lu, T., Pál, D.: Does unlabeled data provably help? worst-case analysis of the sample complexity of semi-supervised learning. In: Proceedings of COLT 2008, pp. 33–44 (2008)
5. Castelli, V., Cover, T.: On the exponential value of labeled samples. Pattern Recognition Letters 16(1), 105–111 (1995)
6. Chapelle, O., Schölkopf, B., Zien, A.: Introduction to semi-supervised learning. In: Semi-Supervised Learning, ch. 1. MIT Press, Cambridge (2006)
7. Chapelle, O., Schölkopf, B., Zien, A.: Semi-Supervised Learning. MIT Press, Cambridge (2006)
8. Cohen, I., Cozman, F., Sebe, N., Cirelo, M., Huang, T.: Semisupervised learning of classifiers: Theory, algorithms, and their application to human-computer interaction. IEEE Transactions on Pattern Analysis and Machine Intelligence pp. 1553–1567 (2004)
9. Cozman, F., Cohen, I.: Risks of semi-supervised learning. In: Semi-Supervised Learning, chap. 4. MIT Press, Cambridge (2006)
10. Dempster, A., Laird, N., Rubin, D.: Maximum likelihood from incomplete data via the em algorithm. Journal of the Royal Statistical Society. Series B (Methodological) 39(1), 1–38 (1977)
11. Duda, R., Hart, P.: Pattern classification and scene analysis. John Wiley & Sons, Chichester (1973)
12. Hastie, T., Buja, A., Tibshirani, R.: Penalized discriminant analysis. The Annals of Statistics 23(1), 73–102 (1995)
13. Lafferty, J., Wasserman, L.: Statistical analysis of semi-supervised regression. In: Advances in Neural Information Processing Systems, vol. 20, pp. 801–808 (2007)
14. Liu, Q., Sung, A., Chen, Z., Liu, J., Huang, X., Deng, Y.: Feature selection and classification of MAQC-II breast cancer and multiple myeloma microarray gene expression data. PLoS ONE 4(12), e8250 (2009)
15. Liu, W., Laitinen, S., Khan, S., Vihinen, M., Kowalski, J., Yu, G., Chen, L., Ewing, C., Eisenberger, M., Carducci, M., Nelson, W., Yegnasubramanian, S., Luo, J., Wang, Y., Xu, J., Isaacs, W., Visakorpi, T., Bova, G.: Copy number analysis indicates monoclonal origin of lethal metastatic prostate cancer. Nature Medicine 15(5), 559–565 (2009)

16. McLachlan, G.: Iterative reclassification procedure for constructing an asymptotically optimal rule of allocation in discriminant analysis. Journal of the American Statistical Association 70(350), 365–369 (1975)
17. McLachlan, G.: Discriminant analysis and statistical pattern recognition. John Wiley & Sons, Chichester (1992)
18. McLachlan, G., Ganesalingam, S.: Updating a discriminant function on the basis of unclassified data. Communications in Statistics - Simulation and Computation 11(6), 753–767 (1982)
19. Nigam, K., McCallum, A., Thrun, S., Mitchell, T.: Learning to classify text from labeled and unlabeled documents. In: Proceedings of the Fifteenth National Conference on Artificial Intelligence, pp. 792–799 (1998)
20. Noguchi, S., Nagasawa, K., Oizumi, J.: The evaluation of the statistical classifier. In: Watanabe, S. (ed.) Methodologies of Pattern Recognition, pp. 437–456. Academic Press, London (1969)
21. Roepman, P., Jassem, J., Smit, E., Muley, T., Niklinski, J., van de Velde, T., Witteveen, A., Rzyman, W., Floore, A., Burgers, S., Giaccone, G., Meister, M., Dienemann, H., Skrzypski, M., Kozlowski, M., Mooi, W., van Zandwijk, N.: An immune response enriched 72-gene prognostic profile for early-stage non-small-cell lung cancer. Clinical Cancer Research 15(1), 284 (2009)
22. Schölkopf, B.: The kernel trick for distances. In: Advances in Neural Information Processing Systems, vol. 13, p. 301. The MIT Press, Cambridge (2001)
23. Seeger, M.: A taxonomy for semi-supervised learning methods. In: Semi-Supervised Learning, ch. 2. MIT Press, Cambridge (2006)
24. Singh, A., Nowak, R., Zhu, X.: Unlabeled data: Now it helps, now it doesn't. In: Advances in Neural Information Processing Systems, vol. 21 (2008)
25. Sokolovska, N., Cappé, O., Yvon, F.: The asymptotics of semi-supervised learning in discriminative probabilistic models. In: Proceedings of the 25th International Conference on Machine Learning, pp. 984–991 (2008)
26. Titterington, D.: Updating a diagnostic system using unconfirmed cases. Journal of the Royal Statistical Society. Series C (Applied Statistics) 25(3), 238–247 (1976)
27. Vittaut, J., Amini, M., Gallinari, P.: Learning classification with both labeled and unlabeled data. In: Elomaa, T., Mannila, H., Toivonen, H. (eds.) ECML 2002. LNCS (LNAI), vol. 2430, pp. 69–78. Springer, Heidelberg (2002)
28. Wessels, L., Reinders, M., Hart, A., Veenman, C., Dai, H., He, Y., Veer, L.: A protocol for building and evaluating predictors of disease state based on microarray data. Bioinformatics 21(19), 3755 (2005)
29. Yarowsky, D.: Unsupervised word sense disambiguation rivaling supervised methods. In: Proceedings of the 33rd annual meeting on Association for Computational Linguistics, pp. 189–196 (1995)
30. Zhu, X., Goldberg, A.: Introduction to Semi-Supervised Learning. Morgan & Claypool Publishers, San Francisco (2009)

Online Learning in Adversarial Lipschitz Environments

Odalric-Ambrym Maillard and Rémi Munos

SequeL Project, INRIA Lille - Nord Europe, France
{odalric.maillard,remi.munos}@inria.fr

Abstract. We consider the problem of online learning in an adversarial environment when the reward functions chosen by the adversary are assumed to be Lipschitz. This setting extends previous works on linear and convex online learning. We provide a class of algorithms with cumulative regret upper bounded by $\tilde{O}(\sqrt{dT\ln(\lambda)})$ where d is the dimension of the search space, T the time horizon, and λ the Lipschitz constant. Efficient numerical implementations using particle methods are discussed. Applications include online supervised learning problems for both full and partial (bandit) information settings, for a large class of non-linear regressors/classifiers, such as neural networks.

Introduction

The adversarial online learning problem is defined as a repeated game between an agent (the learner) and an opponent, where at each round t, simultaneously the agent chooses an action (or decision, or arm, or state) $\theta_t \in \Theta$ (where Θ is a subset of $\mathbb{R}^d$) and the opponent chooses a reward function $f_t : \Theta \mapsto [0,1]$. The agent receives the reward $f_t(\theta_t)$. In this paper we will consider different assumptions about the amount of information received by the agent at each round. In the *full information* case, the full reward function f_t is revealed to the agent after each round, whereas in the case of *bandit information* only the reward corresponding to its own choice $f_t(\theta_t)$ is provided.

The goal of the agent is to allocate its actions $(\theta_t)_{1\leq t\leq T}$ in order to maximize the sum of obtained rewards $F_T \stackrel{\text{def}}{=} \sum_{t=1}^T f_t(\theta_t)$ up to time T and its performance is assessed in terms of the best constant strategy $\theta \in \Theta$ on the same reward functions, i.e. $F_T(\theta) \stackrel{\text{def}}{=} \sum_{t=1}^T f_t(\theta)$. Defining the **cumulative regret**:

$$R_T(\theta) \stackrel{\text{def}}{=} F_T(\theta) - F_T,$$

with respect to (w.r.t.) a strategy θ, the agent aims at minimizing $R_T(\theta)$ for all $\theta \in \Theta$.

In this paper we consider the case when the functions f_t are Lipschitz w.r.t. the decision variable θ (with Lipschitz constant upper bounded by λ).

J.L. Balcázar et al. (Eds.): ECML PKDD 2010, Part II, LNAI 6322, pp. 305–320, 2010.

Previous results. Several works on adversarial online learning include the case of finite action spaces (the so-called *learning from experts* [1] and the *multi-armed bandit problem* [2,3]), countably infinite action spaces [4], and the case of continuous action spaces, where many works have considered strong assumptions on the reward functions, i.e. linearity or convexity.

In the *online linear optimization* (see e.g. [5,6,7,8] in the adversarial case and [9,10] in the stochastic case) where the functions f_t are linear, the resulting upper- and lower-bounds on the regret are of order (up to logarithmic factors) $\sqrt{dT}$ in the case of full information and $d^{3/2}\sqrt{T}$ in the case of bandit information [6] (and in good cases $d\sqrt{T}$ [5]). In *online convex optimization* f_t is assumed to be convex [11] or σ-strongly convex [12], and the resulting upper bounds are of order $C\sqrt{T}$ and $C^2\sigma^{-1}\ln(T)$ (where C is a bound on the gradient of the functions, which implicitly depends on the space dimension). Other extensions have been considered in [13,14,15] and a minimax lower bound analysis in the full information case in [16]. These results hold in bandit information settings where either the value or the gradient of the function is revealed.

To our knowledge, the weaker Lipschitz assumption that we consider here has not been studied in the adversarial optimization literature. However, in the stochastic bandit setting (where noisy evaluations of a fixed function are revealed), the Lipschitz assumption has been previously considered in [17,18], see the discussion in Section 2.3.

Motivations: In many applications (such as the problem of matching ads to web-page contents on the Internet) it is important to be able to consider both large action spaces and general reward functions. The continuous space problem appears naturally in online learning, where a decision point is a classifier in a parametric space of dimension d. Since many non-linear non-convex classifiers/regressors have shown success (such as neural-networks, support vector machines, matching pursuits), we wish to extend the results of online learning to those non-linear non-convex cases. In this paper we consider a Lipschitz assumption (illustrated in the case of neural network architectures) which is much weaker than linearity or convexity.

What we do: We start in Section 1 by describing a general continuous version of the Exponentially Weighted Forecaster and state (Theorem 1) an upper bound on the cumulative regret of $O(\sqrt{dT\ln(d\lambda T)})$ under a non-trivial geometrical property of the action space. The algorithm requires, as a sub-routine, being able to sample actions according to continuous distributions, which may be impossible to do perfectly well in general.

To address the issue of sampling, we may use different sampling techniques, such as uniform grids, random or quasi-random grids, or use adaptive methods such as Monte-Carlo Markov chains (MCMC) or Population Monte-Carlo (PMC).

However, since any sampling technique introduces a sampling bias (compared to an ideal sampling from the continuous distribution), this also impacts the resulting performance of the method in terms of regret. This shows a tradeoff between regret and numerical complexity, which is illustrated by numerical experiments in Section 1.3 where PMC techniques are compared to sampling from uniform grids.

Then in Section 2 we describe several applications to learning problems. In the full information setting (when the desired outputs are revealed after each round), the case of regression is described in Section 2.1 and the case of classification in Section 2.2. Then Section 2.3 considers a classification problem in a bandit setting (i.e. when only the information of whether the prediction is correct or not is revealed). In the later case, we show that the expected number of mistakes does not exceed that of the best classifier by more than $O(\sqrt{dTK\ln(d\lambda T)})$, where K is the number of labels. We detail a possible PMC implementation in this case.

We believe that the work reported in this paper provides arguments that the use of MCMC, PMC, and other adaptive sampling techniques is a promising direction for designing numerically efficient algorithms for online learning in adversarial Lipschitz environments.

1 Adversarial Learning with Full Information

We consider a search space $\Theta \subset \mathbb{R}^d$ equipped with the Lebesgue measure μ. We write $\mu(\Theta) = \int_\Theta 1$. We assume that all reward functions f_t have values in $[0,1]$ and are Lipschitz w.r.t. some norm $||\cdot||$ (e.g. L_1, L_2, or L_∞) with a Lipschitz constant upper bounded by $\lambda > 0$, i.e. for all $t \geq 1$ and $\theta_1, \theta_2 \in \Theta$,

$$|f_t(\theta_1) - f_t(\theta_2)| \leq \lambda ||\theta_1 - \theta_2||.$$

1.1 The ALF Algorithm

We consider the natural extension of the EWF (Exponentially Weighted Forecaster) algorithm [19,20,1] to the continuous action setting. Figure 1 describes this ALF algorithm (for Adversarial Lipschitz Full-information environment).

At each time step, the forecaster samples θ_t from a probability distribution $p_t \stackrel{\text{def}}{=} \frac{w_t}{\int_\Theta w_t}$ with w_t being the weight function defined according to the previously observed reward functions $(f_s)_{s<t}$. The function f_t is then revealed and the weight function is updated. We have $w_{t+1}(\theta) = \exp(\eta F_t(\theta))$, and η is a parameter of the algorithm.

Geometric considerations: The performance of the algorithm depends on the geometry of the space $\Theta \subset \mathbb{R}^d$ (relatively to the chosen norm), and since we want to derive bounds as a function of the dimension d, we now define classes of domains ($(\Theta_d)_{d\geq 0}$ indexed by their dimension) with similar geometrical properties.

Definition 1. *For the class of domains $(\Theta_d)_{d\geq 1}$, we define $\kappa(d) > 1$:*

$$\kappa(d) \stackrel{\text{def}}{=} \sup_{\theta\in\Theta_d, r>0} \frac{\min\left[\mu(B(\theta,r)), \mu(\Theta_d)\right]}{\mu(B(\theta,r)\cap\Theta_d)} \tag{1}$$

Initialization: Set $w_1(\theta) = 1$ for all $\theta \in \Theta$.

For each round $t = 1, 2, \ldots, T$

(1) Simultaneously the adversary chooses the reward function $f_t : \Theta \mapsto [0,1]$, and the learner chooses $\theta_t \overset{iid}{\sim} p_t$, where $p_t(\theta) \overset{\text{def}}{=} \frac{w_t(\theta)}{\int_\Theta w_t(\theta)d\theta}$,
(2) The learner incurs the reward $f_t(\theta_t)$,
(3) The reward function f_t is revealed to the learner. The weight function w_t is updated as:

$$w_{t+1}(\theta) \overset{\text{def}}{=} w_t(\theta)e^{\eta f_t(\theta)}, \text{ for all } \theta \in \Theta$$

Fig. 1. Adversarial Lipschitz learning algorithm in a Full-information setting (ALF algorithm)

Assumption A1 There exists $\kappa > 0$ such that $\kappa(d) \leq \kappa^d$, for all $d \geq 1$, and there exists $\kappa' > 0$ and $\alpha \geq 0$ such that $\mu(B(\theta, r)) \geq (r/(\kappa' d^\alpha))^d$ for all $r > 0$, $d \geq 1$, and $\theta \in \mathbb{R}^d$.

The first part of this assumption says that $\kappa(d)$ scales at most exponentially with the dimension. This is reasonable if we consider domains with similar geometries (i.e. whenever the "angles" of the domains do not go to zero when the dimension d increases). For example, in the domains $\Theta_d = [0,1]^d$, this assumption holds with $\kappa = 2$ for any usual norm (L_1,L_2 and L_∞). The second part of the assumption about the volume of d-balls is a property of the norms and holds naturally for any usual norm: for example, $\kappa' = 1/2$, $\alpha = 0$ for L_∞, and $\kappa' = \sqrt{\pi}/(\sqrt{2}e)$, $\alpha = 3/2$ for any norm L_p, $p \geq 1$, since for L_p norms, $\mu(B(\theta, r)) \geq (2r)^d/d!$ and from Stirling formula, $d! \sim \sqrt{2\pi d}(d/e)^d$, thus $\mu(B(\theta, r)) \geq \left(r/(\frac{\sqrt{2\pi}}{2e} d^{3/2})\right)^d$.

Remark 1. Notice that Assumption A1 makes explicit the required geometry of the domain in order to derive tight regret bounds.

We now provide upper-bounds for the ALF algorithm on the worst expected regret (i.e. $\sup_{\theta\in\Theta} \mathbb{E}R_T(\theta)$) and high probability bounds on the worst regret $\sup_{\theta\in\Theta} R_T(\theta)$.

Theorem 1. *(ALF algorithm) Under Assumption A1, for any $\eta \leq 1$, the expected (w.r.t. the internal randomization of the algorithm) cumulative regret of the ALF algorithm is bounded as:*

$$\sup_{\theta\in\Theta} \mathbb{E}R_T(\theta) \leq T\eta + \frac{1}{\eta}\Big[d \ln(cd^\alpha \eta\lambda T) + \ln(\mu(\Theta))\Big], \tag{2}$$

whenever $(d^\alpha \eta \lambda T)^d \mu(\Theta) \geq 1$, where $c \overset{\text{def}}{=} 2\kappa \max(\kappa', 1)$ is a constant (which depends on the geometry of Θ and the considered norm). Under the same assumptions, with probability $1 - \beta$,

$$\sup_{\theta\in\Theta} R_T(\theta) \le T\eta + \frac{1}{\eta}\Big[d\ln(cd^{\alpha}\eta\lambda T) + \ln(\mu(\Theta))\Big] + \sqrt{2T\ln(\beta^{-1})}. \qquad (3)$$

We deduce that for the choice $\eta = \big(\frac{d}{T}\ln(cd^{\alpha}\lambda T)\big)^{1/2}$, *when* $\eta \le 1$ *and assuming* $\mu(\Theta) = 1$, *we have:*

$$\sup_{\theta\in\Theta} \mathbb{E}R_T(\theta) \le 2\sqrt{dT\ln(cd^{\alpha}\lambda T)},$$

and a similar bound holds in high probability.

The proof is given in Appendix A. Note that the parameter η of the algorithm depends very mildly on the (unknown) Lipschitz constant λ. Actually even if λ was totally unknown, the choice $\eta = \big(\frac{d}{T}\ln(cd^{\alpha}T)\big)^{1/2}$ would yield a bound $\sup_{\theta\in\Theta}\mathbb{E}R_T(\theta) = O(\sqrt{dT\ln(dT)}\ln\lambda)$ which is still logarithmic in λ (instead of linear in the case of the discretization) and enables to consider classes of functions for which λ may be large (and unknown).

Anytime algorithm. Like in the discrete version of EWF (see e.g. [21,22,1]) this algorithms may easily be extended to an anytime algorithm (i.e. providing similar performance even when the time horizon T is not known in advance) by considering a decreasing coefficient $\eta_t = \big(\frac{d}{2t}\ln(cd^{\alpha}\lambda t)\big)^{1/2}$ in the definition of the weight function w_t. We refer to [22] for a description of the methodology.

The issue of sampling. In order to implement the ALF algorithm detailed in Figure 1 one should be able to sample θ_t from the continuous distribution p_t. However it is in general impossible to sample perfectly from arbitrary continuous distributions p_t, thus we need to resort to approximate sampling techniques, such as based on uniform grids, random or quasi-random grids, or adaptive methods such as Monte-Carlo Markov Chain (MCMC) methods or population Monte-Carlo (PMC) methods. If we write p_t^N the distribution from which the samples are actually generated, where N stands for the computational resources (e.g. the number of grid points if we use a grid) used to generate the samples, then the expected regret $\mathbb{E}R_T(\theta)$ will suffer an additional term of at most $\sum_{t=1}^{T}|\int_{\Theta} p_t f_t - \int_{\Theta} p_t^N f_t|$. This shows a tradeoff between the regret (low when N is large, i.e. p_t^N is close to p_t) and numerical complexity and memory requirement (which scales with N). In the next two sub-sections we discuss sampling techniques based on fixed grids and adaptive PMC methods, respectively.

1.2 Uniform Grid over the Unit Hypercube

A first approach consists in setting a uniform grid (say with N grid points) before the learning starts and consider the naive approximation of p_t by sampling at each round one point of the grid, since in that case the distribution has finite support and the sampling is easy.

Actually, in the case when the domain Θ is the unit hypercube $[0,1]^d$, we can easily do the analysis of an Exponentially Weighted Forecaster (EWF) playing on the grid and shows that the total expected regret is small provided that N is large enough. Indeed, let $\Theta_N \stackrel{\text{def}}{=} \{\theta_1, \ldots, \theta_N\}$ be a uniform grid of resolution $h > 0$, i.e. such that for any $\theta \in \Theta$, $\min_{1\leq i\leq N} ||\theta - \theta_i|| \leq h$. This means that at each round t, we select the action $\theta_{I_t} \in \Theta_N$, where $I_t \stackrel{iid}{\sim} p_t^N$ with p_t^N the distribution on $\{1, \ldots, N\}$ defined by $p_t^N(i) \stackrel{\text{def}}{=} w_t(i)/\sum_{j=1}^N w_t(j)$, where the weights are defined as $w_t(i) \stackrel{\text{def}}{=} e^{\eta F_{t-1}(\theta_i)}$ for some appropriate constant $\eta = \sqrt{2\ln N/T}$.

The usual analysis of EWF implies that the regret relatively to any point of the grid is upper bounded as: $\sup_{1\leq i\leq N} \mathbb{E}R_T(\theta_i) \leq \sqrt{2T\ln N}$.

Now, since we consider the unit hypercube $\Theta = [0, 1]^d$, and under the assumption that the functions f_t are λ-Lipschitz with respect to L_∞-norm, we have that $F_T(\theta) \leq \min_{1\leq i\leq N} F_T(\theta_i) + \lambda T h$. We deduce that the expected regret relatively to any $\theta \in \Theta$ is bounded as $\sup_{\theta\in\Theta} \mathbb{E}R_T(\theta) \leq \sqrt{2T\ln N} + \lambda T h$.

Setting $N = h^{-d}$ with the optimal choice of h in the previous bound (up to a logarithmic term) $h = \frac{1}{\lambda}\sqrt{d/T}$ gives the upper bound on the regret: $\sup_{\theta\in\Theta} \mathbb{E}R_T = O(\sqrt{dT\ln(\lambda\sqrt{T})})$.

However this discretized EWF algorithm suffers from severe limitations from a practical point of view:

1. The choice of the best resolution h of the grid depends crucially on the knowledge of the Lipschitz constant λ and has an important impact on the regret bound. However, usually λ is not known exaclty (but an upper-bound may be available, e.g. in the case of neural networks discussed below). If we choose h irrespective of λ (e.g. $h = \sqrt{d/T}$) then the resulting bound on the regret will be of order $O(\lambda\sqrt{dT})$ which is much worst in terms of λ than its optimal order $\sqrt{\ln \lambda}$.
2. The number of grid points (which determines the memory requirement and the numerical complexity of the EWF algorithm) scales exponentially with the dimension d.

Notice that instead of using a uniform grid, one may resort to the use of random (or quasi-random) grids with a given number of points N, which would scale better in high dimensions. However all those method are non-adaptive in the sense that the position of the grid point do not adapt to the actual reward functions f_t observed through time. We would like to sample points according to an "adaptive discretization" that would allocate more points where the cumulative reward function F_t is high. In the next sub-section we consider the ALF algorithm where we use adaptive sampling techniques such as MCMC and PMC which are designed for sampling from (possibly high dimensional) continuous distributions.

1.3 A Population Monte-Carlo Sampling Technique

The idea of sampling techniques such as Metropolis-Hasting (MH) or other MCMC (Monte-Carlo Markov Chain) methods (see e.g. [23,24]) is to build a Markov chain that has p_t as its equilibrium distribution, and starting from an initial distribution, iterates its transition kernel K times so as to approximate p_t. Note that the rate of convergence of the distribution towards p_t is exponential with K (see e.g. [25]): $\delta(k) \le (2\epsilon)^{k/\tau(\epsilon)}$, where $\delta(k)$ is the total variation distance between p_t and the distribution at step k, and $\tau(\epsilon) = \min\{k; \delta(k) \le \epsilon\}$ is the so called mixing time of the Markov Chain ($\epsilon < 1/2$).

Thus sampling $\theta_t \sim p_t$ only requires being able to compute $w_t(\theta)$ at a finite number of points K (the number of transitions of the corresponding Markov chain needed to approximate the stationary distribution p_t). This is possible whenever the reward functions f_t can be stored by using a finite amount of information, which is the case in the applications to learning, described in the next section.

However, using MCMC at each time step to sample from a distribution p_t which is similar to the previous one p_{t-1} (since the cumulative functions F_t do not change much from one iteration to the next) is a waste of MC transitions. The exponential decay of $\delta(k)$ depends on the mixing time $\tau(\epsilon)$ which depends on both the target distribution and the transition kernel, and can be reduced when considering efficient methods based on interacting particles systems. The population Monte-Carlo (PMC) method (see e.g. [26]) approximates p_t by a population of N particles $(x_{t,k}^{1:N})$ which evolve (during $1 \le k \le K$ rounds) according to a transition/selection scheme:

- At round k, the **transition step** generates a successor population $\widetilde{x}_{t,k}^{1:N} \overset{iid}{\sim} g_{t,k}(x_{t,k-1}^{1:N}, \cdot)$ according to a transition kernel $g_{t,k}(\cdot, \cdot)$. Then likelihood ratios are defined as $w_{t,k}^{1:N} = \frac{p_t(\widetilde{x}_{t,k}^{1:N})}{g(x_{t,k-1}^{1:N}, \widetilde{x}_{t,k}^{1:N})}$,
- The **selection step** resamples N particles $x_{t,k}^i = \widetilde{x}_{t,k}^{I_i}$ for $1 \le i \le N$ where the selection indices $(I_i)_{1 \le i \le N}$ are drawn (with replacement) from the set $\{1 \ldots N\}$ according to a multinomial distribution with parameters $(w_{t,k}^i)_{1 \le i \le N}$

At round K, one particle (out of N) is selected uniformly randomly, which defines the sample $\theta_t \sim p_t^N$ that is returned by the sampling technique. Some properties of this approch is that the proposed sample tends to an unbiased independent sample of p_t (when either N or $K \to \infty$). We do not provide additional implementation details about this method here since this is not the goal of this paper, but we refer the interested reader to [26] for discussion about the choice of good kernels $g_{t,k}$ and automatic tuning methods of the parameter K and number of particles N. Note that in [26], the authors prove a Central Limit Theorem showing that the term $\sqrt{N}(\int_\Theta p_t f - \int_\Theta p_t^N f)$ is asymptotically gaussian with explicit variance depending on the previous parameters (that we do not report here for it would require additional specific notations), thus giving the speed of convergence towards 0. We also refer to [27] for known theoretical results of the general PMC theory.

When using this sampling techniques in the ALF algorithm, since the distribution p_{t+1} does not differ much from p_t, we can initialize the particles at round $t+1$ with the particles obtained at the previous round t at the last step of the

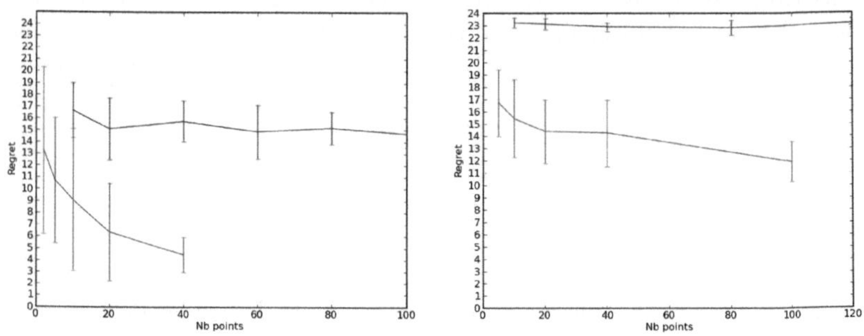

Fig. 2. Regret as a function of N, for dimensions $d = 2$ (left figure) and 20 (right figure). In both figures, the top curve represents the grid sampling and the bottom curve the PMC sampling.

PMC sampling: $x^i_{t+1,1} \stackrel{\text{def}}{=} x^i_{t,K}$, for $1 \leq i \leq N$. In the numerical experiments reported in the next sub-section, this enabled to reduce drastically the number of rounds K per time step (less than 5 in all experiments below).

1.4 Numerical Experiments

For illustation, let us consider the problem defined by: $\Theta = [0,1]^d$, $f_t(\theta) = (1 - ||\theta - \theta_t||/\sqrt{d})^3$ where $\theta_t = t/T\,(1,\ldots,1)'$. The optimal θ^* (i.e. $\arg\max_\theta F_T(\theta)$) is $1/2\,(1,\ldots,1)'$. Figure 2 plots the expected regret $\sup_{\theta\in\Theta} \mathbb{E}R_T(\theta)$ (with $T = 100$, averaged over 10 experiments) as a function of the parameter N (number of sampling points/particles) for two sampling methods: the random grid mentioned in the end of Section 1.2 and the PMC method. We considered two values of the space dimension: $d = 2$ and $d = 20$. Note that the uniform discretization technique is not applicable in the case of dimension $d = 20$ (because of the curse of dimensionality). We used $K = 5$ steps and used a Gaussian centered kernel $g_{t,k}$ of variance $\sigma^2 = 0.1$ for the PMC method.

Since the complexity of sampling from a PMC method with N particles and from a grid of N points is not the same, in order to compare the performance of the two methods both in terms of regret and runtime, we plot in Figure 3 the regret as a function of the CPU time required to do the sampling, for different values of N.

As expected, the PMC method is more efficient since its allocation of points (particles) depends on the cumulative rewards F_t (it thus may be considered as an adaptive algorithm).

2 Applications to Learning Problems

2.1 Online Regression

Consider an online adversarial regression problem defined as follows: at each round t, an opponent selects a couple (x_t, y_t) where $x_t \in \mathcal{X}$ and $y_t \in \mathcal{Y} \subset \mathbb{R}$,

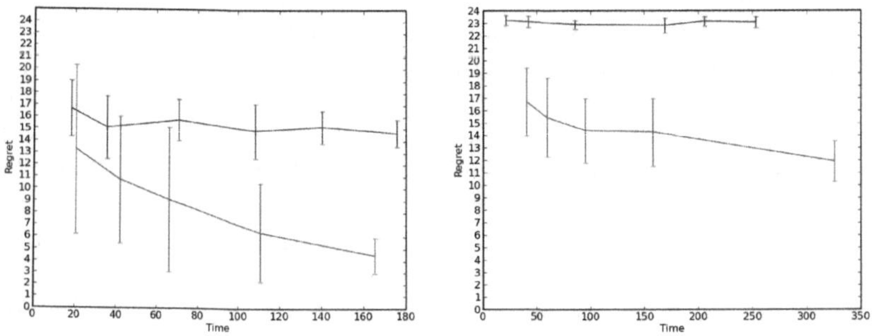

Fig. 3. Regret as a function of the CPU time used for sampling, for dimensions $d = 2$ (left figure) and 20 (right figure). Again, in both figures, the top curve represents the grid sampling and the bottom curve the PMC sampling.

and shows the input x_t to the learner. The learner selects a regression function $g_t \in \mathcal{G}$ and predicts $\hat{y}_t = g_t(x_t)$. Then the output y_t is revealed and the learner incurs the reward (or equivalently a loss) $l(\hat{y}_t, y_t) \in [0, 1]$.

Since the true output is revealed, it is possible to evaluate the reward of any $g \in \mathcal{G}$, which corresponds to the full information case.

Now, consider a parametric space $\mathcal{G} = \{g_\theta, \theta \in \Theta \subset \mathbb{R}^d\}$ of regression functions, and assume that the mapping $\theta \mapsto l(g_\theta(x), y)$ is Lipschitz w.r.t. θ with a uniform (over $x \in \mathcal{X}, y \in \mathcal{Y}$) Lipschitz constant $\lambda < \infty$. This happens for example when $\mathcal{X}$ and $\mathcal{Y}$ are compact domains, the regression $\theta \mapsto g_\theta$ is Lipschitz, and the loss function $(u, v) \mapsto l(u, v)$ is also Lipschitz w.r.t. its first variable (such as for e.g. L_1 or L_2 loss functions) on compact domains.

The online learning problem consists in selecting at each round t a parameter $\theta_t \in \Theta$ such as to optimize the accuracy of the prediction of y_t with $g_{\theta_t}(x_t)$. If we define $f_t(\theta) \stackrel{\text{def}}{=} l(g_\theta(x), y)$, then applying the ALF algorithm described previously (changing rewards into losses by using the transformation $u \mapsto 1 - u$), we obtain directly that the expected cumulative loss of the ALF algorithm is almost as small as that of the best regression function in $\mathcal{G}$, in the sense that:

$$\mathbb{E}\Big[\sum_{t=1}^{T} l_t\Big] - \inf_{g \in \mathcal{G}} \mathbb{E}\Big[\sum_{t=1}^{T} l(g(x_t), y_t)\Big] \leq 2\sqrt{dT \ln(d^\alpha \lambda T)},$$

where $l_t \stackrel{\text{def}}{=} l(g_{\theta_t}(x_t), y_t)$. To illustrate, consider a feedforward neural network (NN) [28] with parameter space Θ (the set of weights of the network) and one hidden layer. Let n and m be the number of input (respectively hidden) neurons. Thus if $x \in \mathcal{X} \subset \mathbb{R}^n$ is the input of the NN, a possible NN architecture would produce the output: $g_\theta(x) \stackrel{\text{def}}{=} \theta^o \cdot \sigma(x)$ with $\sigma(x) \in \mathbb{R}^m$ and $\sigma(x)_l \stackrel{\text{def}}{=} \sigma(\theta_l^i \cdot x)$ (where σ is the sigmoid function) is the output of the l-th hidden neuron. Here $\theta = (\theta^i, \theta^o) \in \Theta \subset \mathbb{R}^d$ the set of (input, output) weights (thus here $d = n \times m + m$).

The Lipschitz constant of the mapping $\theta \mapsto g_\theta(x)$ is upper bounded by $\sup_{x\in\mathcal{X},\theta\in\Theta}||x||_\infty||\theta||_\infty$, thus assuming that the domains $\mathcal{X}$, $\mathcal{Y}$, and Θ are compacts, the assumption that $\theta \mapsto l(g_\theta(x), y)$ is uniformly (over $\mathcal{X}, \mathcal{Y}$) Lipschitz w.r.t. θ holds e.g. for L_1 or L_2 loss functions, and the previous result applies.

Now, as discussed above about the practical aspects of the ALF algorithm, in this online regression problem, the knowledge of the past input-output pairs $(x_s, y_s)_{s<t}$ enables to compute the weight $w_t(\theta) = \exp(\eta \sum_{s=1}^{t-1} l(g_\theta(x_s), y_s))$ for any $\theta \in \Theta$, and thus enables to use a PMC algorithm to sample θ_t from the distribution p_t. Up to our knowledge, we believe this is a first regret bound analysis of online learning for non-linear NN regression, in an adversarial setting.

2.2 Online Classification

Now consider the problem of online classification (i.e. when the set of labels $\mathcal{Y}$ is finite). Here we can no longer make the assumption that the classifier's prediction $g_\theta(x) \in \mathcal{Y}$ is Lipschitz w.r.t. the parameter θ (and neither that the loss function $l(y, y') = \mathbb{I}_{\{y=y'\}}$ is Lipschitz w.r.t. its first variable). One way to circumvent this problem is to consider a class $\mathcal{G} = \{g_\theta, \theta \in \Theta\}$ of stochastic classifiers, so that $g_\theta(y|x)$ represents the probability of predicting label y given input x. The ALF algorithm would apply as follows: at round t, the algorithms chooses $\theta_t \in \Theta$ and samples the prediction $\hat{y}_t$ from the distribution $g_{\theta_t}(\cdot|x_t)$.

When the label y_t is revealed, the loss function $f_t(\theta) \stackrel{\text{def}}{=} g_\theta(y_t|x_t)$ for all classifiers g_θ may be computed. Thus assuming that the mapping $\theta \mapsto g_\theta(y|x)$ is Lipschitz w.r.t. θ with uniform (over $\mathcal{X}\times\mathcal{Y}$) Lipschitz constant λ, then Theorem 1 applies, and we have that

$$\underbrace{\sup_{g\in\mathcal{G}} \mathbb{E}\left\{\sum_{t=1}^{T} g(y_t|x_t)\right\}}_{\substack{\text{Exp. nb. of correct}\\ \text{predictions of best classifier}}} - \underbrace{\mathbb{E}\left\{\sum_{t=1}^{T} g_{\theta_t}(y_t|x_t)\right\}}_{\substack{\text{Exp. nb. of correct}\\ \text{predictions of ALF algo.}}} \leq 2\sqrt{dT\ln(cd^\alpha\lambda T)}$$

which says that the expected number of good predictions of the ALF algorithm is almost as good as that of the best classifier in $\mathcal{G}$. An example of such parametric regression setting is the case of neural networks (parameterized by θ) where the activation of the output neurons (one for each label y of $\mathcal{Y}$), up to some renormalization, define the probability distribution $g_\theta(y|x)$.

2.3 Online Classification with Bandit Information

In the previous section, the information revealed by the opponent enables to compute the reward (or loss) function $f_t(\theta)$ for all $\theta \in \Theta$. In the bandit information case considered now only the reward $f_t(\theta_t)$ of the selected action is revealed. Under our Lipschitz assumption on the functions, the knowledge of f_t at a point θ_t reveals very few information about f_t elsewhere. Thus we cannot expect to derive tight regret bounds in general. However we can obtain interesting bounds in the case when the reward function f_t may actually be coded by a

Initialization: Set $w_1(\theta) = 1$ for all $\theta \in \Theta$.

For each round $t = 1, 2, \dots, T$

(1) The adversary chooses $(x_t, y_t) \in \mathcal{X} \times \mathcal{Y}$ and shows x_t to the learner,

(2) The learner chooses $\theta_t \sim p_t$, where $p_t(\theta) \stackrel{\text{def}}{=} \frac{w_t(\theta)}{\int_\Theta w_t(\theta) d\theta}$, and predicts $\hat{y}_t \sim q_{t,\theta_t}$, where $q_{t,\theta}(y) \stackrel{\text{def}}{=} (1-\gamma) g_{\theta_t}(y|x_t) + \frac{\gamma}{K}$,

(3) The learner sees the (bandit) information $Z_t \stackrel{\text{def}}{=} \mathbb{I}_{\hat{y}_t = y_t}$, from which he defines $\tilde{f}_t(\theta) \stackrel{\text{def}}{=} \frac{g_\theta(\hat{y}_t|x_t)}{q_t(\hat{y}_t)} Z_t$, where $q_t(y) \stackrel{\text{def}}{=} \int_\Theta p_t(\theta) q_{t,\theta}(y) d\theta$, for any $y \in \mathcal{Y}$.

(4) The weight function w_t is updated according to $w_{t+1}(\theta) = w_t(\theta) e^{\eta \tilde{f}_t(\theta)}$, for all $\theta \in \Theta$.

Fig. 4. The Adversarial Lipschitz Bandit Classifier (ALBC algo)

finite amount of information. We illustrate this setting on the online classification problem described in Section 2.2 but with the difference that the true label $y_t \in \mathcal{Y} = \{1, \dots, K\}$ is not revealed at each round: the only available information is $Z_t \stackrel{\text{def}}{=} \mathbb{I}_{\{\hat{y}_t = y_t\}}$, i.e. whether the prediction $\hat{y}_t$ is correct or not. An example of applications is the problem of web advertisement systems, where the user's click is the only received feedback.

Again, we consider a parametric family of stochastic classifiers $\mathcal{G} = \{g_\theta, \theta \in \Theta\}$, where $g_\theta(y|x)$ corresponds to the probability of selecting $y \in \mathcal{Y}$ given the input x. Now, in each round, a classifier g_{θ_t} is selected (by sampling $\theta_t \sim p_t$) and a prediction $\hat{y}_t$ is made. However, in this bandit setting, the feedback information $Z_t = \mathbb{I}_{\{\hat{y}_t = y_t\}}$ does not enable to evaluate the performance $f_t(\theta) \stackrel{\text{def}}{=} g_\theta(y_t|x_t)$ of any classifiers g_θ, $\theta \in \Theta$. Instead, we randomize the prediction by considering a mixture distribution between g_{θ_t} and the uniform distribution: $\hat{y}_t \sim q_{t,\theta_t}$, where $q_{t,\theta}$ is the distribution over the labels $\mathcal{Y}$ defined by $q_{t,\theta}(y) \stackrel{\text{def}}{=} (1-\gamma) g_\theta(y|x_t) + \frac{\gamma}{K}$.

This idea is close to the Exp4 algorithm in [3]. Given the information Z_t, we build an estimate $\tilde{f}_t(\theta)$ of the performance $f_t(\theta)$ of any classifiers g_θ: $\tilde{f}_t(\theta) \stackrel{\text{def}}{=} \frac{g_\theta(\hat{y}_t|x_t)}{q_t(\hat{y}_t)} Z_t$, where $q_t(y) \stackrel{\text{def}}{=} \mathbb{E}_{\theta \sim p_t}[q_{t,\theta}(y)]$, for any $y \in \mathcal{Y}$. This estimate is unbiased since:

$$\mathbb{E}_{\theta_t, \hat{y}_t} \tilde{f}_t(\theta) = \int_\Theta p_t(\theta') \sum_{y \in \mathcal{Y}} \frac{q_{t,\theta'}(y) g_\theta(y|x_t)}{q_t(y)} \mathbb{I}_{\{y = y_t\}} d\theta'$$

$$= \int_\Theta p_t(\theta') \frac{q_{t,\theta'}(y) g_\theta(y_t|x_t)}{q_t(y_t)} d\theta' = g_\theta(y_t|x_t) = f_t(\theta)$$

Figure 4 describes this Adversarial Lipschitz Bandit Classifier (ALBC) algorithm. The next result assesses the expected performance of the ALBC algorithm $\sum_{t=1}^T \mathbb{I}_{\{\hat{y}_t = y_t\}}$ in comparison with the expected performance of the best classifier $g \in \mathcal{G}$, in terms of number of correct predictions. Define the regret:

$$R_T(\theta) \stackrel{\text{def}}{=} \sum_{t=1}^{T} g_\theta(y_t|x_t) - \mathbb{E}\Big[\sum_{t=1}^{T} \mathbb{I}_{\{\hat{y}_t = y_t\}}\Big].$$

The ALBC algorithm has a regret $\sup_{\theta\in\Theta} \mathbb{E}R_T(\theta) \leq 4\sqrt{KdT\ln(cd^\alpha \lambda T)}$ (the proof is omitted from this extended abstract but follows the same lines as the proof of ALF algorithm combined with EXP4 ideas). Notice that like in the multi-armed bandit problem, in this bandit setting, the regret suffers from an additional factor K per round (i.e. $\sqrt{T}$ is replaced by $\sqrt{KT}$ in the bound), compared to the full information case.

A practical algorithm. A practical implementation of the ALBC algorithm requires being able to sample θ_t from p_t. The key difference with the technique detailed in Section 1.3 is that in the ALBC algorithm, the functions $\tilde{f}_t(\theta)$ depend on $q_t(\hat{y}_t)$ which is not directly known. However a refined MCMC or PMC algorithm is possible: at round t, assume that we have kept in memory the information: $H_{<t} \stackrel{\text{def}}{=} \{(x_s, \hat{y}_s, Z_s, q_s(\hat{y}_s))_{s<t}\}$.

We now show that (1) this is possible, and (2) this is sufficient for sampling $\theta_t \sim p_t$. We prove (1) recursively by showing that from $H_{<t}$ we are able to calculate $q_t(\hat{y}_t)$ (the other pieces of information $x_t, \hat{y}_t$, and Z_t are available at the end of round t). Thus we only need to prove that from $H_{<t}$, we can sample $\theta_t \sim p_t$ and compute $q_t(\hat{y}_t)$. But since $q_t(\hat{y}_t)$ is the expectation of $q_{t,\theta}(\hat{y}_t)$ for $\theta \sim p_t$, we may consider a MCMC or PMC method where the same Markov chain (having p_t as stationnary distribution) or particle population serves for both sampling $\theta_t \sim p_t$ and estimating $q_t(\hat{y}_t)$. Finally, this is possible since the pointwise evaluation of w_t (thus of p_t up to a renormalization constant) only depends on information in $H_{<t}$.

3 Conclusion

We have considered the adversarial online learning framework in the case of Lipschitz functions. In the full information case, the bound shows the same rate $\sqrt{dT}$ as for linear functions. This enables to derive similar performance bounds for online regression and classification, thus extending previous results to non-linear parametric approximation, such as neural networks. Our main contribution was to consider a continuous extension of the EWF algorithm (ALF algorithm) for which we provide geometrical conditions for sound regret analysis, and discuss the use of different approximation schemes and especially the use of a PMC sampling method compared to non adaptive sampling methods. We provided experiments showing the benefit of using a PMC sampling method for minimizing regret under computational time constraint compared to naive random grid.

We applied this result to derive bounds for (full information) regression and classification online learning problems and (bandit information) K-classes classification problems where the revealed information is the correctness of the prediction. We derived a regret bound on the expected number of mistakes of order $\sqrt{dTK}$, and illustrate the case of a Neural Networks architecture.

Acknowledgment

This work has been supported by French National Research Agency (ANR) through COSINUS program (project EXPLO-RA number ANR-08-COSI-004).

References

1. Cesa-Bianchi, N., Lugosi, G.: Prediction, Learning, and Games. Cambridge University Press, New York (2006)
2. Auer, P., Cesa-bianchi, N., Freund, Y., Schapire, R.E.: Gambling in a rigged casino: The adversarial multi-armed bandit problem. In: Proceedings of the 36th Annual Symposium on Foundations of Computer Science, pp. 322–331. IEEE Computer Society Press, Los Alamitos (1995)
3. Auer, P., Cesa-bianchi, N., Freund, Y., Schapire, R.E.: The non-stochastic multi-armed bandit problem. SIAM Journal on Computing 32 (2002)
4. Poland, J.: Nonstochastic bandits: Countable decision set, unbounded costs and reactive environments. Theor. Comput. Sci. 397(1-3), 77–93 (2008)
5. Dani, V., Hayes, T., Kakade, S.: The price of bandit information for online optimization. In: Platt, J., Koller, D., Singer, Y., Roweis, S. (eds.) Advances in Neural Information Processing Systems 20, pp. 345–352. MIT Press, Cambridge (2008)
6. Abernethy, J., Hazan, E., Rakhlin, A.: Competing in the dark: An efficient algorithm for bandit linear optimization. In: Servedio, R.A., Zhang, T. (eds.) Conference on Learning Theory, pp. 263–274. Omnipress (2008)
7. Cesa-Bianchi, N., Lugosi, G.: Combinatorial bandits. In: Conference on Learning Theory (2009)
8. Kakade, S.M., Shalev-Shwartz, S., Tewari, A.: Efficient bandit algorithms for online multiclass prediction. In: Proceedings of the 25th International Conference on Machine learning, pp. 440–447. ACM, New York (2008)
9. Auer, P.: Using confidence bounds for exploitation-exploration trade-offs. Journal of Machine Learning Research, 397–422 (2002)
10. Dani, V., Hayes, T.P., Kakade, S.M.: Stochastic linear optimization under bandit feedback (2008) (in submission)
11. Zinkevich, M.: Online convex programming and generalized infinitesimal gradient ascent. In: International Conference on Machine learning, pp. 928–936 (2003)
12. Hazan, E., Agarwal, A., Kale, S.: Logarithmic regret algorithms for online convex optimization. In: Conference on Learning Theory, pp. 499–513 (2006)
13. Bartlett, P., Hazan, E., Rakhlin, A.: Adaptive online gradient descent. In: Platt, J.C., Koller, D., Singer, Y., Roweis, S.T. (eds.) Advances in Neural Information Processing Systems. MIT Press, Cambridge (2007)
14. Shalev-Shwartz, S.: Online Learning: Theory, Algorithms, and Applications. PhD thesis (July 2007)
15. Flaxman, A.D., Kalai, A.T., McMahan, H.B.: Online convex optimization in the bandit setting: gradient descent without a gradient. In: Proceedings of the sixteenth annual ACM-SIAM Symposium on Discrete algorithms, pp. 385–394. SIAM, Philadelphia (2005)

16. Abernethy, J.D., Bartlett, P., Rakhlin, A., Tewari, A.: Optimal strategies and minimax lower bounds for online convex games. Technical Report UCB/EECS-2008-19, EECS Department, University of California, Berkeley (February 2008)
17. Kleinberg, R., Slivkins, A., Upfal, E.: Multi-armed bandit problems in metric spaces. In: Proceedings of the 40th ACM Symposium on Theory of Computing, pp. 681–690 (2008)
18. Bubeck, S., Munos, R., Stoltz, G., Szepesvári, C.: Online optimization of X-armed bandits. In: Advances in Neural Information Processing Systems (2008)
19. Littlestone, N., Warmuth, M.: The weighted majority algorithm. Information and Computation 108, 212–261 (1994)
20. Cesa-Bianchi, N., Freund, Y., Haussler, D., Helmbold, D.P., Shapire, R., Warmuth, M.: How to use expert advice. Journal of the ACM 44(3), 427–485 (1997)
21. Auer, P., Cesa-bianchi, N., Gentile, C.: Adaptive and self-confident on-line learning algorithms. Journal of Computer and System Sciences 64 (2000)
22. Stoltz, G.: Incomplete information and internal regret in prediction of individual sequences. PhD thesis (2005)
23. Gilks, W., Richardson, S., Spiegelhalter, D.: Markov Chain Monte Carlo in Practice. Chapman Hall/CRC, Boca Raton (1996)
24. Andrieu, C., De Freitas, N., Doucet, A., Jordan, M.: An introduction to mcmc for machine learning. Journal of Machine Learning Research 50, 5–43 (2003)
25. Levin, D.A., Peres, Y., Wilmer, E.L.: Markov Chains and Mixing Times. American Mathematical Society, Providence (2008)
26. Douc, R., Guillin, A., Marin, J., Robert, C.: Minimum variance importance sampling via population monte carlo. Esaim P&S 11 (2007)
27. Del Moral, P.: Feynman-Kac formulae: genealogical and interacting particle systems with applications/Pierre Del Moral, p. 555. Springer, Heidelberg (2004)
28. Bishop, C.M.: Pattern Recognition and Machine Learning (Information Science and Statistics). Springer, Heidelberg (2006)
29. Devroye, L., Györfi, L., Lugosi, G.: A Probabilistic Theory of Pattern Recognition. Springer, New York (1996)

A Proof of Theorem 1 (ALF algorithm)

We start by following the usual proof for exponentially weighted forecasting. Define $W_t \stackrel{\text{def}}{=} \int_\Theta w_t$. For any $t \in \{1, \ldots, T\}$, we have:

$$\frac{W_{t+1}}{W_t} = \frac{\int_\Theta \exp(\eta F_t)}{\int_\Theta \exp(\eta F_{t-1})} = \int_\Theta p_t(\theta)\exp(\eta f_t(\theta)).$$

Since $\exp(u) \le 1 + u + u^2$ for $u \le 1$, then, whenever $\eta \le 1$, we have $\frac{W_{t+1}}{W_t} \le 1 + \eta \int_\Theta p_t f_t + \eta^2 \int_\Theta p_t f_t^2$. Moreover, since $W_1 = \mu(\Theta)$, we get:

$$\ln(W_{T+1}) \le \eta \sum_{t=1}^{T} \int_\Theta p_t f_t + T\eta^2 + \ln(\mu(\Theta)). \tag{4}$$

Let us write $h(\theta) \stackrel{\text{def}}{=} \exp(\eta F_T(\theta))$, and $h^* \stackrel{\text{def}}{=} \max_{x\in\Theta} h(\theta)$. We have that

$$\begin{aligned} |h(\theta_1) - h(\theta_2)| &\le \eta |F_T(\theta_1) - F_T(\theta_2)| h^* \\ &\le \eta \lambda T h^* ||\theta_1 - \theta_2||, \end{aligned} \tag{5}$$

since the function F_T is λT-Lipschitz. Let θ^* be any point of maximum of h, and define $\pi(\theta) \stackrel{\text{def}}{=} \max(0, 1 - \eta\lambda T||\theta - \theta^*||)$. Then for all $\theta \in \Theta$,

$$h(\theta) \geq h^* \pi(\theta). \tag{6}$$

Indeed, this holds for any $\theta \notin B(\theta^*, 1/(\eta\lambda T))$ where $B(\theta, r)$ is the ball $\{x', ||x - x'|| \leq r\}$, since in that case, $\pi(\theta) = 0$. Now if there were some $\theta \in B(\theta^*, 1/(\eta\lambda T))$ such that $h(\theta) < h^*\pi(\theta)$, then we would have: $h(\theta^*) - h(\theta) > \eta\lambda T h^* ||x - x^*||$, which would contradict the Lipschitz property (5) of h.

Notice that π is a pyramid function with base $B(\theta^*, 1/(\eta\lambda T))$ and height 1. We now state a Lemma that will enable us to derive a lower bound on $\int_\Theta \pi$.

Lemma 1. *For any $\theta^* \in \Theta$, $r > 0$, let π be the function defined by $\pi(\theta) \stackrel{\text{def}}{=} \max(0, 1 - ||x - x^*||/r)$. Then:*

$$\int_\Theta \pi \geq \frac{1}{(d+1)\kappa(d)} \min\left[\mu(B(\theta^*, r)), \mu(\Theta)\right]$$

Proof.

$$\begin{aligned}
\int_\Theta \pi &= \int_{\mathbb{R}^D} \mathbb{I}_{\theta \in \Theta \cap B(\theta^*, r)} (1 - \frac{||\theta^* - \theta||}{r}) \mu(d\theta) \\
&= \int_{\mathbb{R}^D} \mathbb{I}_{\theta \subset \Theta \cap B(\theta^*, r)} \int_0^1 \mathbb{I}_{||\theta^* - \theta|| \leq \alpha r} d\alpha \mu(d\theta) \\
&= \int_0^1 \int_{\mathbb{R}^D} \mathbb{I}_{\theta \in \Theta \cap B(\theta^*, \alpha r)} \mu(d\theta) d\alpha \\
&= \int_0^1 \mu(\Theta \cap B(\theta^*, \alpha r)) d\alpha
\end{aligned}$$

Now, using the definition of $\kappa(d)$ from (1),

$$\int_\Theta \pi \geq \int_0^1 \frac{1}{\kappa(d)} \min[\alpha^d \mu(B(\theta^*, r)), \mu(\Theta)] d\alpha$$

We deduce that if $\mu(\Theta) \geq \mu(B(\theta^*, r))$ then $\int_\Theta \pi \geq \frac{\mu(B(\theta^*, r))}{(d+1)\kappa(d)}$. And otherwise, $\exists \alpha_0 < 1$ such that $\mu(\Theta) = \alpha_0^d \mu(B(\theta^*, r))$ thus we have $\int_\Theta \pi \geq \frac{\mu(\Theta)}{\kappa(d)}(1 - \alpha_0 + \frac{\alpha_0}{d+1}) \geq \frac{\mu(\Theta)}{(d+1)\kappa(d)}$ and the Lemma is proved.

We apply this Lemma with the π function and $r = 1/\eta\lambda T$ to obtain:

$$\int_\Theta \pi \geq \frac{1}{(d+1)\kappa(d)} \min\left[\mu(B(\theta^*, \frac{1}{\eta\lambda T})), \mu(\Theta)\right]$$

Now using (6) together with the previous bound combined with Assumption A1 (i.e. $\kappa(d) \leq \kappa^d$ and $\mu(B(\theta^*, r)) \geq (r/(\kappa' d^\alpha)^d)$, we derive the lower bound:

$$\int_\Theta h \geq h^* \min\left[\frac{1}{(cd^\alpha \eta\lambda T)^d}, \frac{\mu(\Theta)}{c^d}\right].$$

where we set $c = 2\kappa \max(\kappa', 1)$.

From its definition, $W_{T+1} = \int_\Theta h$, thus

$$\ln(W_{T+1}) \geq \eta \max_{\theta\in\Theta} F_T(\theta) - \ln\left(\max\left[(cd^\alpha\eta\lambda T)^d, \frac{c^d}{\mu(\Theta)}\right]\right),$$

which, together with (4) yields:

$$\sup_{\theta\in\Theta} F_T(\theta) - \sum_{t=1}^{T}\int_\Theta p_t f_t \leq T\eta + \frac{1}{\eta}\max\left[d\ln(cd^\alpha\eta\lambda T) + \ln(\mu(\Theta)), d\ln c\right].$$

Since $\int_\Theta p_t f_t = \mathbb{E}_t[f_t(\theta_t)]$, where $\mathbb{E}_t$ denotes the expectation w.r.t. the choice of $\theta_t \sim p_t$, we deduce that the expected regret (w.r.t. the internal randomization of the learner) of any $\theta \in \Theta$ is bounded according to:

$$\mathbb{E}R_T(\theta) \leq T\eta + \frac{1}{\eta}(d\ln(cd^\alpha\eta\lambda T) + \ln(\mu(\Theta))),$$

whenever $d\ln(d^\alpha\eta\lambda T) \geq -\ln(\mu(\Theta))$.

Now, for the high probability result, if we introduce $Y_t = \int_\Theta p_t f_t - f_t(\theta_t)$ and $\mathcal{F}_{<t}$ the σ-algebra generated by the past random decisions, then $\mathbb{E}[Y_t|\mathcal{F}_{<t}] = 0$, thus $Y_1, ..., Y_T$ is a martingale difference sequence, and since $f_t \in [0,1]$, $|Y_t| \leq 1$ a.s., using Hoeffding-Azuma's inequality (see e.g. [29]), we obtain that with probability at least $1-\beta$,

$$\sum_{t=1}^{T}\int_\Theta p_t f_t \leq F_T + \sqrt{2T\ln(\beta^{-1})},$$

which enables to deduce (3).

Summarising Data by Clustering Items

Michael Mampaey and Jilles Vreeken

Department of Mathematics and Computer Science
Universiteit Antwerpen
{michael.mampaey,jilles.vreeken}@ua.ac.be

Abstract. For a book, the title and abstract provide a good first impression of what to expect from it. For a database, getting a first impression is not so straightforward. While low-order statistics only provide limited insight, mining the data quickly provides too much detail. In this paper we propose a middle ground, and introduce a parameter-free method for constructing high-quality summaries for binary data. Our method builds a summary by grouping items that strongly correlate, and uses the Minimum Description Length principle to identify the best grouping —without requiring a distance measure between items.

Besides offering a practical overview of which attributes interact most strongly, these summaries are also easily-queried surrogates for the data. Experiments show that our method discovers high-quality results: correlated attributes are correctly grouped and the supports of frequent itemsets are closely approximated.

1 Introduction

When handling a book, and wondering about its contents, we can simply start reading it from A to Z. In practice, however, to get a good first impression we usually first refer to the summary. For a book, this can be anything from the title, the abstract, up to simply paging through it. The common denominator here is that a summary quickly provides high-quality and high-level information about the book. A summary may already contain exactly what we were looking for, but in general we expect to get enough insight to judge what the book contains and whether we need to read it further.

When handling a transaction database, and wondering about its content and whether (or how) we should analyse it, it is quite hard to get a good first impression. Of course, one can inspect the schema of the database, and the attribute labels will also convey some information. However, these do not provide an overview of what is *in* the database. To this end, basic statistics can help to a limited extent, e.g. first order statistics tell us which items occur often, and which do not. For binary transaction databases, however, further basic statistics are not readily available. Ironically, this means that while the goal is to get a first impression, we have to analyse the data in detail. For non-trivially sized databases especially, this means investing far more time and effort than we should at this stage of the analysis.

J.L. Balcázar et al. (Eds.): ECML PKDD 2010, Part II, LNAI 6322, pp. 321–336, 2010.

When analysing data, a good first impression of the data is important, as mining data is essentially an iterative process [9], where each step provides extra insight, which allows us to extract increasingly more knowledge. A good summary allows us to make a well-informed decision on what basic assumptions to make and how to mine the data.

Here, we propose a simple and parameter-free method for providing high-quality summary overviews for binary transaction data. The outcome provides insight into which attributes are most correlated and in what value-configurations these occur. They are probabilistic models of the data that can be queried fast and accurately, allowing them to be used instead of the data. Further, by showing which attributes interact most strongly, these summaries can provide insight for selecting and constructing features. In short, like a proper summary, they provide a good first impression and can be used as a surrogate.

To the best of our knowledge, there currently do not exist light-weight data analysis methods that can be easily used for summary purposes. Instead, for binary data the standard approach is to mine for frequent itemsets first, the result of which quickly grows up to many times the size of the original data. Resulting, many proposals exist that focus on summarising sets of frequent patterns. That is, to choose groups of representative itemsets such that the information in the complete pattern set is maintained as well as possible. Here, we do not summarise the outcome of an analysis, i.e. a set of patterns, but instead provide a summary which can be used to decide how to further analyse the data.

Existing proposals for data summarisation, such as KRIMP [17] and Summarization [3], provide highly detailed results. Although this has obvious merit, analysing these summaries consequently also requires significant effort. Our method shares the approach of using compression to find a good summary. However, we do not aim at finding a group of descriptive itemsets. Instead, we view the data symmetrically with regard to 0s and 1s and aim to optimally group those items that interact most strongly. In this regard, our approach is also related to selecting low-entropy sets [10], itemsets that identify strong interactions in the data. An existing proposal to this end, LESS [11], requires a collection of low-entropy sets as input, and the resulting model cannot easily be queried. For a more complete discussion of related work, please refer to Section 5.

The method we propose in this paper groups attributes that interact strongly, i.e. that have low entropy. We identify the best grouping through the Minimum Description Length principle; no parameter needs to be set by the user. No distance measure between attributes is required, and the similarity between clusters is easily calculated. Experiments show that our method discovers high-quality results: correlated attributes are correctly grouped, representative features are identified, and the supports of itemsets are closely approximated.

The roadmap of this paper is as follows. First, we introduce notation and formalise the problem. In Section 3 we introduce our method for finding good summarisations and how information can be extracted from these in Section 4. Related work is discussed in Section 5. We experimentally evaluate our method in Section 6. We round up with a discussion in Section 7 and conclude in Section 8.

2 MDL for Attribute Clustering

In this section we formally introduce our method. We start by covering the preliminaries and notation, then define what an attribute clustering is, and how to use MDL to identify good clusterings.

2.1 Preliminaries

We denote the set of all items by $\mathcal{I} = \{I_1, \ldots, I_n\}$. A dataset $\mathcal{D}$ is a bag of transactions t. A transaction is a binary vector of length n. An *item* is a binary attribute, that is, a pair $(I = v)$, where $I \in \mathcal{I}$ and $v \in \{0,1\}$. Then, an *itemset* is simply a pair $(X = v)$, where $X \subseteq \mathcal{I}$ is a set of items, and $v \in \{0,1\}^{|X|}$ is a binary vector of length $|X|$. Sometimes we will also refer to a set of attributes as an itemset. A transaction t is said to contain an itemset $X = v$, denoted as $X \subset t$, if for all items $x_i \in X$ it holds that $t_i = v_i$. The *support* of $X = v$ is the number of transactions in $\mathcal{D}$ that contain $X = v$, i.e. $supp(X) = |\{t \in \mathcal{D} \mid X \subset t\}|$. The *frequency* of $X = v$ is defined as its support, divided by the size of $\mathcal{D}$, i.e. $freq(X = v) = supp(X = v)/|\mathcal{D}|$. The entropy of an itemset X over $\mathcal{D}$ is defined as $H(X) = -\sum_v freq(X = v) \log freq(X = v)$, where the logarithm is to base 2.

2.2 Definitions

The summaries we use are based on attribute clusterings. Therefore, we first formally introduce the concept of an attribute clustering.

Definition 1. *An attribute clustering $\mathcal{A} = \{A_1, \ldots, A_k\}$ of a set of items $\mathcal{I}$ is a partition of $\mathcal{I}$, where*

1. *each cluster is not empty:* $\forall A_i \in \mathcal{A} : A_i \neq \emptyset$,
2. *all clusters are pairwise disjoint:* $\forall i \neq j : A_i \cap A_j = \emptyset$,
3. *every item belongs to a cluster:* $\bigcup_i A_i = \mathcal{I}$.

Next, we must define what the *best* attribute clustering is. For this, we use the *Minimum Description Length* principle (MDL). This principle [7] can be roughly described as follows. Given a set of models $\mathcal{M}$ for $\mathcal{D}$, the best model $M \in \mathcal{M}$ is the one that minimises

$$L(\mathcal{D} \mid M) + L(M) ,$$

where $L(M)$ is the length, in bits, of the description of the model, and $L(\mathcal{D} \mid M)$ is the length of the description of the data when encoded by the model. To use MDL, we need to define how to encode a model, and how it describes the data.

First, let us determine how to describe the attribute clustering. To begin with, we must state how many items there are, and then, for each item we describe to which cluster it belongs. In this description there is some redundancy, since any permutation of the cluster labels yields an equivalent partition. Taking this into account, the description of the partition requires $\log n + n \log k - \log k!$ bits.

Secondly, we use *code tables* to describe the distribution of each cluster. Let $A_i \in \mathcal{A}$ be an attribute cluster, then the code table CT_i for A_i describes which

CT_i				
a b c	$code(v)$		$freq(v)$	$L(code(v))$
1 1 1	0		50%	1
1 1 0	10		25%	2
1 0 1	110		12.5%	3
0 1 0	1110		6.25%	4
0 0 0	1111		6.25%	4

Fig. 1. Example of a code table CT_i for the cluster $A_i = \{a, b, c\}$. The frequencies are not actually part of the code table, they are merely included as illustration. Moreover, the specific codes are examples—in our computations we are not interested in materialised codes, only in their lengths, $L(code(v))$

itemset values occur in the data with respect to this cluster, together with their codes. That is, a code table for a cluster A_i is a two-column table with value-assignments $v \in \{0,1\}^{|A_i|}$ for A_i on the left-hand side, and the corresponding codes on right-hand side. Figure 1 shows an example of a code table.

The values v can best be described as strings of bits, so their length is simply $|A_i|$. A well-known result from information theory [4] states that the code lengths for $A_i = v$ are optimal when $L(code(v)) = -\log freq(v)$. Note that we are not interested in actual materialised codes (e.g. computed through Huffman coding [4]), but only in their lengths. If a certain v has a frequency of 0, that is, it does not occur in the data, then we do not record it in the code table. Hence, the description length of the code table of a cluster A_i can be computed as $L(CT_i) = \sum_{v;freq(v)\neq 0} |A_i| - \log freq(v)$.

Finally, we need to define how to compute $L(\mathcal{D} \mid \mathcal{A})$, the length of the encoded description of database $\mathcal{D}$ given the clustering $\mathcal{A}$. Each transaction $t \in \mathcal{D}$ is partitioned according to $\mathcal{A}$, and encoded using the optimal codes found in the code tables. Since an itemset $X = v$ is used $|\mathcal{D}| \cdot freq(X = v)$ times, the total encoded size of $\mathcal{D}$ with respect to a single cluster A_i can be written as $L(\mathcal{D}_{A_i} \mid \mathcal{A}) = -\sum_t \log freq(t) = -|\mathcal{D}| \cdot \sum_v freq(v) \log freq(v) = |\mathcal{D}| \cdot H(A_i)$.

Putting this all together, we define the total encoded size $L(\mathcal{A}, \mathcal{D})$ as follows.

Definition 2. *The total description length of a clustering $\mathcal{A} = \{A_i\}_{i=1}^k$ of size k for a dataset $\mathcal{D}$ is:*

$$L(\mathcal{A}, \mathcal{D}) = L(\mathcal{D} \mid \mathcal{A}) + L(\mathcal{A}),$$

where

$$\begin{cases} L(\mathcal{D} \mid \mathcal{A}) & = |\mathcal{D}| \cdot \sum_{i=1}^k H(A_i) \\ L(\mathcal{A}) & = \log n + n \log k - \log k! + \sum_{i=1}^k L(CT_i) \\ L(CT_i) & = \sum_{v;freq(v)\neq 0} |A_i| + L(code(v)) \\ L(code(v)) & = -\log freq(v) \end{cases}$$

2.3 Problem Definition

Our goal is to discover an optimal summary of a transaction database. A summary consists of a partitioning of the attributes of a binary transaction database;

it must be optimal in the sense that the attribute groups should be relatively independent, while the individual clusters should exhibit strong correlations on the data, as clusters with a lot of structure can be described succinctly.

Formally, the problem we address is as follows. Given a transaction database $\mathcal{D}$ over a set of binary attributes $\mathcal{I}$, find the attribute clustering $\mathcal{A}$ that minimises

$$L(\mathcal{A}, \mathcal{D}) = L(\mathcal{D} \mid \mathcal{A}) + L(\mathcal{A}) .$$

With this problem statement we let MDL decide what the optimal number of attribute clusters is, by choosing the clustering that minimises the number of bits required to describe the model and the data. This also ensures that two unrelated groups of attributes will not be combined into one, as it will be far cheaper to describe the two groups separately.

The search space we have to consider for our problem is rather large. The total number of possible partitions of a set of n elements is known as the Bell number B_n, which is at least $\Omega(2^n)$. Therefore we cannot simply enumerate and test all possible partitions for a non-trivial dataset. We must traverse the search space and exploit its structure to arrive at a good clustering. The *refinement* of partitions naturally structures the search space into a lattice. A partition $\mathcal{A}$ refines a partition $\mathcal{A}'$ if for all $A \in \mathcal{A}$ there exists $A' \in \mathcal{A}'$ such that $A \subseteq A'$. The minimal clustering with respect to refinement contains a cluster for each individual item. We will call this the *independence clustering*. The transitive reduction of the refinement relation corresponds to merging two clusters, and this is how we will traverse the search space. Note that $L(\mathcal{A}, \mathcal{D})$ is not (anti-) monotonic with respect to refinement, otherwise the best clustering would simply be $\{\mathcal{I}\}$ or $\{\{I\} \mid I \in \mathcal{I}\}$, respectively. Further, this means there is no structure we can exploit to efficiently find the optimal clustering.

2.4 Measuring Cluster Similarity

Instead of requiring the user to specify a distance metric for individual items, we can derive a similarity measure between clusters from our definition of $L(\mathcal{A}, \mathcal{D})$.

Let $\mathcal{A}$ be an attribute clustering and let $\mathcal{A}'$ be the result of merging the clusters A_i and A_j in $\mathcal{A}$. In other words, $\mathcal{A}$ is a refinement of $\mathcal{A}'$. Then the difference of description lengths defines a similarity measure between A_i and A_j.

Definition 3. *The similarity of two clusters A_i and A_j in $\mathcal{A}$ is defined as*

$$CS_{\mathcal{D}}(A_i, A_j) = L(\mathcal{A}, \mathcal{D}) - L(\mathcal{A}', \mathcal{D}),$$

where $\mathcal{A}' = \mathcal{A} \setminus \{A_i, A_j\} \cup \{A_i \cup A_j\}$.

If A_i and A_j are very similar, i.e. have a low joint entropy, then this merger improves the total description length, meaning $CS_{\mathcal{D}}(A_i, A_j)$ will be positive, otherwise it is negative. Note that this similarity is local, in that it is not influenced by the other clusters in $\mathcal{A}$. This is further supported by the following lemma, which allows us to compute cluster similarity without having to compute the entire cluster description length. For the sake of exposition, we here ignore the cluster description term $\log n + n \log k - \log k!$, which is not a dominating term for the total description length.

Lemma 1. *Let $\mathcal{A}$ be an attribute clustering of $\mathcal{I}$, with $A_i, A_j \in \mathcal{A}$. Then*

$$CS_{\mathcal{D}}(A_i, A_j) = |\mathcal{D}| \cdot I(A_i, A_j) + \Delta L(CT),$$

where $I(A_i, A_j) = H(A_i) + H(A_j) - H(A_i A_j)$ is the mutual information between A_i and A_j, and $\Delta L(CT) = L(CT_i) + L(CT_j) - L(CT_{ij})$.

Lemma 1 shows that we can decompose cluster similarity into a mutual information term, and a term expressing the difference in code table size. Both of these are high when the attributes in A_i and A_j are highly correlated.

3 Mining Attribute Clusterings

As detailed above, the search space we have to consider is extremely large, and there exists no structure we can exploit to find the optimum. Hence, we have to settle for heuristics. In this section we introduce our algorithm, which finds a good attribute clustering $\mathcal{A}$ with a low description length $L(\mathcal{A}, \mathcal{D})$.

Since we do not employ a distance metric between the attributes, the problem is not as easy as simply applying an existing clustering algorithm such as k-means [13]. Instead, we use a greedy bottom-up clustering algorithm, which iteratively merges clusters by selecting those two clusters whose union has the shortest description. This results in a hierarchy of clusters, which can be represented visually as a dendrogram, as shown on the left hand side of Figure 2. At the bottom we have the independence distribution and at the top the joint empirical distribution of the data. An advantage of this approach is that we can so visualise how the clusters were formed.

The pseudo-code is given in Algorithm 1. We start by placing each item in its own cluster (line 1), which corresponds to the independence model. Then, we iteratively find the two clusters with the highest similarity (4), an merge them (5). In other words, in each iteration the algorithm tries to reduce the total description length as much as possible. If a merge reduces the lowest description length seen yet, we remember it (6-7), and finally return the best clustering (10).

The graph on the right hand side of Figure 2 shows how the description length behaves during the course of the algorithm on the *Pen Digits* dataset. Starting at $k = n$, the description length $L(\mathcal{A}, \mathcal{D})$ gradually decreases as similar clusters are being merged. This indicates that there is some definite structure present in the data. It continues to decrease until $k = 5$, which yields the best clustering found for this dataset. After this, the description length of the code tables increases dramatically, which implies that no more structure is present.

3.1 Convexity

Figure 2 seems to suggest that the description length evolves convexly with respect to k. That is, there is a single local minimum, and once $L(\mathcal{A}, \mathcal{D})$ starts to

Algorithm 1. ATTRIBUTECLUSTERING

Input: A transactional dataset $\mathcal{D}$ over a set of items $\mathcal{I}$.
Output: A clustering of the items $\mathcal{A} = \cup_{i=1}^{k} A_i$.

$\mathcal{A} \leftarrow \{\{I\} \mid I \in \mathcal{I}\}$
$\mathcal{A}_{min} \leftarrow \mathcal{A}$
while $|\mathcal{A}| > 1$ **do**
  $A_i, A_j \leftarrow \text{argmax}_{i,j} CS_{\mathcal{D}}(A_i, A_j)$
  $\mathcal{A} \leftarrow \mathcal{A} \setminus \{A_i, A_j\} \cup \{A_i \cup A_j\}$
  if $L(\mathcal{A}, \mathcal{D}) < L(\mathcal{A}_{min}, \mathcal{D})$ **then**
    $\mathcal{A}_{min} \leftarrow \mathcal{A}$
  end if
end while
return $\mathcal{A}_{min}$

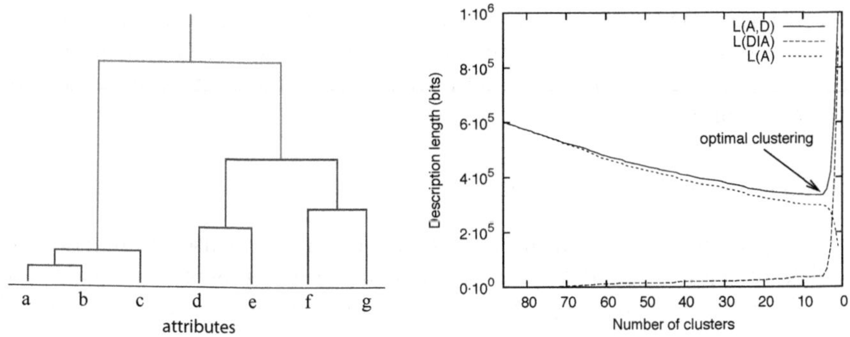

Fig. 2. (left) Example dendrogram. Merges that save bits are depicted in green (dark grey), merges that cost bits are red (light grey). Here the optimal $k = 2$. (right) Evolution of the encoded length $L(\mathcal{A}, \mathcal{D})$ with respect to the number of clusters k, on the *Pen Digits* dataset. The optimum is at $k = 5$.

increase, there are no more steps in which the description length decreases. Naturally, the question arises whether this is the case in general; if so, the algorithm can terminate as soon as a local minimum is detected. Intuitively, we would expect that if the currently best cluster merge increases the total description length, then all other merges are even worse, and we expect the same from all future merges. However, the following example shows that this is not the case.

Consider a dataset $\mathcal{D}$ with $\mathcal{I} = \{a, b, c, d\}$. Let us assume that for the transactions of $\mathcal{D}$ it holds that $d = a \oplus b \oplus c$, where $\oplus$ denotes exclusive or. Now, using this dependency, let $\mathcal{D}$ contain a transaction for every $v \in \{0, 1\}^3$ as values for abc. Then, every pair of clusters whose union contains up to three items (e.g. $A_i = ab$ and $A_j = d$) is independent. It is clear that as the algorithm starts to merge clusters, the entropy remains constant, and the code tables become more complex and thus $L(\mathcal{A}, \mathcal{D})$ increases. Only at the last step, when the two last clusters are merged, the dependency is recognised and the entropy drops, leading to a decrease of the total encoded size. Hence, the total description length $L(\mathcal{A}, \mathcal{D})$ is non-convex with respect to to cluster merges.

Note, however, that the gain in encoded length depends on the number of transactions in the database. In the above example, if every unique transaction occurs 20 times, the complete clustering would be preferred over the independence model. However, if there are fewer transactions (say, every transaction occurs four times), then while the dependencies are the same, the algorithm decides that the best clustering corresponds to the independence model (i.e. there is no significant structure). Intuitively this can be explained by the fact that if there are only few samples then the observed dependencies might be coincidental, but if many transactions follow it, the dependencies are truly present. This is one of the nice properties we get from using MDL to identify the best model.

While this example shows that in general we should not stop the algorithm at a local minimum, it is a very synthetic example with a strong requirement on the number of transactions. For instance, if we generalise the XOR example to 20 attributes, the minimum number of transactions for it to be detectable is already larger than 20 million. Furthermore, in none of our experiments with real data did we encounter a local minimum which was not also a global minimum. Therefore, we can say that in practice it is acceptable to stop the algorithm at a local minimum.

3.2 Algorithmic Complexity

Naturally, a summarisation method should be fast, because of our goal to get a quick overview of the data. For complex data mining algorithms on the other hand, it is often found acceptable that they are exponential. Here we show that our algorithm is polynomial in the number of attributes.

In the first iteration, we compute the description length for each singleton cluster $\{I\}$, and then determine which clusters to merge. To do this, we must compute $O(n^2)$ cluster similarities, where $n = |\mathcal{I}|$. Since we might need some of the similarities later on, we store them in a heap, such that we can easily retrieve the maximum. Now say that in a subsequent iteration k we have just merged A_i and A_j into A_{ij}. Then we delete $2k - 1$ similarities from the heap, and compute and insert $k - 1$ new similarities, i.e. between A_{ij} and the remaining clusters. Since heap insertion and deletion is logarithmic, maintaining the similarities in one iteration takes $O(k \log k)$ time. The computation of the similarities $CS_{\mathcal{D}}(A_i, A_j)$ requires collecting all nonzero frequencies $\mathit{freq}(A_{ij} = v)$, and we do this by simply iterating over all transactions t and computing $A_{ij} \cap t$, which takes $O(n|\mathcal{D}|)$ time. In total, the time complexity of our algorithm is $O(n^2 \log n \times n|\mathcal{D}|)$.

The biggest cost in terms of storage are the cluster similarities, and hence the memory complexity is $O(n^2)$.

4 Querying a Summary

Besides providing a general overview of which attributes interact most strongly, and in which value-assignments they typically occur, our summaries can also be used as surrogates for the data. That is, we can query a summary. For binary data, a query comes down to calculating marginals: counting how often a particular value-assignment occurs, in other words, determining supports.

The frequency of an itemset (or conjunctive query) can be estimated from an attribute clustering, by assuming that the clusters are independent. By MDL, we know this is a safe assumption: if two clusters A_i and A_j were dependent, it would have been far cheaper to combine them into a single cluster A_{ij}.

Let $\mathcal{A} = \{A_i\}_{i=1}^k$ be a clustering of $\mathcal{I}$, and let $X \subset \mathcal{I}$ be an itemset. Then the frequency of X can be estimated as

$$\hat{freq}(X) = \prod_{i=1}^{k} freq(X \cap A_i)$$

As an example, let $\mathcal{I} = \{a, b, c, d, e, f\}$ and $\mathcal{A} = \{abc, de, f\}$, and let $X = \{a, b, e, f\}$, then $\hat{freq}(X) = freq(ab) \cdot freq(e) \cdot freq(f)$.

As each CT_i implicitly contains the frequencies of the value-assignments for A_i, we can use our clustering models as very efficient surrogates for $\mathcal{D}$.

5 Related Work

The main goal of this proposal is to offer a good first impression of the data. For numerical data, averages and correlations can easily be computed, and more importantly, are informative. For binary transaction data such informative statistics are not readily available. As such, our work can be seen as to provide an informative 'average' for binary data; for those attributes that interact strongly, it shows how often the value-assignments occur.

Most existing techniques for summarisation are aimed at giving a succinct representation of a given collection of itemsets. Well-known examples include closed itemsets [15] and non-derivable itemsets [2], which both provide a lossless reduction of the complete collection. A lossy approach that provides a succinct summary of the patterns was proposed by Yan et al. [20]. Experiments show our method provides better frequency estimates, while requiring fewer 'profiles'. Wang and Karypis gave a method [19] for directly mining a summary of the frequent pattern collection for a given *minsup* threshold. Please refer to [8] for a more complete overview of pattern mining and summarisation techniques.

For summarising data fewer proposals exist. Chandola and Kumar [3] propose to induce k transaction templates such that the database can be reconstructed with minimal loss of information. Alternatively, the KRIMP algorithm [17] selects those itemsets that provide the best lossless compression of the database, i.e. the best description. While it only considers the 1s in the data, it provides high-quality and detailed results, which are consequently not as small and easily interpreted as our summaries. Though the KRIMP code tables can generate data virtually indistinguishable from the original [18], they are not probabilistic models and cannot be queried directly, they are no surrogate for the data.

Most related to our method are low-entropy sets [10], itemsets for which the entropy of the data is below a given threshold. As entropy is strongly monotonically increasing, typically very many low-entropy sets are discovered even for low thresholds. Heikinheimo et al. introduced a filtering proposal [11], LESS, to select those low-entropy sets that together describe the data well. Here, instead of filtering, we discover itemsets with low entropy directly on the data.

Orthogonal to our approach, the maximally informative k-itemsets (miki's) by Knobbe and Ho [12] are k items (or patterns) that together split the data optimally, found through exhaustive search. Bringmann and Zimmermann [1] proposed a greedy alternative to this exhaustive method that can consider larger sets of items. We group items together that correlate strongly, so the correlations between groups are weak. As future work, we plan to investigate whether good approximate miki's can be extracted from our summaries.

As our approach employs clustering, the work in this field is not unrelated. However, clustering is foremost concerned with grouping rows together, typically requiring a distance measure between objects. Bi-clustering [16] is a type of clustering in which clusters are detected over both attributes and rows. In our setup we only group attributes, not rows, and do not require a distance measure between items.

6 Experiments

In this section we experimentally evaluate our method and validate the quality of the returned summaries.

6.1 Setup

We implemented our algorithm in C++, and provide the source code for research purposes[1]. All experiments were executed on an quad-core Intel Xeon machine with 6GB of memory, running Linux.

We evaluate our method on three synthetic datasets, as well as on seven publicly available real-world datasets. Their basic characteristics are given in Table 1. The *Independent* data has independent attributes with random frequencies. In *Markov* each item is a copy of the previous one with a random probability. The *DAG* dataset is generated according to a directed acyclic graph among the items. An item depends on a small amount of preceding items, the probabilities in the corresponding contingency table are generated at random.

The *Accidents*, *BMS-Webview-1*, *Chess*, *Connect*, and *Mushroom* datasets were obtained from the FIMI dataset repository[2] and the *Pen Digits* data was obtained from the LUCS-KDD data library[3]. Further, we use the *DNA Amplification* database, which contains data on DNA copy number amplifications. Such copies activate oncogenes and are hallmarks of nearly all advanced tumors [14]. Amplified genes represent targets for therapy, diagnostics and prognostics.

6.2 Evaluation

In Table 1 we are interested in k, the number of clusters our algorithm finds, and what the total compressed size $L(\mathcal{A}, \mathcal{D})$ is, relative to the independence clustering

[1] `http://www.adrem.ua.ac.be/implementations/`
[2] `http://fimi.cs.helsinki.fi/data/`
[3] `http://www.csc.liv.ac.uk/~frans/KDD/`

Table 1. Results of our Attribute Clustering algorithm for 3 synthetic and 7 real datasets. As basic statistics per dataset, shown are the number of binary attributes, and the number of transactions. For the result of our method, shown are the number of identified groups, the attained compression ratio relative to the independence model, and the wall-clock time used to generate the summary.

	Basic Statistics		*Attribute Clustering*		
Dataset	$\|\mathcal{I}\|$	$\|\mathcal{D}\|$	k	$\frac{L(\mathcal{A},\mathcal{D})}{L(\mathcal{I},\mathcal{D})}$	*time*
Independent	50	20000	50	100%	3 s
Markov	50	20000	14	89.6%	5 s
DAG	50	20000	12	95.7%	6 s
Accidents	468	340183	199	64.7%	165 m
BMS-Webview-1	497	59602	150	89.6%	434 s
Chess	75	3196	9	40.8%	2 s
Connect	129	67557	7	43.4%	182 s
DNA Amplification	391	4950	52	42.0%	22 s
Mushroom	119	8124	5	37.9%	14 s
Pen Digits	86	10992	5	55.4%	9 s

$L(\mathcal{I},\mathcal{D})$. A low number of clusters and a short description length indicate that our algorithm models structure present in the data. The algorithm correctly detects 50 clusters in the *Independent* data, even though there might seem some accidental dependencies present due to the randomness of the data generation. For the other datasets we see that the number of clusters k is much lower than the number of items $|\mathcal{I}|$. As such, it is perfectly feasible to inspect these clusters by hand. Many of the datasets are highly structured, which can be seen from the strong compression ratios the clusterings achieve.

In Table 2 we test whether the clustering that our algorithm finds actually reflects true structure in the data, rather then just finding some random artifacts. For each dataset $\mathcal{D}$ we create 1000 swap randomised datasets $\mathcal{D}_S$ [6], and run our algorithm. These datasets have the same row and column margins as the original data, but are random otherwise. Only patterns depending on the margins are therefore retained. We see that in all cases all structure disappears, and the average number of clusters our algorithm returns is very close to the number of attributes, i.e. the best clustering is close to the independence clustering. Furthermore, we also see that the average description length is basically the same as for the independence clustering. For each dataset, we also created 1000 random partitions, $\mathcal{A}_r$, of k groups. The last column in Table 2 shows the average description length compared to $L(\mathcal{I},\mathcal{D})$. We see that while for several of the datasets random partitions can still compress the data better than the independence clustering and hence model *some* structure, the gain is much lower than the compression gain that our algorithm attains.

Next, we investigate the actual clusterings discovered by our attribute clustering algorithm in closer detail.

Table 2. Results for the randomisation experiments. The second and third columns are the averaged results of our algorithm on 1000 swap randomised datasets (100 for *BMS-Webview-1* and 20 for *Accidents*). The number of swaps is equal to the number of ones in the data as suggested in [6]. The fourth column is the average total description length for 1000 random k-partitions.

	Swap Randomisation		*Random k-partition*
Dataset	k_{swap}	$\frac{L(\mathcal{A}_s,\mathcal{D}_s)}{L(\mathcal{I},\mathcal{D}_s)}$	$\frac{L(\mathcal{A}_r,\mathcal{D})}{L(\mathcal{I},\mathcal{D})}$
Independent	49.95 ± 0.22	100.0% ± 0.0	100.0% ± 0.0
Markov	49.93 ± 0.88	100.0% ± 0.0	99.5% ± 0.0
DAG	49.92 ± 0.59	100.0% ± 0.0	100.4% ± 1.1
Accidents	432.7 ± 28.6	99.9% ± 0.2	99.7% ± 0.3
BMS-Webview-1	339.1 ± 4.32	99.6% ± 0.0	100.0 ± 0.0
Chess	73.36 ± 2.00	99.9% ± 0.1	94.5% ± 3.6
Connect	100.8 ± 3.64	100.0% ± 0.2	90.6% ± 2.0
DNA Amplification	348.9 ± 3.18	99.8% ± 0.0	104.9% ± 0.0
Mushroom	114.8 ± 2.13	100.0% ± 0.0	67.7% ± 2.3
Pen Digits	80.47 ± 7.56	100.0% ± 0.0	102.6% ± 3.5

For the synthetic data, we see that the embedded structures are correctly recovered. For *Independent* the algorithm of course returns the independence clustering. The items in the *Markov* dataset form a Markov chain, and the clusters found by our algorithm contain adjacent items, i.e. they are Markov chains themselves. Interestingly, when regarding its dendrogram (not shown), we see that the chains are split up at exactly those places where the dependency between items is low, i.e. the copy probability is close to 50%. Likewise, in the *DAG* dataset, which has attribute dependencies forming a directed acyclic graph, the clusters contain items which form tightly linked groups in the graph.

The *DNA Amplification* dataset is an approximately *banded* dataset [5], that is, the majority of the ones form a staircase pattern, and are located in blocks along the diagonal. In Figure 3, a submatrix of the data is plotted, along with the attribute clustering our algorithm finds. The clustering clearly distinguishes the blocks in the data. In turn, these blocks correspond to related oncogenes.

The *Connect* dataset contains all legal 8-ply positions of the well-known Connect Four game. The game has 7 columns and 6 rows, and for each of the 42 squares, an attribute describes whether it is blank, or which one of the two players has positioned a chip there. Furthermore, a class label describes which player can win or whether the game will result in a draw. The dataset we use is binary, and contains an item for each possible attribute-value pair, as well as an item for each class label. First of all, we see that all attribute-value pairs (items) originating from a single attribute (i.e. location) are grouped into the same cluster. Furthermore, our algorithm discovers 7 clusters. Each one of these clusters correctly corresponds to a column in the game, i.e. the structure found by our algorithm reflects the physical structure of the game. The class label is

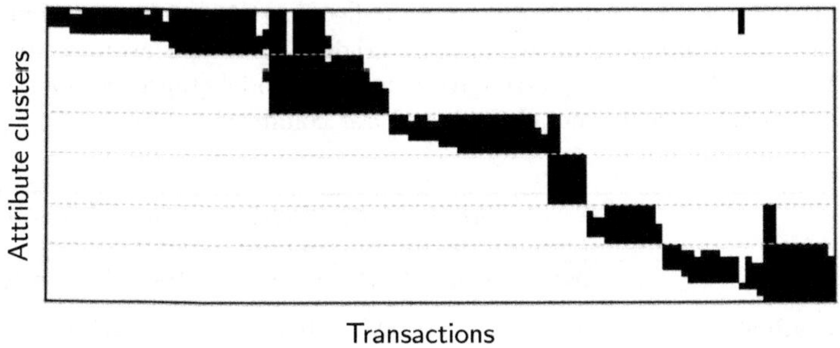

Fig. 3. A (transposed) submatrix of the *DNA Amplification* data and the corresponding discovered attribute clusters, separated by the dotted lines.

placed in the cluster of the middle column; this makes a lot of sense since any horizontal or diagonal line of four must pass through the middle column, and hence this column is key for winning the game.

6.3 Estimating Itemset Frequencies

In this subsection we investigate how well our summaries can be used to estimate itemset frequencies. For each dataset we first mine up to the top-10 000 closed frequent itemsets. Then, for each itemset in this collection, we estimate its frequency according to our model and compute both its absolute and relative error. For comparison the same is done for the independence model, which is equivalent to the singleton clustering. As can be seen from the results in Table 3, the models returned by our algorithm allow for very good frequency estimates; for most datasets the average absolute error is less than 1% and much better than that for the independence model. While for the *BMS-Webview-1*, *DNA Amplification* and *Pen Digits* datasets the average relative error seems rather high (50%), this is explained by the fact that the frequencies for the top-10 000 closed itemsets for these datasets are very low, as can be read from the first column.

In Figure 4 we plot the cumulative probability of the absolute errors for *Connect* and *Mushroom*. For every $\epsilon \in [0, 1]$ we determine the probability $\delta = p(err > \epsilon)$ that the absolute estimation error $|freq(X) - \widehat{freq}(X)|$ is greater than ϵ. For both datasets we see that the best clustering outperforms the independence clustering. For instance, in the *Mushroom* dataset we see that probability of an absolute error larger than 5% is about 50% for the independence model, while for our clustering method this is only 1%.

Lastly, we compare the frequency estimation capabilities of our attribute clusterings with the profile-based summarisation approach by Yan et al. [20]. In short, a profile is a submatrix of the data, in which the items are assumed to be independent. A collection of profiles can be overlapping, and summarises a given set of patterns, rather than being a global model for the data. Even though

Table 3. Results for frequency estimation of the top-10 000 closed frequent itemsets. Depicted are the average frequency in the original data, the average absolute and relative errors of the frequency estimates using our model (third and fourth column) and using the independence model (fifth and last column)

		Attribute Clustering		*Independence Model*	
Dataset	$\overline{freq}$	$\lvert freq - \hat{freq}\rvert$	$\frac{\lvert freq-\hat{freq}\rvert}{freq}$	$\lvert freq - \hat{freq}\rvert$	$\frac{\lvert freq-\hat{freq}\rvert}{freq}$
Independent	29.0%	0.15%	0.54%	0.15%	0.54%
Markov	15.7%	0.30%	2.02%	1.36%	8.47%
DAG	20.9%	0.50%	2.51%	0.92%	4.69%
Accidents	55.8%	1.47%	2.74%	2.89%	5.38%
BMS-Webview-1	0.1%	0.09%	83.1%	0.10%	91.14%
Chess	81.2%	0.93%	1.16%	1.47%	1.83%
Connect	88.8%	0.38%	0.45%	2.56%	2.95%
DNA Amplification	0.5%	0.08%	53.24%	0.46%	92.25%
Mushroom	12.5%	1.30%	13.55%	5.48%	48.46%
Pen Digits	6.1%	2.89%	51.52%	3.86%	67.52%

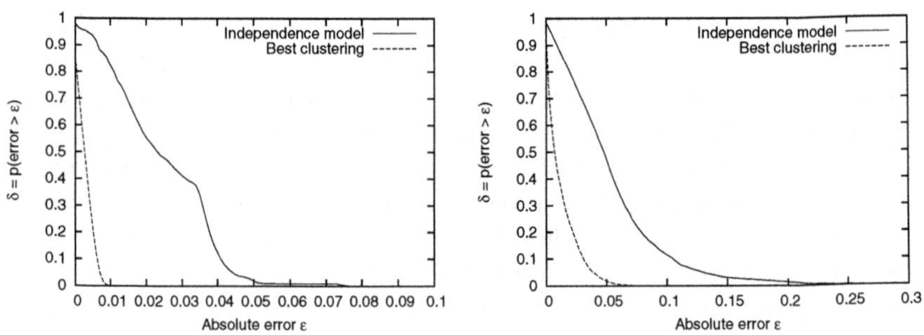

Fig. 4. Probability of an estimation error larger than ϵ for *Connect* (left) and *Mushroom* (right) on the top-10 000 closed frequent itemsets

summarisation with profiles is different from our clustering approach, we can compare the quality of the frequency estimates. We mimic the experiments in [20] on *Mushroom* and *BMS-Webview-1* by comparing the average relative error, also called restoration error. The collection of itemsets contains all frequent closed itemsets for a minsup threshold of 25% and 0.1% respectively. On *Mushroom* we attain a restoration error of 2.31%, which is lower than the results reported in [20] for any number of profiles. For *BMS-Webview-1* our restoration error is 70.4%, which is on par with Yan et al.'s results when using about 100 profiles. Their results improve when increasing the number of profiles; however, the best scores require over a thousand profiles, a number at which it rapidly becomes infeasible to inspect the profile-based summary.

7 Discussion

The experiments show our method discovers high-quality summaries. The high compression ratios for real data, and the inability to compress swap-randomised data, show our models capture the significant structure of the data. The summaries are good surrogates for the data, which can be queried quickly and accurately to approximate the frequencies of itemsets. Inspection of the models showed that correlated attributes are correctly grouped, providing necessary insight when constructing background knowledge to effectively mine the data [9]. Also, this information could be used to select or construct features. Further research into this matter is required, however.

Even while our current implementation is crude, the summaries considered were constructed fast. The implementation can be trivially parallelised, and optimised by using *tid*-lists. We especially regard the development of fast *approximate* summarisation techniques for databases with many transaction and/or items as an important topic for future research, in particular as many data mining techniques cannot consider such datasets directly, but could be made to consider the summary surrogate. Another important open problem is the generation of summaries for data consisting of both numeric and binary attributes.

8 Conclusions

In this paper we introduced a method for getting a good first impression of a binary transaction dataset. Our parameter-free method builds such summaries by grouping items that strongly correlate, and uses the Minimum Description Length principle to identify the best grouping—without requiring a distance measure between items. The result offers an overview of which attributes interact most strongly, and in what value-instantiations these typically occur. Further, as they consider the data symmetrically with regard to 0/1 and form probabilistic models for it, these summaries are good surrogates for the data that can be queried efficiently.

Experiments showed that our method provides high-quality results that correctly identify groups of correlated items, and can be used to obtain close approximations of itemset frequencies.

Acknowledgements

Michael Mampaey is supported by the Institute for the Promotion of Innovation through Science and Technology in Flanders (IWT-Vlaanderen).

References

1. Bringmann, B., Zimmermann, A.: The chosen few: On identifying valuable patterns. In: Perner, P. (ed.) ICDM 2007. LNCS (LNAI), vol. 4597, pp. 63–72. Springer, Heidelberg (2007)

2. Calders, T., Goethals, B.: Mining all non-derivable frequent itemsets. In: Elomaa, T., Mannila, H., Toivonen, H. (eds.) PKDD 2002. LNCS (LNAI), vol. 2431, pp. 74–85. Springer, Heidelberg (2002)
3. Chandola, V., Kumar, V.: Summarization – compressing data into an informative representation. In: Proceedings of ICDM 2005, pp. 98–105 (2005)
4. Cover, T.M., Thomas, J.A.: Elements of Information Theory, 2nd edn. John Wiley and Sons, Chichester (2006)
5. Garriga, G.C., Junttila, E., Mannila, H.: Banded structure in binary matrices. In: Proceedings of KDD 2008, pp. 292–300 (2008)
6. Gionis, A., Mannila, H., Mielikäinen, T., Tsaparas, P.: Assessing data mining results via swap randomization. TKDD, 1(3) (2007)
7. Grünwald, P.D.: The Minimum Description Length Principle. MIT Press, Cambridge (2007)
8. Han, J., Cheng, H., Xin, D., Yan, X.: Frequent pattern mining: Current status and future directions. Data Mining and Knowledge Discovery 15(1), 55–86 (2007)
9. Hanhijärvi, S., Ojala, M., Vuokko, N., Puolamäki, K., Tatti, N., Mannila, H.: Tell me something I don't know: randomization strategies for iterative data mining. In: Proceedings of KDD 2009, pp. 379–388. ACM, New York (2009)
10. Heikinheimo, H., Hinkkanen, E., Mannila, H., Mielikäinen, T., Seppänen, J.K.: Finding low-entropy sets and trees from binary data. In: Proceedings of KDD 2007, pp. 350–359 (2007)
11. Heikinheimo, H., Vreeken, J., Siebes, A., Mannila, H.: Low-entropy set selection. In: Jonker, W., Petković, M. (eds.) Secure Data Management. LNCS, vol. 5776, pp. 569–579. Springer, Heidelberg (2009)
12. Knobbe, A.J., Ho, E.K.Y.: Maximally informative k-itemsets and their efficient discovery. In: Proceedings of KDD 2006, pp. 237–244 (2006)
13. MacQueen, J.: Some methods for classification and analysis of multivariate observations. In: Proceedings of the 5th Symposium on Mathematical Statistics and Probability (1967)
14. Myllykangas, S., Himberg, J., Böhling, T., Nagy, B., Hollmén, J., Knuutila, S.: DNA copy number amplification profiling of human neoplasms. Oncogene 25(55), 7324–7332 (2006)
15. Pasquier, N., Bastide, Y., Taouil, R., Lakhal, L.: Discovering frequent closed itemsets for association rules. In: Beeri, C., Bruneman, P. (eds.) ICDT 1999. LNCS, vol. 1540, pp. 398–416. Springer, Heidelberg (1998)
16. Pensa, R., Robardet, C., Boulicaut, J.-F.: A bi-clustering framework for categorical data. In: Jorge, A.M., Torgo, L., Brazdil, P.B., Camacho, R., Gama, J. (eds.) PKDD 2005. LNCS (LNAI), vol. 3721, pp. 643–650. Springer, Heidelberg (2005)
17. Siebes, A., Vreeken, J., van Leeuwen, M.: Item sets that compress. In: Jonker, W., Petković, M. (eds.) SDM 2006. LNCS, vol. 4165, pp. 393–404. Springer, Heidelberg (2006)
18. Vreeken, J., van Leeuwen, M., Siebes, A.: Preserving privacy through data generation. In: Perner, P. (ed.) ICDM 2007. LNCS (LNAI), vol. 4597, pp. 685–690. Springer, Heidelberg (2007)
19. Wang, J., Karypis, G.: SUMMARY: Efficiently summarizing transactions for clustering. In: Perner, P. (ed.) ICDM 2004. LNCS (LNAI), vol. 3275, pp. 241–248. Springer, Heidelberg (2004)
20. Yan, X., Cheng, H., Han, J., Xin, D.: Summarizing itemset patterns: A profile-based approach. In: Proceedings of KDD 2005, pp. 314–323 (2005)

Classification and Novel Class Detection of Data Streams in a Dynamic Feature Space

Mohammad M. Masud[1], Qing Chen[1], Jing Gao[2], Latifur Khan[1], Jiawei Han[2], and Bhavani Thuraisingham[1]

[1] University of Texas at Dallas
[2] University of Illinois at Urbana Champaign
{mehedy,qingch}@utdallas.edu, jinggao3@uiuc.edu
lkhan@utdallas.edu, hanj@cs.uiuc.edu, bhavani.thuraisingham@utdallas.edu

Abstract. Data stream classification poses many challenges, most of which are not addressed by the state-of-the-art. We present DXMiner, which addresses four major challenges to data stream classification, namely, infinite length, concept-drift, concept-evolution, and feature-evolution. Data streams are assumed to be infinite in length, which necessitates single-pass incremental learning techniques. Concept-drift occurs in a data stream when the underlying concept changes over time. Most existing data stream classification techniques address only the infinite length and concept-drift problems. However, concept-evolution and feature- evolution are also major challenges, and these are ignored by most of the existing approaches. Concept-evolution occurs in the stream when novel classes arrive, and feature-evolution occurs when new features emerge in the stream. Our previous work addresses the concept-evolution problem in addition to addressing the infinite length and concept-drift problems. Most of the existing data stream classification techniques, including our previous work, assume that the feature space of the data points in the stream is static. This assumption may be impractical for some type of data, for example text data. DXMiner considers the dynamic nature of the feature space and provides an elegant solution for classification and novel class detection when the feature space is dynamic. We show that our approach outperforms state-of-the-art stream classification techniques in classifying and detecting novel classes in real data streams.

1 Introduction

The goal of data stream classification is to learn a model from past labeled data, and classify future instances using the model. There are many challenges in data stream classification. First, data streams have *infinite length*, and so, it is impossible to store all the historical data for training. Therefore, traditional learning algorithms that require multiple passes over the whole training data are not directly applicable to data streams. Second, data streams observe *concept-drift*, which occurs when the underlying concept of the data changes over time. A classification model must adapt itself to the most recent concept in order to

J.L. Balcázar et al. (Eds.): ECML PKDD 2010, Part II, LNAI 6322, pp. 337–352, 2010.

cope with concept-drift. Third, novel classes may appear in the stream, which we call *concept-evolution*. In order to cope with concept-evolution, a classification model must be able to automatically detect novel classes.

Finally, the feature space that represents a data point in the stream may change over time. For example, consider a text stream where each data point is a document, and each word is a feature. Since it is impossible to know which words will appear in the future, the complete feature space is unknown. Besides, it is customary to use only a subset of the words as the feature set because most of the words are likely to be redundant for classification. Therefore at any given time, the feature space is defined by the useful words (i.e., features) selected using some selection criteria. Since in the future, new words may become useful and old useful words may become redundant, the feature space changes dynamically. We call this dynamic nature of features as *feature-evolution*. In order to cope with feature-evolution, the classification model should be able to correctly classify a data point having a different feature space than the feature space of the model. Most existing data stream classification techniques address only the infinite length, and concept-drift problems [1, 11, 5, 3, 9]. Our previous work XMiner [6] addresses the concept-evolution problem in addition to the infinite length and concept-drift problems. In this paper, we propose DXMiner, which addresses *feature-evolution* as well as the other three challenges. Dealing with the feature-evolution problem becomes much challenging in the presence of concept-drift and concept-evolution.

DXMiner addresses the infinite length and concept-drift problems by applying a hybrid batch-incremental process [6, 9], which is done as follows. The data stream is divided into equal sized chunks and a classification model is trained from each chunk. An ensemble of L such models is used to classify the unlabeled data. When a new model is trained from a data chunk, it replaces one of the existing models in the ensemble. In this way the ensemble is kept up-to-date. The infinite length problem is addressed by maintaining a fixed sized ensemble, and the concept-drift is addressed by keeping the ensemble up-to-date. DXMiner solves the concept-evolution problem by automatically detecting novel classes in the data stream [6]. In order to detect novel class, it first builds a decision boundary around the training data. During classification of unlabeled data, it first identifies the test data points that are outside the decision boundary. Such data points are called filtered outliers (F-*outlier*s), and they represent data points that are *well separated* from the training data. Then if sufficient number of F-*outlier*s are found that show strong cohesion among themselves (i.e., they are close together), the F-*outlier*s are classified as novel class instances. Finally, DXMiner solves the feature-evolution problem by applying effective feature selection technique and dynamically converting the feature spaces of the classification models and the test instances.

We have several contributions. First, we propose a framework for classifying a data stream that observes infinite-length, concept-drift, concept-evolution, and feature-evolution. To the best of our knowledge, this is the first work that addresses all these challenges in a single framework. Second, we propose a realistic feature extraction and selection technique for data streams, which selects the features for the test instances without knowing their labels. Third, we propose a fast and effective feature space conversion technique to address the feature-evolution problem. In this technique, we convert different heterogeneous feature spaces into one homogeneous space without losing any feature value. The effectiveness of this technique is established both analytically and empirically. Finally, we evaluate our framework on real data streams, such as Twitter messages, and NASA safety aviation reports, and achieve satisfactory performance over existing state-of-the-art data stream classification techniques.

The rest of the paper is organized as follows. Section 2 discusses relevant works in data stream classification. Section 3 describes the proposed framework in details, and Section 4 then explains our feature space conversion technique to cope with dynamic feature space. Section 5 reports the experimental results and analyzes them. Finally, Section 6 concludes with directions to future works.

2 Related Work

The challenges of data stream classification are addressed by different researchers in different ways. These approaches can be divided into three categories. Approaches belonging to the first category address the infinite length and concept-drift problems; approaches belonging to the second category address the infinite length, concept-drift, and feature-evolution problems; and approaches belonging to the third category address the infinite length, concept-drift, and concept-evolution problems.

Most of the existing techniques fall into the first category. There are two different approaches: single model classification, and ensemble classification. The single model classification techniques apply some form of incremental learning to address the infinite length problem, and strive to adapt themselves to the most recent concept to address the concept-drift problem [3, 1, 11]. Ensemble classification techniques [9, 5, 2] maintain a fixed-sized ensemble of models, and use ensemble voting to classify unlabeled instances. These techniques address the infinite length problem by applying a hybrid batch-incremental technique. Here the data stream is divided into equal sized chunks and a classification model is trained from each chunk. This model replaces one of the existing models in the ensemble, keeping the ensemble size constant. The concept-drift problem is addressed by continuously updating the ensemble with newer models, and striving to keep the ensemble consistent with the current concept. DXMiner also applies an ensemble classification technique.

Techniques in the second category address the feature-evolution problem on top of the infinite length and concept-drift problems. Katakis et al. [4] propose a feature selection technique for data streams having dynamic feature space. Their technique consists of an incremental feature ranking method and an incremental learning algorithm. Wenerstrom and Giraud-Carrier [10] propose a technique, called FAE, which also applies incremental feature selection, but their incremental learner is an ensemble of models. Their approach achieves relatively better performance than the approach of Katakis et al [4]. There are several differences in the way that FAE and DXMiner approaches the feature-evolution problem. First, FAE uses the $\mathcal{X}^2$ statistics for feature selection, whereas DXMiner uses deviation weight (section 3.2). Second, in FAE, if a test instance has a different feature space than the classification model, the model uses its own feature space, but the test instance uses only those features that belong to the model's feature space. In other words, FAE uses a Lossy-L conversion, whereas DXMiner uses Lossless converion (see section 4). Furthermore, none of the proposed approaches of the second category detects novel class, but DXMiner does.

Techniques in the third category deal with the concept-evolution problem in addition to addressing the infinite length and concept-drift problems. An unsupervised novel concept detection technique for data streams is proposed in [8], but it is not applicable to multi-class classification. Our previous works MineClass and XMiner [6] address the concept-evolution problem on a multi-class classification framework. They can detect the arrival of a novel class automatically, without being trained with any labeled instances of that class. However, they do not address the feature-evolution problem. On the other hand, DXMiner addresses the more general case where features can evolve dynamically. DXMiner differs from all other data stream classification techniques in that it addresses all four major challenges in a single framework, whereas previous techniques address three or less challenges. Its effectiveness is shown analytically and demonstrated empirically on a number of real data streams.

3 Overview of DXMiner

In this section, we will briefly describe the system architecture of DXMiner (or DECSMiner), which stands for Dynamic feature based Enhanced Classifier for Data Streams with novel class Miner. Before describing the system, we define the concept of *novel class* and *existing class.*

Definition 1. *[Existing class and Novel class] Let M be the current ensemble of classification models. A class c is an existing class if at least one of the models $M_i \in M$ has been trained with class c. Otherwise, c is a novel class.*

3.1 Top Level Description

Algorithm 1 sketches the basic steps of DXMiner. The system consists of an ensemble of L classification models, $\{M_1, ..., M_L\}$. The data stream is divided into equal sized chunks. When the data points of a chunk are labeled by an expert, it is used for training. The initial ensemble is built from first L data chunks (line 1).

Feature extraction and selection: It is applied on the raw data to extract all the features and select the best features for the latest unlabeled data chunk D_u (line 5). The feature selection technique is described in section 3.2. However, if the feature set is pre-determined, then the function (Extract&SelectFeatures) simply returns that feature set.

Algorithm 1. DXMiner

```
1: M ← Build-initial-ensemble()
2: buf ← empty //temporary buffer
3: D_u ← latest chunk of unlabeled instances
4: D_l ← sliding window of last r data chunks
5: F_u ← Extract&Select-Features(D_l,D_u) //Feature set for D_u (section 3.2)
6: Q ⇐ D_u //FIFO queue of data chunks waiting to be labeled
7: while true do
8:    for all x_j ∈ D_u do
9:       M', x'_j ←Convert-Featurespace(M,x_j,F_u) //(section 3.4)
10:      NovelClass-Detection&Classification(M',x'_j,buf) //(section 3.5)
11:   end for
12:   if the instances in Q.front() are now labeled then
13:      D_f ⇐ Q //Dequeue
14:      M ← Train&Update(M,D_f) //(section 3.3)
15:      D_l ← move-window(D_l,D_f) //slide the window to include D_f
16:   end if
17:   D_u ← new chunk of unlabeled data
18:   F_u ← Extract&Select-Features(D_l,D_u) //Feature set for D_u
19:   Q ⇐ D_u //Enqueue
20: end while
```

D_u is enqueued into a queue of unlabeled data chunks waiting to be labeled (line 6). Each instance of the chunk D_u is then classified by the ensemble M (lines 8-11). Before classification, the models in the ensemble, as well as the test instances need to pass through a feature space conversion process.

Feature space conversion (line 9): It is not needed if the feature set for the whole data stream is static. However, if the feature space is dynamic, then we would have different feature sets in different data chunks. As a result, each model in the ensemble would be trained on different feature sets. Besides, the feature space of the test instances would also be different from the feature space of the models. Therefore, we apply a feature space conversion technique to homogenize the feature sets of the models and the test instances. See section 4 for details.

Novel class detection and classification (line 10): After the conversion of feature spaces, the test instance is examined by the ensemble of models to determine whether the instance should be identified as a novel class instance, or as one of the existing class instances. The buffer buf is used to temporarily store potential novel class instances. See section 3.5 for details.

The queue Q is checked to see if the chunk at the front (i.e., oldest chunk) is labeled. If yes, the chunk is dequeued, used to train a model, and the sliding window of labeled chunks is shifted right. By keeping the queue to store unlabeled data, we eliminate the constraint imposed by many approaches (e.g. [10]) that each new data point arriving in the stream should be labeled as soon as it is classified by the existing model.

Training and update(line 14): We learn a model from the training data. We also build a decision boundary around the training data in order to detect novel classes. Each model also saves the set of features with which it is trained. The newly trained model replaces an existing model in the ensemble. The model to be replaced is selected by evaluating each of the models in the ensemble on the training data, and choosing the one with the highest error. See section 3.3 for details. Finally, when a new data chunk arrives, we again select best features for that chunk, and enqueue the chunk into Q.

3.2 Feature Extraction and Selection

The data points in the stream may or may not have a fixed feature set. If they have a fixed feature set, then we simply use that feature set. Otherwise, we apply a feature extraction and feature selection technique. Note that we need to select features for the instances of the test chunk before they can be classified by the existing models, since the classification models require the feature vectors for the test instances. However, since the instances of the test chunk are unlabeled, we cannot use supervised feature selection (e.g. information gain) on that chunk. To solve this problem, we propose two alternatives: *predictive* feature selection, and *informative* feature selection, to be explained shortly. Once the feature set has been selected for a test chunk, the feature values for each instance are computed, and feature vectors are produced. The same feature vector is used during classification (when unlabeled) and training (when labeled).

Predictive feature selection: Here, we predict the features of the test instances without using any of their information, rather we use the past labeled instances to predict the feature set of the test instances. This is done by extracting all features from the last r labeled chunks (D_l in the DXMiner algorithm), and then selecting the best R features using some selection criteria. In our experiments, we use r=3. One such popular selection criterion is information gain. We use another criterion which we call *deviation weight*. The deviation weight for the i-th feature for class c is given by: $dw_i = freq_i * \frac{freq_i^c}{N_c} * \frac{N-N_c}{freq_i - freq_i^c + \epsilon}$, where $freq_i$ is the total frequency of the i-th feature, $freq_i^c$ is the frequency of the i-th feature in class c, N_c is the number of instances of class c, N is the total number of instances, and ϵ is a smoothing constant. A higher value of deviation weight means greater discriminating power. For each class, we choose the top r features having the highest deviation weight. So, if there are total $|C|$ classes, then we select $R = |C|r$ features this way. These features are used as the feature space for the test instances. We use deviation weight instead of information gain in some data streams because this selection criterion achieves better classification accuracy (see section 5.3). Although information gain or deviation weight consider fixed number of classes, this does not affect the novel class detection process since the feature selection is used just to select the best features for the test instances. The test instances are still unlabeled, and therefore, novel class detection mechanism is applicable to them.

Informative feature selection: Here, we use the test chunk to select the features. We extract all possible features from the test chunk (D_u in the DXMiner algorithm), and select the best R features in an unsupervised way. For example, one such unsupervised selection criterion is to choose the R highest frequency features in the chunk. This strategy is very useful in data streams like to the Twitter (see section 5).

3.3 Training and Update

The feature vectors constructed in the previous step (section 3.2) are supplied to the learning algorithm to train a model. In our case, we use a semi-supervised clustering technique to train a K-NN based classifier [7]. We build K clusters with the training data, applying a semi-supervised clustering technique. After building the clusters, we save the cluster summary (mentioned as *pseudopoint*) of each cluster. The summary contains the *centroid*, *radius*, and *frequencies* of data points belonging to each class. The radius of a pseudopoint is defined as the distance between the centroid and the farthest data point in the cluster. The raw data points are discarded after creating the summary. Therefore, each model M_i is a collection of K pseudopoints. A test instance x_j is classified using M_i as follows. We find the pseudopoint $h \in M_i$ whose centroid is nearest from x_j. The predicted class of x_j is the class that has the highest frequency in h. x_j is classified using the ensemble M by taking a majority voting among all classifiers.

Each pseudopoint corresponds to a "hypersphere" in the feature space having center at the centroid, and a radius equal to its radius. Let $\mathcal{S}(h)$ be the *feature space* covered by such a hypersphere of pseudopoint h. The *decision boundary* of a model M_i (or $\mathcal{B}(M_i)$) is the union of the feature spaces (i.e., $\mathcal{S}(h)$) of all pseudopoints $h \in M_i$. The decision boundary of the ensemble M (or $\mathcal{B}(M)$) is the union of the decision boundaries (i.e., $\mathcal{B}(M_i)$) of all models $M_i \in M$.

The ensemble is updated by the newly trained classifier as follows. Each existing model in the ensemble is evaluated on the latest training chunk, and their error rates are obtained. The model having the highest error is replaced with the newly trained model. This ensures that we have exactly L models in the ensemble at any given point of time.

3.4 Feature Space Conversion: Explained in Details in Section 4

3.5 Classification and Novel Class Detection

Each instance in the most recent unlabeled chunk is first examined by the ensemble of models to see if it is outside the decision boundary of the ensemble (i.e., $\mathcal{B}(M)$). If it is inside, then it is classified normally (i.e., using majority voting) using the ensemble of models. Otherwise, it is declared as an *F-outlier*, or filtered outlier. We assume that any class of data has the following property.

Property 1. *A data point should be closer to the data points of its own class (cohesion) and farther apart from the data points of other classes (separation).*

So, if there is a novel class in the stream, instances belonging to the class will be far from the existing class instances and will be close to other novel class instances. Since *F-outliers* are outside $\mathcal{B}(M)$, they are far from the existing class instances. So, the separation property for a novel class is satisfied by the *F-outliers*. Therefore, *F-outliers* are potential novel class instances, and they are temporarily stored in the buffer *buf* (see algorithm 1) to observe whether they also satisfy the cohesion property. We then examine whether there are enough *F-outliers* that are close to each other. This is done by computing the following metric, which we call the q-Neighborhood Silhouette Coefficient, or q-NSC [6] (to be explained shortly).

Definition 2 (λ_c-neighborhood). *The λ_c-neighborhood of an Foutlier x is the set of q-nearest neighbors of x belonging to class c.*

Here q is a user defined parameter. For brevity, we denote the λ_c-neighborhood of an *F-outlier* x as $\lambda_c(x)$. Thus, $\lambda_+(x)$ of an *F-outlier* x is the set of q instances of class c_+, that are closest to the outlier x. Similarly, $\lambda_o(x)$ refers to the set of q *F-outliers* that are closest to x. Let $\bar{D}_{c_{out},q}(x)$ be the mean distance from an *F-outlier* x to its q-nearest *F-outlier* instances (i.e., to its $\lambda_o(x)$ neighborhood), Also, let $\bar{D}_{c_{min},q}(x)$ be the mean distance from x to its closest existing class neighborhood ($\lambda_{c_{min}}(x)$). Then q-NSC of x is given by:

$$q\text{-}NSC(x) = \frac{\bar{D}_{c_{min},q}(x) - \bar{D}_{c_{out},q}(x)}{max(\bar{D}_{c_{min},q}(x), \bar{D}_{c_{out},q}(x))} \tag{1}$$

q-NSC, a unified measure of cohesion and separation, yields a value between -1 and +1. A positive value indicates that x is closer to the *F-outlier* instances (more cohesion) and farther away from existing class instances (more separation), and vice versa. q-NSC(x) of an *F-outlier* x must be computed separately for each classifier $M_i \in M$. We declare a *new class* if there are at least q' ($> q$) *F-outliers* having positive q-NSC for all classifiers $M_i \in M$. In order to reduce the time complexity in computing q-NSC(), we cluster the *F-outliers*, and compute q-NSC() of those clusters only. The q-NSC() of each such cluster is used as the approximate q-NSC() value of each data point in the cluster.

It is worthwhile to mention here that we do not make any assumption about the number of novel classes. If there are two or more novel classes appearing at the same time, all of them will be detected as long as each one of them satisfies property-1 and each of them has $> q$ instances. However, we will tag them simply as "novel class", i.e., no distinction will be made among them. But the distinction will be learned by our model as soon as those instances are labeled by human experts, and a classifier is trained with them.

4 Feature Space Conversion

It is obvious that the data streams that do not have any fixed feature space (such as text stream) will have different feature spaces for different models in the ensemble, since different sets of features would likely be selected for different chunks. Besides, the feature space of test instances is also likely to be different from the feature space of the classification models. Therefore, when we need to classify an instance, we need to come up with a homogeneous feature space for the model and the test instances. There are three possible alternatives: i) Lossy fixed conversion (or *Lossy-F* conversion in short), ii) Lossy local conversion (or *Lossy-L* conversion in short), and iii) Lossless homogenizing conversion (or *Lossless* conversion in short).

4.1 Lossy Fixed (Lossy-F) Conversion

Here we use the same feature set for the entire stream, which had been selected for the first data chunk (or first n data chunks). This will make the feature set fixed, and therefore all the instances in the stream, whether training or testing, will be mapped to this feature set. We call this a lossy conversion because future models and instances may lose important features due to this conversion.

Example: let $\mathcal{F}_S = \{\mathcal{F}_a, \mathcal{F}_b, \mathcal{F}_c\}$ be the features selected in the first n chunks of the stream. With the Lossy-F conversion, all future instances will be mapped to this feature set. That is, suppose the set of features for a future instance x be: $\{\mathcal{F}_a, \mathcal{F}_c, \mathcal{F}_d, \mathcal{F}_e\}$, and the corresponding feature values of x be: $\{x_a, x_c, x_d, x_e\}$. Then after conversion, x will be represented by the following values: $\{x_a, 0, x_c\}$. In other words, any feature of x that is not in F_S (i.e.,$\mathcal{F}_d$ and $\mathcal{F}_e$) will be discarded, and any feature of F_S that is not in x (i.e., $\mathcal{F}_b$) will be assumed to have a zero value. All future models will also be trained using F_S.

4.2 Lossy Local (Lossy-L) Conversion

In this case, each training chunk, as well as the model built from the chunk, will have its own feature set selected using the feature extraction and selection technique. When a test instance is to be classified using a model M_i, the model will use its own feature set as the feature set of the test instance. This conversion is also lossy because the test instance might lose important features as a result of this conversion.

Example: the same example of section 4.1 is applicable here, if we let F_S to be the selected feature set for a model M_i, and let x to be an instance being classified using M_i. Note that for the Lossy-F conversion, F_S is the same over all models, whereas for Lossy-L conversion, F_S is different for different models.

4.3 Lossless Homogenizing (Lossless) Conversion

Here, each model has its own selected set of features. When a test instance x is to be classified using a model M_i, both the model and the instance will convert their feature sets to the union of their feature sets. We call this conversion "lossless homogenizing" since both the model and the test instance preserve their dimensions (i.e., features), and the converted feature space becomes homogeneous for both the model and the test instance. Therefore, no useful features are lost as a result of the conversion.

Example: continuing from the previous example, let $\mathcal{F}_S = \{\mathcal{F}_a, \mathcal{F}_b, \mathcal{F}_c\}$ be the feature set of a model M_i, $\{\mathcal{F}_a, \mathcal{F}_c, \mathcal{F}_d, \mathcal{F}_e\}$ be the feature set of the test instance x, and $\{x_a, x_c, x_d, x_e\}$ be the corresponding feature values of x. Then after conversion, both x and M_i will have the following features: $\{\mathcal{F}_a, \mathcal{F}_b, \mathcal{F}_c, \mathcal{F}_d, \mathcal{F}_e\}$. Also, x will be represented with the following feature values: $\{x_a, 0, x_c, x_d, x_e\}$. In other words, all the features of x will be included in the converted feature set, and any feature of F_S that is not in x (i.e., $\mathcal{F}_b$) will be assumed to be zero.

4.4 Advantage of Lossless Conversion over Lossy Conversions

Lossless conversion is preferred over Lossy conversions because no features are lost due to this conversion. Our main assumption is that Lossless conversion preserves the properties of a novel class. That is, if an instance belongs to a novel class, it remains outside the decision boundary of any model M_i of the ensemble M in the converted feature space. However, this is not true for a Lossy-L conversion, as the following theorem states.

Lemma 1. *If a test point x belongs to a novel class, it will be mis-classified by the ensemble M as an existing class instance under certain conditions when the Lossy-L conversion is used.*

Proof. According to our algorithm, if x remains inside the decision boundary of any model $M_i \in M$, then the ensemble M considers it as an existing class instance. Let $M_i \in M$ be the model under question. Without loss of generality, let M_i and x have m and n features, respectively, l of which are common features. That is, let the features of the model be $\{\mathcal{F}_{i_1}, ..., \mathcal{F}_{i_m}\}$ and the features of x be $\{\mathcal{F}_{j_1}, ..., \mathcal{F}_{j_n}\}$, where $i_k = j_k$ for $0 \le k \le l$. In the boundary case, l=0, i.e., no features are common between M_i and x. Let h be the pseudopoint in M_i that is closest to x, and also, R be the radius of h, and $\mathcal{C}$ be the centroid of h. The Lossless feature space would be the union of the features of M_i and x, which is: $\{\mathcal{F}_{i_1}, ..., \mathcal{F}_{i_l}, \mathcal{F}_{i_{l+1}}, ..., \mathcal{F}_{i_m}, \mathcal{F}_{j_{l+1}}, ..., \mathcal{F}_{j_n}\}$ According to our assumption that the properties of novel class are preserved with the Lossless conversion, x will remain outside the decision boundary of all models $M_i \in M$ in the converted feature space. Therefore, the distance from x to the centroid $\mathcal{C}$ will be greater than R.

Let the feature values of the centroid $\mathcal{C}$ in the original feature space be: $\{y_{i_1}, ..., y_{i_m}\}$, where y_{i_k} is the value of feature $\mathcal{F}_{i_k}$. After Lossless conversion, the feature values of $\mathcal{C}$ in the new feature space would be: $\{y_{i_1}, ..., y_{i_m}, 0, ..., 0\}$. That is, all feature values for the added features $\{\mathcal{F}_{j_{l+1}}, ..., \mathcal{F}_{j_n}\}$ are zeros. Also, let the feature values of x in the original feature space be: $\{x_{j_1}, ..., x_{j_n}\}$. The feature values of x after the Lossless conversion would be:
$\{x_{j_1}, ..., x_{j_l}, 0, ..., 0, x_{j_{l+1}}, ..., x_{j_n}\}$, that is, the feature values for the added features are all zeros. Without loss of generality, let Euclidean distance be the distance metric. Let D be the distance from x to the centroid $\mathcal{C}$. Therefore, we can deduce:

$$D^2 = (\boldsymbol{C} - \boldsymbol{x})^2 > R^2$$
$$\Rightarrow R^2 < \sum_{k=1}^{l}(y_{i_k} - x_{j_k})^2 + \sum_{k=l+1}^{m}(y_{i_k} - 0)^2 + \sum_{k=l+1}^{n}(0 - x_{j_k})^2 \quad (2)$$

Now, let $A^2 = \sum_{k=1}^{l}(y_{i_k} - x_{j_k})^2 + \sum_{k=l+1}^{m}(y_{i_k} - 0)^2$, and $B^2 = \sum_{k=l+1}^{n}(0 - x_{j_k})^2$. Note that with the Lossy-L conversion, the distance from x to $\mathcal{C}$ would be A, since the converted feature space is the same as the original feature space of M_i. So, it follows that:

$$R^2 < A^2 + B^2 \Rightarrow R^2 = A^2 + B^2 - e^2 \quad \text{(letting } e > 0\text{)}$$
$$\Rightarrow A^2 = R^2 + (e^2 - B^2) \Rightarrow A^2 < R^2 \quad \text{(provided that } e^2 - B^2 < 0\text{)}$$

Therefore, in the Lossy-L converted feature space, the distance from x to the centroid $\mathcal{C}$ is less than the radius of the pseudopoint h, meaning, x is inside the region of h, and as a result, x is inside decision boundary of M_i. Therefore, x is mis-classified as an existing class instance by M_i when the Lossy-L conversion is used, under the condition that $e^2 < B^2$ □.

This lemma is supported by our experimental results, which show that Lossy-L conversion mis-classifies most of the novel class instances as existing class. It might appear to the reader that increasing the dimension of the models and the test instances may have an undesirable side effect due to curse of dimensionality. However, it is reasonable to assume that the feature set of the test instances is not dramatically different from the feature sets of the classification models because the models usually represent the most recent concept. Therefore, the converted dimension of the feature space should be almost the same as the original feature spaces. Furthermore, this type of conversion has been proved to be successful in other popular classification techniques such as Support Vector Machines.

5 Experiments

5.1 Dataset

We use four different datasets having different characteristics (see table 1).

Twitter dataset (Twitter): This dataset contains 170,000 Twitter messages (tweets) of seven different trends (classes). These tweets have been retrieved from http://search.twitter.com/trends/weekly.json using a tweets crawling program written in Perl script. The raw data is in free text and we apply preprocessing to get a useful dataset. The preprocessing consists of two steps. First,

Table 1. Summary of the datasets used

Dataset	Concept-drift	Concept-evolution	Feature-evolution	Features	Instances	Classes
Twitter	√	√	√	30	170,000	7
ASRS	X	X	√	50	135,000	13
KDD	√	√	X	34	490,000	22
Forest	X	√	X	54	581,000	7

filtering is performed on the messages to filter out words that match against a stop word list. Examples or stop words are articles ('a', 'an', 'the'), acronyms ('lol', 'btw') etc. Second, we use Wiktionary to retrieve the parts of speech (POS) of the remaining words, and remove all pronouns (e.g., 'I', 'u'), change tense of verbs (e.g. change 'did' and 'done' to 'do'), change plurals to singulars and so on. We apply the *informative feature selection* (section 3.2) technique on the Twitter dataset. Also, we generate the feature vector for each message using the following formula: $w_{ij} = \beta * f(a_i, m_j) \sum_{j=1}^{S} f(a_i, m_j)$ where w_{ij} is the value of the ith feature (a_i) for the jth message in the chunk, $f(a_i, m_j)$ is the frequency of feature a_i in message m_j, and β is a normalizing constant.

NASA Aviation Safety Reporting System dataset (ASRS): This dataset contains around 135,000 text documents. Each document is actually a report corresponding to a flight anomaly. There are 13 different types of anomalies (or classes), such as "aircraft equipment problem : critical", "aircraft equipment problem : less severe". These documents are treated as a data stream by arranging the reports in order of their creation time. The documents are normalized using a software called PLADS, which removes stop words, expands abbreviations, and performs stemming (e.g. changing tense of verbs). The instances in the dataset are multi-label, meaning, an instance may have more than one class labels. We transform the multi-label classification problem into 13 separate binary classification problems, one for each class. When reporting the accuracy, we report the average accuracy of the 13 datasets. We apply the *predictive feature selection* (section 3.2) for the ASRS dataset. We use deviation weight for feature selection, which works better than information gain (see section 5.3). The feature values are produced using the same formula that is used for Twitter dataset.

KDD cup 1999 intrusion detection dataset (KDD) and **Forest cover dataset from UCI repository (Forest):** See [6] for details.

5.2 Experimental Setup

Baseline techniques: ***DXMiner:*** This is the proposed approach with the *Lossless* feature space conversion. ***Lossy-F:*** This approach the same as DXMiner except that the *Lossy-F* feature space conversion is used. ***Lossy-L:*** This is DXMiner with the *Lossy-L* feature space conversion. ***O-F:*** This is a combination of the *OLINDDA* [8] approach with *FAE* [10] approach. We combine these two, because to the best of our knowledge, no other approach can work with dynamic feature vector and detect novel classes in data streams. In this combination, *OLINDDA works as the novel class detector, and FAE performs classification.* This is done as follows: For each chunk, we first detect the novel class instances using OLINDDA. All other instances in the chunk are assumed to be in the existing classes, and they are classified using FAE. FAE uses the Lossy-L conversion of feature spaces. OLINDDA is also adapted to this conversion. For fairness, the underlying learning algorithm for FAE is chosen the same as that of DXMiner. Since OLINDDA assumes that there is only one "normal" class, we build parallel OLINDDA models, one for each class, which evolve simultaneously. Whenever the instances of a novel class appear, we create a new

OLINDDA model for that class. A test instance is declared as novel, if *all the existing class models* identify this instance as novel.

Parameters settings: DXMiner: R (feature set size) = 30 for Twitter and 50 for ASRS. Note that R is only used for data streams having feature-evolution. K (number of pseudopoints per chunk) = 50, S (chunk size) = 1000, L (ensemble size) = 6, and q (minimum number of *F-outliers* required to declare a novel class) = 50. These parameter values are reasonably stable, which are obtained by running DXMiner on a number of real and synthetic datasets. Sensitivity to different parameters are discussed in details in [6].

OLINDDA: Number of data points per cluster (N_{excl}) = 30, least number of normal instances needed to update the existing model = 100, least number of instances needed to build the initial model = 100. FAE: m (maturity) = 200, p (probation time)=4000, f (feature change threshold) =5, r(growth rate)=10, N(number of instances) =1000, M (feature selected) = same as R of DXMiner. These parameters are chosen either according to the default values used in OLINDDA, FAE, or by trial and error to get an overall satisfactory performance.

5.3 Evaluation

Evaluation approach: We use the following performance metrics for evaluation: M_{new} = % of novel class instances Misclassified as existing class, F_{new} = % of existing class instances Falsely identified as novel class, **ERR** = Total misclassification error (%)(including M_{new} and F_{new}). We build the initial models in each method with the first 3 chunks. From the 4th chunk onward, we first evaluate the performances of each method on that chunk, then use that chunk to update the existing models. The performance metrics for each chunk for each method are saved and averaged for producing the summary result.

Figures 1(a),(c) show the ERR rates and total number of missed novel classes respectively, for each approach throughout the stream in Twitter dataset. For example in fugure 1(a), at X axis = 150, the Y values show the average ERR of each approach from the beginning of the stream to chunk 150. At this point, the ERR of DXMiner, Lossy-F, Lossy-L, and O-F are 4.4% 35.0%, 1.3%, and 3.2%, respectively. Figure 1(c) show the total number of novel instances missed for each of the baseline approaches. For example, at the same value of X axis, the Y values show the total novel instances missed (i.e., misclassified as existing class) for each approach from the beginning of the stream to chunk 150. At this point, the number of novel instances missed by DXMiner, Lossy-F, Lossy-L, and O-F are 929, 0, 1731, and 2229 respectively. The total number of novel class instances at this point is 2287, which is also shown in the graph.

Note that although O-F and Lossy-L have lower ERR than DXMiner, they have higher M_{new} rates, as they misses most of the novel class instances. This is because both FAE and Lossy-L use the Lossy-L conversion, which, according to Lemma 1, is likely to mis-classify more novel class instances as existing class instance (i.e., have higher M_{new} rates). On the other hand, Lossy-F has zero

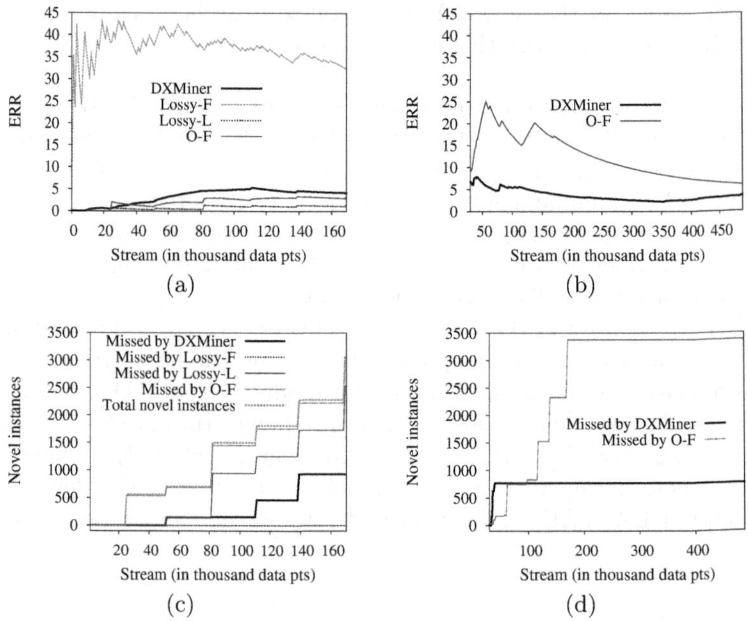

Fig. 1. ERR rates and missed novel classes in Twitter (a,c) and Forest (b,d) datasets

M_{new} rate, but it has very high false positive rate. This is because it wrongly recognizes most of the data points as novel class as a fixed feature vector is used for training the models; although newer and more powerful features evolve often in the stream. Figure 1(b),(d) show the ERR rates and number of novel classes missed, respectively, for Forest dataset. Note that since the feature vector is fixed for this dataset, no feature space conversion is required, and therefore, Lossy-L and Lossy-F are not applicable here. We also generate ROC curves for the Twitter, KDD, and Forest datasets by plotting false novel class detection rate (false positive rate if we consider novel class as positive class and existing classes as negative class) against true novel class detection rate (true positive

Table 2. Summary of the results

Dataset	Method	ERR	M_{new}	F_{new}	AUC	FP	FN
Twitter	DXMiner	4.2	30.5	0.8	**0.887**	-	-
	Lossy-F	32.5	0.0	32.6	0.834	-	-
	Lossy-L	1.6	82.0	0.0	0.764	-	-
	O-F	3.4	96.7	1.6	0.557	-	-
ASRS	DXMiner	**0.02**	-	-	**0.996**	**0.00**	**0.1**
	DXMiner(info-gain)	1.4	-	-	0.967	0.04	10.3
	O-F	3.4	-	-	0.876	0.00	24.7
Forest	DXMiner	**3.6**	**8.4**	1.3	**0.973**	-	-
	O-F	5.9	20.6	1.1	0.743	-	-
KDD	DXMiner	**1.2**	**5.9**	**0.9**	**0.986**	-	-
	O-F	4.7	9.6	4.4	0.967	-	-

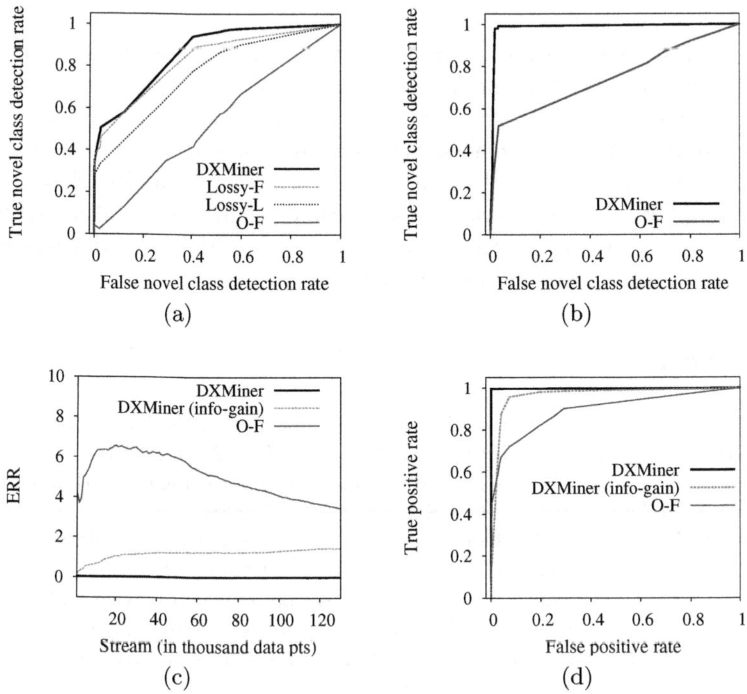

Fig. 2. ROC curves for (a) Twitter, (b) Forest dataset; ERR rates (c) and ROC curves (d) for ASRS dataset

rate). The ROC curves corresponding to Twitter and Forest datasets are shown in figures 2(a,b), and the corresponding AUCs are reported in table 2.

Figure 2(c) shows the ERR rates for ASRS dataset, averaged over all 13 classes. Here DXMiner (with deviation weight feature selection criterion) has the lowest error rate. Figure 2(d) shows the corresponding ROC curves. Each ROC curve is averaged over all 13 classes. Here too, DXMiner has the highest area under the curve (AUC), which is 0.996, whereas O-F has AUC=0.876. Table 2 shows the summary of performances of all approaches in all datasets. Note that for the ASRS we report false positive (FP) and false negative (FN) rates, since ASRS does not have any novel classes. The FP and FN rates are averaged over all 13 classes. For any dataset, DXMiner has the highest AUC.

The running times (training plus classification time per 1,000 data points) of DXMiner and O-F for different datasets are 26.4 and 258 (Twitter), 34.9 and 141 (ASRS), 2.2 and 13.1 (Forest), and 2.6 and 66.7 seconds (KDD), respectively. It is obvious that DXMiner is at least 4 times or more faster than O-F in any dataset. Twitter and ASRS datasets require longer running times than Forest and KDD due to the feature space conversions at runtime. O-F is much slower than DXMiner because $|C|$ OLINDDA models run in parallel, where $|C|$ is the number of classes, making O-F roughly $|C|$ times slower than DXMiner.

6 Conclusion

We have presented a novel technique to detect new classes in concept-drifting data streams having dynamic feature space. Most of the existing data stream classification techniques either cannot detect novel class, or does not consider the dynamic nature of feature spaces. We have analytically demonstrated the effectiveness of our approach, and empirically shown that our approach outperforms the state-of-the art data stream classification techniques in both classification accuracy and processing speed. In the future, we would like to address the multi-label classification problem in data streams.

References

1. Chen, S., Wang, H., Zhou, S., Yu, P.: Stop chasing trends: Discovering high order models in evolving data. In: Proc. ICDE 2008, pp. 923–932 (2008)
2. Fan, W.: Systematic data selection to mine concept-drifting data streams. In: Proc. ACM SIGKDD, Seattle, WA, USA, pp. 128–137 (2004)
3. Hulten, G., Spencer, L., Domingos, P.: Mining time-changing data streams. In: SIGKDD, San Francisco, CA, USA, pp. 97–106 (August 2001)
4. Katakis, I., Tsoumakas, G., Vlahavas, I.: Dynamic feature space and incremental feature selection for the classification of textual data streams. In: Fürnkranz, J., Scheffer, T., Spiliopoulou, M. (eds.) PKDD 2006. LNCS (LNAI), vol. 4213, pp. 102–116. Springer, Heidelberg (2006)
5. Kolter, J., Maloof, M.: Using additive expert ensembles to cope with concept drift. In: ICML, Bonn, Germany, pp. 449–456 (August 2005)
6. Masud, M.M., Gao, J., Khan, L., Han, J., Thuraisingham, B.M.: Integrating novel class detection with classification for concept-drifting data streams. In: Buntine, W., Grobelnik, M., Mladenić, D., Shawe-Taylor, J. (eds.) ECML PKDD 2009. LNCS, vol. 5782, pp. 79–94. Springer, Heidelberg (2009); Extended version is in the preprints, IEEE TKDE, vol. 99 (2010), doi = http://doi.ieeecomputersociety.org/10.1109/TKDE.2010.61
7. Masud, M.M., Gao, J., Khan, L., Han, J., Thuraisingham, B.M.: A practical approach to classify evolving data streams: Training with limited amount of labeled data. In: Perner, P. (ed.) ICDM 2008. LNCS (LNAI), vol. 5077, pp. 929–934. Springer, Heidelberg (2008)
8. Spinosa, E.J., de Leon, A.P., de Carvalho, F., Gama, J.: Cluster-based novel concept detection in data streams applied to intrusion detection in computer networks. In: ACM SAC, pp. 976–980 (2008)
9. Wang, H., Fan, W., Yu, P.S., Han, J.: Mining concept-drifting data streams using ensemble classifiers. In: KDD 2003, pp. 226–235 (2003)
10. Wenerstrom, B., Giraud-Carrier, C.: Temporal data mining in dynamic feature spaces. In: Perner, P. (ed.) ICDM 2006. LNCS (LNAI), vol. 4065, pp. 1141–1145. Springer, Heidelberg (2006)
11. Yang, Y., Wu, X., Zhu, X.: Combining proactive and reactive predictions for data streams. In: Proc. SIGKDD, pp. 710–715 (2005)

Latent Structure Pattern Mining

Andreas Maunz[1], Christoph Helma[2], Tobias Cramer[1], and Stefan Kramer[3]

[1] Freiburg Center for Data Analysis and Modeling (FDM),
Hermann-Herder-Str. 3, D-79104 Freiburg im Breisgau, Germany
maunza@fdm.uni-freiburg.de, cramer@seminar-fr.de
[2] in-silico Toxicology, Altkircherstr. 4, CH-4054 Basel, Switzerland
helma@in-silico.ch
[3] Institut für Informatik/I12, Technische Universität München,
Boltzmannstr. 3, D-85748 Garching bei München, Germany
kramer@in.tum.de

Abstract. Pattern mining methods for graph data have largely been restricted to ground features, such as frequent or correlated subgraphs. Kazius et al. have demonstrated the use of elaborate patterns in the biochemical domain, summarizing several ground features at once. Such patterns bear the potential to reveal latent information not present in any individual ground feature. However, those patterns were handcrafted by chemical experts. In this paper, we present a data-driven bottom-up method for pattern generation that takes advantage of the embedding relationships among individual ground features. The method works fully automatically and does not require data preprocessing (e.g., to introduce abstract node or edge labels). Controlling the process of generating ground features, it is possible to align them canonically and merge (stack) them, yielding a weighted edge graph. In a subsequent step, the subgraph features can further be reduced by singular value decomposition (SVD). Our experiments show that the resulting features enable substantial performance improvements on chemical datasets that have been problematic so far for graph mining approaches.

1 Introduction

Graph mining algorithms have focused almost exclusively on ground features so far, such as frequent or correlated substructures. In the biochemical domain, Kazius *et al.* [6] have demonstrated the use of more elaborate patterns that can represent several ground features at once. Such patterns bear the potential to reveal latent information which is not present in any individual ground feature. To illustrate the concept of non-ground features, Figure 1 shows two molecules, taken from a biochemical study investigating the ability of chemicals to cross the blood-brain barrier, with similar gray fragments in each of them (in fact, due to symmetry of the ring structure, the respective fragment occurs twice in the second molecule). Note that the fragments are not completely identical, but differ in the arrow-marked atom (nitrogen vs. oxygen). However, regardless of this difference, both atoms have a strong electronegativity, resulting in a decreased

J.L. Balcázar et al. (Eds.): ECML PKDD 2010, Part II, LNAI 6322, pp. 353–368, 2010.

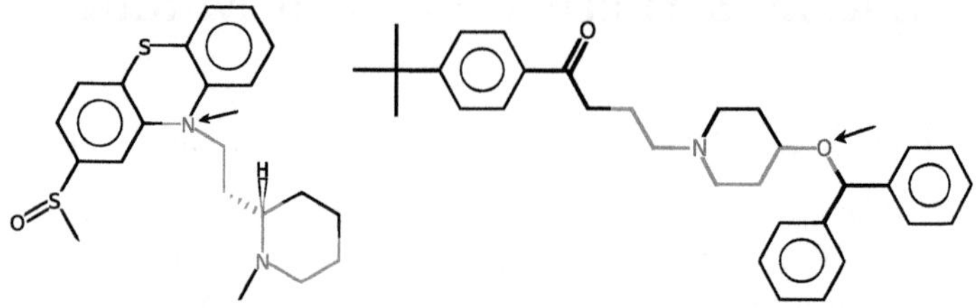

Fig. 1. Two molecules with strong polarity, induced by similar fragments (gray)

ability to cross membranes in the body, such as the blood-brain barrier. So far, the identification of such patterns requires expert knowledge [6] or extensive pre-processing of the data (annotating certain nodes or edges by wildcards or specific labels) [3].

We present a modular graph mining algorithm to identify higher level (latent) and mechanistically interpretable motifs for the first time in a fully automated fashion. Technically, the approach is based on so-called *alignments* of features, i.e. orderings of nodes and edges with fixed positions in the structure. Such alignments may be obtained for features by controlling the feature generating process in a graph mining algorithm with a canonical enumeration strategy. This is feasible, for instance, on top of current a-priori based graph mining algorithms. Subsequently, based on the canonical alignments, ground features can be stacked onto each other, yielding a weighted edge graph that represents the number of occurrences in the fragment set (see the left and middle panel of Figure 2). In a final step, the weighted edge graph is reduced again (in our case by singular value decomposition) to reveal the latent structure of the feature (see the right panel of Figure 2). In summary, we execute a pipeline with the steps (a) align, (b) stack, and (c) compress. A schematic overview of the algorithm, called LAST-PM (Latent Structure Pattern Mining) in the following, is shown in Figure 2 (from left to right).

The goal of LAST-PM is to find chemical substructures that are chemically meaningful (further examples not shown due to lack of space) and ultimately useful for prediction. More specifically, we compare LAST-PM favorably to the

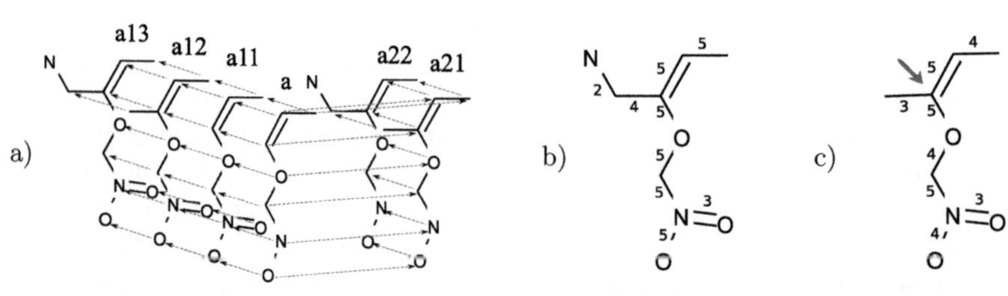

Fig. 2. Illustration of the pipeline with the three steps (a) align, (b) stack, and (c) compress. Left: Aligned ground features in the partial order. Center: Corresponding weighted graph. Right: Latent structure graph.

complete set of ground features from which they were derived in terms of classification accuracy and feature count (baseline comparison), while the tradeoff between runtime and feature count reduction remains advantageous. We also compare accuracy to other state-of-the-art compressed and abstract representations. Finally, we present the results for QSAR endpoints for which data mining approaches have not reached the performance of classical approaches (using physico-chemical properties as features) yet: bioavailability [12] and the ability to cross the blood-brain barrier [7,4]. Our results suggest that graph mining approaches can in fact reach the performance of approaches that require the careful selection of physico-chemical properties on such data.

The remainder of the paper is organized as follows: Section 2 will introduce the graph-theoretic concepts needed to explain the approach. In Section 3, we will present the workflow and basic components (conflict detection, conflict resolution, the stopping criterion and calculating the latent structure graph). Section 4 discusses the algorithm and also briefly the output of the method. Subsequently, we will present experimental results on blood-brain barrier, estrogen receptor binding and bioavailability data, and compare against other types of descriptors. Finally, we will discuss LAST-PM in the context of related work (Section 6) and come to our conclusions (Section 7).

2 Graph Theory and Concepts

We assume a graph database $R = (\mathbf{r}, a)$, where $\mathbf{r}$ is a set of undirected, labeled graphs, and $a : \mathbf{r} \rightarrow \{0, 1\}$ is a function that assigns a class value to every graph (binary classification). Graphs with the same classification are collectively referred to as *target classes*. Every graph is a tuple $r = (V, E, \Sigma, l)$, where $l : V \cup E \rightarrow \Sigma$ is a label function for nodes and edges. An *alignment* of a graph r is a bijection $\phi_r : (V, E) \rightarrow P$, where P is a set of distinct, partially ordered, identifiers of size $n = |V| + |E|$, such as natural numbers. Thus, the alignment function applies to both nodes and edges. We use the usual notion of edge-induced subgraph, denoted by $\subseteq$. If $r' \subseteq r$, then r' is said to *cover* r. This induces a partial order on graphs, the more-general-than relation "$\preceq$", which is commonly used in graph mining: for any graphs r, r', s,

$$r' \preceq r, \text{ if } r \subseteq s \Rightarrow r' \subseteq s. \tag{1}$$

Subgraphs are also referred to as *(ground) features.* The subset of $\mathbf{r}$ that a feature r covers is referred to as the *occurrences* of r, its size as *support* of r in $\mathbf{r}$. A node refinement is an addition of an edge and a node to a feature r. Given a graph r with at least two edges, a *branch* is a node refinement that extends r at a node adjacent to at least two edges. Two (distinct) features obtained by node refinements of a specific parent feature are called *siblings.* Two aligned siblings r and s are called *mutually exclusive*, if they branch at different locations of the parent structure, i.e. let v_i and v_j be the nodes where the corresponding node refinements are attached in the parent structure, then $\phi_r(v_i) \neq \phi_s(v_j)$. Conversely, two siblings r and s are called *conflicting*, if they refine at the same location of the parent structure.

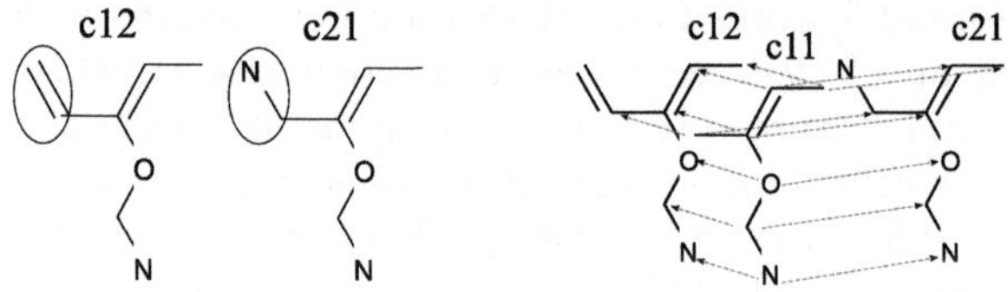

Fig. 3. Left: Conflicting siblings $c12$ and $c21$. Right: Corresponding partial order

For several ground features, alignments can be visualized by overlaying or *stacking* the structures. It is possible to count the occurrences of every component (identified by its position), inducing a weighted graph. Assume a collection of aligned ground features with occurrences significantly skewed towards a single target class, as compared to the overall activity distribution. A "heavy" component in the associated weighted graph is then due to many ground features significant for a specific target class. Assuming correct alignments, the identity of different components is guaranteed, hence multiple adjacent components with equal weight can be considered equivalent in terms of their classification potential.

Figure 2 illustrates the pipeline consisting of the three steps (a) align, (b) stack, and (c) compress, which exploits these relationships. It shows aligned ground features a, $a11$, $a12$, $a13$, $a21$, and $a22$ in the partial order (*search tree*) built by a depth-first algorithm. The aligned features can be stacked onto each other, yielding a weighted edge graph. Subsequently, latent information (such as the main components) can be extracted by SVD. Inspecting the partial order, we note that refining a branches the search due to the sibling pair $a11$ and $a21$. Siblings always induce a branch in the partial order. Note that the algorithm will have to backtrack to the branching positions.

However, in general, the proposed approach is not directly applicable. In contrast to $a11$ and $a21$, which was a mutually exclusive pair, Figure 3 shows a conflicting sibling pair, $c12$ and $c21$, together with their associated part of the partial order (matching elements are drawn on corresponding positions). It is not clear a priori, how conflicting features could be stacked, thus a conflict resolution mechanism is necessary.

The introduced concepts (alignment, conflicts, conflict resolution, and stacking) will now be used in the workflow and algorithm of LAST-PM.

3 Workflow and Basic Steps

In this section, we will elaborate on the main steps of latent structure pattern mining:

1. Ground features are repeatedly stacked, resolving conflicts as they occur. A pattern representing several ground features is created.
2. The process in step 1. is bounded by a criterion to prevent the incorporation of too diverse features.

3. The components with the least information are removed from the structure obtained after step 2. Then the result (*latent structure*) is returned.

In the following, we describe the basic components of the approach in some detail.

3.1 Efficient Conflict Detection

We detect conflicts based primarily on edges and secondarily on nodes. A node list is a vector of nodes, where new nodes are added to the back of the vector during the search. The edge list first enumerates all edges emanating from the first node, then from the second, and so forth. For each specific node, the order of edges is also maintained. Note, that for this implementation of alignment, the ground graph algorithm must fulfill certain conditions, such as partial order on the ground features as well as canonical enumeration (see Section 4). In the following, the *core component* of two siblings denotes their maximum subgraph, i.e. the parent.

Figure 4 shows lists for features $a11$ and $a21$, representing the matching alignment. Underlined entries represent core nodes and adjacent edges. In line with our previous observations, no distinct nodes and no distinct edges have been assigned the same position, so there is no conflict. The node refinement involving node identifier 7 has taken place at different positions. This would be different for the feature pair $c12/c21$.

Due to the monotonic addition of nodes and edges to the lists, conflicts between two ground features become immediately evident through checking corresponding entries in the alignment for inequality. Three cases are observed:

1. Edge lists of f_1 and f_2 do not contain exactly the same elements, but all elements with identical positions, i.e. pairs of ids, are equal. This does not indicate a conflict.
2. There exists an element in each of the lists with the same position that differs in the label. This indicates a conflict.

id	label
0	7
1	6
2	8
3	6
4	6
5	6
6	8
7	8

id1	id2	label
0	1	1
0	6	1
0	7	2
1	2	1
2	3	1
3	4	2
4	5	1

(a) $a11$ node and edge lists

id	label
0	7
1	6
2	8
3	6
4	6
5	6
6	8
7	6

id1	id2	label
0	1	1
0	6	1
1	2	1
2	3	1
3	4	2
3	7	1
4	5	1

(b) $a21$ node and edge lists

Fig. 4. Node and edge lists for conflicting nodes $c12$ and $c21$, sorted by id (position). Underlined entries represent core nodes and adjacent edges.

3. No difference is observed between the edge lists at all. This indicates a conflict, since the difference is in the node list (due to double-free enumeration, there must be a difference).

For siblings $a11$ and $a21$, case 1. applies, and for $c12$ and $c21$, case 2. applies. A conflict is equivalent to a missing maximal feature for two aligned search structures (see Section 3.2). Such conflicts arise through different embeddings of the conflicting features in the database instances. Small differences (e.g., a difference by just one node/edge), however, should be generalized.

3.2 Conflict Resolution

Let r and s be graphs. A *maximum refinement* m of r and s is defined as $(r \preceq m) \wedge (s \preceq m) \wedge (\forall n \succeq r : m \succeq n) \wedge (\forall o \succeq s : m \succeq o)$.

Lemma 1. *Let r and s be two aligned graphs. Then the following two configurations are equivalent:*

1. *There is no maximum refinement m of r and s with alignment ϕ_m induced by ϕ_r and ϕ_s, i.e. $\phi_m \supseteq \phi_r \cup \phi_s$.*
2. *A conflict occurs between r and s, i.e. either*
   (a) $v_i \neq v_j$ for nodes $v_i \in r$ and $v_j \in s$ with $\phi_r(v_i) = \phi_s(v_j)$, or
   (b) $e_i \neq e_j$ for edges $e_i \in r$ and $e_j \in s$ with $\phi_r(e_i) = \phi_s(e_j)$.

Proof. Two directions:
"1. $\Rightarrow$ 2.": Assume the contrary. Then the alignments are compatible, i.e. no unequal nodes $v_i \neq v_j$ or edges $e_i \neq e_j$ are assigned the same position. Thus, there is a common maximum feature m with $\phi_m \supseteq \phi_r \cup \phi_s$.
"1. $\Leftarrow$ 2.": Since ϕ is a bijection, there can be at most one value assigned by ϕ for every node and edge. However, the set $\phi_m \supseteq \phi_r \cup \phi_s$ violates this condition due to the conflict. Thus, there is no m with $\phi_m \supseteq \phi_r \cup \phi_s$.

In Figure 3, the refinements of $c11$ have no maximum element, since they include conflicting ground features $c12$ and $c21$. In contrast, refinements of a in Figure 2 do have a maximum element (namely feature $a13$).

As a consequence of Lemma 1, conflicts prove to be barriers when we wish to merge several features to patterns, especially in case of patterns that stretch beyond the conflict position. A way to resolve conflicts and to incorporate two

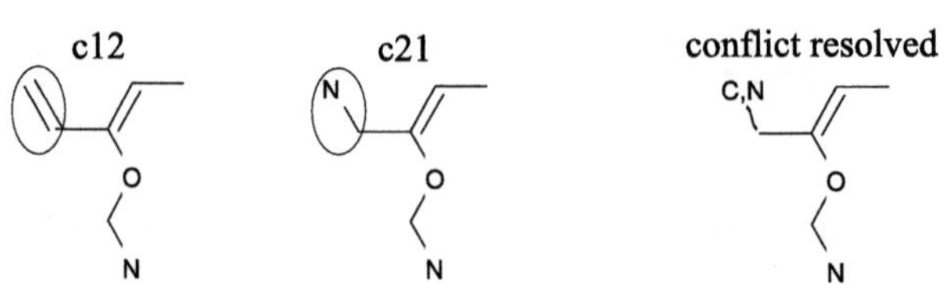

Fig. 5. Conflict resolution by logical OR

conflicting features in a latent feature is by logical OR, i.e. any of the two labels may be present for a match. For instance, $c12$ and $c21$ can be merged by allowing either single or double bond and either node label of $\{N, C\}$ at the conflicting edge and node, as shown in Figure 5, represented by a curly edge and multiple node labels.

Conflicts and mutually exclusive ground features arise from different embeddings of the features in the database, i.e. the anti-monotonic property of diminishing support is lost between pairs of conflicting or mutually exclusive features. This also poses a problem for directly calculating the support of latent patterns.

3.3 Stopping Criterion

Since the alignment, and therefore equal and unequal parts, are induced by the partial order of the mining process, which is in turn a result of the embeddings of ground features in the database, we employ those to mark the boundaries within which merging should take place. Given a ground feature f, its support in the positive class is defined as $y = |\{r \in \mathbf{r} \mid covers(f,r) \wedge a(r) = 1\}|$, its (global) support as x. We use χ^2 values to bound the merging process, since they incorporate a notion of *weight*: a pattern with low (global) support is downweighted, whereas the occurrences of a pattern with high support are similar to the overall distribution. Assuming $n = |\mathbf{r}|$ the number of graphs, define the *weight* of a feature as $w = \frac{x}{n}$. Moreover, assuming $m = \{r \in \mathbf{r} \mid a(r) = 1\}$, define the *expected support* in the positive [negative] class as wm [$w(n-m)$]. The function

$$\chi^2_d(x,y) = \frac{(y-wm)^2}{m} + \frac{(x-y-w(n-m))^2}{w(n-m)} \quad (2)$$

calculates the χ^2 value for the distribution test as the sum of squares of deviation from the expected support for both classes. Values exceeding 3.84 ($\approx$ 95%

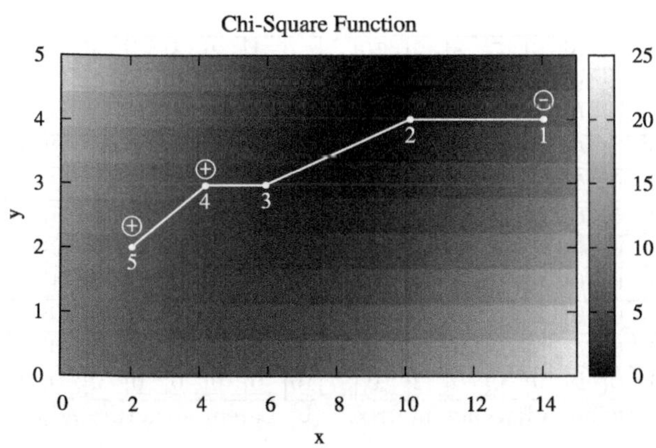

Fig. 6. Contour map of χ^2 values for a balanced class distribution and possible values for a refinement path

significance for $1df$) are considered significant. Here, we consider significance for each target class individually. Thus, a significant feature f is correlated to either (a) the positive class, denoted by $f_\oplus$, if $y > wm$, or (b) the negative class, denoted by $f_\ominus$, if $x - y > w(n - m)$.

Definition 1. *Patch*
Given a graph database $R = \{r, a\}$, a patch P is a set of significant ground features, where for each ground feature f there is a ground feature in P that is either sibling or parent of f, and for each pair of ground features $(f_X, g_Y) : X = Y,\quad X, Y \in \{\oplus, \ominus\}$.

The contour map for equally balanced target classes, a sample size of 20 and occurrence in half of the compounds in Figure 6 illustrates the (well-known) convexity of the χ^2 function and a particular refinement path in the search tree with features partially ordered by χ^2 values as $1_\ominus > 2 < 3 < 4_\oplus < 5_\oplus$.

3.4 Latent Structure Graph Calculation

In order to find the latent (hidden) structures, a "mixture model" for ground features can be used, i.e. elements (nodes and edges) are weighted by the sum of ground features that contain this element. It is obtained by stacking the aligned features of a specific patch, followed by a compression step. To extract the latent information, singular value decomposition (SVD) can be applied. It is recommended by Fukunaga to keep $80\% - 90\%$ of the information [2].

The first step is to count the occurrences of the edges in the ground features and put them in an adjacency table. For instance, Table 7(a) shows the pattern that results from the aligned features $a11$, $a12$, $a13$, $a21$, and $a22$ (see Figure 2). As a specific example, edge $1 - 2$ was present in all five ground features, whereas edge $9 - 10$ occurred in two features only. We applied SVD with 90% to the corresponding matrix and obtained the latent structure graph matrix in Figure 7(b). Here, we removed spurious edges that were introduced by SVD

	1	2	3	4	5	6	7	8	9	10
1	0	5	0	0	0	0	0	0	0	0
2	5	0	5	0	0	0	0	3	0	0
3	0	5	0	5	0	0	0	0	0	0
4	0	0	5	0	5	0	0	0	0	0
5	0	0	0	5	0	5	0	0	4	0
6	0	0	0	0	5	0	5	0	0	0
7	0	0	0	0	0	5	0	0	0	0
8	0	3	0	0	0	0	0	0	0	0
9	0	0	0	0	4	0	0	0	0	2
10	0	0	0	0	0	0	0	0	2	0

(a) Weighted original adjacency matrix.

	1	2	3	4	5	6	7	8	9	10
1	0	4	0	0	0	0	0	0	0	0
2	4	0	5	0	0	0	0	3	0	0
3	0	5	0	4	0	0	0	0	0	0
4	0	0	4	0	5	0	0	0	0	0
5	0	0	0	5	0	5	0	0	3	0
6	0	0	0	0	5	0	4	0	0	0
7	0	0	0	0	0	4	0	0	0	0
8	0	3	0	0	0	0	0	0	0	0
9	0	0	0	0	3	0	0	0	0	0
10	0	0	0	0	0	0	0	0	0	0

(b) Latent structure adjacency matrix.

Fig. 7. Input (left) and output (right) of latent structure graph calculation, obtained by aligning the features $a11 - a22$

(compression artifacts). As can be seen, the edges leading to the two nodes with degree 3 are fully retained, while the peripheral ones are downweighted. In fact, edge $9-10$ is even removed, since it was downweighted to weight 0. In general, SVD downweights weakly interconnected areas, corresponding to a blurred or downsampled picture of the original graph, which has previously proven useful in finding a basic motif in several ground patterns [13].

Definition 2. *Latent Structure Pattern Mining (LAST-PM)*
Given a graph database R, and a user-defined minimum support m, calculate the latent structure graph of all patches in the search space, where for each ground feature f, $supp(f) \geq m$.

4 Algorithm

Given the preliminaries and description of the individual steps, we are now in a position to present a unified approach to latent structure pattern mining, combining alignment, conflict resolution, and component weighting. The method assumes (a) a partial order on ground features (vertical ordering), and (b) canonical representations for ground features, avoiding multiple enumerations of features (horizontal ordering). A depth-first pattern mining algorithm, possibly driven by anti-monotonic constraints, can be used to fulfill these requirements. We follow a strategy to extract latent structures from patches. A latent structure is a graph more general than defined in Section 2: the edges are attributed with weights, and the label function is replaced by a label relation, allowing multiple labels. Since patches stretch horizontally (sibling relation), as well as vertically (parent relation), we need a recursive updating scheme to embed the construction of the latent structure in the ground graph mining algorithm.

We first inspect the horizontal merging: given a specific level of refinement i, we start with an empty latent structure l^i and aggregate siblings from low to high in the lexicographic ordering, starting with empty l^i. For each sibling s and innate l^i, it holds that either

1. s is not significant for any target class, or
2. s is significant for the same target class as l^i, i.e. $X = Y$, for s_X, l^i_Y (if empty, s initializes l^i to its class), or
3. s is significant for the other target class.

In cases 1. and 3., l^i is subjected to latent structure graph calculation and output, and a new, empty latent l^i is created. For case 3., it is additionally initialized with s. For case 2., however, s and l^i are merged, i.e. subjected to conflict resolution, aligning s and l^i, and stacking s onto l^i.

For the vertical or topdown merging, we return l^i to the calling refinement level $i-1$, when all siblings have been processed as described above. Structures l^i and l^{i-1} are merged, if l^i is significant for the same target class as l^{i-1}, i.e. $X = Y$, for l^i_X, l^{i-1}_Y. Also, condition 1. must not be fulfilled for the current sibling on level $i-1$. Otherwise, both l^i and l^{i-1} are subjected to latent structure graph calculation and output, and a new l^{i-1} is created.

Input : Latent structures l_1, l_2; an interval C of core node positions.
Output: Aligned and stacked version of l_1 and l_2, conflicts resolved.

1 **repeat**
2   E.clear() ; E_{l_1}.clear() ; E_{l_2}.clear() ;
3   **for** $j = 0$ *to (size(C)-1)* **do**
4   index = $C[j]$;
5         $I = (l_1$.to[index] $\cap$ l_2.to[index]) ;
6   E.insert($I \setminus C$) ;
7   E_{l_1}.insert(l_2.to[index] $\setminus I$) ;
8   E_{l_2}.insert(l_1.to[index] $\setminus I$) ;
9   **end**
10   **if** $min(E_{l_1}) \leq min(E_{l_2})$ **then** $M_1 = E_{l_1}$ **else** $M_1 = E_{l_2}$;
11   **if** $min(E) < min(M_1)$ **then** $M_2 = E$ **else** $M_2 = M_1$;
12   *core_new*.insert(min(M_2)) ;
13   **if** $M_1 == E_{l_1}$ **then** l_2.add_edge(min(M_1)) **else** l_1.add_edge(min(M_1)) ;
14 **until** *E.size==0* $\wedge$ E_{l_1}*.size==0* $\wedge$ E_{l_2}*.size==0* ;
15 l_1 = stack(l_1, l_2) ;
16 l_1 = alignment(l_1, l_2, *core_new*) ;
17 **return** l_1;

Algorithm 1. Alignment Calculation

Alignment calculation (Algorithm 1) works recursively: In lines 3-9, it extracts mutually exclusive edges leaving core positions to non-core positions, i.e. there is a distinction between edges leaving the core, but are shared by $l1$ and $l2$ (conflicting edges, E), vs. edges that are unique to either l_1 or l_2 (non-conflicting edges, E_{l_1}, E_{l_2}). The overall minimum edge is remembered for the next iteration, ordered by "to"-node position (lines 11-12). The minimum edge of E_{l_1} and E_{l_2} (line 10; in case of equality, E_{l_1} takes precedence) is added to the other structure where it was missing (line 13).

The procedure can be seen as inserting pseudo-edges into the two candidate structures that were only present in the other one before, thus creating a canonical alignment. For instance, in Figure 4, exclusive edge 0-7 from $a11$ would be first inserted into $a21$, *pushing* node 7 to node 8 and edge 3-7 to edge 3-8 in $a21$. Subsequently, vice versa, exclusive edge 3-8 would be inserted into $a11$, leaving no more exclusive edges, i.e. the two structures are aligned.

This process is repeated until no more edges are found, resulting in the alignment of l_1 and l_2. Line 15 then calls the stacking routine, a set-insertion of l_2's node and edge labels into l_1's and the addition of l_2's edge weights to l_1's, and line 16 repeats the process for the next block of core ids. Due to the definition of node and edge lists, the following invariant holds in each iteration: For the node list, core components are always enumerated in a contiguous block, and for each edge e, the core components are always enumerated at the beginning of the partition of the edge list that corresponds to e. For horizontal (vertical) merging, we call Algorithm 1 with $l_1 := l^i, l_2 := s$ ($l_1 := l^{i-1}, l_2 := l^i$). This ensures that l_1 comprises only ground features lower in the canonical ordering than l_2. Thus, Algorithm 1 correctly calculates the alignments (we omit a formal proof due to space constraints).

4.1 Complexity

We modified the graph miner Gaston by Nijssen and Kok [9] to support latent structure pattern mining[1]. It is especially well-suited for our purposes: First, Gaston uses a highly efficient canonical representation for graphs. Specifically, no refinement is enumerated twice. Second, Gaston employs a canonical depth sequence formulation that induces a partial order among trees (we do not consider cycle-closing structures due the complexity of the isomorphism problem for general graphs). Siblings in the partial order can be compared lexicographically.

LAST-PM allows the use of anti-monotonic constraints for pruning the search in the forward direction, such as minimum frequency or upper bounds for convex functions, e.g χ^2. The former is integrated in Gaston. For the latter, we implemented statistical metric pruning using χ^2 upper bound as described in [8]. Obviously, the additional complexity incurred by LAST-PM depends on conflict resolution, alignments, and stacking (see Algorithm 1), as well as weighting (SVD).

- Algorithm 1 for latent structures $l1$, $l2$ takes at most $|l1| + |l2|$ insert operations, i.e. is linear in the number of edges (including conflict resolution).
- For each patch, an SVD of the $m \times n$ latent structure graph is required ($mn^2 - n^3/3$ multiplications).

Thus, the overhead compared to the underlying Gaston algorithm is rather small (see Section 5).

5 Experiments

In the following, we present our experimental results on three chemical datasets with binary class labels from the study by Rückert and Kramer [10]. The nctrer dataset deals with the binding activity of small molecules at the estrogen receptor, the Yoshida dataset classifies molecules according to their bioavailability, and the bloodbarr dataset deals with the degree to which a molecule can cross the blood-brain barrier. For the bloodbarr/ nctrer/ yoshida datasets, the percentage of active molecules is 66.8/ 59.9/ 60.0. For efficiency reasons, we only consider the core chemical structure without hydrogen atoms. Hydrogens attached to fragments can be inferred from matching the fragments back to the training structures. Program code, datasets and examples are provided on the supporting website `http://last-pm.maunz.de`

5.1 Methodology

Given the output XML file of LAST-PM, SMARTS patterns for instantiation are created by parsing patterns depth-first (directed). Focusing on a node, all outgoing edges have weights according to Section 3.4. This forms weight levels of branches with the same weight. We may choose to make some branches optional, based on the size of weight levels, or demand all branches to be attached:

[1] Version 1.1 (with embedding lists), see `http://www.liacs.nl/~snijssen/gaston/`

- *nop*: demand all (**no** **op**tional) branches.
- *msa*: demand number of branches equal to **m**aximum **s**ize of **a**ll levels
- *nls*: demand number of branches equal to highest (**n**ext) **l**evel **s**ize

For example, *nop* would simply disregard weights and require all of the three bonds leaving the arrow-marked atom of Figure 2 (right), while *nls* (here also *msa*) would require any two of the three branches to be attached. With *msa* and *nls*, we hope to better capture combinations of important branches. The two methods allow, besides simple disjunctions of atomic node and edge labels such as in Figure 1, for (nested) optional parts of the structure.[2]

All experimental results were obtained from repeated ten-fold stratified cross-validation (two times with different folds) in the following way: We used edge-induced subgraphs as ground features. For each training set in a crossvalidation, descriptors were calculated using 6% minimum frequency and 95% χ^2-significance on ground features. This ensures that features are selected ignorant of test sets. Atoms were not attributed with aromatic information but only labeled by their atomic number. Edges were attributed as single, double and triple, or as aromatic bond, as inferred from the molecular structure. Features were converted to SMARTS according to the variants *msa*, *nls*, and *nop*, and matched onto training and test instances, yielding instantiation tables. We employed unoptimized linear SVM models and a constant parameter $C = 1$ for each pair of training and test set. The statistics in the tables were derived from pooling the twenty test set results into a global table first. Due to the skewed target class distributions in the datasets (see above), it is easy to obtain relatively high predictive accuracies by predicting the majority class. Thus, the evaluation of a model's performance should be based primarily on a measure that is insensitive to skew. We chose AUROC for that purpose. A 20% SVD compression (percentage of sum of singular value squares) is reported for the LAST-PM features, since this gave the best AUROC values of 10, 15, and 20% in preliminary trials in two out of three times. Significance is determined by the 95% confidence interval.

5.2 Validation Results

We compare the performance of LAST-PM descriptors in Table 1 with

1. ALL ground features from which LAST-PM descriptors were obtained (baseline comparison).
2. BBRC features by Maunz, Helma, and Kramer [8] to relate to structurally diverse and class-correlated ground features.
3. MOSS features by Borgelt and Berthold [3] to see the performance of another type of abstract patterns.
4. SLS features by Rückert and Kramer [10] to see the performance of ground features compressed according to the so-called dispersion score.

[2] Figure 1 is an actual pattern found by LAST-PM in the bloodbarr dataset. See the supporting website at `http://last-pm.maunz.de` for the implementation in SMARTS.

Table 1. Comparative analysis (repeated 10-fold crossvalidation)

Dataset	LAST-PM			ALL	BBRC	MOSS	SLS
	Variant	%Train	%Test	%Test	%Test	%Test	%Test
bloodbarr	nls+nls	84.19	72.20	70.49[a]	68.50[a]	67.49[a]	70.4[b]
nctrer	nls+msa	88.01	80.22	79.13	80.22	77.17[a]	78.4[b]
yoshida	nop+msa	82.43	69.81	65.19[a]	65.96[a]	66.46[a]	63.8[b]

[a] significant difference to LAST-PM.
[b] result from the literature, no significance testing possible

For ALL and BBRC, a minimum frequency of 6% and a significance level of 95% were used. For the MOSS approach, we obtained features with MoSS [3]. This involves cyclic fragments and special labels for aromatic nodes. In order to generalize from ground patterns, ring bonds were distinguished from other bonds. Otherwise (including minimum frequency) default settings were used, yielding only the most specific patterns with the same support (closed features). For SLS, we report the overall best figures for the dispersion score and the SVM model from Table 1 in their paper. As can be seen from Table 1, using the given variants for the first and second fold, respectively, LAST-PM outperforms ALL, BBRC and MOSS significantly for the bloodbarr and yoshida dataset (paired corrected t-test, $n = 20$), as well as MOSS for the nctrer dataset (seven out of nine times in total).

Table 2 relates feature count and runtime of LAST-PM and ALL (median of 20 folds). FCR is the feature count ratio, RTR the runtime ratio between LAST-PM and ALL, as measured for descriptor calculation on our 2.4 GHz Intel Xeon test system with 16GB of RAM, running Linux 2.6. Since $1/FCR$ always exceeds RTR, we conclude that the additional computational effort is justified. Note that nctrer seems to be an especially dense dataset. Profiling showed that most CPU time is spent on alignment calculation, while SVD can be neglected.

In their original paper [12], Yoshida and Topliss report on the prediction on an external test set of 40 compounds with physico-chemical descriptors, in which they achieved a false negative count of 2 and false positive count of 7. We obtained the test set and could reproduce their exact accuracy with 1 false negative and 8 false positives, using LAST-PM features.

Hu and co-workers [7], authors of the bloodbarr dataset study, provided us with the composition of their "external" validation set, which is in fact a subset of

Table 2. Analysis of feature count and runtime

Dataset	LAST-PM	ALL	FCR/RTR
bloodbarr	249 (1.23s)	1613 (0.36s)	0.15 /3.41
nctrer	193 (12.49s)	22942 (0.13s)	0.0084 /96.0769
yoshida	124 (0.28s)	462 (0.09s)	0.27 /3.11

the complete dataset, comprising 64 positive and 32 negative compounds. Their SVM model was based on carefully selected physico-chemical descriptors, and yielded only seven false positives and seven false negatives, an overall accuracy of 85.4%. Using LAST-PM features and our unoptimized polynomial kernel, we predicted only five false positives and two false negatives, an overall accuracy of 91.7%.

We conducted further experiments with another 110 molecule blood-brain barrier dataset (46 active and 64 inactive compounds) by Hou and Xu [4], that we obtained together with pre-computed physico-chemical descriptors. Here, we achieved a AUROC value of 0.78 using LAST-PM features in repeated 10-fold crossvalidation, close to the 0.80 that the authors obtained with the former. However, when combined, both descriptor types give an AUROC of 0.82. In contrast to this, AUROC could not be improved in combination with BBRC instead of LAST-PM descriptors.

6 Related Work

Latent structure pattern mining allows deriving basic motifs within the corresponding ground features that are frequent and significantly correlated with the target classes. The approach falls into the general framework of graph mining. Roughly speaking, the goal of pattern mining approaches to graph mining is to enumerate all interesting subgraphs occurring in a graph database (interestingness defined, e.g., in terms of frequency, class correlation, non-redundancy, structural diversity, ...). Since this ensemble is in general exponentially large, different techniques for selecting representative subgraphs for classification purposes have been proposed, e.g. by Yan [11]. Due to the NP-completeness of the subgraph isomorphism problem, no efficient algorithm is known for general graph mining (i.e. including cyclic structures). For a detailed introduction to the tractable case of non-cyclic graph mining, see the overview by Muntz *et al.* [1], which mostly targets methods with minimum frequency as interestingness criterion. Regarding advanced methods that go beyond the mining of ground features, we relate our method to approaches that provide or require basic motifs in the data, and/or are capable of dealing with conflicts.

Kazius *et al.* [6] created two types of (fixed) high level molecule representations (aromatic and planar) based on expert knowledge. These representations are the basis of graph mining experiments.

Inokuchi [5] proposed a method for mining generalized subgraphs based on a user-defined taxonomy of node labels. Thus, the search extends not only due to structural specialization, but also along the node label hierarchy. The method finds the most specific (closed) patterns at any level of taxonomy and support. Since the exact node and edge label representation is not explicitly given beforehand, the derivation of abstract patterns is semi-automatic.

Hofer, Borgelt and Berthold [3] present a pattern mining approach for ground features with class-specific minimum and maximum frequency constraints, that can be initialized with arbitrary motifs. All solution features are required to contain the seed. Moreover, their algorithm MoSS offers the facility to collapse ring structures into special nodes, to mark ring components with special node and edge labels, or to use wildcard atom types: Under certain conditions (such as if the atom is part of a ring), multiple atom types are allowed for a fixed position. It also mines cyclic structures at the cost of losing double-free enumeration.

All approaches have in common that the (chemical expert) user specifies high-level motifs of interest beforehand via a specific molecule representation. They integrate in different ways user-defined wildcard search into the search tree expansion process, whereas the approach presented here derives abstract patterns automatically by resolving conflicts during backtracking and weighting.

7 Conclusions

In the paper, we introduced a method for generating abstract non-ground features for large databases of molecular graphs. The approach differs from traditional graph mining approaches in several ways: Incorporating several similar features into a larger pattern reveals additional (*latent*) information, e.g., on the most frequently or infrequently incorporated parts, emphasizing a common interesting motif. It can thus be seen as graph mining on subgraphs. In traditional frequent or correlated pattern mining, sets of ground features are returned, including groups of very similar ones with only minor variations of the same interesting basic motif. It is, however, hard and error-prone (or sometimes even impossible) to appropriately select a representative from each group, such that it conveys the basic motif. Latent structure pattern mining can also be regarded as a form of abstraction, which has been shown to be useful for noise handling in many areas. It is, however, new to graph and substructure mining.

The key experimental results were obtained on blood-brain barrier (BBB), estrogen receptor binding and bioavailability data, which have been hard for substructure-based approaches so far. In the experiments, we showed that the non-ground feature sets improve over the set of all ground features from which they were derived, but also over MOSS [3], BBRC [8] and compressed [10] ground feature sets when used with SVM models. In seven out of nine cases, the improvements are statistically significant. We also found a favorable tradeoff between feature count of and runtime for computing LAST-PM descriptors compared to the complete set of frequent and correlated ground features.

We took bioavailability and blood-brain barrier data and QSAR models from the literature and showed that, on three test sets obtained from the original authors, the purely substructure-based approach is on par with or even better than their approach based on physico-chemical properties only. We also showed that LAST-PM features can enhance the performance of solely physico-chemical properties. Therefore, latent structure patterns show some promise to make hard (Q)SAR problems amenable to graph mining approaches.

Acknowledgements. The research was supported by the EU seventh framework programme under contract no Health-F5-2008-200787 (OpenTox).

References

1. Chi, Y., Muntz, R.R., Nijssen, S., Kok, J.N.: Frequent Subtree Mining - An Overview, 2001. Fundamenta Informaticae 66(1-2), 161–198 (2004)
2. Fukunaga, K.: Introduction to Statistical Pattern Recognition, 2nd edn. Academic Press Professional, Inc., San Diego (1990)
3. Hofer, H., Borgelt, C., Berthold, M.R.: Large Scale Mining of Molecular Fragments with Wildcards. Intelligent Data Analysis 8(5), 495–504 (2004)
4. Hou, T.J., Xu, X.J.: ADME Evaluation in Drug Discovery. 3. Modeling Blood-Brain Barrier Partitioning Using Simple Molecular Descriptors. Journal of Chemical Information and Computer Sciences 43(6), 2137–2152 (2003)
5. Inokuchi, A.: Mining Generalized Substructures from a Set of Labeled Graphs. In: IEEE International Conference on Data Mining, pp. 415–418 (2004)
6. Kazius, J., Nijssen, S., Kok, J., Baeck, T., Ijzerman, A.P.: Substructure Mining Using Elaborate Chemical Representation. Journal of Chemical Information and Modeling 46, 597–605 (2006)
7. Li, H., Yap, C.W., Ung, C.Y., Xue, Y., Cao, Z.W., Chen, Y.Z.: Effect of Selection of Molecular Descriptors on the Prediction of Blood-Brain Barrier Penetrating and Nonpenetrating Agents by Statistical Learning Methods. Journal of Chemical Information and Modeling 45(5), 1376–1384 (2005)
8. Maunz, A., Helma, C., Kramer, S.: Large-Scale Graph Mining Using Backbone Refinement Classes. In: KDD 2009: Proceedings of the 15th ACM SIGKDD International Conference on Knowledge Discovery and Data Mining, pp. 617–626. ACM, New York (2009)
9. Nijssen, S., Kok, J.N.: A Quickstart in Frequent Structure Mining can make a Difference. In: KDD 2004: Proceedings of the Tenth ACM SIGKDD International Conference on Knowledge Discovery and Data Mining, pp. 647–652. ACM, New York (2004)
10. Rückert, U., Kramer, S.: Optimizing Feature Sets for Structured Data. In: Kok, J.N., Koronacki, J., Lopez de Mantaras, R., Matwin, S., Mladenič, D., Skowron, A. (eds.) ECML 2007. LNCS (LNAI), vol. 4701, pp. 716–723. Springer, Heidelberg (2007)
11. Yan, X., Cheng, H., Han, J., Yu, P.S.: Mining Significant Graph Patterns by Leap Search. In: SIGMOD 2008: Proceedings of the 2008 ACM SIGMOD International Conference on Management of Data, pp. 433–444. ACM, New York (2008)
12. Yoshida, F., Topliss, J.G.: QSAR Model for Drug Human Oral Bioavailability. Journal of Medicinal Chemistry 43(13), 2575–2585 (2000)
13. Zhu, Q., Wang, X., Keogh, E., Lee, S.-H.: Augmenting the Generalized Hough Transform to Enable the Mining of Petroglyphs. In: KDD 2009: Proceedings of the 15th ACM SIGKDD International Conference on Knowledge Discovery and Data Mining, pp. 1057–1066. ACM, New York (2009)

First-Order Bayes-Ball

Wannes Meert, Nima Taghipour, and Hendrik Blockeel

Katholieke Universiteit Leuven, Department of Computer Science,
Celestijnenlaan 200A, 3001 Heverlee, Belgium

Abstract. Efficient probabilistic inference is key to the success of statistical relational learning. One issue that increases the cost of inference is the presence of irrelevant random variables. The Bayes-ball algorithm can identify the requisite variables in a propositional Bayesian network and thus ignore irrelevant variables. This paper presents a lifted version of Bayes-ball, which works directly on the first-order level, and shows how this algorithm applies to (lifted) inference in directed first-order probabilistic models.

1 Introduction

Probabilistic logic models [2] bring the expressive power of first-order logic to probabilistic models, enabling them to capture both the relational structure and the uncertainty present in data. Efficient inference, however, is a bottleneck in these models, affecting also the cost of learning these models from data. Many attempts have been made to make inference more efficient in these formalisms, including lifted inference methods which try to exploit the symmetries present in the first-order probabilistic model [9,10,8,12]. In this paper, we make use of the first-order structure to focus on the issue of irrelevant random variables in probabilistic logic inference. To answer a specific probabilistic query, there is a minimum set of random variables which are required to be included in the computations. This set is called the *minimal requisite network* (MRN) [11]. Inference becomes more efficient by restricting computations to the MRN.

Bayes-ball [11] is an efficient algorithm that finds the MRN for inference in (propositional) Bayesian networks. The naive way of applying Bayes-ball to a probabilistic logic model is to ground the entire model and apply Bayes-ball on the resulting propositional network. This can be computationally expensive because the grounded network is often large or infinite, and its construction can be a significant part of the total inference cost.

Another way to compute the MRN for a probabilistic logic model consists of two steps: In the first step, all logical proofs for the query and for all evidence atoms are computed (e.g., using SLD resolution [2, Ch.10]), and a network is built using all ground clauses that are used therein. In the second step, the MRN is computed by applying Bayes-ball to this network. The second step is necessary since some atoms encountered in certain proofs of an evidence atom may be D-separated [11] from the query. This method has the disadvantage that it initially computes a Bayesian network that may be larger than the MRN.

J.L. Balcázar et al. (Eds.): ECML PKDD 2010, Part II, LNAI 6322, pp. 369–384, 2010.

None of the mentioned methods take full advantage of the first-order representation of the probabilistic logic model. First-order probabilistic models introduce many random variables which are essentially identical with respect to inference, and hence also share the same status of relevance for a specific probabilistic query. We propose a first-order version of the Bayes-ball algorithm called first-order Bayes-ball (FOBB) that exploits these symmetries to efficiently compute the MRN for probabilistic logic inference. FOBB works directly at the first-order level while building the MRN; there is no need to build the ground network in the beginning. This algorithm treats indistinguishable random variables as one first-order atom and can process them in one single step (performing only one instance of identical operations).

Another contribution of FOBB is in the *first-order representation of the MRN*. This is valuable for lifted inference algorithms, which are designed specifically to make use of the first-order representation. The existing methods for computing the MRN all produce a ground MRN which does not preserve the first-order structure of the original model. FOBB, on the other hand, produces a first-order MRN for each query. This network can be used by lifted inference algorithms and can also make lifted inference more efficient by removing unnecessary but costly operations on the irrelevant parts of the model (such as shattering [8,4]). To the best of our knowledge the issue of relevance has not been addressed in lifted inference settings before.

While FOBB is a general method that applies to several directed probabilistic logic formalisms such as BLPs [3] and CP-logic [6], we illustrate it for the particular case of parametrized Bayesian networks [9], because of the intuitive relationship of this formalism to Bayesian networks.

In section 2 we review the Bayes-ball algorithm and parametrized Bayesian networks. Section 3 introduces the FOBB algorithm. Experiments are presented in Section 4. Section 5 contains the conclusions.

2 Preliminaries

2.1 Bayes-Ball

Bayes-ball [11] identifies the MRN of a Bayesian network for a given set of query and evidence nodes. It is based on the analogy of bouncing balls that travel over the edges of the Bayesian network (Fig. 1). In this analogy a visit from node a to b is compared to passing a ball from a to b. The balls start at the query nodes (each query node receives a ball). Upon reaching each node, a ball may *pass through*, *bounce back* and/or be *blocked*. The action chosen depends on the direction from which it came and on whether the node is probabilistically or deterministically dependent on its parents (based on D-separation). When a ball is to be passed to multiple nodes, it means that a visit to each of those nodes is put in a schedule. At each step a visit is selected from the schedule and processed.

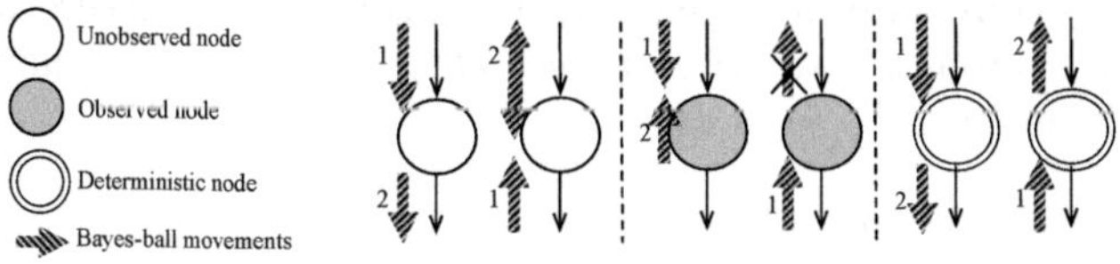

Fig. 1. Different actions performed on the Bayes-ball depending on the type of the node and direction from which the ball comes

The rules by which the balls moves through the network can be summarized as follows:

(a) An unobserved probabilistic node passes balls from parents on, that is, if such a node receives the ball from a parent it passes the ball to all its children. When such a node receives the ball from a child, it passes the ball to both its parents and children.
(b) An observed node bounces balls back from parents, that is, upon receiving the ball from a parent, such a node passes the ball to all its parents. However, an observed node blocks balls from children, that is, the ball is not passed on anymore from this node.
(c) A functional unobserved node always passes balls through, that is, it passes the ball coming from parents to children and vice versa.

Nodes are marked at each visit of a ball, depending on the type of action performed on the ball: when the ball is passed from a node to its parents (children), the node receives a mark on top (bottom). These marks help the algorithm avoid repeating the same action, and guarantee the termination of the algorithm. Having a mark on top (bottom) indicates that there is no need to visit the parents (children) anymore. In the end, these marks indicate the relevance of each node: the MRN consists of all the nodes marked on the top together with the set of evidence atoms visited during the algorithm.

2.2 Parametrized Bayesian Networks

To illustrate FOBB we represent probabilistic logic models as *parametrized Bayesian networks* (PBNs) [9]. In such a model, random variables of a Bayesian network are represented by (ground) atoms, and definite clauses are used to capture the structure of the network. In this way, a first-order atom represents a class of random variables, and a first-order clause represents repeating structures in the Bayesian network. An example PBN is presented in Fig. 2.a.

Each predicate p takes on values from a specific range $Range(p)$, which contains the possible states of the random variables represented by this predicate. Moreover, each argument of a predicate takes values of a specific type. A PBN includes *domain declarations* defining the set of constants of each type, and *functor declarations* assigning a type to each argument of each predicate.

A PBN also consists of a *theory*, which captures the structure of the probabilistic model through a set of *Bayesian clauses.* A Bayesian clause bc is an expression of the form: [1]

$$(bc)\ \forall X_1,\ldots,X_n : C;\ h|b_1,\ldots,b_n$$

with $h, b_1, \ldots, b_n$ first-order atoms with logic variables $X_1, \ldots, X_n$ as arguments. Each logic variable X_i is implicitly assigned to a domain, considering the *functor declarations.* The constraint C restricts all possible groundings of $X_1, \ldots, X_n$ to a subset of the Cartesian product of their respective domains. We define $head(bc) = h$ and $body(bc) = \bigcup_i b_i$. A Bayesian clause specifies that for each substitution $\theta = \{X_1/t_1, \ldots, X_n/t_n\}$ that grounds the clause and is in accordance with C, the random variable $h\theta$ depends on $b_1\theta, \ldots, b_n\theta$.

Each Bayesian clause is associated with a conditional probability distribution (CPD) $P(h|b_1, \ldots, b_n)$, which specifies the same distribution $P(h\theta|b_1\theta, \ldots, b_n\theta)$ for every ground instance of the clause. If a ground atom appears in the head of more than one ground clause, then a *combining rule* is used to obtain the desired probability distribution from the CPDs associated to those clauses. Each predicate has an associated combining rule.

2.3 Equivalent Bayesian Network

Like most directed probabilistic first-order models, a PBN can be transformed to a Bayesian network. Here, we use a transformation to what we call an equivalent Bayesian network (EBN), similar to the EBNs in Meert et al [6]. The EBN is a regular Bayesian network, but we consider it as a bipartite graph containing two types of nodes: *atom nodes* and *factor nodes* (see Fig. 2.b).

Atom nodes correspond to the random variables; there is an atom node in the EBN for each atom in the grounding of the theory.

Factor nodes explicitly capture the factorization declared by the Bayesian clauses. For each clause bc and each grounding substitution θ complying with the constraint C of bc, there is a factor node corresponding to $bc\theta$ in the EBN. We denote factor nodes by atoms too. Each Bayesian clause bc is associated to an n-ary predicate bc', with $n = |Var(bc)|$. This predicate has the same domain as $head(bc)$. The factor node corresponding to ground clause $bc\theta$ can be represented by the atom $bc'\theta$.

There are edges between factor nodes and atom nodes, but no edges between nodes of the same type. The edges of the network can be fully described with this rule: Each factor node $bc'\theta$, corresponding to a ground clause $bc\theta$, has as parents all the atoms in $body(bc\theta)$, and is the parent of the atom $head(bc\theta)$.

The nodes and edges represent the structure of the EBN. In the quantitative component of the EBN, the CPD of each clause bc is associated to the corresponding

[1] We use a slightly different version of Poole's PBN [9]: here each rule is associated with an entire CPD, instead of declaring the probability of one specific combination of values for a node and its parents. This idea is introduced in BLPs [3], from which we also borrow the term *Bayesian clause.*

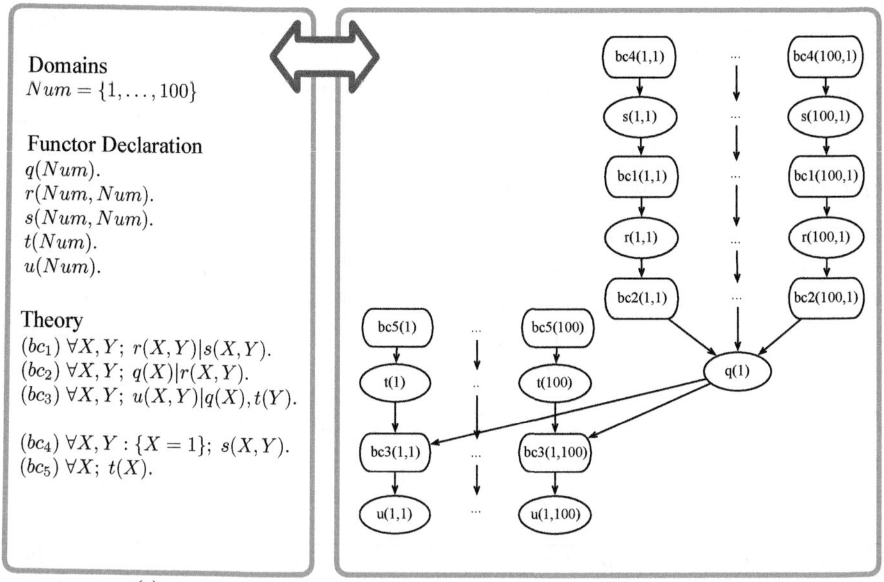

(a) (b)

Fig. 2. (a) Parametrized Bayesian network (b) Equivalent Bayesian network having the same probability distribution as the theory. Bayes-ball can be used on such a Bayesian network to find the MRN given a query and evidence.

factor node bc'. This assignment of CPD to factor nodes is straightforward since by definition factor nodes have the same domain as $head(bc)$. The CPD on the atom nodes is a deterministic function of the parent factor nodes, implementing the effect of the combining rule associated to the atom. In this work, we consider combining rules that represent *independent causation*, such as *noisy-or*, *-and*, *-max*, and *-min*.

3 First-Order Bayes-Ball

FOBB is based on the same principles as Bayes-ball, building upon the transformability of a probabilistic logic model to an EBN. Its main advantage is the possibility to perform some steps at the first-order level. That is, several nodes can be represented by what we call a *first-order node*, and be visited in one single step. After a definition of first-order nodes and related operations on such nodes, we show the main features of the algorithm through an example; this is followed by a more detailed description.

Definition 1 ***(Constraint)*** *Having logic variables* $\mathbf{X} = \{X_1, X_2, \ldots, X_n\}$, *with* $D(X_i)$ *the associated domain of* X_i, *a constraint* C *on* $\mathbf{X}$ *is a relation on* $\mathbf{X}$, *indicating a subset of the Cartesian product* $D(\mathbf{X}) = \times_i D(X_i)$.

There is no restriction on how to represent and store the constraints. The choice of representation, however, affects the efficiency of the algorithm. For example, storing them as ground tuples would cancel the advantages of FOBB over Bayes-ball. For the implementation we opted to store constraints as decision trees with set membership tests in the nodes. This representation is different from that used in [9] for PBNs and other work about lifted inference, where a constraint is a set of (in)equalities involving logic variables and constants. One such conjunction is equivalent to one branch in our decision tree.

Definition 2 ***(First-order node)*** *A first order node $\mathcal{F}$ is a pair (p, C), where $p = a(X_1, \ldots, X_n)$ is a first-order atom, and C is a constraint on logic variables $\mathbf{X} = \{X_1, X_2, \ldots, X_n\}$. Each first-order node $\mathcal{F} = (p, C)$ represents the set of ground random variables $p\theta$, where $\mathbf{X}\theta \in C$. We denote the set of (ground) random variables represented by $\mathcal{F}$ as $RV(\mathcal{F})$.*

For two first-order nodes $\mathcal{F}_1 = (p, C_1)$ and $\mathcal{F}_2 = (p, C_2)$, we define:

1. $\mathcal{F}_1 \subseteq \mathcal{F}_2$ iff $RV(\mathcal{F}_1) \subseteq RV(\mathcal{F}_2)$ iff $C_1 \subseteq C_2$.
2. $\mathcal{F}_1 \Delta \mathcal{F}_2 = \mathcal{F}'$ iff $\mathcal{F}' = (p, C_1 \Delta C_2)$, for $\Delta \in \{\cap, \cup, \setminus\}$.

Definition 3 ***(Splitting)*** *The result of splitting a first-order node $\mathcal{F} = (p, C)$ is a set of first-order nodes $\{\mathcal{F}_1, \ldots, \mathcal{F}_n\}$, where each $\mathcal{F}_i = (p, C_i)$, and such that $\bigcup_i C_i = C$ and $\bigcap_i C_i = \emptyset$.*

FOBB also uses the operation of *projection* on constraints:

Definition 4 ***(Projection)*** *Let C be a constraint on logic variables $\mathbf{X}$. Projection of C on a subset of its variables $\mathbf{Y} \subseteq \mathbf{X}$ is given by the constraint $\pi_Y(C) = \{\mathbf{y} = (y_1, \ldots, y_{|Y|}) | \exists \overline{\mathbf{y}} \in C$, and $\overline{\mathbf{y}}$ is an extension of $\mathbf{y}\}$.*

3.1 Overview

FOBB computes the MRN for a probabilistic query $P(Q|E)$ on the probabilistic logic model M. It is assumed that the query atoms Q are ground and that the theory T of M has a finite grounding. The outer structure of Alg. 1 closely resembles the original Bayes-ball (see [11]), the main differences are that it works with first-order nodes instead of ground nodes, and that it uses the given first-order probabilistic logic model to compute the parents and children of a first-order node.

We use an example to illustrate FOBB and compare it with the original Bayes-ball: Suppose we need to compute $P(Q|E)$ where $Q = \{q(1)\}$ and $E = \{r(1,1), r(1,2), u(1,1), \ldots, u(1,10)\}$, given the theory in Fig. 2.

Similar to Bayes-ball, FOBB schedules the nodes which are to be visited. Instead of scheduling ground nodes to visit, FOBB schedules first-order nodes. Each entry in the schedule is represented by a tuple $\langle \mathcal{F}, \mathit{direction} \rangle$, containing a first-order node $\mathcal{F}$, and the *direction* of the visit ($\mathit{fromChild}$ or $\mathit{fromParent}$). This entry stands for a visit to each node in $RV(\mathcal{F})$, in Bayes-ball. In the beginning, FOBB starts by scheduling the query: $\langle (q(X), \{X{=}1\}), \mathit{fromChild} \rangle$. Next, FOBB retrieves this tuple from the schedule and computes its parents by matching $(q(X), \{X{=}1\})$ to the heads of clauses of T. In this case, only Bayesian clause

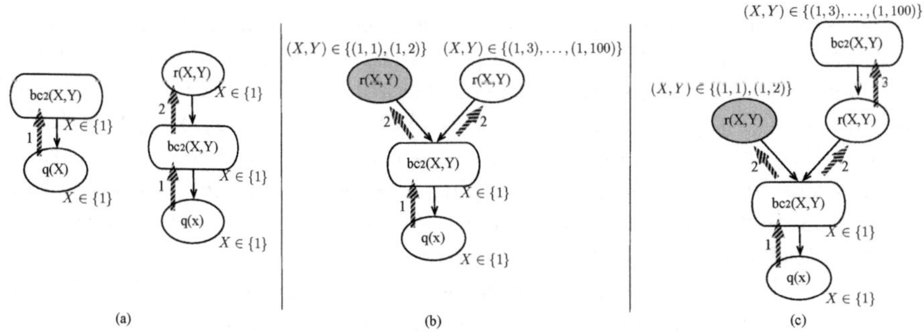

Fig. 3. Illustration of the FOBB algorithm as explained in Sec. 3.1. (a) A ball is passed on from a node to its ancestors. (b) When not all ground nodes represented by a first-order node respond identically to the ball, the node is split up. In this case, part of the nodes represented by the first-order node are observed and do not pass on the ball. (c) The first-order node that represents the ground nodes that are not observed and do not yet have top mark passes on the ball to its ancestors.

bc_2 matches, so FOBB schedules the factor node $\langle(bc_2'(X,Y),\{X{=}1\}), \mathit{fromChild}\rangle$. The constraint $\{X{=}1\}$ makes sure that only the subset of $bc_2'(X,Y)$ that are the parents of the query, are included. We will elaborate on this later. Note that the scheduled first-order node actually represents multiple nodes in the EBN (due to the free variable Y). This is shown in the first step in Fig. 3.

When a subset $\mathcal{F}'$ of the nodes in $\mathcal{F}$ interact differently with the rest of the network (e.g. they are observed variables), we need to separate them from other nodes in $\mathcal{F}$. We call this operation *splitting*, following Poole [9]. Continuing our example, when $\mathcal{F} = (r(X,Y),\{X{=}1\})$ receives the ball (from child $(bc_2'(X,Y),\{X{=}1\})$) it contains nodes $r(1,1)$,$r(1,2)$ which are evidence while the rest of the nodes are not observed. Hence, the algorithm splits the original first-order node $\mathcal{F}$ into $\mathcal{F}_{\neg E} = (r(X,Y),\{X = 1, Y \notin \{1,2\}\})$, consisting of unobserved nodes, and $\mathcal{F}_E = (r(X,Y),\{X = 1, Y \in \{1,2\}\})$, consisting of the evidence. Now, when $\mathcal{F}_{\neg E}$ receives the ball it passes the ball to its parents and children, while $\mathcal{F}_E$, which contains only observed nodes, blocks the ball. The splitting operation will be defined in detail later.

Passing the ball to the children involves similar operations as sending the ball to parents. For example, to find children of $(q(x),\{x = 1\})$, we need to find clauses which have this atom in their body. We see it only appears in the body of clause bc_3, and so its children would be $(bc_3'(X,Y),\{X = 1\})$.

Bayes-ball assigns *top* and/or *bottom* marks to the nodes it visits. FOBB also keeps track of marks, but here these are marks for first-order nodes. It stores the marks as pairs $\langle\mathcal{F},\mathbf{M}\rangle$ in a mark table, with $\mathcal{F}$ being a first-order node, and $\mathbf{M}$ a set of marks. For example, the initial marks for the query atom and for $bc_1'(X,Y)$ are stored as $\langle(q(X),\{X = 1\}),\{\mathit{top}\}\rangle$ and $\langle(bc_1'(X,Y),\{X{=}1\}),\{\mathit{top}\}\rangle$, showing these two first-order nodes have passed the ball to their respective parents.

During the execution of the algorithm, it is possible that we have to split an atom in the marks table. This happens when a subset of the nodes presented by a first-order node need to be assigned additional marks. In our example, $\mathcal{F}_1 = (bc'_3(X,Y), \{X = 1\})$ passes the ball to its children resulting in $\langle \mathcal{F}_1, \{bottom\} \rangle$ being registered in the marks table. Later when the ball is passed up from $(u(X,Y), \{X = 1, Y \in \{1,\ldots,10\}\})$ to $\mathcal{F}'_1 = (bc'_3(X,Y), \{X = 1, Y \in \{1,\ldots,10\}\})$ this node should in turn pass the ball up and receive a mark on top. For this reason we need to first split $\mathcal{F}_1$, keeping $\langle \mathcal{F}_1 \setminus \mathcal{F}'_1, \{bottom\} \rangle$ from the original first-order node and adding $\langle \mathcal{F}'_1, \{top, bottom\} \rangle$ to the marks table, for the subset which passes the ball up.

When FOBB terminates, all the first-order *factor* nodes marked at the top, together with the visited evidence atoms constitute the MRN.

3.2 The Algorithm

The FOBB algorithm uses the same set of rules as Bayes-ball to send balls through the network, visiting the possibly relevant nodes. The main difference is that the balls are not passed between nodes as we know them from Bayesian networks but between first-order nodes that aggregate multiple ground nodes in one higher level node. This way FOBB can perform multiple identical operations on $RV(\mathcal{F})$ in one single step instead of performing $|RV(\mathcal{F})|$ equivalent steps in Bayes-ball.

FOBB schedules visits to a group of ground nodes aggregated in a first-order node, searches for parents and children of such a first-order node and assigns marks to first-order nodes. The aim is to keep the nodes as aggregated as possible, but when a subset of the nodes behave differently it is necessary to split the first-order node and treat those subsets separately. The splitting happens when needed during the execution of the algorithm. Next, we illustrate how the operations in FOBB differ from those in Bayes-ball:

Initialization
In the initialization of the algorithm, all the query atoms are added to the schedule as if they were visited from child. For this we need to represent the query nodes as first-order nodes. This is done by the GETFONODE method that takes as input a ground atom $q(a_1,\ldots,a_n)$ and outputs a first-order node (p, C) with $p = q(X_1,\ldots,X_n)$ and $C = \{X_i = a_i\}$.

Scheduling visits
Where Bayes-ball has a schedule with pairs of nodes and directions to keep track of scheduled visits, FOBB utilizes a schedule containing pairs of first-order nodes and directions. An entry in the schedule containing first-order node $\mathcal{F}$ stands for a set of visits to the ground nodes in $RV(\mathcal{F})$. When a ground node in Bayes-ball receives a ball it will respond according to the rules in Sec. 2.1. In FOBB, however, it is possible that not all ground nodes represented by $\mathcal{F}$ pass the ball in the same way. This happens when some of the nodes are part of the evidence, or when not all the nodes have the same marks. In this case the first-order node $\mathcal{F}$ is split into new first-order nodes representing subsets of the ground nodes in $\mathcal{F}$ that pass the ball identically.

Algorithm 1. FOBB(M, Q, E)

Input: M: probabilistic logic model, Q: set of ground query atoms, E: set of ground evidence atoms
Output: R: requisite network, E_R: requisite evidence

```
S ← ∅, E_R ← ∅
for each q ∈ Q do
    𝒬 = GetFONode(q); S ← S ∪ ⟨𝒬, fromChild⟩
while S ≠ ∅ do
    pick and remove a visit ⟨ℱ, direction⟩ from S
    (ℱ_E, ℱ_¬E) ← SplitOnEvidence(ℱ, E)
    if ℱ_E ≠ ∅ then
        E_R ← E_R ∪ ℱ_E
    if direction = fromChild ∧ ℱ_¬E ≠ ∅ then                    // Backward chaining
        (ℱ_¬E^top, ℱ_¬E^¬top) ← SplitOnMark(ℱ_¬E, top)
        AddMark(ℱ_¬E^¬top, top)
        for each 𝒫𝒜 ∈ GetParents(ℱ_¬E^¬top, T) do
            S ← S ∪ ⟨𝒫𝒜, fromChild⟩
        (ℱ_¬E^btm, ℱ_¬E^¬btm) ← SplitOnMark(ℱ_¬E, bottom)
        if ¬Functional(ℱ) ∧ ℱ_¬E^¬btm ≠ ∅ then
            AddMark(ℱ_¬E^¬btm, bottom)
            for each 𝒞ℋ ∈ GetChildren(ℱ_¬E^¬btm, T) do
                S ← S ∪ ⟨𝒞ℋ, fromParent⟩
    if direction = fromParent then                               // Forward chaining
        if ℱ_E ≠ ∅ then
            E_R ← E_R ∪ ℱ_E
            (ℱ_E^top, ℱ_E^¬top) ← SplitOnMark(ℱ_E, top)
            AddMark(ℱ_E^¬top, top)
            for each 𝒫𝒜 ∈ GetParents(ℱ_E^¬top, T) do
                S ← S ∪ ⟨𝒫𝒜, fromChild⟩
        if ℱ_¬E ≠ ∅ then
            (ℱ_¬E^btm, ℱ_¬E^¬btm) ← SplitOnMark(ℱ_¬E, bottom)
            AddMark(ℱ_¬E^¬btm, bottom)
            for each 𝒞ℋ ∈ GetChildren(ℱ_¬E^¬btm, T) do
                S ← S ∪ ⟨𝒞ℋ, fromParent⟩
R ← {ℛ | HasMark(ℛ, top)}
return (R, E_R)
```

For example, if $\mathcal{F}$ receives a ball from one of its children, those ground nodes in $RV(\mathcal{F})$ that are part of the evidence and those that already have a top mark do not need to pass the ball to their parents, while the other ones do.

In Alg. 1 two methods are used to split up a first-order node. First, SplitOnEvidence uses the *evidence* to split up first-order node $\mathcal{F}$ into $\mathcal{F}_E$ (containing all the evidence nodes in $\mathcal{F}$), and $\mathcal{F}_{\neg E}$ (containing non-evidence nodes of $\mathcal{F}$). All evidence atoms of predicate p can be represented as a first-order node (p, C_E), then $\mathcal{F}_E = (p, C \cap C_E)$ and $\mathcal{F}_{\neg E} = (p, C \setminus C_E)$. After $\mathcal{F}$ is split on evidence, first $\mathcal{F}_E$ receives the ball (and is added to the set of visited evidence atoms), and then $\mathcal{F}_{\neg E}$.

Second, the obtained first-order node $\mathcal{F}_{\neg E}$ is split further by SPLITONMARK. To split $\mathcal{F} = (p, c)$ on mark m, we need to consult the marks table and find entries $\langle (p, C_i), \mathbf{M}_i \rangle$, such that $m \in \mathbf{M}_i$. Using the found entries, FOBB splits $\mathcal{F}$ into first-order node $\mathcal{F}^m = \bigcup_i \mathcal{F}_i = \bigcup_i (p, C \cap C_i)$, which has the mark m, and $\mathcal{F}^{\neg m} = \mathcal{F} \setminus \mathcal{F}^m$, which does not have the mark m.

After a first-order node is split into subsets which perform the same action to the ball, each subset can pass the ball to its parents/children.

Passing the Ball to Parents and Children.
Like in Bayes-ball, passing a ball from a node to its parents or children is done by following the outgoing or incoming links and scheduling a visit to the found nodes. However, since FOBB does not construct the fully grounded Bayesian network, the PBN must be used to find the parents and children of each first-order node. This is done differently for *atom* and *factor* nodes, considering the transformation of PBNs to their EBN.

The parents of each ground *atom* node $a \in RV(\mathcal{F})$ in the EBN are those factor nodes corresponding to the ground clauses which have a in their head. Thus, to find the parents of a first-order *atom* node $\mathcal{F} = (p, C)$, we first find the set of clauses $\{bc_1, \ldots, bc_k\}$, such that there is a (renaming) substitution θ_i where $head(bc_i) = p\theta_i$. Each bc_i can be represented by a first-order *factor* node $\mathcal{B}_i = (bc_i', C_i)$ where bc_i' is the atom associated to clause bc_i and each C_i is the constraint defined on the variables of bc_i in the PBN. Then, the parents of first-order node $\mathcal{F}$ in the EBN can be represented as first-order nodes $\mathcal{PA}_i = (bc_i', C_i')$, where each C_i' restricts $RV(\mathcal{PA}_i)$ to those which are parent of a node in $RV(\mathcal{F})$. Each constraint C_i' is equivalent to the relation acquired from the *natural join* $C_i \bowtie C$ of relations C_i and C (with the variables of C renamed according to substitution θ_i). This way, all the groundings of a clause bc_i that have an atom $a \in RV(F)$ in their head are captured by the first-order node $\mathcal{PA}_i$. Finding the children of an atom node is similar, only there the connected clauses are those which have a in their body.

When $\mathcal{F} = (bc', C)$ is a *factor* node, its parents are found by considering the body of the clause bc in the theory, to which bc' is associated. Let $body(bc) = \{b_1, \ldots, b_n\}$ and C_{bc} be the constraint associated to bc in the theory. Then, the parents of $\mathcal{F}$, are first-order nodes $\mathcal{B}_i = (b_i, C_i)$, where C_i restricts $RV(\mathcal{B}_i)$ to those which are parents of a node in $RV(\mathcal{F})$. Each $C_i = \pi_{X_i}(C \cap C_{bc})$ is the relation acquired from projecting $C \cap C_{bc}$ on variables $\mathbf{X_i} = Var(b_i)$. Similarly, for finding the children the head of the clause bc is considered instead of the body.

Having computed the parents $\mathcal{PA}_i$ (using GETPARENTS), in the end an entry $\langle \mathcal{PA}_i, fromchild \rangle$ is registered in the schedule, for each $\mathcal{PA}_i \neq \emptyset$, to pass the ball to parents of $\mathcal{F}$. Similarly, to pass the ball to children an entry $\langle \mathcal{CH}_i, fromParent) \rangle$ is added to the schedule for the computed children $\mathcal{CH}_i$ (using GETCHILDREN).

Assigning Marks
After passing the ball to the parents (children) of $\mathcal{F}$, FOBB needs to mark $\mathcal{F}$ on top (bottom). This can be naively done by adding $\langle \mathcal{F}, \{top\} \rangle$ to the marks

table. In this way, however, the marks table might include overlapping entries, that is, there might be a $\langle \mathcal{F}', \mathbf{M}' \rangle$ in the marks were $RV(\mathcal{F}) \cap RV(\mathcal{F}') \neq \emptyset$. In this case, $\mathbf{M}'$ contains only the *bottom* mark, since $\mathcal{F}$ is split on the top mark when retrieved from the schedule, guaranteeing that no subset of it has the top mark. Hence, the subset $\mathcal{F} \cap \mathcal{F}'$ should now have both the *top* and *bottom* marks, and should be grouped together. In general, when assigning a mark m to $\mathcal{F} = (p, C)$, if there is an overlapping mark $\mu = \langle \mathcal{F}' = (p, C'), \mathbf{M}' \rangle$ then we need to *split the marks*: First, μ is removed from the marks table, and then the marks $\mu_1 = \langle \mathcal{F} \cap \mathcal{F}', \mathbf{M}' \cup \{m\} \rangle$, $\mu_2 = \langle \mathcal{F} \setminus \mathcal{F}', \{m\} \rangle$, and $\mu_3 = \langle \mathcal{F}' \setminus \mathcal{F}, \mathbf{M}' \rangle$ are assigned instead. (Assigning these marks might result in further splits.) In this manner all the marks $\langle (p, C_i), \mathbf{M}_i \rangle$ form a partition on all the groundings of p which have been visited, such that all the nodes in each $\mathcal{F}_i = (p, C_i)$ have exactly the same marks.

3.3 Extension for Implicit Domains

The semantics of many probabilistic logic languages, such as BLPs [3] and CP-logic [6], declare an implicit domain for their logic variables. Although FOBB requires explicit domains, it can be extended to deduce the domains dynamically. Formally, we want to restrict the random variables to the *least Herbrand model* of the corresponding logic program. Intuitively, the set of ground nodes represented by a probabilistic logic model M are those which have a proof in M. To comply with these semantics, FOBB too needs to identify which random variables have a proof.

Most formalisms use some form of backward-chaining, such as SLD resolution, to find the least Herbrand model. The same idea can be adopted in FOBB. Note that Bayes-ball (and FOBB) effectively forms the backward-chains for each node from which the ball is passed to its parents and then its ancestors. Hence, FOBB is searching for proofs in a similar way to SLD resolution. The backward-chain ends whenever a root node or an evidence node receives the ball from a child. At this point we know whether this node has a proof. By chaining this information forward through the network, the nodes which have a proof can be identified.

In practice an extra mark, called a *proof mark*, is used to indicate what nodes have been proved. The MRN is then constituted by those nodes that have not only the top mark but also the proof mark. Also, the schedule has to give preference to those balls that have been passed on from proved parents.

4 Experiments

In our experiments, we investigated how the size of the domain of logic variables affects the search for the MRN, and how this MRN affects inference for probabilistic logic models. All experiments are performed on an Intel Pentium D CPU 2.80GHz processor with 1GB of memory available. FOBB itself is implemented in C++.

As a first experiment, we take the theory shown in Fig. 2 and compute the conditional probability of the atom $q(1)$ while varying the size of the domain

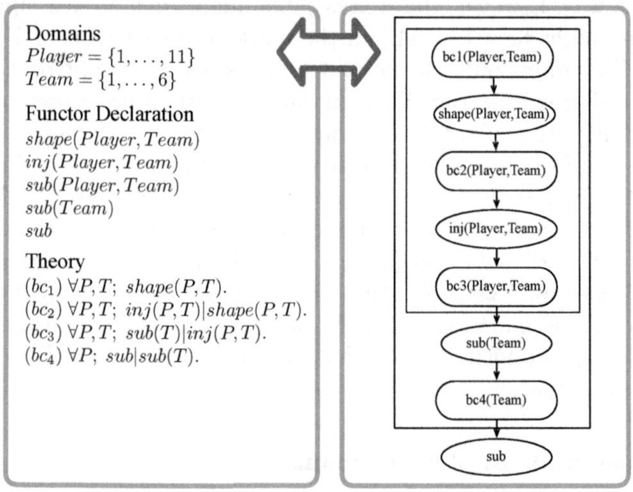

(a) (b)

Fig. 4. (a) First-order model and (b) its equivalent belief network. The scopes of logic variables are indicated by the rectangles (plates).

Num. One third of the ground nodes is chosen at random and considered as observed. For the inference, while any Bayesian network inference could be applied, we used an implementation in C++ performing variable elimination with the optimization proposed in [1] to obtain linear inference for noisy-or nodes. Five approaches were used to obtain a Bayesian network from the original theory: (a) ground the entire network based on the domains; (b) ground the entire network and use Bayes-ball to limit the resulting equivalent Bayesian network to the MRN; (c) ground the network by means of SLD resolution (using Prolog and setting the query and evidence as goals); (d) ground by means of SLD resolution and use Bayes-ball to find the MRN; and (e) use FOBB to find the MRN directly from the theory and ground the MRN. For methods (a) and (b) the theory shown in Fig. 2 was transformed first to a ground theory and afterwards compiled to an equivalent Bayesian network. Methods (c) and (d) required to first ground the facts (clauses with empty body) according to the domains.

Fig. 5 shows the results of the first experiment. The bottom graph shows that FOBB is magnitudes faster in finding the (grounded) MRN than any of the other methods. The top graph shows that the complexity of performing inference grows faster than that of grounding. As a consequence, for large networks the approach used to find the MRN became of less importance in the total inference time. These results, however, confirm the importance of restricting the network to the MRN: The two methods that do not restrict the network, full grounding and SLD resolution, could not handle even the smaller networks and ran out of memory. This result motivates investigating the effect of restricting computations to the MRN in lifted inference, for which no method like Bayes-ball has been proposed to date. We applied FOBB to such a case in our second experiment.

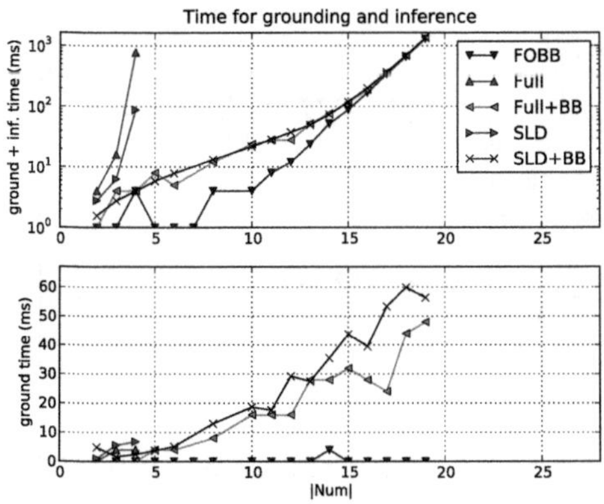

Fig. 5. Performance on the model in example in Fig. 2

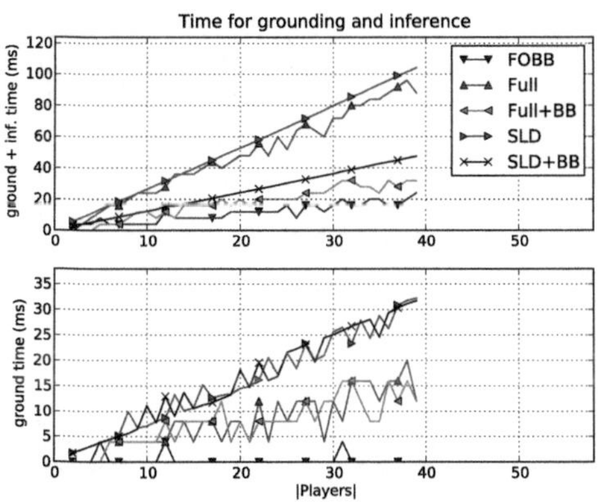

Fig. 6. Performance on the soccer example using propositional inference

In the second experiment we used the theory shown in Fig. 4. This theory is an extension of the theory used in [5] to benchmark lifted inference methods. This theory represents that whether a player in a soccer tournament is substituted during the tournament depends on whether he gets injured. The probability of an injury depends on the physical condition of the player. We compute the conditional probability of *substitution*(1) given that six teams participate, while varying the number of players in a team. For four of the teams there is evidence that some player has been injured.

For the results in Fig. 6 we used the same strategy as for the previous experiment. In addition to propositional inference we also used the lifted inference technique C-FOVE [8] available in BLOG [7] (Java) to calculate the conditional probability of the query. The factors are created based on the optimizations mentioned in [5].

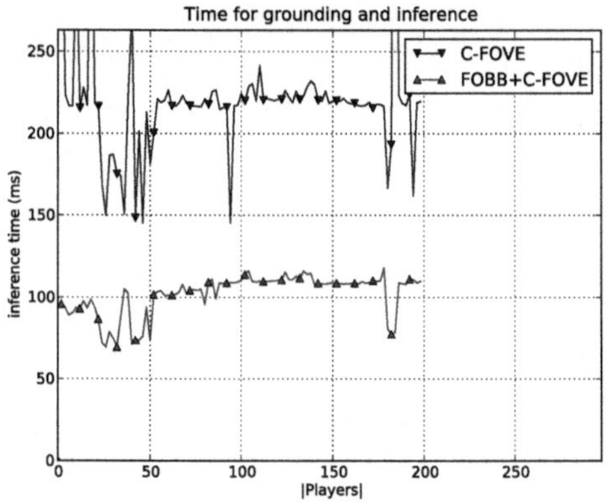

Fig. 7. Performance on the soccer example using lifted inference

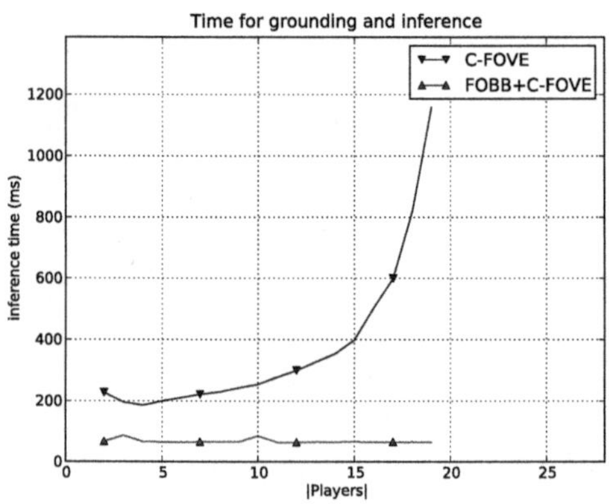

Fig. 8. Performance on the soccer example using lifted inference with a more complex interaction than noisy-or for one extra team

The soccer model is very symmetric and inference is therefore efficient. The results in Fig. 6 show that the complexity of grounding and inference are both linear. In this case the efficiency of grounding has a noticeable influence on the total inference time.

For this model, a lifted inference method can make abstraction of the domain size for performing probabilistic inference, and can therefore calculate the marginal probability of the query in constant time. This is shown in Fig. 7. FOBB allows us to find the MRN in a form that can be interpreted by a lifted inference method. With this combination, we can thus not only make abstraction of the domain size but also ignore non-requisite parts of the first-order probabilistic model. FOBB can have a greater influence on the inference when applied to to more comprehensive models, since it is possible to have non-requisite parts of arbitrary complexity. Such an effect can be observed, for example, when the model contains an extra team that uses a more complex combining rule for *sub* than noisy-or. This causes inference to be exponential on the non-requisite parts. These unnecessary computations are avoided when using FOBB, as shown in Fig. 8.

5 Conclusions and Future Work

In this work, we presented a first-order version of Bayes-ball called FOBB, which finds the minimum relevant network for a given set of query and evidence atoms. The advantages of using FOBB are twofold; first, it is more efficient to find the ground network needed to calculate the probability of the query than current methods. Second, the resulting relevant network is first-order, permitting it to be used as input to lifted inference methods which have shown to offer magnitudes of gain in speed and memory. FOBB resembles the approach by Singla and Domingos [12] in aggregating ground nodes as one unit and building a *lifted network*; major differences are that FOBB is meant for directed graphs instead of undirected graphs, and that it is dependent on the query and not a compilation technique.

In general, empirical evaluations of lifted inference algorithms are done using simple first-order probabilistic models. FOBB is a valuable companion to existing lifted inference methods like the one proposed in [5] to handle more comprehensive and real-life models.

In the future, we want to investigate further the concept of the proof ball. This would make FOBB even more suited for probabilistic logic models that are based on logic programming.

Acknowledgements

Institute for the Promotion of Innovation through Science and Technology in Flanders (IWT-Vlaanderen) to Wannes Meert. GOA/08/008 'Probabilistic Logic Learning' to Nima Taghipour.

References

1. Díez, F.J., Galán, S.F.: Efficient computation for the noisy max. International Journal of Intelligent Systems 18(2), 165–177 (2003)
2. Getoor, L., Taskar, B.: Introduction to Statistical Relational Learning (Adaptive Computation and Machine Learning). The MIT Press, Cambridge (2007)
3. Kersting, K., De Raedt, L.: Bayesian Logic Programming: Theory and Tool. In: Getoor, L., Taskar, B. (eds.) An Introduction to Statistical Relational Learning, pp. 291–322. MIT Press, Cambridge (2007)
4. Kisynski, J., Poole, D.: Constraint processing in lifted probabilistic inference. In: Proceedings of the 25th Conference on Uncertainty in Artificial Intelligence, UAI (2009)
5. Kisynski, J., Poole, D.: Lifted aggregation in directed first-order probabilistic models. In: Proceedings of the 21th International Joint Conference on Artificial Intelligence, IJCAI (2009)

6. Meert, W., Struyf, J., Blockeel, H.: Learning ground CP-logic theories by leveraging Bayesian network learning techniques. Fundamenta Informaticae 89(1), 131–160 (2008)
7. Milch, B.: BLOG (2008), `http://people.csail.mit.edu/milch/blog/`
8. Milch, B., Zettlemoyer, L.S., Kersting, K., Haimes, M., Kaelbling, L.P.: Lifted probabilistic inference with counting formulas. In: Proceedings of the 23rd AAAI Conference on Artificial Intelligence (AAAI), pp. 1062–1608 (2008)
9. Poole, D.: First-order probabilistic inference. In: Proceedings of the 18th International Joint Conference on Artificial Intelligence (IJCAI), pp. 985–991 (2003)
10. Braz, R.d.S., Amir, E., Roth, D.: Lifted first-order probabilistic inference. In: Proceedings of the 19th International Joint Conference on Artificial Intelligence (IJCAI), pp. 1319–1325 (2005)
11. Shachter, R.D.: Bayes-Ball: The rational pastime (for determining irrelevance and requisite information in belief networks and influence diagrams). In: Proceedings of the 14th Conference on Uncertainty in Artificial Intelligence (UAI), pp. 480–487 (1998)
12. Singla, P., Domingos, P.: Lifted first-order belief propagation. In: Proceedings of the 23rd AAAI Conference on Artificial Intelligence (AAAI), pp. 1094–1099 (2008)

Learning from Demonstration Using MDP Induced Metrics*

Francisco S. Melo[1] and Manuel Lopes[2]

[1] INESC-ID/Instituto Superior Técnico
TagusPark - Edifício IST
2780-990 Porto Salvo, Portugal
fmelo@inesc-id.pt
[2] University of Plymouth
Plymouth, Devon, PL4 8AA, UK
manuel.lopes@plymouth.ac.uk

Abstract. In this paper we address the problem of learning a policy from demonstration. Assuming that the policy to be learned is the optimal policy for an underlying MDP, we propose a novel way of leveraging the underlying MDP structure in a kernel-based approach. Our proposed approach rests on the insight that the MDP structure can be encapsulated into an adequate state-space metric. In particular we show that, using MDP metrics, we are able to cast the problem of learning from demonstration as a classification problem and attain similar generalization performance as methods based on inverse reinforcement learning at a much lower online computational cost. Our method is also able to attain superior generalization than other supervised learning methods that fail to consider the MDP structure.

1 Introduction

In this paper we address the problem of learning a policy from demonstration. This problem has garnered particular interest in recent years, as the ability of non-technical users to program complex systems to perform customized tasks rests heavily on fast and efficient solutions to this particular problem. The literature on this topic is too extensive to review here, and we refer to [3, 13] for reviews.

We formalize the problem of learning a policy from demonstration as a standard supervised learning problem: given a demonstration of some target policy consisting of (possibly perturbed) samples thereof, we must recover the target policy from the observed samples. These samples consist of the actions that the agent (learner) should take at specific situations; the learning agent must generalize the observed actions to situations never observed before.

* Work partially supported by the Portuguese Fundação para a Ciência e a Tecnologia (INESC-ID multiannual funding) through the PIDDAC Program funds and by the projects PTDC/EEA-ACR/70174/2006 and Handle (EU-FP7-ICT-231640).

J.L. Balcázar et al. (Eds.): ECML PKDD 2010, Part II, LNAI 6322, pp. 385–401, 2010.

Several works follow a similar approach to the problem of learning from demonstration. Examples go back to the pioneer work of Pomerleau [18], in which an artificial neural network is used to have a vehicle learn how to steer from sample steering motions by a human operator. Other related approaches include [4, 22].

When envisioning real world scenarios – *e.g.,* when a human user must teach an agent some target task, – the algorithm must exhibit good generalization ability, and hence all available information on the structure of the problem should be exploited. Our contribution in this paper addresses the latter issue. We resort to *MDP metrics* [7, 27] as a way to add additional structure to the problem of learning from demonstration. In particular, and unlike other supervised learning approaches to this problem, we assume that the target policy is the optimal policy for some underlying *Markov decision process* (MDP), whose dynamics are known. We thus propose the use of this underlying MDP structure to improve the generalization ability of our learning algorithm.

While many supervised learning methods depend critically on the definition of *features* that capture useful similarities between elements in input-space, MDP metrics naturally achieve this in the class of problems considered in this paper, by identifying two states as "similar" if policies can safely be generalized from one to the other.[1] In this sense, our approach is also close to *inverse reinforcement learning* [17, 19] and related approaches (see, e.g., [1] and references therein). However, our method avoids one bottleneck associated with the IRL-based methods. In fact, IRL-based methods are typically iterative and require solving one different MDP per iteration (see Section 3 for details) rendering these methods computationally expensive in tasks involving large domains.

The remainder of the paper is organized as follows. Section 2 formalizes the problem of learning a policy from demonstration. Section 3 reviews IRL-based and supervised learning approaches to this problem. We introduce our method in Section 4 and illustrate its application in Section 5. Section 6 concludes.

2 Formalizing Learning from Demonstration

We now formalize the problem of learning a policy from demonstration. Throughout the paper, a *policy* is understood as a mapping $\pi : \mathcal{X} \to \Delta(\mathcal{A})$, where $\mathcal{X}$ is some finite set of states, $\mathcal{A}$ is a finite set of actions and $\Delta(\mathcal{A})$ is the set of all probability distributions over $\mathcal{A}$. A policy π thus maps each state $x \in \mathcal{X}$ to a distribution over actions. Intuitively, it can be thought of as a decision rule that an agent/decision-maker can follow to choose the actions in each state. All throughout the paper we focus on problems with finite $\mathcal{X}$, common in scenarios of routing, man-machine interfaces or robot control using high-level skills/options.

[1] This is particularly evident in scenarios in which the input space is discrete. In fact, problems in continuous domains typically assume the input space as some subset of $\mathbb{R}^p$, for which natural metrics exist. The same does not occur in discrete domains, where the notion of two states being close is less immediate.

A *demonstration* is a set $\mathcal{D}$ of state-action pairs, $\mathcal{D} = \{(x_i, a_i), i = 1, \ldots, N\}$, where the states are independently randomly sampled from $\mathcal{X}$, and the corresponding actions are sampled according to the target policy π_{target}. Formally, denoting by $\mathsf{Uni}(\cdot)$ the uniform probability distribution over $\mathcal{X}$, we have that the samples $(x_i, a_i), i = 1, \ldots, N$ are i.i.d. according to

$$\mathbb{P}\left[(X, A) = (x, a)\right] = \mathsf{Uni}(x)\pi_{\text{target}}(x, a),$$

where $(X, A) = (x, a)$ denotes the event that the pair (x, a) is sampled.[2]

We assume that the target policy π_{target} used to generate the demonstration is either the optimal policy for an underlying *Markov decision process* (MDP) or a perturbation thereof. An MDP is herein represented as a tuple $(\mathcal{X}, \mathcal{A}, \mathsf{P}, r, \gamma)$, where $\mathcal{X}$ and $\mathcal{A}$ are as defined above, $\mathsf{P}(x, a, y)$ denotes the probability of moving to state y after choosing action a at state x and $r(x)$ is the reward that the agent receives upon reaching state x. The constant γ is a discount factor. We recall that, given an MDP $\mathcal{M} = (\mathcal{X}, \mathcal{A}, \mathsf{P}, r, \gamma)$, the value associated with a policy π in state $x \in \mathcal{X}$ is

$$V^\pi(x) = \mathbb{E}_\pi\left[\sum_{t=0}^{\infty} \gamma^t r(X_t) \mid X_0 = x\right],$$

where X_t denotes the state of the MDP at time instant t and the expectation $\mathbb{E}_\pi\left[\cdot\right]$ is taken with respect to the distribution over trajectories of the chain induced by π. The optimal policy for an MDP is thus the policy π^* that verifies, for all $x \in \mathcal{X}$ and all π, $V^{\pi^*}(x) \geq V^\pi(x)$. We assume the transition probabilities for the MDP, encapsulated in the matrix P, to be known to the learner. The problem of learning a policy from demonstration can thus be roughly described as that of determining a policy that best approximates the target policy, π_{target}, given a demonstration $\mathcal{D}$ and knowledge of the MDP transition probabilities P.

In our finite action setting, we can treat each action $a \in \mathcal{A}$ as a *class label*, and a policy is essentially a *discriminant function* that assigns class labels to the states in $\mathcal{X}$. In other words, the policy $\hat{\pi}$ computed by our learning algorithm can be interpreted as classifier that assigns each label $a \in \mathcal{A}$ to a state $x \in \mathcal{X}$ according to the probabilities $\mathbb{P}\left[A = a \mid X = x\right] = \hat{\pi}(x, a)$. We henceforth use the designations "classifier" and "policy" interchangeably. We define the 0-1 *loss function* $\ell : \mathcal{A} \times \mathcal{A} \to \{0, 1\}$ as $\ell(a, \hat{a}) = 1 - \delta(a, \hat{a})$, where $\delta(\cdot, \cdot)$ is such that $\delta(a, \hat{a}) = 1$ if $a = \hat{a}$ and 0 otherwise. Our algorithm must then compute the classifier $\hat{\pi}$ that minimizes the *misclassification rate*,

$$\mathcal{E} = \frac{1}{N}\sum_{i=1}^{N} \mathbb{E}_{\hat{\pi}}\left[\ell(a_i, a)\right] = \frac{1}{N}\sum_{i=1}^{N}\sum_{a \in \mathcal{A}} \ell(a_i, a)\hat{\pi}(x_i, a), \tag{1}$$

[2] There may be applications in which the above representation may not be the most adequate (*e.g.*, in some robotic applications). However, this discussion is out of the scope of the paper and we instead refer to [3].

where (x_i, a_i) is the ith sample in the dataset $\mathcal{D}$. To this purpose, we resort to a kernel-based learning algorithm from the literature [14].[3]

Our contribution arises from considering a kernel obtained from a metric over $\mathcal{X}$ that is *induced by the MDP structure.* This *MDP metric* provides a natural way of "injecting" the MDP structure in the learning algorithm, leading to an improved performance when compared to other metrics that ignore the MDP. To put the work of this paper in context, the following section reviews the main ideas behind IRL-based approaches to the problem of learning a policy from demonstration. Also, it shows the supervised learning method used in our experiments to assess the usefulness of MDP metrics in the setting of this paper.

3 IRL and Supervised Learning Approaches

In this section we review some background material on (supervised) learning from demonstration and IRL. As will become apparent from our discussion, the underlying MDP used in IRL approaches is a rich structure that, if used adequately, leads to a significant improvement in estimating the target policy. This observation constitutes the main motivation for the ideas contributed in the paper. The section concludes with a brief revision of MDP metrics.

3.1 Inverse Reinforcement Learning

To describe the main ideas behind IRL-based approaches to the problem of learning a policy from demonstration, we introduce some additional concepts and notation on Markov decision processes. Given a policy π and an MDP $\mathcal{M} = (\mathcal{X}, \mathcal{A}, \mathsf{P}, r, \gamma)$, we have

$$V^\pi(x) = r(x) + \gamma \sum_{y \in \mathcal{X}} \mathsf{P}_\pi(x, y) V^\pi(y)$$

where $\mathsf{P}_\pi(x, y) = \sum_{a \in \mathcal{A}} \pi(x, a) \mathsf{P}(x, a, y)$. For the particular case of the optimal policy π^*,

$$V^*(x) = r(x) + \max_{a \in \mathcal{A}} \gamma \sum_{y \in \mathcal{X}} \mathsf{P}(x, a, y) V^*(y).$$

The Q-function associated with policy π is defined as

$$Q^\pi(x, a) = r(x) + \gamma \sum_{y \in \mathcal{X}} \mathsf{P}(x, a, y) V^\pi(y)$$

[3] We note that, in an MDP, the impact of misclassified actions on the performance of the agent heavily depends on which state such misclassification occurred. If information regarding the "relevance" of states in terms of performance is available, then such information can be integrated into the error $\mathcal{E}$ by weighting differently the loss at different states. In any case, we emphasize that the performance criterion for our learning algorithm is concerned with classification accuracy rather than with a reward-based performance.

Inverse reinforcement learning (IRL) deals with the inverse problem to that of an MDP: *Given a policy* π_{target} *and the model* $(\mathcal{X}, \mathcal{A}, \mathsf{P}, \gamma)$, *IRL seeks to determine a reward function* r^* *such that* π_{target} *is an optimal policy for the MDP* $(\mathcal{X}, \mathcal{A}, \mathsf{P}, r^*, \gamma)$.

IRL was first formalized in the seminal paper by Ng and Russel [17]. Among other things, this paper characterizes the solution space associated with a target policy π_{target} as being the set of rewards $\mathbf{r}$ verifying

$$(\mathsf{P}_\pi - \mathsf{P}_a)(\mathbf{I} - \gamma \mathsf{P}_\pi)^{-1}\mathbf{r} \succeq \mathbf{0}, \tag{2}$$

where we have denoted by $\mathbf{r}$ the column vector with xth component given by $r(x)$. One interesting aspect of the condition (2) is that, although trivially met by some reward functions, it still provides a restriction on the reward space arising solely from considering the structure of the MDP. In other words, by considering the structure of the MDP it is possible to restrict the rewards that are actually compatible with the provided policy.

The sample-based approach to the problem of learning from demonstration considered in this paper, however, is closest to [19,15]. In these works, the demonstration provided to the algorithm consists in perturbed samples from the optimal policy associated with the target reward function. Specifically, in [19], the distribution from which the samples are obtained is used as the *likelihood function* in a Bayesian setting. The paper then estimates the posterior distribution $\mathbb{P}[\mathbf{r} \mid \mathcal{D}]$ using a variant of the Monte-Carlo Markov Chain algorithm. In [15], on the other hand, the authors adopt a gradient approach to recover the reward function that minimizes the empirical mean squared error with respect to the target policy.

For future reference, we review the latter method in more detail. Roughly speaking, the working assumption in [15] is that there is one reward function, r_{target}, that the agent must estimate. Denoting the corresponding optimal Q-function by Q^*_{target}, the paper assumes the demonstrator will choose an action $a \in \mathcal{A}$ in state $x \in \mathcal{X}$ with probability

$$\mathbb{P}[A = a \mid X = x, r_{\text{target}}] = \frac{e^{\eta Q^*_{\text{target}}(x,a)}}{\sum_{b \in \mathcal{A}} e^{\eta Q^*_{\text{target}}(x,b)}}, \tag{3}$$

where η is a non-negative constant, henceforth designated as *confidence parameter*. Now given a demonstration $\mathcal{D} = \{(x_i, a_i), i = 1, \ldots, N\}$ obtained according to the distribution in (3), the algorithm proceeds by estimating the reward function r that minimizes the error

$$\mathcal{E} = \frac{1}{N} \sum_i \left(\tilde{\pi}(x_i, a_i) - \hat{\pi}_r(x_i, a_i)\right)^2,$$

where $\tilde{\pi}(x_i, a_i)$ is the empirical frequency of action a_i in state x_i and $\hat{\pi}_r$ is the policy estimate, corresponding to the optimal policy for the MDP $(\mathcal{X}, \mathcal{A}, \mathsf{P}, r, \gamma)$. Minimization is achieved by (natural) gradient descent, with respect to the parameters of $\hat{\pi}_r$ – the corresponding reward function $\mathbf{r}$. The method thus proceeds by successively updating the reward

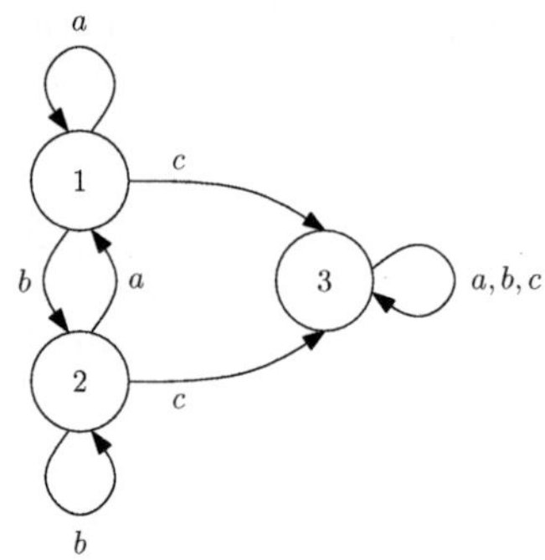

Fig. 1. Transition diagram for a simple MDP with 3 states, 3 actions and deterministic transitions. Even if the reward for this MDP is unknown, the demonstration of an optimal action in state 1, say action c, immediately implies that this action is also optimal in state 2. In fact, if c is optimal in state 1, then $r(3) > r(1)$ and $r(3) > r(2)$ – otherwise c would not be optimal in state 1. But then, the action in state 2 should also be c.

$$r^{(k+1)} = r^{(k)} - \alpha_t \tilde{\nabla}_r \mathcal{E}(r^{(k)}),$$

where $\tilde{\nabla}_r \mathcal{E}(r^{(k)})$ is the natural gradient of $\mathcal{E}$ with respect to r computed at $r^{(k)}$.

We conclude by noting that the computation of the natural gradient $\tilde{\nabla}_r \mathcal{E}(r^{(k)})$ at each iteration k requires solving the corresponding MDP $(\mathcal{X}, \mathcal{A}, \mathsf{P}, r^{(k)}, \gamma)$. This need to solve an MDP at each iteration makes this method computationally expensive for large problems.

3.2 Supervised Learning

There is a significant volume of work on supervised learning approaches to the problem of learning a policy from demonstration. That literature is far too extensive to be reviewed here, and we refer to [3, 13] for more detailed accounts.

A "pure" supervised learning approach addresses the problem of learning a policy from demonstration as a standard classification problem: *Given a set $\mathcal{D}$ of state-action pairs, $\mathcal{D} = \{(x_i, a_i), i = 1, \ldots, N\}$, sampled as detailed in Section 2, we want to compute a discriminant function $\hat{\pi}$ that minimizes the misclassification rate $\mathcal{E}$ in* (1).

Interestingly, even when the target policy is assumed optimal for some MDP, most supervised learning approaches to the problem of learning a policy from demonstration eventually ignore the underlying MDP structure in the learning process (see, for example, [4]). This means that such supervised learning algorithms do not actually require any particular knowledge of the underlying structure of the problem to function. Of course, when such information is available, it can provide valuable information for the learning algorithm and the learning algorithm should be able to accommodate such information.

In Section 4 we propose a principled approach to leverage the MDP structure in a kernel-based learning algorithm. Our proposed approach rests on the insight that the MDP structure can be encapsulated into an *input-space metric* that the learning algorithm can use to generalize along the (implicit) MDP structure. We propose using a metric that "projects" the structure of the MDP in such a way that the target policy in two states that are "close" is likely to be similar. This notion is illustrated in the example of Fig. 1.

We postpone to Section 4 the detailed description of how to encapsulate the MDP structure in the input-space metric and conclude this section by describing the learning algorithm used in our experiments.

Going back to the formulation in Section 2, we are interested in computing a classifier $\hat{\pi}$ (a distribution over the set of class labels – in this case, the action set $\mathcal{A}$) that varies along the state-space $\mathcal{X}$. In particular, we are interested in a learning algorithm that takes advantage of the metric structure in $\mathcal{X}$ to be described next. Possible methods for this include kernel logistic regression [28], beta regression [14] and others. We adopt the trivial extension of the latter to the multi-class case due to its computational efficiency and the possibility of including prior information.

At each state $x \in \mathcal{X}$, the desired classifier $\hat{\pi}$ is simply a multinomial distribution over the set of possible actions. Using a standard Bayesian approach, we compute a posterior distribution over each of the parameters of this multinomial using the corresponding conjugate prior, the Dirichlet distribution. For notational convenience, let us for the moment denote the parameters of the multinomial at a state $x \in \mathcal{X}$ as a vector $\mathbf{p}(x)$ with ath component given by $\mathbf{p}_a(x)$. Using this notation, we want to estimate, for each $x \in \mathcal{X}$, the posterior distribution $\mathbb{P}\left[\mathbf{p}(x) \mid \mathcal{D}\right]$. Let x_i be some state observed in the demonstration $\mathcal{D}$, and let $n_a(x_i)$ denote the number of times that, in the demonstration, the action a was observed in state x_i. Finally, let $n(x_i) = \sum_a n_a(x_i)$. We have

$$\begin{aligned}\mathbb{P}\left[\mathbf{p}(x_i) \mid \mathcal{D}\right] &\propto \mathsf{Multi}(\mathbf{p}_1(x_i), \mathbf{p}_2(x_i), \ldots, \mathbf{p}_{|\mathcal{A}|}(x_i))\mathsf{Dir}(\alpha_1, \alpha_2, \ldots, \alpha_{|\mathcal{A}|})\\ &= \frac{n(x_i)!}{\prod_{a\in\mathcal{A}} n_a(x_i)!}\prod_{a\in\mathcal{A}} \mathbf{p}_a(x_i)^{n_a}\frac{1}{B(\boldsymbol{\alpha})}\prod_{a\in\mathcal{A}} \mathbf{p}_a(x)^{\alpha_a-1}\end{aligned}$$

In other words, the posterior distribution of $\mathbf{p}(x_i)$ is also a Dirichlet distribution with parameters $n_a + \alpha_a, a = 1, \ldots, |\mathcal{A}|$. In order to generalize $\mathbb{P}\left[\mathbf{p}(x) \mid \mathcal{D}\right]$ to unvisited states $x \in \mathcal{X}$, and following the approach in [14], we assume that the parameters of the Dirichlet distribution depend smoothly on x. From the metric structure of $\mathcal{X}$ we define a kernel on $\mathcal{X}$, $\mathbf{k}(\cdot, \cdot)$, and use standard kernel regression to extrapolate the parameters of the Dirichlet from the training dataset to unvisited states [23]. Specifically, for any query point $x^* \in \mathcal{X}$ and all $a \in \mathcal{A}$, we have

$$\hat{n}_a(x^*) = \sum_i \mathbf{k}(x^*, x_i)n_a(x_i) + \alpha_a. \tag{4}$$

Finally, the posterior mean of the distribution over the parameters $\mathbf{p}(x^*)$ – that we will henceforth use as our classifier at x^*, $\hat{\pi}(x^*, \cdot)$ – is given by

$$\hat{\pi}(x^*, a) = \mathbb{E}\left[\mathbf{p}_a(x^*) \mid \mathcal{D}\right] = \frac{\hat{n}_a(x^*)}{\sum_b \hat{n}_b(x^*)}, \tag{5}$$

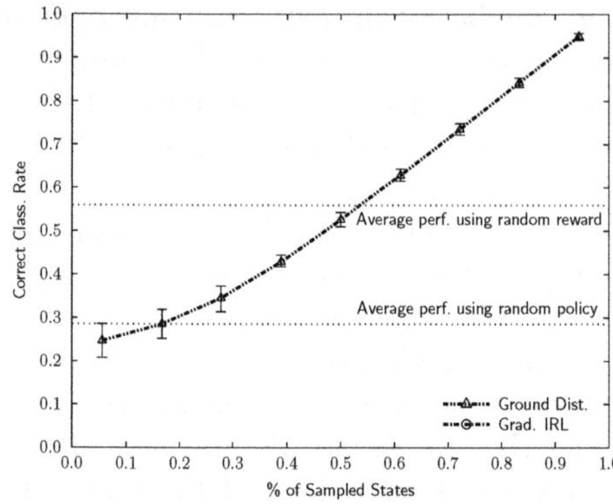

Fig. 2. Classification rate of the method above for different demonstration sizes. The horizontal lines correspond to the performances of a random policy and a policy obtained from the underlying MDP with a random reward.

for all $a \in \mathcal{A}$. In the next section, we introduce our main contribution, proposing a principled way to encapsulate the MDP structure in a suitable metric for $\mathcal{X}$ that can then be used to define the kernel $\mathbf{k}(\cdot, \cdot)$.

To conclude this subsection, we briefly summarize the main differences between IRL-based approaches and "pure" SL-based approaches. IRL-based approaches use the knowledge of the MDP model to search the space of reward functions for one specific reward function that yields an optimal policy that best matches the provided demonstration. For each reward function "tested", these methods solve the corresponding MDP, and so each iteration of IRL-based methods requires solving an MDP. However, the very structure of the MDP significantly reduces the space of possible policies, allowing these methods to exhibit very good generalization.

"Pure" SL-based methods, on the other hand, typically do not assume any underlying MDP model, and as such do not require solving any such model. This renders these methods more computationally efficient in large domains. However, the lack of clear domain knowledge often causes poorer generalization than IRL-based methods.

To illustrating the difference in generalization capabilities of the two classes of methods above, we applied the kernel-based algorithm just described to a problem of learning from demonstration in a grid-like world (a more detailed description of the experimental setting is postponed to Section 5). Figure 2 compares the performance of the algorithm against those of a random policy and a policy obtained by solving the underlying MDP with a random reward. Note that, even using a random reward, the MDP-based solution is able to attain about a policy with 57% of correct actions, clearly establishing the advantage of using the MDP structure.

3.3 Bisimulation and MDP Metrics

Let us start by considering again the MDP depicted in Fig. 1. In this MDP, there is a close relation between states 1 and 2 since their actions and corresponding transitions are similar. In such a scenario, information about the policy, say, in state 1 will typically be also useful in determining the policy in state 2.

This notion of "similarity" has recently been explored in the MDP literature as a means to render solution methods for MDPs more efficient [7,21,27]. In fact, by identifying "similar" states in an MDP $\mathcal{M}$, it may be possible to construct a smaller MDP $\mathcal{M}'$ that can more easily be solved. In this paper we instead use MDP metrics to identify "similar" states and safely generalize the policy observed in the demonstration.

As established in [9], "similarity" between MDP states is best captured by the notion of *bisimulation*. Bisimulation is an equivalence relation $\sim$ on $\mathcal{X}$ in which two states x and y are similar if $r(x) = r(y)$ and

$$\mathbb{P}\left[X_{t+1} \in U \mid X_t = x, A_t = a\right] = \mathbb{P}\left[X_{t+1} \in U \mid X_t = y, A_t = a\right],$$

where U is some set in the partition induced by $\sim$. Lax bisimulation is a generalization of bisimulation that also accounts for action relabeling. Both bisimulation and lax bisimulation led to the development of several *MDP metrics* in which, if the distance between two states x, y is zero, then $x \sim y$ [7,27]. In this paper we adopt one such MDP metric, introduced in [27] and henceforth denoted as $\delta_{d_{\text{MDP}}}$, that is built over an initial metric on $\mathcal{X}$ that we refer as the *ground distance* (see Appendix A for details). We point out, however, that this choice is not unique.

While MDP metrics such as the one above were designed to improve efficiency in MDP solution methods, in this paper we are interested in their use in the problem of learning a policy from demonstration. In this context, MDP metrics arise as a natural way to "embed" the MDP structure in a supervised learning algorithm to improve its generalization performance while avoiding solving multiple MDPs. As will soon become apparent from our results, the use of an MDP metric indeed provides a significant and consistent improvement in performance over other metrics that ignore the MDP structure.

4 A Kernel-Based Approach Using MDP Metrics

We now introduce the main contributions of the paper, namely how to use MDP metrics such as the one discussed above to the particular problem considered in this paper.

The first aspect to consider is that, when learning a policy from demonstration, there is no reward information. While most MDP metrics naturally include a component that is reward-dependent, the particular setting considered here implies that the metric used should not include one such term. Secondly, the metric $\delta_{d_{\text{MDP}}}$ used already implicitly provides the learning algorithm with the necessary information on action relabeling. Therefore, in our algorithm we use the kernel

$$\mathbf{k}\big((x,a),(y,b)\big) = \exp\big(-\delta_{d_{\mathrm{MDP}}}((x,a),(y,b))/\sigma\big),$$

where σ denotes the kernel bandwidth, and (4) and (5) become

$$\hat{n}_a(x^*) = \sum_{i,b} \mathbf{k}\big((x^*,a),(x_i,b)\big)n_b(x_i) + \alpha_a, \tag{6}$$

$$\hat{\pi}(x^*,a) = \frac{\hat{n}_a(x^*)}{\sum_b \hat{n}_b(x^*)}. \tag{7}$$

The complete algorithm is summarized in Algorithm 1.

Algorithm 1. MDP-induced Kernel Classification.

Require: State-action space metric $\delta_{d_{\mathrm{MDP}}}$;
Require: Dataset $\mathcal{D} = \{(x_i, a_i), i = 1, \ldots, N\}$;
1: Given a query point x^*,
2: **for all** $a \in \mathcal{A}$ **do**
3: Compute $\hat{n}_a(x^*)$ using (6)
4: **end for**
5: **for all** $a \in \mathcal{A}$ **do**
6: Compute $\hat{\pi}(x^*, a)$ using (7)
7: **end for**
8: **return** $\hat{\pi}$;

4.1 Active Learning from Demonstration

Algorithm 1 is a Bayesian inference method that computes the posterior distribution over the parameters $\mathbf{p}(x)$ of the multinomial distribution (see Section 3.2). It is possible to use this posterior distribution in a simple active sampling strategy that can lead to a reduction in the sample complexity of our method [24].

One possibility is to compute the variance associated with the posterior distribution over the parameters $\mathbf{p}(x)$ at each state $x \in \mathcal{X}$, choosing as the next sample the state for which this variance is largest. The intuition behind this idea is that states with larger variance were observed less frequently and/or are distant (in the sense of our MDP metric) from other more often sampled states.

Using this active learning strategy, we can reduce the number of samples required for learning, since more informative states will be chosen first. Note also that this active learning approach implicitly takes into account the similarity between the states, in a sense treating similar states as one "aggregated" state and effectively requiring less samples.

5 Results

In this section we discuss several aspects concerning the performance of the proposed approach. Our results feature demonstrations with significantly less samples than the size of the state-space of the underlying MDP, allowing a clear evaluation of the generalization ability of the tested methods. Also, the scenarios used are typically not too large, allowing for easier interpretation of the results.

Our first test aims at verifying the general applicability of our approach. To this purpose, we applied the algorithm in Section 3.2 using the MDP metric described in Section 4 to 50 randomly generated MDPs. The state-space of the MDPs varies between 20 and 60 states, and the action space between 4 and 10 actions. From each state, it is possible to transition to between 20% and 40% of the other states. For each MDP, we randomly sample (without replacement) half of the states in the MDP and provide the learning algorithm with a demonstration consisting of these states and the corresponding optimal actions.

We compare the classification accuracy of the algorithm in Section 3.2 when using the MDP metric against the performance of that same algorithm when using other metrics. In particular, we used the *zero-one metric*, in which each the distance between any two state-action pairs (x, a) and (y, b) is either zero – if $(x, a) = (y, b)$ – or 1, and the *transition distance*, in which the distance between any two state-action pairs (x, a) and (y, b) corresponds to the minimum number of steps that an agent would take to move from state x to state y when taking as a first action the action a. As for the MDP metric, we used the distance $\delta_{d_{\mathrm{MDP}}}$ obtained by using the transition distance just described as ground distance. The bandwidth of the kernel was adjusted experimentally in each case to optimize the performance. The correct classification rate for the different approaches is summarized in Table 1.

Table 1. Average performance over 50 random worlds with between 20 and 50 states and 4 and 10 actions

	Correct Class. Rate (%)
Zero-one Distance	57.95 ± 4.33
Transition Distance	60.67 ± 5.10
$\delta_{d_{\mathrm{MDP}}}$	73.92 ± 6.98

As is clear from the results above, our approach clearly outperforms all other approaches, computing the correct action on 73% of the total number of states. This establishes that, as expected, the use of the MDP metric indeed provides an important boost in the performance of the method.

Our second test aims at comparing the performance of our approach against that of IRL-based algorithm (namely, the algorithm in Section 3.1). We run this comparison in two dimensions. We compare how both methods behave as the size of the demonstration increases and as the noise in the demonstration varies. For this comparison, we used a scenario consisting of eight clusters weakly connected

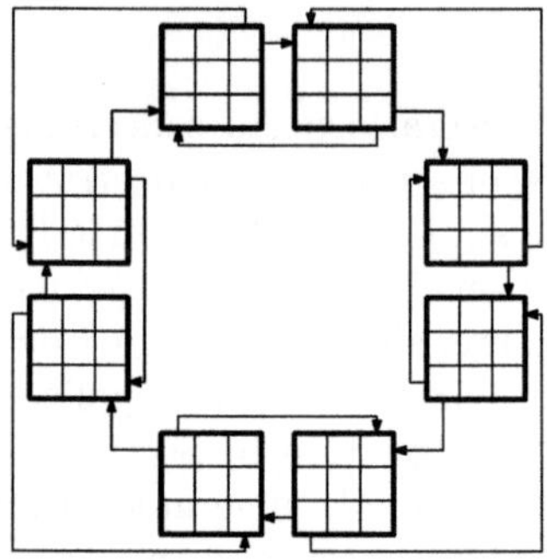

Fig. 3. Cluster-world scenario used in the tests

with one another. Each cluster contains a total of 9 states in a total of 72 states (see Fig. 3). Unlike more standard grid-world scenarios, the weakly connected clusters of states imply that there is a great asymmetry between the states in terms of transitions (unlike what occurs in standard grid-world domains). For this MDP, we considered 5 actions, corresponding to motions in the 4 directions, and the "NoOp" action.

We provided the algorithms with demonstrations of increasing size and evaluated the percentage of correct actions. As before, the demonstration was obtained by sampling (without replacement) a percentage of the total number of states and, for these, providing the corresponding optimal actions according to the target policy. Figure 4(a) compares our approach with the gradient-based IRL approach (GIRL). We also depict the performance of the supervised learning method when using other metrics and the two baselines discussed in Section 3.

The results show that all methods improve as the size of the demonstration increases. For very small demonstrations, GIRL gives better results as its baseline is higher. With a dataset larger than about 30% of the total number of states, our method clearly outperforms all other up to a "full demonstration", when all kernel-based methods perform alike. We also note that the worse performance of GIRL even in large demonstrations is probably due to the existence of local minima.[4]

Figure 4(b) compares the performance of all methods as the noise in the demonstration varies. As in the first experiment, we used a demonstration of half the total number of states in the MDP. The actions were sampled from the distribution in (3) and the noise was adjusted by changing the value of the confidence parameter η. Low values of η correspond to noisy policies (where many suboptimal actions are sampled) while high values of η correspond to near-optimal policies. As expected all methods behave worse as the noise level increases, approaching the baseline performance levels for high levels of noise. As the noise decreases, the performance approaches the levels observed in Fig. 4(a).

[4] The results shown contain only runs that did have an improvement from the initial condition.

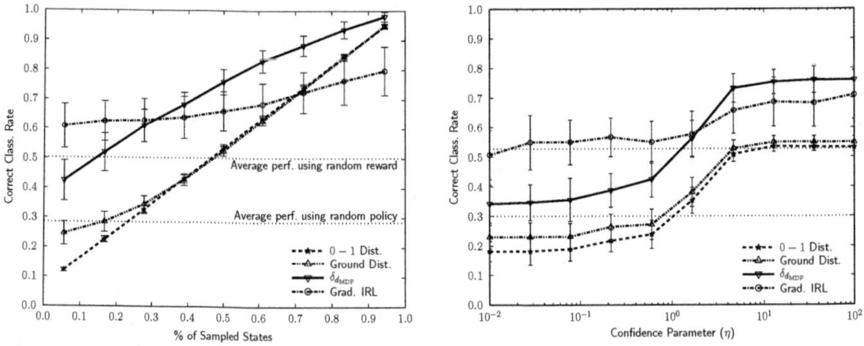

(a) (b)

Fig. 4. Classification rate of the different methods for (a) different demonstration sizes; (b) different confidence levels on the demonstration, with a demonstration size of alf the number of states. The horizontal lines are the baseline references (see main text). The results depicted are averaged over 50 independent Monte-Carlo trials, with the error bars corresponding to sample standard deviation.

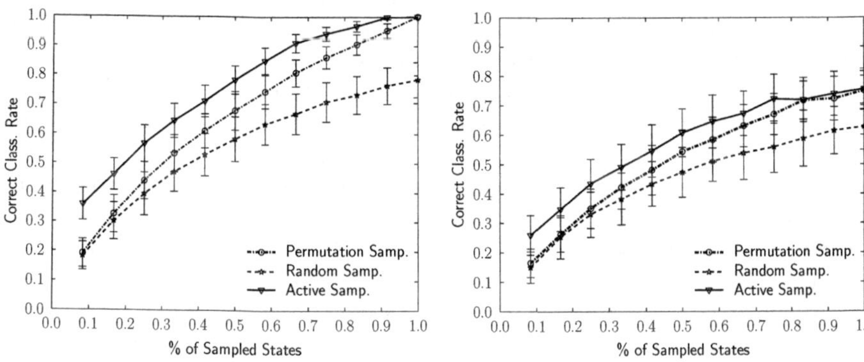

(a) Noise-free demonstration. (b) Noisy demonstration.

Fig. 5. Classification rate of the different exploration methods with the size of the demonstration (as a ration of the number of states). The results depicted are averaged over 50 independent Monte-Carlo trials, with the error bars corresponding to sample standard deviation.

We note, however, that the performance of the kernel-based methods suffers a significant improvement at a certain noise threshold (for $\eta \approx 1$). We also note that larger demonstrations allowing states to be sampled with replacement will lead all methods to better filter out the noise and improve the classification rate. Finally, Fig. 5 shows the results of the active learning approach described in Section 4.1. We compare several possible sampling techniques: *random sampling*, where the states to be demonstrated are sampled uniformly with replacement; permutation sampling, where the states are sampled uniformly without replacement; and *active sampling*, where the states are sampled according to the variance of the posterior distribution over the parameters.

Our results, although specific to the particular scenario considered, illustrates the fact that our active sampling strategy manages to outperform other sampling techniques in terms of sample efficiency. The obtained gain depends on the particular problem, where problems with more symmetries/more clustered states are expected to yield larger gains for the active learning approach.

6 Concluding Remarks

Our results illustrate how the use of the MDP structure indeed provides valuable information when tackling the problem of learning a policy from demonstration. This can be verified easily by comparing, for example, the two baselines in Fig. 2. In these results, using a random policy lead to a correct classification rate of around 30%. However, if we instead use a *random reward function* and solve the corresponding MDP, the correct classification rate goes up to 57%. This clearly indicates that, in this particular example, the MDP structure significantly restricts the set of possible policies. Although these numbers correspond to a rather small environment in which the set of policies is naturally small, some preliminary experiments indicate that this conclusion holds in general (although with different numbers, of course). This conclusion is also supported by the discussion in Section 3.1 about the results in [17].

A second remark is concerned with the computational complexity associated with the particular MDP metric used in our results. As already pointed out by [7], MDP metrics that rely on the Kantorovich metric to evaluate the distance between two distributions $\mathsf{P}(x, a, \cdot)$ and $\mathsf{P}(y, b, \cdot)$ – such as the one used here – are computationally demanding. There are other alternative metrics that can be used (such as the total variation distance [7]) that are significantly more computationally efficient and can thus alleviate this issue. Nevertheless, it is worth pointing out that the MDP metric needs only to be computed once and can then be applied to any demonstration. This point is particularly important when envisioning, for example, robotic applications, since the MDP metric can be computed offline, hard-coded into the robot and used to learn different tasks by demonstration to be able to adapt to different users.

Another important point to make is that there are other methods that share a similar principle to the gradient-based IRL method in Section 3.1 and used for comparison in Section 5. These methods, while not explicitly concerned in recovering a reward description of the demonstrated task, use nevertheless an underlying MDP structure within a supervised learning or optimization setting [2, 20, 26, 25, 29]. Unfortunately, these approaches are close to IRL-based methods in that they still suffer from the need to solve multiple MDPs (see also the discussion in [16] for a more detailed account of the similarities and differences between the methods above). It is also worth noting that most aforementioned methods are designed to run with significantly larger datasets than those used in the experiments [16]. In conclusion they share the same advantages and disadvantages of the GIRL method in Section 3.2.

Finally, we conclude by noting that there is no immediate (theoretical) difficulty in extending the ideas in this paper to continuous scenarios. We did not discuss this extension in the paper since MDP metrics in continuous MDPs require a significantly more evolved machinery to describe [8] that would unnecessarily complicate the presentation. On the otherhand, although some continuous MDP metrics have been identified and shown to have good theoretical properties [8], they are expensive to compute and no computationally efficiently alternatives are known. One important avenue for future research is precisely the identifying continuous MDP metrics that are efficiently computable. It would also be interesting to explore, within the setting of our work, the impact of several recent works that proposed classification-based MDP solution methods [5,6,10,11,12].

Acknowledgements

The authors acknowledge the useful suggestions by Andreas Wichert and the anonymous reviewers.

References

1. Abbeel, P.: Apprenticeship learning and reinforcement learning with application to robotic control. Ph.D. thesis, Dep. Computer Science, Stanford Univ (2008)
2. Abbeel, P., Ng, A.: Apprenticeship learning via inverse reinforcement learning. In: Proc. 21st Int. Conf. Machine Learning, pp. 1–8 (2004)
3. Argall, B., Chernova, S., Veloso, M.: A survey of robot learning from demonstration. Robotics and Autonomous Systems 57(5), 469–483 (2009)
4. Chernova, S., Veloso, M.: Interactive policy learning through confidence-based autonomy. J. Artificial Intelligence Research 34, 1–25 (2009)
5. Fern, A., Yoon, S., Givan, R.: Approximate policy iteration with a policy language bias. In: Adv. Neural Information Proc. Systems 16 (2003)
6. Fern, A., Yoon, S., Givan, R.: Approximate policy iteration with a policy language bias: Solving relational Markov decision processes. J. Artificial Intelligence Research 25, 75–118 (2006)
7. Ferns, N., Panangaden, P., Precup, D.: Metrics for finite Markov decision processes. In: Proc. 20th Conf. Uncertainty in Artificial Intelligence, pp. 162–169 (2004)
8. Ferns, N., Panangaden, P., Precup, D.: Metrics for Markov decision processes with infinite state-spaces. In: Proc. 21st Conf. Uncertainty in Artificial Intelligence, pp. 201–208 (2005)
9. Givan, R., Dean, T., Greig, M.: Equivalence notions and model minimization in Markov Decision Processes. Artificial Intelligence 147, 163–223 (2003)
10. Lagoudakis, M., Parr, R.: Reinforcement learning as classification: Leveraging modern classifiers. In: Proc. 20th Int. Conf. Machine Learning, pp. D424–D431 (2003)
11. Langford, J., Zadrozny, B.: Relating reinforcement learning performance to classification performance. In: Proc. 22nd Int. Conf. Machine Learning, pp. D473–D480 (2005)
12. Lazaric, A., Ghavamzadeh, M., Munos, R.: Analysis of a classification-based policy iteration algorithm. In: Proc. 27th Int. Conf. Machine Learning (to appear, 2010)

13. Lopes, M., Melo, F., Montesano, L., Santos-Victor, J.: Abstraction levels for robotic imitation: Overview and computational approaches. In: From Motor Learning to Interaction Learning in Robots, pp. 313–355 (2010)
14. Montesano, L., Lopes, M.: Learning grasping affordances from local visual descriptors. In: Proc. 8th Int. Conf. Development and Learning, pp. 1–6 (2009)
15. Neu, G., Szepesvári, C.: Apprenticeship learning using inverse reinforcement learning and gradient methods. In: Proc. 23rd Conf. Uncertainty in Artificial Intelligence, pp. 295–302 (2007)
16. Neu, G., Szepesvári, C.: Training parsers by inverse reinforcement learning. Machine Learning (2009) (accepted)
17. Ng, A., Russel, S.: Algorithms for inverse reinforcement learning. In: Proc. 17th Int. Conf. Machine Learning, pp. 663–670 (2000)
18. Pomerleau, D.: Efficient training of artificial neural networks for autonomous navigation. Neural Computation 3(1), 88–97 (1991)
19. Ramachandran, D., Amir, E.: Bayesian inverse reinforcement learning. In: Proc. 20th Int. Joint Conf. Artificial Intelligence, pp. 2586–2591 (2007)
20. Ratliff, N., Bagnell, J., Zinkevich, M.: Maximum margin planning. In: Proc. 23rd Int. Conf. Machine Learning, pp. 729–736 (2006)
21. Ravindran, B., Barto, A.: Approximate homomorphisms: A framework for non-exact minimization in Markov decision processes. In: Proc. 5th Int. Conf. Knowledge-Based Computer Systems (2004)
22. Saunders, J., Nehaniv, C., Dautenhahn, K.: Teaching robots by moulding behavior and scaffolding the environment. In: Proc. 1st Annual Conf. Human-Robot Interaction (2006)
23. Schölkopf, B., Smola, A.: Learning with kernels: Support vector machines, regularization, optimization and beyond. MIT Press, Cambridge (2002)
24. Settles, B.: Active learning literature survey. Tech. Rep. CS Tech. Rep. 1648, Univ. Wisconsin-Maddison (2009)
25. Syed, U., Schapire, R.: A game-theoretic approach to apprenticeship learning. In: Adv. Neural Information Proc. Systems, vol. 20, pp. 1449–1456 (2008)
26. Syed, U., Schapire, R., Bowling, M.: Apprenticeship learning using linear programming. In: Proc. 25th Int. Conf. Machine Learning, pp. 1032–1039 (2008)
27. Taylor, J., Precup, D., Panangaden, P.: Bounding performance loss in approximate MDP homomorphisms. In: Adv. Neural Information Proc. Systems, pp. 1649–1656 (2008)S
28. Zhu, J., Hastie, T.: Kernel logistic regression and the import vector machine. In: Adv. Neural Information Proc. Systems. pp. 1081–1088 (2002)
29. Ziebart, B., Maas, A., Bagnell, J., Dey, A.: Maximum entropy inverse reinforcement learning. In: Proc. 23rd AAAI Conf. Artificial Intelligence, pp. 1433–1438 (2008)

A Description of a Lax-Bisimulation Metric

Let $\mathcal{M} = (\mathcal{X}, \mathcal{A}, \mathsf{P}, r, \gamma)$ be an MDP, where the state-space $\mathcal{X}$ is endowed with a distance d. In other words, $(\mathcal{X}, d)$ is a metric space, and we henceforth refer to d as the *ground distance*. The function d can be any reasonable distance function on $\mathcal{X}$, *e.g.*, the Manhattan distance in the environment of Fig. 1.

Given one such distance d and any two distributions p_1 and p_2 over $\mathcal{X}$, the *Kantorovich distance* (also known as the earth mover's distance) between p_1 and p_2 is defined as the value of the linear program

$$\begin{array}{lll} \max\limits_{\theta_x} & \sum\limits_x (p_1(x) - p_2(x))\theta_x & \\ \text{s.t.} & \theta_x - \theta_y \leq d(x, y), & \text{for all } x, y \in \mathcal{X} \\ & 0 \leq \theta_x \leq 1 & \text{for all } x \in \mathcal{X} \end{array}$$

and denoted as $\mathsf{K}_d(p_1, p_2)$. Now given any two state-action pairs (x, a) and (y, b), we write

$$\delta_d\big((x, a), (y, b)\big) = k_1|r(x) - r(y)| + k_2\mathsf{K}_d\big(\mathsf{P}(x, a, \cdot), \mathsf{P}(y, b, \cdot)\big) \tag{8}$$

where k_1 and k_2 are non-negative constants such that $k_1 + k_2 \leq 1$. The function δ_d is, in a sense, a "one-step distance" between (x, a) and (y, b). It measures how different (x, a) and (y, b) are in terms of immediate reward/next transition. In order to measure differences in terms of long-term behavior, we denote by $\mathsf{H}_d(U, V)$ the Hausdorff distance (associated with d) between two sets U and V,[5] and define the MDP metric d_{MDP} as the fixed point of the operator $\mathbf{F}$ given by

$$\mathbf{F}(d)(x, y) = \mathsf{H}_{\delta_d}(\{x\} \times \mathcal{A}, \{y\} \times \mathcal{A}). \tag{9}$$

As shown in [27], the metric d_{MDP} can be obtained by iterating $\mathbf{F}$ and whenever $d(x, y) = 0$ then x and y are lax-bisimulation equivalent. Also, using the metric d_{MDP} it is possible to relabel the actions at each state to match those of "nearby" states.

We conclude by noting that, as discussed in Section 4, in our results we use $\delta_{d_{\text{MDP}}}$ with $k_1 = 0$ on the right-hand side of (8).

---

[5] Given a metric space $(\mathcal{X}, d)$, the Hausdorff distance between two sets $U, V \subset \mathcal{X}$ is given by

$$\mathsf{H}_d(U, V) = \max\left\{\sup_{x \in U} \inf_{y \in V} d(x, y), \sup_{y \in V} \inf_{x \in U} d(x, y)\right\}.$$

Demand-Driven Tag Recommendation

Guilherme Vale Menezes[1], Jussara M. Almeida[1], Fabiano Belém[1], Marcos André Gonçalves[1], Anísio Lacerda[1], Edleno Silva de Moura[2], Gisele L. Pappa[1], Adriano Veloso[1], and Nivio Ziviani[1]

[1] Universidade Federal de Minas Gerais, Belo Horizonte, Brazil
[2] Universidade Federal do Amazonas, Manaus, Brazil

Abstract. Collaborative tagging allows users to assign arbitrary keywords (or tags) describing the content of objects, which facilitates navigation and improves searching without dependence on pre-configured categories. In large-scale tag-based systems, tag recommendation services can assist a user in the assignment of tags to objects and help consolidate the vocabulary of tags across users. A promising approach for tag recommendation is to exploit the co-occurrence of tags. However, these methods are challenged by the huge size of the tag vocabulary, either because (1) the computational complexity may increase exponentially with the number of tags or (2) the score associated with each tag may become distorted since different tags may operate in different scales and the scores are not directly comparable. In this paper we propose a novel method that recommends tags on a demand-driven basis according to an initial set of tags applied to an object. It reduces the space of possible solutions, so that its complexity increases polynomially with the size of the tag vocabulary. Further, the score of each tag is calibrated using an entropy minimization approach which corrects possible distortions and provides more precise recommendations. We conducted a systematic evaluation of the proposed method using three types of media: audio, bookmarks and video. The experimental results show that the proposed method is fast and boosts recommendation quality on different experimental scenarios. For instance, in the case of a popular audio site it provides improvements in precision (p@5) ranging from 6.4% to 46.7% (depending on the number of tags given as input), outperforming a recently proposed co-occurrence based tag recommendation method.

1 Introduction

The act of associating keywords (tags) with objects is referred to as tagging. It has become a popular activity with the advent of Web 2.0 applications, which facilitated and stimulated end users to contribute with content created by themselves. This content, typically multimedia (e.g. audio, image, video), brings challenges to current information retrieval (IR) methods, not only due to the scale of the collections and speed of update, but also due to the (usually) poor quality of user-generated material. Tags offer a good alternative for personal or community-based organization, dissemination, and retrieval of Web 2.0 content. In fact, recent studies have demonstrated that tags are among the best textual features to be exploited by IR services, such as automatic classification [5].

In this context, tag recommendation services aims at improving the description of an object by suggesting tags that would correctly and more completely describe its

J.L. Balcázar et al. (Eds.): ECML PKDD 2010, Part II, LNAI 6322, pp. 402–417, 2010.

content. Possible sources of information for this recommendation task could be: (1) tags previously associated with objects in the collection; and (2) the textual content of other features (e.g., title, description, user comments) associated with the object for which the recommendation is expected. While in case of (2) there could be more input for the sake of recommendation, problems such as the lack of standardization in content format and the presence of noisy content (e.g., non-existing words [23]) benefit the use of recommendation methods that exploit solely tag co-occurrence information [6,20].

In this paper we present a new co-occurrence based algorithm to recommend tags on a demand-driven fashion according to an initial set of tags applied to an object. We address the tag recommendation problem from the community-oriented perspective, in the sense that the set of recommended tags is useful both to the user and to a larger community of users. Typically, an initial set of tags $\mathcal{I}_o$ associated with an object o is provided to the recommendation method, which outputs a set of related tags $\mathcal{C}_o$, where $\mathcal{I}_o \cap \mathcal{C}_o = \emptyset$.

Since there is an unlimited number of ways to describe an object by choosing arbitrary keywords, an object may have several different tags associated with it. Our strategy is to treat each possible tag already existent in the system as a class for the object, modeling the problem of recommending tags as a *multi-label classification* problem. This approach is challenging since the vocabulary of tags in systems such as *YouTube*[1], *Delicious*[2] and *LastFM*[3] is large and current automatic classifiers cannot deal well with problems with many thousands of classes.

We present a Lazy Associative Tag REcommender, referred to as LATRE from now on, which has been developed to deal with large-scale problems with thousands of tags (or classes). LATRE exploits co-occurrence of tags by extracting association rules on a demand-driven basis by. These rules are the basic components of the classification model produced by LATRE. In this case, rules have the form $X \rightarrow y$, where X is a set of tags and y is the predicted tag.

Rule extraction is a major issue for co-occurrence based recommendation methods, such as [7,20], since the number of extracted rules may increase exponentially with the number of tags. LATRE, on the other hand, extracts rules from the training data on the fly, at recommendation time. The algorithm projects the search space for rules according to qualitative information present in each test instance, allowing the generation of more elaborated rules with efficiency. In other words, LATRE projects/filters the training data according to the tags in $\mathcal{I}_o$, and extracts rules from this projected data. This ensures that only rules that carry information about object o (i.e., a test object) are extracted from the training data, drastically bounding the number of possible rules. In fact, the computational complexity of LATRE is shown to increase polynomially with the number of tags in the vocabulary. This efficiency enables LATRE to explore portions of the rule space that could not be feasibly explored by other methods (i.e., more "complex" rules).

After a set of rules is extracted for object o, LATRE uses them to sort/rank candidate tags that are more likely to be correctly associated with this object. Each extracted rule

[1] www.youtube.com
[2] www.delicious.com
[3] www.lastfm.com

$X \xrightarrow{\theta} y$ is interpreted as a vote given for tag y and the weight of the vote is given by θ, which is the conditional probability of object o being associated with tag y given that o contains all tags in X. Weighted votes for each tag are added, and tags that scored higher are placed on the beginning of the ranking. Usually, there are many candidate tags, and thus properly ranking them is also a difficult issue, since different candidate tags may operate in different scales (i.e., a popular tag may receive a large number of "weak" votes, and this tag is likely to be placed before a specific tag which received a small number of "strong" votes). In order to enforce all tags to operate in the same scale, so that they can be directly compared, we employed an entropy-minimization calibration approach to correct possible distortions in the scores of candidate tags.

Experimental results obtained from collections crawled from Delicious, LastFM and YouTube show that LATRE recommends tags with a significantly higher precision in all collections when compared to a recent co-occurrence based baseline proposed in [20]. The study of the effectiveness of LATRE on three different collections corresponding to different media types (i.e. Web pages, audio and video) is also an important contribution of our work, as most of the methods in the literature are tested only with one collection and one media type. Depending on the number of tags provided as input to LATRE, it obtained gains in precision (p@5) ranging from 10.7% to 23.9% for Delicious, from 6.4% to 46.7% for LastFM, and from 16.2% to 33.1% for YouTube.

This paper is organized as follows. In Section 2 we cover related work. In Section 3 we provide an in-depth description of our proposed method, whereas in Section 4 we describe our collections and discuss our experiments and results. Finally, in Section 5 we offer conclusions and possible directions for future work.

2 Related Work

Previous work has used co-occurrence information to expand an initial set of tags I_o associated with an object o with related tags [6,7,9,20,25,26]. Heymann et al. [7] use association rules to expand a set of tags of an object. A confidence threshold is set to limit the rules used in expansion. Experiments have shown that the expanded set of tags can increase the recall of their tag recommendation method by 50% while keeping a high precision. In [20], authors use conditional probability and the Jaccard coefficient to recommend tags on Flickr. They aggregate these co-occurrence measures for all tags in I_o to obtain the set of related tags. The problem they studied is very similar to ours, i.e., output a ranking of tags using only community knowledge (not personal information). Another related work is presented in [6]. They study the problem of making personal recommendations using the history of all tags a user has applied in the past. The authors use a Naive Bayes classifier to learn a model specific for each user. They concluded that adding personal history can improve the effectiveness of co-occurrence tag-recommendation. Krestel et al. [9] use Latent Dirichlet Allocation to expand the set of tags of objects annotated by only a few users (the cold start problem). After a comparison with [7] they concluded that their method is more accurate and yields more specific recommendations. In [25], Wu et al. model tag recommendation in Flickr as a learn to rank problem using RankBoost. They use tag co-occurrence measures and image content information as features for the learn to rank algorithm.

A content-based approach to the expansion of a tag set is described in [19]. The authors use overlapping and duplicated content in videos to create links in a graph and propagate tags between similar videos. They use the expanded set for video clustering and classification, obtaining significant improvements.

Tag co-occurrence has also been used in contexts different from tag expansion. For example, the tag recommendation algorithm described in [26] uses co-occurrence information to select a small set of informative tags from the tags collectively used to describe an object. They give higher values to tags that have been used together by the same user (complementary tags) and lower value to different tags that have been used by different users to describe the same object (tags that describe the same concept). Another example is the identification of ambiguous tags using co-occurrence distributions. The method in [24] suggests tags that help to disambiguate the set of tags previously assigned to an object. The key observation is that very different distributions of co-occurring tags arise after adding each ambiguous tag. A third example is tag ranking [11], in which the authors used a random walk process based on tag co-occurrence information to generate a ranking, which is shown to improve image search, tag recommendation and group recommendation. Finally, tag translation using tag co-occurrence is described in [19]. The authors created a tag co-occurrence graph and used network similarity measures to find candidates for translation.

Classification algorithms have been used in tag recommendation in the past. In [7], the authors use a SVM classifier to predict whether a tag t is associated with an object o based on the textual content of o. Their approach is not scalable to many thousands of tags since they need to build a binary classifier for each tag. In their experiments they use only the top 100 tags of their Delicious collection. A second approach is used by [22]. They group objects into clusters using a graph partitioning algorithm, and they train a Naive Bayes classifier using the generated clusters as classes. When a new object arrives, their method classifies the object into one of the clusters, and use the tags of the cluster to generate a ranking of candidate tags. A third way is to consider each tag as a class, and model tag recommendation as a multi-label classification problem. In this case, tags are used as both features and labels, and the classification algorithm must be able to deal with a very large number of classes. This approach is discussed in [6,21] and used in this work. While [6] uses a Naive Bayes classifier and [21] proposes a multi-label sparse Gaussian process classification to model tag recommendation, our work is based on associative classification. Between the many applications of social tags we can cite page clustering [12,15], enhancing item search [3,4,16], enhancing item recommendation [8,17,18], uncovering user-induced relations among items [2], clustering users in communities [10] and automatic ontology generation [13].

3 Associative Recommendation

We have essentially modeled the tag recommendation task as a multi-label classification problem. In this case, we have as input the *training data* (referred to as $\mathcal{D}$), which consists of objects of the form $d = <\mathcal{I}_d, \mathcal{Y}_d>$, where both $\mathcal{I}_d$ and $\mathcal{Y}_d$ are sets of tags, and initially, $\mathcal{I}_d$ contains all tags that are associated with object d, while $\mathcal{Y}_d$ is empty. The *test set* (referred to as $\mathcal{T}$) consists of objects of the form $t = <\mathcal{I}_t, \mathcal{Y}_t>$, where both

$\mathcal{I}_t$ and $\mathcal{Y}_t$ are sets of tags associated with object t. However, while tags in $\mathcal{I}_t$ are known in advance, tags in $\mathcal{Y}_t$ are unknown, and functions learned from $\mathcal{D}$ are used to predict (or recommend) tags that are likely to be in $\mathcal{Y}_t$ based on tags in $\mathcal{I}_t$. We developed the LATRE method within this model. Recommendation functions produced by LATRE exploit the co-occurrence of tags in $\mathcal{D}$, which are represented by association rules [1].

Definition 1. *An association rule is an implication $\mathcal{X} \xrightarrow{\theta} y$, where the antecedent $\mathcal{X}$ is a set of tags, and the consequent y is the predicted tag. The domain for $\mathcal{X}$ is denoted as $\mathcal{I}=\{\mathcal{I}_1 \cup \mathcal{I}_2 \cup \ldots \cup \mathcal{I}_m\}$ (i.e., $\mathcal{X} \subseteq \mathcal{I}$), where $m=|\mathcal{D}| + |\mathcal{T}|$. The domain for y is $\mathcal{Y}=\{\mathcal{Y}_1 \cup \mathcal{Y}_2 \cup \ldots \cup \mathcal{Y}_m\}$ (i.e., $y \in \mathcal{Y}$). The size of rule $\mathcal{X} \rightarrow y$ is given by the number of tags in the antecedent, that is $|\mathcal{X}|$. The strength of the association between $\mathcal{X}$ and y is given by θ, which is simply the conditional probability of y being in $\mathcal{Y}_o$ given that $\mathcal{X} \subseteq \mathcal{I}_o$.*

We denote as $\mathcal{R}$ a rule-set composed of rules $\mathcal{X} \xrightarrow{\theta} y$. Next we present the major steps of LATRE: rule extraction, tag ranking, and calibration.

3.1 Demand-Driven Rule Extraction

The search space for rules is huge. Existing co-occurrence based recommendation methods, such as the ones proposed in [20], impose computational cost restrictions during rule extraction. A typical strategy to restrict the search space for rules is to prune rules that are not sufficiently frequent (i.e., minimum support). This strategy, however, leads to serious problems because the vast majority of the tags are usually not frequent enough. An alternate strategy is to extract only rules $\mathcal{X} \rightarrow y$ such that $|\mathcal{X}| \leq \alpha_{max}$, where α_{max} is a pre-specified threshold which limits the maximum size of the extracted rules. However, in actual application scenarios, methods such as [20] are only able to efficiently explore the search space for rules if α_{max}=1. When the value of α_{max} is increased, the number of rules extracted from $\mathcal{D}$ increases at a much faster pace (i.e., there is a combinatorial explosion).

The obvious drawback of this approach (i.e., α_{max}=1) is that more complex rules (i.e., rules with $|\mathcal{X}|$>1) will be not included in $\mathcal{R}$. These complex rules may provide important information for the sake of recommendation, and thus, the immediate question is how to efficiently extract rules from $\mathcal{D}$ using arbitrary values of α_{max}. One possible solution for this question is to extract rules on a demand-driven basis, but before discussing this solution we need to present the definition of useful association rules.

Definition 2. *A rule $\{\mathcal{X} \rightarrow y\} \in \mathcal{R}$ is said to be useful for object $t = <\mathcal{I}_t, \mathcal{Y}_t>$ if $\mathcal{X} \subseteq \mathcal{I}_t$. That is, rule $\{\mathcal{X} \rightarrow y\} \in \mathcal{R}$ can only be used to predict tags for object $t \in \mathcal{T}$ if all tags in $\mathcal{X}$ are included in $\mathcal{I}_t$.*

The idea behind demand-driven rule extraction is to extract only those rules that are useful for objects in $\mathcal{T}$. In this case, rule extraction is delayed until an object $t = <\mathcal{I}_t, \mathcal{Y}_t>$ is informed. Then, tags in $\mathcal{I}_t$ are used as a filter which configures $\mathcal{D}$ in a way that only rules that are useful for object t can be extracted. This filtering process produces a projected training data, $\mathcal{D}_t$, which is composed of objects of the form $d^t = < \mathcal{I}^t_d, \mathcal{Y}^t_d >$, where $\mathcal{I}^t_d=\{\mathcal{I}_d \cap \mathcal{I}_t\}$ and $\mathcal{Y}^t_d=\{\mathcal{I}_d - \mathcal{I}^t_d\}$.

Table 1. Training data and test set

		$\mathcal{I}$	$\mathcal{Y}$
	d_1	unicef children un united nations	$\emptyset$
	d_2	un climatechange summit environment	$\emptyset$
$\mathcal{D}$	d_3	climatechange islands environment	$\emptyset$
	d_4	children games education math	$\emptyset$
	d_5	education children unicef job	$\emptyset$
$\mathcal{T}$	t_1	unicef education haiti	?

Table 2. Projected training data for object t_1

		$\mathcal{I}^t$	$\mathcal{Y}^t$
	$d_1^{t_1}$	unicef	children un united nations
$\mathcal{D}_{t_1}$	$d_4^{t_1}$	education	children games math
	$d_5^{t_1}$	unicef education	children job

The process is illustrated in Tables 1 and 2. There are 5 objects $\{<\mathcal{I}_{d_1}, \mathcal{Y}_{d_1}>; <\mathcal{I}_{d_2}, \mathcal{Y}_{d_2}>; <\mathcal{I}_{d_3}, \mathcal{Y}_{d_3}>; <\mathcal{I}_{d_4}, \mathcal{Y}_{d_4}>; <\mathcal{I}_{d_5}, \mathcal{Y}_{d_5}>\}$ in $\mathcal{D}$, and one object $<\mathcal{I}_{t_1}, \mathcal{Y}_{t_1}>$ in $\mathcal{T}$. Table 2 shows $\mathcal{D}$ after being projected according to $\mathcal{I}_{t_1}$. In this case, $\mathcal{I}_{d_1}^{t_1}$=$\{\mathcal{I}_{t_1} \cap \mathcal{I}_{d_1}\}$=$\{$unicef$\}$, and $\mathcal{Y}_{d_1}^{t_1}$=$\{\mathcal{I}_{d_1} - \mathcal{I}_{d_1}^{t_1}\}$=$\{$children, un, united, nations$\}$. The same procedure is repeated for the remaining objects in $\mathcal{D}$, so that $\mathcal{D}^{t_1}$ is finally obtained. For an arbitrary object $t \in \mathcal{T}$, we denote as $\mathcal{R}_t$ the rule-set extracted from $\mathcal{D}_t$.

Lemma 1. *All rules in $\mathcal{R}_t$ are useful for object t =$<\mathcal{I}_t, \mathcal{Y}_t>$.*

Proof. Let $\mathcal{X} \rightarrow y$ be an arbitrary rule in $\mathcal{R}_t$. In this case, $\mathcal{X} \subseteq \mathcal{I}_t$. Thus, according to Definition 2, this rule must be useful for object t. □

Lemma 1 states that any rule extracted from $\mathcal{D}_{t_1}$ (i.e., Table 2) is useful for object t_1. Examples of rules extracted from $\mathcal{D}_{t_1}$ include:

- unicef $\xrightarrow{\theta=1.00}$ children
- $\{$unicef$\wedge$education$\}$ $\xrightarrow{\theta=1.00}$ children
- education $\xrightarrow{\theta=0.50}$ math

Since $\{$unicef,education$\}$ $\subseteq \mathcal{I}_{t_1}$, all these rules are useful for object t_1. An example of rule that is useless for object t_1 is "climatechange $\xrightarrow{\theta=1.00}$ environment", and it is easy to see that this rule cannot be extracted from $\mathcal{D}_{t_1}$, since tag "climatechange" is not present in $\mathcal{D}_{t_1}$.

The next theorem states that LATRE efficiently extracts rules from $\mathcal{D}$. The key intuition is that LATRE works only on tags that are known to be associated to each other, drastically narrowing down the search space for rules.

Theorem 1. *The complexity of LATRE increases polynomially with the number of tags in the vocabulary.*

Proof. Let n be the number of tags in the vocabulary. Obviously, the number of possible association rules that can be extracted from $\mathcal{D}$ is 2^n. Also, let $t = <\mathcal{I}_t, \mathcal{Y}_t>$ be an arbitrary object in $\mathcal{T}$. Since $\mathcal{I}_t$ contains at most k tags (with $k \ll n$), any rule useful for object t can have at most k tags in its antecedent. Therefore, the number of possible rules that are useful for object t is $(n\text{-}k) \times (k + \binom{k}{2} + \ldots + \binom{k}{k}) = O(n^k)$ (since $k \ll n$), and thus, the number of useful rules increases polynomially in n. Since, according to Lemma 1, LATRE extracts only useful rules for objects in $\mathcal{T}$, then the complexity of LATRE also increases polynomially in n. □

An important practical aspect of LATRE is that the projection of the dataset (as shown in the examples in Tables 1 and 2) greatly reduces the size of both n and k, since we only consider candidate tags that co-occur at least once with any tag in the test object (n is reduced), while the size of the test object is small in practice (k is reduced). For instance, in Tables 1 and 2 we have $k_1 = 1$, $k_2 = 0$, $k_3 = 0$, $k_4 = 1$ and $k_5 = 2$ for the projected dataset. Therefore, the average k per object is $(1+0+0+1+2)/5 = 4/5 = 0.8$. This number is much smaller than the upper bound in the number of tags in an object, which is 5 in the example.

3.2 Tag Ranking

In order to select candidate tags that are more likely to be associated with object $t \in \mathcal{T}$, it is necessary to sort tags by combining rules in $\mathcal{R}_t$. In this case, LATRE interprets $\mathcal{R}_t$ as a poll, in which each rule $\mathcal{X} \xrightarrow{\theta} y \in \mathcal{R}_t$ is a vote given by tags in $\mathcal{X}$ for candidate tag y. Votes have different weights, depending on the strength of the association they represent (i.e., θ). The weighted votes for each tag y are summed, giving the score for tag y with regard to object t, as shown in Equation 1 (where y_i is the i-th candidate tag, and $\theta(\mathcal{X} \rightarrow y_i)$ is the value θ assumes for rule $\mathcal{X} \rightarrow y_i$):

$$s(t, y_i) = \sum \theta(\mathcal{X} \rightarrow y_i), \text{ where } \mathcal{X} \subseteq \mathcal{I}_t \tag{1}$$

Thus, for an object t, the score associated with tag y_i is obtained by summing the θ values of the rules predicting y_i in $\mathcal{R}_t$. The likelihood of t being associated with tag y_i is obtained by normalizing the scores, as expressed by the function $\hat{p}(y_i|t)$, shown in Equation 2:

$$\hat{p}(y_i|t) = \frac{s(t, y_i)}{\sum_{j=0}^{n} s(t, y_j)} \tag{2}$$

Candidate tags for object t are sorted according to Equation 2, and tags appearing first in the ranking are finally recommended.

3.3 Calibration

According to Equation 1, the score associated with a tag is impacted by two characteristics: (1) the number of votes it receives, and also (2) the strength of these votes. While both characteristics are intuitively important to estimate the likelihood of association between tags and objects, it may be difficult to decide which one is more important. In some cases, the scores associated with different tags cannot be directly compared, because they operate in different scales (i.e., the score associated with popular tags are likely to be higher than the scores associated with specific tags, simply because they receive a large number of votes). This means that the same value of score can be considered either high or low, depending on the tag.

An approach for this problem would be to inspect the expected likelihood $\hat{p}(y|t)$, in order to make scores associated with different tags directly comparable. The obvious problem with this approach is that the correct value for $\hat{p}(y|t)$ is not known in advance, since $t \in \mathcal{T}$, and thus we cannot verify if $y \in \mathcal{Y}_t$. An alternative is to use a validation set (denoted as $\mathcal{V}$), which is composed of objects of the form $v = <\mathcal{I}_v, \mathcal{Y}_v>$, where both $\mathcal{I}_v$ and $\mathcal{Y}_v$ are sets of tags associated with object v, and $\{\mathcal{I}_v \cap \mathcal{Y}_v\}=\emptyset$. That is, the validation set essentially mimics the test set, in the sense that $\mathcal{Y}_v$ is not used for the sake of producing rules, but only to find possible distortions in the value of $\hat{p}(y|v)$.

The key intuition of our approach is to contrast values of $\hat{p}(y|v)$ for which $y \notin \mathcal{Y}_v$, and values of $\hat{p}(y|v)$ for which $y \in \mathcal{Y}_v$. In an ideal case, for a given tag y, there is a value f_y such that:

- if $\hat{p}(y|v) \leq f_y$, then $y \notin \mathcal{Y}_v$
- if $\hat{p}(y|v) > f_y$, then $y \in \mathcal{Y}_v$

Once f_y is calculated, it can be used to determine whether a certain value of $\hat{p}(y|v)$ is low or high, so that the score associated with different tags can be directly compared. However, more difficult cases exist, for which it is not possible to obtain a perfect separation in the space of values for $\hat{p}(y|v)$. Thus, we propose a more general approach to calculate f_y. The basic idea is that any value for f_y induces two partitions over the space of values for $\hat{p}(y|v)$ (i.e., one partition with values that are lower than f_y, and another partition with values that are higher than f_y). Our approach is to set f_y with the value which minimizes the average entropy of these two partitions. In the following we present the basic definitions in order to detail this approach.

Definition 3. *Let y be an arbitrary tag, and let $v = <\mathcal{I}_v, \mathcal{Y}_v>$ be an arbitrary object in $\mathcal{V}$. In this case, let $o(y, v)$ be a binary function such that:*

$$o(y,v) = \begin{cases} 1 \text{ if } y \in \mathcal{Y}_v \\ 0 \text{ otherwise} \end{cases}$$

Definition 4. *Consider $\mathcal{O}(y)$ a list of pairs $<o(y,v), \hat{p}(y|v)>$, sorted in increasing order of $\hat{p}(y|v)$. That is, $\mathcal{O}(y)$=$\{\ldots, <o(y,v_i), \hat{p}(y|v_i)>, <o(y,v_j), \hat{p}(y|v_j)>, \ldots\}$, such that $\hat{p}(y|v_i) \leq \hat{p}(y|v_j)$. Also, consider c a candidate value for f_y. In this case, $\mathcal{O}_c(y, \leq)$ is a sub-list of $\mathcal{O}(y)$, that is, $\mathcal{O}_c(y, \leq)$=$\{\ldots, <o(y,v), \hat{p}(y|v)>, \ldots\}$, such that for all pairs in $\mathcal{O}_c(y, \leq)$, $\hat{p}(y|v) \leq c$. Similarly, $\mathcal{O}_c(y, >)$=$\{\ldots, <o(y,v), \hat{p}(y|v)>, \ldots\}$, such that for all pairs in $\mathcal{O}_c(y, >)$, $\hat{p}(y|v) > c$. In other words, $\mathcal{O}_c(y, \leq)$ and $\mathcal{O}_c(y, >)$ are two partitions of $\mathcal{O}(y)$ induced by c.*

Definition 5. *Consider $N_0(\mathcal{O}(y))$ the number of elements in $\mathcal{O}(y)$ for which $o(y, v)$=0. Similarly, consider $N_1(\mathcal{O}(y))$ the number of elements in $\mathcal{O}(y)$ for which $o(y, v)$=1.*

The first step of our entropy-minimization calibration approach is to calculate the entropy of tag y in $\mathcal{O}(y)$, as shown in Equation 3.

$$E(\mathcal{O}(y)) = -\left(\frac{\mathrm{N}_0(\mathcal{O}(y))}{|\mathcal{O}(y)|} \times \log \frac{\mathrm{N}_0(\mathcal{O}(y))}{|\mathcal{O}(y)|}\right) - \left(\frac{\mathrm{N}_1(\mathcal{O}(y))}{|\mathcal{O}(y)|} \times \log \frac{\mathrm{N}_1(\mathcal{O}(y))}{|\mathcal{O}(y)|}\right) \quad (3)$$

The second step is to calculate the sum of the entropies of tag y in each partition induced by c, according to Equation 4.

$$E(\mathcal{O}(y), c) = \frac{|\mathcal{O}_c(y, \leq)|}{|\mathcal{O}(y)|} \times E(\mathcal{O}_c(y, \leq)) + \frac{|\mathcal{O}_c(y, >)|}{|\mathcal{O}(y)|} \times E(\mathcal{O}_c(y, >)) \quad (4)$$

The third step is to set f_y to the value of c which minimizes the difference $E(\mathcal{O}(y)) - E(\mathcal{O}(y), c)$. Now, the final step is to calibrate each $\hat{p}(y|t)$ (note that $t \in \mathcal{T}$) using the corresponding f_y (which was obtained using the validation set). The intuition is that f_y separates values of $\hat{p}(y|t)$ that should be considered high (i.e., $\hat{p}(y|t) > f_y$) from those that should be considered low (i.e., $\hat{p}(y|t) \leq f_y$). Thus, a natural way to calibrate $\hat{p}(y|t)$ is to calculate how many times $\hat{p}(y|t)$ is greater than f_y. This can be easily done as shown in Equation 5. The values of $\hat{c}(y|t)$ are directly comparable, since the corresponding values of $\hat{p}(y|t)$ were normalized by f_y. Thus, $\hat{c}(y|t)$ is used to sort candidate tags that are more likely to be associated with object t:

$$\hat{c}(y|t) = \frac{\hat{p}(y|t)}{f_y} \quad (5)$$

In some cases, calibration may drastically improves recommendation performance, as we will show in the next section.

4 Experimental Evaluation

In this section we empirically analyze the recommendation performance of LATRE. We employ as the basic evaluation measures precision at x (p@x), which measures the proportion of relevant tags in the x first positions in the tag ranking, and MRR [6,20], which shows the capacity of a method to return relevant tags early in the tag ranking. We first present the baseline and collections employed in the evaluation, and then we discuss the recommendation performance of LATRE on these collections.

4.1 Baseline

The baseline method used for comparison is described in [20]. It is a co-occurrence method which also employs association rules. It gives less weight to candidate tags that are either too popular or too rare. Furthermore, its algorithm gives more weight to candidate tags that are higher in each candidate tag list with the goal of smoothening the co-occurrence values decay. Other related methods in Section 2 were not considered as baselines because they use additional information, such as the user history [6], image features [25] and the page content [7,19].

4.2 Collections

Differently from related work that present results restricted to a single collection [6,7,20], in this paper we experiment with several collections, namely, Delicious, LastFM and YouTube.

Delicious

Delicious is a popular social bookmarking application that permits users to store, share and discover bookmarks. Users can categorize their bookmarks using tags, which serve as personal indexes so that a user can retrieve its stored pages. The assignment of tags in Delicious is collaborative, i.e., the set of tags associated with a page is generated by many users in collaboration.

For the Delicious crawl we used its "Recent Bookmarks" page, which is a public timeline that shows a subset of the most recently bookmarked objects. We collected unique bookmarked page entries and extracted the set of most frequently used bookmarks for a page (i.e., its "top bookmarks"). Delicious makes available as many as 30 "top bookmarks". Therefore, this is the maximum number of tags per object in our crawled dataset.

The crawl was performed in October 2009. We obtained 560,033 unique object entries and 872,502 unique tags. The mean number of tags per object is 18.27. The number of tags per object ranged from 1 to 30. The first, second and third quartile of the distribution of tags per page is, respectively, 8, 17 and 27.

LastFM

LastFM is a Web 2.0 music website and Internet radio. It allows users to collaboratively contribute with tags to categorize and describe the characteristics of artists, such as the music genre. LastFM was crawled using a snowball approach, which collects a set of seed artists and follows links to related artists. The artists used as seeds are the ones associated with the system most popular tags.

The crawl was performed in October 2008. We obtained 99,161 unique object entries and 109,443 unique tags. The mean number of tags per object is 26.88. The number of tags per object ranged from 1 to 210. The first quartile of the distribution of tags per object is 7, the second quartile is 14, and the third quartile is 31.

YouTube

YouTube is the largest video sharing application on the Web. Users that upload videos to YouTube can provide a set of tags that describe them for indexing purposes. YouTube is different from Delicious and LastFM in that it is non-collaborative, that is, only the video uploader can provide tags.

YouTube was crawled using a snowball approach, following links between related videos. The all-time most popular videos were used as seeds. Our sample was obtained in July 2008. We obtained 180,778 unique objects, and they were described by 132,694 unique tags. The mean number of tags per object is 9.98. The number of tags per object ranged from 1 to 91. The first quartile of the distribution of tags per object is 6, the second quartile is 9 and the third quartile is 14.

Table 3. We divided each collection into three subsets according to the number of tags per object. We show the number of objects in each of these subsets and the average number of tags per object (and its standard deviation) in each subset. Note that we excluded all objects associated with a single tag, since we need at least one tag in $\mathcal{I}_t$ and one tag in $\mathcal{Y}_t$.

Collection	Range	# Objects	Avg. # Tags
Delicious	2 to 6 tags/object	188,173	3.94 ± 1.38
	7 to 12 tags/object	167,613	9.50 ± 1.73
	13 to 30 tags/object	170,708	15.92 ± 2.53
LastFM	2 to 6 tags/object	29,622	3.96 ± 1.39
	7 to 16 tags/object	30,215	10.55 ± 2.77
	17 to 152 tags/object	31,492	44.99 ± 25.30
YouTube	2 to 5 tags/object	56,721	3.63 ± 1.09
	6 to 9 tags/object	53,284	7.39 ± 1.11
	10 to 74 tags/object	59,285	13.60 ± 5.02

4.3 Pre-processing Steps and Setup

In order to assess the recommendation performance of the evaluated methods, we equally divided the tags associated with test object $t = <\mathcal{I}_t, \mathcal{Y}_t>$: half of the tags is included in $\mathcal{I}_t$, and the other half is included in $\mathcal{Y}_t$ and used to assess the performance. This division is made by shuffling the tags and including the first half in $\mathcal{I}_t$ and the last half in $\mathcal{Y}_t$. A similar approach has been adopted in [6] and [20]. We applied Porter's stemming algorithm [14] to avoid trivial recommendations such as plurals and small variations of the same input word.

We split each collection into three subsets: the first subset is composed of objects with a large number of tags, the second subset is composed of objects with a moderate number of tags, and the third subset is composed of objects with a small number of tags. The range of tags per object was selected in a way that the corresponding subsets have approximately the same number of objects. These subsets are shown in Table 3.

Then, we randomly selected 20,000 objects from each of the subsets. We divided each group of selected objects into 5 partitions of 4,000 objects each, and we used 5-fold cross validation to assess recommendation performance. We use the validation set to find the best parameters for each evaluated method.

4.4 Results

All experiments were performed on a Linux PC with an Intel Core 2 Duo 2.20GHz and 4GBytes RAM. In the following subsections we discuss the effectiveness and the computational efficiency of LATRE.

4.5 Precision

Tables 4, 5 and 6 show the results for p@x for each subset of the three collections. We varied x from 1 to 5, following the analysis performed in previous work [6,20]. The reason is that we are interested in the performance of the methods on the top of the ranking (i.e. the first 5 recommendations), since in tag recommendation the user is not likely to scan a large number of tags before choosing which ones are relevant. Furthermore, it is better to recommend good tags earlier in the ranking (e.g. p@1), so that the user has to scan fewer tags. We executed three algorithms over the subsets: the baseline, LATRE without calibration (referred to as LATNC) and LATRE.

Statistical tests have shown that LATRE performs significantly better ($p < 0.05$) than the baseline in all scenarios we experimented with. LATRE has shown gains in p@5 from 6.4% in LastFM to 23.9% in Delicious if we consider only the lower ranges; considering only the middle ranges, LATRE has shown gains in p@5 from 10.7% in Delicious to 28.9% in YouTube; and considering only the upper ranges, LATRE has shown gains in p@5 from 17.2% in Delicious to 46.7% in LastFM. It is important to note that the absolute precision values shown in this paper are underestimated, since there may be additional tags that are relevant to the user and that were not used by he/she to describe the object (and thus are not in $\mathcal{Y}_t$), as discussed in [6].

One interesting conclusion we could draw from the experiments is that calibration has its best performance in the lower ranges, indicating that the distortions described in Section 3 have a more damaging effect in these ranges. It is specially difficult to perform well in the lower ranges since there is very little information to work with, e.g., when there are two tags associated with an object, only one tag can be used as input and only one tag can be considered to be the correct answer. Several applications that use tag co-occurrence could benefit from calibration in these cases, such as in tag expansion for the cold start problem [9,7] (see Section 2).

As the number of tags per object increases, the benefit of using more elaborated rules becomes clearer. The gains in precision in the middle and upper ranges are mainly due to LATNC (i.e., LATRE without calibration), and the reason is that there are more opportunities for producing complex rules in these ranges, i.e., there is more information available. Applications that use tag co-occurrence in objects with a large number of tags could benefit from these elaborated rules, such as tag expansion for index enrichment [20,7], tag ranking [11] or tag translation [19].

It is interesting to notice that in the range 17-152 of LastFM, the baseline has achieved a recommendation performance which is lower than its performance in the range 7-16 (p@1=0.40 vs. p@1=0.54). The baseline does not perform well in range 17-152 of LastFM because this subset has a high number of tags per object (see Table 3), and the algorithm tends to recommend tags that are too general, such as "music" and "listen".

Furthermore, Tables 4, 5 and 6 show that the absolute precision values for Delicious are lower than the corresponding values in LastFM and YouTube. The reason is that Delicious has a much more diverse set of objects, since Web pages can contain or refer to any kind of data, information or media. As an example, YouTube and LastFM pages can also be stored as a bookmark in Delicious. In fact, the number of distinct tags in Delicious in the five partitions used in our experiment (20,000 objects) is higher than in YouTube and LastFM. In the lower ranges, Delicious has 13,247 unique tags, while LastFM and YouTube have 5,164 and 6,860, respectively. The same relative proportions were found in the middle and higher ranges.

Table 4. Delicious: results for p@1, p@3, p@5 and MRR. Statistically significant differences ($p < 0.05$) are shown in (1) bold face and (2) marked with an asterisk (*), representing respectivelly (1) cases in which LATNC and/or LATRE performs better than the baseline and (2) cases in which LATNC performs worse than the baseline. LATRE relative gains are shown in the last row.

	2-6 tags/object				7-12 tags/object				13-30 tags/object			
	p@1	p@3	p@5	MRR	p@1	p@3	p@5	MRR	p@1	p@3	p@5	MRR
Baseline	.106	.063	.046	.158	.307	.196	.150	.417	.506	.362	.285	.627
LATNC	.105	.059*	.045*	.154	**.328**	**.214**	**.160**	**.426**	**.544**	**.414**	**.335**	**.659**
LATRE	**.129**	**.077**	**.057**	**.188**	**.330**	**.218**	**.166**	**.435**	**.543**	**.413**	**.334**	**.659**
[% gain]	**21.7**	**22.2**	**23.9**	**19.0**	**7.5**	**11.2**	**10.7**	**4.3**	**7.3**	**14.1**	**17.2**	**5.1**

Table 5. LastFM: results for p@1, p@3, p@5 and MRR

	2-6 tags/object				7-16 tags/object				17-152 tags/object			
	p@1	p@3	p@5	MRR	p@1	p@3	p@5	MRR	p@1	p@3	p@5	MRR
Baseline	.313	.174	.125	.403	.539	.366	.279	.646	.400	.328	.289	.560
LATNC	.320	**.180**	.128	.409	**.574**	**.410**	**.314**	**.670**	**.564**	**.476**	**.425**	**.695**
LATRE	**.327**	**.187**	**.133**	**.418**	**.575**	**.411**	**.316**	**.672**	**.564**	**.475**	**.424**	**.695**
[% gain]	**4.5**	**7.5**	**6.4**	**3.7**	**6.7**	**12.3**	**13.3**	**4.0**	**41.0**	**44.8**	**46.7**	**24.1**

Table 6. YouTube: results for p@1, p@3, p@5 and MRR

	2-5 tags/object				6-9 tags/object				10-74 tags/object			
	p@1	p@3	p@5	MRR	p@1	p@3	p@5	MRR	p@1	p@3	p@5	MRR
Baseline	.288	.153	.105	.356	.405	.273	.204	.646	.534	.419	.347	.560
LATNC	**.328**	**.171**	**.113**	**.385**	**.485**	**.356**	**.254**	**.556**	**.626**	**.528**	**.459**	**.704**
LATRE	**.350**	**.184**	**.122**	**.411**	**.491**	**.365**	**.263**	**.567**	**.627**	**.530**	**.462**	**.706**
[% gain]	**21.5**	**20.3**	**16.2**	**15.5**	**21.2**	**33.7**	**28.9**	**14.1**	**17.4**	**26.5**	**33.1**	**10.7**

Mean Reciprocal Rank (MRR)

Tables 4, 5 and 6 also show the values of MRR for all ranges. MRR shows the capacity of a method to return relevant tags early in the tag ranking. Statistical significance tests were performed ($p < 0.05$) and results that are statistically different from the baseline are shown in bold face. We can see that LATRE has results significantly better than the baseline for all ranges in all datasets.

Computational Efficiency

We evaluated LATRE efficiency by measuring the average execution time per object. Table 7 shows the results for each subset. For subsets with few tags per object, the average and maximum tagging time are in the order of few milliseconds. As expected,

Table 7. Tagging time in seconds. We also show the standard deviation of the average tagging time.

Collection	Subset	Avg. Time	Max. Time
	2-6	0.0023 ± 0.0019	0.016
Delicious	7-12	0.067 ± 0.047	0.33
	13-30	0.47 ± 0.24	1.23
	2-6	0.0062 ± 0.0055	0.039
LastFM	7-16	0.20 ± 0.15	1.18
	17-152	1.77 ± 0.34	2.44
	2-5	0.0023 ± 0.0023	0.037
YouTube	6-9	0.027 ± 0.026	0.32
	10-74	0.31 ± 0.27	1.56

Table 8. Number of extracted rules for each subset of LastFM

α_{max}	2-6 tags/object		7-16 tags/object		17-152 tags/object	
	Baseline	LATRE	Baseline	LATRE	Baseline	LATRE
1	2.10^7	3.10^6	5.10^7	2.10^7	5.10^7	5.10^7
2	1.10^{11}	3.10^6	3.10^{11}	3.10^7	4.10^{11}	6.10^7
3	5.10^{14}	3.10^6	2.10^{15}	4.10^7	3.10^{15}	6.10^7

the average tagging time increases with the ratio of tags per object. However, even for objects that are associated with many tags, the average time spent per object is never greater than 1.8 seconds. This makes LATRE specially well-suited for real-time tag recommendation [22].

The last set of experiments aims at verifying the increase in the number of extracted rules as a function of α_{max}. According to Theorem 1, the number of rules extracted by LATRE increases polynomially. Table 8 contrasts the number of rules extracted by LATRE with the number of rules that would be extracted by the baseline. We only show the results for subsets of the LastFM collection, but the same trends are also observed for the subsets of the other two collections. Clearly, the number of rules extracted by the baseline increases exponentially, while the number of rules extracted by LATRE increases at a much slower pace.

5 Conclusions

In this paper we have introduced LATRE, a novel co-occurrence based tag recommendation method. LATRE extracts association rules from the training data on a demand-driven basis. The method projects the search space for rules according to qualitative information in test objects, allowing an efficient extraction of more elaborate rules. LATRE interprets each extracted rule as a vote for a candidate tag. After all votes are summed, tags are sorted according to their scores. Finally, LATRE calibrates the scores in order to correct possible distortions in the final ranked list of tags.

Our experiments involve objects belonging to different media types, namely Web pages from Delicious, audio from LastFM, and videos from YouTube. We have shown that LATRE recommends tags with a significantly higher precision in all subsets when compared against the baseline. While our proposed calibration mechanism has its best performance in subsets with a few number of tags, the use of more elaborate rules improves precision in subsets with a larger number of tags. We have proved that LATRE is able to efficiently extract such elaborate rules from the training data. LATRE achieved improvements in precision (p@5) from 10.7% to 23.9% for Delicious, from 6.4% to 46.7% for LastFM, and from 16.2% to 33.1% for YouTube.

As future work, we will investigate other textual features of the Web 2.0, and how these features may improve tag recommendation. Further, we will extend LATRE, so that recommended tags will be used as additional information which can be exploited to improve recommendation effectiveness.

Acknowledgements

We thank the partial support given by the Brazilian National Institute of Science and Technology for the Web (grant MCT/CNPq 573871/2008-6), Project InfoWeb (grant MCT/CNPq/CT-INFO 550874/2007-0), and authors' individual grants and scholarships from CNPq and FAPEMIG.

References

1. Agrawal, R., Imielinski, T., Swami, A.N.: Mining association rules between sets of items in large databases. In: Proc. of the ACM SIGMOD International Conference on Management of Data, pp. 207–216 (1993)
2. Au Yeung, C.-M., Gibbins, N., Shadbolt, N.: User-induced links in collaborative tagging systems. In: CIKM 2009: Proc. of the 18th ACM Conference on Information and Knowledge Management, pp. 787–796 (2009)
3. Bischoff, K., Firan, C.S., Nejdl, W., Paiu, R.: Can all tags be used for search. In: CIKM 2008: Proc. of the 17th ACM Conference on Information and Knowledge Management, pp. 193–202 (2008)
4. Carman, M.J., Baillie, M., Gwadera, R., Crestani, F.: A statistical comparison of tag and query logs. In: SIGIR 2009: Proc. of the 32nd International ACM SIGIR Conference on Research and Development in Information Retrieval, pp. 123–130 (2009)
5. Figueiredo, F., Belém, F., Pinto, H., Almeida, J., Gonçalves, M., Fernandes, D., Moura, E., Cristo, M.: Evidence of quality of textual features on the web 2.0. In: CIKM 2009: Proc. of the 18th ACM Conference on Information and Knowledge Management, pp. 909–918 (2009)
6. Garg, N., Weber, I.: Personalized, interactive tag recommendation for flickr. In: RecSys 2008: Proc. of the 2008 ACM Conference on Recommender Systems, pp. 67–74 (2008)
7. Heymann, P., Ramage, D., Garcia-Molina, H.: Social tag prediction. In: SIGIR 2008. Proc. of the 31nd International ACM SIGIR Conference on Research and Development in Information Retrieval, pp. 531–538 (2008)
8. Konstas, I., Stathopoulos, V., Jose, J.M.: On social networks and collaborative recommendation. In: SIGIR 2009: Proc. of the 32nd International ACM SIGIR Conference on Research and Development in Information Retrieval, pp. 195–202 (2009)

9. Krestel, R., Fankhauser, P., Nejdl, W.: Latent dirichlet allocation for tag recommendation. In: RecSys 2009: Proc. of the 2009 ACM conference on Recommender systems, pp. 61–68 (2009)
10. Li, X., Guo, L., Zhao, Y.E.: Tag-based social interest discovery. In: WWW 2008: Proc. of the 17th International Conference on World Wide Web, pp. 675–684 (2008)
11. Liu, D., Hua, X.-S., Yang, L., Wang, M., Zhang, H.-J.: Tag ranking. In: WWW 2009: Proc. of the 18th International Conference on World Wide Web, pp. 351–360 (2009)
12. Lu, C., Chen, X., Park, E.K.: Exploit the tripartite network of social tagging for web clustering. In: CIKM 2009: Proc. of the 18th ACM Conference on Information and Knowledge Management, pp. 1545–1548 (2009)
13. Plangprasopchok, A., Lerman, K.: Constructing folksonomies from user-specified relations on flickr. In: WWW 2009: Proc. of the 18th International Conference on World Wide Web, pp. 781–790 (2009)
14. Porter, M.F.: An algorithm for suffix stripping. Program 14(3), 130–137 (1980)
15. Ramage, D., Heymann, P., Manning, C.D., Garcia-Molina, H.: Clustering the tagged web. In: WSDM 2009: Proc. of the Second ACM International Conference on Web Search and Data Mining, pp. 54–63 (2009)
16. Schenkel, R., Crecelius, T., Kacimi, M., Michel, S., Neumann, T., Parreira, J.X., Weikum, G.: Efficient top-k querying over social-tagging networks. In: SIGIR 2008: Proc. of the 31st International ACM SIGIR conference on Research and development in information retrieval, pp. 523–530 (2008)
17. Sen, S., Vig, J., Riedl, J.: Tagommenders: connecting users to items through tags. In: WWW 2009: Proc. of the 18th International Conference on World Wide Web, pp. 671–680 (2009)
18. Shepitsen, A., Gemmell, J., Mobasher, B., Burke, R.: Personalized recommendation in social tagging systems using hierarchical clustering. In: RecSys 2008: Proc. of the 2008 ACM conference on Recommender systems, pp. 259–266 (2008)
19. Siersdorfer, S., San Pedro, J., Sanderson, M.: Automatic video tagging using content redundancy. In: SIGIR 2009: Proc. of the 32nd International ACM SIGIR Conference on Research and Development in Information Retrieval, pp. 395–402 (2009)
20. Sigurbjörnsson, B., van Zwol, R.: Flickr tag recommendation based on collective knowledge. In: WWW 2008: Proc. of the 17th International Conference on World Wide Web, pp. 327–336 (2008)
21. Song, Y., Zhang, L., Giles, C.L.: A sparse gaussian processes classification framework for fast tag suggestions. In: CIKM 2008: Proc. of the 17th ACM Conference on Information and Knowledge Management, pp. 93–102 (2008)
22. Song, Y., Zhuang, Z., Li, H., Zhao, Q., Li, J., Lee, W.-C., Giles, C.L.: Real-time automatic tag recommendation. In: SIGIR 2008: Proc. of the 31nd International ACM SIGIR Conference on Research and Development in Information Retrieval, pp. 515–522 (2008)
23. Suchanek, F.M., Vojnovic, M., Gunawardena, D.: Social tags: meaning and suggestions. In: CIKM 2008: Proc. of the 17th ACM Conference on Information and Knowledge Management, pp. 223–232 (2008)
24. Weinberger, K.Q., Slaney, M., Van Zwol, R.: Resolving tag ambiguity. In: MM 2008: Proc. of the 16th ACM International Conference on Multimedia, pp. 111–120 (2008)
25. Wu, L., Yang, L., Yu, N., Hua, X.-S.: Learning to tag. In: WWW 2009: Proc. of the 18th International Conference on World Wide Web, pp. 361–370 (2009)
26. Xu, Z., Fu, Y., Mao, J., Su, D.: Towards the semantic web: Collaborative tag suggestions. In: WWW 2006: Proc. of the Collaborative Web Tagging Workshop (2006)

Solving Structured Sparsity Regularization with Proximal Methods

Sofia Mosci[1], Lorenzo Rosasco[3,4], Matteo Santoro[1], Alessandro Verri[1], and Silvia Villa[2]

[1] Università degli Studi di Genova - DISI
Via Dodecaneso 35, Genova, Italy
[2] Università degli Studi di Genova - DIMA
Via Dodecaneso 35, Genova, Italy
[3] Istituto Italiano di Tecnologia,
Via Morego, 30 16163 Genova, Italy
[4] CBCL, Massachusetts Institute of Technology
Cambridge, MA 02139 - USA

Abstract. Proximal methods have recently been shown to provide effective optimization procedures to solve the variational problems defining the ℓ_1 regularization algorithms. The goal of the paper is twofold. First we discuss how proximal methods can be applied to solve a large class of machine learning algorithms which can be seen as extensions of ℓ_1 regularization, namely structured sparsity regularization. For all these algorithms, it is possible to derive an optimization procedure which corresponds to an iterative projection algorithm. Second, we discuss the effect of a preconditioning of the optimization procedure achieved by adding a strictly convex functional to the objective function. Structured sparsity algorithms are usually based on minimizing a convex (not strictly convex) objective function and this might lead to undesired unstable behavior. We show that by perturbing the objective function by a small strictly convex term we often reduce substantially the number of required computations without affecting the prediction performance of the obtained solution.

1 Introduction

In this paper we show how proximal methods can be profitably used to study a variety of machine learning algorithms. Recently, methods such as the lasso [22] – based on ℓ_1 regularization – received considerable attention for their property of providing sparse solutions. Sparsity has become a popular way to deal with small samples of high dimensional data and, in a broad sense, refers to the possibility of writing the solution in terms of a few building blocks. The success of ℓ_1 regularization motivated exploring different kinds of sparsity enforcing penalties for linear models as well as kernel methods [13, 14, 18, 25–27]. A common feature of this class of penalties is that they can be often written as suitable sums of euclidean (or Hilbertian) norms.

J.L. Balcázar et al. (Eds.): ECML PKDD 2010, Part II, LNAI 6322, pp. 418–433, 2010.

On the other hand, proximal methods have recently been shown to provide effective optimization procedures to solve the variational problems defining the ℓ_1 regularization algorithms, see [3, 4, 6, 7, 19] and [11] in the specific context of machine learning. In the following we discuss how proximal methods can be applied to solve the class of machine learning algorithms which can be seen as extensions of ℓ_1 regularization, namely structured sparsity regularization. For all these algorithms, it is possible to derive an optimization procedure that corresponds to an efficient iterative projection algorithm which can be easily implemented. Depending on the considered learning algorithm, the projection can be either computed in closed form or approximated by another proximal algorithm. A second contribution of our work is to study the effect of a preconditioning of the optimization procedure achieved by adding a strictly convex functional to the objective function. Indeed, structured sparsity algorithms are usually based on minimizing a convex (not strictly convex) objective function and this might lead to undesired unstable behavior. We show that by perturbing the objective function with a small strictly convex term it is possible to reduce substantially the number of required computations without affecting the prediction property of the obtained solution.

The paper is organized as follows. In Section 2, we begin by setting the notation, necessary to state all the mathematical and algorithmic results presented in Section 3. In Section 4, in order to show the wide applicability of our work, we apply the results to several learning schemes, and in Section 5 we describe the experimental results. An extended version of this work can be found in [21], where the interested reader can find all the proofs and some more detailed discussions.

2 Setting and Assumptions

In this section we describe the setting of structured sparsity regularization, in which a central role is played by the following variational problem. Given a Reproducing Kernel Hilbert Space (RKHS) $\mathcal{H}$, and two fixed positive numbers τ and μ, we consider the problem of computing:

$$f^* = \operatorname*{argmin}_{f\in\mathcal{H}} \mathcal{E}_{\tau,\mu}(f) = \operatorname*{argmin}_{f\in\mathcal{H}}\{F(f) + 2\tau J(f) + \mu \|f\|^2_{\mathcal{H}}\}, \tag{1}$$

where $F : \mathcal{H} \to \mathbb{R}$, $J :\to \mathbb{R}\cup\{+\infty\}$ represent the data and penalty terms, respectively, and $\mu\|f\|^2_{\mathcal{H}}$ is a perturbation discussed below. Note that the choice of RKHS recovers also the case of generalized linear models where f can be written as $f(x) = \sum_{j=1}^d \beta_j\psi_j(x)$ for a given dictionary $(\psi_j)_{j=1}^d$, as well as more general models (see Section 4). In the following, F is assumed to be differentiable and convex. In particular we are interested in the case where the first term is the empirical risk associated to a training set $\{(x_i, y_i)_{i=1}^n\} \subseteq (X\times[-C,C])^n$, and a cost function $\ell : \mathbb{R}\times[-C,C]\to\mathbb{R}^+$,

$$F(f) = \frac{1}{n}\sum_{i=1}^n \ell(f(x_i), y_i). \tag{2}$$

and specifically in the case where $\ell(y, f(x))$ is the square loss $(y - f(x))^2$. (other losses – e.g. the logistic loss – would also fit our framework).

In the following we require J to be lower semicontinuous, convex, and one-homogeneous, $J(\lambda f) = \lambda J(f)$, for all $f \in \mathcal{H}$ and $\lambda \in \mathbb{R}^+$. Indeed these technical assumptions are all satisfied by the vast majority of penalties commonly used in the recent literature of sparse learning. The main examples for the functional J are penalties which are sum of norms in distinct Hilbert spaces $(\mathcal{G}_k, \|\cdot\|_k)$:

$$J(f) = \sum_{k=1}^{M} ||\mathcal{J}_k(f)||_k, \tag{3}$$

where, for all k, $\mathcal{J}_k : \mathcal{H} \to \mathcal{G}_k$ is a bounded linear operator bounded from below. This class of penalties have recently received attention since they allow one to enforce more complex sparsity patterns than the simple ℓ_1 regularization [13, 26]. The regularization methods induced by the above penalties are often referred to as structured sparsity regularization algorithms.

Before describing how proximal methods can be used to compute the regularized solution of structured sparsity methods we note that in general, if we choose F, J as in (2) and (3), when $\mu = 0$, the functional (1) will be convex but not strictly convex. Then the regularized solution is in general not unique. On the other hand by setting $\mu > 0$, strict convexity, hence uniqueness of the solution, is guaranteed. As we discuss in the following, this can be seen as a preconditioning of the problem, and, if μ is small enough, one can see empirically that the solution does not change.

3 General Iterative Algorithm

In this section we describe the general iterative procedure for computing the solution f^* of the convex minimization problem (1).

Let K denote the subdifferential, $\partial J(0)$, of J at the origin, which is a convex and closed subset of $\mathcal{H}$. For any $\lambda \in \mathbb{R}^+$, we call $\pi_{\lambda K} : \mathcal{H} \to \mathcal{H}$ the projection on $\lambda K \subset \mathcal{H}$. The optimization scheme we derive is summarized in Algorithm 1, the parameter σ can be seen as a step-size, which choice is crucial to ensure convergence and is discussed in the following subsection. In general, approaches based on proximal methods decouple the contributions of the two functionals J

Algorithm 1. General Algorithm

Require: $\bar{f} \in \mathcal{H}$, $\sigma, \tau, \mu > 0$
Initialize: $f^0 = \bar{f}$
while `convergence not reached` **do**
  $p := p + 1$

$$f^p = \left(I - \pi_{\frac{\tau}{\sigma} K}\right)\left(\left(1 - \frac{\mu}{\sigma}\right) f^{p-1} - \frac{1}{2\sigma}\nabla F(f^{p-1})\right) \tag{4}$$

end while
return f^p

---

and F, since, at each iteration, the projection $\pi_{\tau/\sigma K}$– which is entirely characterized by J – is applied to a term that depends only on F. The derivation of the above iterative procedure for a general non differentiable J relies on a well-known result in convex optimization (see [6] for details), that has been used in the context of supervised learning by [11]. We recall it for completeness and because it is illustrative for studying the effect of the perturbation term $\mu \|f\|_{\mathcal{H}}^2$ in the context of structured sparsity regularization.

Theorem 1. *Given $\tau, \mu > 0$, $F : \mathcal{H} \to \mathbb{R}$ convex and differentiable and $J : \mathcal{H} \to \mathbb{R} \cup \{+\infty\}$ lower semicontinuous and convex, for all $\sigma > 0$ the minimizer f^* of $\mathcal{E}_{\tau,\mu}$ is the unique fixed point of the map $\mathcal{T}_\sigma : \mathcal{H} \to \mathcal{H}$ defined by*

$$\mathcal{T}_\sigma(f) = \mathrm{prox}_{\frac{\tau}{\sigma} J}\left(\left(1 - \frac{\mu}{\sigma}\right) f - \frac{1}{2\sigma}\nabla F(f)\right),$$

where $\mathrm{prox}_{\frac{\tau}{\sigma} J}(f) = \mathrm{argmin}\left\{\frac{\tau}{\sigma} J(g) + \frac{1}{2}\|f - g\|^2\right\}$.

For suitable choices of σ the map $\mathcal{T}_\sigma$ is a contraction, thus convergence of the iteration is ensured by Banach fixed point theorem and convergence rates can be easily obtained- see next section. The case $\mu = 0$ already received a lot of attention, see for example [3, 4, 6, 7, 19] and references therein.

Here we are interested in the setting of supervised learning when the penalty term is one-homogeneous and, as said before, enforces some structured sparsity property. In [21] we show that such assumption guarantees that

$$\mathrm{prox}_{\frac{\tau}{\sigma} J} = \left(I - \pi_{\frac{\tau}{\sigma} K}\right).$$

In the following subsection we discuss the role of the perturbation term $\mu \|f\|_{\mathcal{H}}^2$.

3.1 Convergence and the Role of the Strictly Convex Perturbation

The effect of $\mu > 0$ is clear if we look at convergence rates for the map $\mathcal{T}_\sigma$. In fact it can be shown ([21]) that a suitable a priori choice of σ is given by

$$\sigma = \frac{1}{4}(aL_{max} + bL_{min}) + \mu,$$

where a and b denote the largest and smallest eigenvalues of the kernel matrix, $[\mathrm{K}]_{i,j} = k(x_i, x_j)$, $i, j = 1, \ldots, n$, with k the kernel function of the RKHS $\mathcal{H}$, and $0 \le L_{min} \le \ell''(w, y) \le L_{max}$, $\forall w \in \mathbb{R}, y \in Y$, where ℓ'' denotes the second derivative of ℓ with respect to w. With such a choice the convergence rate is linear, i.e.

$$\|f^* - f^p\| \le \frac{L_\sigma^p}{1 - L_\sigma}\|f^1 - f^0\|, \qquad \text{with} \quad L_\sigma = \frac{aL_{max} - bL_{min}}{aL_{max} + bL_{min} + 4\mu}. \tag{5}$$

Typical examples of loss functions are the square loss and the exponential loss. In these cases suitable step sizes are $\sigma = \frac{1}{2}(a + b + 2\mu)$, and $\sigma = \frac{1}{4}(aC^2 e^{C^2}) + \mu$, respectively.

The above result highlights the role of the μ-term, $\mu \|\cdot\|_{\mathcal{H}}^2$, as a natural preconditioning of the algorithm. In fact, in general, for a strictly convex F, if the smallest eigenvalue of the second derivative is not uniformly bounded from below by a strictly positive constant, when $\mu = 0$, it might not be possible to choose σ so that $L_\sigma < 1$. One can also argue that, if μ is chosen small enough, the solution is expected not to change and in fact converges to a precise minimizer of $F + 2\tau J$. Infact, the quadratic term performs a further regularization that allows to select, as μ approaches 0, the minimizer of $F + 2\tau J$ having minimal norm (see for instance [10]).

3.2 Computing the Projection

In order to compute the proximity operator associated to a functional J as in (3), it is useful to define the operator $\mathcal{J} : \mathcal{H} \to \prod \mathcal{G}_k$ as $\mathcal{J}(f) = (\mathcal{J}_1(f), \ldots, \mathcal{J}_M(f))$. With this definition the projection of an element $f \in \mathcal{H}$ on the set $\lambda K := \lambda \partial J(0)$ is given by $\mathcal{J}^T \bar{v}$, where

$$\bar{v} \in \operatorname*{argmin}_{v \in \mathcal{G}} \|\mathcal{J}^* v - f\|_{\mathcal{H}}^2 + I_{\lambda B}(v), \quad \text{with } I_{\lambda B}(v) = \begin{cases} 0 & \text{if } \|v_k\|_k \leq \lambda \forall k \\ +\infty & \text{otherwise.} \end{cases} \tag{6}$$

The computation of the solution of the above equation is different in two distinct cases. In the first case $\mathcal{H}_k = \mathcal{G}_k, \mathcal{H} = \prod \mathcal{H}_k$, and $\mathcal{J}_k$ is the weighted projection operator on the k-th component, i.e. $\mathcal{J}_k(v) = v_k \in \mathcal{H}_k$ with $w_k > 0, \forall k$, and $\bar{v}$ can be computed exactly as $\bar{v} = \pi_{\lambda B}(f)$, where $\bar{v} = \pi_{\lambda B}(f)$ is simply the projection on the cartesian product of k balls of radius λw_k

$$(\pi_{\lambda B})_k(f_k) = \min\left\{1, \frac{\lambda w_k}{\|f_k\|_k}\right\} f_k.$$

In this case $(I - \pi_{\lambda K})$ coincides with the block-wise soft-thresholding operator:

$$(I - \pi_{\lambda K})_k(f_k) = (\|f_k\|_k - \lambda w_k)_+ f_k := S_\lambda(f_k) \tag{7}$$

which reduces to the well-known component-wise soft-thresholding operator when $\mathcal{H}_k = \mathbb{R}$ for all k.

On the other hand, when $\mathcal{J}$ is not a blockwise weighted projection operator, $\pi_{\lambda K}(f)$ cannot be computed in closed form. In this case we can again resort to proximal methods since equation (6) amounts to the minimization of a functional which is sum of a differential term $\|\mathcal{J}^* v - f\|_{\mathcal{H}}^2$ and a non-differential one $I_{\lambda B}(v)$. We can therefore apply Theorem 1 in order to compute $\bar{v}$ which is the fixed point of the map $\mathcal{T}_\eta$ defined as

$$\mathcal{T}_\eta(v) = \pi_{\lambda B}\left(v - (\eta)^{-1}\mathcal{J}(\mathcal{J}^* v - f)\right) \quad \text{for all } \eta > 0 \tag{8}$$

where $\mathrm{prox}_{I_{\lambda B}} = \pi_{\lambda B}$. We can therefore evaluate it iteratively as $v^q = \mathcal{T}_\eta(v^{q-1})$.

3.3 Some Relevant Algorithmic Issues

Adaptive Step-Size Choice. In the previous sections we proposed a general scheme as well as a parameter set-up ensuring convergence of the proposed procedure. Here, we discuss some heuristics that were observed to consistently speed up the convergence of the iterative procedure. In particular, we mention the Barzilai-Borwein methods – see for example [15, 16, 23] for references. More precisely in the following we will consider

$$\sigma_p = \langle s^p, r^p \rangle / \|s^p\|^2, \qquad \text{or} \quad \sigma_t = \|r^p\|^2 / \langle s^p, r^p \rangle.$$

where $s^p = f^p - f^{p-1}$ and $r^p = \nabla F(f^p) - \nabla F(f^{p-1})$.

Continuation Strategies and Regularization Path. Finally, we recall the *continuation* strategy proposed in [12] to efficiently compute the solutions corresponding to different values of the regularization parameter $\tau_1 > \tau_2 > \cdots > \tau_T$, often called *regularization path.* The idea is that the solution corresponding to the larger value τ_1 can be usually computed in a fast way since it is very sparse. Then, for τ_k one proceeds using the previously computed solution as the starting point of the corresponding procedure. It can be observed that with this *warm starting* much fewer iterations are typically required to achieve convergence.

4 Examples

In this section we illustrate the specialization of the framework described in the previous sections to a number structured sparsity regularization schemes.

4.1 Lasso and Elastic Net Regularization

We start considering the following functional

$$\mathcal{E}_{\tau,\mu}^{(\ell_1 \ell_2)}(\beta) = \|\Psi\beta - y\|^2 + \mu \sum_{j=1}^{d} \beta_j^2 + 2\tau \sum_{j=1}^{d} w_j |\beta_j|, \tag{9}$$

where Ψ is a $n \times d$ matrix, β, y are the vectors of coefficients and measurements respectively, and $(w_j)_{j=1}^d$ are positive weights. The matrix Ψ is given by the features ψ_j in the dictionary evaluated at some points $x_1, \ldots, x_n$.

Minimization of the above functional corresponds to the so called elastic net regularization, or ℓ_1-ℓ_2 regularization, proposed in [27], and reduces to the lasso algorithm [22] if we set $\mu = 0$. Using the notation introduced in the previous sections, we set $F(\beta) = \|\Psi\beta - y\|^2$ and $J(\beta) = \sum_{j=1}^d w_j |\beta_j|$. Moreover we denote by $\mathbf{S}_{\tau/\sigma}$ the soft-thresholding operator defined component-wise as in (7). The minimizer of (9) can be computed via the iterative update in Algorithm 2.

Note that the iteration in Algorithm 2 with $\mu = 0$ leads to the iterated soft-thresholding studied in [7] (see also [24] and references therein). When $\mu > 0$, the same iteration becomes the damped iterated soft-thresholding proposed in [8]. In the former case, the operator $\mathcal{T}_\sigma$ introduced in Theorem 1 is not contractive but only non-expansive, nonetheless convergence is still ensured [7].

Algorithm 2. Iterative Soft thresholding

Require: $\tau, \sigma > 0$
Initialize: $\beta^0 = 0$
while convergence not reached **do**
  $p := p + 1$

$$\beta^p = \mathbf{S}_{\frac{\tau}{\sigma}}\left((1 - \frac{\mu}{\sigma})\beta^{p-1} + \frac{1}{\sigma}\Psi^T(y - \Psi\beta^{p-1})\right)$$

end while
return β^p

Algorithm 3. Group lasso Algorithm

Require: $\tau, \sigma > 0$
Initialize: $\beta^0 = 0$
while convergence not reached **do**
  $p := p + 1$

$$\beta^p = \tilde{\mathbf{S}}_{\frac{\tau}{\sigma}}\left((1 - \frac{\mu}{\sigma})\beta^{p-1} + \frac{1}{\sigma}\Psi^T(y - \Psi\beta^{p-1})\right)$$

end while
return β^p

4.2 Group Lasso

We consider a variation of the above algorithms where the features are assumed to be composed in *blocks*. This latter assumption is used in [25] to define the so called group lasso, which amounts to minimizing

$$\mathcal{E}^{(grLasso)}_{\tau,\mu}(\beta) = \|\Psi\beta - y\|^2 + \mu\|\beta\|^2 + 2\tau\sum_{k=1}^{M} w_k \sqrt{\sum_{j\in\mathcal{I}_k}\beta_j^2} \qquad (10)$$

for $\mu = 0$, where $(\psi_j)_{j\in\mathcal{I}_k}$ for $k = 1, \ldots, M$ is a block partition of the feature set $(\psi_j)_{j\in\mathcal{I}}$. If we define $\beta^{(k)} \in \mathbb{R}^{|\mathcal{I}_k|}$ the vector built with the components of $\beta \in \mathbb{R}^{|\mathcal{I}|}$ corresponding to the elements $(\psi_j)_{j\in\mathcal{I}_k}$, then the nonlinear operation $(I - \pi_{\lambda K})$ – denoted by $\tilde{\mathbf{S}}_{\tau/\sigma}$ – acts on each block as (7), and the minimizer of (10) can hence be computed through Algorithm 3.

4.3 Composite Absolute Penalties

In [26], the authors propose a novel penalty, named Composite Absolute Penalty (CAP), based on assuming possibly overlapping groups of features. Given $\gamma_k \in \mathbb{R}^+$, for $k = 0, 1, \ldots, M$, the penalty is defined as:

$$J(\beta) = \sum_{k=1}^{M}(\sum_{j\in\mathcal{I}_k}\beta_j^{\gamma_k})^{\frac{\gamma_0}{\gamma_k}},$$

where $(\psi_j)_{j\in\mathcal{I}_k}$ for $k = 1, \ldots, M$ is not necessarily a block partition of the feature set $(\psi_j)_{j\in\mathcal{I}}$. This formulation allows to incorporate in the model not only groupings, but also hierarchical structures present within the features, for

Algorithm 4. CAP Algorithm

Require: $\tau, \sigma > 0$
Initialize: $\beta^0 = 0, p = 0, \bar{v}^0 = 0$
while convergence not reached **do**
  $p := p + 1$

$$\tilde{\beta} = (1 - \frac{\mu}{\sigma})\beta^{p-1} + \frac{1}{\sigma}\Psi^T(y - \Psi\beta^{p-1})$$

  set $v^0 = \bar{v}_{p-1}$
  while convergence not reached **do**
    $q := q + 1$
    for k=1,...,M **do**

$$v_k^q = (\pi_{\frac{\tau}{\eta}B})_k \left(v_k^{q-1} - \frac{1}{\eta}\mathcal{J}_k(\mathcal{J}^T v^{q-1} - \tilde{\beta}) \right)$$

    end for
    $\bar{v} = v^q$
  end while

$$\beta^p = \tilde{\beta} - \frac{\tau}{\sigma}\mathcal{J}^T \bar{v}_p$$

end while
return β^p

instance by setting $\mathcal{I}_k \subset \mathcal{I}_{k-1}$. For $\gamma_0 = 1$, the CAP penalty is one-homogeneous and the solution can be computed through Algorithm 1. Furthermore, when $\gamma_k = 2$ for all $k = 1, \ldots, M$, it can be regarded as a particular case of (3), with $\|\mathcal{J}_k(\beta)\|^2 = \sum_{j=1}^d \beta_j^2 \mathbf{1}_{\mathcal{I}_k}(j)$, with $\mathcal{J}_k : \mathbb{R}^{|\mathcal{I}|} \to \mathbb{R}^{|\mathcal{I}_k|}$. Considering the least square error, we study the minimization of the functional

$$\mathcal{E}_{\tau,\mu}^{(CAP)}(\beta) = \|\Psi\beta - y\|^2 + \mu\|\beta\|^2 + 2\tau\sum_{k=1}^M w_k \sqrt{\sum_{j\in\mathcal{I}_k} \beta_j^2}, \tag{11}$$

which is a CAP functional when $\mu = 0$. Note that, due to the overlapping structure of the features groups, the minimizer of (11) cannot be computed blockwise as in Algorithm 3, and therefore need to combine it with the iterative update (8) for the projection, thus obtaining Algorithm 4.

4.4 Multiple Kernel Learning

Multiple kernel learning (MKL) [2, 18] is the process of finding an optimal kernel from a prescribed (convex) set $\mathcal{K}$ of basis kernels, for learning a real-valued function by regularization. In the following we consider the case where the set $\mathcal{K}$ is the convex hull of a finite number of kernels $k_1, \ldots, k_M$, and the loss function is the square loss. It is possible to show [17] that the problem of multiple kernel learning corresponds to find f^* solving

$$\operatorname*{argmin}_{f\in\mathcal{H}} \left\{ \frac{1}{n}\sum_{i=1}^n (\sum_{j=1}^M f_j(x_i) - y_i)^2 + \mu\sum_{j=1}^M \|f_j\|_{\mathcal{H}_j}^2 + 2\tau\sum_{j=1}^M \|f_j\|_{\mathcal{H}_j} \right\}, \tag{12}$$

for $\mu = 0$, with $\mathcal{H} = \mathcal{H}_{k_1} \oplus \cdots \oplus \mathcal{H}_{k_M}$.

Algorithm 5. MKL Algorithm

set $\alpha^0 = 0$
while convergence not reached **do**
$p := p + 1$

$$\alpha^p = \hat{\mathbf{S}}_{\tau/\sigma}\left(K, \left((1-\frac{\mu}{\sigma})\alpha^{p-1} - \frac{1}{\sigma n}(\mathrm{K}\alpha^{p-1} - \mathbf{y})\right)\right)$$

end while
return $(\alpha^p)^T \mathbf{k}$.

Note that our general hypotheses on the penalty term J are clearly satisfied. Though the space of functions is infinite dimensional, thanks to a generalization of the representer theorem, the minimizer of the above functional (12) can be shown to have the finite representation $f_j^*(\cdot) = \sum_{i=1}^n \alpha_{j,i} k_j(x_i, \cdot)$ for all $j = 1, \ldots, M$. Furthermore, introducing the following notation:

$$\begin{aligned}
&\alpha = (\alpha_1, \ldots, \alpha_M)^T \text{ with } \alpha_j = (\alpha_{j,1}, \ldots, \alpha_{j,n})^T, \\
&\mathbf{k}(x) = (\mathbf{k}_1(x), \ldots, \mathbf{k}_M(x))^T \qquad \text{with } \mathbf{k}_j(x) = (k_j(x_1, x), \ldots, k_j(x_n, x)), \\
&\mathrm{K} = \left.\begin{pmatrix} \mathrm{K}_1 & \ldots & \mathrm{K}_M \\ \vdots & \ddots & \vdots \\ \mathrm{K}_1 & \ldots & \mathrm{K}_M \end{pmatrix}\right\} M \text{ times} \qquad \text{with } [\mathrm{K}_j]_{ii'} = k_j(x_i, x_i'), \\
&\mathbf{y} = (\underbrace{y^T, \ldots, y^T}_{M \text{ times}})^T
\end{aligned}$$

we can write the solution of (12) as $f^*(x) = \alpha_1^T \mathbf{k}_1(x) + \cdots + \alpha_M^T \mathbf{k}_M(x)$, which coefficients can be computed using Algorithm 5, where the soft-thresholding operator $\hat{\mathbf{S}}_\lambda(\mathrm{K}, \alpha)$ acts on the components α_j as

$$\hat{\mathbf{S}}_\lambda(\mathrm{K}, \alpha)_j = \frac{\alpha_j^T}{\sqrt{\alpha_j^T \mathrm{K}_j \alpha_j}} (\sqrt{\alpha_j^T \mathrm{K}_j \alpha_j} - \lambda)_+.$$

4.5 Multitask Learning

Learning multiple tasks simultaneously has been shown to improve performance relative to learning each task independently, when the tasks are related in the sense that they all share a small set of features (see for example [1, 14, 20] and references therein). In particular, given T tasks modeled as $f_t(x) = \sum_{j=1}^d \beta_{j,t} \psi_j(x)$ for $t = 1, \ldots, T$, according to [20], regularized multi-task learning amounts to the minimization of the functional

$$\mathcal{E}_{\tau,\mu}^{(MT)}(\beta) = \sum_{t=1}^T \frac{1}{n_t} \sum_{i=1}^{n_t} (\psi(x_{t,i})\beta_t - y_{t,i})^2 + \mu \sum_{t=1}^T \sum_{j=1}^d \beta_{t,j}^2 + 2\tau \sum_{j=1}^d \sqrt{\sum_{t=1}^T \beta_{t,j}^2}. \tag{13}$$

Algorithm 6. Multi-Task Learning Algorithm

set $\beta^0 = 0$
while convergence not reached **do**
  $p := p + 1$

$$\beta^p = \tilde{\mathbf{S}}_{\frac{\tau}{\sigma}} \left((1 - \frac{\mu}{\sigma})\beta^{p-1} + \frac{1}{\sigma}\Psi^T \mathbf{N}(y - \Psi\beta^{p-1}) \right)$$

end while
return β^p

The last term combines the tasks and ensures that common features will be selected across them. Functional (13) is a particular case of (1), and, defining

$$\begin{aligned}
&\beta = (\beta_1^T, \ldots, \beta_T^T)^T, \\
&\Psi = \mathrm{diag}(\Psi_1, \ldots, \Psi_T), \qquad [\Psi_t]_{ij} = \psi_j(x_{t,i}), \\
&y = (y_1^T, \ldots, y_T^T)^T, \\
&\mathbf{N} = \mathrm{diag}(\underbrace{1/n_1, \ldots, 1/n_1}_{n_1 \text{times}}, \underbrace{1/n_2, \ldots, 1/n_2}_{n_2 \text{times}}, \ldots, \underbrace{1/n_T, \ldots, 1/n_T}_{n_T \text{times}}).
\end{aligned}$$

its minimizer can be computed through Algorithm 6. The Soft-thresholding operator $\tilde{\mathbf{S}}_\lambda$ is applied task-wise, that is it acts simultaneously on the regression coefficients relative to the same variable in all the tasks.

5 Experiments and Discussions

In this section we describe several experiments aimed at testing some features of the proposed method. In particular, we investigate the effect of adding the term $\mu \|f\|_{\mathcal{H}}^2$ to the original functional in terms of
-*prediction*: do different values of μ modify the prediction error of the estimator?
-*selection*: does μ increase/decrease the sparsity level of the estimator?
-*running time*: is there a computational improvement due to the use of $\mu > 0$?

We discuss the above questions for the multi-task scheme proposed in [20]. and show results which are consistent with those reported in [9] for the elastic-net estimator. These two methods are only two special cases of our framework, but indeed we expect that, due to the common structure of the penalty terms, all the other learning algorithms considered in this paper share the same properties.

We note that a computational comparison of different optimization approaches is cumbersome since we consider many different learning schemes and is beyond the scope of this paper. Extensive analysis of different approaches to solve ℓ_1 regularization can be found in [12] and [15], where the authors show that projected gradient methods compare favorably to state of the art methods. We expect that similar results will hold for learning schemes other than ℓ_1 regularization.

5.1 Validation Protocol and Simulated Data

In this section, we briefly present the set-up used in the experiments. We considered simulated data to test the properties of the proposed method in a controlled scenario. More precisely, we considered T regression tasks

$$y = x \cdot \beta_t + \varepsilon \quad t = 1, \ldots, T$$

where x is uniformly drawn from $[0,1]^d$, ε is drawn from the zero-mean Gaussian distribution with $\sigma = 0.1$ and the regression vectors are

$$\beta_t^\dagger = (\beta_{t,1}^\dagger, ..., \beta_{t,r}^\dagger, 0, 0, ..., 0).$$

with $\beta_{t,j}^\dagger$ uniformly drawn from $[-1, 1]$ for $t \leq r$, so that the relevant variables are the first r.

Following [5, 23], we consider a debiasing step after running the sparsity based procedure. This last step is a post-processing and corresponds to training a regularized least square (RLS) estimator[1] with parameter λ on the selected components to avoid an undesired shrinkage of the corresponding coefficients.

In order to obtain a fully data driven procedure we use cross validation to choose the regularization parameters τ, λ. After re-training with the optimal regularization parameters, a test error is computed on an independent set of data. Each validation protocol is replicated 20 times by resampling both the input data and the regression coefficients, $\beta_t^\dagger$, in order to assess the stability of the results.

5.2 Role of the Strictly Convex Penalty

We investigate the impact of adding the perturbation $\mu > 0$. We consider $T = 2$, $r = 3$, $d = 10, 100, 1000$, and $n = 8, 16, 32, 64, 128$. For each data set, that is for fixed d and n, we apply the validation protocol described in Subsection 5.1 for increasing values of μ. The number of samples in the validation and test sets is 1000. Error bars are omitted in order to increase the readability of the Figures.

We preliminary discuss an observation suggesting a useful way to vary μ. As a consequence of (5), when $\mu = 0$ and $b = 0$, the Lipschitz constant, L_σ, of the map T_σ in Theorem 1 is 1 so that T_σ is not a contraction. By choosing $\mu = \frac{1}{4} \left\|\nabla^2 F\right\| \alpha$ with $\alpha > 0$, the Lipschitz constant becomes $L_\sigma = (1+\alpha)^{-1} < 1$, and the map T_σ induced by F_μ is a contraction. In particular in multiple task learning with linear features (see Section 4.5) $X = \Psi$, so that $\nabla^2 F = 2X^T X/n$ and $\left\|\nabla^2 F\right\| = 2a/n$, where a is the largest eigenvalue of the symmetric matrix $X^T X$. We therefore let $\mu = \frac{a}{2n}\alpha$ and vary the absolute parameter α as $\alpha = 0, 0.001, 0.01, 0.1$. We then compare the results obtained for different values of α, and analyze in the details the outcome of our results in terms of the three aspects raised at the beginning of this section.

-*prediction* The test errors associated to different values of μ are essentially overlapping, meaning that the perturbation term does not impact the prediction performance of the algorithm when the τ parameter is accurately tuned. This result is consistent with the theoretical results for elastic net – see [8].

-*selection* In principle the presence of the perturbation term tends to reduce the sparsity of the solution in the presence of very small samples. In practice

[1] A simple ordinary least square is often sufficient and here a little regularization is used to avoid possible unstable behaviors especially in the presence of small samples.

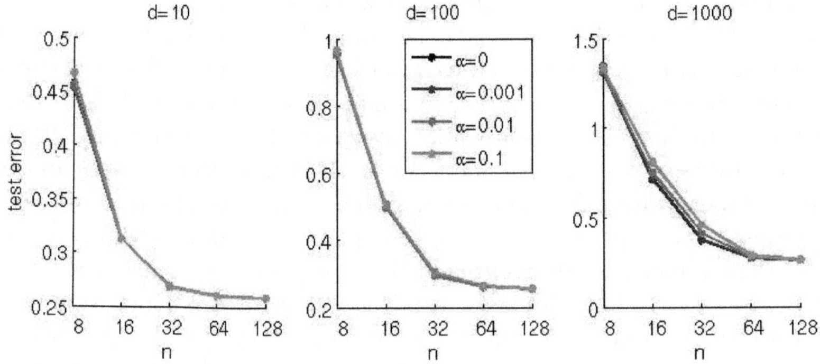

(a) Prediction

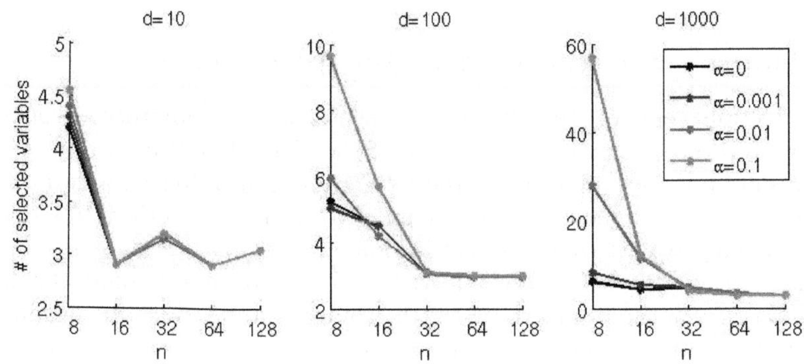

(b) Selection

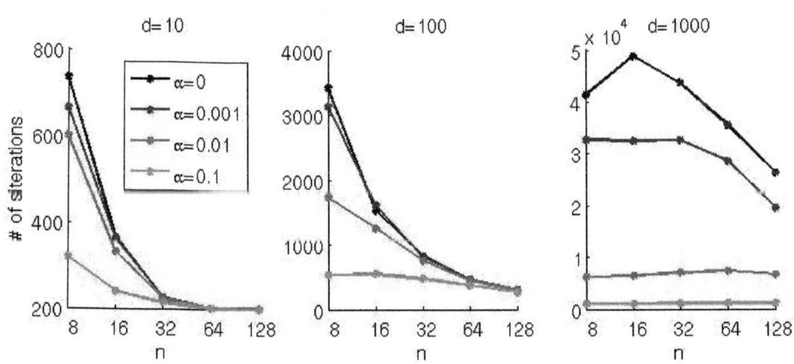

(c) Running time

Fig. 1. Results obtained in the experiments varying the size of the training set and the number of input variables. The properties of the algorithms are evaluated in terms of the prediction error, the ability of selecting the *true* relevant variables, and finally the number of iteration required for the convergence of the algorithm.

one can see that such an effect decreases when the number of input points n increases and is essentially negligible even when $n << d$.

- *running time* From the computational point of view we expect larger values of μ (or equivalently α) to correspond to fewer iterations. This effect is clear in our experiments. Interestingly when $n << d$ small values of μ allow to substantially reduce the computational burden while preserving the prediction property of the algorithm (compare α=0 and α=0.001 for d=1000). Moreover, one can observe that the number of iterations decreases as the number of points increases. This result might seem surprising, but can be explained recalling that the condition number of the underlying problem is likely to improve as n increases.

Finally, we can see that adding the small strictly convex perturbation with $\mu > 0$, has a preconditioning effect on the iterative procedure and can substantially reduce the number of required computations without affecting the prediction property of the obtained solution.

5.3 Impact of Choosing the Step-Size Adaptively

In this section we assess the effectiveness of the adaptive approach proposed in section 3.3 to speed up the convergence of the algorithm. Specifically, we show some results obtained by running the iterative optimization with two different choices of the step-size, namely the one fixed a-priori – as described in section 3.1 – and the adaptive alternative of subsection 3.3.

The experiments have been conducted by first drawing randomly the dataset and finding the optimal solution using the complete validation scheme, and then running two further experiments using, in both cases, the optimal regularization parameters but the two different strategies for the step-size.

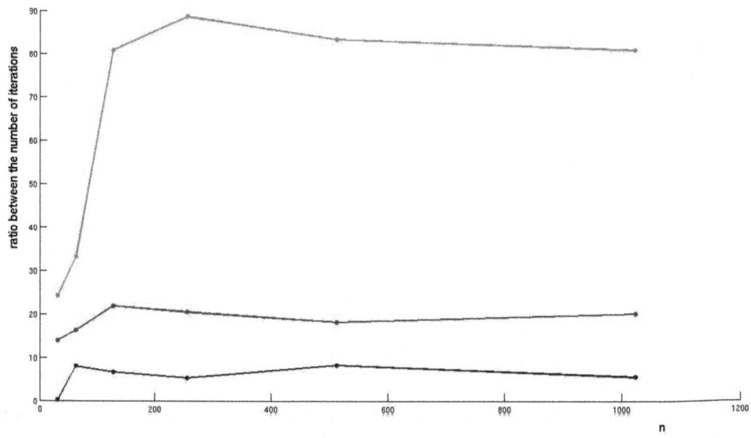

Fig. 2. Comparison of the number of iterations required to compute the regression function using the fixed and the adaptive step-size. The blue plot refers to the experiments using $d = 10$, the red plot to $d = 100$, while the green plot to $d = 500$.

We compared the number of iterations necessary to compute the solution and looked at the ratio between those required by the fixed and the adaptive strategies respectively. In Figure 2, it is easy to note that such ratio is always greater than one, and actually it ranges from the order of tens to the order of hundreds. Moreover, the effectiveness of using an adaptive strategy becomes more and more evident as the number of input variables increases. Finally, for a fixed input dimension, the number of iterations required for both choices of the step-size decreases when the number of training samples increases, in a way that the ratio tends to either remain approximately constant or decrease slightly.

6 Conclusions

This paper shows that many algorithms based on regularization with convex non differentiable penalties can be described within a common framework. This allows to derive a general optimization procedure based on proximal methods whose convergence is guaranteed. The proposed procedure highlights and separates the roles played by the loss terms and the penalty terms, in fact, it corresponds to the iterative projection of the gradient of the loss on a set defined by the penalty. The projection has a simple characterization in the setting we consider: in many cases it can be written in closed form and corresponds to a soft-thresholding operator, in all the other cases it can be iteratively calculated by resorting again to proximal methods. The obtained procedure is simple and its convergence proof is straightforward in the strictly convex case. One can always force such a condition considering a suitable perturbation of the original functional. Interestingly if such a perturbation is small it will act as a preconditioning of the problem and lead to the better computational performances without changing the properties of the solution.

Acknowledgments

This work has been partially supported by the FIRB project LEAP RBIN04PARL, the EU Integrated Project Health-e-Child IST-2004-027749. Matteo Santoro and Sofia Mosci are partially supported by Compagnia di San Paolo. This report describes research done at the Center for Biological & Computational Learning, which is in the McGovern Institute for Brain Research at MIT, as well as in the Dept. of Brain & Cognitive Sciences, and which is affiliated with the Computer Sciences & Artificial Intelligence Laboratory (CSAIL). This research was sponsored by grants from DARPA (IPTO and DSO), National Science Foundation (NSF-0640097, NSF-0827427.) Additional support was provided by: Adobe, Honda Research Institute USA, King Abdullah University Science and Technology grant to B. DeVore, NEC, Sony and especially by the Eugene McDermott Foundation.

References

[1] Argyriou, A., Hauser, R., Micchelli, C.A., Pontil, M.: A dc-programming algorithm for kernel selection. In: Proceedings of the Twenty-Third International Conference on Machine Learning (2006)

[2] Bach, F.R., Lanckriet, G., Jordan, M.I.: Multiple kernel learning, conic duality, and the smo algorithm. In: ICML. ACM International Conference Proceeding Series, vol. 69 (2004)

[3] Beck, A., Teboulle, M.: A fast iterative shrinkage-thresholding algorithm for linear inverse problems. SIAM J. Imaging Sci. 2(1), 183–202 (2009)

[4] Becker, S., Bobin, J., Candes, E.: Nesta: A fast and accurate first-order method for sparse recovery (2009)

[5] Candès, E., Tao, T.: The Dantzig selector: statistical estimation when p is much larger than n. Ann. Statist. 35(6), 2313–2351 (2005)

[6] Combettes, P.L., Wajs, V.R.: Signal recovery by proximal forward-backward splitting. Multiscale Model. Simul. 4(4), 1168–1200 (2005)

[7] Daubechies, I., Defrise, M., De Mol, C.: An iterative thresholding algorithm for linear inverse problems with a sparsity constraint. Communications on Pure and Applied Mathematics 57, 1413–1457 (2004)

[8] De Mol, C., De Vito, E., Rosasco, L.: Elastic-net regularization in learning theory (2009)

[9] De Mol, C., Mosci, S., Traskine, M., Verri, A.: A regularized method for selecting nested groups of relevant genes from microarray data. Journal of Computational Biology, 16 (2009)

[10] Dontchev, A.L., Zolezzi, T.: Well-posed optimization problems. Lecture Notes in Mathematics, vol. 1543. Springer, Heidelberg (1993)

[11] Duchi, J., Singer, Y.: Efficient online and batch learning using forward backward splitting. Journal of Machine Learning Research 10, 2899–2934 (2009)

[12] Hale, E.T., Yin, W., Zhang, Y.: Fixed-point continuation for l1-minimization: Methodology and convergence. SIOPT 19(3), 1107–1130 (2008)

[13] Jenatton, R., Audibert, J.-Y., Bach, F.: Structured variable selection with sparsity-inducing norms. Technical report, INRIA (2009)

[14] Kubota, R.A., Zhang, T.: A framework for learning predictive structures from multiple tasks and unlabeled data. J. Mach. Learn. Res. 6, 1817–1853 (2005)

[15] Loris, I.: On the performance of algorithms for the minimization of l_1-penalized functionals. Inverse Problems 25(3) 035008, 16 (2009)

[16] Loris, I., Bertero, M., De Mol, C., Zanella, R., Zanni, L.: Accelerating gradient projection methods for ℓ_1-constrained signal recovery by steplength selection rules (2009)

[17] Micchelli, C.A., Pontil, M.: Learning the kernel function via regularization. J. Mach. Learn. Res. 6, 1099–1125 (2005)

[18] Micchelli, C.A., Pontil, M.: Feature space perspectives for learning the kernel. Mach. Learn. 66(2-3), 297–319 (2007)

[19] Nesterov, Y.: Smooth minimization of non-smooth functions. Math. Program. 103(1), 127–152 (2005)

[20] Obozinski, G., Taskar, B., Jordan, M.I.: Multi-task feature selection. Technical report, Dept. of Statistics, UC Berkeley (June 2006)

[21] Rosasco, L., Mosci, S., Santoro, A., Verri, M., Villa, S.: Iterative projection methods for structured sparsity regularization. Technical Report MIT-CSAIL-TR-2009-050 CBCL-282 (October 2009)

[22] Tibshirani, R.: Regression shrinkage and selection via the lasso. Journal of the Royal Statistical Society, Series B 56, 267–288 (1996)
[23] Wright, S.J., Nowak, R.D., Figueiredo, M.A.T.: Sparse reconstruction by separable approximation. IEEE Trans. Image Process (2009)
[24] Yin, W., Osher, S., Goldfarb, D., Darbon, J.: Bregman iterative algorithms for ℓ^1-minimization with applications to compressed sensing. SIAM J. Imaging Sciences 1(1), 143–168 (2008)
[25] Yuan, M., Lin, Y.: Model selection and estimation in regression with grouped variables. Journal of the Royal Statistical Society, Series B 68(1), 49–67 (2006)
[26] Zhao, P., Rocha, G., Yu, B.: The composite absolute penalties family for grouped and hierarchical variable selection. Annals of Statistics 37(6A), 3468–3497 (2009)
[27] Zou, Z., Hastie, T.: Regularization and variable selection via the elastic net. Journal of the Royal Statistical Society, Series B 67, 301–320 (2005)

Exploiting Causal Independence in Markov Logic Networks: Combining Undirected and Directed Models

Sriraam Natarajan[1], Tushar Khot[1], Daniel Lowd[2], Prasad Tadepalli[3], Kristian Kersting[4], and Jude Shavlik[1]

[1] University of Wisconsin-Madison
[2] University of Oregon
[3] Oregon State University
[4] Fraunhofer IAIS

Abstract. A new method is proposed for compiling causal independencies into Markov logic networks (MLNs). An MLN can be viewed as compactly representing a factorization of a joint probability into the product of a set of factors guided by logical formulas. We present a notion of causal independence that enables one to further factorize the factors into a combination of even smaller factors and consequently obtain a finer-grain factorization of the joint probability. The causal independence lets us specify the factor in terms of weighted, directed clauses and operators, such as "or", "sum" or "max", on the contribution of the variables involved in the factors, hence combining both undirected and directed knowledge. Our experimental evaluations shows that making use of the finer-grain factorization provided by causal independence can improve quality of parameter learning in MLNs.

1 Introduction

Most traditional AI methods are based on one of the two approaches: *first-order logic*, which excels at capturing the rich relationships among many objects, or *statistical* representations, which handle uncertain environments and noisy observations. Statistical relational learning (SRL) [5], an area of growing interest, seeks to unify these approaches in order to handle problems that are both complex and uncertain.

The principal attraction of SRL models is that they are more succinct than their propositional counterparts, leading to easier specification of their structure by domain experts and faster learning of their parameters. However, different proposed models are good at expressing different kinds of knowledge, making it difficult to compare their empirical performance and to simultaneously exploit the strengths of each. The largest divide is between directed and undirected representations.

One of the primary advantages of the directed graphical models is the notion of "Independence of Causal Influence" (ICI) [6,14], a.k.a "causal independence," i.e., there may be multiple independent causes for a target variable. Directed models can learn conditional distributions due to each of the causes separately and combine them using a (possibly stochastic) function, thus making the process of learning easier. This notion of ICI has been extended to directed SRL models in two different ways: while PRMs [3] use aggregators such as `max`, `min`, and `average` to combine the influences due

J.L. Balcázar et al. (Eds.): ECML PKDD 2010, Part II, LNAI 6322, pp. 434–450, 2010.

to several parents, other formalisms such as BLPs [10] and RBNs [8] use combination functions such as `Noisy-OR`, `mean`, or `weighted mean` to combine distributions.

One weakness of the directed models is the need to keep the graph acyclic while preserving sparsity. This problem is avoided by undirected models such as Markov logic networks (MLNs) [1], which are based on Markov networks. Undirected models do not consider local models (i.e., do not treat each cause as independent from others) and hence do not model the notion of ICI explicitly. Bayesian Networks with tabular CPDs (one parameter for each configuration of the parent variables) can be directly translated to MLNs by introducing one formula for each BN parameter[1]. What is less clear is how combination functions from relational models can best be represented in an MLN. We consider a subset of combination functions called *decomposable combining rules* and derive a representation of MLNs that captures these rules. The important aspect of this representation is that we do not use the "ground" Bayesian network and instead use a "lifted" representation that avoids the grounding of the clauses, since grounding can produce an exponentially large number of (variable-free) clauses.

Representing combining rules using MLNs is a key step towards unifying directed and undirected SRL approaches. Such a unified view on SRL is not only of theoretical interest – it actually has many important practical implications such as more natural model specification and development of specialized, highly efficient inference and learning techniques that can be applied differently to different pieces of the model.

In this work, we make several major contributions: (1) A provable linear representation of decomposable combining functions within MLNs; (2) Explicit examples of average-based and noisy combination functions; (3) A formal description of the algorithm for converting from directed models with combining rules to MLNs; (4) A macro-definition that allows for succinct specification of the resulting MLNs; (5) Empirical proof that combining rules can improve the learning of MLNs when the domain knowledge available is minimal.

We proceed as follows. After introducing the necessary background, we derive the MLN clauses for representing decomposable combining rules and provide the clauses for two common cases of combining rules. Next, we derive a bound on the number of clauses and provide the pseudo-code for the compilation. Before the conclusion, we present empirical results in real-world tasks.

2 MLNs and Directed Models

A Bayesian Network (BN) compactly represents a joint probability distribution over a set of variables $X = \{X_1, \ldots, X_n\}$ as a directed, acyclic graph and a set of conditional probability distributions (CPDs). The graph contains one node for each variable, and encodes the assertion that each variable is independent of its non-descendants given its parents in the graph. These conditional independence assertions allow us to represent the joint probability distribution as the product of the conditional probability of each variable, X_i, given its parents, parents(X_i): $P(X) = \prod_i P(X_i|\text{parents}(X_i))$.

A Markov network (MN) (also called a Markov random field) specifies independencies using an undirected graph. The graph encodes the assertion that each variable

[1] `http://alchemy.cs.washington.edu/faq/index.html`

is independent of all others given its neighbors in the graph. This set of independencies guarantees that the probability distribution can be factored into a set of potentials functions defined over cliques in the graph. Unlike BNs, these factors are not constrained to be conditional probabilities. Instead, a potential function is allowed to take on any non-negative value. The joint probability distribution is therefore defined as follows:$P(X = x) = \frac{1}{Z} \prod_j \phi_j(\mathbf{D}_j)$, where ϕ_j is the jth potential function, $\mathbf{D}_j$ is the set of variables over which ϕ_j is defined, and Z is a normalization constant. MNs are often written as log-linear models, where the potential functions are replaced by a set of weighted features.

One of the most popular and general SRL representations is *Markov logic networks (MLNs)* [1]. An MLN consists of a set of formulas in first-order logic and their real-valued weights, $\{(w_i, f_i)\}$. Together with a set of constants, we can instantiate an MLN as a Markov network with a node for each ground predicate (atom) and a feature for each ground formula. All groundings of the same formula are assigned the same weight, leading to the following joint probability distribution over all atoms:$P(X = x) = \frac{1}{Z} \exp\left(\sum_i w_i n_i(x)\right)$, where $n_i(x)$ is the number of times the ith formula is satisfied by possible world x and Z is a normalization constant (as in Markov networks). Intuitively, a possible world where formula f_i is true one more time than different possible world is e^{w_i} times as probable, all other things being equal.

Directed Models with Combining Rules: Our work does not assume any representation for the directed models. We merely use an abstract syntax called as First-Order Conditional Influence (FOCI) statements [13] to present the semantics of the directed models. We had earlier used this syntax to derive learning algorithms [13] and showed how most directed models such as BLPs [10], RBNs[7], PRMs[3], probabilistic relational language [4] and logical Bayes nets [2] can be represented using this syntax. The goal of this work is not to convert from FOCI statements to MLNs but to show that the knowledge captured by directed models can be represented using MLNs. FOCI statements merely facilitate this conversion. We could replace the FOCI statements with any of the above directed models and still get the same result as all these models share the common syntax as shown in [13].

Each statement has the form: *If* $\langle condition \rangle$ *then* $\langle qualitative\ influence \rangle$, where *condition* is a set of literals, each literal being a predicate symbol applied to the appropriate number of variables. The set of literals is treated as a conjunction. A $\langle qualitative$ *influence*$\rangle$ is of the form $X_1, \ldots, X_k$ *Qinf* Y, where the X_i and Y are of the form $V.a$, and V is a variable that occurs in *condition* and a is an object attribute. Associated with each statement is a *conditional probability distribution* that specifies a probability distribution of the resultant conditioned on the influents, e.g. $P(Y|X_1, \ldots, X_k)$ for the above statement.

```
CR2{
  If {student(S), course(C), takes(T,S,C)} then
    T.grade Qinf (CR1) S.satisfaction.
  If {student(S),paper(P,S)} then
    P.quality Qinf (CR1) S.satisfaction.}
```

The first rule specifies that the grade that a student obtains in a course influences his/her satisfaction. The CPD $P(S.\text{satisfaction} \mid T.\text{grade})$ associated with the first statement (partially) captures the quantitative relationships between the attributes. The second states that if the student has authored a paper, then its quality influences the satisfaction of the student. The distributions due to multiple instantiations of the respective rules (the different course grades or the different paper qualities) are combined using the CR1 combining rule and the distributions due to different rules are combined using CR2 combining rule. We assume discrete values and hence the CPDs are represented using conditional probability tables (CPTs).

Note that there are two levels of combination functions - one for combining multiple instances of the same rule and the other for combining different rules. This idea of two-level combining rules is sufficient to capture the notion of ICI in SRL models and hence we address the 2-level combining rule in this work. The use of combining rules make learning in directed SRL models easier: multiple instances of the same rule share the same CPT and hence can be treated as individual examples while learning the CPTs. Similarly, the different CPTs can be learned *independently* of each other thus exploiting the notion of causal independence. Yet another advantage of the combining rules is that they allow for richer combination of probability distributions. MLNs in their default representation use an exponentiated weighted count as an (indirect) combination function of the different clauses. To express complex functions, a straightforward method would be to construct the grounded BN for each rule and then construct the equivalent Markov Network. Unfortunately, this leads to an exponential number of clauses in the MLN, making the twin problems of learning and inference computationally expensive. Instead we resort to a "lifted" method that avoids unrolling (grounding) all the clauses to create the MLN.

3 Combination Functions Using MLNs

In this section, we present the equivalent MLN representation for decomposable combining functions.

3.1 Decomposable Combining Functions

In this work, we extend the definition of decomposable causal independence due to [6] to the relational setting. To understand the notion of decomposable combining rules, consider Figure 1 where y is the target and x's are the influents. The t_i^j's are the temporary y-values due to each instantiation of x_i^j. Note that there are n rules to predict y. The y's are deterministic nodes obtained using functions f's for the first level (that combines instances of the same rule) and g's for the second level (that combines different rules). The observed nodes are shown using solid circles while the dotted circles correspond to the hidden nodes. These hidden nodes are created when the functions are applied successively. In the figure, we present two levels of combining rules (f_i and g). Let us consider the first level combining rule. For simplicity, consider the set of functions f_1^i. The result y_1 can be represented as:

$$y_1 = f_{1,\sigma}^m(t_{1,\sigma}^m, f_{1,\sigma}^{m-1}(t_{1,\sigma}^{m-1}, ...f_{1,\sigma}^1(t_{1,\sigma}^1, p))) \tag{1}$$

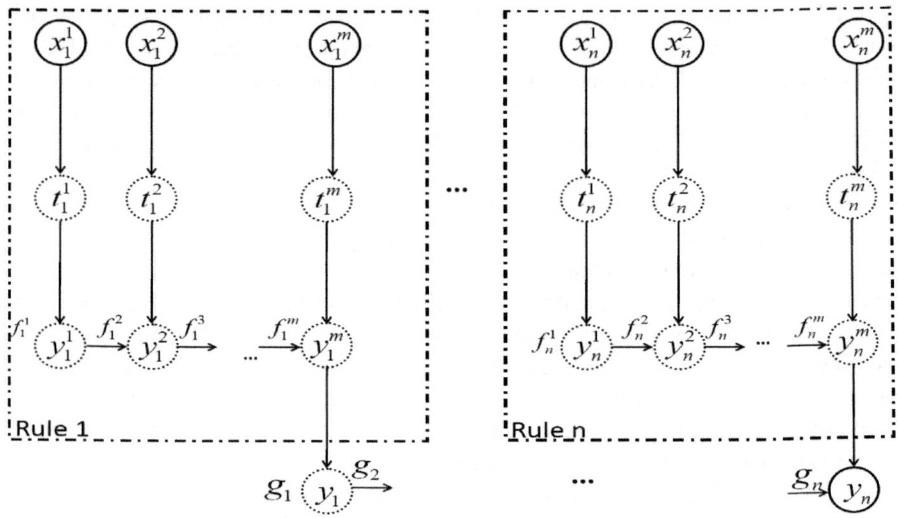

Fig. 1. Decomposable Combining Rules

for the given ordering σ of the different x_1^i. p is some prior on the value of y for rule j. Equation 1 can be written as

$$y_1 = f_{1,\sigma}(t_{1,\sigma}^m, t_{1,\sigma}^{m-1}, ...t_{1,\sigma}^1, p) \tag{2}$$

where f_1 is a combining rule operating over all t_1^i.

Definition: A combining rule is called as *decomposable* if it satisfies equation 2 for **all** orderings (i.e., $\forall\sigma$). This is to say that for every possible ordering of the inputs, the combining rule can be decomposed into a set of functions that yield the same distribution.

In our setting, we require that the combining rules at both the levels are *decomposable*. i.e.,

$$y = g_{n,\sigma}(y_{n,\sigma}, g_{n-1,\sigma}(y_{n-1,\sigma}, ...g_{1,\sigma}(y_{1,\sigma}, p)))\forall\sigma \tag{3}$$

The above condition specifies that the rules can themselves be ordered differently but the resulting distribution always remains the same. Note that most common combination functions used in the literature [12,13,8,6,14] such as: *Noisy-Or*, *Noisy-And*, *Noisy-Existential*, average-based combination functions such as *mean*, *weighted-mean* and *context specific independence*(CSI) can be represented using the above definition. In this work, we show how multi-level combining rules (rules that combine instances of the same clause and the ones that combine the distributions due to different clauses) can be represented and learned using MLNs.

3.2 Decomposable Combining Functions Using MLNs

The notion of decomposability is crucial to deriving the representation of combining rules using MLNs. This allows us to consider the combining rules as multiplexers (the f_i^j and g_i in the figure above) on Bayesian Networks. *The key idea in our work is to*

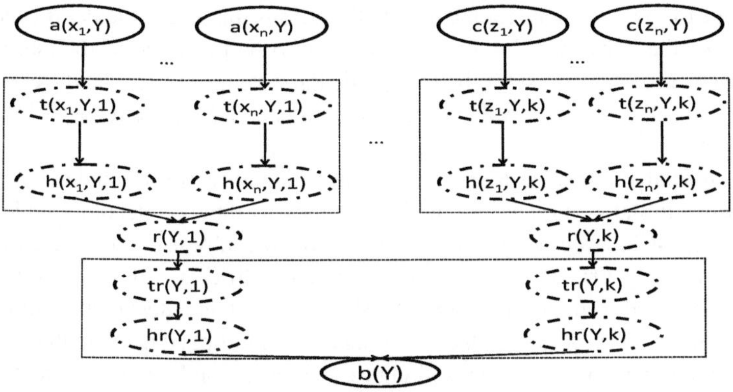

Fig. 2. Understanding combining rules using multiplexers. The dashed nodes are the hidden nodes and the multiplexer nodes, while the solid nodes are observed in the data.

view the combination function as choosing a value among several values proposed by the parents. For instance, taking the average of distributions corresponds to choosing the target value using an uniform distribution among the values proposed by the parents. Weighted mean can be understood as choosing a value based on the distribution given by the weights. The BN representation is used only for the sake of presentation and is not used in its full form during translation. The translation occurs at the logical level (i.e., on the variables rather than the groundings of these variables).

Consider the following two FOCI statements:

$$a(X,Y) \;\; Qinf \;\; b(Y)$$
$$c(Z,Y) \;\; Qinf \;\; b(Y)$$

where $\langle a, b, c\rangle$ are predicates and $\langle X, Y, Z\rangle$ are variables. Associated with each clause is a conditional probability distribution $P(b(y)|\text{parent(b)})$ where the parent for the first statement is $a(x, y)$ and the second is $c(z, y)$. Note that there could be several possible instantiations for X and Z in the above rules. For simplicity, let us assume that the distributions due to the different instances of the same rule are combined using CR_1 and the resulting distributions due to the different rules are combined using CR_2.

Consider the BN presented in Figure 2. For ease of explanation, assume that there are n instantiations of each rule and k such rules (we present only two of them for brevity). In addition to the a, b and c predicates, we introduce two more types of predicates indicated using dashed nodes: *hidden* (temporary) value predicates (t and tr) and *multiplexer* predicates (h and hr). Since there are two levels of combining functions, there are two different sets of multiplexers and hidden nodes represented by two different boxes in the figure. The first box corresponds to choosing a value from a single rule (given by $r(y, i)$, where i is the rule index) and in the next level the final value of the target is chosen from one among the different r-values. We now explain the multiplexers inside the same rule (the top box) and the same idea is extended for different rules (bottom box).

The hidden predicates t's can be understood as choosing a value of the target given the instantiation of the parent based on the CPD. The multiplexers (h-nodes) serve to choose one of the n t-values for the target. The idea is that if a particular h is activated, the value of the corresponding t node is chosen to be the value of the target for the current rule (i.e, $r(y,i)$ is set to be that particular t-value). Given the different values of $r(y,I)$ for all I, the final value of the target b is chosen using the next level of the multiplexer.

In our formalism, there is no restriction on the equality of CR_1 and CR_2, i.e., they need not be similar combination functions as long as they are decomposable. For instance, it is possible to use a mean combining rule to combine the instances of a single rule while a Noisy-Or could be used to combine the different rules themselves. It can be easily observed from our translation to MLNs (presented later) that the only change for the different cases would be the encoding of the multiplexers. Note that it is possible to imagine developing specialized clauses for each combination of the combining functions. In this work, we aim to derive a general representation that covers all decomposable combining functions. Our translation consists for four different kinds of clauses:

1. **CPT Clauses:** This follows the standard translation of Bayesian networks to MLNs. Each independent parameter in the CPT of the Bayes net becomes a clause in the MLN. An example of such a clause is

$$\begin{aligned} w_i^1 &: a(X,Y) \Rightarrow t(X,Y,i) \\ w_i^0 &: \neg a(X,Y) \Rightarrow t(X,Y,i) \end{aligned} \tag{4}$$

where $w_i^j = log\frac{p_i^j}{(1-p_i^j)}$, $p_i^j = P(b(Y) = 1|a(X,Y) = j)$.[2] Hence, for each independent parameter of the original CPT in the directed model, there is a clause in the MLN with the weight as a function of the parameter. In general, the set of arguments in the temporary predicate t is the union of all the arguments in the body of the clause and an argument for the rule index.

2. **Multiplexer Clauses:** These are the clauses that choose a particular value of the target given a set of parent values. For the first-level multiplexer (h in the figure), this set corresponds to the set of values due to different instantiations of the same rule. For the second-level multiplexer, this set consists of the values due to different rules. For the first level, the MLN clauses are of the form

$$\infty : h(X,Y,I) \Rightarrow (t(X,Y,I) \Leftrightarrow r(Y,I)) \tag{5}$$

The above clause is a hard clause (i.e., infinite weight) that specifies that for a particular value of X, if $h(X,Y,i)$ is true for a rule i, then the value of the target for that rule ($r(Y,i)$) must be chosen to be the corresponding $t(X,Y,i)$. Note that the multiplexer always has the same number of variables as that of t. Similarly, for the next level, the multiplexer clause would be,

$$\infty : hr(Y,I) \Rightarrow (tr(Y,I) \Leftrightarrow b(Y)) \tag{6}$$

[2] The CPT clauses are defined for rule 1 that uses predicate a. All the other rules will have similar clauses.

3. **Stochastic Function Clauses:** These are the clauses that specify the stochastic function to be employed on the values. These are essentially the "prior" on the h predicates. For mean, the idea is to choose a target value from the set of h-values uniformly. In the case of Noisy-Or, the target is chosen from using an Or function over the hidden variables (t and tr). We present the stochastic function clauses for two different cases later in the section.

4. **Integrity Constraints:** These are the constraints that are used to specify that among the different multiplexer nodes, only **one** of them can be true for any particular example. These are of the form:

$$\begin{aligned} \infty &: h(X_1, Y, I) \wedge h(X_2, Y, I) \Rightarrow (X_1 = X_2) \\ \infty &: \qquad \exists X.h(X, Y, I) \end{aligned} \tag{7}$$

The above set of clauses specifies that if h is true for 2 values of X, they should be identical and there exists a grounding of X to make h-true. These constraints are exactly similar for the second level as well. [3]

4 Transformation of Combining Rules

As mentioned earlier, the Bayesian network representation is used **only** to explain the translation by the use of multiplexers. The translation itself is independent of the number of groundings (note that all the predicates in the clauses are variablized and not grounded). We now present two most common types of combination functions from literature: (1) *average*-based and (2) *noisy* combination functions. Let us consider just a single clause $a(X, Y) \Rightarrow b(Y)$ for ease of explanation. Associated with this clause is a conditional probability distribution $P(b|a)$ (we use a and b as shorthand notations for the predicates). As we mentioned, the differences between different combination functions lie mainly in the stochastic function clauses. For each case, we first present the translation and prove the correctness of the resulting distribution. We then present the worked example corresponding to the student satisfaction rules presented earlier.

4.1 Average-Based Combining Rules

Assume that the different instantiations of the above rule are combined using the weighted-mean combining rule. Then the posterior over the target b given the different sets of parents is given by

$$P(b|a_1, ..., a_n) = \frac{1}{\sum w_i} \sum_i w_i \times P(b|a_i) \tag{8}$$

where a_i denotes $a(x_i, y)$. For the case of mean, all $w_i = 1$ (note that the w's are not the weights of the MLN clauses, they are the weights of the combining rule). The CPT clauses will be of the form presented in the earlier section, where the weights are

3 Alchemy supports constraints of this form using the syntactic sugar "!" However, we ran into issues when learning weights with "!" and hence explicitly present the constraints.

log functions of the CPT parameters ($log\frac{p_i}{(1-p_i)}$) . The multiplexer clause is again a hard clause that specifies the value of the target based on the value of the multiplexer ($h(X, Y, I)$). The integrity constraints are also the same as the ones presented above. The stochastic function is the weighted-mean. This specifies the prior on the multiplexer nodes i.e., defines the prior probability with which each multiplexer node is true. Hence, they are of the form: $u_i : h(x_i, y, i)$, where $u_i = \log(w_i)$ is the log-odds of the given x_i. Actually, any weight of the form $u_i = \log(const \times w_i) = \log(const) + \log(w_i)$ would work. For mean, the log-odds would imply $u_i = \log(1/n)$, where n is the number of instantiations. From the previous equation, it follows that any u_i are acceptable as long as they are constant for all i. The intuition is that each $t(X, Y)$ chooses the value of the target based on the CPT, and the final value of the target is chosen from the different t's using the multiplexer nodes. The multiplexer is activated such that it takes only one value given by the stochastic function (mean or weighted-mean).

Proposition: The given representation of MLNs exactly captures the distribution given by equation 8.

Proof Sketch: For simplicity, consider only 2 instantiations of the rule presented above and we are interested in $P(b|a_1, a_2)$ which is given by equation 8 for $i = 2$. There will correspondingly be 4 different cases: both t_1 and t_2 (hidden variables) are true and one of h_1 or h_2 (multiplexers) is true (2 cases) and 2 cases where only the multiplexer h_i and the corresponding t_i are true. i.e.,

$$P(b|a_1, a_2) =$$
$$P(b, t_1 = 1, t_2 = 1, h_1 = 1, h_2 = 0|a_1, a_2) + P(b, t_1 = 1, t_2 = 1, h_1 = 0, h_2 = 1|a_1, a_2)+$$
$$P(b, t_1 = 1, t_2 = 0, h_1 = 1, h_2 = 0|a_1, a_2) + P(b, t_1 = 0, t_2 = 1, h_1 = 0, h_2 = 1|a_1, a_2)$$

$$= \frac{1}{Z}\left(e^{\theta_1+\theta_2+\log(w_1)} + e^{\theta_1+\theta_2+log(w_2)} + e^{\theta_1+\log(w_1)} + e^{\theta_2+\log(w_2)}\right), \text{where } \theta_i = \log\frac{p_i}{1-p_i}$$
$$= \frac{1}{Z}\frac{w_1p_1+w_2p_2}{(1-p_1)(1-p_2)} = \frac{w_1p_1+w_2p_2}{w_1+w_2}$$

where Z can be shown to be $\frac{w_1+w_2}{(1-p_1)(1-p_2)}$ by summing over the two values of b i.e., over $(0, 1)$. We omit the calculation of Z for brevity. Thus we can show that the final distribution due to these MLNs is equal to the distribution presented in equation 8. The same proof can be extended for multi-level combining rules as well.

Worked Example: Consider the FOCI statements about student satisfaction presented earlier. We now present the case where $CR1$ is mean while $CR2$ is weighted-mean. We show the translation to MLNs below. First, consider the CPT clauses. Since the grade of the student can be any of say, $\langle A, B, C, D, F\rangle$, we use $+G$ in Alchemy that uses all possible groundings of G. The CPT clauses for each rule are as follows:

$$w_G : student(S), course(C), takes(T, S, C), grade(T, +G) \Rightarrow t1(S, T, C, +G)$$
$$w_Q : student(S), paper(P, S), quality(P, +Q) \Rightarrow t2(S, P, +Q)$$

where each w_G and w_Q are the log-odds for each grade (G) and quality (Q) respectively. We refer to the Alchemy manual for a detailed discussion on $+$. For the purposes of this paper, it suffices to say that for each *grade, student* and *course* combination, there will be a clause corresponding to the CPT entry. Next, we present the multiplexer clauses.

$$\begin{aligned}\infty : h1(S,T,C,G) &\Rightarrow t1(S,T,C,G) \Leftrightarrow r(S,1)\\ \infty : h2(S,P,Q) &\Rightarrow t2(S,P,Q) \Leftrightarrow r(S,2)\\ \infty : hr(S,R) &\Rightarrow r(S,R) \Leftrightarrow satisfaction(S)\end{aligned} \tag{9}$$

The first two clauses serve to choose the intermediate values of r corresponding to rules 1 and 2. The third rule then chooses the final value of satisfaction from the two intermediate values. The stochastic function clauses are given by:

$$\log(w_1) : hr(S,1)$$
$$\log(w_2) : hr(S,2)$$

The above clauses specify the prior over the intermediate nodes as a function of their weights w_i.[4] At the first level, the value of the intermediate node is chosen according to an uniform distribution (mean combining rule) and hence the weights of the MLN clauses are 0 and are not presented here. Finally, we present the integrity constraints that restrict the multiplexer to choose only one value from among a set of possible values

$$\begin{aligned}&\infty : h1(S,T1,C1,G1) \wedge h1(S,T2,C2,G2) \Rightarrow (T1 = T2 \wedge C1 = C2 \wedge G1 = G2)\\ &\infty : \text{ Exists T,C,G } h1(S,T,C,G)\\ &\infty : h2(S1,P1,Q1) \wedge h2(S2,P2,Q2) \Rightarrow (P1 = P2 \wedge Q1 = Q2)\\ &\infty : \text{ Exists P,Q } h2(S,P,Q)\\ &\infty : hr(S,R1) \wedge hr(S,R2) \Rightarrow R1 = R2\\ &\infty : \text{ Exists R } hr(S,R)\end{aligned} \tag{10}$$

4.2 Noisy Functions

For this case, let us assume a single rule and that the different instantiations of that rule are combined using a noisy function. For *Noisy-Or*, the marginal is computed as,

$$P(b = \text{T}|a_1,\ldots,a_n) = 1 - \prod_{i=1}^{n} f_i^{a_i} \tag{11}$$

where f_i's represent the probability that a present (Boolean-valued) cause, a_i, fails to make the result b true. When converting these to MLNs, the transformation is mostly similar to the earlier case. Though the CPT clauses are constructed similarly, we present them for clarity. They are of the form:

$$\begin{aligned}\infty &: \neg a(X,Y) \Rightarrow \neg t(X,Y,1).\\ w_i &: \ a(x_i,Y) \Rightarrow t(x_i,Y,1).\end{aligned}$$

where, $w_i = \log((1 - f_i)/f_i)$. As can be seen, if $a(X,Y)$ is false for a particular value of X, $t(X,Y,1)$ will always be false while if a is true, t can be false due to some noise. The multiplexer and integrity clauses are similar to the average case. A careful reader will note that the multiplexer and integrity clauses are redundant for this case as they derive the r-values directly from t-values as shown below. The stochastic function (deterministic here) is given by,

[4] These weights are the weights of the combining function and must not be confused with the weight of the MLN clauses.

$$\infty : r(Y, I) \Leftrightarrow \exists X.t(X, Y, I) \tag{12}$$

This asserts that $r(Y, i)$ is true if and only if some $t(X, Y, i)$ is true, which is effectively deterministic Or applied to noisy versions of the inputs. It can be shown that this set of clauses exactly capture the distribution given by equation 11. We omit the proof as it is a trivial mathematical exercise similar to the weighted-mean case.

Noisy existentials can be constructed similarly, except that we have tied weights. When constructing noisy-and, the noise adds a probability of success instead of a probability of failure:

$$\begin{aligned} w_i &: \neg a(X, Y, 1) \Rightarrow \neg t(X, Y, 1) \\ \infty &: \; a(x_i, Y, 1) \Rightarrow t(x_i, Y, 1). \end{aligned}$$

The multiplexer and the stochastic functions are also modified accordingly to reflect the *And* function.

Also, note that any MLN can be seen as a noisy-and in which the target $b(Y)$ is known to be true and each $a(x_i, Y)$ is a clause from the original MLN. Because of the infinite-weight conjunction, all t_i must be true. Since t_i is true, we can simplify each implication $\neg a(x_i, Y) \Rightarrow \neg t(x_i, Y)$ to $a(x_i, Y)$. The final, simplified MLN is therefore just the weighted clauses from the original MLN: $w_i : a(x_i, Y)$.

Worked Example: We now present the rules for the satisfaction example where $CR2$ is *Or* while $CR1$ is *Noisy-Or* with q_i as inhibition probability. The CPT clauses are

$$\begin{aligned} &\log(\tfrac{1-q_1}{q_1}) : student(S), course(C), takes(T, S, C), grade(T, G) \Rightarrow t1(S, T, C, G) \\ &\log(\tfrac{1-q_2}{q_2}) : student(S), paper(P, S), quality(P, Q) \Rightarrow t2(S, P, Q) \end{aligned}$$

Note that the CPT parameters are a function of the noise (inhibition) for the two rules. The multiplexer clauses can be constructed similar to the weighted mean case given in Equation 9. The stochastic function clauses are created according to the following clauses. Note that the stochastic function clauses state that there is an Or function at each level (the noise at the first-level is captured in the CPT clauses).

$$\begin{aligned} &\infty : r(S, 1) \Leftrightarrow \text{Exists T,C,G } t1(S, T, C, G) \\ &\infty : r(S, 2) \Leftrightarrow \text{Exists P,Q } t2(S, P, Q) \\ &\infty : satisfaction(S) \Leftrightarrow \text{Exists R } tr(S, R) \end{aligned}$$

The integrity constraints are similar to the earlier case (Equation 10). The example here combines *Noisy-Or* with the Or combining rule. We can similarly imagine combining different decomposable combining rules at the different levels. We are not presenting all the combinations in this work but note that the same templates can be used to construct the different sets of combining functions.

4.3 Algorithm for Creating MLN Clauses from Decomposable Combining Rules

This formulation allows for arbitrary nesting of combining rules. The combining rules used for combining different instantiations of different rules could be different. For instance, we can imagine a situation such as $NoisyAnd(w_1\ A, w_2\ B, w_3\ NoisyOr(w_4\ C, w_5\ Mean(w_6\ D, w_7\ E, w_8\ F)))$ where we have both Noisy-Or and Mean inside the Noisy-And function. w_i's are the weights while A through F are first-order logic Formulae. Such a representation is a significant generalization of MLNs.

1: MLNClauses clauseList = [ ];
2: // Each FOCI Statement s_i has CPT, Predicates, CR_1^i
3: For Each FOCI statement $s_i \in S$
4:     For Each Independent parameter θ_i^j in s_i.CPT
5:         Add one CPT clause to clauseList, e.g. as in Eqn 4
6:     Add to clauseList based on 1st level combining rule CR_1^i:
7:         One multiplexer clause as in Eqn. 5
8:         One stochastic function clause, e.g. as in Eqn. 12
9:         Two integrity constraint clauses as in Eqn. 7
10: For Each FOCI statement $s_i \in S$
11:     Add to clauseList based on the 2nd level combining rule CR_2^i:
12:         One multiplexer clause as in Eqn. 6
13:         One stochastic function clause, e.g. as in Eqn. 12
14: Two integrity constraint clauses as in Eqn. 7

Fig. 3. CreateMLNClauses (FOCI Statements S, CR_2)

Figure 3 describes the pseudocode for constructing MLNs from a set of FOCI statements combined using combining rule CR_2. Each statement s_i has its own 1st level combining rule CR_1^i. Lines 3 through 9 present the methods for constructing the clauses corresponding to s_i and its combining rule. For each independent parameter in the CPT of s_i, a clause is created. Also for each s_i, one multiplexer clause, one stochastic function and two integrity constraints are created. Once all the 1st level combining rules are considered, the clauses corresponding to CR_2 are constructed in lines 10 through 14. We note that it requires $O(1)$ to construct each clause. We now provide a bound on the number of clauses required by such an MLN. In particular, we consider the general SRL case of multi-level combining rules where the instantiations of a single rule are combined using CR_1 and different rules are combined using CR_2.

Theorem 1. *For any joint distribution which can be represented by* n *FOCI statements combined with nested decomposable combining rules, and* k *independent parameters, there exists an equivalent MLN of* O(nk) *rules which can be constructed in* O(nk) *time.*

Proof. (sketch) The proof of equivalence is straightforward from the definition of the various clauses. Let n be the number of FOCI statements and k be the number of independent CPT parameters. From the algorithm in Figure 3, for each rule, there are k CPT clauses, one multiplexer clause, one stochastic function clause and two integrity constraints yielding $k + 4$ clauses. Hence the total number of clauses created in lines $(3 - 9)$ is $n(k + 4)$. For lines $10 - 14$ of the algorithm, the number of clauses is $n(1+1)+2 = 2(n+1)$. Hence the total number of clauses is $n(k+6)+2 = O(nk)$.

Since each clause can be constructed in constant time given the FOCI statement and the combining rule, the resulting MLN can be constructed in $O(nk)$ time. Note that the minimal number of clauses required to model FOCI statements using MLN is $O(nk)$ as we need a clause for every parameter. Hence, our translation creates a model that is no more complex than the minimal MLN. ■

4.4 MLN Macros

While theoretically MLNs can represent most of the distributions that we considered, it seems impractical to expect a domain expert to come up with these rules. Firstly, the domain expert has to understand the underlying distribution and the combining rules as operating on values as against distributions (i.e, multiplexers). Secondly, the domain expert needs to be an MLN expert as well and has to understand the translation. In this section, we present a macro that can be used to construct MLNs given the domain expert's statements. The key idea is to remove the burden of specifying the MLNs from the user and allow our "translator" to create the MLNs corresponding to the true distribution. We now present the structure of the macro:

$\texttt{CR}\{\ CR_1:\ X_1^1 \wedge ... \wedge X_{n_1}^1 \Rightarrow Y$

$CR_2:\ X_1^2 \wedge ... \wedge X_{n_2}^2 \Rightarrow Y \ldots\ \}$

The above macro can be interpreted as: CR, CR_1 and CR_2 are the combination functions – *And, Or, Noisy-Or, Noisy-And etc.* While CR_i combines the multiple instantiations of clause i, CR combines the multiple clauses. X_i^j and Y are predicates. The first clauses specifies n_1 causes for the target predicate Y. X_1^1 is the first cause of Y in rule 1 and so on. Instead of writing 2^n different clauses, the user specifies a single clause that is then unrolled into the different clauses by the translator. The user can specify several clauses that can be combined.

The translator then converts these macros to the MLN clausal representation. A natural question now is: where do the weights come from? A simple solution would be to construct the clauses and allow the underlying MLN package (in our case *Alchemy* [11]), to learn the weights. We could hold the weights of the hard clauses (integrity constraints and some multiplexer clausers) and instruct Alchemy to learn the weights of only the "soft clauses" leading to a more efficient learning. In cases where the training data is not available, we allow the conditional probabilities to be specified as 2^n array corresponding to the different configurations of the predicates in the body of the clause. Hence the clauses are now of the form, $\langle p_1, p_2, ..., p_{2^n}\rangle X_1 \wedge ... \wedge X_n \Rightarrow Y$, where $p_i = P(Y = T|Config(X_1, ..., X_n) = i)$ is the conditional probability of the target being T given that the truth value of the predicates in the body form the ith configuration. Similarly, the parameters of the combining rules (ex. weights of weighted mean) can also be specified as $CR\langle w_1, ..., w_m\rangle$ for m clauses. Based on the combining rule used, the translator then computes the weights of the different clauses based on the probabilities and assigns the weights to the corresponding clauses.

5 Experiments

In the following experiments, we used the Alchemy system[5] to learn the weights and/or perform inference. The same settings were used for both MLNs with combining rules (denoted by MLN^+) and the default MLNs(MLN^*). The clauses of the MLN^* are the parent configurations of the CPT of each rule. Hence, for each independent parameter of the CPT, there exists a clause in MLN^*. MLN^* was chosen so that it had the same number of parameters as that of a directed model to make a fair comparison. The clauses of MLN^+ consist of the CPT clauses and the multiplexer, stochastic function and integrity clauses. For both MLN^+ and MLN^*, we used the same settings for the learning and inference algorithms (i.e., used the same number of iterations, discriminative learning, same number of MCMC steps, MC-SAT for inference etc.).

We present our learning results in two real-world domains: *Cora* and *UW-CSE*. The goal of the experiment is: given minimal domain knowledge (typically 2 rules to predict the target), will the structure imposed by combining rules be useful in learning a good model? For the UW-dataset, the goal was to predict the *advisedBy* relationship between a student and a professor. The rules that we used were:

```
N-Or {
N-Or: student(S) ∧ professor(P) ∧ course(C) ∧ taughtBy(P,C,Q) ∧
ta(S,C,Q) ⇒ advisedBy(S,P).
N-Or: student(S) ∧ professor(P) ∧ publication(P,W) ∧
publication(S,W) ⇒ advisedBy(S,P).}
```

MLN^* used all the combinations of the predicates in the head of the clauses and learned weights for each of them. For MLN^+, we used Noisy-Or as the combining rule at both levels. We learned the weights using Alchemy and used MC-SAT for performing inference. We trained the algorithms on the *AI* group data that consisted of 35 positive instances of the *advisedBy* relation. We present the average likelihood (i.e., $\frac{1}{n}\sum_i P(y_i = \hat{y}_i)$, where n is the number of examples, $\hat{y}$ is the predicted label and y is the true label) of the test set in the last column of Table 1. Note that since we are in the relational setting, the test set will mostly consist of negatives. Hence, an algorithm that always predicts *false* will have a reasonably high likelihood. To avoid this situation, we forced the test set to contain 50% negative examples by sampling the negative examples randomly. This way a likelihood of 0.5 would mean that everything is either predicted *true* or as *false*. There were a total of 80 test examples with 40 positives.

We also compare the area under curve for the ROC and PR curves. MLN^* was not able to learn reasonable weights with a small number of rules and hence predicted everything as 0. In a test-set with 50% positive examples, this yielded a likelihood of 0.5. On the other hand, with MLN^+, we were able to learn a more reasonable model that had a higher likelihood. More importantly, MLN^+ did not predict every query predicate as 0 or 1 and instead had a reasonable distribution over the target. When we added more rules to MLN^* (7 more rules from Alchemy that were earlier used in other MLN experiments to predict *advisedBy*) the average likelihood increased to 0.63. The values of AUC for ROC and PR for MLN^+ are significantly higher than MLN^*. This

[5] http://alchemy.cs.washington.edu/

Table 1. Results on real world domains

Domain	Algorithm	AUC-ROC	AUC-PR	Likelihood
Cora	MLN^+	1.0	1.0	0.987
	MLN^*	1.0	1.0	0.963
UW	MLN^+	0.560	0.672	0.611
	MLN^*	0.472	0.523	0.5

demonstrates that in this domain the use of more complex combining functions seem to improve the performance of MLN learning.

The results were far more impressive for *Cora* dataset where the goal is to predict whether *two* citations refer to the same one. The training set consisted of about 7500 examples (about 70% of them were negative examples) and the test set consisted of 100 examples (out of which 50% were negatives). The two rules that we used were:

```
N-Or {
N-Or : Author(bc1,a1) ∧ Author(bc2,a2) ∧ SameAuthor(a1,a2)
⇒ SameBib(bc1,bc2).
N-Or : Title(bc1,t1) ∧ Title(bc2,t2) ∧ SameTitle(t1,t2)
⇒ SameBib(bc1,bc2).}
```

As can be seen from the table, MLN^+ learned nearly the perfect model for the domain and had a very high likelihood and AUC values. This clearly showed that with just two rules, given some more knowledge (as hard constraints of the combining rules), MLN^+ was able to learn a highly predictive model. While MLN^* with exactly the same setting as MLN^+ (i.e., discriminative, rules for all the combinations of the predicates etc.) predicted all the test examples as 0 and thus had a lower likelihood and AUC values (very similar to the UW data set). To improve the performance of MLN^*, we changed the settings (to generative learning, dropped some seemingly irrelevant clauses that had a large number of groundings). With these changes, we were able to get MLN^* to perform comparably with MLN^+.

Specialized inference algorithms. Admittedly, the presence of hidden predicates increased the running time of Alchemy, but this motivates the need for learning algorithms that exploit the special structure efficiently (as we used the default EM learning algorithm of Alchemy to learn weights for MLN^+).

We implemented an inference algorithm that exploits the special structure of these clauses. We do not present the algorithm in detail in this work as the goal of this work is to show that the combining rules can be captured in MLNs and motivate the need for specialized algorithms that can exploit the local models similar to the ones presented in [13]. Initial results indicate that the time taken for MLN^+ in the UW-data set is 4 seconds while that of MLN^* is 30 seconds to obtain the same results presented in Table 1. As can be seen, there is a drastic improvement when the inference algorithm exploits the knowledge of the structure of the clauses. The modified inference algorithm assumes that the structure of the MLN is the one presented in Figure 2 and performs sampling on this network. Hence, it exploits the knowledge about the structure of the network and the multiplexers in the network. We expect that learning localized models in MLNs will enable efficient learning and inference. We are currently working on formalizing the details of the learning algorithm that uses this inference method. Our hypothesis is that since learning requires inference in its inner loop, the specialized inference algorithm will yield faster learning of the parameters.

6 Conclusions

Combining rules capture the notion of causal independence for SRL models. We have presented an algorithm for representing a class of combining rules (decomposable combining rules) in an undirected model (MLN). We derived the equivalent clauses and provided a bound on the number of clauses required for the representation. Our experiments demonstrated that for a small number of clauses, combining functions are useful in learning more accurate models. The structure imposed by these functions help in guiding the learning algorithms towards reasonable weights. Jaeger [9], showed that RBNs can capture MLNs and pointed out to the reverse as an open problem. We take an important step in that direction by showing how MLNs can capture combination functions of the directed models and in turn, most of the features of directed models.

However, this translation from combining rules to MLNs is not without its cost. We found that the inference in the resulting MLN is 4-5 times slower than the one that does not use the combining rules. The problem is that while the declarative knowledge embedded in the combining rules can be encoded into clauses and given to MLNs, there are no effective means to exploit the causal independence for controlling inference. To be effective, the inference engine has to essentially rediscover the hidden structure that is naturally exploited by the directed models. One possible future direction is to develop specialized inference algorithms that can detect structure in MLNs and exploit it for efficiency. We have taken the first step in this direction, but are still working on the details of the learning algorithm that will exploit the structure. A more general and important direction is to develop hybrid models that allow us to specify different parts of the model differently and combine them using a decomposable structure. This should allow the application of specialized learning algorithms inside each module, and combine the results in an efficient manner.

Acknowledgements

SN, TK and JS gratefully acknowledges the support of DARPA via Air Force Research Laboratory (AFRL) under prime contract no. FA8750-09-C-0181. PT greatfully acknowledges the support of DARPA grant FA8750-09-C-0179. KK was supported by the Fraunhofer ATTRACT fellowship STREAM and by the Europen Commission under contract number FP7-248258-First-MM. DL was supported by the University of Oregon Department of Computer and Information Science. Any opinions, findings, and conclusion or recommendations expressed in this material are those of the author(s) and do not necessarily reflect the view of the Air Force Research Laboratory (AFRL), US government or DARPA.

References

1. Domingos, P., Lowd, D.: Markov Logic: An Interface Layer for AI. Morgan & Claypool, San Rafael (2009)
2. Fierens, D., Blockeel, H., Bruynooghe, M., Ramon, J.: Logical Bayesian networks and their relation to other probabilistic logical models. In: Kramer, S., Pfahringer, B. (eds.) ILP 2005. LNCS (LNAI), vol. 3625, pp. 121–135. Springer, Heidelberg (2005)
3. Getoor, L., Friedman, N., Koller, D., Pfeffer, A.: Learning probabilistic relational models. In: Dzeroski, S., Lavrac, N. (eds.) Relational Data Mining (2001)
4. Getoor, L., Grant, J.: PRL: A probabilistic relational language. Mach. Learn. 62(1-2), 7–31 (2006)
5. Getoor, L., Taskar, B.: Introduction to Statistical Relational Learning. MIT Press, Cambridge (2007)
6. Heckerman, D., Breese, J.: A new look at causal independence. In: UAI (1994)
7. Jaeger, M.: Relational Bayesian networks. In: Proceedings of UAI (1997)
8. Jaeger, M.: Parameter learning for Relational Bayesian networks. In: ICML (2007)
9. Jaeger, M.: Model-theoretic expressivity analysis. In: De Raedt, L., Frasconi, P., Kersting, K., Muggleton, S.H. (eds.) Probabilistic Inductive Logic Programming. LNCS (LNAI), vol. 4911, pp. 325–339. Springer, Heidelberg (2008)
10. Kersting, K., De Raedt, L.: Bayesian logic programming: Theory and tool. An Introduction to Statistical Relational Learning (2007)
11. Kok, S., Sumner, M., Richardson, M., Singla, P., Poon, H., Lowd, D., Domingos, P.: The Alchemy system for statistical relational AI. Technical report, Department of Computer Science and Engineering, University of Washington, Seattle, WA (2007)
12. Koller, D., Pfeffer, A.: Learning probabilities for noisy first-order rules. In: IJCAI (1997)
13. Natarajan, S., Tadepalli, P., Dietterich, T.G., Fern, A.: Learning first-order probabilistic models with combining rules. Special Issue on Probabilistic Relational Learning, AMAI (2009)
14. Zhang, N., Poole, D.: Exploiting causal independence in Bayesian network inference. JAIR 5, 301–328 (1996)

Improved MinMax Cut Graph Clustering with Nonnegative Relaxation

Feiping Nie, Chris Ding, Dijun Luo, and Heng Huang

Department of Computer Science and Engineering,
University of Texas, Arlington, America
{feipingnie,dijun.luo}@gmail.com,{chqding,heng}@uta.edu

Abstract. In graph clustering methods, MinMax Cut tends to provide more balanced clusters as compared to Ratio Cut and Normalized Cut. The traditional approach used spectral relaxation to solve the graph cut problem. The main disadvantage of this approach is that the obtained spectral solution has mixed signs, which could severely deviate from the true solution and have to resort to other clustering methods, such as K-means, to obtain final clusters. In this paper, we propose to apply additional nonnegative constraint into MinMax Cut graph clustering and introduce novel algorithms to optimize the new objective. With the explicit nonnegative constraint, our solutions are very close to the ideal class indicator matrix and can directly assign clusters to data points. We present efficient algorithms to solve the new problem with the nonnegative constraint rigorously. Experimental results show that our new algorithm always converges and significantly outperforms the traditional spectral relaxation approach on ratio cut and normalized cut.

Keywords: Spectral clustering, Normalized cut, MinMax cut, Nonnegative relaxation, cluster balance, random graphs.

1 Introduction

Clustering is an important task in machine learning and data mining areas. In the past decades, many clustering algorithms have been proposed such as K-means clustering, spectral clustering and its variants [1,2,3], support vector clustering [4], and maximum margin clustering [5,6,7]. Among them, the use of manifold information in graph cut clustering has shown the state-of-the-art clustering performance and been widely applied into many applications, such as image segmentation [8], white matter fiber tracking in biomedical image [9], and protein sequence clustering [10].

MinMax Cut was proposed in [11] and showed more compact and balanced clustering results than Ratio Cut [12] and Normalized Cut [8]. Because, in MinMax Cut method, the within-cluster similarities are explicitly maximized. Solving the graph cut clustering problem is a nontrivial task. The main difficulty of the graph clustering problem lies in the constraints on the solution. In order to make the problem tractable, the constraints should be relaxed. Traditional

J.L. Balcázar et al. (Eds.): ECML PKDD 2010, Part II, LNAI 6322, pp. 451–466, 2010.

approach used spectral relaxation to solve this problem. But the main disadvantage of this approach is that the obtained spectral solution has mixed signs, which could severely deviate from the true solution and have to resort to other clustering methods, such as K-means, to obtain final cluster results.

In order to solve this notorious problem, in this paper, we propose a new method to optimize the MinMax Cut graph clustering with additional nonnegative constraint. With the explicit nonnegative constraint, the solutions are very close to the ideal class indicator matrix and can be directly used to assign cluster labels to data points. We propose efficient algorithms to solve this problem with the nonnegative constraint rigorously. Experimental results show that our algorithm always converges and the performance is significantly improved in comparisons with the traditional spectral relaxation approach on Ratio Cut and Normalized Cut.

The rest of this paper is organized as follows. Section 2 reviews the MinMax Cut problem. Our proposed nonnegative relaxation approaches to solve the MinMax Cut clustering problem are introduced in Section 3. Experimental results on real-world data sets are reported in Section 4. Finally, we conclude our work in Section 5.

2 MinMax Cut for Clustering

Suppose we have n data points $\{x_1, x_2, \cdots, x_n\}$, and construct a graph using the data with weight matrix $W \in \mathbb{R}^{n \times n}$. The multi-way MinMax Cut graph clustering objective function is (we also show Min Cut, Ratio Cut and Normalized Cut for comparisons):

$$J = \sum_{1 \le p < q \le K} \frac{s(C_p, C_q)}{\rho(C_p)} + \frac{s(C_p, C_q)}{\rho(C_q)} = \sum_{k=1}^{K} \frac{s(C_k, \bar{C}_k)}{\rho(C_k)}, \tag{1}$$

$$\rho(C_k) = \begin{cases} 1 & \text{for Min Cut} \\ |C_k| & \text{for Ratio Cut} \\ \sum_{i \in C_k} d_i & \text{for Normalized Cut} \\ s(C_k, C_k) & \text{for MinMax Cut} \end{cases} \tag{2}$$

where K is the number of clusters, C_k is the k-th cluster (subgraph in graph G), $\bar{C}_k$ is the complement of subset C_k in graph G, and $s(A, B) = \sum_{i \in A} \sum_{j \in B} W_{ij}$, $d_i = \sum_j W_{ij}$.

Let $\mathbf{q}_k (k = 1, 2, \cdots, K)$ be the cluster indicators where the i-th element of $\mathbf{q}_k$ is 1 if the i-th data point x_i belongs to cluster k, and 0 otherwise. For example, if data points within each cluster are adjacent, then

$$\mathbf{q}_k = (0, \cdots, 0, \overbrace{1, \cdots, 1}^{n_k}, 0, \cdots, 0)^T. \tag{3}$$

We can easily see that $s(C_k, \bar{C}_k) = \sum_{i \in C_k} \sum_{j \in \bar{C}_k} W_{ij} = \mathbf{q}_k^T (D - W)\mathbf{q}_k$, $\sum_{i \in C_k} d_i = \mathbf{q}_k^T D \mathbf{q}_k$, $s(C_k, C_k) = \mathbf{q}_k^T W \mathbf{q}_k$, where D is a diagonal matrix with the i-th diagonal element as d_i. We rewrite the objective functions of these four methods as:

$$J_{\text{mincut}} = \sum_{k=1}^{K} \mathbf{q}_k^T (D - W)\mathbf{q}_k, \quad J_{\text{rcut}} = \sum_{k=1}^{K} \frac{\mathbf{q}_k^T (D - W)\mathbf{q}_k}{\mathbf{q}_k^T \mathbf{q}_k}, \tag{4}$$

$$J_{\text{ncut}} = \sum_{k=1}^{K} \frac{\mathbf{q}_k^T (D - W)\mathbf{q}_k}{\mathbf{q}_k^T D \mathbf{q}_k}, \quad J_{\text{MMC}} = \sum_{k=1}^{K} \frac{\mathbf{q}_k^T (D - W)\mathbf{q}_k}{\mathbf{q}_k^T W \mathbf{q}_k}. \tag{5}$$

2.1 Cluster Balance Analysis on Random Graphs

One important advantage of the `MinMaxCut` method is that it tends to produce balanced clusters in graph clustering, *i.e.*, the resulting subgraphs will have similar size. Here we study the clustering solutions on two popular random graphs: (1) Erdos-Renyi (ER) random graph model [13,14] and (2) Expected degree sequence (EDS) random graph [15].

Erdos-Renyi Random Graph. The ER random graph model is perhaps the mostly wide used random graph model. This is a uniformly distributed random graph with n nodes, where two nodes are connected with probability $p, 0 < p < 1$. Considering the four objective functions, `MINcut`, `Rcut`, `Ncut`, and `MinMaxCut`, we have the following result.

Theorem 1. *For random graphs,* `MinCut` *favors highly skewed cuts.* `MinMaxCut` *favors balanced cut, i.e., both subgraphs have the same size.* `RatioCut` *and* `NormCut` *show no size preferences, i.e., each subgraph could have arbitrary size.*

Proof. We compute the object functions for the partition of G into A and B. Note that the number of edges between A and B are $p|A||B|$ on average. For `MINcut`, we have $J_{\text{mincut}}(A, B) = p|A||B|$. Clearly, `MinCut` favors either $|A| = n-1$ and $|B| = 1$, or $|B| = n - 1$ and $|A| = 1$; both are skewed cuts. For `MinMaxCut`, we have

$$J_{\text{MMC}}(A, B) = \frac{|B|}{|A| - 1} + \frac{|A|}{|B| - 1}.$$

Minimizing $J_{\text{MMC}}(A, B)$, we obtain a balanced cut: $|A| = |B| = n/2$. For `Rcut`, we have

$$J_{\text{rcut}}(A, B) = \frac{p|A||B|}{|A|} + \frac{p|A||B|}{|B|} = np.$$

For `Ncut`, because all nodes have the same degree $(n - 1)p$,

$$J_{\text{ncut}}(A, B) = \frac{p|A||B|}{p|A|(n - 1)} + \frac{p|A||B|}{p|B|(n - 1)} = \frac{n}{n - 1}.$$

Both Rcut and Ncut objectives have no size dependency and no size preference. This random graph model shows that `MinMaxCut` has the tendency of produce a balanced clustering. □

Expected Degree Sequence Random Graph
The degree of a node on a graph is the sum of edge connecting to it. The distribution of the n node degrees of a graph is a critical property. The ER random graph has a degree distribution much like a Gaussian center around the average degree $\bar{d} = np$.

However, much of biological, social and information networks/graphs have a power-law degree distribution The expected degree sequence (EDS) random graph is a graph model for these networks/graphs. In this model, the degrees of each nodes $(d_1 \cdots d_n)$ are pre-specified. The edges are are then randomly distributed, with the constraints that the degrees of nodes satisfy the given fixed degree sequence.

The EDS graph model is a generalization of the ER random graph model, which can be seen as an special case of the EDS random graph by setting $d_i = np$ for all nodes.

For the EDS random graph model,

$$P(W_{ij} = 1) = d_i d_j / M, \ M = \sum_{ij} W_{ij}.$$

[in contrast for ER model, $P(W_{ij} = 1) = p$]. Here, to make things precise, we study a graph whose edge weights are the **average** of the probabilistic distribution: to make things precise:

$$\widehat{W}_{ij} = 1 * P(W_{ij} = 1) + 0 * P(W_{ij} = 0) = P(W_{ij} = 1) = d_i d_j / M, \qquad (6)$$

For notational simplicity, we replace $\widehat{W}$ by W. We have the following

Theorem 2. *For a EDS random graph, Rcut produces highly skewed cuts, MinMaxCutfavors balanced cut, while Ncut has no unique solution and thus shows no size preferences.*

Proof. Suppose we cut the graph into A, B. Then $S(A, B) = \sum_{i \in A} \sum_{j \in B} W_{ij}/M = \sum_{i \in A} \sum_{j \in B} d_i d_j / M = D(A)D(B)/M$. Thus for Ncut,

$$J_{\text{ncut}} = S(A, B)/D(A) + S(A, B)/D(B) + = D(B)/M + D(A)/M = 1,$$

showing no dependence on A, B. Thus Ncut has no unique solution and shows no size preference.

For Rcut and MinMaxCut, we sort the degrees in increasing order and assuming $d_1 < d_2 < d_3 < \cdots < d_n$ (we assume the degrees are different for simplicity). It is easy to see that if $|A| = k, |B| = n - k$ for fixed k, the cut $S(A, B) = D(A)D(B)/M$ is minimized when the graph $G = (v_1 \cdots v_n)$ is cut into $A = \{v_1, \cdots, v_k\}$, $B = \{v_{k+1}, \cdots, v_n\}$. Thus the optimal clustering solution is obtained by searching the minima in the range $k = 1, \cdots, n-1$. The Rcut objective is

$$J_{\text{rcut}}(k) = \frac{1}{M^2} \left[\sum_{i=1}^{k} d_i \right] \left[\sum_{r=k+1}^{n} d_r \right] \left[\frac{1}{k} + \frac{1}{n-k} \right].$$

The clustering solution is found as we search the minima in $k = 1, \cdots, n-1$. Normally, the optimal k^* is very small, $k^* \ll n$, implying a skewed cut. For MinMaxCut, $S(A, A) = \sum_{i\in A}\sum_{j\in A} W_{ij}/M = D(A)^2/M$ and $S(B, B) = D(B)^2/M$. The MinMaxCutobjective becomes

$$J_{\text{MMC}} = \frac{D(A)D(B)}{D(A)^2} + \frac{D(A)D(B)}{D(B)^2} = \frac{D(B)}{D(A)} + \frac{D(A)}{D(B)}.$$

This is minimized when $D(A) = D(B)$. The solution is generally balanced. □

Theorems 1 and 2 show the general tendency regarding to cluster balance for Rcut, Ncut and MinMaxCut. on random graphs. For pure random graphs, there is no true clusters and the clusters obtained are not meaningful. In real applications, graphs are not random. But the general tendency regarding to cluster balance are expected to be similar to Theorems 9 and 10. The following graph examples illustrate these tendencies.

We give two examples to illustrate these tendency in Figure 1, from which we can clearly see that MinMaxCutfavors more balanced cut than Rcut and Ncut.

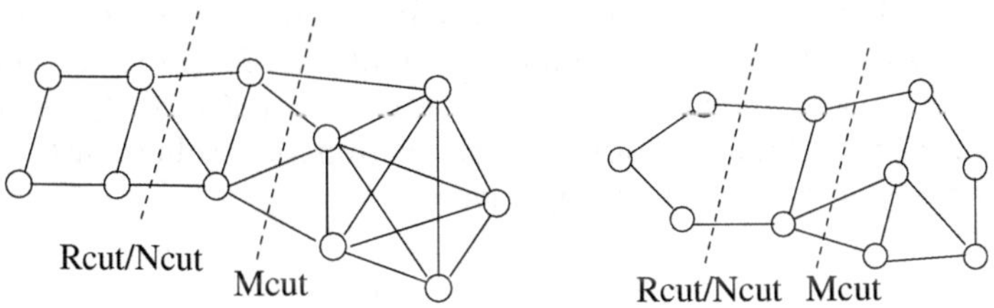

Fig. 1. Two examples that Rcut and Ncut lead to unbalanced clusters, whereas MinMaxCutgives out balanced clusters

2.2 Optimization of MinMax Cut with Spectral Relaxation

We rewrite the MinMaxCutclustering objective J_{MMC} of Eq. (5) by defining

$$Z = (\mathbf{z}_1, \cdots, \mathbf{z}_K),\ \mathbf{z}_k = \frac{\mathbf{q}_k}{\|D^{1/2}\mathbf{q}_k\|}, \tag{7}$$

the MinMaxCutclustering optimization becomes

$$\min_Z J_{\text{MMC}} = \sum_{\ell=1}^{K} \frac{1}{\mathbf{z}_\ell W \mathbf{z}_\ell} - K,\ s.t.\ Z^T D Z = I,\ Z \geq 0. \tag{8}$$

Ignoring the nonnegative constraints $Z \geq 0$ here, we derive the spectral solution.

Using Lagrangian multiplier $\Gamma = \Gamma^T$ to enforce $Z^T D Z = I$, we minimize $\mathcal{L} = J_{\text{MMC}} + \text{Tr}\, \Gamma(Z^T D Z - I)$. Setting $\partial\mathcal{L}/\partial\mathbf{z}_k = 0$, we obtain

$$\frac{W\mathbf{z}_k}{(\mathbf{z}_k^T W \mathbf{z}_k)^2} = \sum_{l=1}^{K} D\mathbf{z}_l \Gamma_{lk} \tag{9}$$

Multiply $\mathbf{z}_l^T$ from the left, we obtain

$$\Gamma_{lk} = \mathbf{z}_l^T W \mathbf{z}_k / (\mathbf{z}_k^T W \mathbf{z}_k)^2, \tag{10}$$

which is not symmetric: $\Gamma_{lk} \neq \Gamma_{kl}$. By definition Γ_{lk} must be symmetric. This implies either (1) Γ is diagonal: $\Gamma_{lk} = \delta_{kl}\Gamma_{kk}$ or (2) $\mathbf{z}_k^T W \mathbf{z}_k = \mathbf{z}_l^T W \mathbf{z}_l, k \neq l$. Condition (2) would render the objective function J_{MMC} of Eq. (8) a constant and thus is impossible. Thus we are left with the only possibility that Γ is diagonal, which in turn implies

$$\mathbf{z}_l^T W \mathbf{z}_k = \delta_{lk}\gamma_k, \tag{11}$$

and $\Gamma_{lk} = \delta_{kl}\Gamma_{kk} = \frac{\delta_{kl}\gamma_k}{\gamma_k^2} = \delta_{kl}\gamma_k^{-1}$. Now with Lagrangian multiplier Γ solved, Eq. (9) becomes

$$W\mathbf{z}_k = \gamma_k^{-1} D\mathbf{z}_k. \tag{12}$$

which is identical to the generalized eigensystem of

$$(D - W)\mathbf{z}_k = \zeta_k D\mathbf{z}_k, \tag{13}$$

where $\zeta_k = 1 - 1/\gamma_k$. Thus the solutions are given by the eigenvectors $(\mathbf{z}_1, \cdots, \mathbf{z}_K)$ of generalized Laplacian as same as Normalized Cut. Since the solution is an relaxed solution of the original minimization problem of Eq. (5), *i.e.*, an optimal solution with enlarged domain from vigorous cluster indicators Q to continuous mixed sign Z, the obtained optimal objective function value must be a lower bound for the true MinMax Cut objective

$$\sum_{k=1}^{K} \frac{1}{1-\zeta_k} - K \leq J_{\text{MMC}}. \tag{14}$$

3 Optimization of MinMax Cut with Nonnegative Relaxation

To solve MinMax Cut problem, the traditional spectral relaxation approach relaxes the solution from binary value to real value. However, this relaxation could make the solution severely deviate from the true solution. Moreover, under this relaxation, the obtained spectral solution cannot be directly used to assign cluster labels for data points. To perform clustering, a commonly used post-processing method is to apply K-means to the space of the spectral solution to obtain clusters.

In this section, we will explicitly constrain the solution $\mathbf{q}_k$ to be nonnegative, and propose efficient algorithms to optimize the MinMax Cut clustering objective function with the nonnegative constraint on $\mathbf{q}_k$ rigorously.

3.1 Orthonormal and Nonnegative Constraints

The main difficulty of the graph clustering problem lies in the constraints of the class indicator matrix Q. The constraints should be relaxed to make the problem solvable. From the definition of the class indicator matrix we can see that only one element is one and others are zeros in each row of Q. Thus the columns of Q are orthogonal and the orthogonality should be preserved in a relaxation of class indicator matrix. Note that the objective of the graph cuts is invariant to the scale of the columns of Q, so traditional spectral relaxation approach relaxes the constraints of Q to the orthonormal constraints:

$$Q^T Q = I. \tag{15}$$

Such relaxation makes the problem to be easily solved, but the obtained solution has mixed signs, which deviates from the class indicator matrix largely.

Note that the class indicator matrix is a nonnegative matrix, a more accurate relaxation is adding the nonnegative constraints on the Q:

$$Q^T Q = I, \; Q \geq 0. \tag{16}$$

One can see that when orthonormal and nonnegative constraints are satisfied simultaneously, only one element is positive and others are zeros in each row of Q, which is very close to the ideal class indicator matrix, and can be used directly to assign cluster labels to data points. This motivates our nonnegative relaxation approach for the MinMax Cut clustering problem, and to solve the following optimization problem:

$$\min_{Q} \sum_{k=1}^{K} \frac{\mathbf{q}_k^T D \mathbf{q}_k}{\mathbf{q}_k^T W \mathbf{q}_k}, \quad s.t. \; Q^T Q = I, \; Q \geq 0. \tag{17}$$

We are going to introduce two efficient algorithms to solve this problem in next subsections. The first algorithm iteratively optimizes the objective function with good performance, and the second one is more concise and also has comparable clustering results.

3.2 An Iterative Algorithm to Solve the Problem with Orthonormal and Nonnegative Constraints

In some cases, minimizing an objective might result in numerical instability [16]. Thus we turn to solve the following identical problem:

$$\max_{Q} J(Q), \quad s.t. \; Q^T Q = I, \; Q \geq 0, \tag{18}$$

where

$$J(Q) = \rho \mathrm{Tr}\, Q^T Q - \sum_{k=1}^{K} \frac{\mathbf{q}_k^T D \mathbf{q}_k}{\mathbf{q}_k^T W \mathbf{q}_k}, \tag{19}$$

ρ is an appropriate positive value such that $\rho W - (D - W)$ is positive semi-definite. Specifically, we set $\rho = \frac{\lambda_{max}(D-W)}{\lambda_{max}(W)}$ in this work, where $\lambda_{max}(W)$ and $\lambda_{max}(D - W)$ denotes the largest eigenvalue of W and $D - W$, respectively.

We begin with the Lagrangian function

$$\mathcal{L} = J(Q) - \mathrm{Tr}\Lambda(Q^TQ - I) - \mathrm{Tr}\Sigma^TQ, \tag{20}$$

where the Lagrange multiplier Λ enforces the orthogonality condition $Q^TQ = I$ and the Lagrange multiplier Σ enforces the nonnegativity condition $Q \geq 0$. Using the KKT complementary slackness condition, we have

$$(\frac{\partial J}{\partial Q} - 2Q\Lambda)_{ik}Q_{ik} = 0. \tag{21}$$

Summing over k, we obtain $(\frac{1}{2}Q^T\frac{\partial J}{\partial Q})_{ii} = (Q^TQ\Lambda)_{ii} = \Lambda_{ii}$. This gives the diagonal elements of Λ. To find the off-diagonal elements of Λ, we temporarily ignore the nonnegativity condition, which gives $(\frac{\partial J}{\partial Q} - 2Q\Lambda)_{ik} = 0$. Left multiplying by $Q_{i'k}$ and summing over k, we obtain $(\frac{1}{2}Q^T\frac{\partial J}{\partial Q})_{i'i} = \Lambda_{i'i}$ for the off-diagonal elements of Λ. Combining these two results yields

$$\Lambda = \frac{1}{2}Q^T\frac{\partial J}{\partial Q}. \tag{22}$$

Note that

$$\frac{1}{2}\frac{\partial J}{\partial Q} = \rho Q - DQ_\alpha + WQ_\beta, \tag{23}$$

where

$$Q_\alpha = \left[\frac{1}{\mathbf{q}_1^TW\mathbf{q}_1}\mathbf{q}_1, \cdots, \frac{1}{\mathbf{q}_K^TW\mathbf{q}_K}\mathbf{q}_K\right], \tag{24}$$

$$Q_\beta = \left[\frac{(\mathbf{q}_1^TD\mathbf{q}_1)}{(\mathbf{q}_1^TW\mathbf{q}_1)^2}\mathbf{q}_1, \cdots, \frac{(\mathbf{q}_K^TD\mathbf{q}_K)}{(\mathbf{q}_K^TW\mathbf{q}_K)^2}\mathbf{q}_K\right], \tag{25}$$

then we have

$$\Lambda = \rho Q^TQ - Q^TDQ_\alpha + Q^TWQ_\beta. \tag{26}$$

Decomposing Λ into positive part and negative part as

$$\Lambda = \Lambda^+ - \Lambda^-, \tag{27}$$

where $\Lambda^+ = (|\Lambda| + \Lambda)/2$ and $\Lambda^- = (|\Lambda| - \Lambda)/2$. Now concentrating on the variable Q while ignoring constant terms in $\mathcal{L}$, we have

$$\begin{aligned}\frac{1}{2}\frac{\partial(J - \mathrm{Tr}\Lambda Q^TQ)}{\partial Q} &= \rho Q - DQ_\alpha + WQ_\beta - Q\Lambda \\ &= \rho Q - DQ_\alpha + WQ_\beta - Q\Lambda^+ + Q\Lambda^- \\ &= (\rho Q + WQ_\beta + Q\Lambda^-) - (DQ_\alpha + Q\Lambda^+). \end{aligned} \tag{28}$$

As in Nonnegative Matrix Factorization (NMF) [17,18], Eq. (28) leads to the following multiplicative update formula:

$$Q_{ik} \leftarrow Q_{ik}\sqrt{\frac{(\rho Q + WQ_\beta + Q\Lambda^-)_{ik}}{(DQ_\alpha + Q\Lambda^+)_{ik}}}. \tag{29}$$

We can see that using this update, Q_{ik} will increase when the corresponding element of the gradient in Eq. (28) is larger than zero, and will decrease otherwise. Therefore, the update direction is consistent to the update direction in the gradient ascent method. Our extensive experiments show that the iterative algorithm presented here always converges and monotonically increases the objective in each iteration. The computational cost in each iteration is of $O(n^2)$, which is similar to traditional spectral clustering algorithm.

As mentioned before, the solution is very close to the ideal class indicator matrix due to the orthonormal and nonnegative constraints. Thus the solution Q can be directly used to assign cluster labels to data points. Specifically, the i-th data point x_i is assigned to cluster label c_i as $c_i = \arg\max_k Q_{ik}$.

3.3 Initialization for the Iterative Algorithm

From the update formula in Eq. (29), we can see that if the initialization of Q is nonnegative, then Q will preserve nonnegative in the update process, and hence the nonnegative constraint of the solution is naturally satisfied. As the spectral relaxation of MinMax Cut problem is identical to the spectral relaxation of Normalized Cut problem, we initialize Q by $Q_0 + 0.2$, where Q_0 is obtained by spectral relaxation of Normalized Cut followed by K-means clustering in the eigenspace. Note that Q_0 is a cluster indicator matrix, and the initialization should not be a cluster indicator matrix (otherwise the values won't change during the iteration), thus we plus 0.2 in practice. It is worth noting that the initialization is not very sensitive, we can also use random initialization as well but the result would be more stable if using the initialization suggested here.

3.4 A New Concise Algorithm

In this section, we propose a more concise NMF algorithm to solve the MinMaxCutproblem. We start with the Eqs. (7, 8) formulation. Using Lagrangian multiplier $\Omega = \Omega^T$ to enforce $Z^T DZ = I$, we minimize

$$\mathcal{L}(Z) = J_{\text{MMC}} + \text{Tr}\ \Omega(Z^T DZ - I). \tag{30}$$

The KKT complementary slackness condition for the nonnegativity condition $Z \geq 0$ gives [noting $Z_{ik} = (\mathbf{z}_k)_i$]

$$0 = \frac{\partial \mathcal{L}}{\partial Z_{ik}} Z_{ik} = \left(-\frac{(W\mathbf{z}_k)_i}{(\mathbf{z}_k^T W \mathbf{z}_k)^2} + (DZ\Omega)_{ik} \right) Z_{ik}. \tag{31}$$

Summing over i, we obtain

$$\Omega_{kk} = \frac{1}{\mathbf{z}_k^T W \mathbf{z}_k} = \frac{1}{(Z^T WZ)_{kk}} \tag{32}$$

To find the off-diagonal elements of Ω, we look at the Lagrangian multiplier for the case where the nonnegativity constraint is ignored, which is given in Eq. (10).

From Eq. (10). we propose three strategies to obtain a symmetrized Ω as follows:

$$S_1: \quad \Omega_{lk} = \frac{(Z^TWZ)_{lk}}{(Z^TWZ)_{kk}(Z^TWZ)_{ll}} \tag{33}$$

$$S_2: \quad \Omega_{lk} = \frac{(Z^TWZ)_{lk}}{2(Z^TWZ)_{kk}^2} + \frac{(Z^TWZ)_{lk}}{2(Z^TWZ)_{ll}^2} \tag{34}$$

$$S_3: \quad \Omega_{lk} = \frac{(Z^TWZ)_{lk}}{\frac{1}{2}(Z^TWZ)_{kk}^2 + \frac{1}{2}(Z^TWZ)_{ll}^2}. \tag{35}$$

Note that all these formulas reduce to Eq. (32) when $l = k$. The gradient decent algorithm is

$$Z_{ik} \leftarrow Z_{ik} - \eta_{ik}\frac{\partial \mathcal{L}}{\partial Z_{ik}} = Z_{ik} - \eta_{ik}\left(-\frac{(WZ)_{ik}}{(Z^TWZ)_{kk}^2} + (DZ\Omega)_{ik}\right) \tag{36}$$

Setting $\eta_{ik} = Z_{ik}/(DZ\Omega)_{ik}$ leads to the update formula

$$Z_{ik} \leftarrow Z_{ik}\sqrt{\frac{(WZ)_{ik}}{(DZ\Omega)_{ik}(Z^TWZ)_{kk}^2}}. \tag{37}$$

Our extensive experiment results show that the iterative algorithm using any one of three above symmetrization strategies always converges and monotonically decreases the objective $\mathcal{L}(Z)$ in each iteration.

4 Experimental Results

In this section, we will evaluate the effectiveness of the proposed nonnegative relaxation algorithms for MinMax Cut graph clustering on eight benchmark data sets. We also compare the clustering performance of our algorithms to the traditional spectral relaxation algorithm for Ratio Cut [12] and for Normalized Cut [8] graph clustering, respectively.

4.1 Experimental Setup

Eight benchmark data sets are used in our experiments, including two UCI data sets [1] (Ecoli and Vehicle), one character data set, Binalpha, one object data set, COIL-20 [19], and four face image data sets, Yale, AT&T [20], Umist [21], and YaleB [22]. Some data sets are resized, and Table 1 summarizes the details of all data sets used in the experiments.

We use Gaussian function to construct the weight matrix W. The weight W_{ij} is defined as

$$W_{ij} = \begin{cases} \exp\left(-\frac{\|x_i - x_j\|^2}{\sigma^2}\right) & x_i \text{ and } x_j \text{ are neighbors;} \\ 0 & \text{otherwise.} \end{cases} \tag{38}$$

[1] http://www.ics.uci.edu/~mlearn/MLRepository.html

Table 1. Data set Descriptions

Data set	Size	Dimensions	Classes	Data set	Size	Dimensions	Classes
Ecoli	336	343	8	AT&T	400	644	40
Vehicle	846	18	4	Umist	575	644	20
Binalpha	1404	320	36	Coil20	1440	1024	20
Yale	165	3456	15	YaleB	2414	1024	38

Table 2. The cluster balance and clustering accuracy of Ratio Cut, Normalized Cut, and the proposed Nonnegative MinMax Cut in Section 3.2. The values after '±' are the standard deviations.

Data set	Ratio Cut		Normalized Cut		MinMax Cut	
	Balance	Accuracy	Balance	Accuracy	Balance	Accuracy
Ecoli	8.41	57.59 ± 2.31 %	6.17	55.95 ± 1.75 %	**5.31**	**58.30 ± 0.85** %
Vehicle	49.51	40.66 ± 0.57 %	56.14	43.74 ± 1.22 %	**49.13**	**44.40 ± 0.91** %
Binalpha	16.29	44.85 ± 1.96 %	24.56	44.29 ± 1.26 %	**9.26**	**46.32 ± 1.32** %
Yale	13.25	65.39 ± 3.96 %	1.57	70.06 ± 0.31 %	**1.13**	**71.52 ± 0.00** %
AT&T	11.15	70.75 ± 2.19 %	3.69	75.92 ± 1.17 %	**2.14**	**79.88 ± 1.13** %
Umist	5.82	60.00 ± 2.86 %	5.78	60.59 ± 1.14 %	**4.17**	**62.92 ± 0.93** %
Coil20	11.35	71.13 ± 4.88 %	6.11	78.19 ± 1.76 %	**5.02**	**79.09 ± 2.18** %
YaleB	49.18	38.55 ± 0.98 %	68.38	39.66 ± 1.35 %	**41.71**	**45.08 ± 1.32** %

The number of neighbors and the parameter σ should be predefined by user. In our experiments, we set the number of neighbors to be 5 (that is a commonly used number) in all data sets, and use the self-tune spectral clustering [23] method to determine the parameter σ.

4.2 Evaluation Metrics

We use the following two standard evaluation metrics to evaluate the performance for the three graph cut clustering algorithms.

Cluster Balance is defined as:

$$CB = \frac{N_{max} - N_{min}}{N_{min}},$$

where N_{max} is the number of data points in the cluster with largest size, and N_{min} is the number of data points in the cluster with smallest size. A smaller CB indicates a more balanced clustering.

Clustering Accuracy is calculated by:

$$ACC = \frac{\sum_{i=1}^{n} \delta(l_i, map(c_i))}{n},$$

where l_i is the true class label and c_i is the obtained cluster label of x_i, $\delta(x, y)$ is the delta function, and $map(\cdot)$ is the best mapping function. Note $\delta(x, y) = 1$,

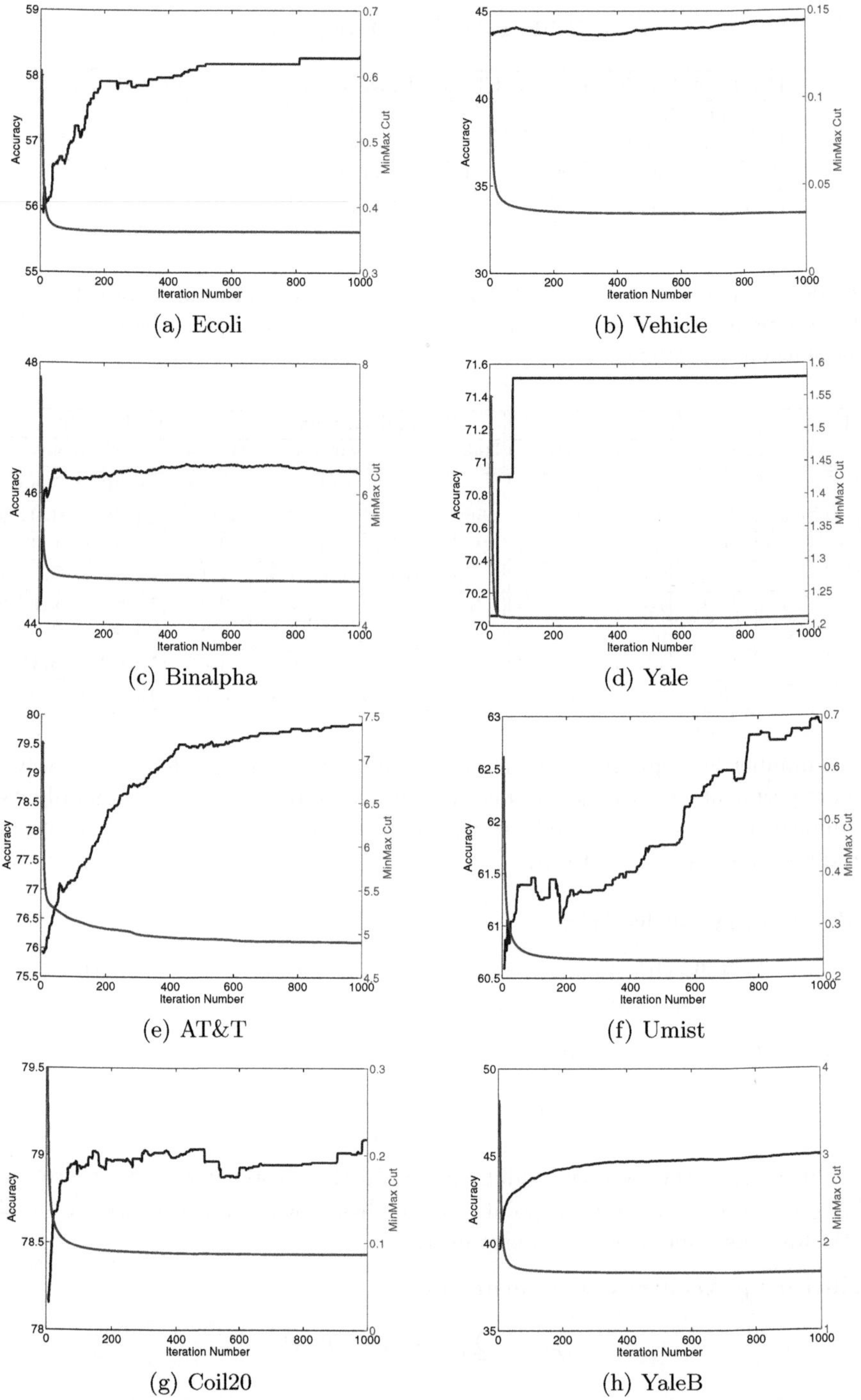

Fig. 2. The variation process of the clustering accuracy and the MinMax Cut objective value *w.r.t.* iteration number for the algorithm proposed in Section 3.2

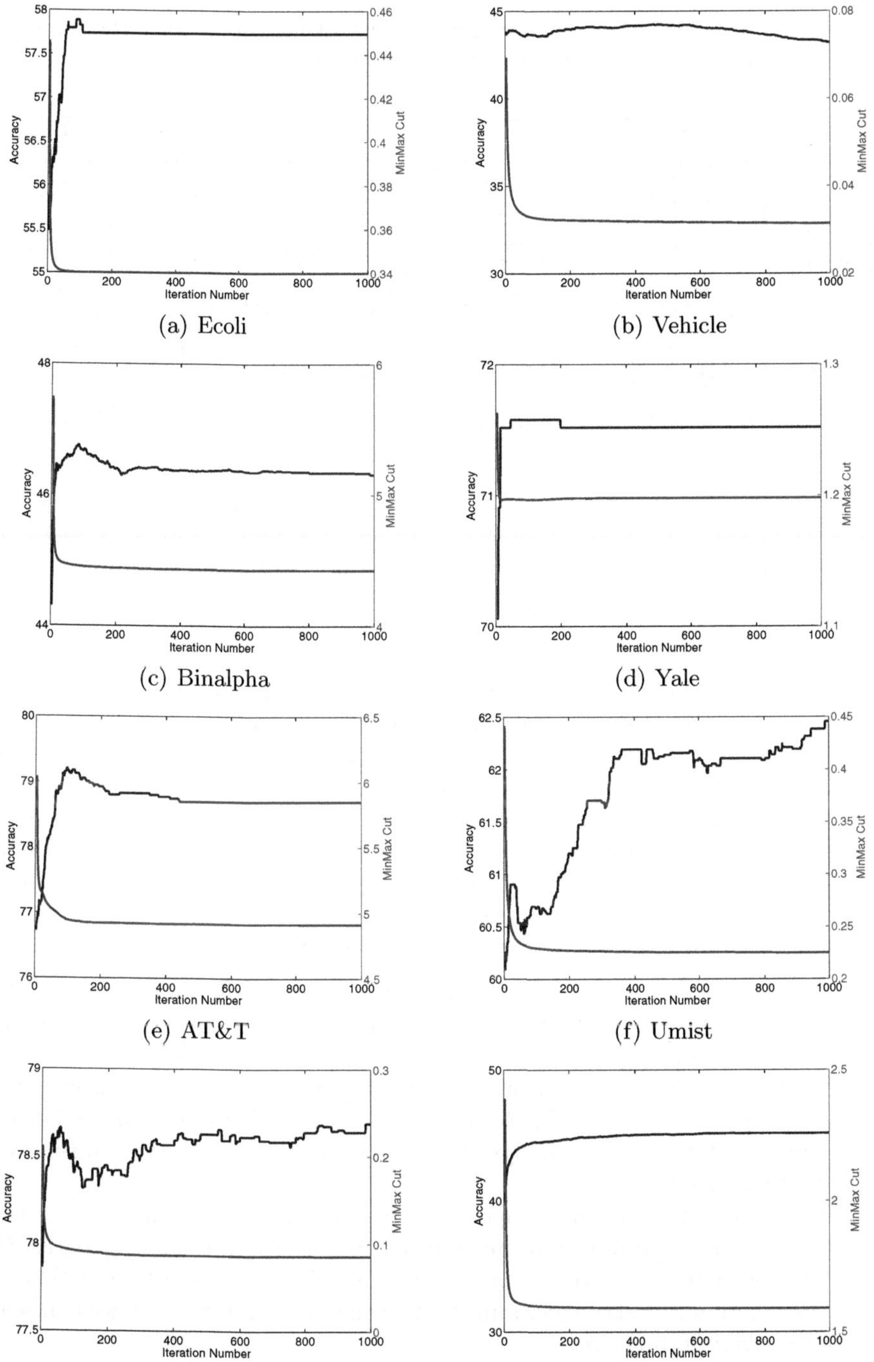

Fig. 3. The variation process of the clustering accuracy and the MinMax Cut objective value *w.r.t.* iteration number for the concise algorithm proposed in Section 3.4 with the first symmetrization strategy

Table 3. Cluster balance and clustering accuracy of the concise algorithm proposed in Section 3.4 with the three symmetrization strategies

Data set	Strategy 1		Strategy 2		Strategy 3	
	Balance	Accuracy	Balance	Accuracy	Balance	Accuracy
Ecoli	5.10	57.74 ± 0.00 %	5.10	57.74 ± 0.00 %	5.10	57.74 ± 0.00 %
Vehicle	47.63	43.20 ± 1.04 %	47.63	43.20 ± 1.04 %	47.63	43.20 ± 1.04 %
Binalpha	7.83	46.31 ± 1.89 %	7.83	46.31 ± 1.89 %	7.72	46.31 ± 1.89 %
Yale	1.13	71.52 ± 0.00 %	1.13	71.52 ± 0.00 %	1.13	71.52 ± 0.00 %
ORL	2.54	78.70 ± 1.81 %	2.54	78.70 ± 1.81 %	2.54	78.70 ± 1.81 %
Umist	4.97	62.45 ± 0.52 %	4.97	62.45 ± 0.52 %	4.97	62.45 ± 0.52 %
Coil20	5.08	78.69 ± 2.21 %	5.08	78.69 ± 2.21 %	5.08	78.69 ± 2.21 %
YaleB	37.73	45.14 ± 1.21 %	37.73	45.14 ± 1.21 %	37.73	45.14 ± 1.21 %

if $x = y$; $\delta(x, y) = 0$, otherwise. The mapping function $map(\cdot)$ matches the true class label and the obtained cluster label and the best mapping is solved by Kuhn-Munkres algorithm [24]. A larger ACC indicates a better performance.

4.3 Evaluation Results

The results of all clustering algorithms depend on the initialization. To reduce statistical variation, we run the Ratio Cut algorithm and the Normalized Cut algorithm with the same 1000 random initializations. Ten results corresponding to the 10 best objective values are selected from the 1000 runs. Then we run the proposed nonnegative MinMax Cut algorithm proposed in Section 3.2 and 3.4 using the 10 results of Normalized Cut as initialization and also obtain 10 new results. We record all the ten results and the mean results are reported in the experiments. The clustering results from three graph cut methods are reported in Table 2 and 3. From the results, we have three following observations:

1) The Normalized Cut frequently, but not always, yields more balanced clustering than Ratio Cut. The Nonnegative MinMax Cut consistently yield more balanced clustering than both Normalized Cut and Ratio Cut.
2) The Normalized Cut frequently, but not always, outperforms Ratio Cut in term of clustering accuracy. The Nonnegative MinMax Cut outperforms Normalized Cut and Ratio Cut on all eight benchmark data sets, and the improvement is significant in some cases.
3) The results of algorithm proposed in Section 3.2 are slightly better than those of algorithm proposed in Section 3.4, but the latter one is simpler and does not need to calculate the ρ in Eq. (19). We can also observe that the three symmetrization strategies almost yield the same results, thus we can select anyone of them in practice.

To evaluate the convergency and effectiveness of our iterative algorithms, we plot the variation process of the clustering accuracy and the MinMax Cut objective value defined in Eq. (30) *w.r.t.* iteration number. Figure 2 shows the variation process for the algorithm proposed in Section 3.2. From the figures, we can see that our algorithm always converges on all eight data sets, and the MinMax Cut objective value is monotonically decreased in each iteration, theoretically proving it is an interesting issue in the future work. On the other hand, the clustering accuracy tends to increase in the iteration, which indicates that the MinMax Cut objective value is consistent to the clustering accuracy, and hence is a reasonable objective for clustering problem.

Figure 3 shows the variation process for the concise algorithm proposed in Section 3.4 (as the results of the three strategies are very similar, we only show the results of the first symmetrization strategy here). From all figures, we can see that the simpler algorithms also converge on all eight data sets. The MinMax Cut objective value is monotonically decreased in each iteration, and the clustering accuracy tends to increase in each iteration.

5 Conclusions

In this paper, we proposed the nonnegative relaxation to solve the MinMax Cut graph clustering problem, and introduced efficient algorithms to solve this problem with explicit nonnegative constraint rigorously. Differing from the traditional spectral relaxation approach, the proposed nonnegative relaxation approach makes the solution very close to the ideal class indicator matrix, and can be directly used to assign cluster labels to data points. Extensive experimental results on eight benchmark data sets show that the proposed algorithms always converge and the performance is significantly improved in comparisons with the traditional spectral relaxation approach on ratio cut and normalized cut.

Acknowledgments. This research is supported by NSF-CCF 0830780, NSF-CCF 0939187, NSF-CCF 0917274, NSF-DMS 0915228, NSF-CNS 0923494.

References

1. Ng, A.Y., Jordan, M.I., Weiss, Y.: On spectral clustering: Analysis and an algorithm. In: Advances in Neural Information Processing Systems (NIPS), vol. 14, pp. 849–856 (2002)
2. Nie, F., Xu, D., Tsang, I.W., Zhang, C.: Spectral embedded clustering. In: IJCAI, pp. 1181–1186 (2009)
3. Yang, Y., Xu, D., Nie, F., Yan, S., Zhuang, Y.: Image clustering using local discriminant models and global integration. IEEE Transactions on Image Process (2010)
4. Ben-Hur, A., Horn, D., Siegelmann, H., Vapnik, V.: Support vector clustering 2, 125–137 (2001)
5. Xu, L., Neufeld, J., Larson, B., Schuurmans, D.: Maximum margin clustering. MIT Press, Cambridge (2005)
6. Zhang, K., Tsang, I., Kwok, J.: Maximum margin clustering made practical. In: ICML, Corvallis, Oregon, USA (2007)

7. Li, Y., Tsang, I., Kwok, J.T., Zhou, Z.: Tighter and convex maximum margin clustering. In: AISTATS (2009)
8. Shi, J., Malik, J.: Normalized cuts and image segmentation. IEEE Transactions on PAMI 22(8), 888–905 (2000)
9. Brun, A., Park, H.J., Shenton, M.E.: Clustering fiber traces using normalized cuts. In: Barillot, C., Haynor, D.R., Hellier, P. (eds.) MICCAI 2004. LNCS, vol. 3216, pp. 368–375. Springer, Heidelberg (2004)
10. Pentney, W., Meila, M.: Spectral clustering of biological sequence data. In: AAAI (2005)
11. Ding, C.H.Q., He, X., Zha, H., Gu, M., Simon, H.D.: A min-max cut algorithm for graph partitioning and data clustering. In: ICDM, pp. 107–114 (2001)
12. Chan, P.K., Schlag, M.D.F., Zien, J.Y.: Spectral k-way ratio-cut partitioning and clustering. IEEE Trans. on CAD of Integrated Circuits and Systems 13(9), 1088–1096 (1994)
13. Cheng, C.K., Wei, Y.C.A.: An improved two-way partitioning algorithm with stable performance [vlsi]. IEEE Trans. on CAD of Integrated Circuits and Systems 10(12), 1502–1511 (1991)
14. Bollobas, B.: Random graphs (1985)
15. Chung, F., Lu, L.: Complex Graphs and Networks. Amer. Math. Society, Providence (2006)
16. Hou, C., Zhang, C., Wu, Y., Jiao, Y.: Stable local dimensionality reduction approaches. Pattern Recognition 42(9), 2054–2066 (2009)
17. Lee, D.D., Seung, H.S.: Algorithms for non-negative matrix factorization. In: NIPS, pp. 556–562. MIT Press, Cambridge (2001)
18. Li, T., Ding, C.H.Q.: The relationships among various nonnegative matrix factorization methods for clustering. In: Perner, P. (ed.) ICDM 2006. LNCS (LNAI), vol. 4065, pp. 362–371. Springer, Heidelberg (2006)
19. Nene, S.A., Nayar, S.K., Murase, H.: Columbia object image library (COIL-20), Technical Report CUCS-005-96, Columbia University (1996)
20. Samaria, F.S., Harter, A.C.: Parameterisation of a stochastic model for human face identification. In: 2nd IEEE Workshop on Applications of Computer Vision, pp. 138–142 (1994)
21. Graham, D.B., Allinson, N.M.: Characterizing virtual eigensignatures for general purpose face recognition. in face recognition: From theory to applications. NATO ASI Series F, Computer and Systems Sciences 163, 446–456 (1998)
22. Georghiades, A., Belhumeur, P., Kriegman, D.: From few to many: Illumination cone models for face recognition under variable lighting and pose. IEEE Transactions on PAMI 23(6), 643–660 (2001)
23. Zelnik-Manor, L., Perona, P.: Self-tuning spectral clustering. In: NIPS (2004)
24. Lovász, L., Plummer, M.: Matching theory. Akadémiai Kiadó, Budapest (1986)

Integrating Constraint Programming and Itemset Mining

Siegfried Nijssen and Tias Guns

K.U. Leuven, Celestijnenlaan 200A, B-3001 Leuven, Belgium
{Siegfried.Nijssen,Tias.Guns}@cs.kuleuven.be

Abstract. Over the years many pattern mining tasks and algorithms have been proposed. Traditionally, the focus of these studies was on the efficiency of the computation and the scalability towards very large databases. Little research has however been done on a general framework that encompasses several of these problems. In earlier work we showed how constraint programming (CP) can offer such a general framework; unfortunately, however, we also found that out-of-the-box CP solvers lack the efficiency and scalability achieved by specialized itemset mining systems, which could discourage their use. Here we study the question whether a framework can be built that inherits the generality of CP systems and the efficiency of specialized algorithms. We propose a CP-based framework for pattern mining that avoids the redundant representations and propagations found in existing CP systems. We show experimentally that an implementation of this framework performs comparable to specialized itemset mining systems; furthermore, under certain conditions it lists itemsets with polynomial delay, which demonstrates that it also is a promising approach for analyzing pattern mining tasks from more theoretical perspectives. This is illustrated on a graph mining problem.

1 Introduction

Constraint-based pattern mining is a topic that has been studied extensively. Popular examples are frequent and closed itemset mining [1,16,11,9,7,15], but also more general constraints have been studied [10,6,4]. Patterns can be used directly, or may serve as an intermediate step when building classifiers or compressing data, in which case additional constraints can be imposed.

The main focus of most studies on pattern mining is on how to efficiently compute the set of solutions, given a specific set of constraints. The frequent itemset mining implementation challenge (FIMI [9]) was organized to fairly compare the runtime efficiency and memory consumption of different algorithms that were proposed over the years. An overwhelming number of aspects have been studied, ranging from algorithmic aspects, such as breadth-first or depth-first search; data structures, such as using row-oriented (horizontal), column-oriented (vertical), or tree-based representations; as well as implementation aspects, such as the most efficient way to represent sets and count their cardinality.

An issue which received considerably less attention is how *generally applicable* some of the algorithms and their proposed optimizations are. Indeed, even

J.L. Balcázar et al. (Eds.): ECML PKDD 2010, Part II, LNAI 6322, pp. 467–482, 2010.

though a number of systems have been developed which support multiple constraints (see [10,6,7,4], for instance), usually the constraint language is limited to a small number of primitives that are hard-coded in the algorithm; no support is provided for adding constraints other than post-processing the patterns. We proposed an alternative approach in previous work [8], which relies on the use of existing *constraint programming* systems. These are generally applicable constraint satisfaction solvers developed by the artificial intelligence community [13,14]. They are commonly used to solve complex problems such as planning, scheduling and resource allocation. We showed that several well-known pattern mining problems can be modeled using constraint programming primitives. At the same time, however, we found that state-of-the-art CP systems often perform an order of magnitude worse than well-known itemset mining algorithms on popular itemset mining tasks such as closed and frequent itemset mining. CP systems were only competitive on tasks characterized by a large number of restrictive constraints. This performance gap could become troublesome for very large datasets, as well as when using very low frequency thresholds. To make constraint programming a viable alternative to specialized algorithms, we believe that it is desirable that the performance gap is reduced.

In this paper we study this problem. We perform an analysis which shows that existing CP systems represent data in a highly redundant way and propagation —the key computational mechanism in CP systems— is more often performed than necessary. To address this problem, we propose several key changes in CP systems: (1) we propose to use representations of data common in data mining in CP; (2) we propose that propagators share these representations; (3) we propose a mechanism through which propagators can share inferred constraints with each other; this allows to eliminate redundant computations by the propagators.

The resulting system maintains the principles of CP systems while being far more efficient than out-of-the-box constraint solvers. In particular, we will show that our CP system also achieves polynomial delay on mining problems that were only recently shown to have such complexity [3,2]; we show that CP provides an alternative framework for deriving algorithms of low computational complexity and illustrate this on a problem in graph mining. This observation is of interest as it shows that CP can be used as fundamental methodology for reasoning about data mining problems, including their computational complexity.

The paper is organized as follows. In Section 2 we summarize the problems of frequent itemset mining and constraint programming and their basic principles. In Section 3 we summarize our earlier work on modeling itemset mining in a CP framework and we identify the bottlenecks in current CP solvers by comparing them with traditional mining algorithms. In Section 4 we present our integrated approach to solving these problems. Section 5 analyses the theoretical complexity of our system while Section 6 studies it experimentally; Section 7 concludes.

2 Itemset Miners and CP Systems

Before we study how to integrate itemset miners and constraint programming systems, let us briefly summarize their basic principles.

2.1 Frequent Itemset Mining

Problem Formulation. Let $\mathcal{I} = \{1, \dots, m\}$ be a set of items, and $\mathcal{T} = \{1, \dots, n\}$ a set of transactions; an itemset database $\mathcal{D}$ is a binary matrix of size $n \times m$. Given such a database, we can define a function $\varphi : 2^{\mathcal{I}} \to 2^{\mathcal{T}}$ which maps an itemset I to a set of transactions as follows:

$$\varphi(I) = \{t \in \mathcal{T} | \forall i \in I : \mathcal{D}_{ti} = 1\}. \tag{1}$$

This set is called a *tid-set*, denoted by T. Using the above function, the frequent itemset mining problem can be formulated as the problem of finding the set:

$$\{(I, T) \mid I \subseteq \mathcal{I}, T \subseteq \mathcal{T}, q(I, T)\}$$

where

$$q(I, T) \Leftrightarrow covers(I, T) \wedge frequent(I, T) \tag{2}$$

$$covers(I, T) \Leftrightarrow (T = \varphi(I)) \tag{3}$$

$$frequent(I, T) \Leftrightarrow (|T| \geq \theta) \tag{4}$$

and θ is the *minimum frequency* parameter of the problem.

Variations of the frequent itemset mining problem include closed itemset mining, maximal itemset mining and itemset mining under monotonic or fault-tolerance constraints. In general, these can be thought of as alternative definitions of $q(I, T)$ that tuples (I, T) should satisfy.

Existing Algorithms. Many algorithms have been proposed for the frequent itemset mining problem. We can distinguish three algorithmic choices: the search strategy, the representation of the data and of the sets.

Search strategy: the most well-known algorithm is the breadth-first APRIORI algorithm [1]; an alternative search strategy is a depth-first search in which each node in the search tree corresponds to an itemset. The latter strategy is taken in most of the recent algorithms for reasons of memory efficiency.

Representation of the data: the itemset database $\mathcal{D}$ can be represented in three equivalent ways:

- as a binary matrix of size $n \times m$, having entries $\mathcal{D}_{ti}$;
- as a bag of n itemsets $\mathcal{D}_t$, each of which represents a transaction with $\mathcal{D}_t = \{i \in \mathcal{I} \mid \mathcal{D}_{ti} = 1\}$. This is referred to as a *horizontal* representation of the binary matrix.
- as a bag of m tid-sets $\mathcal{D}_i^{\mathsf{T}}$ (one for each an item), where $\mathcal{D}^{\mathsf{T}}$ is the transpose of matrix $\mathcal{D}$. Each $\mathcal{D}_i^{\mathsf{T}}$ contains a set of transaction identifiers such that $\mathcal{D}_i^{\mathsf{T}} = \{t \in \mathcal{T} \mid \mathcal{D}_{ti} = 1\}$. This is referred to as a *vertical representation* of the binary matrix.

More complex representations also exist: FP-Growth [11], for instance, represents the itemset database more compactly in a prefix-tree. We do not consider this representation here.

Algorithm 1. Eclat(I,T,I_{pos},$\mathcal{D}$)

Output I
$I'_{pos} = \emptyset$
for all $i \in I_{pos}, i > \max(I)$ **do**
  if $|T \cap \mathcal{D}_i^{\mathsf{T}}| \geq \theta$ **then**
    $I'_{pos} := I'_{pos} \cup \{i\}$
  end if
end for
for all $i \in I'_{pos}$ **do**
  Eclat($I \cup \{i\}$,$T \cap \mathcal{D}_i^{\mathsf{T}}$,$I'_{pos}$,$\mathcal{D}$)
end for

Algorithm 2. Constraint-Search(D)

D :=propagate(D)
if D is a false domain **then**
  return
end if
if $\exists v \in \mathcal{V} : |D(v)| > 1$ **then**
  $v := \arg\min_{v \in \mathcal{V}, D(v)>1} f(v)$
  $D_p :=$ split($D(v)$)
  Constraint-Search($D \cup \{v \mapsto D_p\}$)
  Constraint-Search($D \cup \{v \mapsto D - D_p\}$)
else
  Output solution
end if

Representation of sets: tid-sets such as T and $\mathcal{D}_i^{\mathsf{T}}$ can be represented in multiple ways. One is a *sparse representation*, consisting of a list of transaction identifiers included in the set (which is called a *positive representation*) or not included in the set (which is called a *negative representation*); the size of this representation changes as the number of elements in the set changes. The other is a *dense representation*, in which case the set is represented as a boolean array. Large sets may require less space in this representation.

A representative depth-first algorithm is the ECLAT algorithm [16], which uses a vertical representation of the data; it can be implemented both for sparse and dense representations. The main observation which is used in ECLAT is the following:

$$\varphi(I) = \bigcap_{i \in I} \varphi(\{i\}) = \bigcap_{i \in I} \mathcal{D}_i^{\mathsf{T}}. \tag{5}$$

This allows for a search strategy depicted in Algorithm 1. The initial call of this algorithm is for $I = \emptyset$, $T = \mathcal{T}$ and $I_{pos} = \mathcal{I}$. The depth-first search strategy is based on the following properties. (1) If an itemset I is frequent, but itemset $I \cup \{i\}$ is infrequent, all sets $J \supseteq I \cup \{i\}$ are also infrequent. In I_{pos} we maintain those items which can be added to I and yield a frequent itemset. (2) If we add an item to an itemset, we can calculate the tid-set of the resulting itemset incrementally from the tid-set of the original itemset as follows: $\varphi(I \cup \{i\}) = \varphi(I) \cap \mathcal{D}_i^{\mathsf{T}}$. (3) To avoid an itemset I from being generated multiple times, an order is imposed on the items. We do not add items to an itemset I which are lower than the highest item already in the set.

2.2 Constraint Programming

Problem Formulation Constraint programming is a declarative programming paradigm for solving constraint satisfaction (CSP) and optimization problems. A CSP $\mathcal{P} = (\mathcal{V}, D, \mathcal{C})$ is specified by

- a finite set of variables $\mathcal{V}$;
- an initial domain $D(V)$ for every variable $V \in \mathcal{V}$;
- a finite set of constraints $\mathcal{Q}$.

A *constraint* $q(V_1, \ldots, V_k) \in \mathcal{Q}$ is a boolean function from variables $\{V_1, \ldots, V_k\} \subseteq \mathcal{V}$. A domain usually consists of a finite set of values. A domain D' is called *stronger* than the initial domain D if $D'(V) \subseteq D(V)$ for all $V \in \mathcal{V}$; a variable $V \in \mathcal{V}$ is called *fixed* if $|D(V)| = 1$. A solution to a CSP is a domain D' that (1) fixes all variables ($\forall V \in \mathcal{V} : |D'(V)| = 1$) (2) satisfies all constraints ($\forall q(V_1, \ldots, V_k) \in \mathcal{Q} : q(D'(V_1), \ldots, D'(V_k)) = 1$); (3) is stronger than original domain D (guaranteeing that every variable has a value from its initial domain).

Existing Algorithms. Most constraint programming systems perform depth-first search. A general outline is given in Algorithm 2 [14]. Branches of a node of the search tree are obtained by splitting the domain of a variable in two parts (line 7); for boolean variables, split($\{0, 1\}$)=$\{0\}$ or split($\{0, 1\}$)=$\{1\}$. The search backtracks when a violation of constraints is found (line 2). The search is further optimized by carefully choosing the variable that is fixed next (line 6); a function $f(V)$ scores each variable; the highest ranked is branched on. For instance, $f(V)$ may count the number of constraints the variable is involved in.

The main concept used to speed-up the search is constraint propagation (line 1). Propagation reduces the domains of variables such that the domain remains consistent. In a consistent domain a value d does not occur in the domain of a variable V if it can be determined that there is no solution D' in which $D'(V) = \{d\}$. This way, propagation effectively reduces the size of the search tree, avoiding backtracking as much as possible and hence speeding up the search.

To maintain consistencies *propagators* are used (sometimes also called propagation or filtering rules). A propagator takes as input a domain and outputs a stronger domain. For instance, for a constraint $V < W$ with $D(V) = \{1, 2\}$ and $D(W) = \{1, 4\}$, the propagator may output $D(V) = \{1, 2\}$ and $D(w) = \{4\}$. Propagation continues until a *fixed point* is reached in which the domain does not change any more.

A key ingredient of CP systems is that propagators are evaluated independently of each other; all communication between propagators occurs through the variables and, in some systems, by the insertion of (derived) constraints in the constraint store. This is what allows constraints to be combined and reused across different models.

There are many different CP systems; we identify the following differences:

Types of variables: the variables are at the core of the solver. Many solvers implement integer variables, of which the domains can be represented in two ways: representing every element in the domain separately or saving only the *bounds* of the domain, namely the minimum and maximum value the variable can still take.

Supported constraints: related to the types of variables, the supported constraints determine the problems that can be modeled.

Propagator activation: a constraint is defined on a set of variables. When one of the variables' domains changes, a propagator for the constraint needs to be activated. A common strategy is to tell the propagator which domain changed (in CP, this strategy is known as AC-3); another strategy is to also tell the propagator how the domain changes (this strategy is known as AC-5 [13]). The latter strategy is useful to avoid activating propagators that cannot propagate certain variable assignments.

3 Frequent Itemset Mining in CP Systems

We briefly summarize the CP formulation presented in our previous work [8]. We then study how CP solvers and itemset mining algorithms differ in the properties introduced in the previous section. In the next section we will use the best of both worlds to develop a new algorithm.

Problem Formulation. To model itemset mining in a CP system, we use two sets of boolean variables:

- a variable I_i for each item i, which is 1 if the item is included in the solution and 0 otherwise. The vector $I = (I_1, \ldots, I_m)$ represents an itemset.
- a variable T_t for each transaction t, which is 1 if the transaction is in the solution and 0 otherwise. The vector $T = (T_1, \ldots, T_n)$ represents a tid-set.

A solution hence represents one itemset I with corresponding tid-set T. To find all frequent itemsets, we need to iterate over all solutions satisfying the following constraints:

$$covers(I,T) \Leftrightarrow \forall t \in \mathcal{T} : T_t = 1 \leftrightarrow \sum_{i \in \mathcal{I}} I_i(1 - \mathcal{D}_{ti}) = 0. \tag{6}$$

$$frequent(I,T) \Leftrightarrow \forall i \in \mathcal{I} : I_i = 1 \rightarrow \sum_{t \in \mathcal{T}} T_t \mathcal{D}_{ti} \geq \theta. \tag{7}$$

Constraint (6) is a reformulation of the constraint in equation (4); constraint (7) is derived from a combination of the original coverage and frequency constraints, as follows. We observe that in a solution (I,T): $\forall i \in I : |T| = |T \cap \varphi(\{i\})|$, as $T = \varphi(I) \subseteq \varphi(\{i\})$, and therefore that

$$\forall i \in I : |T \cap \varphi(\{i\})| \geq \theta \Longleftrightarrow \forall i \in \mathcal{I} : I_i = 1 \rightarrow \sum_{t \in \mathcal{T}} T_t \mathcal{D}_{ti} \geq \theta.$$

Other well-known itemset mining problems can be formalized in a similar way. An overview is provided in Table 1; the correctness of these formulas was proved in [8]. The general type of constraint that we used is of the following form:

$$\forall x \in \mathcal{X} : X_x = b \leftarrow \sum_{y \in \mathcal{Y}} Y_y d_{xy} \lesseqgtr \theta. \tag{8}$$

where $b, d_{xy} \in \{0,1\}$ are constants, $\lesseqgtr \in \{\leq, =, \geq\}$ are comparison operators, and each X_x and Y_y is a boolean variable in the CP model. This constraint is called a *reified summation constraint*. Reified summation constraints are available in most CP systems.

Table 1. Formalizations of the primitives of common itemset mining problems in CP; frequent closed itemset mining is for instance formulated by $covers(I,T) \wedge closed(I,T) \wedge frequent(I,T)$; maximal frequent itemset mining by $covers(I,T) \wedge maximal(I,T)$.

Constraint	Reified sums	Matrix notation
$covers(I,T)$	$\forall t \in \mathcal{T} : T_t = 1 \leftrightarrow \sum_{i \in \mathcal{I}} I_i(1 - \mathcal{D}_{ti}) = 0$	$\boldsymbol{T} \leq \mathbf{1}_{=0}((1-\mathcal{D})\boldsymbol{I})$ ❶ and $\boldsymbol{T} \geq \mathbf{1}_{=0}((1-\mathcal{D})\boldsymbol{I})$ ❷
$frequent(I,T)$	$\forall i \in \mathcal{I} : I_i = 1 \rightarrow \sum_{t \in \mathcal{T}} T_t \mathcal{D}_{ti} \geq \theta$	$\boldsymbol{I} \leq \mathbf{1}_{\geq \theta}(\mathcal{D}^{\mathsf{T}}\boldsymbol{T})$ ❸
$closed(I,T)$	$\forall i \in \mathcal{I} : I_i = 1 \leftrightarrow \sum_{t \in \mathcal{T}} T_t(1 - \mathcal{D}_{ti}) = 0$	$\boldsymbol{I} \leq \mathbf{1}_{=0}((1-\mathcal{D})\boldsymbol{T})$ ❹ and $\boldsymbol{I} \geq \mathbf{1}_{=0}((1-\mathcal{D})\boldsymbol{T})$ ❺
$\delta - closed(I,T)$	$\forall i \in \mathcal{I} : I_i = 1 \leftrightarrow \sum_{t \in \mathcal{T}} T_t(1 - \delta - \mathcal{D}_{ti}) \leq 0$	$\boldsymbol{I} = \mathbf{1}_{\leq 0}((1-\delta-\mathcal{D})\boldsymbol{T})$
$maximal(I,T)$	$\forall i \in \mathcal{I} : I_i = 1 \leftrightarrow \sum_{t \in \mathcal{T}} T_t \mathcal{D}_{ti} \geq \theta$	$\boldsymbol{I} \geq \mathbf{1}_{\geq \theta}(\mathcal{D}^{\mathsf{T}}\boldsymbol{T})$ ❻ and ❸
$minsize(I,T)$	$\forall t \in \mathcal{T} : T_t = 1 \rightarrow \sum_{i \in \mathcal{I}} I_i \mathcal{D}_{ti} \geq \theta'$	$\boldsymbol{T} \leq \mathbf{1}_{\geq \theta'}(\mathcal{D}\boldsymbol{I})$ ❼
$mincost(I,T)$	$\forall t \in \mathcal{T} : T_t = 1 \rightarrow \sum_{i \in \mathcal{I}} I_i \mathcal{D}_{ti} c_i \geq \theta'$	$\boldsymbol{T} \leq \mathbf{1}_{\geq \theta'}((\mathcal{D}C)\boldsymbol{I})$ ❼

CP compared to itemset mining algorithms We use the Gecode constraint programming system [14] as a representative CP system. It is one of the most prominent open and extendable constraint programming systems, and is known to be very efficient. We will again use the ECLAT algorithm as representative itemset mining algorithm. To recapitulate from Section 2.1, key choices for itemset mining algorithms are the search strategy and the representation of the data.

In both Gecode and ECLAT search happens depth-first, but in Gecode the search tree is binary: every node corresponds to setting a boolean variable to 0 or 1; in ECLAT on the other hand, every node in the search tree corresponds to one itemset, and the search only adds items to the set.

The database is not explicitly represented in the CP model; instead rows and columns are spread over the constraints. Studying constraints (6) and (7) in detail, we observe that for every transaction there is a reified summation constraint containing all the items not in this transaction; for every item there is a constraint containing all transactions containing this item. Furthermore, to activate a propagator when the domain of a variable is changed, for every item and transaction variable there is a list containing propagators depending on it. Overall, the data is hence stored 4 times, both in horizontal and vertical representations, and in positive and negative representations. In ECLAT a database is only stored in one representation, which can be tuned to the type of data at hand (for instance, using positive or negative representations). Although in terms of worst case space complexity the performance is hence the same, the amount of overhead in the CP system is much higher.

Finally, we wish to point out that in CP, all constraints are independent. The frequency summation constraints check for every item whether it is still frequent, even if it is already included in the itemset. As the constraints are independent, this redundant computation cannot be avoided.

Itemset mining compared to CP solvers. In Section 2.2 we identified types of variables, supported constraints and propagator activation as key differences between solvers. In Gecode, every item and every transaction is represented by an individual boolean variable. The representation of variables is hence dense. The representation of the constraints, on the other hand, is sparse. Itemset mining algorithms like ECLAT are more flexible; the choice of representation for sets is

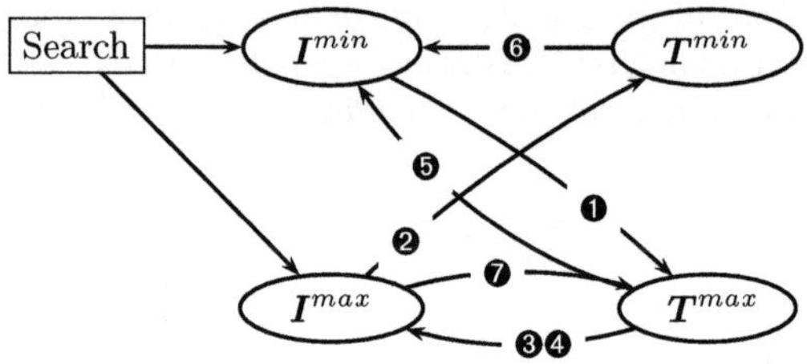

Fig. 1. Propagators for itemset mining as functions between bounds of variables; the search is assumed to branch over items

often left open. Considering supported constraints, the original ECLAT algorithm deals mainly with one type of constraint: the minimum frequency constraint. For other types of constraints, such as closedness constraints, the algorithm needs to be (and has been) extended. Gecode on the other hand supports many different constraints out-of-the-box, and imposes no restriction on which ones to use, or how to combine them.

To compare propagator activation in ECLAT and Gecode, let us consider when the change of a variable is propagated. Figure 1 visualizes the propagation that is necessary for the constraints presented in Table 1. In the case of frequent itemset mining, only propagations ❶ and ❸ are strictly needed; propagation ❷ is useless as the lower-bound of the transaction variables (representing transactions that are certainly covered) is never used later on. Indeed ECLAT only performs propagation ❶ and ❸. However, Gecode activates propagators more often (to no possible beneficial effect in itemset mining). The reason is that in Gecode a propagator is activated whenever a variable's domain is changed, independent of whether the upper-bound or lower-bound changed (similar to AC-3). Hence, if an item's domain changes after frequency propagation, the coverage propagator will be triggered again already due to the presence of constraint ❶ [1].

Conclusion. Overall, we can observe that the search procedures are similar at a high level, but that the constraint programming system faces significant overhead in data representation, data maintenance, and constraint activation.

4 An Integrated Approach

We propose an integrated approach that builds on the generality of CP, but aims to avoid the overhead of existing systems. Our algorithm implements the basic constraint search algorithm depicted in Algorithm 2, augmented with the following features: (1) a boolean vector as basic variable type; (2) support for multiple data representations; (3) a general matrix constraint that encompasses itemset mining constraints; (4) an auxiliary store in which facts are shared; (5) efficient propagators for matrix constraints. To the best of our knowledge, there is currently no CP system that implements these features. In particular, our use

[1] If we would enable propagation ❷, the domain of T may change; this would lead to even further (useless) propagation towards the item variables.

of an auxiliary store that all constraints can access is uncommon in existing CP systems and mainly motivated by our itemset mining requirements.

Boolean vectors Our first choice is to use boolean vectors as basic variable types; such a vector can be interpreted as a subset of a finite set of elements, where the ith bit indicates if element i is part of the set. Constraints will be posted on the set of items I and the set of transactions T as a whole. We put no restriction on whether to implement the boolean vectors in a sparse or dense representation. The domain of a boolean vector B is represented by its bounds in two boolean vectors, B^{min} and B^{max}, as in [6]. We split a domain on one boolean.

Data representation. When posting constraints, we support multiple matrix representations, such as vertical, horizontal, positive and negative representations. Constraints will operate on all representations, but more efficient propagators will be provided for some. Note that certain matrices can be seen as *views* on other matrices: for instance, $\mathcal{D}^{\mathsf{T}}$ is a *horizontal* view on a *vertically* represented matrix $\mathcal{D}$. Doing this avoids that different constraints need to maintain their own representations of the data, and hence reduces the amount of redundancy.

General matrix constraint We now reformulate the reified summation constraint of Equation (8) on boolean vectors. The general form of the reified matrix constraint is the following:

$$\boldsymbol{X} \geq_1 \mathbf{1}_{\geq_2 \theta}(\mathcal{A} \cdot \boldsymbol{Y}); \tag{9}$$

both the first $\geq_1$ and second $\geq_2$ can be replaced by a $\leq$; $\boldsymbol{X}$ and $\boldsymbol{Y}$ are boolean column vectors; $\mathcal{A}$ is a matrix; $\cdot$ denotes the traditional matrix product; $\mathbf{1}_{\geq\theta}$ is an *indicator function* which is applied element-wise to vectors: in this case, if the ith component of $\mathcal{A} \cdot \boldsymbol{Y}$ exceeds threshold θ, the ith component of the result is 1; otherwise it is 0. In comparing two vectors $\boldsymbol{X}_1 \geq \boldsymbol{X}_2$ it is required that every component of $\boldsymbol{X}_1$ is not lower than the corresponding component of $\boldsymbol{Y}_2$.

For instance, the frequency constraint can now be formalized as follows:

$$\boldsymbol{I} \leq \mathbf{1}_{\geq\theta}(\mathcal{D}^{\mathsf{T}} \cdot \boldsymbol{T}), \tag{10}$$

which expresses that only if the vector product of the transaction vector $\boldsymbol{T}$ with column vector i of the data exceeds threshold θ, the ith component of vector $\boldsymbol{I}$ can be 1.

We can reformulate many itemset mining problems using this notation, as shown in Table 1. In this notation we assume that $(x - \mathcal{A})$ yields a matrix $\mathcal{A}'$ in which $\mathcal{A}'_{ij} = x - \mathcal{A}_{ij}$; furthermore, C represents a diagonal matrix in which the costs of items are on the diagonal.

As with any CP system, other constraints and even other variable types can be added; for the typical itemset mining settings discussed in this paper no other constraints are needed.

Propagation framework. A propagator for the general matrix constraint above can be thought of as a function taking boolean vectors (representing bounds) as input, and producing a boolean vector (representing another bound) as output.

This is similar to how binary constraints are dealt with in the AC-5 propagation strategy. For example, a propagator for the frequency constraint in Equation 10 is essentially a function that derives $\boldsymbol{I}^{max}$ from $\boldsymbol{T}^{max}$. Figure 1 lists all these relationships for the constraints in Table 1. In our system, we activate the propagator of a constraint only if the bound(s) it takes as input have changed, avoiding useless propagation calls.

Auxiliary store. The main idea is to store a set of simple, redundant constraints that are known to hold given the constraints and the current variable assignments. These constraints are of the form $\boldsymbol{X} \subseteq \mathcal{A}_i$, where $\subseteq$ may be replaced by other set operators, $\boldsymbol{X}$ is a vector variable in the constraint program, and $\mathcal{A}_i$ may also be a column in the data. Propagators may insert such constraints in the store and may use them later on to avoid data access and the subsequent computation. Note that the worst case size of this store is $O(m+n)$.

Efficient Propagators. The general propagator for one particular choice for the inequality directions in the matrix constraint, $\boldsymbol{X} \leq \mathbf{1}_{\geq\theta}(\mathcal{A} \cdot \boldsymbol{Y})$, is the following:

$$\boldsymbol{X}^{max} \leftarrow \min(\boldsymbol{X}^{max}, \mathbf{1}_{\geq\theta}(\mathcal{A} \cdot \boldsymbol{Y}^{max})), \tag{11}$$

where we presume $\mathcal{A}$ only contains non-negative entries. This propagator takes as input vector $\boldsymbol{Y}^{max}$ and computes $\boldsymbol{X}^{max}$ as output, possibly exploiting the value $\boldsymbol{X}^{max}$ currently has. Upon completion of the propagator, it should be checked if $\boldsymbol{X}^{min} \leq \boldsymbol{X}^{max}$; otherwise a contradiction is found and the search will backtrack.

We provide the specialized propagators of the coverage and frequency constraint on a vertical representation of the data below. We lack the space to provide details for other propagators.

Coverage: The coverage constraint is $\boldsymbol{T} \leq \mathbf{1}_{\leq 0}(\mathcal{A} \cdot \boldsymbol{I})$, where $\mathcal{A} = 1 - \mathcal{D}$, i.e. a binary matrix represented vertically in a negative representation where $\mathcal{D}$ is the positive representation. The propagator should evaluate $\boldsymbol{T}^{max} \leftarrow \min(\boldsymbol{T}^{max}, \mathbf{1}_{\leq 0}(\mathcal{A} \cdot \boldsymbol{I}^{min}))$. Taking inspiration from itemset mining (Equation (5)), we can also evaluate this propagator as follows:

$$T^{max} \leftarrow T^{max} \cap \Big(\bigcap_{i \in I^{min}} \mathcal{D}_i^{\mathsf{T}} \Big). \tag{12}$$

Observe that the propagator uses the vertical matrix representation $\mathcal{D}^{\mathsf{T}}$ directly, without needing to compute the negative matrix $\mathcal{A}$. The propagator skips columns which are not currently included in the itemset ($i \in I^{min}$). After having executed this propagator, we know for a number of items $i \in \mathcal{I}$ that $T \subseteq \mathcal{D}_i^{\mathsf{T}}$. We will store this knowledge in the auxiliary constraint store to avoid recomputing it. The propagator is given in Algorithm 3.

Frequency: The minimum frequency constraint is $\boldsymbol{I} \leq \mathbf{1}_{\geq\theta}(\mathcal{A} \cdot \boldsymbol{T})$, where $\mathcal{A}$ is here a binary matrix horizontally represented; hence $\mathcal{A} = \mathcal{D}^{\mathsf{T}}$, for the vertically represented matrix $\mathcal{D}$. The propagator should evaluate:

$$\boldsymbol{I}^{max} \leftarrow \min(\boldsymbol{I}^{max}, \mathbf{1}_{\geq\theta}(\mathcal{A} \cdot \boldsymbol{T}^{max})).$$

Algorithm 3. PropCoverage(P,D)

Let $\boldsymbol{I}^{min}$ point to input of P in D
Let $\boldsymbol{T}^{max}$ point to output of P in D
for all $i \in \boldsymbol{I}^{min}$ **do**
  if $(\boldsymbol{T}^{max} \subseteq \mathcal{D}_i^{\mathsf{T}}) \notin$ store **then**
    $\boldsymbol{T}^{max} := \boldsymbol{T}^{max} \cap \mathcal{D}_i^{\mathsf{T}}$
    Add $(\boldsymbol{T}^{max} \subseteq \mathcal{D}_i^{\mathsf{T}})$ to store
  end if
end for

Algorithm 4. PropFrequency(P,D)

Let $\boldsymbol{T}^{max}$ point to input of P in D
Let $\boldsymbol{I}^{max}$ point to output of P in D
$F := |\boldsymbol{T}^{max}|$
for all $i \in \boldsymbol{I}^{max}$ **do**
  if $(\boldsymbol{T}^{max} \subseteq \mathcal{D}_i^{\mathsf{T}}) \notin$ store **then**
    $F' := |\mathcal{D}_i^{\mathsf{T}} \boldsymbol{T}^{max}|$
    if $F' < \theta$ **then** $I_i^{max} := 0$
    if $F' = F$ **then**
      Add $(\boldsymbol{T}^{max} \subseteq \mathcal{D}_i^{\mathsf{T}})$ to store
  end if
end for

We can evaluate this constraint by computing for every $i \in I^{max}$ the size of the vector product $|\mathcal{D}_i^{\mathsf{T}} \cdot T^{max}|$. If this number is lower than θ, we can set $I_i^{max} = 0$.

We can speed up this computation using the auxiliary constraint store. If for an item i we know that $T^{max} \subseteq \mathcal{D}_i^{\mathsf{T}}$, then $|\mathcal{D}_i^{\mathsf{T}} \cdot T^{max}| = |T^{max}|$ so we only have to compute $|T^{max}| \geq \theta$. The propagator can use the auxiliary store for this; see Algorithm 4. This is the same information that the coverage constraint uses; hence it can happen at some point during the search that the propagators do not need to access the data any longer. Storing and maintaining the additional information in the auxiliary constraint store will require additional $O(m+n)$ time and space for each itemset (as we need to store an additional transaction set or itemset for each itemset); the potential gain in performance in the frequency propagators is $O(mn)$, as we can potentially avoid having to consider all elements of the data again.

5 Analysis

In this section we study the complexity of our constraint programming system on itemset mining tasks. Assuming that propagators are evaluated in polynomial space, the space complexity of the approach is polynomial, as the search proceeds depth-first and no information is passed from one branch of the search tree to another. In general, polynomial time complexity is not to be expected since CP systems can be used to solve NP complete problems.

For certain tasks, however, we can prove *polynomial delay* complexity, i.e. before and after each solution is printed, the computation time is polynomial in the size of the input. In particular, the CP-based approach provides an alternative to the recent theory of accessible set systems [2], which was used to prove that certain closed itemset mining problems can be solved with polynomial delay.

Consider propagation of constraints in our system, if k is the sum of the lengths of the boolean vector variables in the constraint program for a problem, a fixed point will be reached in $O(k)$ iterations, as this is the maximum number of bits that may change in all iterations of a propagation phase together. If each propagator can be evaluated in polynomial time, then a call to Propagate(D) will execute in polynomial time. Using this we can prove the following.

Theorem 1. *If all propagators and the variable selection function f can be evaluated in time polynomial in the input, and each failing node in the search tree has a non-failing sibling, solutions to the constraint program will be listed with polynomial delay by Algorithm 2.*

Proof. Essentially we need to show that the number of failing leaves between two successive non-failing leaves of the search tree is polynomial in the size of the input. Assume we have an internal node, then given our assumption, independent of the order in which we consider its children, we will reach a succeeding leaf after considering $O(k)$ nodes in the depth-first search. Consider the successive non-failing leaf, we will need at most $O(k)$ steps to reach the ancestor for which we need to consider the next child. From the ancestor we will reach the non-failing leaf after considering $O(k)$ nodes.

From this theorem follows that frequent and frequent closed itemsets can be listed with polynomial delay: considering a model without the useless propagator ❷ (see Section 3), in a node's child either an item is set to 1 or to 0, changing either I^{min} or I^{max}. When I^{max} changes, no propagation will happen, which provides us one branch that does not fail, as required by the theorem.

The same procedure can also be used for more complex itemset mining problems. We illustrate this here for a simple graph mining problem introduced in [3], which illustrates at the same time how our approach can be extended to other mining problems. In addition to a traditional itemset database, in this problem setting a graph G is given in which items constitute nodes and edges connect nodes. An itemset I satisfies constraint *connected*(I) if the subgraph induced in G by the nodes in I is connected. An itemset (I, T) satisfies constraint *connectedClosed*(I, T) iff I corresponds to a complete connected component in the (possibly unconnected) subgraph induced in G by the nodes in $\psi(T) = \cap_{t \in T} \mathcal{D}_t$. We can solve this problem by adding the following elements in the CP system:

- the coverage and frequency constraint as in standard itemset mining;
- a propagator for *connected*(I), which takes $\boldsymbol{I}^{max}$ as input and $\boldsymbol{I}^{max}$ as output; given one arbitrary variable for which $I_i^{min} = 1$, it determines the connected component induced by $\boldsymbol{I}^{max}$ that i is included in, and changes the domain $I_i^{max} = 0$ for all items i not in this component. It fails if this means that for some i': $I_{i'}^{min} > I_{i'}^{max}$;
- a propagator for *connectedClosed*(I, T), which for a given set $\boldsymbol{T}^{max}$ calculates $\psi(\boldsymbol{T}^{max})$, and sets $I_i^{min} = 1$ for all items in the component the current items in $\boldsymbol{I}^{min}$ are included in, possibly leading to a failure;
- a variable selection function f which as next item to split on selects an item which is connected in G to an item for which $I_i^{min} = 1$.

The effect of the variable selection function is that the propagator for *connected*(I) will never fail in practice. Given the absence of another propagator depending on $\boldsymbol{I}^{max}$, the requirements for the theorem are fulfilled.

Table 2. Characteristics of the data used in the experiments

Name	$\lvert\mathcal{T}\rvert$	$\lvert\mathcal{I}\rvert$	Density	# Freq Patterns at $\theta = 25\%$	Pattern Rich?	Source
T10I4D100K	100000	1000	1%	1	Poor	[9]
Splice	3190	290	21%	118	Poor	[12]
Mushroom	8124	120	19%	5546	Rich	[9]
Ionosphere	351	99	50%	184990186	Rich	[12]

6 Experiments

In our experiments we aim to determine how much our specialized framework can contribute to reducing the performance gap between CP systems and itemset miners. As CP is particularly of interest when dealing with many different types of constraints, our main interest is achieving competitive behavior across multiple tasks.

Fairly comparing itemset miners is however a daunting task, as witnessed by the large number of results presented in the FIMI competition [9]. We report a limited number of settings here and refer to our website for more information, including the source code of our system[2].

We choose to restrict our comparison to the problems of frequent, closed and maximal itemset mining, as well as frequent itemset mining under a minimum size constraint. The reason is that there are a relatively large number of systems supporting all these tasks, some of which were initially developed for the FIMI competition, but have since been extended to deal with additional constraints. This makes these systems a good test case for how easily and efficiently specialized algorithms can be extended to deal with additional constraints. In particular, we used these systems: DMCP: our new constraint programming system, in which we always used a dense representation of sets and used a static order of items; FIMCP: our implementation based on Gecode [8]; PATTERNIST: the engine underneath the ConQuEst constraint-based itemset mining system [4]; LCM: an algorithm from the FIMI competition, in two versions [15]; ECLAT, ECLAT_NOR, FPGrowth and APRIORI, as implemented by Borgelt [5]. ECLAT checks maximality of an itemset by searching for related itemsets in a repository of already determined maximal itemsets; ECLAT_NOR checks the maximality of an itemset in the data. Unless mentioned otherwise, the algorithms were run with default parameters; output was requested, but written to `/dev/null`. The algorithms were run on machines with Intel Q9550 processors and 8GB of RAM. Experiments were timed out after 1800s.

Characteristics of the data sets are given in Table 2. We can make a distinction between *pattern-poor* datasets in which the number of frequent itemsets is small (the number of frequent itemsets at a minimum support of 25% is smaller than the number of items in the database; here post-processing a pre-computed set of patterns is an option for most constraints), and *pattern-rich* datasets for which

[2] `http://dtai.cs.kuleuven.be/CP4IM/`

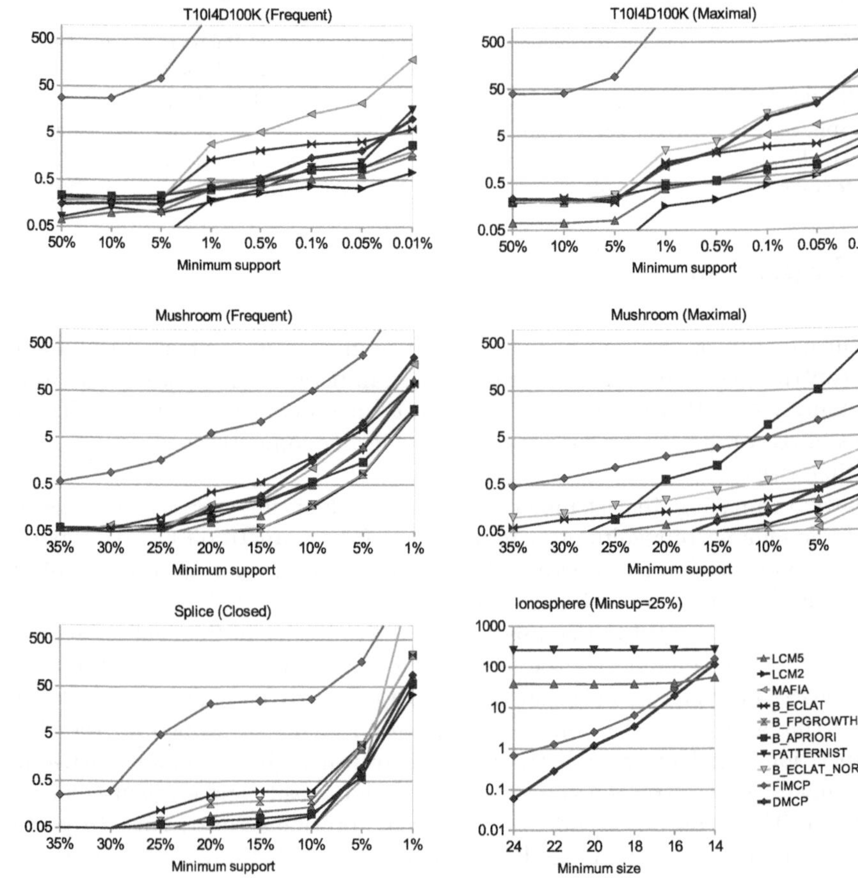

Fig. 2. Run times in seconds of several algorithms; see the main text for a discussion

the number of frequent itemsets is large. We are interested in the behavior of our system in both settings. All results are given in Figure 2.

Pattern-rich Data. On data with a large number of frequent itemsets, the use of constraints is necessary to make the mining feasible and useful. Compared to frequent itemset mining, maximal frequent itemset mining imposes more constraints; hence we would expect constraint programming to be beneficial in this setting. Indeed, on the Mushroom data we can observe that DMCP outperforms approaches such as B_APRIORI and B_ECLAT_NOR, which in this case are filtering frequent itemsets[3]. The CP-based FIMCP does not sufficiently profit from the propagation to overcome its redundant data representation.

Investigating the use of constraints further, we apply a number of minimum size constraints on the Ionosphere data. A minimum size constraint is a common

[3] The difference between B_APRIORI and B_ECLAT_NOR is mainly caused by *perfect extension pruning* [5], which we did not disable.

test case in constraint-based itemset mining, as it has a monotonicity property which is opposite to that of the frequency constraint. Some itemset mining implementations, such as LCM, have added this constraint as an option to their systems. The results show that on this type of data, LCM and the PATTERNIST system, which was designed for constraint-based mining, are (almost) not affected by this constraint; the CP based approaches, on the other hand, can effectively prune the search space as the constraint becomes more restrictive, where DMCP is again superior to FIMCP.

Pattern-poor Data. We illustrate the behavior of the systems on pattern-poor data using the T10I4D100K dataset for frequent and maximal itemset mining, and using Splice for closed itemset mining. In all cases, FIMCP suffers from its redundant representation of the sparse data. For most systems, the run time is not affected by the maximality constraint. The reason is that the number of maximal frequent itemsets is very similar to that of the total number of frequent itemsets. In particular in systems such as B_ECLAT and B_FPGROWTH, which use a repository, the maximality check gives minimal overhead. If we disable the repository (B_ECLAT_NOR), Eclat's performance is very similar to that of the DMCP system, as both systems are essentially filtering itemsets by checking maximality in the data. Similar behavior is observed for the splice data, where the difference between closed and non-closed itemsets is small. In all figures it is clear that our system operates in the same ballpark as other itemset mining systems.

7 Conclusions

In this paper we studied the differences in performance between general CP solvers and specialized mining algorithms. We focused our investigation on the representation of the data and the activation of propagators. This provided insights allowing us to create a new algorithm based on the ideas of AC-5 constraint propagation; it uses boolean vectors as basic type, supports general matrix constraints, enables multiple data representations and uses an auxiliary store inspired by the efficient constraint evaluations of itemset mining algorithms. Additionally, we demonstrated how the framework can be used for complexity analysis of mining tasks and illustrated this on a problem in graph mining. We showed experimentally that our system overcomes most performance differences.

Many questions have still been left unanswered. At the moment, we implemented the optimized propagators in a new CP system which does not support the wide range of constraints general systems such as Gecode do. Whether it is possible to include the same optimizations in Gecode is an open question. Another question is which other itemset mining strategies can be incorporated in a general constraint programming setting. An essential component of our current approach is that constraints express a relationship between items and transactions. However, other itemset mining systems, of which FP-Growth [11] is the most well-known example, represent the data more compactly by merging identical transactions during the search; as our current approach builds on transaction identifiers, this approach faces problems.

Finally, the use and possible extension of constraint programming for dealing with other data mining problems than itemset mining is the largest challenge we are currently working on.

Acknowledgements. This work was supported by a Postdoc and a project grant from the Research Foundation—Flanders, project "Principles of Patternset Mining", as well as a grant from the Institute for the Promotion and Innovation through Science and Technology in Flanders (IWT-Vlaanderen).

References

1. Agrawal, R., Mannila, H., Srikant, R., Toivonen, H., Verkamo, A.I.: Fast discovery of association rules. In: Advances in Knowledge Discovery and Data Mining, pp. 307–328. AAAI Press, Menlo Park (1996)
2. Arimura, H., Uno, T.: Polynomial-delay and polynomial-space algorithms for mining closed sequences, graphs, and pictures in accessible set systems. In: SDM, pp. 1087–1098. SIAM, Philadelphia (2009)
3. Boley, M., Horváth, T., Poigné, A., Wrobel, S.: Efficient closed pattern mining in strongly accessible set systems. In: Kok, J.N., Koronacki, J., Lopez de Mantaras, R., Matwin, S., Mladenič, D., Skowron, A. (eds.) PKDD 2007. LNCS (LNAI), vol. 4702, pp. 382–389. Springer, Heidelberg (2007)
4. Bonchi, F., Giannotti, F., Lucchese, C., Orlando, S., Perego, R., Trasarti, R.: A constraint-based querying system for exploratory pattern discovery. Inf. Syst. 34(1), 3–27 (2009)
5. Borgelt, C.: Efficient implementations of Apriori and Eclat. In: Workshop of Frequent Item Set Mining Implementations, FIMI (2003)
6. Bucila, C., Gehrke, J., Kifer, D., White, W.M.: Dualminer: A dual-pruning algorithm for itemsets with constraints. Data Min. Knowl. Discov. 7(3), 241–272 (2003)
7. Burdick, D., Calimlim, M., Flannick, J., Gehrke, J., Yiu, T.: MAFIA: A maximal frequent itemset algorithm. IEEE TKDE 17(11), 1490–1504 (2005)
8. De Raedt, L., Guns, T., Nijssen, S.: Constraint programming for itemset mining. In: KDD, pp. 204–212 (2008)
9. Goethals, B., Zaki, M.J.: Advances in frequent itemset mining implementations: report on FIMI 2003. In: SIGKDD Explorations, volume 6, pp. 109–117 (2004)
10. Han, J., Lakshmanan, L.V.S., Ng, R.T.: Constraint-based multidimensional data mining. IEEE Computer 32(8), 46–50 (1999)
11. Han, J., Pei, J., Yin, Y.: Mining frequent patterns without candidate generation. In: Proceedings of the ACM SIGMOD International Conference on Management of Data, pp. 1–12 (2000)
12. Nijssen, S., Guns, T., De Raedt, L.: Correlated itemset mining in ROC space: a constraint programming approach. In: KDD, pp. 647–656 (2009)
13. Rossi, F., van Beek, P., Walsh, T.: Handbook of Constraint Programming (Foundations of Artificial Intelligence). Elsevier Science Inc., Amsterdam (2006)
14. Schulte, C., Stuckey, P.J.: Efficient constraint propagation engines. Transactions on Programming Languages and Systems 31(1) (2008)
15. Uno, T., Kiyomi, M., Arimura, H.: LCM ver.3: collaboration of array, bitmap and prefix tree for frequent itemset mining. In: OSDM 2005: Proceedings of the 1st International Workshop on Open Source Data Mining, pp. 77–86 (2005)
16. Zaki, M.J., Parthasarathy, S., Ogihara, M., Li, W.: New algorithms for fast discovery of association rules. In: KDD, pp. 283–286 (1997)

Topic Modeling for Personalized Recommendation of Volatile Items

Maks Ovsjanikov[1] and Ye Chen[2]

[1] Stanford University
maks@stanford.edu
[2] Microsoft Corporation
yec@microsoft.com

Abstract. One of the major strengths of probabilistic topic modeling is the ability to reveal hidden relations via the analysis of co-occurrence patterns on dyadic observations, such as document-term pairs. However, in many practical settings, the extreme sparsity and volatility of co-occurrence patterns within the data, when the majority of terms appear in a single document, limits the applicability of topic models. In this paper, we propose an efficient topic modeling framework in the presence of volatile dyadic observations when direct topic modeling is infeasible. We show both theoretically and empirically that often-available unstructured and semantically-rich meta-data can serve as a link between dyadic sets, and can allow accurate and efficient inference. Our approach is general and can work with most latent variable models, which rely on stable dyadic data, such as pLSI, LDA, and GaP. Using transactional data from a major e-commerce site, we demonstrate the effectiveness as well as the applicability of our method in a personalized recommendation system for volatile items. Our experiments show that the proposed learning method outperforms the traditional LDA by capturing more persistent relations between dyadic sets of wide and practical significance.

1 Introduction

Probabilistic topic models have emerged as a natural, statistically sound method for inferring hidden semantic relations between terms in large collections of documents, e.g., [4,12,11]. Most topic-based models start by assuming that each term in a given document is generated from a hidden topic, and a document can be characterized as a probability distribution over the set of topics. Thus, learning high level semantic relations between documents and terms can be reduced to learning the topic models from a large corpus of documents. At the core of many learning methods for topic-based models lies the idea that terms that often occur together are likely to be explained by the same topic. This is a powerful idea that has been successfully applied in a wide range of fields including computer vision, e.g., [9], and shape retrieval [14].

One of the limitations of direct topic modeling, however, is that co-occurrence patterns can be very sparse and are often volatile. For example, terms cannot

J.L. Balcázar et al. (Eds.): ECML PKDD 2010, Part II, LNAI 6322, pp. 483–498, 2010.

be readily substituted by images and documents by websites in the LDA model, since the vast majority of images will only occur in a single website. Nevertheless, it is often possible to obtain large amounts of *unstructured meta-data*, annotating the input. In this paper, we show that accurate topic modeling can still be performed in these settings, and occurrence probabilities can be computed despite extreme sparsity of the input data.

Our motivating application is a personalized recommendation system of volatile items, which aims to exploit past behavior of users to estimate their preferences for a large collection of items. Our data consists of eBay buyers and their behavioral history. The set of items at eBay is large, constantly evolving (over 9 million items added every day as of August 2009), and each item is only loosely categorized by eBay sellers, while obtaining an accurate catalog for all items is a very difficult task. Therefore, the overlap between purchase and even browsing history of individual users is minimal, which greatly complicates obtaining an accurate topic model for users and items [6].

In this paper we address these challenges by leveraging the unstructured meta-data that accompanies the user-item interactions. Specifically, we map both users and items to a common latent topic space by analyzing the search queries issued by users, and then obtain user and item profiles using statistical inference on the latent model in real time. Thus, the search queries act as a link between users and items, and, remarkably, allow us to build a useful latent model without considering direct user-item interaction. We also show how to use this decomposition to efficiently update occurrence probabilities when new items or users arrive.

2 Related Work

Our work lies on the crossroads between probabilistic topic modeling and personalized recommendation (see e.g., [17] and [1] for surveys of these two fields). The primary objective of probabilistic topic models [4,12,11] and their variants, is to capture latent topical information from a large collection of discrete, often textual, data. These methods are very popular due to their relative simplicity, efficiency and the results which are often easy to interpret. However, most topic models require a relatively stable vocabulary and a significant overlap of terms across documents, which is often difficult to enforce in practice.

Personalized recommendation systems aim to recommend relevant items to users based primarily on their observed behavior, e.g., search personalization [15], Google News personalization [8], and Yahoo! behavioral targeting [7] among others. Most personalization techniques are supervised or semi-supervised, and fit the model parameters to some known user preference values. More recently, several techniques have been proposed to leverage available meta-data to build topic models, which are then used for recommendation, e.g., [2,18]. In most cases, however, users and items are coupled during topic modeling. In our setting, this can lead to overfitting in the presence of sparse data and inefficiencies if the model parameters need to be recomputed whenever a new user or item enter the system. Our method overcomes these issues by decoupling users and items when building the topic model. This not only allows us to obtain more stable topic models, but much more importantly allows us to model new users and items in real time.

Finally, our personalized recommendation system is similar, in spirit, to recently proposed Polylingual Topic Models [16], that aim to learn a consistent topic model for a set of related documents in different languages. In this work, the lack of overlap of terms across documents in different languages is countered by the fact that similar documents will have similar topic distributions, and thus the topic model can be inferred consistently across languages. This technique, however, requires pre-clustering of documents into similar sets, whereas our goal is to infer *personalized* topic models for a set of independent users without a clear cluster structure.

3 Topic Models with Triadic Observations

Although our approach is general, for clarity of exposition we concentrate on inferring unobserved user preferences for a large set of highly volatile items. In this section we describe the setting in detail, and give an overview of our method.

3.1 Motivation and Overview

Our primary goal is to infer the probability that a user u is interested in some item i. In the text setting, this is similar to obtaining the probability that a document will contain a given term. However, the set of items is extremely large and volatile, implying that the overlap between individual users' item browsing or purchasing histories is minimal. This prohibits standard user modeling as well as collaborative filtering techniques based either on finding similar users or factorizing the user-item matrix. Note, however, that in addition to the direct user-item interactions, in the majority of cases, users issue search queries to arrive at the desired items. This means that a single user-item observation can be augmented to a user-query-item triple. This augmentation only exacerbates the sparsity issue, however, since the set of triples is at least as sparse as the original user-item matrix. Therefore, we project this tensor onto the subspace of users and queries. In this space, topic modeling can be performed using standard techniques. Furthermore, since each item is equipped with the set of queries that were used to retrieve it, we use statistical inference on the set of item-query pairs to learn the probability that a given item is characterized by each latent topic. Finally, since the users and items are now modeled in a common reduced latent topic space, the user-item preferences can be obtained by combining the user and item distributions over the latent topics. Intuitively, the search queries serve as a link, which allows us to model both users and items in a common latent topic space. Our approach is therefore applicable whenever there exist meta-data drawn from a relatively stable vocabulary which allows linking two possibly heterogeneous and ephemerally coupled sets.

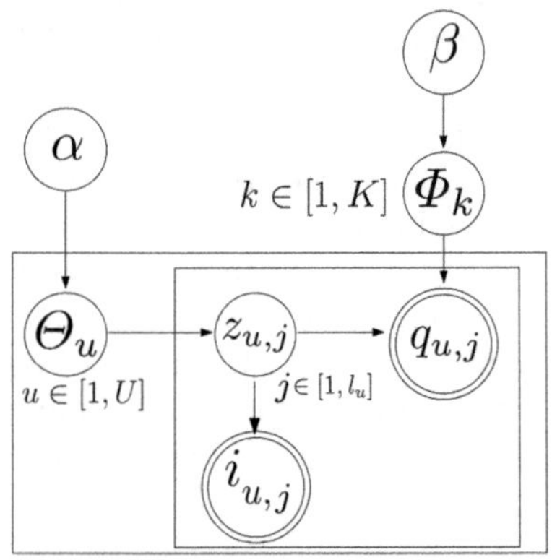

Fig. 1. Graphical representation of our Generative Model

Note that our approach can work with most state-of-the-art latent variable models (e.g., pLSI [12], LDA [4] or GaP [5]). We describe it using the structure of LDA, since it is a fully generative model, natural for our extension.

3.2 Generative Model

Our generative model starts by assuming a hidden set of user intentions that are responsible for generating both queries and items. In other words, a user searching for an item is guided by a certain interest, characterized by the latent topic, which results in a search query and an observed item. Formally, let i, q, u and k represent an item, query, user, and latent topic respectively, then $P(q, i|u)$ is the probability of user u observing a query-item pair (q, i). The simplest way to capture this probability is by assuming conditional independence of items and queries given the underlying latent topic:

$$\begin{aligned} P(q, i|u) &= \sum_k P(q|k, i, u)P(k, i|u) \\ &= \sum_k P(q|k)P(i|k)P(k|u). \end{aligned} \tag{1}$$

Following LDA, we assume that $P(k|u)$ is a multinomial probability distribution Θ_u over a fixed set of K topics. We interpret each element k of Θ_u as the interest of user u in topic k. We also assume that $P(q|k)$ is a multinomial distribution over the set of V distinct *search queries*, which can be represented by a vector Φ_k. Note that unlike the standard unigram model in LDA, a search query can contain multiple words or special characters. This is crucial in our setting since in a web commerce site, queries often contain brand names, having multiple words with conceptually different meanings (e.g. "Banana Republic"). On the other hand, due to high specialization of the web commerce domain the size of the vocabulary of queries is comparable with the size as individual terms.

Thus dealing with queries directly avoids the otherwise lossy n-gram extraction. Similarly to LDA, we introduce Dirichlet priors on Θ_u and Φ_k with specified symmetric hyperparameters α and β respectively. See Figure 1 for a graphical representation of our generative model.

Given this generative model, the marginal probability of a user u generating a particular query q is:

$$P(q|u) = \sum_k P(k|u)P(q|k) = \sum_k \Theta_u(k)\Phi_k(q). \tag{2}$$

Thus, similarly to LDA, our generative model assumes that each search query $q_{u,j}$ by user u corresponds to a latent topic $z_{u,j}$, sampled independently according to $P(z_{u,j} = k) = \Theta_u(k)$. Then, the search query is drawn according to Φ_k, i.e., $P(q_{u,j} = q|z_{u,j} = k) = \Phi_k(q)$.

In the following, we interpret $P(q|u)$ as the user's preference for a query q. Ultimately, we are interested in the user's preference for an item $P(i|u)$. Note that a straightforward extension of LDA would entail learning the topic probabilities over items $P(i|k)$, and using them to estimate $P(i|u)$. This, however, would require jointly modeling user and item topic models. Instead, we integrate, over all users, the topic model with triadic observations as in Eq. (1):

$$\begin{aligned} P(q,i) &= \int_u \sum_k P(q|k)P(k,i|u)P(u)du \\ &= \sum_k P(q|k)P(k|i)P(i), \end{aligned} \tag{3}$$

where the probability $P(k|i)$ can be interpreted as the probability that the latent topic k is responsible for generating an instance of item i. We encode these probabilities by a multinomial distribution *over the set of topics*, or a vector $\Psi_i \in \mathbb{R}^k$. Following LDA, we impose a Dirichlet prior γ on Ψ_i.

We stress that modeling $P(k|i)$ rather than $P(i|k)$ is crucial in allowing our method to decouple user and item behavior as well as to efficiently learn the model parameters for new users and items without recomputing the topic model. This decoupling is only made possible by the presence of textual meta-data (search queries) that links the users and items.

Estimating the model parameters from a corpus of queries generated by a population of users can then be regarded as a Bayesian approach to collaborative filtering, since it leverages co-occurrence information from multiple users to learn user preferences for unseen data. We also discuss the relation of our method to traditional collaborative filtering in the next section.

4 Personalized Recommendation

Recall that the motivating application of our proposed learning method was to derive a statistical formulation for recommendation of volatile items. In this section, we carry out the derivation using the generative model described above.

We also establish a theoretical connection between our method and matrix factorization methods in collaborative filtering.

Intuitively, a generative model is well-suited for a recommendation system since it allows mimicking the user behavior after learning the model parameters. In other words, once the model parameters have been learned, one can generate queries as well as items for a particular user automatically by simply following the generative process. The key property that translates this observation into a recommendation system, is that probabilistic topic models allow to assign meaningful probabilities to unobserved pairs, through the hidden layer of topics. Thus, a recommendation system based on latent topic modeling is unlikely to result in the same item that the user has already observed. However, the recommended query and item will have a high probability of belonging to the latent topic that the user has shown preference for in the past.

4.1 Query Recommendation

Given the user preference vector for latent topics Θ_u and the latent topic distributions $\Phi_k, k \in 1..K$, we recommend search queries to the user u by simulating the generative process described in Section 3.2. Namely, to generate a query q, first sample a latent topic z with $P(z = k) = \Theta_u(k)$, and then pick a query q from Φ_z, s.t. $P(q = t) = \Phi_z(t)$. Note that unlike traditional recommendation systems, which suggest queries or items that the user is most likely interested in, this process is randomized. This stochastic delivery mechanism allows us to diversify the recommendations while maintaining relevancy to the user, and yield a 100% recall of the entire inventory in an asymptotic sense.

4.2 Item Recommendation

Note that the process described above also allows us to recommend items to users. To do this, we issue the recommended queries to the search engine and use the retrieved results as the recommended items. In this way, the queries used for recommendation will correspond to likely queries in the latent topic k, for which the user has shown interest. This means, in particular, that the queries used for recommendation will rarely come from the set of queries issued by the user in question. Instead, they will be the most representative queries in the user's topics of interest. Note that this recommendation step is possible only because we perform topic modeling on full search queries, rather than individual terms, since reconstructing a query from a set of possibly unrelated search terms is a difficult task.

In real time to recommend N items, we can generate N distinct search queries, by first picking N latent topics from Θ_u *without replacement*, and use the top item provided by the search engine for each of these queries. This allows us to significantly diversify the recommended set while keeping it relevant to the user.

The preference $P(i|u)$ of user u for item i is given by:

$$P(i|u) = \sum_k P(i|k)P(k|u) = \sum_k \frac{P(i)}{P(k)} P(k|i)P(k|u)$$

Note that a symmetric Dirichlet prior, with parameters $\alpha_i = \alpha_j \; \forall \; i, j$, is invariant to permutations of the variables. In other words: $p(\Theta) = p(\Pi\Theta)$, for any permutation matrix Π. Moreover, if $\Theta' = \Pi\Theta$, the Jacobian determinant of this transformation is always 1, so $d\Theta' = d\Theta$, and the marginal:

$$P(k) = \int_{\Theta} \Theta_k p(\Theta) d\Theta = \int_{\Theta'} \Theta'_j p(\Theta') d\Theta' = P(j) \; \forall \; k, j$$

Thus, under symmetric Dirichlet prior on Θ_u, the marginal distribution $P(k)$ of latent topics is uniform and can be factored out, meaning $P(i|u) \propto P(i)\Theta_u^T \Psi_i$, where $P(i)$ is the likelihood of item i.

4.3 Relation to Latent Factor Models

We also point out that the generative model presented above is similar in spirit, to latent factor models in collaborative filtering, e.g., [13,3]. These methods map each user and each item to points in a common Euclidean space, where the prediction for user u's preference for item i is approximated by: $\hat{r}_{ui} = p_u^T q_i$, with a possible addition of a baseline predictor [13].

Note that in our model, $\hat{r}_{ui} \propto P(i)\Theta_u^T \Psi_i$, and therefore, traditional latent factor models can be thought of imposing a uniform prior on the space of items. Also note that unlike standard latent factor models, our topic model is derived via the relations between users and search queries, and is independent of user-item interactions. As we show in the next section, this allows us to efficiently infer model parameters for new users and items and perform recommendation.

5 Learning and Inference

In this section we describe the learning and inference procedures for our generative model. As mentioned earlier, our emphasis is on decoupling user and item interactions. Moreover, since the number of items exceeds the number of users by orders of magnitude, we aim to first learn the latent topic parameters by considering user behavior, and then adopt a fast inference procedure to obtain the topic distributions for each item.

5.1 Model Fitting

The marginal distribution of users and search queries described in Section 3.2 mirrors the generative model for documents and terms in LDA. Therefore, estimating Φ_k from data, can be done in a similar way as for LDA. Here, we adopt the Gibbs sampling approach [11], which has been shown to yield accurate results efficiently in practice.

The input to our user-based model estimation is a set of users, with a list of search queries that each user issued in a fixed period of time, where repetitions are naturally allowed. The goal of the Gibbs sampler is to determine for each query in the dataset the latent topic that this query was generated by. Then, the model parameters Φ_k can be computed as statistics on these topic assignments. The main idea of Gibbs sampling for LDA is to derive the distribution:

$$p_k = P(z_{u,j} = k \mid \boldsymbol{z}_{\neg u,j}, \boldsymbol{u}, \boldsymbol{q}), \tag{4}$$

where $z_{u,j}$ is the topic responsible for generating query j of user u, $\boldsymbol{z}_{\neg u,j}$ are topic assignments for all other queries, while $\boldsymbol{u}$ and $\boldsymbol{q}$ are users and queries respectively. Once this distribution is established, we run the Gibbs sampler with a random initialization of $\boldsymbol{z}$ until convergence. In our model, as in LDA:

$$\begin{aligned} p_k &= \frac{P(\boldsymbol{z}, \boldsymbol{u}, \boldsymbol{q})}{P(\boldsymbol{z}_{\neg u,j}, \boldsymbol{u}, \boldsymbol{q})} \\ &= \frac{P(\boldsymbol{q}|\boldsymbol{z})}{P(q_{u,j})P(\boldsymbol{q}_{\neg u,j}|\boldsymbol{z}_{\neg u,j})} \frac{P(\boldsymbol{z}|\boldsymbol{u})}{P(\boldsymbol{z}_{\neg u,j}|\boldsymbol{u})}. \end{aligned} \tag{5}$$

Both $P(\boldsymbol{q}|\boldsymbol{z})$ and $P(\boldsymbol{z}|\boldsymbol{u})$ can easily be derived by integrating over the parameter space, since e.g.:

$$\begin{aligned} P(\boldsymbol{q}|\boldsymbol{z}) &= \int_{\boldsymbol{\Phi}} P(\boldsymbol{q}|\boldsymbol{z}, \boldsymbol{\Phi}) P(\boldsymbol{\Phi}|\beta) d\boldsymbol{\Phi} \\ &= \prod_{k=1}^{K} \frac{1}{\mathrm{B}(\beta)} \mathrm{B}(\boldsymbol{z}^k + \beta), \end{aligned} \tag{6}$$

where $\boldsymbol{z}^k(q)$ is the number of times query q is assigned to topic k in the topic assignment vector $\boldsymbol{z}$ and B is the multivariate beta function. The final distribution for the Gibbs sampler is:

$$p_k \propto \frac{\boldsymbol{z}^k_{\neg u,j}(q_{u,j}) + \beta}{\sum\limits_{w=1}^{V} \left(\boldsymbol{z}^k_{\neg u,j}(w) + \beta\right)} \frac{\boldsymbol{z}^k_{\neg u,j}(u) + \alpha}{\sum\limits_{j=1}^{K} \left(\boldsymbol{z}^j_{\neg u,j}(u) + \alpha\right)}, \tag{7}$$

where $\boldsymbol{z}^k_{\neg u,j}(q)$ is the number of times query q is assigned to topic k in $\boldsymbol{z}_{\neg u,j}$, and $\boldsymbol{z}^k_{\neg u,j}(u)$ is the number of times topic k is assigned to a query by user u. Once the Gibbs sampler converges to a stationary distribution, the model parameters can be computed as:

$$\Phi_k(q) = \frac{\boldsymbol{z}^k(q) + \beta}{\sum_{w=1}^{V} \boldsymbol{z}^k(w) + \beta}. \tag{8}$$

We approximate $P(i)$ by measuring the fraction of times that item i was observed in the corpus of user-item pairs.

5.2 Inference

Once the model parameters Φ_k have been estimated, the inferential task consists of learning Θ_u and Ψ_i, for a user u or item i.

First, suppose the data set consists of a single user u with all the search queries issued by this user in a fixed time period, and our goal is to estimate the user preference vector Θ_u. The Gibbs sampling procedure described above allows us to perform inference in a straightforward fashion. Again, for every query $q_{u,j}$ in the new dataset, we aim to determine which latent topic z is responsible for generating this query. For this, we run the Gibbs sampler in the same fashion as above, while *only iterating over queries of user u*. The probability distribution for the Gibbs sampler is nearly identical to Eq. (7):

$$
\begin{aligned}
p_k \propto & \frac{\boldsymbol{n}^k(q_{u,j}) + \boldsymbol{z}^k_{\neg u,j}(q_{u,j}) + \beta}{\sum_q^V \left(\boldsymbol{n}^k(q_{u,j}) + \boldsymbol{z}^k_{\neg u,j}(q) + \beta\right)} \\
& \cdot \frac{\boldsymbol{z}^k_{\neg u,j}(u) + \alpha}{\sum_j^K \left(\boldsymbol{z}^j_{\neg u,j}(u) + \alpha\right)},
\end{aligned}
\tag{9}
$$

where $\boldsymbol{z}$ is the assignment of latent topics to queries *only of user u*, while $\boldsymbol{n}^k(q_{u,j})$ is the number of times query $q_{u,j}$ was assigned to topic k *in the model fitting step*. Note that $\boldsymbol{n}$ is the only part of the distribution that depends on the model fitting, and can be seen as a precomputed sparse $V \times K$ matrix of counts. After convergence, the user preference vector Θ_u is:

$$
\Theta_u(k) = \frac{\boldsymbol{z}^k(u) + \alpha}{\sum_{j=1}^K \boldsymbol{z}^j(u) + \alpha}. \tag{10}
$$

To derive the inferential procedure for topic probability vector Ψ_i, given an item i, we assume that the data consists of all queries used by all users to arrive at item i. Then, our goal is to determine $z_{i,j}$: the latent topic responsible for generating query j used to arrive at item i. Note that in this case, Eq. (5) can be written as:

$$
\begin{aligned}
p_k =& P(z_{i,j} = k | \boldsymbol{z}_{\neg i,j}, \boldsymbol{i}, \boldsymbol{q}) \\
=& \frac{P(\boldsymbol{q}|\boldsymbol{i}, \boldsymbol{z})P(\boldsymbol{z}|\boldsymbol{i})P(\boldsymbol{i})}{P(\boldsymbol{q}|\boldsymbol{i}, \boldsymbol{z}_{\neg i,j})P(\boldsymbol{z}_{\neg i,j}|\boldsymbol{i})P(\boldsymbol{i})} \\
=& \frac{P(\boldsymbol{q}|\boldsymbol{z})P(\boldsymbol{z}|\boldsymbol{i})}{P(\boldsymbol{q}|\boldsymbol{z}_{\neg i,j})P(\boldsymbol{z}_{\neg i,j}|\boldsymbol{i})},
\end{aligned}
\tag{11}
$$

where the third equality follows from the conditional independence of items $\boldsymbol{i}$ and queries $\boldsymbol{q}$ given $\boldsymbol{z}$ assumed in our generative model. Note that the final expression for p_k has the same form as Eq. (5). In addition, the Dirichlet prior γ assumed on Ψ_i, where $\Psi_i(k) = P(k|i)$, forces the distribution of topics given an item to have the same form as the distribution of topics given a user. Thus, we can run the Gibbs sampler in the identical way as for Θ_u, to get Ψ_i.

Note that if the user u issued N search queries (or the item i was observed through N search queries), one sampling iteration of the Gibbs sampler will require only N topic assignments. Convergence is usually achieved after several Gibbs iterations, and inference is very fast. Therefore, learning model parameters Θ_u and Ψ_i for new users and items through inference is very efficient and does not require extensive computations, unlike conventional matrix factorization models.

6 Experiments

Our dataset consists of the search queries entered by the eBay users over a two-month period in 2009. We only consider queries, for which the search engine returned at least one item in the "Clothing, Shoes & Accessories" (CSA) meta-category. Furthermore, during the model fitting step we remove casual users, who did not purchase any items in this time period. This greatly reduces the data set, and leaves 2.6M users, with an average of 2.5 queries per user per day. We also limit the vocabulary of queries to those entered by at least 30 users to reduce the complexity of the Gibbs sampler. This reduces the data by an additional 8 percent, so that our final vocabulary consists of approximately 50K search queries.

Figures 2(a) and 2(b) show two trends observed in the data. In particular, Figure 2(a) shows that over 50 percent of the users who bought items in the CSA meta-category of eBay during two months, only bought one item in this meta-category (while the average number of purchased items is 3, which is explained by the heavy tailed nature of the distribution). On the other hand, Figure 2(b) shows that the median number of subcategories (as defined by the current eBay taxonomy) that individual users looked at is 5, with the mean 10.7. This shows that users at eBay are willing to explore products in a variety of categories. In particular, this means that recommendations based purely on purchasing behavior may not be able to capture the wide range of interests that users have.

6.1 Choice of Dirichlet Priors α and β

One of the principal advantages of the LDA model over pLSI [12] is the flexibility of prior parameters α and β. Indeed, pLSI is a *maximum a posteriori* LDA model estimated with a uniform Dirichlet prior $\alpha = 1$ [10]. To demonstrate the

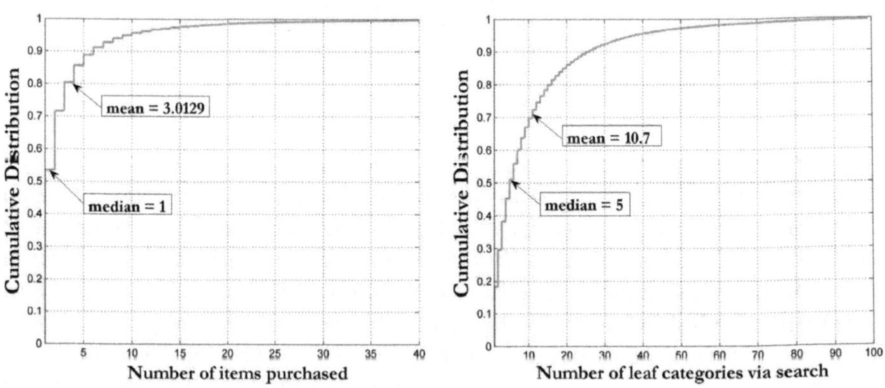

(a) (b)

Fig. 2. (a) Number of items purchased by users (b) Number of subcategories of CSA explored by users

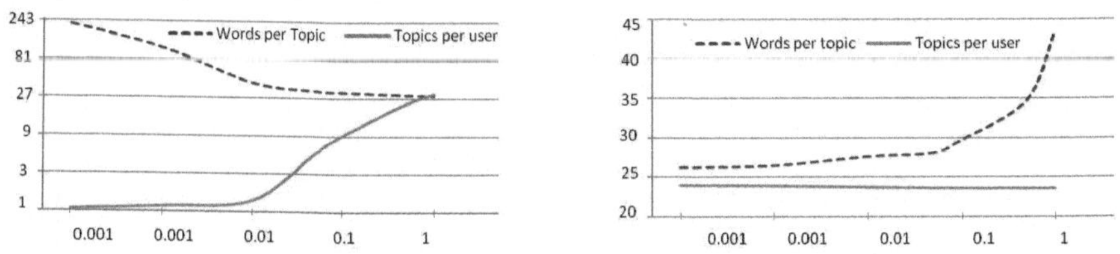

(a) Uniform prior α (b) Uniform prior β

Fig. 3. (a): Influence of α, at $\beta = 0.1$, (b): influence of β, at $\alpha = 0.5$

importance of this distinction, which holds even for a large dataset, we considered a sample of queries issued by 400K users, and estimated the model for 100 topics and various values of α and β. Figures 3(a) and 3(b) show the dependencies of the average number of topics per user and the average number of queries per topic on the priors α and β. For each user u we consider the median of Θ_u: the minimum number of topics, with cumulative distribution in Θ_u at least 0.5, and similarly, for each topic k, we consider the median of Φ_k.

Note that as α approaches 1, the average number of topics per user grows, which means that on average, users' interests become diffused over more topics. However, as a result of this, fewer queries are necessary to explain each topic, and Φ_k becomes more and more concentrated. The effect of β is less pronounced on the average number of topics per user, whereas the median number of queries per topic grows. This is consistent with the intuition that β controls how concentrated each topic is. Since our ultimate goal is to recommend items to users, we would like to have highly concentrated topics with the most relevant queries having a high probability. Furthermore, since the majority of users only explore a small number of subcategories, we expect each user to be interested in a small number of latent topics. Therefore, we use $\beta = 0.1, \alpha = 0.05$, so that on average, the median of Θ_u is 5.

6.2 Personalized versus Global Models

To evaluate the quality of our method, we compute the log-likelihood of unseen data given the model. For this, we first compute the model parameters Φ_k and Θ_u for the two-month data described above. This allows us to predict the preference $P(q|u)$ of user u for a particular query q using Eq. (2). The log-likelihood of a set of queries for a user is given simply as $\sum_j \log(P(q_j|u))$. Thus, to evaluate the quality of our approach to personalized recommendation, we evaluate the log-likelihood of the search queries issued by the same users for which the model was estimated, but in the four days following the training period. A better predictive model would result in a smaller absolute value of the log-likelihood.

As a baseline predictor we use a global model, which is oblivious to individual user's preferences. This corresponds to setting the number of topics to 1, since in this case, under uniform Dirichlet priors, the probability $P(q|u)$ is simply the

Query	$\Phi_k(q)$
golf	0.2975
golf+shoes	0.0684
nike+golf	0.0652
tiger+woods	0.0571
callaway	0.0527
golf+shirts	0.0477
scotty+cameron	0.0375
greg+norman	0.0369
adidas+golf	0.0351
nike+golf+shirt	0.0339
golf+shirt	0.0334
puma+golf	0.0329
footjoy+classics	0.0299

(a)

Query	$\Phi_k(q)$
sunglasses	0.3168
ray+ban	0.1390
ray+ban+sunglasses	0.0888
mens+sunglasses	0.0459
mephisto	0.0388
fishing	0.0364
rayban	0.0341
aviator+sunglasses	0.0336
polarized+sunglasses	0.0331
ray+ban+aviator	0.0303
sun+glasses	0.0268
eyeglasses	0.0251
rayban+sunglasses	0.0245

(b)

Query	$\Phi_k(q)$
oakley	0.3337
oakley+sunglasses	0.2075
oakley+juliet	0.0561
oakley+gascan	0.0469
oakley+half+jacket	0.0405
oakley+m+frame	0.0387
oakley+radar	0.0342
oakley+frogskins	0.0293
oakley+flak+jacket	0.0281
oakley+rare	0.0206
oakley+display	0.0205
oakley+oil+rig	0.0197
oakley+romeo	0.0170

(c)

Query	$\Phi_k(q)$
gucci	0.2616
prada	0.1685
armani	0.1207
dolce+&+gabbana	0.0707
versace	0.0663
ferragamo	0.0651
dolce	0.0444
dolce+gabbana	0.0366
cavalli	0.0331
d&g	0.0171
roberto+cavalli	0.0169
bally	0.0155
dior	0.0142

(d)

Fig. 4. Sample topics inferred using $\alpha = 0.05, \beta = 0.1$, and $K = 100$ along with queries having maximum probabilities $\Phi_k(q)$

fraction of times this query was used in the training data independent of u. In other words, each query is characterized by its *overall popularity* in the training data. Figure 5(a) shows the dependence of the absolute log-likelihood of the testing data on the number of topics K. Note that the global user-oblivious model with 1 topic results in 36% higher absolute log-likelihood than the model with 500 topics and 31% higher than the model with 200 topics. Interestingly, we do not observe any over-saturation phenomena, and the absolute log-likelihood decreases as far as 500 latent topics. However, the rate of the decrease slows down beyond 200 topics. Figure 5(b) shows that the improvement over the global model is significantly more pronounced for the users who issued many queries (in the training set). Thus, we achieve over 50 percent improvement in absolute log-likelihood using a model with 200 topics for users who entered over 400 search queries. This shows, not only that topical preferences are learned more accurately for users who have a longer search history, but also that the topics themselves represent persistent structure in the data.

Figure 4 shows a few sample topics inferred using $K = 100$ together with queries having maximum probabilities $\Phi_k(q)$. Note the highly specific nature of each topic, which will allow us to achieve personalized recommendations. Thus, a user whose history contains many queries related to e.g. golf will be presented with a recommendation that corresponds to a likely query-item pair within this category (Figure 4(a)). Moreover, remark the flexibility provided by the topic modeling framework, where an individual topic can correspond to an activity (Figure 4(a)), a product (Figure 4(b)), a brand (Figure 4(c)), or even a set of conceptually related brands (Figure 4(d)). Finally, note that individual queries often contain multiple terms, and performing topic modeling on the query rather than term level is essential for accurate personalized recommendation.

6.3 Triadic versus Dyadic Models

Finally, we compare our method with standard LDA, to validate the advantage of capturing the hidden structure from a relatively persistent yet often-available intermediate layer, e.g., search queries in our case. Given a set of training data

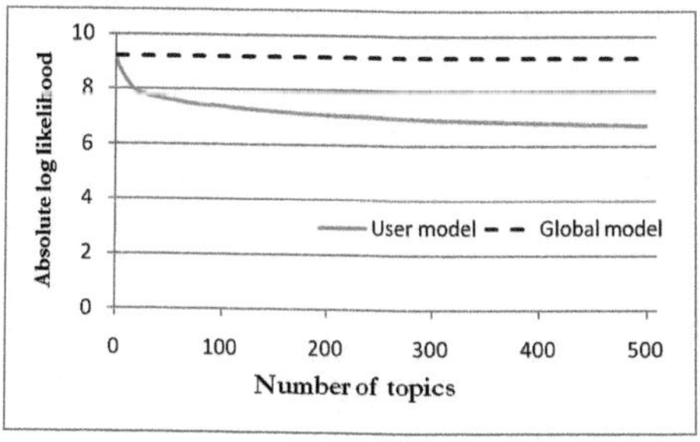

(a) Absolute log-likelihood of the test data given as a function of number of topics.

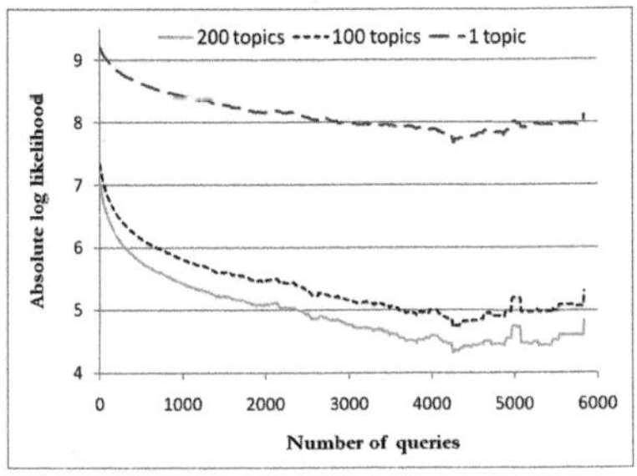

(b) Absolute log-likelihood of the test data given as a function of number of queries.

Fig. 5. Absolute log-likelihood of the test data for different choices of K (a) and for users with different search history sizes (b)

in the form of triadic observations user-query-item, where each example triple means that a user issues a query and then conducts some transactions (click, watch, bid or purchase event) on an item, we first trained a topic model with triadic observations as described in Section 3. To leverage the semantically-rich query layer, the training set of triples was projected to a user-item matrix, while the item dimension uses queries (converting to corresponding items) as descriptors. In other words, the vocabulary consists of converting queries. On the other hand, a standard LDA was trained on a direct projection of user-item matrix by ignoring the query layer, where the vocabulary consists of all unique item ids. Now the difference in volatility between queries and item ids shall be appreciated, and this setting applies to many practical domains such as web images, graphic ads, and page URLs.

Let *word* w denote query q or item i for triadic and dyadic models respectively. After training and inference of both models, we have the posterior Dirichlet parameter given each user $P(k|u)$, and word multinomials conditioned on each topic $P(w|k)$. Given a test data set in the same triple form and collected after the training period, we then can compare the log likelihoods of user-item conversions under both models:

$$\ell = \sum_u \sum_w \log \left(\sum_k P(w|k)p(k|u) \right) . \tag{12}$$

It is important to note that although w denotes different objects for different models, the log likelihood reflects the prediction quality against a same ground truth, which is the user-item conversions in future. Personalization is achieved by using different posterior per-user topical mixture $P(k|u)$ for each user. Online prediction can easily follow. Under triadic topic models, to predict and rank a candidate set of items, one needs to augment items with their historically converting queries.

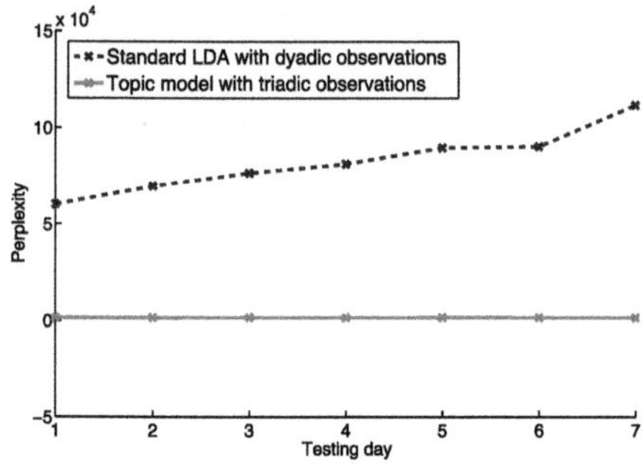

Fig. 6. Daily perplexity on one-week test data. Triadic model outperforms standard LDA, with the gap widening as the testing day is further away from the training day.

We trained both models using one-day worth of user-query-item conversion data, and evaluated both models against the following one-week test data on a daily basis. The training data contains 5.3M triples from 500K users, which gives a vocabulary size $V=$ 539,092 queries for the triadic model and $V=$ 2,186,472 item ids for the standard LDA. We set $\alpha = 0.05$, $\beta = 0.1$, and $K = 100$ as separately tuned in Section 6.1. For a normalized comparison, we report perplexity (two to the power of minus per-word test-set log likelihood in our case), as plotted in Figure 6. The results show that our proposed topic model with triadic observations consistently outperforms the standard LDA with dyadic observations, with the edge widening as testing day farther away from the training day. On day one right after the training date, the per-word log likelihood of the triadic model is -15.88, while a LDA yields -10.68. The over 30% log likelihood improvement translates to over 30 folds decrease in perplexity, with $PP^{\text{triadic}}_{\text{day 1}} = 1.6\text{K}$ and $PP^{\text{lda}}_{\text{day 1}} = 60\text{K}$. As the testing day moves farther away from the training day, the prediction performance of our approach stays stable, while that of a standard LDA deteriorates drastically. On day seven the perplexity gain of our approach over the conventional LDA becomes over 60 times, with $PP^{\text{triadic}}_{\text{day 7}} = 1.7\text{K}$ and $PP^{\text{lda}}_{\text{day 7}} = 112\text{K}$. This observation exactly shows the volatility of items and the persistency of queries, thus motivating the idea of modeling ephemeral relations through a persistent layer. The purpose of comparing our approach with the standard LDA exposed to a sparse dyadic setting is not to claim a better statistical method, but rather to propose a general framework that can help realize the full potential of topic modeling.

7 Conclusion and Future Work

In this paper, we presented a method for building reliable topic models by using unstructured meta-data when direct topic modeling is infeasible due to the sparsity and volatility of co-occurrence patterns. We show both how to use the meta-data to densify the original input and to build a better topic model. Furthermore, we show how to efficiently estimate the model parameters in practice by decoupling the interactions between the two original input sets. Most notably, this allows us to efficiently compute model parameters in the presence of new data without recomputing all model parameters. We demonstrate the usefulness of our method by deriving a statistical formulation of a novel personalized recommendation system for volatile items. Our data is challenging due to extreme sparsity of the input which prohibits the use of standard topic modeling and collaborative filtering techniques. We show that by using the search queries, we can still build a reliable topic model, which can then be used for efficient recommendation.

In the future, we would like to add a temporal aspect to our method to reflect the evolution of the latent topics and user preferences. This could include modeling short-term effects, such as topics that become prominent because of external events (e.g., Super Bowl), cyclic topic shifts that arise in certain seasons, as well as long term topic changes that appear, e.g., from new products introduced in the market. Moreover, we are planning to compare the efficiency and effectiveness of the various item recommendation methods. Finally, it is desired to benchmark our approach with standard LDA with a larger scale data set, e.g., three-month training data and one-month daily evaluation. We expect to further strengthen the advantages of our method.

Acknowledgments

The authors would like to thank Neel Sundaresan (eBay Inc.), as well as Yanen Li (University of Illinois at Urbana-Champaign) and Qing Xu (Vanderbilt University) for the help with acquiring and processing of the data, as well as for the many valuable comments and suggestions.

References

1. Adomavicius, G., Tuzhilin, A.: Toward the next generation of recommender systems: A survey of the state-of-the-art and possible extensions. IEEE Tr. on Knowl. and Data Eng. 17(6), 734–749 (2005)
2. Agarwal, D., Chen, B.: fLDA: Matrix factorization through latent Dirichlet allocation. In: Proc. WSDM (2010)
3. Bell, R., Koren, Y., Volinsky, C.: Modeling relationships at multiple scales to improve accuracy of large recommender systems. In: Proc. KDD, pp. 95–104 (2007)
4. Blei, D.M., Ng, A.Y., Jordan, M.I.: Latent Dirichlet allocation. J. Mach. Learn. Res. 3, 993–1022 (2003)
5. Canny, J.: GaP: a factor model for discrete data. In: Proc. SIGIR, pp. 122–129 (2004)
6. Chen, Y., Canny, J.: Probabilistic clustering of an item. U.S. Patent Application 12/694,885, filed with eBay (2010)
7. Chen, Y., Pavlov, D., Canny, J.: Large-scale behavioral targeting. In: Proc. KDD, pp. 209–218 (2009)
8. Das, A.S., Datar, M., Garg, A., Rajaram, S.: Google news personalization: scalable online collaborative filtering. In: Proc. WWW, pp. 271–280 (2007)

9. Fei-Fei, L., Perona, P.: A Bayesian hierarchical model for learning natural scene categories. In: CVPR, pp. 524–531 (2005)
10. Girolami, M., Kabán, A.: On an equivalence between pLSI and LDA. In: Proc. SIGIR, pp. 433–434 (2003)
11. Griffiths, T.L., Steyvers, M.: Finding scientific topics. Proc. Natl. Acad. Sci. U.S. 101, 5228–5235 (2004)
12. Hofmann, T.: Probabilistic latent semantic analysis. In: Proc. UAI, pp. 289–296 (1999)
13. Koren, Y.: Factorization meets the neighborhood: a multifaceted collaborative filtering model. In: Proc. KDD, pp. 426–434 (2008)
14. Liu, Y., Zha, H., Qin, H.: Shape topics: A compact representation and new algorithms for 3d partial shape retrieval. In: Proc. CVPR, vol. 2, pp. 2025–2032 (2006)
15. Micarelli, A., Gasparetti, F., Sciarrone, F., Gauch, S.: Personalized search on the World Wide Web. In: The Adaptive Web, pp. 195–230 (2007)
16. Mimno, D., Wallach, H.M., Naradowsky, J., Smith, D.A., McCallum, A.: Polylingual topic models. In: Proc. EMNLP, pp. 880–889 (2009)
17. Steyvers, M., Griffiths, T.: Probabilistic topic models. In: Handbook of Latent Semantic Analysis, pp. 424–440 (2007)
18. Wetzker, R., Umbrath, W., Said, A.: A hybrid approach to item recommendation in folksonomies. In: Proc. WSDM, pp. 25–29. ACM, New York (2009)

Conditional Ranking on Relational Data

Tapio Pahikkala[1], Willem Waegeman[2], Antti Airola[1], Tapio Salakoski[1], and Bernard De Baets[2]

[1] University of Turku and Turku Centre for Computer Science, Joukahaisenkatu 3-5 B, FIN-20520, Turku, Finland
forname.surname@utu.fi

[2] Department of Applied Mathematics, Biometrics and Process Control, Ghent University, Coupure links 653, B-9000 Ghent, Belgium
forname.surname@UGent.be

Abstract. In domains like bioinformatics, information retrieval and social network analysis, one can find learning tasks where the goal consists of inferring a ranking of objects, conditioned on a particular target object. We present a general kernel framework for learning conditional rankings from various types of relational data, where rankings can be conditioned on unseen data objects. Conditional ranking from symmetric or reciprocal relations can in this framework be treated as two important special cases. Furthermore, we propose an efficient algorithm for conditional ranking by optimizing a squared ranking loss function. Experiments on synthetic and real-world data illustrate that such an approach delivers state-of-the-art performance in terms of predictive power and computational complexity. Moreover, we also show empirically that incorporating domain knowledge in the model about the underlying relations can improve the generalization performance.

1 Introduction

Let us start with two introductory examples to explain the problem setting of conditional ranking. Firstly, suppose that a number of persons are playing an online computer game. For many people it is always more fun to play against someone with similar skills, so players might be interested in receiving a ranking of other players, ranging from extremely difficult to beat to novice players with no experience at all. Unfortunately, pairwise strategies of players in many games – not only in computer games but also in board or sports games – tend to exhibit a rock-paper-scissors type of relationship [1], in the sense that player A beats with a high probability player B, who on his term beats with a high probability person C, while player A has a high chance of losing from the same player C. Mathematically speaking, the relation between players is not transitive, leading to a cyclic relationship and implying that no global (consistent) ranking of skills exists, yet a conditional ranking can always be obtained for a specific player [2].

As a second introductory example, let us consider the supervised inference of biological networks, like protein-protein interaction networks, where the goal usually consists of predicting new interactions from a set of highly-confident

J.L. Balcázar et al. (Eds.): ECML PKDD 2010, Part II, LNAI 6322, pp. 499–514, 2010.

interactions [3]. Similarly, one can also define a conditional ranking task in such a context, as predicting a ranking of all proteins in the network that are likely to interact with a given target protein [4]. However, this conditional ranking task differs from the previous one because (a) rankings are computed from symmetric relations instead of reciprocal ones and (b) the values of the relations are here usually not continuous but discrete.

Applications for conditional ranking tasks arise in many domains where relational information between objects is observed, such as relations between persons in preference modelling, social network analysis and game theory, links between documents or websites in information retrieval and text mining, interactions between genes or proteins in bioinformatics, etc. When approaching conditional ranking from a graph inference point of view, the goal consists of returning a ranking of all nodes given a particular target node, in which the nodes provide information in terms of features and edges in terms of labels or relations. At least two properties of graphs play a key role in such a setting. Firstly, the type of information stored in the edges defines the learning task: binary-valued edge labels lead to bipartite ranking tasks [5], ordinal-valued edge labels to multipartite or layered ranking tasks [6,7] and continuous labels result in rankings that are nothing more than total orders (when no ties occur). Secondly, the relations that are represented by the edges might have interesting properties, namely symmetry or reciprocity, for which conditional ranking can be interpreted differently.

We present in this article a kernel framework for conditional ranking, which covers all above situations. Unlike existing single-task or multi-task ranking algorithms, where the conditioning is respectively ignored or only happening for training objects, our approach also allows to condition on new data objects that are not known during the training phase. Thus, in light of Figure 1 that will be explained below, the algorithm is not only able to predict conditional rankings for objects A to E, but also for objects F and G that do not contribute to the training dataset. One might argue that existing ranking methods like RankSVM [8] or RankRLS [9,10] can learn conditional rankings, by interpreting the conditioning object as a query. We will show in Section 4 that these methods become computationally intractable in many applications, unlike our approach, which is much more efficient due to exploiting the knowledge that the objects to be ranked and the queries (here objects too) are sampled from the same domain.

As a second difference, information retrieval methods that could be interpreted as conditional ranking predominantly use similarity as underlying relation, often in a pure intuitive manner, as a nearest neighbor type of learning. Think in this context at the example of protein ranking given above [4] or algorithms like *query by document* [11]. These methods simply look for rankings where the most similar objects w.r.t. the conditioning object appear on top, contrary to our approach, which should be considered as much more general, since we learn rankings from any type of binary relation. Nonetheless, similarity relations will of course still occupy a prominent place in our framework as an important special case. We will demonstrate below that domain knowledge about the underlying relations can be easily incorporated in our framework.

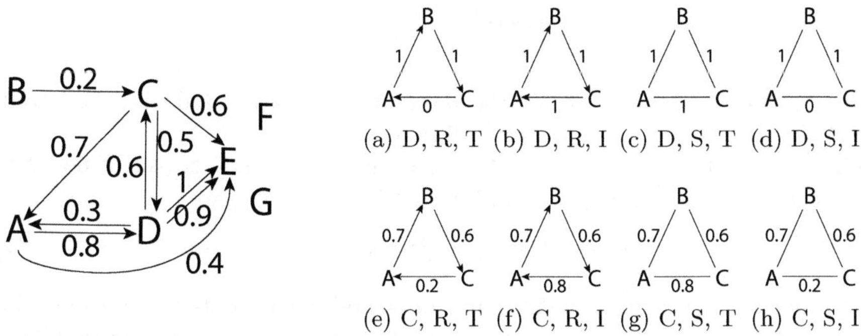

Fig. 1. Left: example of a multi-graph representing the most general case, where no additional properties of relations are assumed. Right: examples of eight different types of relations in a graph of cardinality three. The following relational properties are illustrated: (D) discrete, (C) continuous, (R) reciprocal, (S) symmetric, (T) transitive and (I) intransitive.

2 General Framework

Let us start with introducing some notations. We consider ranking of data structured as a graph $G = (\mathcal{V}, \mathcal{E}, Q)$, where $\mathcal{V}$ corresponds to the set of nodes and $\mathcal{E} \subseteq \mathcal{V}^2$ represents the set of edges e, for which training labels are provided in terms of relations. Moreover, these relations are represented by training weights y_e on the edges and these relations are generated from an unknown underlying relation $Q : \mathcal{V}^2 \to [0,1]$. Relations are required to take values in the interval $[0,1]$ because some properties that we need are historically defined for such relations, but an extension to real-valued relations $h : \mathcal{V}^2 \to \mathbb{R}$ can always be realized with a simple monotonic mapping $g : \mathbb{R} \to [0,1]$ such that

$$Q(v, v') = g(h(v, v')) , \quad \forall (v, v') \in \mathcal{V}^2 . \tag{1}$$

Following the standard notations for kernel methods, we formulate our learning problem as the selection of a suitable function $h \in \mathcal{H}$, with $\mathcal{H}$ a certain hypothesis space, in particular a reproducing kernel Hilbert space (RKHS). Hypotheses $h : \mathcal{V}^2 \to \mathbb{R}$ are usually denoted as $h(e) = \langle \mathbf{w}, \Phi(e) \rangle$ with $\mathbf{w}$ a vector of parameters that needs to be estimated based on training data. Let us denote a training dataset of cardinality $q = |\mathcal{E}|$ as a sequence $T = \{(e, y_e) \mid e \in \mathcal{E}\}$ of input-label pairs, then we formally consider the following variational problem in which we select an appropriate hypothesis h from $\mathcal{H}$ for training data T. Namely, we consider an algorithm

$$\mathcal{A}(T) = \underset{h \in \mathcal{H}}{\operatorname{argmin}} \, \mathcal{L}(h, T) + \lambda \|h\|^2_{\mathcal{H}} \tag{2}$$

with $\mathcal{L}$ a given loss function and $\lambda > 0$ a regularization parameter. According to the representer theorem [12], any minimizer $h \in \mathcal{H}$ of (2) admits a dual representation of the following form:

$$h(\overline{e}) = \langle \mathbf{w}, \Phi(\overline{e}) \rangle = \sum_{e \in \mathcal{E}} a_e K^{\Phi}(e, \overline{e}), \tag{3}$$

with $a_e \in \mathbb{R}$ dual parameters, K^{Φ} the kernel function associated with the RKHS and Φ the feature mapping corresponding to K^{Φ}.

Given two relations $Q(v, v')$ and $Q(v, v'')$ defined on any triplet of nodes in $\mathcal{V}$, we compose the ranking of v' and v'' conditioned on v as

$$v' \succeq_v v'' \Leftrightarrow Q(v, v') \geq Q(v, v''). \tag{4}$$

Let the number of correctly ranked pairs for all nodes in the dataset serve as evaluation criterion for verifying (4), then one aims to minimize the following empirical loss when computing the loss over all conditional rankings simultaneously:

$$\mathcal{L}(h, T) = \sum_{v \in \mathcal{V}} \sum_{e, \overline{e} \in \mathcal{E}_v : y_e < y_{\overline{e}}} I(h(e) - h(\overline{e})), \tag{5}$$

with I the Heaviside function returning one when its argument is strictly positive, returning 1/2 when its argument is exactly zero and returning zero otherwise. Importantly, $\mathcal{E}_v$ denotes the set of all edges starting from or ending at the node v, depending on the specific task. For example, concerning the relation "trust" in a social network, the former loss would correspond to ranking the persons in the network who *are trusted by* a specific person, while the latter loss corresponds to ranking the persons who *trust* that person. So, taking Figure 1 into account, we would in such an application respectively use the rankings $A \succ_C E \succ_C D$ (outgoing edges) and $D \succ_C B$ (incoming edges) as training info for node C.

Since (5) is neither convex nor differentiable, one should look for a differentiable and convex approximation of it. Let us to this end start by considering the following squared loss function over the q observed edges in the training set:

$$\mathcal{L}(h, T) = \sum_{e \in \mathcal{E}} (y_e - h(e))^2. \tag{6}$$

Such a setting would correspond to directly learning the labels on the edges in a regression or classification setting. For the latter case, optimizing (6) instead of the more conventional hinge loss has the advantage that the solution can be found by simply solving a system of linear equations [13]. However, when doing conditional ranking, the simple squared loss might not be optimal. Consider for example that we have a node v and we aim to learn to predict which of the two other nodes, v' or v'', would be closer to it. Let us denote $e = (v, v')$ and $\overline{e} = (v, v'')$, and let y_e and $y_{\overline{e}}$ denote the relation between v and v' and between v and v'', respectively. Then, it would be beneficial for the regression function to have a minimal squared difference $(y_e - y_{\overline{e}} - h(e) + h(\overline{e}))^2$, leading to the following loss function:

$$\mathcal{L}(h, T) = \sum_{v \in \mathcal{V}} \sum_{e, \overline{e} \in \mathcal{E}_v} (y_e - y_{\overline{e}} - h(e) + h(\overline{e}))^2, \tag{7}$$

which can be interpreted as a differentiable and convex approximation of (5).

When no further restrictions on the underlying relation can be specified, then the following Kronecker product feature mapping is used to express pairwise interactions between features of nodes:

$$\Phi(e) = \Phi(v, v') = \phi(v) \otimes \phi(v'),$$

where ϕ represents the feature mapping for individual nodes. As shown in [14], such a pairwise feature mapping yields the tensor product pairwise kernel in the dual model:

$$K^{\Phi}_{\otimes}(e, \overline{e}) = K^{\Phi}_{\otimes}(v, v', \overline{v}, \overline{v}') = K^{\phi}(v, \overline{v})K^{\phi}(v', \overline{v}'),$$

with K^{ϕ} the kernel corresponding to ϕ. With an appropriate choice for K^{ϕ}, such as the Gaussian RBF kernel, the kernel K^{Φ} generates a class $\mathcal{H}$ of universally approximating functions for learning any type of relation (formal proof omitted).

3 Special Relations

In the above framework, we generally allow that for any pair of nodes in the graph several edges can exist, in which an edge in one direction not necessarily imposes constraints on the edge in the opposite direction and multiple edges in the same direction can connect two nodes, leading to a multi-graph like Figure 1, where two different edges in the same direction connect nodes D and E. This construction is required to allow repeated measurements. However, within the context of conditional ranking, two particular cases deserve further attention: symmetric relations and reciprocal relations

A binary relation $Q : \mathcal{V}^2 \rightarrow [0, 1]$ is called a symmetric relation if for all $(v, v') \in \mathcal{V}^2$ it holds that $Q(v, v') = Q(v', v)$. For symmetric relations, edges in multi-graphs like Figure 1 become undirected. Applications arise in many domains and metric learning or learning similarity measures can be seen as special cases. If the relation becomes discrete as $Q : \mathcal{V}^2 \rightarrow \{0, 1\}$, then we end up with a classification setting instead of a regression setting.

A binary relation $Q : \mathcal{V}^2 \rightarrow [0, 1]$ is called a reciprocal relation if for all $(v, v') \in \mathcal{V}^2$ it holds that $Q(v, v') = 1 - Q(v', v)$. For reciprocal relations, every edge $e = (v, v')$ in a multi-graph like Figure 1 induces an unobserved invisible edge $e_R = (v', v)$ with appropriate weight in the opposite direction. Applications arise here in domains such as preference learning, game theory and bioinformatics for representing preference relations, choice probabilities, winning probabilities, gene regulation, etc. The weight on the edge defines the real direction of such an edge. If the weight on the edge $e = (v, v')$ is higher than 0.5, then the direction is from v to v', but when the weight is lower than 0.5, then the direction should be interpreted as inverted, for example, the edges from A to C in Figures 1 (a) and (e) should be interpreted as edges starting from A instead of C. If the relation becomes discrete as $Q : \mathcal{V}^2 \rightarrow \{0, 1/2, 1\}$, then we end up with a three-class ordinal regression setting instead of an ordinary regression setting.

Both for reciprocal relations and symmetric relations, additional properties can be assumed. Among these properties, we will further only use transitivity, which is differently defined for reciprocal relations and symmetric relations. Due to lack of space, we omit a formal discussion here, but Figure 1 shows with examples what transitivity means for symmetric and reciprocal relations that are discrete and continuous. Symmetry and reciprocity can be easily incorporated in our framework. Let Ψ be a feature mapping on $\mathcal{V}^2$, let $g : \mathbb{R} \rightarrow [0,1]$ be a monotonically increasing mapping and let h be a hypothesis defined by (3), then the relation Q of type (1) is

1. reciprocal, if Φ is given by $\Phi_R(e) = \Phi_R(v,v') = \Psi(v,v') - \Psi(v',v)$ and g satisfies $g(1/2) = 0$ and for all $x \in \mathbb{R}$, $g(x) = 1 - g(-x)$.
2. symmetric, if Φ is given by $\Phi_S(e) = \Phi_S(v,v') = \Psi(v,v') + \Psi(v',v)$.

In addition, one can easily show that reciprocity and symmetry as domain knowledge can be enforced in the dual formulation. Let us in the least restrictive form now consider the Kronecker product for Ψ, then one obtains for Φ_R and Φ_S respectively the kernels $K^{\Phi}_{\otimes R}$ and $K^{\Phi}_{\otimes S}$ given by

$$K^{\Phi}_{\otimes R}(e,\overline{e}) = 2\big(K^{\phi}(v,\overline{v})K^{\phi}(v',\overline{v}') - K^{\phi}(v,\overline{v}')K^{\phi}(v',\overline{v})\big),$$
$$K^{\Phi}_{\otimes S}(e,\overline{e}) = 2\big(K^{\phi}(v,\overline{v})K^{\phi}(v',\overline{v}') + K^{\phi}(v,\overline{v}')K^{\phi}(v',\overline{v})\big).$$

The former kernel has been proposed for learning preference relations where transitivity not necessarily holds [2], while the latter one is used for predicting protein-protein interactions in bioinformatics [14]. Unlike many existing kernel-based methods for pairwise data, the models obtained with these kernels are able to represent any reciprocal or symmetric relation respectively, without imposing additional transitivity properties of the relations.

For arbitrary relations, the rankings obtained when conditioning on the first or second argument of Q are not necessarily consistent – recall *trusts* versus *is trusted by*. Yet, for symmetric relations, the two conditional rankings obtained from $Q(v,\cdot)$ and $Q(\cdot,v)$ are identical since for all $v,v',v'' \in \mathcal{V}$ it holds that

$$Q(v,v') \geq Q(v,v'') \Leftrightarrow Q(v',v) \geq Q(v'',v)\,. \tag{8}$$

Conversely, for reciprocal relations, the first ranking corresponds to reversing the second one and vice versa, since then for all $v,v',v'' \in \mathcal{V}$ it holds that

$$Q(v,v') \geq Q(v,v'') \Leftrightarrow Q(v',v) \leq Q(v'',v)\,. \tag{9}$$

In the examples in Figure 1, if we condition on the first or the last node, respectively, then we get for the symmetric graph (g) twice $C \succ_A B$, but the reciprocal graph (e) yields $C \succ_A B$ versus $B \succ_A C$.

In addition to enforcing symmetry or reciprocity as domain knowledge in the kernel, one can incorporate the extra information available for this type of relations in the loss function. Remark that in (5), each edge in the training set is considered only once, in the ranking conditioned either on the first or the second node of the edge. For reciprocal or symmetric relations, each observed edge can be twice taken into account, in the rankings conditioned on both nodes, as (9) and (8) respectively hold.

4 Links with Existing Ranking Methods

Examining the pairwise loss (5) reveals that there exists a quite straightforward mapping from the task of conditional ranking to that of traditional ranking. Relation graph edges are in this mapping explicitly used for training and prediction. In recent years, several algorithms for learning to rank have been proposed, which can be used for conditional ranking, by interpreting the conditioning node as a query (see e.g. [15,8,5,10,16]). The main application has been in information retrieval, where the examples are joint feature representations of queries and documents, and preferences are induced only between documents connected to the same query. One of the earliest and most successful of these methods is the ranking support vector machine RankSVM [8], which optimizes the pairwise hinge loss. Even much more closely related is the ranking regularized least-squares method RankRLS [9,10], previously proposed by some of the present authors. The method is based on minimizing the pairwise regularized squared loss and becomes equivalent to the algorithms proposed in this article, if it is trained directly on the relation graph edges.

What this in practice means is that when the training relation graph is sparse enough, say consisting of only a few thousand edges, existing methods for learning to rank can be used to train conditional ranking models. In fact this is how we perform the rock-paper-scissors experiments, as discussed in Section 6.1. However, if the training graph is dense, existing methods for learning to rank are of quite limited use.

Let us assume a training graph that has p nodes. Further, we assume that most of the edges in the graph are connected, meaning that the number of edges is of the order p^2. Using a learning algorithm that explicitly calculates the kernel matrix for the edges would thus need to construct and store a $p^2 * p^2$ matrix, which is intractable already when p is less than thousand. When the standard Kronecker kernel is used together with a linear kernel for the nodes, primal training algorithms could be used without forming the kernel matrix. Assuming on average d non-zero features per node, this would result in having to form a data matrix with $p^2 * d^2$ non-zero entries. Again, this would be both memorywise and computationally infeasible for relatively modest values of p and d.

Thus, building practical algorithms for solving the conditional ranking task requires computational shortcuts to avoid the above-mentioned space and time complexities. The methods presented in this article are based on such shortcuts, because queries and objects come from the same domain, resulting in a special structure of the Kronecker product kernel and a closed-form solution for the minimizer of the pairwise regularized squared loss.

5 Algorithmic Aspects

Let p and q respectively represent the number of nodes and edges in T. Let $\mathbf{K} \in \mathbb{R}^{p \times p}$ be the kernel matrix of K^{ϕ}, containing similarities for all nodes in T, then $\overline{\mathbf{K}} = \mathbf{K} \otimes \mathbf{K} \in \mathbb{R}^{p^2 \times p^2}$ is the kernel matrix of $K^{\Phi}_{\otimes}$, computed on all possible couples of nodes in T. Moreover, let $\mathbf{B} \in \{0,1\}^{q \times p^2}$ be a bookkeeping matrix of the training data, that is, its rows and columns are indexed by the edges in the training set and the set of all possible pairs of nodes, respectively. Each row of $\mathbf{B}$ contains a single nonzero entry indicating to which pair of nodes the corresponding edge is connected.

Now, we show how the loss function (7) can be represented in a matrix form. This representation is similar to the RankRLS loss introduced by [9,10]. Let

$$\mathbf{L}_l = \mathbf{I} - \frac{1}{l}\mathbf{1}\mathbf{1}^{\mathrm{T}} \tag{10}$$

be the $l \times l$-centering matrix with $l \in \mathbb{N}$. The matrix $\mathbf{L}$ is an idempotent matrix and multiplying it with a vector removes the mean of the vector entries from all elements of the vector. Moreover, the following equality can be shown

$$\frac{1}{2l^2}\sum_{i,j=1}^{l}(c_i - c_j)^2 = \frac{1}{l}\mathbf{c}^{\mathrm{T}}\mathbf{L}_l\mathbf{c},$$

where c_i are the entries of any vector $\mathbf{c}$. Now, let us consider the following quasi-diagonal matrix:

$$\mathbf{L} = \begin{pmatrix} \mathbf{L}_{l_1} & & \\ & \ddots & \\ & & \mathbf{L}_{l_p} \end{pmatrix}, \tag{11}$$

where $l_i = |\mathcal{E}_{v_i}|$ for $i \in \{1, \ldots, p\}$. Up to multiplication with a constant, the loss function (7) can be represented in a matrix form as

$$\mathcal{L} = (\mathbf{y} - \mathbf{B}\overline{\mathbf{K}}\mathbf{a})^{\mathrm{T}}\mathbf{L}(\mathbf{y} - \mathbf{B}\overline{\mathbf{K}}\mathbf{a}), \tag{12}$$

provided that the entries of $\mathbf{y}$ and $\mathbf{B}$ are ordered in a way compatible with the entries of $\mathbf{L}$, that is, the training edges are arranged according to their starting nodes. Note that if we use an identity matrix instead of $\mathbf{L}$ in (12), the loss becomes the regression loss (6). In our experiments in Section 6, the conditional ranking and regression approaches are empirically compared.

Next, we will consider how the kernel matrices corresponding to the reciprocal kernel $K^{\Phi}_{\otimes R}$ and the symmetric kernel $K^{\Phi}_{\otimes S}$ can be represented in a matrix notation. Below, we assume that $\mathbf{M}, \mathbf{N} \in \mathbb{R}^{r \times r}$. Let us consider the $r^2 \times r^2$-matrix defined as

$$\mathbf{P} = \sum_{i=1}^{r}\sum_{j=1}^{r}\mathbf{e}_{(i-1)r+j}\mathbf{e}^{\mathrm{T}}_{(j-1)r+i},$$

where $\mathbf{e}_i$ are the standard basis vectors of $\mathbb{R}^{r^2}$. In the literature, $\mathbf{P}$ is called the commutation matrix by [17]. For $\mathbf{P}$, we have the following properties. Firstly, $\mathbf{PP} = \mathbf{I}$, since $\mathbf{P}$ is a symmetric permutation matrix. Moreover, we have $\mathbf{P}\mathrm{vec}(\mathbf{M}) = \mathrm{vec}(\mathbf{M}^{\mathrm{T}})$. Furthermore, we have $\mathbf{P}(\mathbf{M} \otimes \mathbf{N}) = (\mathbf{N} \otimes \mathbf{M})\mathbf{P}$. Next, we consider the matrices

$$\mathbf{S} = \frac{1}{2}(\mathbf{I} + \mathbf{P}),\ \mathbf{A} = \frac{1}{2}(\mathbf{I} - \mathbf{P}),$$

where $\mathbf{I}$ is the identity matrix. In the literature, $\mathbf{S}$ and $\mathbf{A}$ are known as the symmetrizer and skew-symmetrizer matrix, respectively (see e.g. [17]). From the properties of the commutation matrix, it is straightforward to determine the following properties of $\mathbf{S}$ and $\mathbf{A}$. Firstly, $\mathbf{S}\text{vec}(\mathbf{M}) = \frac{1}{2}\text{vec}(\mathbf{M} + \mathbf{M}^{\mathrm{T}})$ and $\mathbf{A}\text{vec}(\mathbf{M}) = \frac{1}{2}\text{vec}(\mathbf{M} - \mathbf{M}^{\mathrm{T}})$. Secondly, the matrices $\mathbf{S}$ and $\mathbf{A}$ are idempotent. Finally, the matrices $\mathbf{S}$ and $\mathbf{A}$ commute with the matrix $\mathbf{M} \otimes \mathbf{M}$.

It can be shown that the symmetry and reciprocity of the pairwise kernels can be enforced by using the matrices $\mathbf{S}$ and $\mathbf{A}$ of the same size as $\mathbf{K}\otimes\mathbf{K}$. Namely, the symmetrized and skew-symmetrized Kronecker kernel matrices for the couples are, up to multiplication with a constant, $\overline{\mathbf{K}} = \mathbf{S}(\mathbf{K} \otimes \mathbf{K})$ and $\overline{\mathbf{K}} = \mathbf{A}(\mathbf{K} \otimes \mathbf{K})$, respectively. Subsequently, we consider different approaches for solving the least-squares problem with loss (12). As already discussed in Section 4, the existing ranking algorithms can be used but this is feasible only in limited cases.

Alternatively, we can also use iterative training algorithms for solving the system of linear equations providing a solution to the learning task:

$$(\mathbf{B}^{\mathrm{T}}\mathbf{L}\mathbf{B}(\mathbf{K} \otimes \mathbf{K})\mathbf{S} + \lambda\mathbf{I})\mathbf{a} = \mathbf{B}^{\mathrm{T}}\mathbf{L}\mathbf{y},$$

where the symmetrizer matrix $\mathbf{S}$ can be replaced with the skew-symmetrizer or identity matrix, depending on the prior knowledge we have about the underlying relation. Here, we consider an approach based on conjugate gradient type of methods, which take advantage of the special structure of the kernel matrices and the loss function. The Kronecker product $(\mathbf{K}\otimes\mathbf{K})\mathbf{v}$ can be written as $\text{vec}(\mathbf{KVK})$, where $\mathbf{v} = \text{vec}(\mathbf{V}) \in \mathbb{R}^{p^2}$, $\mathbf{V} \in \mathbb{R}^{p\times p}$, and vec is the column vectorizing operator, which stacks the columns of a matrix in a column vector. Computing this product is cubic in the number of nodes. Moreover, multiplying a vector with the matrices $\mathbf{S}$ or $\mathbf{A}$ does not increase the computational complexity, since they are sparse. Finally, multiplying a vector with matrices $\mathbf{L}$ or $\mathbf{B}$ can be performed in $O(q)$ time, since the multiplication with the former is equivalent to performing a series of centering operations on the vector and the latter has only q non-zero elements. Conjugate gradient methods require, in the worst case, $O(p^4)$ iterations in order to solve the system of linear equations under consideration. However, the number of iterations required in practice is a small constant, as we show in the experiments. In addition, since using early stopping with the gradient-based methods has a regularizing effect on the learning process (see e.g. [18]), this approach can be used instead of the quadratic regularizer.

As a third approach, we note that in the special case in which $\mathbf{B}$ is the identity matrix – meaning that our training set consists of a graph having exactly two edges between each node, one for both directions – and in which the ordinary Kronecker kernel is used, the training of the conditional ranking method admits the following closed-form solution.

Proposition 1. *If* $\mathbf{B}$ *is the identity matrix and* $\overline{\mathbf{K}} = \mathbf{K}\otimes\mathbf{K}$, *the dual parameters corresponding to the minimizer of (2) with the loss (12) can be obtained from*

$$\mathbf{a} = \text{vec}(\mathbf{U}(\mathbf{C} \odot (\mathbf{U}^{-1}\mathbf{L}_p\mathbf{Y}\mathbf{V}))\mathbf{V}^{\mathrm{T}}) \qquad (13)$$

where $\odot$ *is the Hadamard product,* $\mathbf{y} = \mathrm{vec}(\mathbf{Y})$,

$$\mathrm{diag}(\mathrm{vec}(\mathbf{C})) = (\mathbf{\Lambda} \otimes \mathbf{\Sigma} + \lambda \mathbf{I})^{-1},$$

diag *is the operator that maps vectors to diagonal matrices, and* $\mathbf{V}\mathbf{\Lambda}\mathbf{V}^{\mathrm{T}}$ *and* $\mathbf{U}\mathbf{\Sigma}\mathbf{U}^{-1}$ *are the eigen decompositions of* $\mathbf{K}$ *and* $\mathbf{L}_p\mathbf{K}$, *respectively.*

The proof consists of standard Kronecker product algebra. Since the eigen decompositions and matrix products in (13) can be performed in $O(p^3)$ time, this is also the time complexity of solving the above special case.

To conclude, we have the following three approaches for solving conditional ranking problems: (a) off-the-shelf ranking algorithms can be used when they can be computationally afforded, i.e., when the number of edges in the training set is small; (b) the above-presented approach based on the conjugate gradient method with early stopping and taking advantage of the special matrix structures is recommended when using off-the-shelf methods becomes intractable; (c) the closed-form solution presented in Proposition 1 is recommended if its requirements are fulfilled, since its computational complexity is equivalent to that of a single iteration of the conjugate gradient method.

6 Experiments

In the experiments we consider conditional ranking tasks on both synthetic and real-world data, illustrating different aspects of the generality of our approach. The first experiment is run on the synthetic rock-paper-scissors data set, in which the underlying relation is both reciprocal and intransitive. The task is to learn a model for ranking players according to their likelihood of winning against any other player on whom the ranking is conditioned. In the second experiment, run on the 20-newsgroups data set, the task is to rank documents according to their similarity to any other document, on which the ranking is conditioned.

In all the experiments, we run both the conditional ranker that minimizes the convex edgewise ranking loss approximation (7) and the method that minimizes the regression loss (6) over the edges. Further, in the rock-paper-scissors experiment we also train a conditional ranker with RankSVM. For the 20-newsgroups data this is not possible due to the large amount of edges present in the relational graph, resulting in too high memory requirements and computational costs for RankSVM training. We use the Kronecker kernel $K_{\otimes}^{\Phi}$ for edges in all the experiments, and also test the effects of enforcing domain knowledge by applying the reciprocal kernel $K_{\otimes R}^{\Phi}$ to the rock-paper-scissors dataset, and applying the symmetric kernel $K_{\otimes S}^{\Phi}$ in the 20-newsgroups experiments. The linear kernel is used for individual nodes (thus, for K^{ϕ}).

In all the experiments, performance is measured using the ranking loss (5) on the test set. As a test of statistical significance for comparing the performance differences between the learning methods, we use the paired Wilcoxon-signed-rank test with significance level 0.05. For the rock-paper-scissors dataset, the test error is calculated over 100 repetitions of the experiments and for the 20-newsgroups dataset the test error is calculated separately for each test node.

We use a variety of approaches for minimizing the squared conditional ranking and regression losses, depending on the characteristics of the task. All the used solvers are written in the Python programming language, and rely on the Numpy and Scipy libraries. In the rock-paper-scissors experiment, we train the methods directly on the feature representations of the edges using the standard RankRLS and RLS solvers available in the RLScore software package[1]. The majority of the 20-newsgroups experiments are run by solving the closed-form solution of the conditional ranker presented in Proposition 1, and the analogous solution for the conditional regressor, using standard matrix operations. For the experiment where the training is performed iteratively, we apply the biconjugate gradient stabilized method (BGSM)[19]. The RankSVM conditional ranker is trained with the SVM^{rank} software[2] described in [20].

6.1 Rock-Paper-Scissors

The synthetic benchmark data, whose generation process is described in detail in [2], consists of simulated games of the well-known game of rock-paper-scissors between pairs of players. The training set contains the outcomes of 1000 games played between 100 players, the outcomes are labeled according to which of the players won. The test set consists of another group of 100 players, and for each pair of players the probability of the first player winning against the second one. Different players differ in how often they play each of the three possible moves in the game. The data set can be considered as a directed graph where players are nodes and edges played games, the true underlying relation generating the data is in this case reciprocal. Moreover, the relation is intransitive. It represents the probability that one player wins against another player. Thus, it is not meaningful to try to construct a global ranking of the players. The task of conditional ranking, where players are ranked according to their estimated probability of winning against a given player, however is a sensible task.

We experiment with three different variations of the data set, the w1, w10 and w100 sets. These data sets differ in how balanced the strategies played by the players are. In w1 all the players have close to equal probability of playing any of the three available moves, while in w100 each of the players has a favorite strategy he/she will use much more often than the other strategies. Both the training and test sets in the three cases are generated one hundred times and the hundred ranking results are averaged for each of the three cases and for every tested learning method.

Since the training set consists of only one thousand games, it is feasible to adapt existing ranking algorithm implementations for solving the conditional ranking task. Each game is represented as two edges, labeled as +1 if the edge starts from the winner, and as −1 if the edge starts from the loser. Each node has only 3 features, and thus the explicit feature representation where the Kronecker kernel is used together with a linear kernel results in 9 product features for each

[1] Available at `http://www.tucs.fi/RLScore`

[2] Available at `http://www.cs.cornell.edu/People/tj/svm_light/svm_rank.html`

Method	RLS	RLS (rec)	RankRLS	RankRLS (rec)	RankSVM	RankSVM (rec)
$w = 1$	0.4875	0.4868	0.4876	0.4880	0.4984	0.4930
$w = 10$	0.04172	0.04145	0.04519	0.04291	0.04724	0.04273
$w = 100$	0.001384	0.001370	0.001428	0.001358	0.007408	0.006123

Fig. 2. Overview of the results for rock-paper-scissors. With the abbreviation rec we refer to the use of a reciprocal Kronecker kernel.

edge. In addition, we generate an analogous feature representation for the reciprocal Kronecker kernel. We use these generated feature representations for the edges to train three algorithms. RLS regresses directly the edge scores, RankRLS minimizes pairwise regularized squared loss on the edges, and RankSVM minimizes pairwise hinge loss on the edges. For RankRLS and RankSVM, pairwise preferences are generated only between edges starting from the same node.

In initial preliminary experiments we found out that on this data set regularization is harmful. All the considered methods reach their optimal performance with close to zero regularization parameter values, while the performance almost monotonically decreases with increased regularization. Since some regularization is necessary to guarantee the numeric stability and convergence of the solvers, we set the regularization parameter close to zero for all the considered methods.

The results of the experiments are presented in Figure 2. Clearly the methods are successful in learning conditional ranking models, and the easier the problem is made, the better is the performance. The one hundred repetitions are used to calculate the statistical significances between the performance differences of the methods and the general trends indicated by the tests are discussed next.

For w1, the results are very close to random and there are no statistical significances between the performances of the different methods. For w10, the pairwise ranking methods using the ordinary Kronecker kernel are statistically significantly worse than the pairwise ranking methods using the reciprocal Kronecker kernel. They are also significantly worse than both regression methods. However, there are no significant differences between the pairwise ranking methods using the reciprocal Kronecker kernel and the regression methods. The results for w100 are analogous to those for w10 except that RankSVM performs worse than the other methods. A possible reason for this is that for small values of the regularization parameter λ the training algorithms for optimizing least-squares types of losses are much more stable than the cutting plane training algorithm for RankSVM. The observation that regression can yield as low as, or even lower ranking error than using a pairwise loss is compatible with earlier results in the literature. For example the results presented in [10] show that for some data sets the regression approach may work as well as pairwise models.

In conclusion, we have in this section shown that highly intransitive relations can be modeled and successfully learned in the conditional ranking setting. Moreover, we have shown that when the relation graph of the training set is sparse enough, existing ranking algorithms can be applied by explicitly using the edges of the graph as training examples. Further, both pairwise ranking methods benefit from the use of the reciprocal Kronecker kernels instead of the ordinary Kronecker kernels, while it does not have an effect on the regression method. Finally, with this data, it appears that a regression-based approach performs as well as the pairwise ranking methods.

6.2 20-Newsgroups

In the second set of experiments we aim to learn to rank newsgroup documents according to their similarity with respect to a document the ranking is conditioned on. We use the publicly available 20-newsgroups data set[3] for the experiments. The data set consists of documents from 20 newsgroups, each containing approximately 1000 documents, the document features are word frequencies. Some of the newsgroups are considered to have similar topics, such as the rec.sport.baseball, and rec.sport.hockey newsgroups, which both contain messages about sports. We define a three-level conditional ranking task. Given a document, documents from the same newsgroup should be ranked highest, documents from similar newsgroups next, and documents from unrelated newsgroups last. Thus, we aim to learn the conditional ranking model from an undirected graph, and the underlying similarity relation is a symmetric relation. The setup is similar to that of [21], the difference is that we aim to learn a model for conditional ranking instead of just ranking documents against a fixed newsgroup.

Since the training relation graph is fully connected, the number of edges grows quadratically with the number of nodes. For 5000 training nodes, as considered in one of the experiments, this results already in a graph of approximately 25 million edges, with $1.25 * 10^{11}$ pairwise preferences. Thus unlike in the previous rock-paper-scissors experiment, training a ranking algorithm directly on the edges of the graph is no longer feasible. Instead, we solve the closed-form presented in Proposition 1. At the end of this section we also present experimental results for the iterative BGSM training algorithm, as this allows us to examine the effects of early stopping, and enforcing symmetry on the prediction function.

In the first two experiments, where the closed form solution is applied, we assume a setting where the set of available newsgroups is not static, but rather over time old newsgroups may wither and die out, or new groups may be added. Thus we cannot assume, when seeing new examples, that we have seen documents from the same newsgroup already when training our model. We simulate this by selecting different newsgroups for testing than for training. We form two disjoint sets of newsgroups. Set 1 contains the messages from the newsgroups rec.autos, rec.sport.baseball, comp.sys.ibm.pc.hardware and comp.windows.x, set 2 the messages from the newsgroups rec.motorcycles, rec.sport.hockey, comp.graphics, comp.os.ms-windows.misc and comp.sys.mac.hardware. Thus the graph formed by set 1 consists of approximately 4000 nodes and 16 million edges, and the graph formed by set 2 contains approximately 5000 nodes and 25 million edges. In the first experiment, set 1 is used for training and set 2 for testing. In the second experiment, set 2 is used for training and set 1 for testing. The regularization

[3] Available at: `http://people.csail.mit.edu/jrennie/20Newsgroups/`

Method	Exp. 1	Exp. 2
c.rank	**0.2562**	**0.2895**
c.reg	0.3685	0.3967
p-value	< 0.001	< 0.001

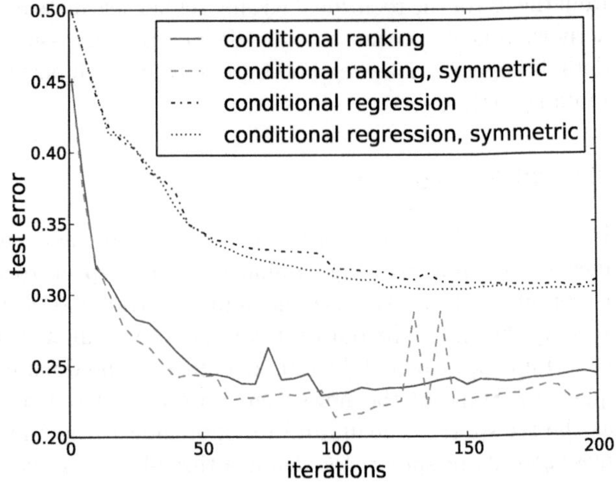

Fig. 3. Experimental results on the newsgroup data. Results for the large-scale experiments with closed-form solution (left table). Results for the small-scale experiment with BGSM and early stopping (right image).

parameter is selected by using half of the training newsgroups as a holdout set against which the parameters are tested. When training the final model all the training data is combined back together.

The results for the closed form solution experiments are presented in Figure 3. Both methods are successful in learning a conditional ranking model that generalizes to new newsgroups which were not seen during the training phase. The method optimizing a ranking based loss over the pairs outperforms the one regressing the values for the pairwise relations in a statistically significant way.

Finally, we study whether enforcing the prior knowledge about the underlying relation being symmetric is beneficial. In this final experiment we use the iterative BGSM method, as it is compatible with the symmetric Kronecker kernel, unlike the solution of Proposition 1. The change in setup results in an increased computational cost, since each iteration of the BGSM method costs as much as using Proposition 1 to calculate the solution. Therefore, we simplify the previous experimental setup by sampling a training set of 1000 nodes, and a test set of 500 nodes from 5 newsgroups. The task is now easier than before, since the training and test sets have the same distribution. All the methods are trained for 200 iterations, and test error is plotted. We do not apply any regularization, but rather rely on the regularizing effect of early stopping, as discussed in Section 5.

Figure 3 contains the performance curves. Again, we see that the pairwise ranking loss quite clearly outperforms the regression loss. Using prior knowledge about the learned relation by enforcing symmetry leads to increased performance, most notably for the ranking loss. The performance curves flatten out within the 200 iterations, demonstrating the feasibility of early stopping.

In conclusion, we have demonstrated various characteristics of our approach in the newsgroups experiments. First, we showed that the introduced methods scale to training graphs that consist of tens of millions of edges, each having a high-dimensional feature representation. Second, we showed the generality of our approach, as it is possible to learn conditional ranking models even when the test newsgroups are not represented in the training data, as long as data from similar newsgroups is available. Unlike the earlier experiments on the rock-paper-scissors data, the pairwise loss yields a dramatic improvement in performance compared to a regression based loss. Finally, enforcing prior knowledge about the type of the underlying relation with kernels was shown to be advantageous.

7 Conclusion

We presented in this article a general framework for conditional ranking from various types of relational data, where rankings can be conditioned on unseen objects and reciprocal or symmetric relations can be treated as two important special cases. We proposed in addition an efficient least-squares algorithm that optimizes a ranking-based loss function of type (5). Experimental results on a synthetic and a real-world dataset confirm that such an approach can lead to statistically significant improvements in performance when the task consists of conditional ranking instead of just trying to predict the underlying relations. Moreover, we also showed empirically that incorporating domain knowledge about the underlying relations can boost the generalization performance.

Acknowledgments

We would like to thank the anonymous reviewers for their insightful comments. T.P. is supported for this work by the Academy of Finland and W.W. by the Research Foundation of Flanders.

References

1. Fisher, L.: Rock, Paper, Scissors: Game Theory in Everyday Life. Basic Books (2008)
2. Pahikkala, T., Waegeman, W., Tsivtsivadze, E., Salakoski, T., De Baets, B.: Learning intransitive reciprocal relations with kernel methods. European Journal of Operational Research 206(3), 676–685 (2010)
3. Yamanishi, Y., Vert, J., Kanehisa, M.: Protein network inference from multiple genomic data: a supervised approach. Bioinformatics 20, 1363–1370 (2004)
4. Weston, J., Eliseeff, A., Zhou, D., Leslie, C., Noble, W.S.: Protein ranking: from local to global structure in the protein similarity network. Proceedings of the National Academy of Sciences of the United States of America 101(17), 6559–6563 (2004)
5. Freund, Y., Yier, R., Schapire, R., Singer, Y.: An efficient boosting algorithm for combining preferences. Journal of Machine Learning Research 4, 933–969 (2003)

6. Waegeman, W., De Baets, B., Boullart, L.: Learning layered ranking functions with structured support vector machines. Neural Networks 21(10), 1511–1523 (2008)
7. Fürnkranz, J., Hüllermeier, E., Vanderlooy, S.: Binary decomposition methods for multipartite ranking. In: Buntine, W., Grobelnik, M., Mladenić, D., Shawe-Taylor, J. (eds.) ECML PKDD 2009. LNCS, vol. 5781, pp. 359–374. Springer, Heidelberg (2009)
8. Joachims, T.: Optimizing search engines using clickthrough data. In: Hand, D., Keim, D., Ng, R. (eds.) Proceedings of the 8th ACM SIGKDD Conference on Knowledge Discovery and Data Mining (KDD 2002), pp. 133–142. ACM Press, New York (2002)
9. Pahikkala, T., Tsivtsivadze, E., Airola, A., Boberg, J., Salakoski, T.: Learning to rank with pairwise regularized least-squares. In: Joachims, T., Li, H., Liu, T.Y., Zhai, C. (eds.) SIGIR 2007 Workshop on Learning to Rank for Information Retrieval, pp. 27–33 (2007)
10. Pahikkala, T., Tsivtsivadze, E., Airola, A., Järvinen, J., Boberg, J.: An efficient algorithm for learning to rank from preference graphs. Machine Learning 75(1), 129–165 (2009)
11. Yang, Y., Bansal, N., Dakka, W., Ipeirotis, P., Koudas, N., Papadias, D.: Query by document. In: Baeza-Yates, R.A., Boldi, P., Ribeiro-Neto, B.A., Cambazoglu, B.B. (eds.) Proceedings of the 2nd International Conference on Web Search and Data Mining, pp. 34–43. ACM Press, New York (2009)
12. Schölkopf, B., Smola, A.: Learning with Kernels, Support Vector Machines, Regularisation, Optimization and Beyond. MIT Press, Cambridge (2002)
13. Suykens, J., Van Gestel, T., De Brabanter, J., De Moor, B., Vandewalle, J.: Least Squares Support Vector Machines. World Scientific Pub. Co., Singapore (2002)
14. Ben-Hur, A., Noble, W.: Kernel methods for predicting protein-protein interactions. Bioinformatics 21(suppl. 1), 38–46 (2005)
15. Burges, C., Shaked, T., Renshaw, E., Lazier, A., Deeds, M., Hamilton, N., Hullender, G.: Learning to rank using gradient descent. In: De Raedt, L., Wrobel, S. (eds.) Proceedings of the 22nd international conference on Machine learning. ACM International Conference Proceeding Series, vol. 119, pp. 89–96. ACM Press, New York (2005)
16. Rendle, S., Freudenthaler, C., Gantner, Z., Schmidt-Thieme, L.: Bpr: Bayesian personalized ranking from implicit feedback. In: UAI 2009: Proceedings of the 25th Conference on Uncertainty in Artificial Intelligence. AUAI Press (2009)
17. Abadir, M., Magnus, J.: Matrix Algebra. Cambridge University Press, Cambridge (2005)
18. Engl, H.W., Hanke, M., Neubauer, A.: Regularization of Inverse Problems. Mathematics and Its Applications, vol. 375. Kluwer Academic Publishers, Dordrecht (1996)
19. van der Vorst, H.A.: BI-CGSTAB: a fast and smoothly converging variant of BI-CG for the solution of nonsymmetric linear systems. SIAM Journal on Scientific and Statistical Computing 13(2), 631–644 (1992)
20. Joachims, T.: Training linear SVMs in linear time. In: Eliassi-Rad, T., Ungar, L.H., Craven, M., Gunopulos, D. (eds.) Proceedings of the 12th ACM SIGKDD international conference on Knowledge discovery and data mining, pp. 217–226. ACM Press, New York (2006)
21. Agarwal, S.: Ranking on graph data. In: Cohen, W.W., Moore, A. (eds.) Proceedings of the 23rd International Conference on Machine Learning. ACM International Conference Proceeding Series, vol. 148, pp. 25–32. ACM Press, New York (2006)

Author Index

GPSR Compliance

The European Union's (EU) General Product Safety Regulation (GPSR) is a set of rules that requires consumer products to be safe and our obligations to ensure this.

If you have any concerns about our products, you can contact us on ProductSafety@springernature.com

In case Publisher is established outside the EU, the EU authorized representative is:

Springer Nature Customer Service Center GmbH
Europaplatz 3
69115 Heidelberg, Germany

Batch number: 10704693

Printed by Printforce, the Netherlands